CONTENTS

PART *III*

RETAIL MATHEMATICS

PART *IV*

MATHEMATICS OF FINANCE

PREFACE

In its 25th year of publication, *Business Mathematics: A Collegiate Approach* continues to meet the needs of business students throughout the nation. Over the years, the text has evolved into its current highly accepted position. The success of the book confirms that it has accomplished its primary objective—to familiarize students with a wide range of business procedures that require the use of mathematics.

Business Mathematics is a comprehensive textbook. A wide range of mathematical procedures exposes students to various business applications. The teaching methodology used in *Business Mathematics* has proved to be successful with many types of students and in varied academic environments. The text is used extensively in the business programs of two-year community and technical colleges, as well as in the lower-level programs at four-year colleges and universities. The content is sufficient for a full year's course, or selected topics/chapters may be used for a one-term course.

The authors' overriding concern has always been for the student. Extreme care is taken to present each topic in a clear and logical manner, with all steps included to facilitate understanding. In addition, concise discussions describe the business applications of each topic, so that the student can appreciate its relevance. Undoubtedly, this emphasis on the student's needs has contributed to the enduring acceptance of the text.

Business Mathematics sharpens the mathematical skills of students preparing to enter business employment, as it simultaneously provides an introduction to accounting, finance, insurance, statistics, taxation, and other math-related subjects. If anything, the widespread use of calculators and computers in business has raised the expectation that employees will be knowledgable in mathematical procedures. The fact that machines perform the final calculations in no way lessens the necessity that employees understand what needs to be done and the sequence of operations required. Thus, students anticipating a career in business should see this course as valuable preparation for a wide range of career opportunities.

Even those students who plan nonbusiness careers will receive valuable consumer education. The text provides an understanding of consumer issues such as bank reconciliations, discounts, markups and markdowns, as well as installment purchases and simple and compound interest.

Among the aspects that make *Business Mathematics* especially student oriented are the following features:

- The terminology and business applications of each topic are presented in a precise, readable discussion that is as concise as possible without sacrificing content. The first use of each important term is highlighted through the use of **boldface.** Moreover, each chapter concludes with a glossary that defines all the key terms of that chapter.

- An unusually large number of examples covers every type of problem that the student may be assigned. The thorough explanations include every step, to ensure that the student will understand the entire solution. In addition, illustrations of many business forms, notes, and statements are featured.

- Over 1,000 problems are included in this text. Each exercise set contains two types of problems: tabular problems designed for quick mastery of technique, followed by written "word" problems to ensure full understanding of the topic under discussion. All odd/ even problems form a pair—providing one problem for classwork and a similar one for homework—and are arranged in order of increasing difficulty. Special care has been taken to use compatible number combinations, so as to emphasize the inherent business procedures and avoid tedious exercises in arithmetic.

- Student learning objectives introduce each chapter and describe the skills to be developed. Each objective is completely cross-referenced to the text, identifying the section and examples where the related information is explained, as well as which problems should be studied to accomplish that objective.

- Brief reviews of arithmetic, equations, and percent are included for the benefit of students whose math skills need refreshing. This essential review is made more relevant by using review problems that are also business-oriented, which contributes to the overall goals of the course.

- Elementary equation-solving is used throughout the text in all topics where it is logical and natural to do so. Students can thus apply the same techniques to many topics and avoid trying to memorize several variations of each formula (which inexperienced math students find nearly impossible to do).

- The use of hand-held LCD calculators, which is allowed (or even required) at many institutions, is introduced in a topic that covers basic functionality and gives students general practice. Throughout the text, highlighted "Calculator Techniques" illustrate how to apply specific calculator operations to the current section's examples, in order to help students compute the problems correctly and efficiently.

Prior to this seventh edition, business mathematics instructors at a number of colleges were surveyed to determine (a) whether any existing topics were considered obsolete and (b) whether any specific new topics were recommended. Consensus strongly reinforced the text's existing subject matter, with appropriate updating. Although not all suggestions could be incorporated into the seventh edition, the authors are grateful to those instructors who responded. Among the recommended ideas incorporated in this edition are the following:

- Over 50 percent of the student problems are new in this edition. The new problems update the text to reflect business trends, while providing instructors with fresh assignment selections.

- Chapter 8, Wages and Payrolls, reflects changes in Social Security, Medicare, and federal income tax deductions. Examples and problems incorporate 1995 tax tables, rates, and forms. Included are illustrations of Form W–4, the Employee's Withholding Allowance Certificate; Form W–2, Wage and Tax Statement; and a quarterly individual earnings summary; as well as illustrations and problems for Form 941, the Employer Quarterly Tax Return.

- To facilitate international transactions in the global market, a new appendix on metrics and foreign exchange conversions has been prepared.

- Extensive tables with ten decimal places have been retained within the text, as requested by a majority of survey participants. Thus, instructors and students can determine the appropriate number of decimals for text problems and personal computations.

- Interest rates, prices, and wages in the examples and problems have been adjusted to reflect current trends.

- The same sequence of topics as presented in previous editions has been maintained. The survey respondents recommended the retention of the first three chapters of basic mathematics review, preferring it in the body of the text.

- A new section entitled "How to Study Business Mathematics" has been added.

An extensive *Instructor's Manual* is available upon adoption of the text. For the instructor's convenience, complete solutions are included for all problems in the text. In addition, suggestions for organizing the course and teaching each chapter are offered, as well as sample quizzes and a final examination. Reflecting the wide acceptance of *Business Mathematics*, an extensive array of supplementary materials is available for instructors and/ or students, as follows:

- Test Item File
- Student Interactive Software
- Achievement Tests
- Prentice Hall Custom Test File
- Instructor's Overhead Calculator (available upon adoption)
- Solutions Manual with Transparency Masters

The content of this seventh edition is due in large measure to the contributions of the following people, to whom we express special appreciation:

- The instructors who participated in the prerevision survey for this seventh edition. The time you took to respond is greatly appreciated, for your suggestions ensure a comprehensiveness and consensus that would be impossible to achieve were the modifications derived only from the authors. Our thanks to the following survey participants: Prof. Russell Baker, Howard Community College; Dr. Mark Bambach, Community College of Philadelphia; Prof. Bettie Davis, Community College of Philadelphia; Prof. James J. Dougher, Devry Technical Institute of Technology, Woodbridge, N.J.; Prof. Cecil Green, Riverside Community College; Dr. Luther Guynes, Los Angeles Community College; Prof. Kathleen K. Hodges, Jamestown Community College; Prof. Nancy Jones, Dundalk Community College; Prof. Paul L. McClellan, Volunteer State Community College; Prof. Richard B. Monbrod, Devry Institute of Technology, Chicago; Prof. J. E. Nassimos, SUNY, Morrisville, N.Y.; Prof. Richard A. Paradiso, Thomas Nelson Community College; Prof. Isaac Pirian, Devry Institute of Technology, Chicago; Prof. Daniel L. Timmons, Alamance Community College; Prof. Thomas J. Witten, Jr., Southwest Virginia Community College; and Dr. Steve Yang, Western International College.
- The following instructors at the Alexandria campus of the Northern Virginia Community College for sharing their expertise in various technical areas of business: Prof. Thomas Atwater, Prof. Pamela Carter, Prof. James Gale, Prof. Ann Marie Klinko, Prof. James Lock, Prof. William T. McDaniel, Prof. Michael Palguta, and Prof. Joyce Wood; and also Mr. William Dudley for assistance with the manuscript.
- Prof. James V. Gray, who provided federal tax information and forms for Chapter 8, and Prof. Lynn Pape, who served as the

quantitative editor, offering helpful suggestions in procedures and steps for examples and problems, as well as checking the mathematical solutions in the *Instructor's Manual*. Our special thanks to Mr. Gray and Mr. Pape, both also of the Alexandria campus of the Northern Virginia Community College.

- Ms. Susanne Stevenson, a former business mathematics student, colleague, and friend, who word-processed the *Instructor's Manual*, a marathon project in itself.

- Prof. David C. Mayne, Anne Arundel Community College, who wrote the Achievement Tests and the Test Item File.

- Many experts outside the academic community who provided important information concerning their area of expertise: Ms. Jennifer E. Bashaw of Lockheed Martin Corporation; Mr. Robert G. Cappello of International Business Machines Corporation; Dr. James Duggan of the U.S. Department of Treasury; Ms. Verenda Smith of the Federation of Tax Administrators; Mr. Gary Hess of Bankers Systems, Inc.; Mr. Ralph A. Richter of the U.S. Department of Commerce; and Ms. Suzanne Stemnock of the American Council of Life Insurance.

- Mr. Craig Campanella, editorial assistant at Prentice Hall, who cheerfully assisted the authors in numerous ways.

- Production editor Ms. Linda Zuk, who provided dedicated support and enthusiasm.

- Ms. Judy Casillo, supplements editor, who coordinated the activities of many different people.

- Special thanks to our husbands, Dan Taylor and Edwin Graves, and daughters, Meg and Sarah Graves. Their encouragement and understanding provided the supportive environment to complete this revision.

It is our sincere hope that all who teach or study *Business Mathematics* will find it to be a very worthwhile experience.

NELDA W. ROUECHE
Los Gatos, California

VIRGINIA H. GRAVES
Alexandria, Virginia

HOW TO STUDY BUSINESS MATH

The authors remember their time on "the other side of the desk" as students. We clearly remember wishing that a math book had more examples, or trying to determine what happened in the steps that were left out, or wondering what some vague explanation was supposed to mean. So when we began *Business Mathematics: A Collegiate Approach*, we designed it not for the instructor but especially for you, the student.

Successful completion of business math requires only two things—time and effort. If you are willing to supply these, you can be assured that your efforts will be rewarded. And you will find that it is well worth the effort required, for business math is a very practical course. Regardless of what position you may later accept, almost any mathematics required by the job will have been introduced in business math. (The text will serve as a good on-the-job reference book after you finish the course.)

By following these suggestions, we think you will be successful and enjoy the course.

- The first step seems so easy, and yet we are amazed at how many students do not do it—***read the book.*** Read it carefully and thoroughly. If something seems unclear when you first read it, read it again. The explanations will familiarize you with the business applications of each topic and the special terminology associated with it. For every problem in the text, there is an example that includes the steps to be followed to find the solution. Try working out the example itself with paper and pencil. Make a practice of reading each section before it is discussed in class; this will enable you to better understand the instructor's discussions and examples.

- ***Keep a notebook*** for taking notes in class. Try to take very complete notes on both the instructor's lecture and examples given. Copy instructions and examples written on the board or overhead projector.

Also, write down questions you encounter in reading the text, in your instructor's lecture, and in your attempts at solving problems.

- *Ask questions*—never hesitate to ask questions in class or after class. If you go on to the next section with a question still unanswered, you may find yourself unable to understand the topic discussed there. (Besides, when one student has a question, others usually have the same question.)

- *Work assigned problems* using a pencil (everybody makes a few careless mistakes), and don't skip steps, unless you are quite adept at math. If a set of problems requires a formula, write it with each problem. Always start by listing the information that is given and indicating what must be found, the unknown. Allow plenty of space for each problem; many mistakes are made because the problem becomes so crowded that the student cannot read his/her numbers or determine what the sequence of steps has been. If you absolutely cannot solve a problem, ask for assistance from your instructor.

- Learn to *use a calculator.* Section 2 of Chapter 1 provides general instructions for calculator usage. Throughout the text, "Calculator Techniques" show steps to be taken to solve numerous examples. Work through these exercises with your calculator. Also, read the instruction manual that comes with your calculator to fine-tune your skills.

- Before working a problem, *estimate the answer.* It may be difficult to determine in your head the answer to 25% of 496, but you should be able to estimate the answer to be approximately 125 (¼ of 500). *Compare your computed answer* to your estimate for reasonableness.

- *Keep homework papers organized* to study before quizzes and exams. Test yourself on the problems and briefly write down the steps or formulas that would be required, or simply set up the problem. Then check what you have written against a similar problem that you have worked previously. *Review* the chapter learning objectives as well as the glossaries.

- During a quiz or exam, *read the problem through* completely before attempting to solve it. Work as quickly as possible without being careless. *Double-check your calculations.* When quizzes are returned to you, *correct your errors and save the papers* as study material for the final exam.

As indicated above, this study plan will require substantial time and effort, but you will learn much useful information in the process. Best wishes for success in business math, and good luck in the business world.

NELDA W. ROUECHE
VIRGINIA H. GRAVES

Basics

REVIEW OF OPERATIONS

OBJECTIVES

Upon completion of Chapter 1, you will be able to:

1. Define and use correctly the terminology associated with each topic.

2. Multiply efficiently using (Section 1):
   a. Numbers containing zero (example: $307 \times 1,400$) (Examples 1, 2; Problems 1, 2)
   b. Decimal numbers with fractions (example: $15 \times 3.5\frac{1}{4}$) (Example 3; Problems 1, 2)
   c. Numbers with exponents (example: 3^4) (Example 4; Problems 1, 2).

3. Simplify fractions containing (Section 1: Example 5; Problems 3, 4):
   a. Decimal parts $\left(\text{example: } \dfrac{6}{1.5}\right)$ b. Fractional parts $\left(\text{example: } \dfrac{6}{1\frac{1}{2}}\right)$.

4. Find negative differences (deficits) (Section 1: Example 6; Problems 3, 4).

5. Simplify expressions containing parentheses of the type $M(1 - dt)$ or $P(1 + rt)$ (Section 1: Examples 7–9; Problems 5, 6).

6. Round off numbers to a specified decimal place (Section 1: Example 10; Problems 7, 8).

7. Round a multiplication product to an accurate number of digits, based on the digits of the original numbers (Section 1: Example 11; Problems 9–12).

8. Using a hand-held LCD calculator (battery or solar powered), perform computations for (Section 2):
   a. Basic arithmetic (Example 1; Problems 1, 2)
   b. Percent calculations (Example 2; Problems 3, 4)
   c. Memory operations (Example 3; Problems 5, 6).

1

The normal day-to-day operations of most businesses require frequent computations with numbers. Many of these computations are done in modern business by calculators and computers. Some computations, however, are still performed manually. And even those processes whose final calculations will be done automatically must first be set up correctly, so that the proper numbers will be fed into the machines and the proper operation performed. Section 1 covers these basic arithmetic techniques. In Section 2, you will have an opportunity to practice using a calculator for basic operations.

SECTION 1
ARITHMETIC TECHNIQUES

The following topics are presented to eliminate some of the weaknesses that many students have in working with specific types of numbers. A more comprehensive review is included as Appendix A, Arithmetic, for those students who need further practice.

MULTIPLICATION

Whenever a number is to be multiplied by some *number ending in zeros,* the zeros should be written to the right of the actual problem; the zeros are then brought down and affixed to the right of the product without actually entering into the **operation.** If either number contains a decimal, this also does not affect the problem until the operation has been completed.

Example 1   (a) 125×40 (b) $2.13 \times 1,500$ (c) $13,000 \times 18$

$$
\begin{array}{r}
125 \\
\times \quad 4\,|0 \\
\hline
5,00\,|0
\end{array}
\qquad
\begin{array}{r}
2.13 \\
\times \quad 15\ |00 \\
\hline
1\ 065 \quad | \\
2\ 13 \qquad | \\
\hline
3,195.\,|00
\end{array}
\qquad
\begin{array}{r}
18 \\
\times \quad 13,\,|000 \\
\hline
54 \qquad \\
18 \qquad \\
\hline
234,\,|000
\end{array}
$$

When multiplying by a *number containing inner zeros,* students often write whole rows of zeros unnecessarily to assure themselves that the other digits will be aligned correctly. The useless zeros can be eliminated if you will remember the following rule: On each line of multiplication, the first digit written down goes directly underneath the digit that was used for multiplying.

Example 2   (a) $2{,}145 \times 307$ (b) $1{,}005 \times 7{,}208$

	Right		_Inefficient_

<pre>
        Right   Inefficient   7,208
        2,145   2,145   ×   1,005
    ×   307   ×   307   3ϕ 040
      15 ϕ15   15 015   7 208
      643 5   00 00   7,244,040
     ───────   643 5
     658,515   ───────
                             658,515
</pre>

When multiplying a **whole number** by a **mixed number** (a mixed number is a whole number plus a fraction), you should first multiply by the fraction and then multiply by the other numbers in the usual manner. If a decimal is involved, the decimal point is marked off in the usual way—the fraction does not increase the number of decimal places.

Example 3   (a) $24 \times 15\frac{3}{4}$ (b) $35 \times 1.3\frac{3}{7}$

<pre>
          24   3 5
     × 15¾   × 1.3³⁄₇
        1ϕ   (¾ × 24 = 18)   1ϕ
        120   10 5
         24   35
       ─────   ──────
        378   47.0
</pre>

There are several different ways to indicate that multiplication is required. The common symbols are the "times" sign ("×") and the raised dot ("·"). Many formulas contain **variables,** which are letters or symbols used to represent numerical values. Often variables are written together; this indicates that the numbers that these variables represent are to be multiplied. A number written beside a variable indicates multiplication of the number and variable. Numbers or variables within parentheses written together should also be multiplied. A number or variable written adjoining parentheses should be multiplied by the expression within the parentheses. Thus,

$$3 \cdot 5 = 3 \times 5$$

$$Prt = P \times r \times t$$

$$4k = 4 \times k$$

$$(2.5)(4)(6.8) = 2.5 \times 4 \times 6.8$$

$$7(12) = 7 \times 12$$

Some few problems require the use of exponents. An **exponent** is merely a number which, when written as a superscript to the right of another number, called the **base,** indicates how many times the base is to be written in repeated multiplication times itself.

Example 4

(a) $x^2 = x \cdot x$ (exponent, base)

(b) $5^3 = 5 \cdot 5 \cdot 5 = 125$ (exponent, base)

(c) $2^4 = 2 \cdot 2 \cdot 2 \cdot 2$
$= 16$

(d) $(1.02)^3 = (1.02)(1.02)(1.02)$
$= 1.061208$

DIVISION

Recall that a fraction indicates division—the numerator of the fraction (above the line) is to be divided by the denominator (below the line). Thus, if the denominator contains a fraction, it must be inverted and multiplied by the numerator *of the entire fraction.* If the denominator contains a decimal, it must be moved to the end of the number and the decimal in the numerator moved a corresponding number of places.

Example 5

(a) $\dfrac{3}{\frac{3}{4}} = \dfrac{3}{1} \div \dfrac{3}{4} = \dfrac{\overset{1}{\cancel{3}}}{1} \times \dfrac{4}{\cancel{3}} = \dfrac{4}{1} = 4$

(b) $\dfrac{6}{1\frac{1}{2}} = \dfrac{6}{\frac{3}{2}} = \dfrac{6}{1} \div \dfrac{3}{2} = \dfrac{\overset{2}{\cancel{6}}}{1} \times \dfrac{2}{\cancel{3}} = \dfrac{4}{1} = 4$

(c) $\dfrac{4}{0.5} = \dfrac{4.0}{0.5} = \dfrac{40}{5} = 8$

(d) $\dfrac{5.2}{0.13} = \dfrac{5.20}{0.13} = \dfrac{520}{13} = 40$

(e) $\dfrac{2.31}{0.3} = \dfrac{2.31}{0.3} = \dfrac{23.1}{3} = 7.7$

SUBTRACTION

Unfortunately, business expenses sometimes exceed the funds budgeted for them. In this case, the **deficit** is called a **negative difference.** (It is also common to say that the account is "in the red.") A negative difference is found by taking the numerical difference between the larger and smaller numbers. The result is indicated as being a deficit either by placing a minus sign before the number or by placing the result in parentheses (the method used on most financial statements).

Example 6

(a)
Bank balance	$598.00
Checks written	− 650.00
Deficit	−$ 52.00

(b)
Profit earned	$3,500.00
Salary owed	− 4,700.00
Negative difference	($1,200.00)

PARENTHESES

Several formulas used in finding simple interest and discount contain parentheses. You should be familiar with the correct procedure to follow in working with parentheses: If parentheses contain both multiplication and addition, or multiplication and subtraction, the multiplication should be performed first and the addition or subtraction last. This procedure simply follows the standard rule for order of operations: Multiplication and division should always be performed before addition and subtraction. Thus, $3 \times 4 - 8 = 12 - 8 = 4$; and $9 + 16 \div 8 = 9 + 2 = 11$.

Example 7

(a) $(1 + rt) = \left(1 + \dfrac{\overset{3}{\cancel{6}}}{100} \cdot \dfrac{1}{\cancel{2}}\right)$ $\quad \left(\text{where } r = \dfrac{6}{100} \text{ and } t = \dfrac{1}{2}\right)$

$= \left(1 + \dfrac{3}{100}\right)$

$= \left(\dfrac{100}{100} + \dfrac{3}{100}\right)$

$= \left(\dfrac{103}{100}\right)$

(b) $(1 - dt) = \left(1 - \dfrac{\overset{2}{\cancel{8}}}{100} \cdot \dfrac{1}{\cancel{4}}\right)$ $\quad \left(\text{where } d = \dfrac{8}{100} \text{ and } t = \dfrac{1}{4}\right)$

$= \left(1 - \dfrac{2}{100}\right)$

$= \left(\dfrac{100}{100} - \dfrac{2}{100}\right)$

$= \left(\dfrac{98}{100}\right)$

(c) $(1 + rt) = \left(1 + \dfrac{\overset{2}{\cancel{8}}}{100}\cdot\dfrac{7}{\underset{3}{\cancel{12}}}\right)$ $\left(\text{where } r = \dfrac{8}{100} \text{ and } t = \dfrac{7}{12}\right)$

$$= \left(1 + \dfrac{14}{300}\right)$$

$$= \left(\dfrac{300}{300} + \dfrac{14}{300}\right)$$

$$= \left(\dfrac{314}{300}\right)$$

Note. Parts (a) through (c) are left "unreduced" intentionally, since each fraction represents part of a larger problem.

If a parenthesis is preceded by a number, that number must be multiplied times the whole expression within the parentheses. That is, the terms within the parentheses should be consolidated into a single number or fraction, if possible, *before* multiplying. Consider the formula $P(1 + rt)$ when $P = \$600$, $r = \frac{5}{100}$, and $t = \frac{1}{3}$:

Example 8

	Right	*Wrong*

$P(1 + rt) = \$600\left(1 + \dfrac{5}{100}\cdot\dfrac{1}{3}\right)$ $P(1 + rt) = \$600\left(1 + \dfrac{5}{100}\cdot\dfrac{1}{3}\right)$

$= 600\left(1 + \dfrac{5}{300}\right)$ $= 600\left(1 + \dfrac{5}{300}\right)$

$= 600\left(\dfrac{300}{300} + \dfrac{5}{300}\right)$ $= \overset{2}{\cancel{600}}\left(1 + \dfrac{5}{\cancel{300}}\right)$

$= \overset{2}{\cancel{600}}\left(\dfrac{305}{\cancel{300}}\right)$ $= 2(6)$

$= \$610$ $= \$12$

If it is impossible to consolidate the terms within the parentheses into a single number or fraction, then the number in front of the parentheses must be multiplied times *each* separate term within the parentheses. (Separate terms may be identified by the fact that a plus or minus sign always appears between them.)

Example 9

(a) $P(1 + rt) = P{\cdot}1 \;+\; P{\cdot}rt = P + Prt$

(b) $M(1 - dt) = M{\cdot}1 \;-\; M{\cdot}dt = M - Mdt$

(c) $M(1 - dt)$ with $M = \$400$ and $d = \frac{6}{100}$ becomes

$$M(1 - dt) = \$400\left(1 - \frac{6}{100}t\right)$$

$$= \$400{\cdot}1 - \overset{4}{\cancel{400}}{\cdot}\frac{6}{\cancel{100}}\,t$$

$$= \$400 - 24t$$

ROUNDING OFF DECIMALS

The general rule for rounding off decimals is as follows:

If the last decimal place that you wish to include is followed by any digit from 0 through 4, the digit in question remains unchanged. If followed by any digit from 5 through 9, the digit in question is increased by one.

Example 10

(a) Rounding off to tenths (to one decimal place):
14.3)274 = 14.3

(b) Rounding off to hundredths (to two decimal places):
5.37)812 = 5.38

(c) Rounding off to thousandths (to three decimal places):
0.032)569 = 0.033

(d) Rounding off 24.14759:
to tenths: 24.1
to hundredths: 24.15
to thousandths: 24.148

Note.   A special case sometimes arises when the last digit that you wish to use is followed by a single 5 (with no other succeeding digits). In this case, particularly when working with numbers that should total 100 percent, you may use the following rule:

If the last digit to be included is an odd number, round it off to the next higher number; if the digit in question is even, leave it unchanged.

The following example illustrates this case:

Example 10
(cont.)

(e)   Round to the nearest percent.

$$
\begin{array}{r}
62.5\% = 62\% \\
+ \ \underline{37.5\%} = \underline{38\%} \\
100.0\% = 100\%
\end{array}
$$

ACCURACY OF COMPUTATION

Business, engineering, manufacturing, and the like all demand careful accuracy in their calculations, processes, and products. Computational accuracy is dependent on the number of significant digits used. **Significant digits** are digits obtained by precise measurement rather than by rough approximation. For example, 4,283′ has four significant digits; 4,300′ has only two significant digits if it was measured only to the nearest hundred feet. Similarly, 8.75 has three significant digits; 9 has one significant digit. There are two significant digits in 9.0 if a measurement made to the nearest tenth falls between 8.95 and 9.05.

The concept of significant digits requires that the following rule be observed: *The result obtained from a mathematical calculation can never be more accurate than the least accurate figure used in making the calculation.* Specifically, additional accuracy (more significant digits) cannot be created artificially by the simple use of a mathematical operation (such as multiplication).

This rule requires an initial calculation to be rounded off in many instances. The answer must at least be rounded off so as to contain no more than the number of decimal places contained in the original figure which had the *least number of decimal places.* To be more restrictive, the answer should contain no more significant digits than the original number which had the *fewest significant digits.*

Example 11

(a)   Find the area of a room 17.27 feet in length and 13.6 feet in width.

$A = lw$

$ = 17.27 \times 13.6$

$ = 234.872$

$A = 234.9$ square feet   (since 13.6 had only one decimal place)

or

$A = 235$ square feet   (since 13.6 had three significant digits)

This rule is of particular importance in computing amounts invested at compound interest, since tables containing many decimal places are used.

Students will want to minimize work by using no more of these decimal places than are necessary—but at the same time, obtain an answer that is correct to the nearest cent.

To do this, you should first estimate the answer to determine the number of digits it will contain (including the cent's place). This number plus one more (to ensure absolute accuracy) will determine the number of digits you need to copy from the table.

It should be noted that any known exact amount (such as $350) is considered accurate for any number of decimal places desired.

Example 11
(cont.)

(b) Suppose an interest problem requires that $200 be multiplied by the table value 1.48594740. We wish to use only enough digits from the table to ensure that our answer is correct to the nearest cent.

First, estimate the value of the tabular number: It is approximately 1.5. Therefore, the answer we obtain will be approximately $200 × 1.5 = $300.

The number $300.00 contains 5 digits; thus we must copy 5 + 1 = 6 digits from the table. (The sixth digit from the table will be rounded off using the previously discussed rules for rounding off decimals.) The number 1.48594740, rounded to six digits, equals 1.48595.

Thus, $200 × 1.48595 = $297.19000 = $297.19 is the solution correct to the nearest cent.

(c) Find $1,500 times 1.86102237 correct to the nearest cent, using no more digits than necessary.

1.86102237 equals approximately 2.
2 × $1,500 = $3,000, which contains 6 digits (including the cents digits).
6 + 1 = 7 digits are required from the table.
$1,500 × 1.861022 = $2,791.533000 = $2,791.53, correct to cents.

(d) Find $800 × 0.50752126 correct to cents, using no more digits than necessary.

0.50752126 equals approximately 0.5.
$800 × 0.5 = $400, which has 5 digits (to the nearest cent).
5 + 1 = 6 digits are required from the table.
$800 × 0.507521 = $406.016800 = $406.02.

Note.   The zero before the decimal in the number 0.50752126 does not qualify as a significant digit (that is, as a digit which should be counted), because the zero could have been omitted without changing the value of the number.

Another suggestion that minimizes work is to substitute fractional equivalents for percents (such as $\frac{1}{3}$ for $33\frac{1}{3}\%$, or $\frac{3}{7}$ for $42\frac{6}{7}\%$) in computations. Such equivalents usually simplify calculations done manually and often provide more accurate results than the more tedious computations performed with rounded decimal equivalents. For this reason, even students using calculators are often advised to use fraction equivalents (a two-step process). Thus, all students should familiarize themselves with the percents having convenient fractional equivalents. A table of these equivalents appears on page 53.

PROBLEM SOLVING

Students often read a problem, decide that they do not know how to solve it, and simply "give up" without writing anything at all. By attacking problems in an organized manner, however, you can succeed with many problems for which the solution process was not at all obvious on first reading.

1. Make a list of all information that was given. Name the term that each given amount represents and/or associate each with a variable that can be used in an equation or formula. (For instance, overhead = $3,000, or $r = 5\%$.)

2. Based on the question asked in the problem, indicate what needs to be found. (For example, markup = ?% of selling price, or P = ?$.)

3. Ask yourself how this unknown quantity could logically be found. (A formula, such as $I = Prt$? An algebraic equation based on the question in the problem? Some process, such as using a table to determine annual percentage rate for monthly payments?)

4. After deciding on a logical method, reexamine your list of given information. Do you already have everything required to use this method? If not, could your given information be used to compute the additional information required by this method?

5. Attempt some calculations using the method you selected. Examine your result to decide whether it really answers the question that was asked and whether that answer is reasonable. (If necessary, go back to step 3 and consider another method. For instance, the unknown quantity may be included in several different formulas and could perhaps be found using a different formula than was first tried.)

The foregoing approach may not be foolproof, of course, but it certainly increases your chances for success. You have no doubt heard the expression "If all else fails, read the directions." In this case, the "directions" are the examples in the text, and the authors assume that they will be studied before any problems are attempted!

SECTION 1   PROBLEMS

Find the product.

1. a. 15×60
   b. 20×312
   c. 400×37
   d. $27 \times 1,400$
   e. $2,500 \times 114$
   f. $1,425 \times 504$
   g. 208×729
   h. $1,006 \times 24,304$
   i. $2,050 \times 354$
   j. $28 \times 2.3\frac{1}{4}$
   k. $1.75 \times 14\frac{2}{5}$
   l. $15\frac{3}{8} \times 0.24$
   m. $3.6 \times 14\frac{2}{9}$
   n. $2.7 \times 1.2\frac{1}{3}$
   o. 8^4
   p. 3^5
   q. 6^3
   r. 1.04^2

2. a. 83×30
   b. 50×256
   c. 183×600
   d. $1,300 \times 44$
   e. $3,600 \times 118$
   f. $1,641 \times 302$
   g. $405 \times 1,765$
   h. $2,009 \times 13,202$
   i. $1,070 \times 423$
   j. $34 \times 6.4\frac{1}{2}$
   k. $5.2 \times 18\frac{1}{4}$
   l. $28\frac{1}{3} \times 0.48$
   m. $5.4 \times 32\frac{1}{6}$
   n. $2.4 \times 3.5\frac{3}{4}$
   o. 5^4
   p. 4^8
   q. 7^3
   r. 2.01^2

Divide or subtract, as indicated.

3. a. $\dfrac{9}{\frac{3}{5}}$
   b. $\dfrac{14}{2\frac{1}{3}}$
   c. $\dfrac{21}{0.3}$
   d. $\dfrac{9.8}{0.14}$
   e. $\dfrac{2.34}{0.6}$

f. Checkbook balance $127.13
   Checks written $-\ \underline{184.94}$

g. Gross profit $72,089
   Operating expenses $-\ \underline{84,256}$

h. Total handling cost $4,115
   Selling price $-\ \underline{4,300}$

i. Insurance coverage $1,000
   Doctor's charge $-\ \underline{3,225}$

4. a. $\dfrac{48}{\frac{6}{7}}$
   b. $\dfrac{72}{4\frac{1}{2}}$
   c. $\dfrac{20}{0.5}$
   d. $\dfrac{3.6}{0.12}$
   e. $\dfrac{1.36}{0.8}$

f. Net sales $66,708
   Cost of goods sold $-\ \underline{68,134}$

g. Travel allowance $300
   Travel expenses $-\ \underline{475}$

h. Net income $ 82,500
   Partners' salaries $-\ \underline{100,000}$

i. Escrow for taxes $2,575
   Taxes assessed $-\ \underline{3,100}$

Find the value of each expression.

5. a. $\left(1 + \dfrac{6}{100}\cdot\dfrac{7}{36}\right)$
   b. $\left(1 + \dfrac{4}{100}\cdot\dfrac{30}{40}\right)$
   c. $\left(1 - \dfrac{9}{100}\cdot\dfrac{4}{5}\right)$

d. $\left(1 - \dfrac{8}{100}\cdot\dfrac{6}{42}\right)$
   e. $1,200\left(1 + \dfrac{14}{100}\cdot\dfrac{5}{12}\right)$
   f. $2,400\left(1 + \dfrac{10}{100}\cdot\dfrac{5}{12}\right)$

g. $800\left(1 - \dfrac{9}{100}\cdot\dfrac{5}{6}\right)$
   h. $200\left(1 - \dfrac{7y}{100}\right)$
   i. $400\left(1 - \dfrac{3a}{100}\right)$

j. $c(de + f)$
   k. $s(1 - tu)$

6. a. $\left(1 + \dfrac{16}{100}\cdot\dfrac{1}{4}\right)$

b. $\left(1 + \dfrac{7}{100}\cdot\dfrac{1}{3}\right)$

c. $\left(1 - \dfrac{3}{100}\cdot\dfrac{1}{6}\right)$

d. $\left(1 - \dfrac{4}{100}\cdot\dfrac{1}{2}\right)$

e. $5{,}000\left(1 - \dfrac{18}{100}\cdot\dfrac{4}{9}\right)$

f. $1{,}500\left(1 + \dfrac{28}{100}\cdot\dfrac{2}{7}\right)$

g. $1{,}000\left(1 - \dfrac{6}{100}\cdot\dfrac{2}{5}\right)$

h. $500\left(1 - \dfrac{12g}{100}\right)$

i. $300\left(1 + \dfrac{9b}{100}\right)$

j. $j(kl + m)$

k. $w(1 - xy)$

Round off each number as indicated.

7. a. To tenths:	b. To hundredths:	c. To thousandths:	d. To tenths, hundredths, and thousandths:
66.6666	84.6717	86.45451	8.1875
367.9810	488.9261	410.01463	14.5266
1,542.3425	591.5544	217.67528	

8. a. To tenths:	b. To hundredths:	c. To thousandths:	d. To tenths, hundredths, and thousandths:
43.258	8.9426	18.92453	5.08473
156.643	26.4453	0.56641	23.67521
1,680.952	160.0639	337.00894	

Compute each product. Round first according to the least accurate decimal, and then round according to the least number of significant digits.

9. a. 8.61×13.5　　b. 1.5×4.03

c. 3.45×7.002　　d. 50.6×0.32

10. a. 14.2×12.35　　b. 4.56×7.3

c. 5.8×7.83　　d. 1.111×3.85

Compute each product, correct to the nearest cent. Use no more digits than are necessary.

11. a. $\$600 \times 1.50252492$　　b. $\$900 \times 2.0015973$

c. $\$2{,}000 \times 0.40324726$　　d. $\$40 \times 16.09689554$

12. a. $\$300 \times 1.91301845$　　b. $\$500 \times 1.52161826$

c. $\$4{,}000 \times 0.37440925$　　d. $\$20 \times 19.08162643$

SECTION 2
USING A CALCULATOR

Calculators are allowed (or required) for the business mathematics courses at many colleges, usually beginning with Chapter 4 after the Basics unit is complete. This current section teaches some basic techniques for using a calculator.

Then, beginning with Chapter 4, various topics throughout the text will contain boxes that illustrate calculator techniques applicable to those corresponding problems.

Different types of calculators require slightly different methods of operation. However, the techniques described here apply for the basic calculators used by most students: hand-held LCD (liquid crystal display) calculators allowing a maximum of eight digits (or seven digits after the decimal point). Only basic key-functions are assumed; thus, if your calculator contains special keys not described here, you should consult the instructions that came with your calculator to learn their uses.

If needed, press the $\boxed{\text{ON/C}}$ key to activate your calculator. (Some light/solar-powered calculators have no "on" key but activate as soon as they are uncovered.) After each problem, press $\boxed{\text{ON/C}}$ again to "clear" the calculator, resetting the display to zero.

ARITHMETIC OPERATIONS

Arithmetic operations compute in the order you enter them into the calculator. The first number is "known" by the calculator as soon as you press the digits; thus, the first operation (or function) symbol that you press determines what process happens with the second number. In the examples, square boxes such as "$\boxed{+}$" indicate that you should press the key containing that symbol. An arrow "→" points to the solution that should display on your calculator.

Example 1   This series of additions and subtractions is typical of calculations you would use for a checkbook, where you make deposits (add) and write checks (subtract).

(a)   Find $550 - 75 - 80 + 100 - 125 = 370$.

$$550 \boxed{-} 75 \boxed{-} 80 \boxed{+} 100 \boxed{-} 125 \boxed{=} \longrightarrow 370$$

Notice as you press the operation symbol before the third number (and similarly before all subsequent numbers) that the current subtotal temporarily displays, just before you enter the number itself. This subtotal could be used to update your checkbook balance after each transaction.

Hints.   If you accidentally press the wrong *operation* key (such as $\boxed{+}$ when you meant $\boxed{-}$), simply press the correct key immediately. It will cancel the previous operation and take effect itself. That is, $550 \boxed{+} \boxed{-} 75$ computes as $550 \boxed{-} 75$. If you enter a *number* incorrectly, immediately press the $\boxed{\text{CE}}$ (cancel entry) key; then enter the correct number and continue with the calculation.

Multiplication and division are required for many business mathematics computations. Problems containing fractions also make use of the same techniques.

Example 1
(cont.)

(b) Find $3,600 \div 90 \times 15 = 600$.

$$3,600 \boxed{\div} 90 \boxed{\times} 15 \boxed{=} \longrightarrow 600$$

(c) Compute $\dfrac{3}{5}(750) = 450$.

$$3 \boxed{\times} 750 \boxed{\div} 5 \boxed{=} \longrightarrow 450$$

Observe in this case that $3 \boxed{\div} 5 \boxed{\times} 750 \boxed{=} 450$, as previously calculated. On a calculator, however, reversing the order for a fraction does not always produce exactly the same result, as demonstrated by the next example.

(d) Multiply $\dfrac{3}{7} \times 280 = 120$.

$$3 \boxed{\times} 280 \boxed{\div} 7 \boxed{=} \longrightarrow 120 \qquad \text{(preferred method)}$$

$$3 \boxed{\div} 7 \boxed{\times} 280 \boxed{=} \longrightarrow 119.99999$$

Thus, you should make a habit of multiplying by the numerator first and then dividing by the denominator. (Most problems in this text are designed for convenient calculations by students without calculators; thus denominators usually "cancel" into the second number, but they may not divide evenly into the numerator, as shown in this example.)

PERCENT CALCULATIONS

A large portion of business mathematics calculations involve percent. The $\boxed{\%}$ key produces an immediate result on your calculator; it is not necessary to press the $\boxed{=}$ key for multiplication problems that involve percent. However, this dictates the order in which numbers must be entered, as follows.

Example 2

(a) Find $720 \times 35\% = 252$.

$$720 \boxed{\times} 35 \boxed{\%} \longrightarrow 252$$

(b) Multiply $48\% \times 250 = 120$.

$$250 \boxed{\times} 48 \boxed{\%} \longrightarrow 120$$

Notice from part (b) that the percent (here, 48%) must be entered *last*, although it appears first in the problem.

Memory Operations

The memory capability of calculators allows you to save values that you will use again—either for repeated, similar calculations or else for a later portion of the current computation. For instance, you might store a tax rate or a Social Security rate and use that rate in several similar calculations. Or, given a complicated fraction, you would determine the denominator first and save it in memory, to use after the numerator has been computed.

The keys $\boxed{M+}$ and $\boxed{M-}$ are used to add or subtract a value into memory, combining it with any value already existing in memory. If a multiple-step computation is in progress, the $\boxed{M+}$ or $\boxed{M-}$ key can be used without pressing $\boxed{=}$; that is, the memory key will simultaneously determine the solution and store it into memory. An "M" will appear at one side of your display, indicating that memory currently contains a nonzero value. You can use the memory value simply by pressing the memory recall key, $\boxed{MR}$ (sometimes labeled $\boxed{M_C^R}$), at the corresponding point where that value is used in a calculation.

Example 3

(a) Place into memory the sales tax rate 6.25%. Then find the sales tax on purchases of $20 and $50.

$$.0625 \;\boxed{M+} \longrightarrow (\text{"M" appears at one side of the display})$$

$$20 \;\boxed{\times}\; \boxed{MR}\; \boxed{=} \longrightarrow 1.25$$

$$\boxed{MR}\; \boxed{\times}\; 50 \;\boxed{=} \longrightarrow 3.125$$

Notice that the $\boxed{MR}$ key can be used in either order in a multiplication problem; for instance, the last line could just as easily be computed as $50 \;\boxed{\times}\; \boxed{MR}$.

Hint. If you are unsure what decimal value is equivalent to a percent, multiply 1 times the percent to enter its decimal value. For instance, the first line above could be: $1 \;\boxed{\times}\; 6.25 \;\boxed{\%}\; \boxed{M+} \rightarrow 0.0625$.

Before proceeding, you need to "clear" the current value from memory. This procedure varies according to the type of key your calculator has: (1) If your calculator has an $\boxed{MC}$ (or memory clear) key, press that key. The "M" will then disappear, although the last solution remains in the display. (2) If there is no $\boxed{MC}$ key, you could clear the memory simply by pressing $\boxed{OFF}$ and then again pressing $\boxed{ON/C}$. Alternatively, you can press $\boxed{MR}$ to recall the memory value, then press $\boxed{M-}$ to subtract it from itself, which leaves the memory value at zero. As before, the "M" immediately disappears from your display, indicating that memory has cleared, although the number itself remains as the displayed value. (3) If you have an $\boxed{AC}$ (or all clear) key, this will clear both memory and the current display simultaneously.

Example 3
(cont.)

(b) Upon completing part (a), clear 6.25% from the calculator memory.

If you have an "MC" key: $\boxed{\text{MC}} \longrightarrow$ ("M" disappears; 3.125 remains)

Without an "MC" key: $\boxed{\text{MR}} \longrightarrow 0.0625$
$\boxed{\text{M}-} \longrightarrow$ ("M" disappears; 0.0625 remains)

If you have an "AC" key: $\boxed{\text{AC}} \longrightarrow 0.$ ("M" disappears; zero appears)

When using fractions that have multiple parts in the denominator, evaluate the denominator first and save it into memory, before entering the numerator. Similarly, multiple values inside parentheses should be computed first and stored into memory, before you multiply by the value preceding the parentheses.

Example 3
(cont.)

(c) Calculate the value of $\dfrac{2{,}000 + 47.5}{1 - 0.025} = 2{,}100.$

$1 \boxed{-} .025 \boxed{\text{M}+} \longrightarrow 0.975$ ("M" appears and remains until you clear memory)

$2000 \boxed{+} 47.5 \boxed{=} \longrightarrow 2047.5 \boxed{\div} \boxed{\text{MR}} \boxed{=} \longrightarrow 2100$

Note. The last line can be computed correctly as $2000 \boxed{+} 47.5 \boxed{\div} \boxed{\text{MR}} \boxed{=} \rightarrow 2100$. However, most students feel more comfortable including the first $\boxed{=}$, in order to see the 2047.5 total value of the numerator.

Hint. Remember to clear memory before each new problem. For assurance, you may also wish to clear the display by pressing $\boxed{\text{ON/C}}$ (although that is not essential if the display contains the result of a previous calculation).

Business mathematics requires many financial formulas like the next example, where a "rate × time" calculation (the last two items in parentheses below) must be added/subtracted into memory as a separate step. The "1 $\boxed{\text{M}+}$" step can be done either before or after you place the "rate × time" value into memory; here, it is shown as the first step.

(d) Calculate $1{,}200 \left(1 - 0.06 \times \dfrac{5}{12} \right) = 1{,}170.$

$1 \boxed{\text{M}+} \longrightarrow$ ("M" appears; 1 remains)

$.06 \boxed{\times} 5 \boxed{\div} 12 \boxed{\text{M}-} \longrightarrow 0.025$

$1200 \boxed{\times} \boxed{\text{MR}} \boxed{=} \longrightarrow 1170$

As you press $\boxed{\text{MR}}$, notice that this briefly displays (recalls) the memory value 0.975, which is the computed value within parentheses.

Note. On some calculators, the second row above can also be computed as: 5 $\boxed{\times}$ 6 $\boxed{\%}$ $\boxed{\div}$ 12 $\boxed{\text{M}-}$, to give the same result in exact dollars. In most real-world calculations, however, your final result must be rounded to the nearest cent. Observe that the $\boxed{\text{M}-}$ key is used here because a "−" appears before the "0.06 × $\frac{5}{12}$." Had a "+" appeared there, $\boxed{\text{M}+}$ would be used to add the "rate × time" value into memory.

SECTION 2   PROBLEMS

Perform the following operations using a calculator.

1. a. 1,550 + 145 − 104 − 91   b. 462 − 13 + 88 − 200   c. 600 × 8 ÷ 20

    d. 120 ÷ 12 × 8   e. $\frac{340}{5} \times 4$ f. $\frac{2}{9}(720)$

    g. $\frac{5}{6}(54)$ h. $\frac{1,000}{40} \times 16$

2. a. 3,120 − 48 + 188 − 251   b. 964 − 410 + 17 + 8   c. 76 × 9 ÷ 4

    d. 12 ÷ 8 × 46   e. $\frac{3}{8} \times 400$ f. $\frac{4}{9}(108)$

    g. $\frac{2}{7}(420)$ h. $\frac{608}{8} \times 12$

3. a. 520 × 43%   b. 640 × 12 × 10%

    c. 3.5% × 1,600   d. 75% × 80% × 1,200

4. a. 136 × 22%   b. 582 × 30 × 5%

    c. 6.2% × 405   d. 90% × 60% × 1,500

Use the calculator memory for the following problems. Clear the memory between parts.

5. a. 45% of $90; of $160; of $300   b. 6.5% × 48; × 96; × 133

    c. $\frac{208 + 642}{1 - 0.60}$ d. $\frac{48.315 + 32.093}{3.3 + 0.70}$

    e. 500(1 + 0.12 × 18)   f. $1,500\left(1 - \frac{1}{2} \times 14\%\right)$

    g. $830\left(1 + \frac{1}{4} \times 10\%\right)$ h. $\frac{48}{25 \times 6\%}$

6. a. 12.25% of 32; of 60; of 180   b. 14% × 12; × 25; × 1,400

    c. $\frac{335 + 785}{1 - 0.80}$ d. $\frac{874 - 56.42 - 21.26}{12.34 + 4.25}$

e. $4,000(1 + 0.08 \times 24)$

f. $1,600\left(1 - \dfrac{3}{4} \times 9\%\right)$

g. $720\left(1 + \dfrac{5}{8} \times 5\%\right)$

h. $\dfrac{126.9}{47 \times 15\%}$

CHAPTER 1 GLOSSARY

Base.   A number that is to be used in multiplication times itself.

Deficit.   The result when a larger number is subtracted from a smaller number.

Exponent.   A superscript that indicates how many times a base is to be written in repeated multiplication times itself. (Example: The exponent 3 denotes that $5^3 = 5 \cdot 5 \cdot 5 = 125$.)

Mixed number.   A number that combines a whole number and a fraction. (Example: $12\frac{1}{2}$)

Negative difference.   (See "Deficit.")

Operation.   Any of the arithmetic processes of addition, subtraction, multiplication, and division.

Significant digits.   Digits obtained by precise measurement rather than by rough approximation (or artificially by computation).

Variable.   A letter or symbol used to represent a number.

Whole number.   A number from the set

$$\{0, 1, 2, 3, \ldots\}$$

which has no decimal or fractional part.

2

USING EQUATIONS

OBJECTIVES

Upon completion of Chapter 2, you will be able to:

1. Define and use correctly the terminology associated with each topic.
2. Solve basic equations that require only (Section 1: Examples 1–11; Problems 1–36):

    a. Combining the similar terms of the equation; and/or

    b. Using the operations of addition, subtraction, multiplication, and division.

    c. Examples:

    $$x - 12 = 16, \qquad 5y - 4 = 6 - 3y, \qquad \text{or} \qquad \frac{3x}{4} + 5 = 23$$

3. Express any written problem in a concise sentence which provides the structure for the equation that solves the problem (Section 2: Examples 1–8; Problems 1–36).

4. **a.** Express numbers in ratios (Section 3: Example 1; Problems 1, 2, 5–8).

    b. Use proportions to find numerical amounts (Section 3: Examples 2–4; Problems 3, 4, 9–28).

The procedure required to solve many problems in business mathematics is much more obvious if you have a basic knowledge of equation-solving techniques. The equations we will consider, such as $x - 7 = 5$ or $\frac{2}{3}y + 5 = y + 2$, are **first-degree equations in one variable.** "First-degree" means that the variable, or letter, in each equation has an exponent of 1 (although this exponent is not usually written; that is, x means x^1). Notice that the "one variable" may appear more than one time, but there will only be one distinct variable in any given equation.

It is important to note the difference between an expression and an equation. An **expression** is any indicated mathematical operation(s) written with no equal sign. An **equation** is a mathematical statement that two expressions (or an expression and a number) are equal. For instance, "$2x - 4$" is an expression, whereas "$2x - 4 = 10$" is an equation.

Before you start to solve any actual equations, a discussion of some characteristics of equations may prove helpful.

SECTION 1
BASIC EQUATIONS

An equation may be compared to an old-fashioned balancing scale. The "equals" sign is the center post of the scale, and the two sides of the equation balance each other as do the pans of the scale.

We know that you can either add weights or remove them from the pans of a scale, and as long as you make the same changes in both pans, the scales remain in balance. The same is true of equations: In solving an equation, you may perform any operation (addition, subtraction, multiplication, or division), and as long as the *same operation* is performed with the *same numbers* on *both sides* of the equation, the balance of the equation will not be upset.

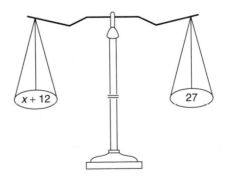

The object in solving any equation with unknowns in it is to determine what value of the unknown quantity (the *variable*) will make the equation a true statement. Basically, this is accomplished by isolating the variable **term**

or terms on one side of the equation, with the ordinary numbers on the other side, as follows:

$$\text{Variables} = \text{Numbers}$$

or

$$\text{Numbers} = \text{Variables}$$

For instance, given $5p - 3 = 2p + 15$, we isolate the variable terms on one side and the numbers on the other to obtain $5p - 2p = 15 + 3$. The **similar terms** (those that include the same variable) are isolated by performing exactly the same operation(s) on both sides of the equation, as demonstrated in the examples that follow.

There are only four basic equation forms—those whose solution requires a single operation of either addition, subtraction, multiplication, or division. All other equations are variations or combinations of these four forms. We shall now consider several elementary equations, beginning with the four basic forms.

In the basic equation forms to follow, there will be only one number associated with the unknown (variable) on the same side of the equation. You should first determine which operation is involved with the number. The solution of the equation will then require that the *opposite* operation be performed on both sides of the equation.

Example 1   Solve for x in the equation: $x + 5 = 38$

To isolate the variable x on the left side of the equation, we need to remove the 5. In the original equation, the 5 is added to x; therefore, we must perform the opposite of addition, or subtract 5 from both sides.

$$x + 5 = 38$$

$$\text{(Subtract 5 from both sides)} \quad x + 5 - 5 = 38 - 5$$

$$(5 - 5 = 0, \text{ and } x + 0 = x) \quad \underbrace{x + 5 - 5}_{} = \underbrace{38 - 5}_{}$$

$$x = 33$$

Note. Numbers or variables in a given equation may be added or subtracted only if they are on the *same side* of the equation; thus, in Example 1, equation-solving techniques were used to get 38 and 5 on the same side of the equation before the numbers could be combined. Also, students who have had little experience in solving equations should develop the following habit: On each succeeding line of the problem, first copy the adjusted equation resulting from the preceding step; then perform the next step.

Example 2   Solve for r: $r - 26 = 47$

      The 26 is subtracted from r; therefore, we must *add* 26 to both sides of the equation.

$$r - 26 = 47$$

(Add 26 to both sides) $r - 26 + 26 = 47 + 26$

($-26 + 26 = 0$) $r - 26 + 26 = 47 + 26$

$$r = 73$$

Example 3   Solve for y: $6y = 72$

      The expression $6y$ means "6 times y" (or $6 \times y$). Any number that multiplies a variable is called a **coefficient;** in this equation, 6 is the coefficient of y.

      Because 6 multiplies y, we must perform the opposite of multiplication, or divide both sides of the equation by 6.

$$6y = 72$$

(Divide both sides by 6) $\dfrac{6y}{6} = \dfrac{72}{6}$

$\left(\dfrac{6}{6} = 1; 1y = y \right)$ $\dfrac{\overset{1}{\cancel{6}}y}{\cancel{6}} = \dfrac{72}{6}$

$$y = 12$$

Example 4   Solve for k: $\dfrac{k}{3} = 17$

      Since k is divided by 3, we *multiply* both sides of the equation by 3.

$$\frac{k}{3} = 17$$

(Multiply both sides by 3) $\dfrac{3}{1} \cdot \dfrac{k}{3} = 17 \cdot 3$

$\dfrac{\overset{1}{\cancel{3}}}{1} \cdot \dfrac{k}{\cancel{3}} = 17 \cdot 3$

$$k = 51$$

Now let us consider some variations and combinations of the basic equation forms.

Example 5   Solve for t: $\dfrac{2t}{3} = 36$

Just as $\dfrac{2 \times 12}{3}$ gives the same result as $\dfrac{2}{3} \times 12$, so $\dfrac{2t}{3}$ is the same as $\dfrac{2}{3}t$. In order to solve the equation for t, we need to remove the coefficient $\frac{2}{3}$ and obtain a coefficient of 1. This can be accomplished easily by multiplying by the reciprocal of $\frac{2}{3}$. The **reciprocal** of any number is the result obtained when the number is inverted (that is, the numerator and denominator are interchanged). Thus, the reciprocal of $\frac{2}{3}$ is $\frac{3}{2}$.

$$\frac{2t}{3} = 36$$

$\left(\text{Multiply by } \dfrac{3}{2}, \text{ the reciprocal of } \dfrac{2}{3}\right)$ $\dfrac{\cancel{3}^{1}}{\cancel{2}_{1}} \cdot \dfrac{\cancel{2}^{1}}{\cancel{3}_{1}} t = \cancel{36}^{18} \cdot \dfrac{3}{\cancel{2}_{1}}$

$\left(\dfrac{3}{2} \cdot \dfrac{2}{3} = 1; 1t = t\right)$

$$t = 54$$

Example 6   Solve for p: $6p + p - 3p - 48$

We must first combine the three p terms into a single p term. This is done by adding and/or subtracting the coefficients, whichever the signs indicate. (Recall that $p = 1p$.)

$$6p + p - 3p = 48$$

(Combine the coefficients: $6 + 1 - 3 = 4$) $4p = 48$

(Divide both sides by 4) $\dfrac{\cancel{4}^{1}p}{\cancel{4}} = \dfrac{48}{4}$

$$p = 12$$

Example 7   Solve for m: $4m + 5 = 33$

When one side of an equation contains both a variable term and a number term, it is customary to work with the number term first, in order to obtain an altered equation of the type

$$\text{Variable term} = \text{Numbers}$$

$$4m + 5 = 33$$

(Subtract 5 from both sides) $\qquad 4m + \cancel{5 - 5} = 33 - 5$

$$4m = 28$$

(Divide both sides by 4) $\qquad \dfrac{\overset{1}{\cancel{4}}m}{\cancel{4}} = \dfrac{28}{4}$

$$m = 7$$

Example 8   Solve for n: $\qquad 6 = \dfrac{n}{7} - 3$

We must first isolate the variable term so that

$$\text{Numbers} = \text{Variable term}$$

$$6 = \frac{n}{7} - 3$$

(Add 3 to both sides) $\qquad 6 + 3 = \dfrac{n}{7} - \cancel{3 + 3}$

$$9 = \frac{n}{7}$$

(Multiply both sides by 7) $\qquad 7 \cdot 9 = \dfrac{n}{\cancel{7}} \cdot \overset{1}{\cancel{7}}$

$$63 = n$$

Example 9   Solve for s: $\qquad \dfrac{3s}{5} - 4 = 14$

$$\frac{3s}{5} - 4 = 14$$

(Add 4) $\qquad \dfrac{3s}{5} - \cancel{4 + 4} = 14 + 4$

$$\frac{3s}{5} = 18$$

$$\left(\text{Multiply by } \frac{5}{3}\right) \qquad \frac{\overset{1}{\cancel{5}}}{\cancel{3}} \cdot \frac{\overset{1}{\cancel{3}}s}{\cancel{5}} = \overset{6}{\cancel{18}} \cdot \frac{5}{\cancel{3}}$$

$$s = 30$$

Example 10 Solve for z: $9(z - 3) = 45$

Recall that the coefficient 9 must multiply times each term within the parentheses.

$$9(z - 3) = 45$$

$$9z - 27 = 45$$

(Add 27) $9z - \underline{27 + 27} = 45 + 27$

$$9z = 72$$

(Divide by 9) $$\frac{\overset{1}{\cancel{9}}z}{\cancel{9}} = \frac{72}{9}$$

$$z = 8$$

Example 11 Solve for d: $8d + 2 = 5d + 17$

Remember that variable terms or number terms can be added and/or subtracted only when they appear on the same side of the equation. Therefore, the first steps will involve getting all the variable terms on one side of the equation and all the number terms on the other. When the solution of an equation requires that both of these steps be done, it is customary—although not essential—to work with the variable terms first.

$$8d + 2 = 5d + 17$$

(Subtract 5d from both sides) $8d - 5d + 2 = \cancel{5d} - \cancel{5d} + 17$

$$3d + 2 = 17$$

(Subtract 2 from both sides) $3d + \cancel{2} - \cancel{2} = 17 - 2$

$$3d = 15$$

(Divide both sides by 3) $$\frac{\overset{1}{\cancel{3}}d}{\cancel{3}} = \frac{15}{3}$$

$$d = 5$$

SECTION 1   PROBLEMS

Solve the following equations.

1. $x + 14 = 61$
2. $x + 23 = 88$
3. $x - 5 = 7$
4. $x - 9 = 21$
5. $8t = 72$
6. $4y = 112$
7. $9z = 180$
8. $12b = 108$
9. $2x - 5 = 15$
10. $8x - 6 = 18$
11. $4p + 12 = 156$
12. $7c + 48 = 125$
13. $5y - 6 = 44$
14. $26n + 3 = 601$
15. $14v + 20 = 4v + 100$
16. $8z - 8 = 2z + 52$
17. $\dfrac{4s}{7} = 36$
18. $\dfrac{3a}{8} = 48$
19. $10b = 6b + 24$
20. $7q = 5q + 16$
21. $8(x - 2) = x + 47$
22. $6(y + 3) = y + 128$
23. $7c - 16 = 2c + 79$
24. $8f + 22 = f + 57$
25. $12x - 11 = 3x + 34$
26. $15x - 6 = 7x + 10$
27. $13z - 10 = 8 - 5z$
28. $2h - 10 = 32 - 4h$
29. $7(d - 3) = 2d - 6$
30. $6(x - 2) = 4x - 2$
31. $\dfrac{h}{5} - 4 = 11$
32. $\dfrac{r}{7} - 6 = 12$
33. $\dfrac{5m}{7} + 14 = 59$
34. $\dfrac{3x}{5} + 8 = 98$
35. $\dfrac{4b}{9} - 1 = 35$
36. $47 = \dfrac{3d}{4} - 22$

SECTION 2
WRITTEN PROBLEMS

Many business problems of a mathematical nature can have no predetermined formula applied. In these cases, the business person must be able to set the facts of the situation into an original equation to obtain the solution.

An equation is simply a mathematical sentence (a statement of equality). If you can express the facts of a mathematical problem in a clear, concise, English sentence, the mathematical sentence (equation) will follow in exactly the same pattern. (By the same token, a student who cannot express the facts in a clear, English sentence probably does not understand the situation well enough to be able to obtain a mathematical solution.)

One fact that helps immensely in converting English sentences into mathematical sentences is that the *verb* of the English sentence corresponds to the *equal sign* of the equation. The mathematical equivalents of several other words will be pointed out in the following examples.

Example 1   What number increased by 15 gives 68?

$$\begin{array}{cccc}
\underline{\text{What number}} & \underline{\text{increased by}} & \underline{15} & \underline{\text{gives}} & \underline{68?} \\
\downarrow & \downarrow & \downarrow & \downarrow & \downarrow \\
n & + & 15 & = & 68?
\end{array}$$

Thus,

$$n + 15 = 68$$

$$n + 15 - 15 = 68 - 15$$

$$n = 53$$

Example 2   Advertising expense last month was $300 less than utilities expense. If $745 was spent for advertising, how much were utilities?

(subtracted from)

$$\begin{array}{ccccc}
\underline{\text{Advertising}} & \underline{\text{was}} & \underline{\$300} & \underline{\text{less than}} & \underline{\text{utilities}} \\
\downarrow & \downarrow & & \downarrow & \\
a & = & u & - & \$300 \\
\$745 & = & u & - & \$300 \\
745 + 300 & = & u & - & 300 + 300 \\
\$1{,}045 & = & u
\end{array}$$

Example 3   About $\frac{1}{6}$ of a family's net (after-tax) monthly income is budgeted for food. What is their net monthly income if food bills average $105 per week? (Assume that 4 weeks equal 1 month.)

$$\begin{array}{ccccc}
\underline{\text{One-sixth}} & \underline{\text{of}} & \underline{\text{net income}} & \underline{\text{is}} & \underline{\text{food expense}} \\
\downarrow & \downarrow & \downarrow & \downarrow & \downarrow \\
\frac{1}{6} & \times & I & = & (4 \times \$105) \\
& & \frac{I}{6} & = & 420 \\
& & (\cancel{6})\frac{I}{\cancel{6}} & = & 420(6) \\
& & I & = & \$2{,}520
\end{array}$$

Example 4   A training workshop held by Field Sales Corp. cost the firm $4,800. If 75 salespeople attended the workshop, what was the cost per participant?

$$
\begin{array}{ccccc}
\underline{\text{Number of}} & \text{times} & \underline{\text{cost per}} & \text{equals} & \underline{\text{total}} \\
\text{participants} & & \text{participant} & & \\
75 & \times & c & = & \$4{,}800 \\
& & 75c & = & 4{,}800 \\
& & \dfrac{\cancel{75}c}{\cancel{75}} & = & \dfrac{4{,}800}{75} \\
& & c & = & \$64
\end{array}
$$

Example 5   The state welfare department pays $\frac{2}{3}$ of the cost of day care for indigent children. The remaining balance is financed by Whitworth County. If the state paid $1,400 for day care, what was the total cost of the day-care project?

$$
\begin{array}{ccccc}
\underline{\text{Two-thirds}} & \underline{\text{of}} & \underline{\text{total cost}} & \underline{\text{was}} & \underline{\text{state's share}} \\
\dfrac{2}{3} & \times & c & = & \$1{,}400 \\
& & \dfrac{2c}{3} & = & 1{,}400 \\
& & \dfrac{3}{2} \times \dfrac{2c}{3} & = & \overset{700}{\cancel{1{,}400}} \times \dfrac{3}{\cancel{2}} \\
& & c & = & \$2{,}100
\end{array}
$$

Example 6   The number of television sets sold at Entertainment Associates was three times the combined total of compact disc players and radios. If their sales included 72 TV sets and 8 radios, how many CD players were sold?

$$
\begin{array}{ccccc}
\underline{\text{TVs}} & \underline{\text{were}} & \underline{\text{three times}} & & \underline{\text{CDs and radios combined}} \\
t & = & 3 & \times & (c + r) \\
72 & = & 3 & \times & (c + 8) \\
72 & = & 3(c + 8) & & \\
72 & = & 3c + 24 & & \\
72 - 24 & = & 3c + 24 - 24 & & \\
48 & = & 3c & & \\
\dfrac{48}{3} & = & \dfrac{\cancel{3}c}{\cancel{3}} & & \\
16 & = & c & & \text{(CD players)}
\end{array}
$$

(handwritten notes)
$T = 3(C + R)$
$72 = 3(C + 8)$
$72 = 3C + 24$
$\dfrac{48}{3} = \dfrac{3C}{3}$
$16 = C$

Example 7   Harris and Smith together sold 36 new insurance policies. If Harris sold three times as many policies as Smith, how many sales did each make?

Note.   When a problem involves two amounts, it is usually easier to let the variable of your equation represent the smaller quantity.

Since Harris sold three times as many policies as Smith, then Smith is the smaller quantity:

$$\begin{array}{ccccc}\underline{\text{Harris}} & \underline{\text{sold}} & \underline{\text{three}} & \underline{\text{times}} & \underline{\text{Smith}} \\ H & = & 3 & \times & S\end{array}$$

Then,

$$\begin{array}{cccc}\underline{\text{Harris}} \ \underline{\text{and}} \ \underline{\text{Smith}} \ \underline{\text{together}} & \underline{\text{sold}} & \underline{\text{36 policies}} \\ H \ + \ S & = & 36 \\ \downarrow \\ 3S \ + \ S & = & 36 \\ \\ \dfrac{\cancel{4}S}{\cancel{4}} & = & \dfrac{36}{4} \\ \\ \text{(Smith's sales)} \quad S & = & 9 \text{ policies} \\ \\ \text{(Harris's sales)} \quad H & = & 3S \\ \\ & = & 3 \times 9 \\ \\ H & = & 27 \text{ policies}\end{array}$$

Example 8   The Texas Barbeque charges $9 for a barbeque dinner and $15 for a steak dinner. A recent dinner for 250 people totaled $2,850. How many of each type of dinner were served?

We know that a total of 250 dinners were served, some barbeque and some steak. So, suppose we knew there were 50 barbeque dinners; how would we find the number of steak dinners? (Steak = 250 − 50.) Regardless of the actual number of each type of dinners,

$$\text{Barbeque} + \text{Steak} = 250$$

$$b \ + \ s \ = 250$$

$$b \ = 250 - s$$

Now,

Cost of barbeque	plus	cost of steak	totals	$2,850
$9 × b$	+	$15 × s$	=	$2,850$

$$\downarrow$$

$$9(250 - s) \quad + \quad 15s \quad = \quad 2,850$$

$$2,250 - 9s \quad + \quad 15s \quad = \quad 2,850$$

$$2,250 \quad + \quad 6s \quad = \quad 2,850$$

$$2,250 - 2,250 \quad + \quad 6s \quad = \quad 2,850 - 2,250$$

$$6s \quad = \quad 600$$

$$\frac{\cancel{6}s}{\cancel{6}} \quad = \quad \frac{600}{6}$$

(Steak dinners) $\quad s \quad = \quad 100$ dinners

(Barbeque dinners) $\quad b \quad = \quad 250 - s$

$$= \quad 250 - 100$$

$$b \quad = \quad 150 \text{ dinners}$$

The following list is a review of words and their mathematical equivalents. Knowing the mathematical equivalents of these words will be helpful in solving equations.

ENGLISH WORD	SYMBOL
Is, are, was, were, gives, sold	=
Increased by, more than, combined, together, sum	+
Decreased by, less than, fewer than	−
Of, times, product of	×
Per, out of	÷

SECTION 2 PROBLEMS

"Translate" the following expressions into algebraic symbols.

1. a. A number decreased by 12
   b. Dick and Jane together

2. a. 6 times x
   b. Apples and oranges combined

c. 3 more than b

d. $15 less than x

e. ¼ of c

f. 10 less than ⅔ of k

g. 4 times the sum of c and d

h. w made $18 less than z

i. b totals 8 times j and k combined

j. d costs 3.8 times as much as e

k. h equals 5 less than ½ of m

l. $30 per office visit

c. A number increased by 10

d. A number decreased by 18

e. ⅔ of c

f. 5 more than ¼ of p

g. Twice the sum of r and s

h. g costs $4 less than h

i. d totals 2 times a and b together

j. b costs 8.5 times as much as f

k. m equals 9 less than ⅓ of n

l. $10 per barrel

Express each of the following as both an English equation and a math equation, and solve.

3. What number increased by 14 yields 79?   N+14

4. What number decreased by 26 yields 56?

5. Last year's price of a lamp, increased by $12, gives this year's price of $55. What was last year's price?

6. The Corner Market charges $1.50 less than Wilson's Mart for the same-size package of dog food. If the Corner Market's price is $12, what does Wilson's Mart charge?

7. Bradlee's charges $6 more than Arnold Discount Center for a sweater. If Bradlee's price is $32, what does Arnold charge?

8. The Scott Shop charges $15 more than the Garcia Co. for a winter jacket. If the Scott Shop's price is $77, how much does the Garcia Co. charge?

9. Two-thirds of a firm's expenditures are related to payroll expenses. If the total expenditures are $42,000, how much are the payroll expenses?

10. Three-fifths of a sports shop's sales were charge sales. What were their charge sales, if the total sales were $3,000?

11. One-twelfth of Ace Co.'s gross sales were returned last year. If $9,000 of merchandise was returned last year, what were the gross sales?

12. One-fourth of Happy Ed's Auto sales last year were repeat customers. If 800 cars were sold to repeat customers, how many cars were sold?

13. A sporting goods store found that 4 less than 4/7 of its sales were for men's sportswear. If 80 items of men's sportswear were sold last month, how many total sales were made?

14. Eight less than ⅖ of the employees at a manufacturing plant took no sick leave during January. If there were 40 employees without an absence, how many people does the plant employ?

15. A California law office handled 3.5 times as many cases as its Georgia counterpart last year. If the California office handled 17,500 cases, how many cases did the Georgia office handle in the year?

16. Utility expenses for February were 1.2 times March utility expenses for the Dawson family. If February utilities totaled $192, how much were the expenses for March?

17. On a recent business trip, an employee was reimbursed for car expenses. If the car was driven 1,900 miles and the employee received $342, what was the reimbursement per mile?

18. Depreciation amounting to $1,176 was claimed on robotic equipment by an auto manufacturer. The equipment was used 4,900 hours last year. What was the depreciation per hour?

19. The total handling cost of merchandise consists of the wholesale cost plus the overhead. In August, the total handling cost was $25,000. If the wholesale cost ran 3 times as much as the overhead, how much was each?

20. Salaries for managers are 1.8 times the salaries for staff. If the total salaries were $280,000, how much was each?

21. A professional painter purchased semigloss latex paint and flat latex paint for an office building. He spent $236 for 21 gallons of paint. If the semigloss cost $10 a gallon and the flat cost $12 a gallon, how many gallons of each kind did he purchase?

22. An electronic store's September sales of cellular phones and pagers totaled $1,900. Cellular phones sold for $50 and pagers sold for $60. If 35 items were sold, how many of each were sold?

23. A college freshman budgets a monthly income of $500 for food, clothing, and entertainment. Food purchases are 4 times his clothing purchases, and he spends $20 more on entertainment than he does on clothing. How much does he budget for each item?

24. A homeowner pays $1,005 per month for mortgage, taxes, and home insurance combined. The mortgage payment is 9 times the insurance payment, and the taxes are $15 more than the insurance. How much is spent for each item?

25. At its grand opening sale, The Eye Store sold 4.6 times as many eye glasses as it sold contact lenses. If 224 sales were made altogether, how many of each were sold?

26. A store sells three brands of watches. The number of Brand A watches sold was 3 times as many as Brands B and C together. Find the number of Brand C watches sold, if 129 Brand A and 25 Brand B watches were sold.

27. A young woman paid $8 for shampoo, hair conditioner, and hair spray. The bottle of conditioner costs twice as much as the shampoo, and the hair spray was $1 less than the shampoo. How much did each bottle cost?

28. A man purchased a jacket, a pair of slacks, and a flannel shirt for $137. The jacket cost 2.5 times as much as the slacks, and the shirt cost $7 less than the slacks. How much did each item cost?

29. In a recent order, The Mother Goose Shop purchased CDs and cassette tapes. There were 50 total items included in the shipment, and the total cost was $220. The CDs cost $5 each, whereas the tapes cost $4 each. How many of each were shipped?

30. A gym shop sold 54 leotards for a total of $900. The cotton leotards sold for $15 each, and the nylon leotards sold for $18 each. How many of each were sold?

31. To raise money for the city council campaign, a candidate sold campaign buttons and bumper stickers. The price for a button was $3.50, and a sticker cost $2.00. One afternoon, 110 items were sold, raising $265. How many of each were sold?

32. One style of athletic shoes costs $20 wholesale in canvas and $36 in leather. An invoice for $776 accompanied an order for 26 pairs of shoes. How many pairs of each were bought?

33. A pastry shop ran a special sale on glazed donuts and jelly-filled donuts. The shop sold 780 donuts and collected $225. If the glazed donuts cost $0.20 each and the jelly-filled donuts cost $0.35 each, how many of each were sold?

34. A theater sold tickets for a total of $7,050. Friday night tickets were priced at $30 each, while Saturday matinee tickets were $22 each. If 275 tickets were sold, how many of each were sold?

35. A jewelry store ordered the same bracelet in silver and gold. The total order was for 12 bracelets and the total invoice price was $275. If the silver bracelets cost $20 each and the gold ones cost $25 each, how many of each were ordered?

36. The Lamplighter sold 200 lamps during a two-week special sale. Desk lamps sold for $19 each and floor lamps sold for $40 each. If $5,165 was received from the sale of the 200 lamps, how many of each were sold?

SECTION 3
RATIO AND PROPORTION

A **ratio** is a way of using division or fractions to compare numbers. When two numbers are being compared, the ratio may be written in any of three ways: for example, (1) 3 to 5, (2) 3:5, or (3) $\frac{3}{5}$. Method 3 indicates that common fractions (which are **rational numbers**) are also ratios. (Notice that "rational" and "ratio" are variations of the same word.) If more than two numbers are being compared, the ratio is written in either of the first two ways: as 3 to 2 to 4, or 3:2:4.

Ratios are often used instead of percent in order to compare items of expense, particularly when the percent would have exceeded 100%. In this case it is customary to reduce the ratio to a comparison to 1, such as 1.3 to 1 or as $\frac{2.47}{1}$. (A ratio is "reduced" by expressing it as a common fraction and reducing the fraction. A ratio is reduced to "a comparison to 1" by dividing the denominator into the numerator.) Ratios may also be used as a basis for dividing expenses among several categories, which we will consider when apportioning overhead in a later chapter.

Example 1   Determine the following ratios.

(a)   There are 20 doctors and 48 registered nurses on the staff of Neuman County Hospital. What is the ratio of doctors to nurses?

$$\text{Doctors to registered nurses} = \frac{\text{Doctors}}{\text{Nurses}}$$

$$= \frac{20}{48}$$

$$= \frac{5}{12}$$

The ratio of doctors to registered nurses is 5 to 12, or 5:12, or $\frac{5}{12}$.

(b) Net sales of Warner Corp. were $105,000 this year, compared to $84,000 last year. What is the ratio of this year's sales to last year's?

$$\text{Current sales to previous} = \frac{\text{Current sales}}{\text{Previous sales}}$$

$$= \frac{\$105,000}{\$84,000}$$

$$= \frac{5}{4}$$

$$= 1.25 \quad \text{or} \quad \frac{1.25}{1}$$

The ratio may be given correctly in any of the following forms: 5 to 4 or 1.25 to 1; 5:4 or 1.25:1; $\frac{5}{4}$ or $\frac{1.25}{1}$. Since most actual business figures would not reduce to a common fraction like $\frac{5}{4}$, business ratios are usually computed by ordinary division, without attempting to reduce the fraction. Thus,

$$\frac{\$105,000}{\$84,000} = 1.25 = \frac{1.25}{1} \quad \text{or} \quad 1.25 \text{ to } 1 \quad \text{or} \quad 1.25:1$$

The term *proportion* is frequently used in connection with ratio. A **proportion** is simply a mathematical statement that two ratios are equal. Thus, $\frac{6}{10} = \frac{3}{5}$ is a proportion. The same proportion could also be indicated as 6:10::3:5, which is read "six is to ten as three is to five."

A proportion may be used to find an unknown amount if we first know the ratio that exists between this unknown and another, known amount. This method is particularly useful when relationships (ratios) are given as percent. Such proportions are then solved like ordinary equations.

Example 2 The net profit of Shaw Hardware was 7% of its net sales. What was the net profit if the firm had $60,000 in net sales?

Since 7% $= \frac{7}{100}$, then

$$\frac{\text{Profit}}{\text{Net sales}} = \frac{7}{100}$$

$$\frac{P}{\$60,000} = \frac{7}{100}$$

$$(\cancel{60,000}^{1}) \frac{P}{\cancel{60,000}} = \frac{7}{\cancel{100}} (\cancel{60,000}^{600})$$

$$P = \$4,200$$

For any given proportion, the **cross products** are always equal. Given $\frac{6}{10} = \frac{3}{5}$, the cross products are computed as

$$\frac{6}{10} \diagup\!\!\!\!\diagdown \frac{3}{5}$$

$$6 \cdot 5 = 3 \cdot 10$$

$$30 = 30$$

Cross products* are useful in solving any proportion where the unknown appears in the denominator, as follows.

Example 3   Lien and Pham, who are sales representatives for Nguyen & Co., received orders last week in the ratio of $\frac{2}{3}$. How many orders did Pham sell, if Lien sold 56 orders?

$$\frac{\text{Lien sales}}{\text{Pham sales}} = \frac{2}{3}$$

$$\frac{56}{P} = \frac{2}{3}$$

$$(3)56 = 2P$$

$$168 = 2P$$

$$\frac{168}{2} = \frac{\cancel{2}P}{\cancel{2}}$$

$$84 = P$$

Example 4   Citizens of San Rio paid \$39 for 700 kilowatt-hours (kWh) of electricity. What was the bill for a family that used 1,400 kWh?

$$\frac{39}{700} = \frac{?}{1400}$$

---

*Cross products are a shortcut which produces the same result that is obtained when both sides of an equation are multiplied by the product of the two given denominators. Given $\frac{6}{10} = \frac{3}{5}$, then $50(\frac{6}{10}) = 50(\frac{3}{5})$, which gives $30 = 30$, as before.

$$\frac{\text{Cost}}{\text{kWh}} = \frac{\$39}{700}$$

$$\frac{c}{1,400} = \frac{39}{700}$$

$$(\cancel{1,400})\frac{c}{\cancel{1,400}} = \frac{39}{\cancel{700}}(\overset{2}{\cancel{1,400}})$$

$$c = \$78$$

SECTION 3   PROBLEMS

Reduce the following ratios and express in each of the three forms.

1. a. 12 to 72
   b. 90 to 120
   c. $2,700 to $8,100
   d. $560 to $480
   e. 54,000 to 9,000

2. a. 8 to 21
   b. 150 to 450
   c. $1,000 to $2,500
   d. $400 to $640
   e. 36,000 to 16,000

Find the missing element in each proportion.

3. a. $\dfrac{x}{36} = \dfrac{3}{4}$
   b. $\dfrac{7}{21} = \dfrac{d}{60}$
   c. $\dfrac{14}{a} = \dfrac{8}{12}$
   d. $\dfrac{7}{4} = \dfrac{14}{m}$

4. a. $\dfrac{c}{14} = \dfrac{2}{7}$
   b. $\dfrac{3}{8} = \dfrac{r}{16}$
   c. $\dfrac{3}{g} = \dfrac{9}{24}$
   d. $\dfrac{15}{10} = \dfrac{9}{z}$

Use ratio or proportion to solve the following problems.

5. John made 20 sales calls to merchants last week, whereas Tom made 36. What was the ratio of calls for John to Tom?

6. Carlita read 8 books over spring break while Susanne read only 2. What was the ratio of books read by Carlita to books read by Susanne?

7. Systems Inc. had $120,000 in net sales and $144,000 in total sales for the year. What was the ratio of net sales to total sales?

8. In a beginning word processing class, 22 people had completed a previous keyboarding class whereas 4 had not. What was the ratio of students with previous keyboarding skills to those who had none?

9. The production department this month completed 96% of the number of jobs it completed last month. Last month's production was 1,600 jobs. How many jobs were completed this month?

10. The operating expenses of Randolph Motors were 28% of their net sales last month. Find the operating expenses if net sales were $66,000.

11. In a comparison test of two cars, Car A averaged 60% of the miles per gallon (mpg) that Car B averaged. How many mpg does Car B get if Car A averaged 18 mpg?

12. Fifteen percent of Pam's gross wages is deducted for an annuity. If her annuity deduction was $33, what was the amount of her gross wages?

13. The ratio of passing to failing grades received in a business statistics course was $5:2$. Find the number of passing grades if 6 failing grades were assigned.

14. The ratio of men to women in an aerobic program was $3:8$. If 6 men were enrolled, how many women were enrolled?

15. During the spring subscription compaign for a local public ratio station, the ratio of renewal subscribers to new subscribers was $8:5$. If 130 new subscriptions were received, how many renewals were received?

16. The ratio of business administration majors to marketing majors at a local community college was 5 to 2. If there were 240 business administration majors, how many marketing majors were there?

17. The ratio of fall to spring registrants for Alexandria City soccer was $3:4$. How many fall registrants were there if 488 people signed up for the spring?

18. The ratio of team sales to individual sales for warm-up suits was $2:7$ for Gray Sports Shop. If sales to individuals totaled 1,260, how many were sold to teams?

19. The Boles Construction Co. finds that the ratio of repair work to new construction is $7:4$. If there were 602 repairs during the past six-month period, how many new construction jobs did they have?

20. A computer store sold 54 packages of Brand X software. The ratio of sales of Brand X to Brand Y was $3:2$. How many Brand Y packages were sold?

21. A typist can type 3 pages of straight copy in 15 minutes. How many pages can she type in 1½ hours?

22. A secretary transcribes 900 words in 30 minutes. How many words can she transcribe in 3½ hours?

23. A doctor sees 2 patients in 30 minutes. At this rate, how many people can he see in a 7-hour day?

24. George packs 10 boxes in 45 minutes. How many boxes can he pack in 3 hours?

25. The laser printer just purchased by Graff & Associates will print 40 pages in 5 minutes. How many pages will it print in 2 hours?

26. A nurse counted 18 heartbeats in 15 seconds. At this rate, how many times will the heart beat in 1 minute?

27. After 6 hours, a truck driver had covered 312 miles. At this rate, how long will it take to complete his 728-mile trip?

28. On a vacation, a family covered a distance of 165 miles in 3 hours. At this rate, how long did it take them to complete their 1,100-mile trip?

CHAPTER 2 GLOSSARY

Coefficient.   A number that multiplies a variable or a quantity in a set of parentheses. [Example: 3 is a coefficient in $3x$ or in $3(x + z)$.]

Cross products.   The equal products obtained when each numerator of a proportion is multiplied by the opposite denominator. (Given $\frac{3}{4} = \frac{9}{12}$, the cross products $3 \cdot 12$ and $4 \cdot 9$ both equal 36.)

Equation.   A mathematical statement that two expressions (or an expression and a number) are equal.

Expression. Any indicated mathematical operation(s) written without an equal sign. $\left(\text{Examples: } 2x + 3; \quad \dfrac{3t}{5} - 1\right)$

First-degree equation in one variable. An equation with one distinct variable whose highest exponent (degree) is 1. (Example: $2x + 7 = x + 12$)

Proportion. A mathematical statement that two ratios are equal. (Example: $\frac{3}{4} = \frac{9}{12}$)

Ratio. A comparison of two (or more) numbers, frequently indicated by a common fraction. (Examples: 4 to 7;   4:7; $\frac{4}{7}$)

Rational number. A number that can be expressed as a common fraction (the quotient of two integers).

Reciprocal. The result obtained when a number is inverted by interchanging the numerator and the denominator. Or, that value which, when multiplied by the original value, yields 1. (Example: The reciprocal of $\frac{3}{4}$ is $\frac{4}{3}$.)

Similar terms. Terms that include the same variable. (Example: The expression $5x + x - 2x$ contains three similar terms.)

Terms. The products, quotients, or numbers in a mathematical expression that are separated by plus or minus signs. $\left(\text{Example: The expression } 4x - \dfrac{3x}{5} + 7 \text{ contains three terms.}\right)$

REVIEW OF PERCENT

OBJECTIVES

Upon completion of Chapter 3, you will be able to:

1. Define and use correctly the terminology associated with each topic.

2. **a.** Change a percent to its equivalent decimal or fraction (Section 1: Examples 1, 2; Problems 1–32).

   b. Change a decimal or fraction to its equivalent percent (Section 1: Examples 3, 4; Problems 33–64).

3. Use an equation to find the missing element in a percentage relationship (examples: What percent of 30 is 25?   60 is 120% of what number?) (Section 2: Example 1; Problems 1–36).

4. Use the basic equation form "__% of Original = Change?" to find the percent of change (increase or decrease) (example: $33 is what percent more than $27?) (Section 2: Example 2; Problems 37–48).

5. Use an equation to find the original number when the percent of change (increase or decrease) and the result are both known (example: What number increased by 25% of itself gives 30?) (Section 2: Example 3; Problems 49–60).

6. Given a word problem containing percents, express the problem in a concise sentence which translates into an equation that solves the problem (Section 3: Examples 1–5; Problems 1–50).

Percent is a fundamental topic with which everyone has some familiarity. We use a percent like a ratio to make a comparison. A ratio is often expressed as a fraction, such as $\frac{18}{24}$ or $\frac{3}{4}$, whereas a percent would express a relationship using 100 as the denominator of the fraction, such as $\frac{75}{100}$. Percent compares the number of parts out of 100 to which the fraction is equivalent.

Because percent is one of the most frequently applied mathematical concepts in all areas of business, it is extremely important for anyone entering business to be capable of accurate percent calculations.

For a further discussion of percents, fractions, and decimals, see Appendix A.

SECTION 1
BASIC PERCENT

The use of percent will be easier if you remember that **percent** means **hundredths.** That is, the expression "46%" may just as correctly be read "46 hundredths." The word "hundredths" denotes either a common fraction with 100 as denominator or a decimal fraction of two decimal places. Thus,

$$46\% = 46 \text{ hundredths} = \frac{46}{100} \quad \text{or} \quad 0.46$$

A percent must first be changed to either a fractional or decimal form before that percent of any number can be found. These conversions are simplified by applying the fact that "percent" means "hundredths."

A. CHANGING A PERCENT TO A DECIMAL

When people think of percent, they normally think of the most common percents, those between 1% and 99%. These are the percents that occupy the first two places in the decimal representation of a percent. (For example, 99% = 99 hundredths = 0.99, and 1% = 1 hundredth = 0.01.)

> **A.1** *When converting percents to decimals, write the whole percents between* 1% *and* 99% *in the first two decimal places.*

When this rule is followed, the other digits will then naturally fall into their correct places, as demonstrated next.

Example 1   **% to Decimal**

Express each of the following percents as a decimal.

(a)   5% = 5 hundredths = 0.05

(b)   86% = 86 hundredths = 0.86

(c)   16.3% = 0.163   (since 16% = 0.16)

(d)   131% = 1.31   (since 31% = 0.31)

(e) 122.75% = 1.2275   (since 22% = 0.22)

(f) 6.09% = 0.0609   (since 6% = 0.06)

(g) 0.4% = 0.004   (since 0% = 0.00)

(h) 0.03% = 0.0003   (since 0% = 0.00)

Some percents are written using common fractions. To change a fraction to a decimal, you must divide the numerator by the denominator.

$$\frac{n}{d}) \qquad \text{or} \qquad d\overline{)n}$$

A.2 *When a fractional percent is to be changed to a decimal, first convert the fractional percent to a decimal percent, and then convert the decimal percent to an ordinary decimal.*

(Machine calculations of the decimal percent may be rounded to the third decimal place.)

**Example 1
(cont.)**

Fractional % to Decimal

Express each of the following percents as a decimal.

(i) $\frac{7}{8}$%: Recall that $\frac{7}{8}$ means $8\overline{)7.000} = 0.875$ giving $.875$

And since $\frac{7}{8} = 0.875$

then $\frac{7}{8}\% = 0.875\% = 0.00875$

(j) $\frac{4}{7}$%: First, $\frac{4}{7}$ means $7\overline{)4.00} = 0.57\frac{1}{7}$ giving $.57\frac{1}{7}$ (or 0.571)

Then $\frac{4}{7} = 0.57\frac{1}{7}$

implies that $\frac{4}{7}\% = 0.57\frac{1}{7}\% = 0.0057\frac{1}{7}$ (or 0.00571)

A fractional percent always indicates a percent less than 1% (or it indicates that fractional part of 1%). Thus, $\frac{1}{4}\% = \frac{1}{4}$ of 1%; $\frac{3}{5}\% = \frac{3}{5}$ of 1%.

When the given fraction is a common one for which the decimal equivalent is known, the process of *converting a fractional percent to a decimal* may be shortened:

A.3 *Write two zeros to the right of the decimal point (0.00) to indicate that the percent is less than 1%, and then write the digits that are normally used to denote the decimal equivalent of the fraction.*

**Example 1
(cont.)**

(k) $\frac{1}{4}\%$: $\left(\frac{1}{4}\% = \frac{1}{4} \text{ of } 1\%\right)$

To show that the percent is less than 1%, write: 0.00

To indicate $\frac{1}{4}$, affix the digits "25" $\left(\frac{1}{4} = 25 \text{ hundredths}\right)$: 0.0025

Thus, $\frac{1}{4}\% = 0.0025$.

(This may be easily verified, since $4 \times 0.0025 = 0.0100 = 1\%$.)

(l) $\frac{2}{3}\%$: $\left(\frac{2}{3}\% = \frac{2}{3} \text{ of } 1\%\right)$

Indicate a percent less than 1%: 0.00

Affix the digits $66\frac{2}{3}$: $0.0066\frac{2}{3}$

B. CHANGING A PERCENT TO A FRACTION

The procedure for changing a percent to a fraction is summarized as follows:

B.1 *After an ordinary percent has been changed to hundredths (by dropping the percent sign and placing the number over a denominator of 100), the resulting fraction should be reduced to lowest terms.*

An improper fraction—one in which the numerator is greater than the denominator—is considered to be in lowest terms provided that it cannot be further reduced.

Example 2 **% to Fraction**

Convert each percent to its fractional equivalent in lowest terms.

(a) $45\% = 45 \text{ hundredths} = \frac{45}{100} = \frac{9}{20}$

(b) $8\% = 8 \text{ hundredths} = \frac{8}{100} = \frac{2}{25}$

(c) $175\% = 175 \text{ hundredths} = \frac{175}{100} = \frac{7}{4}$

Percents containing fractions may be converted to their ordinary fractional equivalents by altering the above procedure slightly. The fact that "percent" means "hundredths" may also be shown as

$$\% = \text{hundredth} = \frac{1}{100}$$

B.2 *Percents containing fractions are changed to ordinary fractions by substituting $\frac{1}{100}$ for the % sign and multiplying.*

Now let us apply this fact to change some fractional percents to common fractions.

Example 2
(cont.)

% with Fraction to Fraction

(d) $12\frac{1}{2}\% = 12\frac{1}{2} \times \frac{1}{100} = \frac{25}{2} \times \frac{1}{100} = \frac{25}{200} = \frac{1}{8}$

(e) $55\frac{5}{9}\% = 55\frac{5}{9} \times \frac{1}{100} = \frac{500}{9} \times \frac{1}{100} = \frac{500}{900} = \frac{5}{9}$

(f) $\frac{1}{3}\% = \frac{1}{3} \times \frac{1}{100} = \frac{1}{300}$

B.3 *Decimal percents may be converted to fractions by applying the following procedure:*

1. *Change the percent to its decimal equivalent.*
2. *Pronounce the value of the decimal; then write this number as a fraction and reduce it.*

Example 2
(cont.)

Decimal % to Fraction

(g) 22.5%:
   1. Since 22% = 0.22, then 22.5% = 0.225.
   2. $0.225 = 225$ thousandths $= \frac{225}{1,000} = \frac{9}{40}$.

(h) 0.56%: Since 0% = 0.00, then

$$0.56\% = 0.0056$$

$$= 56 \text{ ten-thousandths}$$

$$= \frac{56}{10,000}$$

$$= \frac{7}{1,250}$$

(i) 0.6%:

$$0.6\% = 0.006$$

$$= 6 \text{ thousandths}$$

$$= \frac{6}{1,000}$$

$$= \frac{3}{500}$$

C. CHANGING A DECIMAL TO A PERCENT

Just as "percent" means "hundredths," so is the opposite also true—that is, *hundredths = percent.* Rule C states an important concept that should be remembered when working with decimals and percents.

> **C.** *When a decimal number is to be changed to a percent, the hundredths places of the decimal indicate the whole "percents" between 1% and 99% (0.01 = 1%, and 0.99 = 99%).*

By first isolating the whole percents between 1% and 99%, you will have no difficulty in placing the other digits correctly.

If there is to be a decimal point in the percent, it will come after the hundredth's place of the decimal number. (This is why it is sometimes said: "To change a decimal to a percent, move the decimal point two places to the right and add a percent sign.")

Example 3   **Decimal to %**

Express each of these decimals as a percent.

(a) $0.67 = 67$ hundredths $= 67\%$
(b) $0.82\frac{1}{2} = 82\frac{1}{2}$ hundredths $= 82\frac{1}{2}\%$
(c) $0.08 = 8$ hundredths $= 8\%$
(d) 0.721:   Since $0.72 = 72\%$, then $0.721 = 72.1\%$.

It may prove helpful to circle the first two decimal places, as an aid in identifying the percents between 1% and 99%.

(e) 1.④⑤9:   Given 1.459, the .45 indicates 45%.

Thus, $1.459 = 145.9\%$.

(f)   0.$\textcircled{9}$:   Given 0.9, the .9 or .90 denotes 90%.

$$\text{So, } 0.9 = 90\%.$$

(g)   0.$\textcircled{00}$26:   Given 0.0026, the 0.00 represents 0%.

$$\text{Hence, } 0.0026 = 0.26\%.$$

D. CHANGING A FRACTION TO A PERCENT

The procedure for changing a fraction to a percent is summarized in Rule D.

D. *To change a fraction to a percent, apply the following steps:*

1. *Convert the fraction to its decimal equivalent by dividing the numerator by the denominator:*

$$\frac{n}{d}\overline{)}\qquad or \qquad d\overline{)n}$$

   (Machine calculations may be rounded to the third decimal place.)

2. *Then change the decimal to a percent, as illustrated in the preceding example.*

Example 4   **Fraction to %**

Change each fraction to its equivalent percent.

(a) $\dfrac{5}{8} = \dfrac{.625}{8\overline{)5.000}} = 0.625 = 62.5\%$

(b) $\dfrac{7}{18} = \dfrac{.38\frac{16}{18}}{18\overline{)7.00}} = 0.38\frac{8}{9} = 38\frac{8}{9}\%$ (or 38.9%)

(c) $\dfrac{5}{4} = \dfrac{1.25}{4\overline{)5.00}} = 1.25 = 125\%$

(d) $1\dfrac{4}{5}\left(\text{since }\dfrac{4}{5} = \dfrac{.80}{5\overline{)4.00}} = 0.80\right) = 1.80 = 180\%$

   or $1\dfrac{4}{5} = \dfrac{9}{5} = \dfrac{1.80}{5\overline{)9.00}} = 1.80 = 180\%$

(e) $2\dfrac{3}{4}\left(\text{since }\dfrac{3}{4} = 0.75\right) = 2.75 = 275\%$

   or $2\dfrac{3}{4} = \dfrac{11}{4} = 2.75 = 275\%$

SECTION 1   PROBLEMS

Express each percent both as a decimal and as a fraction in lowest terms.

1. 35%	**2.** 11%	**3.** 16%	**4.** 32%
5. 3%	**6.** 2%	**7.** 52.25%	**8.** 30.5%
9. 4.5%	**10.** 6.25%	**11.** 250%	**12.** 174%
13. 137.5%	**14.** 128.4%	**15.** 0.8%	**16.** 0.48%
17. 1.75%	**18.** 1.3%	**19.** 145%	**20.** 105%
21. ¾%	**22.** ⅛%	**23.** ⅖%	**24.** 1/20%
25. ⅜%	**26.** ⅘%	**27.** 1⅗%	**28.** 2¼%
29. 87½%	**30.** 16⅗%	**31.** 12½%	**32.** 8⅕%

Express each of the following as a percent.

33. 0.07	**34.** 0.09	**35.** 0.35	**36.** 0.41
37. 0.36	**38.** 0.52	**39.** 0.165	**40.** 0.232
41. 2.11	**42.** 1.45	**43.** 1.06	**44.** 4.38
45. 0.001	**46.** 0.022	**47.** 0.5	**48.** 0.8
49. 0.005	**50.** 0.008	**51.** 3.1	**52.** 2.88
53. 4.6	**54.** 5.5	**55.** 0.03	**56.** 0.01
57. 4/9	**58.** ⅚	**59.** ⅗	**60.** 5/12
61. 1¾	**62.** 3½	**63.** 4⅘	**64.** 1⅛

SECTION 2

PERCENT EQUATION FORMS

All problems that involve the use of percent are some variation of the basic percent form: "Some percent of one number equals another number." This basic **percent equation form** may be abbreviated as:

$$\underline{\quad}\% \text{ of } \underline{\quad} = \underline{\quad}$$

Given in reverse order, the equation form is:

$$\underline{\quad} = \underline{\quad}\% \text{ of } \underline{\quad}$$

Another way of expressing the preceding formula is:

$$\text{rate} \times \text{base} = \text{percentage} \qquad (r \cdot b = p)$$

Because there can be only one unknown in an elementary equation, there are only three variations of the basic equation form: The unknown can be either (1) the percent (or **rate**), (2) the first number (or **base**), or (3) the second number (the **percentage**). For instance, consider

$$70\% \text{ of } 52 = 36.4$$

Here, 70% is the rate, 52 is the base, and 36.4 is the percentage. (This is somewhat confusing because the percentage is not a percent; the rate is the percent. This confusion, however, is avoided by using the percent equation form above.)

Equation-solving procedures that were studied earlier will be applied to solve percentage problems. *Recall that before a percent of any number can be computed, the percent must first be changed to either a fraction or a decimal.* Conversely, if the unknown represents a percent, the solution to the equation will be a decimal or fraction that must then be converted to the percent.

Example 1

(a)   16% of 45 is what number?

$$0.16 \times 45 = n$$

$$7.2 = n$$

(b)   What percent of 30 is 6?

$$r \times 30 = 6$$

$$30r = 6$$

$$\frac{30r}{30} = \frac{6}{30}$$

$$r = \frac{1}{5}$$

$$r = 20\%$$

(c) $66\frac{2}{3}\%$ of what number is 12?

Because $66\frac{2}{3}\%$ is exactly $\frac{2}{3}$, we will use $\frac{2}{3}$:

$$\frac{2}{3} \times n = 12$$

$$\frac{2n}{3} = 12$$

$$\frac{3}{2} \times \frac{2n}{3} = 12 \times \frac{3}{2}$$

$$n = 18$$

Many of the formulas used to solve mathematical problems in business are nothing more than the basic percent equation form with different words (or variables) substituted for the percent and the first and second numbers. The methods used to solve the formulas are identical to those of Example 1.

One such formula is used to find **percent of change** (that is, percent of increase or decrease). In words, the formula can be stated, "What percent of the original number is the change?" In more abbreviated form it is

$$\_\_\_\_\% \text{ of Original} = \text{Change?}$$

Example 2

(a)   What percent more than 24 is 33?

1.   *Original number.*   To have "more than 24" implies that we originally had 24; thus, 24 is the original number. In general, it can be said that the "original number" is the number that follows the words "more than" or "less than" in the stated problem.

2.   *Change.*   The "change" is the amount of increase or decrease that occurred. That is, the change is the numerical difference between the two numbers in the problem. (The fact that it may be a negative difference is not important to this formula.) Thus, the change in this example is $33 - 24 = 9$.

$$\_\_\_\_\% \text{ of Original} = \text{Change}$$

$$\_\_\_\_\% \text{ of } \qquad 24 = 9$$

$$24x = 9$$

$$\frac{24x}{24} = \frac{9}{24}$$

$$x = \frac{3}{8}$$

$$x = 37\tfrac{1}{2}\%$$

Thus, 24 increased by $37\tfrac{1}{2}\%$ of itself gives 33.

(b)   12 is what percent less than 18?

Original = 18	$\_\_\_\_\% \text{ of Original} = \text{Change}$
Change = 18 − 12	$18r = 6$
= 6	$\frac{18r}{18} = \frac{6}{18}$
	$r = \tfrac{1}{3}$
	$r = 33\tfrac{1}{3}\%$

Thus, 18 decreased by $33\tfrac{1}{3}\%$ of itself is 12.

This formula is often applied to changes in the prices of merchandise or stocks and to changes in volume of business from one year to the next. Other applications include increase in cost of living or unemployment and decrease in expenses or net profit.

A third type of problem is a variation of the "percent of change" type. In this application we know the percent of change (percent increase or decrease) and the result, and we compute the amount of the original number. The original number is either increased or decreased by a percent of itself. The number is expressed as 100%n or 1n so that the variables can be combined.

Example 3   (a)   What number decreased by 12% of itself gives 66?

$$\underline{\text{What number}} \text{ decreased by } \underline{(12\% \text{ of itself})} \text{ gives } \underline{66}?$$

$$n \qquad - \qquad (0.12 \times n) \quad = \quad 66$$

$$n* \qquad - \qquad 0.12n \quad = \quad 66$$

$$0.88n \quad = \quad 66$$

$$\frac{\cancel{0.88}n}{\cancel{0.88}} \quad = \quad \frac{66.00}{0.88}$$

$$n \quad = \quad 75$$

(b)   What number increased by $33\frac{1}{3}\%$ of itself gives 28?

When the percent contains a repeating decimal, it is easier to work with the fraction equivalent instead of the decimal equivalent.

$$\frac{3}{3}n + \frac{1n}{3} = 28$$

$$\frac{4n}{3} = 28$$

$$\frac{\cancel{3}}{\cancel{4}} \times \frac{\cancel{4}n}{\cancel{3}} = \overset{7}{\cancel{28}} \times \frac{3}{\cancel{4}}$$

$$n = 21$$

*Recall that $n = 1n$; thus,

$$n - 0.12n = 1n - 0.12n$$

$$= 1.00n - 0.12n$$

$$= 0.88n$$

SECTION 2 PROBLEMS

Use equations to obtain the following solutions. Express percent remainders either as fractions or rounded to hundredths (example: $28\frac{4}{7}$% or 28.57%).

Part One

1. 5% of 120 is what amount?
2. 8% of 700 is what number?
3. 12% of 900 is what number?
4. 20% of 640 is how much?
5. 25% of 78 is how much?
6. 42% of 90 is what amount?
7. What is $\frac{3}{5}$% of 37,000?
8. What is $\frac{3}{4}$% of 6,000?
9. What is $37\frac{1}{2}$% of 88?
10. How much is $12\frac{1}{2}$% of 560?
11. What is $4\frac{1}{2}$% of 600?
12. $\frac{1}{4}$% of 14,000 is what amount?
13. What percent of 48 is 12?
14. What percent of 90 is 63?
15. 18 is what percent of 96?
16. What percent of 100 is 13?
17. 28 is what percent of 80?
18. 240 is what percent of 4,000?
19. 5.4 is what percent of 180?
20. 45 is what percent of 900?
21. What percent of 9 is 2?
22. What percent of 18 is 4.5?
23. 15% of what number is 48?
24. 120% of what amount is 78?
25. 44% of what number is 33?
26. $11\frac{1}{2}$% of what number is 115?
27. 18 is $2\frac{1}{4}$% of what amount?
28. 90 is $33\frac{1}{3}$% of what amount?
29. $42\frac{6}{7}$% of what number is 9?
30. $16\frac{2}{3}$% of what amount is 45?
31. 84% of what amount is 105?
32. 100 is $1\frac{1}{4}$% of what amount?
33. 49.5 is 5.5% of what number?
34. 3.5 is 0.7% of what number?
35. 4.5 is 0.6% of what amount?
36. 48 is $22\frac{2}{9}$% of what amount?

Part Two

37. What percent more than 5 is 7?
38. What percent more than 150 is 180?
39. 44 is what percent less than 55?
40. What percent less than 150 is 105?
41. 130 is what percent less than 325?
42. 195 is what percent less than 260?
43. 132 is what percent more than 108?
44. 300 is what percent more than 200?
45 $54 is what percent more than $42?
46. $210 is what percent more than $180?
47. What percent less than $600 is $597?
48. What percent less than $800 is $794?

Part Three

49. What number increased by 30% of itself gives 78?
50. What number increased by 25% of itself gives 90?
51. What amount decreased by 25% of itself gives 36?
52. What amount decreased by $11\frac{1}{9}$% of itself gives 48?
53. What number increased by 50% of itself gives 108?
54. What number increased by 26.5% of itself gives 253?
55. What amount decreased by $16\frac{2}{3}$% of itself gives 540?
56. What amount decreased by $33\frac{1}{3}$% of itself gives 150?
57. What number increased by $83\frac{1}{3}$% of itself gives 121?
58. What number increased by 0.4% of itself gives 1,004?
59. What amount decreased by $11\frac{1}{9}$% gives 56?
60. What amount decreased by $37\frac{1}{2}$% gives 120?

WORD PROBLEMS USING PERCENTS

Three primary types of percent applications are illustrated by the following five examples. These types are:

1. The basic percent equation form: ____% of ____ = ____ (Examples 1–3).
2. Finding *percent* of change (increase or decrease), using ____% of Original = Change (Example 4).
3. Finding an *amount* that has been increased or decreased by a percent of itself (Example 5).

Example 1 A direct-mail campaign by a magazine produced $16,000 in renewals and new subscriptions. If renewals totaled $10,000, what percent of the subscriptions were renewals?

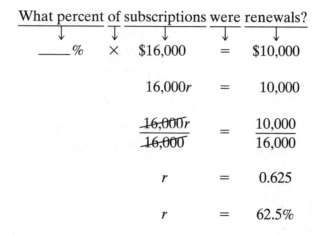

$$\text{What percent of subscriptions were renewals?}$$
$$\underline{\quad}\% \ \times \ \$16,000 \ = \ \$10,000$$
$$16,000r \ = \ 10,000$$
$$\frac{\cancel{16,000}r}{\cancel{16,000}} \ = \ \frac{10,000}{16,000}$$
$$r \ = \ 0.625$$
$$r \ = \ 62.5\%$$

Example 2 Thirty-five percent of a payment made to a partnership land purchase was tax deductible. If a partner receives a notice that he qualifies for a $700 tax deduction, how much was his total payment?

$$\text{35\% of payment was tax deduction}$$
$$35\% \times \quad p \quad = \quad \$700$$
$$0.35p \ = \ 700$$
$$\frac{\cancel{0.35}p}{\cancel{0.35}} \ = \ \frac{700.00}{0.35}$$
$$p \ = \ \$2,000$$

Example 3   A quarterback completed 40% of his attempted passes in a professional football game. If he attempted 65 passes, how many did he complete?

$$\underline{40\% \text{ of }} \underline{\text{attempted passes}} \underline{\text{ were }} \underline{\text{completed passes}}$$

$$0.4 \times \quad 65 \quad = \quad c$$

$$0.4(65) \quad = \quad c$$

$$26 \quad = \quad c$$

Example 4   Last year's sales were $45,000; sales for this year totaled $48,600. What was the percent of increase in sales?

Original = $45,000   _____% of Original = Change?

Change = $48,600 − $45,000   _____% of $45,000 = $3,600

$\quad\quad = \$3,600$ $45,000r = 3,600$

$$\frac{\cancel{45,000}r}{\cancel{45,000}} = \frac{3,600}{45,000}$$

$$r = 0.08$$

$$r = 8\%$$

Example 5   Labor expenses (wages) at a construction project increased by 10% this month. If wages totaled $13,200 this month, how much were last month's wages?

$$\underline{\text{Old wages}} \underline{\text{ increased by }} \underline{10\% \text{ (of itself)}} \underline{\text{ equals }} \underline{\text{new wages}}$$

$$w \quad + \quad (10\% \times w) \quad = \quad \$13,200$$

$$w \quad + \quad 0.1w \quad = \quad 13,200$$

$$\frac{\cancel{1.1}w}{\cancel{1.1}} \quad = \quad \frac{13,200.0}{1.1}$$

$$w \quad = \quad \$12,000$$

Note.   Percents are never used just by themselves. In an equation, any percent that is included must be a percent *of* ("times") some other number or variable.

For convenience, the following problems are divided into two parts, each of which covers all the given examples. Part 1 covers the three types of problems presented: (1) Examples 1–3, see Problems 1–16; (2) Example 4, see Problems 17–22; (3) Example 5, see Problems 23–32. Additional practice is provided in Part 2, but the order is scrambled.

PERCENT EQUIVALENTS OF COMMON FRACTIONS

The following percents (many with fractional remainders) will be given in many problems so that students without calculators can conveniently use the exact fractional equivalents. (Students using calculators may also use the fractional equivalents by entering the numerators and denominators separately.)

$$\frac{1}{2} = 50\%$$

$$\frac{1}{3} = 33\frac{1}{3}\% \qquad \frac{2}{3} = 66\frac{2}{3}\%$$

$$\frac{1}{4} = 25\% \qquad \frac{3}{4} = 75\%$$

$$\frac{1}{5} = 20\% \qquad \frac{2}{5} = 40\% \qquad \frac{3}{5} = 60\% \qquad \frac{4}{5} = 80\%$$

$$\frac{1}{6} = 16\frac{2}{3}\% \qquad \frac{5}{6} = 83\frac{1}{3}\%$$

$$\frac{1}{7} = 14\frac{2}{7}\% \qquad \frac{2}{7} = 28\frac{4}{7}\% \qquad \frac{3}{7} = 42\frac{6}{7}\%$$

$$\frac{1}{8} = 12\frac{1}{2}\% \qquad \frac{3}{8} = 37\frac{1}{2}\% \qquad \frac{5}{8} = 62\frac{1}{2}\% \qquad \frac{7}{8} = 87\frac{1}{2}\%$$

$$\frac{1}{9} = 11\frac{1}{9}\% \qquad \frac{2}{9} = 22\frac{2}{9}\% \qquad \frac{4}{9} = 44\frac{4}{9}\% \qquad \frac{5}{9} = 55\frac{5}{9}\%$$

$$\frac{1}{10} = 10\% \qquad \frac{3}{10} = 30\% \qquad \frac{7}{10} = 70\% \qquad \frac{9}{10} = 90\%$$

$$\frac{1}{12} = 8\frac{1}{3}\%$$

SECTION 3   PROBLEMS

Part One

1. A real estate agent earned a commission of $5,760 on a house sale of $144,000. What percent of the house sale was the commission?

2. A yogurt store's overhead last quarter was $6,750 while sales were $15,000. The store's overhead was what percent of its sales?

3. An office manager finds that she spends 3 hours each day on personnel problems. If she works an 8-hour day, what percent of her day is spent on personnel problems?

4. Out of its total budget of $60,000, the human resources department spent $51,600 on wages. What percent of the budget was spent on wages?

5. A salesclerk makes a 7% commission on his net sales. If his net sales last week were $5,618, what was his commission?

6. Income from advertisements placed in the Montana Ledger newspaper amounted to 75% of total revenue. What were the receipts from advertising if revenues totaled $150,000 last month?

7. A taxpayer pays 1½% of the assessed value of his real estate for real property tax. If the assessed value of the property is $75,000, how much must he pay in property tax?

8. Jana spends about 12½% of her weekly wages for lunches. If she earns $360 a week, how much does she spend on average for lunches?

9. A candidate in a state senate race won 52% of the votes cast. If the candidate received 64,272 votes, how many votes were cast for all candidates in this race?

10. Charles receives 20% of his income from investments. What was his total income, if his investment income was $10,200?

11. There are 16 employees at Cranshaw Co. who chose the group health insurance plan. This number represents 33⅓% of all Cranshaw employees. How many people are employed by the company?

12. Inventory represents 60.25% of a hardware store's total assets. What are the total assets if the inventory is valued at $180,750?

13. Eight percent of a stock's cost was paid in dividends last year. If the stock cost $85, what dividend did it yield?

14. A collection agency charges 33⅓% of accounts receivable collections as a fee. The agency collected $186,000 last month. What fee did it earn for these collections?

15. Thirty percent of the selling price of a battery charger is the markup. If the markup is $7.50, how much is the selling price?

16. The raw material cost to manufacture a pair of safety boots is $10. The raw materials run 22⅔% of the total production cost. What is the total cost?

17. The population of Jackson City in 1980 was 54,000. By 1990, the population had jumped to 78,000. What percent increase does this represent?

18. Monthly cable charges last year were $36 while the same charges this year amount to $37.98. What percent increase does this represent?

19. Accounting majors who were hired after graduating last spring received an average of $27,000. Accounting majors this year received an average of $27,810. What percent increase does this represent?

20. Jack made 12 account calls on Monday and 8 calls on Tuesday. What percent decrease in calls is this?

21. At a holiday sale, the price of a dress was $76.80. Before the sale, it had a list price of $96. What was the percent decrease in price?

22. Sandy's travel expense vouchers totaled $570 last month compared to $595.65 this month. What is the percent increase in her travel expenses?

23. What number increased by 25% of itself yields 35?

24. What number increased by 40% of itself yields 4,200?

25. What amount decreased by 16⅔% of itself leaves 25?

26. What amount decreased by 2% of itself equals $19.60?

27. It is estimated that by the year 2000 there will be 39 million Americans over the age of 65. This number represents a 30% increase from the year 1990. How many people were over 65 in 1990?

28. Within the past two years, the average price of a haircut has increased by 11⅛%. If a haircut costs $20 today, what was the cost two years ago?

29. A local union's membership decreased by 12% over a 5-year period. At the end of this period, there were 5,632 members. How many members were there at the beginning of the period?

30. During a special promotion, the price of a multimedia system was reduced by 30%. What was the price before the reduction if the sales price was $2,100?

31. The current selling price of a blender is 15% higher than it was last year. If this year's price is $41.40, what was the price last year?

32. This year, the Simmons Co. showed $93,960 in current liabilities on its annual balance sheet. This was an 8% increase from the previous year. What was the total of the current liabilities for the previous year?

Part Two

33. Thrifty Stores Inc. estimates that ½% of its credit sales of $243,000 will be uncollectible. How much is its estimated uncollectible accounts expense?

34. At its grand opening, Rossbach Co. estimated that 16⅔% of its cutomers were teenagers. Nine hundred and thirty customers visited the store during this time. How many were teenagers?

35. The average price for a new home in Crystal City decreased from $120,000 to $112,800. What was the percent decrease in the price of a home?

36. The cost of coffee to a grocery store has increased from $200 per case to $206 per case over the past six months. What is the percent increase in cost?

37. The price of a photocopier was reduced by 25%, making the sales price $600. What was the price before the reduction?

38. Monthly expenses for courier service at Campanella Co. have increased by 4½% since last year. The current month's expense is $209. How much was the courier expense during the same month last year?

39. Twelve percent of merchandise sold was returned to the store. If $5,520 of goods was returned, how much were the total sales?

40. Seventeen percent of recent graduates of Parker College had secured jobs before graduation. If 68 people had jobs before graduation, how many graduates were there?

41. Enrollment in a tax seminar decreased from 130 to 117 over a 2-year period. What percent change does this represent?

SUMMARY OF CONVERSIONS

Given:	To Convert to:	
PERCENT	A. DECIMAL	B. FRACTION
Basic % (1% to 99%)	Basic percents become the first two (hundredths) decimal places. (Other digits align accordingly.) (Example: 76% = 76 hundredths = 0.76)	Percent sign is dropped and given number is placed over a denominator of 100; reduce. (Example: 76% = 76 hundredths = $\frac{76}{100}$ = $\frac{19}{25}$)
Decimal % (125.73%; 0.45%)	Digits representing percents greater than 99% or less than 1% are aligned around the first two (hundredths) decimal places (see also Section 1, "Basic Percent"). (Example: 0.45% = 0.0045)	Convert the percent to a decimal (as indicated at left). Pronounce the decimal value; then write this as a fraction and reduce. (Example: 0.45% = 0.0045 = 45 ten-thousandths = $\frac{45}{10,000}$ = $\frac{9}{2,000}$)
Fractional % ($\frac{1}{4}$%; $\frac{7}{5}$%; $1\frac{2}{3}$%)	Convert the fractional percent to a decimal percent (by dividing the numerator by the denominator). Then convert this decimal percent to an ordinary decimal (as shown above). (Example: $\frac{7}{5}$% = 1.4% = 0.014)	Percent sign is replaced by $\frac{1}{100}$; multiply and reduce. (Example: $\frac{7}{5}$% = $\frac{7}{5} \times \frac{1}{100}$ = $\frac{7}{500}$)

	To Convert to Percent
C. *Decimals* (0.473; 1.2; 0.0015)	Isolate the first two (hundredths) decimal places to identify the basic percent (1% to 99%). Other digits will then be aligned correctly. (Example: 0.473 = 0.④⑦3 = 47.3%)
D. *Fractions* ($\frac{2}{9}$; $\frac{7}{3}$; $1\frac{1}{2}$)	Convert the fraction to a decimal (by dividing the numerator by the denominator). Then follow the above procedure for converting a decimal to percent. (Example: $\frac{2}{9}$ = 0.22$\frac{2}{9}$ = 0.㉒$\frac{2}{9}$ = 22$\frac{2}{9}$%)

42. The price of an oriental rug dropped from $5,400 to $3,294. What percent decrease is this?

43. The value of a store's inventory increased from $50,000 to $62,000. By what percent did the value increase?

44. Orders for a statistics textbook increased from 4,800 to 5,760 at a publishing company this year. What was the percent increase in textbooks ordered?

45. A car dealership saw a 3.8% increase in sales of new cars in May over April. If 519 new cars were sold in May, how many were sold in April?

46. The price of the latest edition of a word-processing software package increased 8⅓% over a previous edition. The new software costs $104. What was the previous edition's price?

47. The owner of a small business has total assets of $840,000 while her net worth is $336,000. What percent of her total assets is her net worth?

48. Office equipment purchased by an accounting firm for $6,000 has been depreciated by $4,000. What percent of the purchase price has been depreciated?

49. Find a newspaper or magazine article that includes percents, and bring it to class for discussion. Does the use of percents clarify or enhance the information in the article? Why or why not? Are the percents presented in a way that would enable you to compute additional data not specifically stated in the article?

50. Continue with the newspaper or magazine article from Problem 49. Create a word problem using information from the article, and solve the problem using techniques you have used for earlier problems in this assignment. Repeat this with a second word problem.

CHAPTER 3   GLOSSARY

Base.   The number that a percent multiplies: the first number in the basic percent equation form. (The base represents 100%.)

Hundredths.   Equals percent; a value associated with the first two decimal places or with a common fraction having a denominator of 100.

Percent.   Means hundredths; indicates a common fraction with a denominator of 100 or a decimal fraction of two decimal places.

Percentage.   A part of the base (a number or an amount) that is determined by multiplying the given percent (rate) times the base. (See the accompanying figure.)

Percent equation form.   The basic equation, ____% of ____ = ____; in words, what % of a first number (base) is a second number (percentage)?

Percent of change.   The equation, ____% of original = change; a variation of the basic equation form, used to find a percent of increase or decrease.

Rate.   The percent that multiplies a base. (See the accompanying figure.)

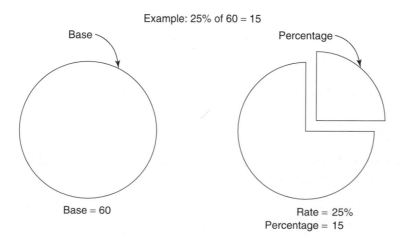

Example: 25% of 60 = 15

Base

Base = 60

Percentage

Rate = 25%
Percentage = 15

Accounting Mathematics

BASIC STATISTICS AND GRAPHS

OBJECTIVES

Upon completion of Chapter 4, you will be able to:

1. Define and use correctly the terminology associated with each topic.

2. Compute the three simple averages—mean, median, and mode (Section 1: Examples 1–7; Problems 1–18).

3. Use a frequency distribution to determine mean, median, and modal class for grouped data (Section 2: Examples 1, 2; Problems 1–8).

4. Using data that assume a normal curve distribution (Section 3):

   a. Determine the percents associated with various combinations of intervals $\pm 1\sigma$, $\pm 2\sigma$, and $\pm 3\sigma$ (Example 1; Problems 1–14)

   b. Compute standard deviation for ungrouped data, using the formula

   $$\sigma = \sqrt{\frac{\sum d^2}{n}}$$

   (Example 2; Problems 5–14).

5. Interpret index numbers and compute simple price relatives (Section 4: Examples 1, 2; Problems 1–6).

6. Construct the common types of graphs—bar, line, and circle (Section 5: Examples 1, 2; Problems 1–12).

Decisions about what course a business may take are rarely made without first consulting records indicating what past experience has been. When numerical records are involved, these data must be organized before they can present meaningful information. The process of collecting, organizing, tabulating, and interpreting numerical data is called **statistics.** The information presented as a result of this process is also known, collectively, as **statistics.** Statistics are invaluable to business in making comparisons, indicating trends, and analyzing facts affecting operations, production, profits, and so forth. However, the significance of statistics can often be realized more easily when the information is displayed in graph form. Thus, graphs are an integral part of statistics.

SECTION 1
AVERAGES

If anyone attempted to analyze a large group of numbers simultaneously, it would usually be impossible to obtain much useful information. Thus, it is customary to condense such a group into a single number representative of the entire group. This representative number is generally some central value around which the other numbers seem to cluster; for this reason, it is called a **measure of central tendency** or an **average.** The three most common averages are the **arithmetic mean** (or just the "mean"), the **median,** and the **mode,** all of which we will consider briefly.

MEAN

The **arithmetic mean** is the common average with which you are already familiar. It is the average referred to in everyday expressions about average temperature, average cost, the average grade on a quiz, average age, and so on. The mean is found by adding the various numbers and dividing this sum by the number of items.

Example 1   The five departments at Smith Sporting Goods made sales of the following amounts: A, $1,050; B, $1,200; C, $1,350; D, $1,800; E, $1,500. What were the average (mean) sales per department?

$$\frac{\$1,050 + \$1,200 + \$1,350 + \$1,800 + \$1,500}{5} = \frac{\$6,900}{5} = \$1,380$$

The mean sales per department were $1,380. Notice that there was not a single department that had sales of exactly $1,380. It is quite common for the "average" of a group of numbers to be a value not actually contained in the group. However, this mean is still representative of the entire group in that it is similar to most of the values in the group—some are somewhat larger and others somewhat smaller.

Further observe that if each department had made sales of $1,380, the total sales would also have been $6,900. This kind of observation applies to any situation in which an arithmetic mean is utilized.

WEIGHTED MEAN

The ordinary mean will present a true picture only if the data involved are of equal importance. If they are not, a weighted mean is needed. A **weighted arithmetic mean** is found by "weighting" or multiplying each number according to its importance. These products are then added and that sum is divided by the total number of weights.

A weighted average of particular importance to students is the **grade-point average.**

Example 2   Listed below are Carol Dickerson's course credit hours and term grades for the spring quarter at West Ridge Community College. Each A carries 4 quality points; a B, 3 points; a C, 2 points; and a D, 1 point. Compute Carol's grade-poing average (average per credit hour) for the quarter.

CREDIT HOURS	FINAL GRADE	QUALITY POINTS
4	A	4
3	C	2
3	B	3
3	C	2
2	C	2

If we simply computed the average of Carol's quality points,

$$\frac{4 + 2 + 3 + 2 + 2}{5} = \frac{13}{5} = 2.6$$

the result would be misleading, because her courses are not of equal importance from a credit standpoint. We must therefore weight the quality points for each course according to the number of credit hours each course carries, as follows:

CREDIT HOURS		QUALITY POINTS		
4	×	4	=	16
3	×	2	=	6
3	×	3	=	9
3	×	2	=	6
+ 2	×	2	=	4
15				41

$$\frac{41}{15} = 2.733\ldots = 2.73$$

Carol's grade-point average for the quarter is thus 2.73. This is somewhat better than the 2.6 obtained by the ordinary averaging method.

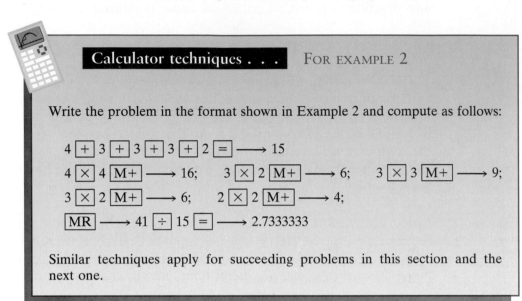

Calculator techniques . . .   FOR EXAMPLE 2

Write the problem in the format shown in Example 2 and compute as follows:

4 $+$ 3 $+$ 3 $+$ 3 $+$ 2 $=$ ⟶ 15

4 $\times$ 4 M+ ⟶ 16;   3 $\times$ 2 M+ ⟶ 6;   3 $\times$ 3 M+ ⟶ 9;

3 $\times$ 2 M+ ⟶ 6;   2 $\times$ 2 M+ ⟶ 4;

MR ⟶ 41 $\div$ 15 $=$ ⟶ 2.7333333

Similar techniques apply for succeeding problems in this section and the next one.

Example 3   The same model of microwave oven is available from several appliance stores. Prices and number of ovens sold are shown below. What is (a) the average price per store, and (b) the average price per oven sold?

STORE	NUMBER SOLD	PRICE
1	40	$450
2	60	400
3	120	350
4	80	375

It may help to first analyze the question. The word "per" indicates the division line of a fraction. The word following "per" tells which total number of objects to divide by. Hence, average price "per store" is found by dividing by the total number of stores, and average price "per oven" is found by dividing by the total number of ovens.

(a)   Average price per store $= \dfrac{\text{Sum of prices at all stores}}{\text{Total number of stores}}$

$$\frac{\$450 + \$400 + \$350 + \$375}{4} = \frac{\$1{,}575}{4}$$

$$= \$393.75 \qquad \text{Average price per store}$$

(b)   Average price per oven = $\dfrac{\text{Sum of prices of all ovens}}{\text{Total number of ovens}}$

For this average, the various prices must be weighted according to the number of ovens sold at each price.

NUMBER		PRICE		
40	×	$450	=	$ 18,000
60	×	400	=	24,000
120	×	350	=	42,000
+ 80	×	375	=	30,000
300				$114,000

$$\frac{\$114,000}{300} = \$380 \qquad \text{Average price per oven}$$

Example 4   David Cohen invested $5,000 in a business on January 1. On May 1, he withdrew $200; on July 1, he reinvested $1,000; and on August 1, he invested another $1,400. What was his average investment during the year?

$$\text{Average investment} = \frac{\text{Sum of investments of all months}}{\text{Total number of months}}$$

Before we can find the average investment, we must first determine the new balance after each change. Each succeeding balance must then be multiplied (weighted) by the number of months it remained invested.

DATE	CHANGE	AMOUNT OF INVESTMENT		MONTHS INVESTED		
January 1		$5,000	×	4	=	$20,000
May 1	− $200	4,800	×	2	=	9,600
July 1	+1,000	5,800	×	1	=	5,800
August 1	+1,400	7,200	×	5	=	36,000
				12		$71,400

$$\frac{\$71,400}{12} = \$5,950 \qquad \text{Average investment}$$

MEDIAN

Even when a group of numbers are of equal importance, the arithmetic mean may not convey an accurate description if the group contains a few values which differ greatly from the others. In such a case, the median may be the better average (or measure of central tendency). The **median** is the midpoint (or middle number) of a group of numbers. It is the central value of an array (or listing of data).

In order to find the median, we must arrange the numbers in order of size (such an arrangement is called an **ordered array**). If the group contains an odd number of entries, the median is the exact middle number. Its position can be determined quickly by first dividing the number of entries by 2. Then, the next whole number larger than the quotient indicates the position held by the median when the numbers are ordered in a series from smallest to largest.

When a group contains an even number of values, the median is again found by first dividing the number of entries by 2. The median is then the arithmetic mean (average) of the number in this position and the next number, when the numbers are ordered from smallest to largest.

Example 5   The Wingate Insurance Agency has seven agents with the following respective years of experience: 5, 3, 22, 8, 6, 10, and 30. What is the median years of experience of these agents?

Notice that in this case the mean years of experience,

$$\frac{5 + 3 + 22 + 8 + 6 + 10 + 30}{7} = \frac{84}{7} = 12 \text{ years}$$

does not give a very good indication of central tendency, since only two of the seven agents have had at least that much experience. The median is more descriptive:

Since there are seven numbers in the group, divide 7 by 2: $\frac{7}{2} = 3+$

The median is the next larger (or fourth) number when the numbers are arranged in increasing order of size. Hence, in the series, 3, 5, 6, $\underline{8}$, 10, 22, 30, the median is 8 years of experience.

Example 6   Wilcox & Co., a clothing store, employs six people. Their monthly salaries are $1,240, $1,350, $1,160, $1,400, $1,450, and $2,400. What is the median salary at the store?

$2,400
1,450
1,400⎫
1,350⎭
1,240
1,160

Since there are six employees, there is no exact middle number. The median is the arithmetic mean of the third ($\frac{6}{2} = 3$) and fourth salaries:

$$\frac{\$1,400 + \$1,350}{2} = \frac{\$2,750}{2} = \$1,375$$

The median salary is $1,375. (The mean salary, however, is $1,500.)

A major difference between the mean and the median is that all the given values have an influence on the mean, whereas the median is determined basically by position. The median is often an appropriate measure of central tendency where the items can be ranked. However, if the number of items is quite large, it may be difficult to prepare the array necessary for determining the median. An alternative measure sometimes used in such circumstances is the **modified mean,** in which the mean is computed in the usual manner after the extremely high and/or low scores have been deleted. If each highest and lowest score were deleted in Examples 5 and 6, the modified means would be 10.2 years of experience and $1,360 per month, both of which are much nearer to the median scores than were the ordinary means.

MODE

The **mode** of a group of numbers is the value (or values) that occurs most frequently. If all the values in a group are different, there is no mode. A set of numbers may have two or more modes when these all occur an equal number of times (and more often than the other values).

The mode may be quite different from the mean and the median, and it does not necessarily indicate anything about the quality of the group. Like the median, the mode is not influenced by all of the given values, and it may be very erratic. Unlike the other averages, however, the mode always represents actual items from the given data. The mode's principal application to business, from a production or sales standpoint, is in indicating the most popular items.

Example 7   The employees of Noland Construction Co. worked the following numbers of hours last week: 40, 48, 25, 36, 43, 48, 45, 60, 52, 43, 56, 54, 48, 33, and 53. What is the modal number of hours worked?

The numbers of hours worked, arranged in numerical order, are

60
56
54
53
52
48 ⎫
48 ⎬   There are three employees who worked exactly 48 hours
48 ⎭   last week; hence, 48 is the mode. If there were only two
45   employees with 48-hour weeks, then 48 and 43 would both
43   be modes, since these numbers would each be worked by
43   two employees.
40
36
33
25

SECTION 1   PROBLEMS

Find (1) the arithmetic mean, (2) the median, and (3) the mode for each of the following groups of numbers.

1. a. 86, 71, 29, 59, 71, 88, 44, 71, 70, 65, 29, 37, 47
   b. 1050, 1120, 956, 1000, 1050, 981, 986, 1040, 1200, 990, 1250, 1050, 995, 989, 981, 1114

2. a. 10, 22, 29, 22, 45, 30, 43, 18, 39, 28, 59, 41, 42, 40, 12
   b. 170, 154, 120, 133, 142, 161, 115, 121, 130, 154

3. By the end of the semester, Janice earned the following grades. She earns 4 quality points for each A, 3 points for each B, 2 points for each C, and 1 point for each D.
   a. What were her average (mean) quality points per class?
   b. What was her quality-point average for the semester (her average per credit hour)?

CREDIT HOURS	GRADES
5	C
4	A
4	B
3	C
1	A

4. The credit hours and final grades for Julio's four courses are listed below. He receives 4 quality points for each A, 3 points for each B, 2 points for each C, and 1 point for each D.
   a. What are Julio's average (mean) quality points per class?
   b. What is his quality-point average for the semester (his average per credit hour)?

CREDIT HOURS	GRADES
3	D
3	C
4	A
5	B

5. Jefferson Hardware employs 15 people. Their jobs and wages are listed below.
   a. What was the mean wage per job?
   b. What was the mean wage per employee?

JOB	NUMBER IN JOB	WEEKLY WAGES
Manager	3	$1,000
Sales clerk	7	500
Clerical	5	600

6. Scott's Heating and Air Conditioning Service employs 20 people. Their jobs and wages are shown below.
   a. What was the mean wage per job?
   b. What was the mean wage per employee?

JOB	NUMBER IN JOB	WEEKLY WAGES
Manager	1	$950
Sales representative	4	250
Technician	10	500
Office support	5	325

[handwritten: 950, 1000, 5000, 1625 } 8575 ÷ 20 = 428.75]

[handwritten: mean · 20 ÷ 4 = 5   2025 ÷ 4 = 506.25]

7. The Pants Place sold 14 pairs of women's jeans yesterday at the prices shown below.

   a. What was the mean price per brand?

   b. What was the mean price per pair sold?

BRAND	PRICE	NUMBER SOLD
A	$38	1
B	36	2
C	35	4
D	31	2
E	30	5

[handwritten: = 38, = 72, 140, 62, 150]

[handwritten: 170 = $34   14   462]

8. Six stores in a city sell the same brand and model of vacuum cleaner. The price and number sold at each store are listed below.

   a. What is the average price per store?

   b. What is the mean price per vacuum cleaner sold?

STORE	PRICE	NUMBER SOLD
A	$ 85	12
B	135	8
C	96	22
D	89	20
E	112	9
F	107	9

9. Ed Laser invested $10,000 in the King Associates Partnership on January 1. On May 1, he invested another $2,000. On August 1, he withdrew $800 for personal reasons. On October 1, he reinvested $1,000. His investment then remained the same through the end of the year. What was Ed's average (mean) investment during the year?

10. Hope Powers invested $5,000 in the Aerobic Connection Partnership on January 1. On March 1, she withdrew $200. She reinvested $400 on August 1, and withdrew $600 on September 1. No further change occurred during the year. Find Hope's average (mean) investment during the year.

11. Determine the median-priced charcoal grill sold last month at a local department store if grills were sold for the following prices: $39, $199, $48, $45, $39, $118, $48, $42, $45, $85, $48.

12. Determine the median salary earned by 7 college presidents throughout the state last year if the salaries earned were: $151,000, $133,000, $108,000, 130,000, $138,000, $110,000, $140,000.

13. The Dickens Co. sold 13 recliner chairs during a 2-week period. The selling prices were: $250, $268, $260, $255, $250, $300, $320, $264, $250, $318, $268, $264, $250. What was the most popularly priced chair sold (the modal price)?

14. Determine the most popularly priced CD/cassette player sold (the modal price) at Johnson Electronics last week if the selling prices were: $130, $86, $82, $94, $82, $79, $124, $82, $79, $86, $158.

15. The 20 students in a management class were the following ages: 18, 19, 18, 21, 20, 50, 37, 35, 19, 28, 19, 31, 26, 24, 25, 19, 22, 24, 23, 22. Find the following:
    a. Mean age
    b. Median age
    c. Modal age

16. The junior partners and law clerks in Jones, Hecht, and Isaac worked the following hours last week: 48, 50, 54, 60, 48, 54, 56, 60, 59, 68, 52, 60, 51, 60, 45, 47. Find the following:
    a. Mean number of hours worked
    b. Median number of hours worked
    c. Most frequent (modal) number of hours worked

17. During a sales promotion at Shank Furniture, nine sales representatives sold the following number of sleep sofas: 2, 6, 5, 11, 6, 12, 8, 10, 3. Determine the following:
    a. Mean number sold
    b. Median number sold
    c. Modal number sold

18. Employees at Armez Textiles work on a production basis. They produced the following numbers of units last month: 1,600, 1,630, 1,640, 1,630, 1,620, 1,640, 1,560, 1,640, 1,620.
    a. What was the mean number of units?
    b. What was the median number of units?
    c. What was the modal number of units?

SECTION 2

AVERAGES (GROUPED DATA)

Business statistics often involve hundreds or thousands of values. If a computer is not available, it would be extremely time consuming to arrange each value in an array. It is easier to handle large numbers of values by organizing them into groups (or **classes**). The difference between the lower limit of one class and the lower limit of the next class is the **interval** of the class. (All classes of a problem usually have the same-sized interval.) Each class is handled by first determining the number of values within the class (the **frequency**) and then using the **midpoint** of the class as a basis for calculation. The end result of this classification and tabulation is called a **frequency distribution.**

The first step in making a frequency distribution is to decide what the class intervals shall be. For the sake of convenience, the intervals are usually round numbers. A tally is then made as each given value is classified; the total tally in each class is the frequency f of the class. After the midpoint of each class is computed, the frequency is multiplied by the midpoint value to obtain the final column of the frequency distribution. Example 1 shows how to compute the interval midpoints and the arithmetic mean of the frequency distribution.

Example 1   Lady Fair Mills is a textile firm producing ladies' lingerie. Using a frequency distribution, determine (a) the mean and (b) the median number of garments processed per employee last week in the sewing department. Production per employee (in dozens) was as follows:

368	363	352	333
340	347	380	321
386	342	308	357
320	382	341	374
355	335	364	391
372	325	378	348

The smallest number is 308, and the largest is 391; thus, a table from 300 to 400 will include all entries. The difference between the upper and lower limits (400 − 300) gives a **range** of 100. We can conveniently divide this range into five classes.* Since $\frac{100}{5} = 20$, each class interval will be 20.

CLASS INTERVALS	TALLY	f	MIDPOINT	$f \times$ MIDPOINT
380–399	||||	4	390	1,560
360–379	卌 |	6	370	2,220
340–359	卌 |||	8	350	2,800
320–339	卌	5	330	1,650
300–319	|	+ 1	310	+ 310
		24		8,540

The midpoint of each interval is the arithmetic mean of the lower limit of that class and the lower limit of the next class.†

$$\text{Example:} \quad \frac{320 + 340}{2} = \frac{660}{2} = 330$$

(a)   The arithmetic mean of the frequency distribution is found by dividing the $f \times$ midpoint grand total by the total number of items

---

*The choice of 5 classes is arbitrary. Four classes with an interval of $\frac{100}{4} = 25$ could be chosen, as could 10 classes with an interval of $\frac{100}{10} = 10$.

†Adherence to strict statistical methods requires a midpoint that is the mean of the lower and upper boundaries of the interval. For instance, $\frac{320 + 339}{2} = 329.5$, rather than the 330 midpoint determined above. The simplicity of the method in Example 1 seems preferable for introductory business mathematics; however, the instructor may specify the standard statistical midpoint if desired. [The group mean of part (a) would then be $355\frac{1}{3}$ rather than $355\frac{5}{6}$.] Further discussion appears in the Instructor's Manual.

(the frequency total). Thus,

$$\frac{8,540}{24} = 355\frac{5}{6} \quad \text{or} \quad 356$$

The mean production per employee in the sewing department was 356 dozen garments, to the nearest whole dozen.

MEDIAN FOR GROUPED DATA

The *median for grouped data* is found as follows:

1. Divide the total number of items (total frequency) by 2. This quotient represents the position which the median holds in the tally.
2. Counting from the smallest number, find the class where this item is located, and determine how many values within this class must be counted to reach this position.
3. Using the number of values that must be counted (step 2) as a numerator and the *total* number within the class as a denominator, form a fraction to be multiplied times the class interval.
4. Add this result to the lower limit of the interval to determine the median.

Example 1 (cont.)

(b)   The median for this example is thus determined as follows:

(1) $\frac{24}{2} = 12$; we are concerned with the 12th item from the bottom.

(2)   The bottom classes contain only $1 + 5 = 6$ items. We must therefore include 6 more items from the class 340–359.

(3)   There are 8 values in the class, and the class interval is 20. Hence,

$$\frac{6}{8} \times 20 = \frac{3}{\cancel{4}} \times \cancel{20}^{5} = 15$$

(4) $340 + 15 = 355$

The median of 355 is thus the dividing point where half of the employees obtained higher production and half had lower.

As a point of interest, compare the mean and median for Example 1 when the data are computed individually (without a frequency distribution). The mean is then 353.4 (as compared to the grouped data mean of 356), and the median is 353.5 (compared to the 355 obtained using a frequency distribution). The differences between these means and medians are small, even though we used a small number of items (24). For large numbers of items, the results obtained by the individual and the grouped-data methods are almost identical.

Since a frequency distribution does not contain individual values, there is no single mode. However, the distribution may contain a *modal class* (the class containing the largest number of frequencies). The class interval 340–359 is the modal class of Example 1. This indicates that the most common production per employee was between 340 and 359 dozen garments.

Example 2   What is the median production if a frequency distribution contains 25 items and is distributed as follows?

CLASS INTERVALS	f
2,000–2,399	3
1,600–1,999	4
1,200–1,599	10
800–1,199	7
400–799	+ 1
	25

The steps for solution (see the preceding rules) are as follows:

(1) $\dfrac{25}{2} = 12.5$; we must locate the class containing the imaginary item numbered 12.5.

(2) The two bottom classes contain only $1 + 7 = 8$ entries. Thus 12.5 is in the next class, 1,200–1,599, and 4.5 more values from this class complete the required 12.5.

(3) The class 1,200–1,599 contains 10 items, and the interval is 400. Thus,

$$\frac{4.5}{\cancel{10}} \times \cancel{400}^{\,40} = 4.5 \times 40 = 180$$

(4) The median production is $1,200 + 180 = 1,380$ pieces.

Note.   The median can also be found by counting *down from the top* to the required item, multiplying the obtained fraction times the interval in the usual manner, and then *subtracting* this product from the *lower limit* of the adjacent class. For example, since the top two classes in Example 2 contain $3 + 4 = 7$ items, 5.5 more values from the class 1,200–1,599 must be included. Now, $\frac{5.5}{10} \times 400 = 220$; and $1,600 - 220$ gives 1,380, as before.

SECTION 2   PROBLEMS

Use a frequency distribution to determine (a) the mean, (b) the median, and (c) the modal class for the following sets of data.

1. Use intervals of 0–9, 10–19, etc.

24	15	21	17	8
6	25	16	23	18
3	48	19	30	26
28	17	35	38	10
20	27	43	22	

2. Use intervals of 0–19, 20–39, etc.

82	22	28	36	39
68	23	61	38	12
96	44	10	21	45
73	78	51	40	26
14	30	17	65	56

3. Use intervals of $50, starting with $100.

$100	$240	$210	$220	$190
340	270	360	120	110
300	230	260	240	150
380	230	360	170	245

4. Use intervals of $100, starting with $500.

$ 570	$ 640	$590	$ 960	$980	$ 800
670	940	660	730	820	1,050
710	860	850	740	770	680
970	1,020	890	1,080	820	760
1,000	780	950	870	500	840

5. The following are salaries of managers at different levels of a corporation. Using intervals of $10,000 and beginning with $30,000, construct a frequency distribution to determine (a) the mean, (b) the median, and (c) the modal class.

$77,000	$34,000	$53,000	$48,000	$32,000	$61,000
55,000	38,000	59,000	66,000	74,000	65,000
73,000	49,000	79,000	36,000	45,000	75,000
64,000	56,000	47,000	64,000	72,000	81,000
88,000	72,000				

6. During a President's Day Sale, the 20 sales associates at Stephen's Electronics registered sales in the amounts shown below. Make a frequency distribution using intervals of $100 (starting with $50), and determine (a) the mean, (b) the median, and (c) the most typical sales interval (modal class).

$155	$ 95	$ 75	$385	$190
420	160	50	260	330
170	50	180	175	240
140	250	300	80	200

7. The Security Insurance Co. last month sold life insurance policies in the amounts shown below. Using $5,000 intervals, starting with the lowest amount, prepare a frequency distribution. Compute the (a) mean, (b) median, and (c) mode.

$33,000	$10,000	$24,000	$15,000	$38,000	$12,000
40,000	35,000	22,000	36,000	42,000	54,000
17,000	33,000	25,000	26,000	18,000	31,000
33,000	54,000	46,000	28,000	30,000	32,000

8. Wexford Laboratories employs 26 lab technicians whose years of experience are listed below. Use 3-year intervals in completing a frequency distribution, and compute (a) the mean, (b) the median, and (c) the most common interval (modal class) of experience.

15	13	6	3	3
12	16	4	8	6
8	7	10	0	6
7	18	6	22	14
6	1	12	20	5
3				

SECTION 3
STANDARD DEVIATION

Measures of central tendency—the arithmetic mean, median, and mode—are used to identify some typical value(s) around which the other values cluster. It is frequently also helpful to know something about how spread out the data are: Are many (or most) of the given amounts close in value to the measure of central tendency? Or do the amounts scatter out without concentrating near any particular value? To provide some indication of this scatteredness (or **variability**), we compute **measures of dispersion.** One simple measure of dispersion which has already been mentioned is the **range**—the difference between the upper and lower limits of a frequency distribution or between the highest and lowest values in a group of numbers. Probably the most widely used measure of dispersion is the **standard deviation,** which we will briefly consider.

The tally column of a frequency distribution provides a visual illustration of the central tendency and of the dispersion of a set of data. If a tally column were turned sideways and a smooth line drawn outlining the tallies, we could obtain a simple graph of the data, as shown below. This example—with a heavy concentration of values in the middle and quickly tapering off symmetrically on both sides—is typical of the distribution for many sets of data. For example, the graphs of many human characteristics (such as height, weight, intelligence, athletic ability) would form such a graph, with most people obtaining some average, central score and those with higher and lower scores distributed evenly on both sides.

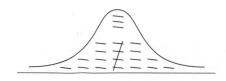

Many sets of business-related data would also form a similar graph. For example, on quality-control tests of the chemical content of a medicine or on tests of the sizing accuracy of a mechanical part, most test results would match the desired standard, and variations higher and lower should taper off quickly. In business and industry, however, it is vital to know exactly how much variation exists.

When statistical data are collected on large groups of people who are typical of the entire population (rather than just some specialized group), the graph of these data would approach the perfect mathematical model known as the **normal curve** (or **bell curve**), shown in the following diagram. In this normal distribution, the mean, median, and mode all have the same value (M)—the value (on the horizontal line) which corresponds to the highest point on the curve. The range of values is evenly divided into six equal intervals, each of which represents one standard deviation, σ. The six intervals are marked off symmetrically around the mean at $\pm 1\sigma$, $\pm 2\sigma$, and $\pm 3\sigma$.*

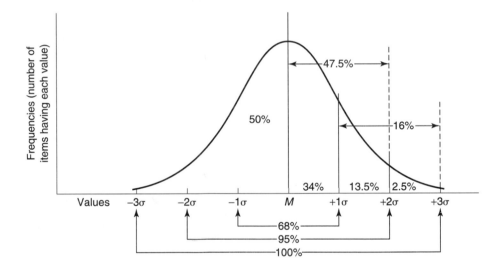

The standard deviation is significant because we know what percent of the total items (people) fall within each interval or combination of intervals. These percent categories are indicated on the normal curve above (and apply symmetrically on each side of the mean). Thus, as shown below, if the mean height of American males is 5′10″ and one standard deivation $\sigma = 2″$, then 68% of all men have heights between 5′8″ and 6′0″ tall; 95% of the men fall between 5′6″ and 6′2″; and 100% of the men fall between 5′4″ and 6′4″. (Obviously, there are men shorter than 5′4″ and taller than 6′4″. However, since less than

---

*Technically, the normal curve continues infinitely in both directions, so that additional higher and lower values are possible. However, we assume for practical purposes that the given range of values lies entirely within $\pm 3\sigma$, since the area within $\pm 3\sigma$ includes 99.73% of all items depicted by the normal curve.

0.3% of the men would fall in either category, we assume for convenience that 100% have been included.)

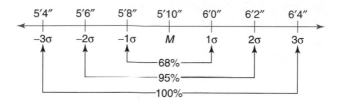

Since the standard deviation for a set of data can be computed, industries can test their products to determine how accurate their manufacturing processes are and what percent of their products fall into various tolerance categories. Clothing manufacturers or other firms whose products are sized may use the percents associated with standard deviation as a guide in deciding how many items of each size to produce.

Example 1   Assume that the mean size of women's dresses is size 12, with a standard deviation of 4, as shown below. Use the percent categories of the normal curve to answer each question.

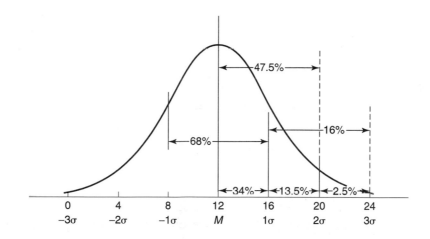

(a)   What percent of the women are size 12 or larger?   50%

(b)   What percent of the women are size 16 or larger?   16%

(c)   What percent are 24 or smaller?   100%

(d)   What percent are smaller than size 4?   2.5%

(e)   What percent are smaller than size 4 *or* larger than size 20?   2.5% + 2.5% = 5.0%

(f)   If a dress company plans to manufacture 5,000 dresses, how many of those should be in sizes 8 through 16?   68% × 5,000 = 3,400

The computation of standard deviation from a set of grouped data in a frequency distribution requires more time and explanation than can be included in introductory business math. However, we can compute the standard deviation for a set of ungrouped values. Using this type of data, the formula for standard deviation is

$$\sigma = \sqrt{\frac{\sum d^2}{n}}$$

where the symbols are as follows:

σ = standard deviation

$\sum$ = "the sum of"

d^2 = the square of the deviations from the mean (for each given value)

n = the number of values given

$\sqrt{}$ = "the square root of"

This procedure for finding standard deviation is outlined as follows.

STANDARD DEVIATION PROCEDURE

1. Find the mean: Add the given data, and divide by n.

2. Find each deviation, d: Subtract the mean from each given value, obtaining either a positive or negative difference. (The sum of these deviations must be 0.)

3. Square each deviation, and find their sum, $\sum d^2$.

4. Apply the standard deviation formula, $\sigma = \sqrt{\dfrac{\sum d^2}{n}}$: Divide $\sum d^2$ by n, and take the square root. (If the square root is not a whole number, let the standard deviation remain in $\sqrt{}$ form.)

Example 2   Find the standard deviation of the following values: 61, 62, 65, 67, 68, and 73.

MEAN	DEVIATION, d	d^2	STANDARD DEVIATION
61	$61 - 66 = -5$	25	$\sigma = \sqrt{\dfrac{\sum d^2}{n}}$
62	$62 - 66 = -4$	16	
65	$65 - 66 = -1$	1	
67	$67 - 66 = 1$	1	$= \sqrt{\dfrac{96}{6}}$
68	$68 - 66 = 2$	4	
+ 73	$73 - 66 = 7$	49	$= \sqrt{16}$
$6\overline{)396} = 66$	0	$\sum d^2 = 96$	$\sigma = 4$

The standard deviation is 4. Thus, the values associated with each interval would appear on a line graph as shown below. The dots represent the six values given above. For a normal curve distribution, 68% of the values should fall within $\pm 1\sigma$; that is, 68% of the 6 numbers (or 4 numbers) should fall between 62 and 70. Four of our given numbers (62, 65, 67, and 68) fall within $\pm 1\sigma$. Within $\pm 2\sigma$ should be 95% of the 6 numbers, which equals 6 values to the nearest whole unit. All 6 of our given values do fall within $\pm 2\sigma$, which includes values from 58 through 74.

Our given set of data above contains too few numbers to include values in the two outside intervals under the normal curve. Sets of data containing many more values, however, would closely match the normal distribution. Industries using standard deviation for quality control would thus use larger groups of data in order to obtain dependable statistical results.

Calculator techniques . . .   FOR EXAMPLE 2

Set up the problem using the format shown in the example. Then,

61 ⊞ 62 ⊞ 65 ⊞ 67 ⊞ 68 ⊞ 73 ⊟ ⟶ 396 ⊡ 6 ⟶ 66

Compute the Deviation and d^2 columns mentally; then calculate as follows to find their sum and take the square root (using the $\boxed{\sqrt{x}}$ key):

25 ⊞ 16 ⊞ 1 ⊞ 1 ⊞ 4 ⊞ 49 ⊟ ⟶ 96 ⊡ 6 ⟶ 16 $\boxed{\sqrt{x}}$ ⟶ 4

Complete the line graph by hand.

SECTION 3   PROBLEMS

Use the percent categories associated with the normal curve in order to answer the following questions.

1. Suppose that the mean neck size for a man's shirt is 15½, with a standard deviation of ½ size, as shown below.

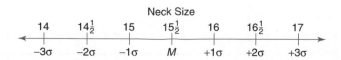

   a. What percent of the men are size 15 or larger?

   b. What percent are size 16½ or larger?

   c. What percent are between sizes 14½ and 16½?

   d. What percent are larger than size 16?

   e. If a shirt company manufactures 80,000 shirts, how many should be size 15½ or larger?

   f. How many of the 80,000 shirts should be sizes 15½ to 16½?

2. Assume that 75 was the mean test score on a finance exam, with a standard deviation of 5 points, as shown below.

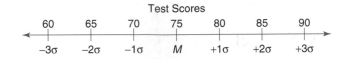

   a. What percent of students had scores of 75 or higher?

   b. What percent had scores of 85 or higher?

   c. What percent had scores between 65 and 75?

   d. What percent had scores of 65 or higher?

   e. If 50 people took the exam, how many had scores between 60 and 70?

   f. How many of the 50 people scored over 70?

3. The average chicken potpie contains 11 grams of protein, with a standard deviation of 3 grams. Prepare a line graph (as in Problems 1 and 2) to show the grams associated with each deviation $\pm 1\sigma$, $\pm 2\sigma$, and $\pm 3\sigma$.

   a. What percent of the pies have between 2 and 20 grams of protein?

   b. What percent have less than 8 grams?

   c. What percent have more than 5 grams?

   d. What percent have between 8 and 14 grams?

   e. If 5,000 pies are manufactured, how many have between 8 and 14 grams of protein?

   f. How many of the 5,000 pies have between 14 and 17 grams?

   g. How many have 11 grams or higher?

4. Suppose that the mean high temperature in Jacksonville for the month of August was 92°, with a standard deviation of 2°. Prepare a line graph to show the temperatures associated with each deviation $\pm 1\sigma$, $\pm 2\sigma$, and $\pm 3\sigma$.

   a. What pecent of the days have temperatures above 88°?

   b. What percent of the days have temperatures below 92°?

   c. What percent of the days have temperatures between 90° and 94°?

   d. What percent of the days have temperatures above 94°?

   e. For the 31 days in August, how many days were above 94°?

   f. How many days were below 92°?

   g. How many days were between 90° and 94°?

     For the following sets of data:

       (a) Compute each standard deviation.

       (b) Prepare a line graph showing the value associated with each deviation $\pm 1\sigma$, $\pm 2\sigma$, and $\pm 3\sigma$.

(c) *Check each interval (within ±1σ, ±2σ, and ±3σ) to determine whether it contains the expected number of given values. If not, what does this mean?*

5. 25, 27, 29, 33, 36

6. 8, 10, 12, 16, 19

7. 40, 42, 45, 51, 55, 55

8. 28, 34, 35, 35, 40, 44

9. 48, 51, 51, 52, 52, 53, 54, 55

10. 31, 32, 35, 41, 44, 44, 46, 47

11. 37, 43, 44, 45, 45, 46, 48, 52

12. 23, 24, 25, 31, 31, 34, 36, 36

13. 57, 63, 64, 65, 65, 66, 68, 72

14. 14, 15, 16, 18, 20, 20, 23, 23, 25, 26

SECTION 4
INDEX NUMBERS

An **index number** is similar to a percent in both appearance and purpose. Like percent, index numbers use a scale of 100 in order to show the relative size between certain costs or to indicate change in costs over a period of time, rather than showing the actual, numerical differences. Whereas the actual dollars-and-cents changes in costs often could not be readily analyzed, both the index number and percent allow the reader to make comparisons between items more easily. Index numbers are often used in business to depict long-range trends in prices or volume (of sales, production, and so forth).

The simplest index number that can be computed involves a single item. Such a simple price index is sometimes called a **price relative.** This index is simply the ratio of a given year's price to the base year's price, multiplied by 100:

$$\text{Index number (or price relative)} = \frac{\text{Given year's price}}{\text{Base year's price}} \times 100$$

Example 1 Suppose that a half-gallon of milk cost $1.50 in 1990, $1.45 in 1994, and $1.58 in 1996. Compute the price relatives (or index numbers) for the milk using 1990 as the base year.

YEAR	PRICE		PRICE INDEX (1990 = 100.0)
1990	$1.50	Base year	100.0
1994	1.45	$\frac{1.45}{1.50} \times 100 =$	96.7
1996	1.58	$\frac{1.58}{1.50} \times 100 =$	105.3

The student should realize that this index is nothing more than a simple percent, but written without the percent sign. The question "What percent of the 1990 price was the 1996 price?" would be answered "105.3%."

An excellent example of the use of index numbers is the Consumer Price Index, published monthly by the U.S. Department of Labor (Bureau of Labor Statistics). The costs of a large number of goods and services are indexed for the United States as a whole and for selected cities and urban areas individually.

| TABLE 4-1 | CONSUMER PRICE INDEX OF SELECTED EXPENSES IN SELECTED URBAN AREAS (1982–84 = 100) |

RISING MORE Rapidly (handwritten)

EXPENSE	U.S. URBAN AVERAGE	ATLANTA	CHICAGO	DALLAS	LOS ANGELES	NEW YORK	SEATTLE
Food	147.2	141.1	147.0	142.2	148.5	151.9	146.9
Housing	145.4	140.1	144.8	129.0	151.0	159.9	147.9
Utilities	122.0	132.2	110.7	126.3	143.1	112.4	112.7
Transportation	137.1	123.8	130.3	134.5	140.5	141.8	135.0
Medical care	215.3	227.0	213.2	205.6	215.2	217.6	199.8

Source: U.S. Department of Labor.

then & now same place (handwritten)

Consider the example given in Table 4-1. (Assume that this index is for the present year.)

Example 2

(a) Indicating rise of price in each city separately

The index numbers show the "current" year-end costs of the major consumer expenses as compared with the costs of these same expenses in *that* locality during the base years, 1982–84. (These three years, 1982, 1983, and 1984, were averaged to give a composite base year.) The same food that cost $100 in Dallas in 1982–84 would cost $142.20 at the present time. Similarly, housing that cost $100 in Seattle in 1982–84 has now increased to $147.90. (These indexes also may be read, "Transportation that cost $10 in Chicago in 1982–84 would now cost $13.03.")

Various index numbers cannot be used for comparison among themselves unless they refer to the same base amount. For instance, Table 4-1 gives the price index for medical care in Atlanta as 227.0 and in New York as 217.6. This means that medical care that cost $100 in Atlanta in 1982–84 would now cost $227 and that medical care that cost $100 in New York in 1982–84 would now cost $217.60. It does *not* mean that $100 in 1982–84 would have purchased the same medical services and drugs in both Atlanta and New York.

Thus, Table 4-1 does not necessarily indicate that medical care in Atlanta costs more today than similar care in New York; it simply indicates that the cost of medical care in Atlanta has *increased* more rapidly since 1982–84 than have medical expenses in New York (227.0 versus 217.6). If a medical procedure in Atlanta had cost substantially less in 1982–84 than did the same procedure in New York, then the Atlanta medical procedure could still be less expensive than the New York procedure (although Atlanta medical fees have increased faster than have the New York fees).

Example 2 (cont.)

(b) Comparing actual $ prices between cities

As part (a) implies, the actual cost of living in Los Angeles as compared with the cost of living in Chicago (or any other locality) cannot be determined from the Consumer Price Index, but would have to be obtained from some other statistical source that contains an index of *comparative* living costs. Such an index, released periodically by the American Chamber of Commerce Researchers Association, measures price levels for consumer goods and services. The national average is set at 100, and each city's index is read as a percentage of the national average.

A typical index of this type is given in Table 4-2, based on the U.S. urban average as 100. This comparative index enables us to determine that, for every $100 that the average urban family spends, the Seattle family spends $106.20, the Dallas family spends $102.00, and the Chicago family $103.50—all to maintain the same intermediate standard of living. As a matter of interest, notice on the comparative index of transportation expenses that transportation in Chicago (108.0) is more expensive than in Los Angeles (106.8), despite the fact that the Consumer Price Index showed transportation costs have increased more rapidly in Los Angeles (140.5) than in Chicago (130.3).

Example 1 illustrated the computation of price relatives for only a single item—milk. Actually, the term "index number" is often reserved for a number derived by combining several related items, each weighted according to its relative importance. For instance, the index numbers for food expenses in Example 2 are a composite of many different grocery items, each weighted according to its portion of the total food budget in a typical family (as one consideration). There are many methods for constructing such composite index numbers, all of which require more statistical knowledge than can be acquired in basic business math. Even for the same data, these composite index numbers vary significantly according to the method used; hence, the best method to use depends on careful analysis of the purpose that the index is intended to serve. Obtaining an accurate and meaningful composite index number is not a simple matter. Our index number problems will therefore be limited to simple price relatives.

TABLE 4-2 COMPARATIVE INDEX OF SELECTED EXPENSES IN SELECTED URBAN AREAS AT INTERMEDIATE STANDARD OF LIVING

U.S. Urban Average = 100

EXPENSE	ATLANTA	CHICAGO	DALLAS	LOS ANGELES	NEW YORK	SEATTLE
Food	98.6	99.3	98.1	110.3	112.8	97.3
Housing	86.8	107.5	93.8	147.9	105.6	132.8
Utilities	110.4	113.2	124.0	91.0	131.0	57.1
Transportation	98.3	108.0	103.2	106.8	95.5	94.7
Medical care	110.4	105.8	113.8	145.5	106.2	124.8
Combined index	97.0	103.5	102.0	121.1	106.0	106.2

Source: American Chamber of Commerce Research Association.

SECTION 4   PROBLEMS

Base your answers for Problems 1 and 2 on the Consumer Price Index in Table 4-1 and the Comparative Index in Table 4-2. (Include the appropriate index number with each answer.)

1. a. How much would a Seattle resident pay now for housing that cost $1,000 in the composite base period, 1982–84?
   b. How much would a Chicago resident pay now for food that cost $100 in 1982–84?
   c. How much does the average U.S. citizen pay today for transportation that cost $10 in 1982–84?
   d. In which city are utility costs the highest?
   e. In which city have utility costs risen the fastest?
   f. In which city have housing costs increased the slowest?
   g. In which city have food costs increased the least?
   h. In which city are food costs the most expensive?
   i. In which city have transportation costs increased most rapidly?

2. a. How much would a Los Angeles resident spend today for housing that cost $1,000 in the composite base period, 1982–84?
   b. If a medical fee was $10 in 1982–84, how much should a Dallas resident expect to pay now for the same service?
   c. How much does the average U.S. citizen pay today for utilities that cost $100 in 1982–84?
   d. In which area have utility costs risen most slowly?
   e. Where are food prices least expensive?
   f. In which area are housing costs least?
   g. In which area are transportation costs least?
   h. In which area have medical costs risen most slowly?
   i. In which area is medical care least expensive?

3. Determine the price index for a can of corn using 1991 as the base year.

YEAR	PRICE	INDEX (1991 = 100.0)
1991	$0.50	
1992	0.54	
1993	0.60	
1994	0.49	
1995	0.51	

4. Calculate the price relative for a popular magazine, using 1992 as the base year.

YEAR	PRICE	INDEX (1992 = 100.0)
1992	$1.50	
1993	1.47	
1994	1.65	
1995	1.80	
1996	1.95	

5. Determine the price index for a candy bar using 1993 as the base year.

YEAR	PRICE	INDEX (1993 = 100.0)
1993	$0.80	
1994	0.84	
1995	0.72	
1996	0.76	

6. Determine the price index for a can of soft drink, using 1980 as the base year.

YEAR	PRICE	INDEX (1980 = 100.0)
1980	$0.40	
1985	0.38	
1990	0.55	
1995	0.60	

SECTION 5
GRAPHS

Many people read business statistics without fully realizing their significance. Statistics usually make a much stronger impression if they can be visualized. This is particularly true when the data are used to make comparisons among several items or when the statistics give the various parts that comprise a whole. **Graphs** are visual representations of numerical data. The purpose of business graphs is to present data in a manner that will be both accurate and easy to interpret. Today, businesses assemble, tabulate, and graph information through computer programs. These computer-generated graphs make a presentation visually appealing and interesting. For instance, a spreadsheet is used to assemble and tabulate a firm's cash flow over several months. From the same spreadsheet, the results can be presented in a variety of graph formats. The most common graphs are *bar graphs, line graphs,* and *circle graphs.*

All graphs have certain elements in common, whether they are computer generated or hand drawn. Every graph should have a title so that the reader will know what facts are presented. Every graph should show the origin (zero) and also some low value on the scale to help establish the scale for the reader. Very large figures may be rounded off before graphing.

Breaks in the scale, to eliminate unused space on the graph, should be used with care. The difference between amounts may appear much greater proportionally than they actually are; thus a break can cause a distortion in the graph. The accompanying bar graphs both illustrate the same information. The break in the left bar graph creates an overemphasis on the difference between the net sales for the two years. The difference appears greater than it actually is; thus a distortion is created. The vertical scale in the bar graph on the right corrects this distortion by eliminating the break and by using a lower value as the beginning number.

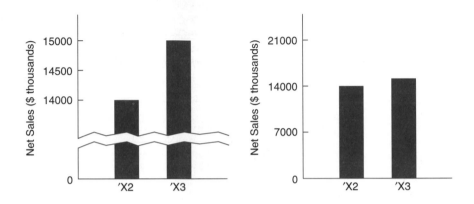

You are undoubtedly familiar with bar graphs, since they are so widely used. For this topic, we will therefore consider in detail only line graphs and circle graphs.

Line Graphs

The line graph usually records change (in sales, production, income, and so forth) over a period of time. The only basic difference between it and the vertical bar graph is that, instead of having bars that come up to the various values, the values are connected by a line. The line may be broken or unbroken, straight or curved. Two or more related statistics are often presented on the same line graph, using additional lines.

Line graphs should ordinarily be drawn on graph paper. After the scale of values has been labeled (time should be on the horizontal scale), locate the appropriate values and mark each with a dot. Finally, connect the dots with a line. When the graph contains more than one line, be sure to make each line distinct and to identify each.

Example 1   Koenig Building Distributors, Inc. experienced the following totals of net sales, cost of goods sold, and net profit during a 5-year period. Depict these data on a line graph.

	19X1	19X2	19X3	19X4	19X5
Net sales	$180,000	$210,000	$160,000	$220,000	$237,000
Cost of goods sold	105,000	110,000	80,000	112,000	95,000
Net profit	25,000	32,000	8,000	10,000	42,000

9. The total floor space at Simpson Co. is utilized as shown below. Prepare a circle graph to compare this utilization.

Receiving	4,000 sq. ft.
Cutting/sewing	8,000 sq. ft.
Inspecting	2,000 sq. ft.
Shipping/storage	5,000 sq. ft.
Office	1,000 sq. ft.

10. Wadge Industries had the following operating expenses last year. Construct a circle graph to compare these expenses.

Wages	$126,000
Utilities	32,000
Depreciation	20,000
Supplies	15,000
Taxes	5,000
Other	2,000

11. Find examples of bar, line, and circle graphs in recent periodicals. What information is depicted in the graphs? Using the data in one of the graphs, describe in narrative form the information that is shown.

12. Choose a graph from a periodical. Using the same data, construct a graph different from that depicted in the periodical. For instance, if the periodical depicts a bar graph, use the same data to construct a line or circle graph.

CHAPTER 4 GLOSSARY

Arithmetic mean. (See "Mean.")

Array, ordered. A listing of numerical data in order of size.

Average. A typical value or measure of central tendency for a group of numbers. (Examples: Mean, median, and mode.)

Central tendency, measure of. A representative number around which the other numbers in a group seem to cluster; an average.

Classes. The groups (or intervals) into which given values are categorized in a frequency distribution.

Dispersion, measure of. A value that gives some idea of how spread out (or how scattered from the mean) a group of numbers are. (Examples: Range and standard deviation.)

Frequency. The number of values within a class interval of a frequency distribution.

Frequency distribution. A process for classifying large numbers of values and tabulating the data in groups (classes) rather than individually; used to compute measures of central tendency and/or dispersion.

Grade-point average. A weighted scholastic average where each letter grade is assigned a quality-point value that is multiplied by the number of credit hours that the course earns. This total, divided by the total hours, gives the grade-point average (or average per credit hour).

Graph. A visual representation of numerical data.

Index number. A number based on a scale of 100 which shows the relative size between certain costs or indicates change in costs over a period of time. (Example: If 1987 = 100 and the current index is 115, then $115 is now required to purchase an item that cost $100 in 1987.)

Interval. On a frequency distribution, the difference between the lower limit of one class and the lower limit of the next class.

Mean, arithmetic. The most common average; the sum of all given values divided by the number of values.
(Example: $\dfrac{7+5+6}{3} = \dfrac{18}{3} = 6$)

Median. The midpoint (or middle value) of a group of numbers; the central value of an array. (Examples: Given 1, 2, 3, 7, and 9, the median is 3; given 1, 2, 3, 7, 9, and 10, the median is $\dfrac{3+7}{2} = \dfrac{10}{2} = 5$.)

Midpoint. Of a class in a frequency distribution, the arithmetic mean of the lower limit of that class and the lower limit of the next class.

Mode. The value in a set of data that occurs most frequently. (Example: Given 1, 3, 3, 5, 5, 5, 7, 9, and 10, the mode is 5.)

Modified mean. An arithmetic mean computed after the extreme high and low values have been deleted.

Normal (or bell) curve. A symmetrical, bell-shaped graph associated with the expected distribution for many types of

data; the graph for which standard deviation applies.

Price relative. A simple price index for one item:
$$\left(= \frac{\text{current price}}{\text{base price}} \times 100 \right)$$

Range. The difference between the highest and lowest values in a group of numbers (or between the upper and lower limits in a frequency distribution); a simple measure of dispersion.

Standard deviation (σ). For a normal curve distribution, the deviation (difference) between the mean and two crucial points symmetrical to the mean, within which interval fall 68% of all the given values.

For ungrouped data, $\sigma = \sqrt{\dfrac{\sum d^2}{n}}$.

Statistics. Informally, the process of collecting, organizing, tabulating, and interpreting numerical data; also, the information presented as a result of this process.

Variability. Dispersion (or scatteredness) of data.

Weighted mean. A mean computed for data whose importance differs and where each value is weighted (or multiplied) according to its relative importance.

CHAPTER 5

TAXES

OBJECTIVES

Upon completion of Chapter 5, you will be able to:

1. Define and use correctly the terminology associated with each topic.
2. Compute sales tax (Section 1):
   a. Using a table (Example 1)
   b. Based on a percentage rate (Example 2; Problems 1–6, 11–24).
3. Find the marked price of an item, given the sales (or other) tax rate and the total tax or the total price (Section 1: Examples 2–4; Problems 1, 2, 7–22).
4. Express a given (decimal) property tax rate (Section 2: Examples 1, 2; Problems 1–6):
   a. As a percent
   b. As a rate per $100
   c. As a rate per $1,000
   d. In mills.
5. Calculate the property tax rate of a tax district (Section 2: Example 1; Problems 5, 6).
6. Using the property tax formula $R \cdot V = T$ and given two of the values, determine the unknown (Section 2: Examples 3–7; Problems 7–22):
   a. Property tax
   b. Tax rate
   c. Assessed value of property.
7. Determine the change in the property tax rate or the assessed value from one year to another (Section 2: Example 7; Problems 23–26).

Increased government activity (at the federal, state, and local levels) has made taxes—personal as well as business—an integral part of modern life. Two of the most common taxes are the sales tax and the property tax. The sales tax has become an everyday part of commerce. Property owners, including business organizations, must pay a property tax on their possessions. The following sections present the calculation of sales tax and property tax. (A third important tax is income tax. However, any significant study of income tax would require an entire course by itself and could not be accomplished in one unit of a business math course. Chapter 8, "Wages and Payrolls," discusses the taxes that must be deducted from payrolls.)

SECTION 1

SALES TAX

A large majority of state legislatures have approved **taxes on retail sales** as a means of securing revenue. Retail merchants are responsible for collecting these taxes at the time of each sale and relaying them to the state. In some states the legislatures have also allowed their cities the option of adopting a municipal sales tax in addition to the state tax in order to obtain local funds. Sales tax is computed as a specified percent of the selling price, with the percent usually varying from 4% to 8%.

Sales taxes historically have applied only to sales made within the jurisdiction of the taxing body. Thus, merchandise sold outside a state (such as by mail or by telephone) has not usually been subject to that state's sales tax. However, merchandise brought into another state for use in the state is usually subject to a **use tax.** For instance, merchandise sold in Maryland by telephone to a Virginia resident is not subject to the Maryland state sales tax but is subject to the Virginia use tax. All states imposing sales taxes also impose use

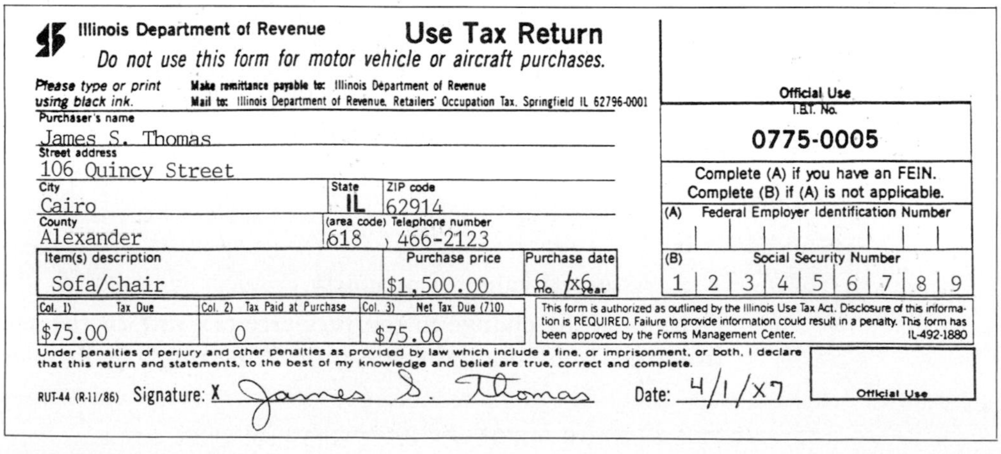
Illinois Department of Revenue
Use Tax Return
Do not use this form for motor vehicle or aircraft purchases.

Please type or print using black ink.
Make remittance payable to: Illinois Department of Revenue
Mail to: Illinois Department of Revenue, Retailers' Occupation Tax, Springfield IL 62796-0001

Official Use
I.B.T. No.
0775-0005

Purchaser's name: James S. Thomas
Street address: 106 Quincy Street
City: Cairo | State: IL | ZIP code: 62914
County: Alexander | (area code) Telephone number: 618) 466-2123

Complete (A) if you have an FEIN.
Complete (B) if (A) is not applicable.
(A) Federal Employer Identification Number
(B) Social Security Number: 1 2 3 4 5 6 7 8 9

Item(s) description	Purchase price	Purchase date
Sofa/chair	$1,500.00	6 mo. / x6 year

Col. 1) Tax Due	Col. 2) Tax Paid at Purchase	Col. 3) Net Tax Due (710)
$75.00	0	$75.00

This form is authorized as outlined by the Illinois Use Tax Act. Disclosure of this information is REQUIRED. Failure to provide information could result in a penalty. This form has been approved by the Forms Management Center. IL-492-1880

Under penalties of perjury and other penalties as provided by law which include a fine, or imprisonment, or both, I declare that this return and statements, to the best of my knowledge and belief are true, correct and complete.

RUT-44 (R-11/86) Signature: X James S. Thomas Date: 4/1/x7

Official Use

FIGURE 5-1

TABLE 5-1	SALES TAX INTERVALS
MARKED PRICE	6% SALES TAX
$40.00–40.08	$2.40
40.09–40.24	2.41
40.25–40.41	2.42
40.42–40.58	2.43
40.59–40.74	2.44
40.75–40.91	2.45
40.92–41.08	2.46
41.09–41.24	2.47
41.25–41.41	2.48

taxes. Credit is given for sales tax paid to other states, although there are a few states that do not provide credit for sales tax on motor vehicle purchases in other states. Many consumers are unaware of the use tax—they think their out-of-state purchase simply is not taxed. The state collects the use tax from the out-of-state vendor or from the purchaser. Because the purchase is difficult to track—and thus the use tax is difficult to collect—many states do lose tax revenue. That is one of the reasons many state legislatures are considering passage of sales taxes on all retail purchases regardless of the state of residence of the buyer. The Urban Institute estimates that only ten percent of the combined sales and use tax revenue comes from the use tax. Figure 5-1 shows a state's use tax return.

In areas subject to sales tax, charts are available so that the cashier may simply "look up" the tax. These charts eliminate the problem of rounding off the tax to the nearest cent by listing consecutive price intervals and indicating the tax to be charged in each interval. Table 5-1 illustrates a partial sales tax table.

Example 1 Watts Plumbing Equipment Inc. made the following sales, subject to a 6% tax rate: (a) $40.10, (b) $40.50, and (c) $41.30. Use Table 5-1 to determine the sales tax and the total amount collected for each sale.

(a) $40.10 falls in the $40.09–40.24 interval. The tax from the table is $2.41.

The total amount due is:

Marked price	$40.10
Sales tax	+ 2.41
Total price	$42.51

(b) $40.50 falls in the $40.42–40.58 interval.

The tax from the table is $2.43.

The total amount due is:

Marked price	$40.50
Sales tax	+ 2.43
Total price	$42.93

(c) $41.30 falls in the $41.25–41.41 interval. The tax from the table is $2.48.

The total amount due is:

Marked price	$41.30
Sales tax	+ 2.48
Total price	$43.78

Example 2   A 5% sales tax is charged on all retail sales made by Sun Country Pool Supplies. (a) How much tax must be charged on a $34.85 sale? (b) The sales tax on a case of pool chlorine was $2.85. What was the price of the case of chlorine to the nearest dollar?

(a)   We will use the basic percent formula, ____% of ____ = ____.

$$5\% \times \text{Price} = \text{Tax}$$

$$0.05 \times \$34.85 = t$$

$$\$1.74 = t$$

The sales tax due is $1.74.

(b)   The same formula is used to find the price when the sales tax is known. Since the tax was $2.85,

$$5\% \times \text{Price} = \text{Tax}$$

$$5\% \times p = \$2.85$$

$$0.05p = 2.85$$

$$p = \frac{2.85}{0.05}$$

$$p = \$57$$

The case of chlorine was priced at $57; thus, the total price, including the sales tax, was $59.85.

Note. Because sales tax rounds to the nearest cent (with half a cent rounding up to the next higher cent), $2.85 tax would be due on all sales in the range $56.90 to $57.09. Our procedure finds the price at the midpoint of this range.

Example 3 The total price of a camera, including a 6% sales tax, was $137.80. What was the marked price of the camera?

$$\text{Marked price} + \text{Sales tax} = \text{Total price}$$

$$\text{Price} + 6\% \text{ of price} = \text{Total price}$$

$$p + 0.06p = \$137.80$$

$$1.06p = 137.80$$

$$p = \frac{137.80}{1.06}$$

$$p = \$130$$

The marked price of the camera was $130.

Note. Observe that the sales tax is 6% of the marked price, *not* 6% of the total price. Therefore, problems of this type *cannot* be solved by taking 6% of the total price and subtracting it to find the marked price. The results would be close, but they would differ by a few cents; hence, they would be wrong. (You are invited to try a few examples if you are still skeptical! The only exception is for small amounts under $2.00.)

Excise taxes are levied on items to help limit the purchase of potentially harmful products such as firearms, or to help pay for products or services used only by certain people, such as highways financed by gasoline taxes. Other examples of excise taxes include taxes on automobiles, air transportation, telephone service, sporting and recreational equipment, alcoholic beverages, tobacco, entertainment, and business licenses.

A similar tax levied by the federal government is a **customs duty, import tax,** or **tariff.** These are taxes levied on products brought into the United States from outside the country. The purpose of these taxes is to protect American business against unfair foreign competition. Over the years, bilateral and multi-lateral agreements such as NAFTA (Northern American Free Trade Agreement) and GATT (General Agreement on Tariffs and Trade) have eliminated or reduced many of these taxes.

Another federal tax is the **luxury tax.** This 10% tax is levied on automobiles whose market price exceeds a certain amount. The threshold amount, which began at $32,000 in 1993, is adjusted annually for inflation. This tax also had been applied to the purchase of furs, jewelry, boats, and airplanes, but Congress repealed the tax on these items in 1993. Luxury taxes will not be included in the problems in this text.

Example 4   The $52.65 total price of a target pistol includes a 7% sales tax and a 10% federal excise tax. What was the marked price of the pistol?

$$\text{Marked price} + \text{Sales tax} + \text{Excise tax} = \text{Total price}$$

$$\text{Price} + 7\% \text{ of price} + 10\% \text{ of price} = \$52.65$$

$$p + 0.07p + 0.10p = 52.65$$

$$1.17p = 52.65$$

$$p = \frac{52.65}{1.17}$$

$$p = \$45$$

SECTION 1   PROBLEMS

Compute the missing items.

	MARKED PRICE	SALES TAX RATE	SALES TAX	TOTAL PRICE
1. a.	$23.00	5%		
b.	45.00	8		
c.		6	$4.68	
d.		7	6.44	
e.		8	4.32	
f.		5		$19.95
g.		6		51.94
h.		7		86.67
2. a.	$15.75	4%		
b.	66.00	8		
c.		6	$1.95	
d.		5	2.05	
e.		6	3.48	
f.		7		$25.68
g.		8		38.88
h.		5		91.35

3. Determine the sales tax and the total price to be charged for a backpack marked $14.60, if a 5% sales tax is charged.

4. Determine the sales tax and the total price to be charged for an answering machine marked $29, if a 6% sales tax is required.

5. What is the amount of sales tax and the total price paid for a videotape marked $9.90, if a 7% sales tax is to be collected?

6. If a 9% sales tax is to be collected, determine the sales tax and the total price on a five-piece dining set marked $310.

7. If a sales tax of 5% amounted to $6.20,
   a. What was the marked price?
   b. What was the total price including the sales tax?

8. A 6% sales tax on a humidifier was $5.34.
   a. What was the marked price?
   b. What was the total price including the sales tax?

9. The excise tax on a pair of gold earrings was $4.32. If a 12% tax was levied on the jewelry,
   a. What was the marked price?
   b. What was the total price?

10. The import duty on a crate of citrus fruit was $7.50. If a 15% duty was charged,
    a. What was the crate's declared value?
    b. What was the total price including the duty?

11. The total cost of a briefcase was $95.04. If the tax rate was 8%,
    a. What was the marked price of the case?
    b. What was the sales tax?

12. The total cost of a tent was $198.55. If the tax rate was 4½%,
    a. What was the marked price of the tent?
    b. What was the sales tax?

13. The total cost of a word-processing package was $299.60. If the tax rate was 7%,
    a. What was the marked price?
    b. What was the sales tax?

14. The total cost of a hammock was $93.45. If the tax rate was 5%,
    a. What was the marked price?
    b. What was the sales tax?

15. The total price of a pair of canvas shoes was $40.56 after a 4% sales tax was added to the marked price.
    a. What was the marked price of the shoes?
    b. How much was the sales tax?

16. After a 6½% sales tax was added, the total cost of a braided rug was $66.03.
    a. How much was the marked price of the rug?
    b. How much sales tax was paid?

17. A beach towel was purchased for a total price of $18.55. If the sales tax rate was 6%,
    a. What was the marked price of the towel?
    b. What was the sales tax on the towel?

18. The price of a camera was $93.96, which included an 8% sales tax.
    a. What was the marked price of the camera?
    b. How much was the sales tax?

19. A ticket to a professional basketball game cost $23.20, which included a 6% sales tax and a 10% excise tax.
    a. What was the price of the ticket before the taxes were added?
    b. What was the amount of the sales tax?
    c. How much was the excise tax?

20. The gasoline bill for a car was $15.60, which included a 5% sales tax and a 15% excise tax.
    a. What was the basic charge for the gasoline?
    b. How much sales tax was paid?
    c. How much excise tax was paid?

21. A diamond ring had a total price of $2,340. This price included a 5% city/state sales tax and a 12% federal excise tax.
    a. What was the marked price of the ring?
    b. What was the amount of the sales tax?
    c. How much federal excise tax was included?

22. The price of a five-piece place setting of imported china was $84. This price included a 7% city/state sales tax and a 13% import tax.
    a. What was the marked price of the china?
    b. What was the amount of the sales tax?
    c. How much import tax was included?

23. An automobile dealership advertised an imported car for $22,000. In addition, the buyer must pay a 5% state sales tax and a 10% import duty.
    a. Determine the total price of the automobile.
    b. Determine the sales tax.
    c. Determine the import duty.

24. Imported photographic equipment had a price tag of $25,000. This price was subject to a 6% state sales tax and a 12% import tax.
    a. What was the total price of the equipment?
    b. What was the sales tax?
    c. What was the import tax?

SECTION 2
PROPERTY TAX

The typical means by which counties, cities, and independent public school boards secure revenue is the **property tax.** This tax applies to **real estate** (land and the building improvements on it), as well as to **personal property** (cash, cars, household furnishings, appliances, jewelry, and so on). Often, the real-property tax and personal-property tax are combined into a single assessment. Two things affect the amount of tax that will be paid: (1) the assessed value of the property and (2) the tax rate.

Property tax is normally paid on the **assessed value** of property rather than on its market value. Assessed value in any given locality is usually determined by taking a percent of the estimated fair market value of property. The percent used varies greatly from locality to locality, although assessed valuations between 50% and 100% are probably most common.

Thus, the tax rate alone is not a dependable indication of how high taxes are in a community; one must also know how the assessed value of property is determined. Even though two towns have the same tax rate, taxes in one of the towns might be much higher if the property there is assessed at a higher rate. Similarly, a community with a higher tax rate than surrounding areas might actually have lower taxes if property there is assessed at a much lower value. The recent trend has been for property to be assessed at higher percents of market value and for tax rates to be reduced somewhat—with the net result that property taxes continue to increase. Assessed valuations supposedly equal to market value are becoming more common, although even then the property would probably bring somewhat more if actually sold.

The tax *rate* of an area is usually expressed in one of three ways: as a percent, as an amount per $100, or as an amount per $1,000. The latter method is less common than percent or per $100 rates. A variation of the amount per $1,000 is to express the rate in mills. **Mills** indicates "thousandths."

The county commissioners, city council members, or other board responsible for establishing the tax rate use the following procedure: After the budget (total tax needed) for the coming year has been determined and the total assessed value of property in the area is known, the rate is computed:

$$\text{Rate} = \frac{\text{Total tax needed}}{\text{Total assessed valuation}}$$

Example 1 Markly County has an annual budget of $4,800,000. If the taxable property in the county is assessed as $300,000,000, what will be the tax rate for next year? Express this rate (a) as a percent, (b) as an amount per $100, (c) as an amount per $1,000, and (d) in mills.

$$\text{Rate} = \frac{\text{Total tax needed}}{\text{Total assessed value}}$$

$$= \frac{\$4,800,000}{\$300,000,000}$$

$$\text{Rate} = 0.016$$

(a) The property tax rate is 1.6% of the assessed value of property.

(b) When the rate is expressed per $100, the decimal point is marked off after the hundredths place. Thus, the rate is $1.60 per C (or per $100).

(c) The rate per $1,000 is found by marking off the decimal point after the thousandths place. Hence, the rate is $16.00 per M (or per $1,000).

(d)   The rate in mills is found the same way as in part (c), but the $ sign is not used (16 mills).

In most instances, the tax rate does not divide out to an even quotient as in Example 1. When a tax rate does not come out even, the final digit to be used is always rounded up to the next higher digit, regardless of the size of the next digit in the remainder.

Example 2   The tax rate initially computed for the Richland Independent School District is 0.0152437+. Express this rate (a) as a percent to two decimal places, (b) per $100 of valuation (correct to cents), (c) per $1,000 of valuation (correct to cents), and (d) in whole mills.

(a)   1.53%   (b) $1.53 per C   (c) $15.25 per M   (d)   16 mills

Property tax is found using the formula

$$\text{Rate} \times \text{Valuation} = \text{Tax}$$

The same basic formula is also used to find the rate or the valuation (assessed value) of property when either one is unknown.

Example 3   How much property tax will be due on a home assessed at $75,600 if the tax rate is 2%?

$$R = 2\% \qquad\qquad R \times V = T$$

$$V = \$75,600 \qquad 2\% \times \$75,600 = T$$

$$T = ? \qquad\qquad\qquad \$1,512 = T$$

The real property tax on this home is $1,512.

Example 4   A duplex apartment in Orange County is assessed at $79,500 and is subject to tax at the rate of $1.60 per C. Calculate the real property tax.

$$R = \$1.60 \text{ per C} \qquad R \times V = T$$

$$V = \$79,500 \qquad 1.60 \times \$795 = T$$

$$= \$795 \text{ hundred} \qquad \$1,272 = T$$

$$T = ?$$

A property tax of $1,272 is due on the duplex.

Note.   When the tax formula is solved for valuation (V), the quotient obtained represents that many hundreds or thousands of dollars of valuation, according to whether the rate is per $100 or per $1,000. The original quotient must be converted to a complete dollar valuation. (For example, a solution of "$721" means "$721 hundred" when the rate is per $100; hence, the assessed value must be rewritten $72,100.)

Example 5   A property tax of $1,120 was paid in Brazos City, where the tax rate is $17.50 per M. What is the assessed value of the property?

$$R = \$17.50 \text{ per M} \qquad R \times V = T$$

$$V = ?$$

$$\frac{17.50V}{17.50} = \frac{\$1,120}{17.50}$$

$$T = \$1,120$$

$$V = \frac{1,120}{17.50}$$

$$= 64 \text{ thousand}$$

$$V = \$64,000$$

The property is assessed at $64,000.

Example 6   The office building of Wesson & White, Inc., which is assessed at $86,500, was taxed $1,557. Determine the property tax rate per $100.

$$R = ? \text{ per C} \qquad R \times V = T$$

$$V = \$86,500 \qquad R \times \$865 = \$1,557$$

$$= \$865 \text{ hundred} \qquad 865R = 1,557$$

$$T = \$1,557$$

$$R = \frac{1,557}{865}$$

$$R = \$1.80 \text{ per C}$$

The tax rate is $1.80 per $100 of assessed valuation.

Example 7   The property tax was $792 on a home when the rate was 1.65%. During the succeeding year, the assessed value of the home was increased by $3,000, and the property tax amounted to $867. Determine the amount of increase in the tax rate.

$$R = 1.65\% \qquad R \times V = T$$

$$V = ? \qquad 0.0165V = \$792$$

$$T = \$792 \qquad V = \frac{792}{0.0165}$$

$$V = \$48,000$$

The assessed value of the home for the first year was $48,000. The assessed value of the home for the second year was $48,000 + $3,000 = $51,000.

$$R = ? \qquad R \times V \quad = T$$

$$V = \$51,000 \qquad R(\$51,000) = \$867$$

$$T = \$867 \qquad R = \frac{867}{51,000}$$

$$= 0.017$$

$$R = 1.7\%$$

$$\begin{array}{r} 1.70\% \\ -\ 1.65 \\ \hline 0.05\% \end{array} \quad \text{Increase in rate}$$

SECTION 2 PROBLEMS

In the following problems, express each tax rate as indicated.

1. 0.0163102+
  a. As a percent with three decimal places
  b. As an amount per $100, correct to cents
  c. As an amount per $1,000, correct to cents
  d. In whole mills

2. 0.0184402+
  a. As a percent with two decimal places
  b. As an amount per $100, correct to cents
  c. As an amount per $1,000, correct to cents
  d. In whole mills

3. 0.0215335+
  a. As a percent with one decimal place
  b. As an amount per $100, correct to cents
  c. As an amount per $1,000, correct to cents
  d. In whole mills

4. 0.0192356+

    a. As a percent with two decimal places

    b. As an amount per $100, correct to cents

    c. As an amount per $1,000, correct to cents

    d. In whole mills

For Problems 5 and 6, determine the tax rate (a) as a percent with one decimal place, (b) as an amount per $100, correct to cents, (c) as an amount per $1,000, correct to cents, and (d) in whole mills.

5. The total assessed valuation of all real property in Beauregard City was $540,000,000. The city council voted an annual budget of $17,000,000. Express the tax rate in the four variations.

6. In Queens County, the county commissioners voted a budget of $24,000,000. The total assessed value of all real estate in the county amounted to $980,000,000. Express the tax rate in the four variations.

Find each missing factor relating to property tax.

	RATE	ASSESSED VALUE	TAX
7. a.	1.4%	$64,000	896
b.	$1.15 per C	72,000	828
c.	$10.20 per M	55,000	561
d.	3.7%	46080	$1,702
e.	$2.05 per C	98000	2,009
f.	$16.20 per M	85000	1,377
g.	?%	63,000	1,638
h.	? per C	92,000	1,288
i.	? per M	80,000	1,232
8. a.	2.8%	$94,000	
b.	$1.25 per C	59,200	
c.	$12.40 per M	75,000	
d.	2.1%		$1,050
e.	$2.60 per C		1,768
f.	$18.80 per M		1,222
g.	?%	82,000	1,148
h.	? per C	74,500	2,384
i.	? per M	95,000	912

9. Property owners in Auburn pay a tax rate of 2.6%. Determine the tax due on real property assessed at $74,000.

10. Determine the tax due on real estate in Charles County assessed at $35,000 if the tax rate is 1.8%.

11. The tax rate on real property in Littleville is $1.85 per C. How much tax is due on a residence assessed at $40,000?

12. The property tax rate in Marshall City is $2.05 per C. A commercial building has been assessed at $90,000. How much tax is due?

13. A resident of Alexandria paid $1,764 in property tax last year. What was the assessed value of the property if the tax rate was $1.26 per C?

14. A property tax of $429 was paid on undeveloped land in Denver. What was the assessed value of the property if the tax rate was $1.95 per C?

15. Determine the assessed value of a home in Fairfax County if the tax rate is $10.40 per $1,000 and the property taxes due are $468.

16. Determine the assessed value of a home in Wadley if the tax rate is $12.75 per M and the property taxes are $918.

17. What percent is the tax rate in Montgomery County if a property owner paid $6,240 in property taxes on a home assessed at $240,000?

18. A resident paid $2,016 in property taxes on a home assessed at $84,000. What percent is the tax rate?

19. Property taxes in Edgewood cost a homeowner $715. The assessed value of the house was $55,000. What was the tax rate expressed in whole mills?

20. A homeowner in Hope paid $525 in real estate taxes on a home assessed at $35,000. What was the tax rate expressed in whole mills?

21. Property tax in Albany cost a homeowner $1,785. If the home was assessed at $105,000, what was the tax rate per $100?

22. Property taxes cost a small business $3,960. If the real property was assessed at $180,000, what was the tax rate per $100?

23. The property tax on real property was $1,184 when the assessed value was $64,000. The following year, the tax rate increased by $0.20 per $100 and the property tax was $1,230. Determine how much the assessed value had decreased. (Hint: Solve each year separately.)

24. The property tax was $1,349 on commercial property when the assessed value was $95,000. During the succeeding year, the tax rate was increased by $0.08 per $100, and the property tax amounted to $1,434. Determine the amount of increase in the assessed value. (Hint: Solve each year separately.)

25. The property tax rate in Cropwell was 1.15% and the property tax on a home was $897. During the next year, the assessed value of the home was increased by $2,000, and the property tax amounted to $984. Determine the tax rate for the second year.

26. The first-year property tax on a new home was $390, and the property tax rate was 1.3%. During the second year, the assessed value of the home increased by $4,000, and the property tax amounted to $544. What was the tax rate for the second year?

CHAPTER 5 GLOSSARY

Assessed value. A property value (usually computed as some percent of true market value) which is used as a basis for levying property taxes.

Customs duty (import tax, tariff). A tax levied on products brought into this country.

Excise tax. A tax levied on the manufacture or sale of certain products and services, usually luxury items, or to finance projects related to the item taxed; paid in addition to sales tax.

Luxury tax.   A tax on automobiles imposed on that portion of the marked price that exceeds a certain cutoff point.

Mills.   A property tax rate that denotes "thousandths." (Example: 16 mills indicates a rate of 0.016.)

Personal property.   Possessions (such as cars, cash, household furnishings, appliances, jewelry, etc.) which together with real property are subject to property tax.

Property tax.   A tax levied on real property and personal property; the chief means by which counties, cities, and independent school districts obtain revenue.

Real property.   Real estate (land and buildings) which together with personal property is subject to property tax.

Sales tax.   Tax levied by states (and sometimes by cities and counties) on the sale of merchandise and services.

Use tax.   A tax levied on merchandise brought into a state after its purchase in another state.

INSURANCE

OBJECTIVES

Upon completion of Chapter 6, you will be able to:

1. Define and use correctly the terminology associated with each topic.

2.   **a.** Compute fire insurance premiums (Section 1: Example 1; Problems 1–4, 11, 12)

   b. Compute the premium refund due when a policy is canceled (1) by the insured or (2) by the carrier (Examples 2, 3; Problems 3, 4, 13, 14).

3.   **a.** Determine compensation due following a fire loss if a coinsurance clause requirement is not met (Section 1: Example 4; Problems 5, 6, 9, 10, 15–20)

   b. Find each company's share of a loss when full coverage is divided among several companies (Example 5; Problems 7, 8, 21–24)

   c. Find each company's share of a loss when a coinsurance requirement is not met (Example 6; Problems 9, 10, 25–28).

4.   **a.** Compute the annual premium for (1) automobile liability and medical payment insurance and (2) comprehensive/collision insurance (Section 2: Examples 1, 3; Problems 1–12)

   b. Find the compensation due following an accident (Examples 2, 4; Problems 13–20).

5. Determine premiums for the following life insurance policies: (a) term, (b) whole life, (c) 20-payment life, and (d) variable universal life (Section 3: Examples 1–3; Problems 1, 2, 7–14).

6. Calculate the following nonforfeiture values: (a) cash value, (b) paid-up insurance, and (c) extended term insurance (Section 3: Example 4; Problems 3, 4, 13–18).

7. Compare the monthly (and total) annuities that a beneficiary could receive under various settlement options (Section 3: Examples 5–7; Problems 5, 6, 19–26).

All insurance is carried to provide financial compensation should some undesirable event occur. Insurance operates on the principle of shared risk. Many persons purchase insurance protection and their payments are pooled in order to provide funds with which to pay those who experience losses. Thus, rates are lower when many different people purchase the same protection and divide the risk; rates would be much higher if only a few people had to pay in enough money to provide reimbursement for a major loss.

The risk that an insurance company assumes by insuring a person, organization, or event always influences how high the rates will be. The company determines the mathematical probability that the insured entity will incur losses and then sets its rates accordingly. The insurance company does not know which of its clients will suffer a loss, but it knows from experience how many losses to expect. The insurance rates must be high enough to pay the expected losses, to pay the expenses of operating the company, and to provide a reserve fund should losses exceed the expected amounts during any period.

The payment made to purchase insurance coverage is called a **premium.** Insurance premiums are usually paid either annually, semiannually, quarterly, or montly. The total cost of insurance is slightly higher when payments are made more frequently, in order to cover the cost of increased bookkeeping. An insurance contract, or **policy,** defines in detail the provisions and limitations of the coverage. The amount of insurance specified by the policy is the **face value** of the policy. Insurance premiums are paid in advance, and the **term** of the policy is the time period for which the policy will remain in effect. The insurance company is often referred to as the **insurer,** the **underwriter,** or the **carrier.** The person or business that purchases insurance is called the **insured** or the **policyholder.** When an insured loss occurs, the payment that the insurance company makes to reimburse the policyholder (or another designated person) is often called an **indemnity.**

As indicated above, insurance is sold in order to provide financial protection to the policyholder; it is not intended to be a profit-making proposition when a loss occurs. Thus, insurance settlements are always limited to either the actual amount of loss or the face amount of the policy, whichever is smaller.

Individual states have insurance commissions that regulate premium structure and insurance requirements (such as required automobile insurance) passed by their legislatures. Thus, insurance practices and premium computation vary widely from state to state.

SECTION *1*
BUSINESS INSURANCE

Certain standard forms of insurance are carried by almost all businesses; these types of insurance are discussed on the following pages. Because of its importance to private citizens as well as to business organizations, fire insurance receives particular attention.

FIRE INSURANCE

Basic fire insurance provides financial protection against property damage resulting from fire or lightning. (Insured property may be buildings and their contents, vehicles, building sites, agricultural crops, forest crops, and so forth.) Further damage resulting from smoke or from attempts to extinguish or contain a fire may be as costly as the fire damage itself. Therefore, **extended coverage** is usually included to provide protection against smoke, water, and chemical damage, as well as physical destruction caused by firefighters having to break into the property or taking measures to keep a fire from spreading. Extended coverage also covers riot, civil commotion, hurricane, wind, and hail damage.

Homeowners often purchase a **comprehensive** "package" **policy,** which, in addition to the coverage above, also provides protection against loss due to theft, vandalism, and malicious mischief; protection against accidental loss or damage; and liability coverage in case of personal injury or property damage to visitors on the property. Similar protection may also be purchased by business firms, although it is sometimes obtained in separate policies.

PREMIUMS

Historically, fire insurance policies were written for several years in advance and reduced long-term rates were given for premium prepayments. (For instance, a prepaid 3-year policy might have cost only 2.7 times as much as a 1-year policy.) In recent years, however, spiraling costs for reimbursing fire losses have forced insurance companies to discontinue these practices. Most fire insurance policies are now written for either 1 or 3 years, and the prepaid cost of a 3-year policy is three times the 1-year premium. If the 3-year premium is paid in three annual payments, interest may also be charged for the latter 2 years. It is probable that this total cost would still be cheaper than three 1-year policies, however, since rates have been increasing so rapidly.

Each state has a fire-rating bureau that establishes fire insurance rates. A representative of the bureau inspects every commercial property and determines the rate to be paid by each. These rates vary greatly, because they are influenced by many different factors. The primary factors that the inspector must consider are (1) the construction of the building, (2) the contents stored or manufactured therein, (3) the quality of fire protection available, (4) the location of the buildings, and (5) past experience regarding the probability of a fire.

Thus, a building of fireproof construction (all steel and concrete, and with a fire-resistive roof) would have a much lower insurance rate than a frame building with a flammable roof. Similarly, property located near fire hydrants and served by a well-trained fire department in an area that receives abundant rainfall would require much lower premiums than property located in an arid region and some distance from a fire station. (The installation of fire extinguishers or a sprinkler system would reduce the rate somewhat.) A dress shop would normally have a lower rate than a paint store, because of the combustible nature of painting supplies; if the dress shop relocated to a site beside a gasoline storage area, however, its fire insurance rate would increase greatly.

Fire insurance rates are quoted as an amount per $100 of insurance coverage. Separate rates are charged for insurance on the building itself, on the furniture and fixtures, and on the stock or merchandise—normally in that order of increasing premium expense. Table 6-1 and Example 1 illustrate the way in which the premiums may be determined.

Example 1 Using Table 6-1, determine the fire insurance premium due for a $125,000 home policy in Territory 1, Class B category, with contents valued at $30,000.

Fire insurance premiums are computed using the procedure

$$\text{Premium} = \text{Rate} \times \text{Value of policy}$$

or simply

$$P = R \times V$$

Thus, the premium for the preceding home would be computed as follows:

$R = \$0.56 \text{ per } \$100$ $\qquad$ $P = R \times V$

$V = \$125,000$ $\qquad\qquad\qquad = \$0.56 \times \$1,250$

$\qquad = \$1,250 \text{ hundreds}$ $\qquad P = \$700 \text{ (annual premium)}$

$P = ?$

The annual premium on the building itself is $700. The premium to insure the contents would be separately computed in a similar manner.

$R = \$0.79 \text{ per } \$100$ $\qquad$ $P = R \times V$

$V = \$30,000$ $\qquad\qquad\qquad = \$0.79 \times \$300$

$\qquad = \$300 \text{ hundreds}$ $\qquad P = \$237$

$P = ?$

TABLE 6-1	ANNUAL FIRE INSURANCE PREMIUMS PER $100 OF FACE VALUE							
	STRUCTURE CLASS				CONTENTS CLASS			
TERRITORY	A	B	C	D	A	B	C	D
1	$0.45	$0.56	$0.66	$0.75	$0.51	$0.79	$0.79	$0.98
2	0.58	0.64	0.78	0.80	0.68	0.86	0.91	1.03
3	0.63	0.72	0.85	0.93	0.71	0.89	0.95	1.12

The total premium for the building and the contents would be:

$$P = \$700 + \$237$$

$$P = \$937$$

Note.   Fire insurance premiums are rounded to the nearest whole dollar.

SHORT-TERM POLICIES

Sometimes a business may wish to purchase fire insurance for only a limited time in order to obtain coverage on merchandise that will soon be sold. Also, policies are sometimes canceled by the policyholders if they sell the insured property, if they move, and so on. If, for any reason, a policy is in effect for less than 1 year, it is considered a **short-term policy.** The premium to be charged for such a policy is computed according to a short-rate table, such as Table 6-2. A very short policy is relatively expensive because of the cost to the insurance company of selling, writing, and processing the policy.

Example 2   Grant Avery owned a home on which he paid $800 for fire insurance. He moved after 8 months and canceled the insurance policy. How much refund will Avery receive?

For an 8-month premium, 80% of the annual premium will be charged:

Annual premium	$800	Annual premium	$800
×	.80	−	640
8-month premium	$640	Refund due	$160

Occasionally, it becomes necessary for an insurance company to cancel a policy. This might happen because of excessive claims or a refusal of the policyholder to accept higher rates when the fire risk has increased. The company ordinarily must give advance notice that the policy will be canceled, thus allowing the policyholder time to purchase another policy before the coverage ceases. When a company cancels, it may keep only that fraction of

TABLE 6-2	SHORT-TERM RATES INCLUDING CANCELLATION BY THE INSURED		
MONTHS OF COVERAGE	PERCENT OF ANNUAL PREMIUM CHARGED	MONTHS OF COVERAGE	PERCENT OF ANNUAL PREMIUM CHARGED
1	20%	7	75%
2	30	8	80
3	40	9	85
4	50	10	90
5	60	11	95
6	70	12	100

the annual premium equivalent to the fraction of the year which has elapsed. Many companies compute this time precisely, using $\dfrac{\text{exact days}}{365}$; however, we shall use approximate time for our calculations. The insurance company is not entitled to short-term rates when it cancels the policy.

Example 3   A $640,000 commercial fire policy, sold at the rate of $0.75 per $100, is canceled by the underwriter after 5 months. How much refund will the policyholder receive?

The annual premium is determined in the usual manner:

$R = \$0.75$ per $100   Annual premium $= R \times V$

$V = \$640,000$ $= \$0.75 \times \$6,400$

$\quad = \$6,400$ hundreds $P = \$4,800$

$P = ?$

Because the underwriter canceled after 5 months, only ⁵⁄₁₂ of the annual premium may be retained:

5-month premium $= \frac{5}{12} \times \$4,800$ Annual premium $\$4,800$
$\qquad\qquad\qquad = \$2,000$ 5-month premium $-$ $\underline{2,000}$
                                                                      Refund due $\$2,800$

Calculator techniques . . .   FOR EXAMPLE 3

Set up the problem as demonstrated in Example 3. In the first step, recall that $\boxed{\text{M+}}$ after an operation includes the $\boxed{=}$. Parentheses below mean to use the displayed number without re-entering it.

0.75 $\boxed{\times}$ 6,400 $\boxed{\text{M+}}$ ⟶ 4,800

(4,800) $\boxed{\times}$ 5 $\boxed{\div}$ 12 $\boxed{=}$ ⟶ 2,000

$\boxed{\text{MR}}$ ⟶ 4,800 $\boxed{-}$ 2,000 $\boxed{=}$ ⟶ 2,800

COMPENSATION FOR LOSS; COINSURANCE CLAUSE

Experience indicates that most fires result in the loss of only a small portion of the total value of the property. Realizing that the total destruction is relatively rare, many people would be tempted to economize by purchasing only partial coverage. However, to pay the full amount of these damages would be extremely expensive to an insurance company if most policyholders were paying premiums on only the partial value of their property. Therefore, to maintain an economically sound business operation, as well as to encourage property owners to purchase full insurance coverage, many policies contain a coinsurance clause. A **coinsurance clause** guarantees full payment after a loss (up to the face amount of the policy) only if the policyholder purchases coverage which is at least a specified percent of the total value of the property at the time the fire occurs. For example, if a house would cost $125,000 to replace, then this figure represents the total value, regardless of how much less money the house may have cost originally.

Many policies contain coinsurance clauses requiring 80% coverage, although lower coverage is sometimes allowed, and higher (particularly for extended coverage policies) may often be necessary. If the required coverage is not carried, the company will pay damages equivalent to the ratio of the required coverage that is carried; the policyholder must then bear the remaining loss. (In this respect, the policyholder "coinsures" the property. The term "coinsurance" implies that the insurance company and the policyholder each assume part of the total risk.) Most policies on homes do provide full coverage; businesses are more likely than homeowners to partially coinsure themselves.

A property owner may purchase more insurance than the minimum requirement, of course. However, in no case will payment for damages exceed the actual value of the property destroyed, no matter how much insurance is carried. The maximum payment is always the amount of damages or the face value of the policy, whichever is smaller. When a fire occurs, the policyholder must contact the insurance company, which sends out an *insurance adjuster* to inspect the damage and determine the indemnity to be paid. (The policyholder may also have to obtain repair or reconstruction estimates from several companies.)

Example 4   Centex Distributors, Inc. operates in a building valued at $750,000. The building is insured for $500,000 under a fire policy containing an 80% coinsurance clause. (a) What part of a fire loss will the company pay? (b) How much will the company pay on a $120,000 loss? (c) On a $420,000 loss? (d) On a $720,000 loss?

(a)   According to the coinsurance clause, 80% of the property value must be insured:

$$\$750,000 \times 80\% = \$600,000 \text{ insurance required}$$

$$\frac{\text{Insurance carried}}{\text{Insurance required}} = \frac{\$500,000}{\$600,000} = \frac{5}{6}$$

Centex Distributors carries only $\frac{5}{6}$ of the insurance required by the coinsurance clause; therefore, the insurance company will pay only

$\frac{5}{6}$ of any loss, up to the face value of the policy. By their failure to carry the required coverage, Centex has indirectly agreed to coinsure the remaining portion of any fire loss themselves.

(b) $\frac{5}{6} \times \$120,000 = \$100,000$ insurance reimbursement

(c) $\frac{5}{6} \times \$420,000 = \$350,000$ indemnity

(d) $\frac{5}{6} \times \$720,000 = \$600,000$

Preliminary calculations indicate the insurance company's share of a $720,000 loss to be $600,000. However, this exceeds the face value of Centex's policy; therefore, the insurance company's payment is limited to the full value of the policy, or $500,000.

MULTIPLE CARRIERS

Insurance coverage on property may often be divided among several companies. This may happen because the value of the property is so high that no single company can afford to assume the entire risk, because the owner wishes to distribute his business for public relations purposes, or simply because coverage on various portions of the property was purchased separately over a period of years.

When property is insured with multiple carriers, each company pays damages in the ratio that its policy bears to the total insurance coverage. As noted previously, the total amount paid by all carriers will never exceed the value of the damage, nor will any company's payment exceed the face value of its policy. Multiple policies may also contain coinsurance clauses.

Example 5   **Coinsurance Met**

Farnsworth Equipment suffered a $180,000 fire. Farnsworth was insured under the following policies: $150,000 with Company A, $100,000 with Company B, and $50,000 with Company C. What indemnity will each company pay? (Assume that Farnsworth meets the coinsurance requirement necessary to receive full coverage on a loss.)

Policies

$150,000	A pays $\dfrac{\$150,000}{\$300,000}$	or $\dfrac{1}{2}$ of the loss
100,000	B pays $\dfrac{\$100,000}{\$300,000}$	or $\dfrac{1}{3}$ of the loss
$\underline{50,000}$	C pays $\dfrac{\$\ 50,000}{\$300,000}$	or $\dfrac{1}{6}$ of the loss
$300,000		

The full $180,000 loss is insured; therefore, each company's payment is as follows:

Company A: $\frac{1}{2} \times \$180,000 = \$ 90,000$

Company B: $\frac{1}{3} \times \$180,000 = 60,000$

Company C: $\frac{1}{6} \times \$180,000 = \underline{\quad 30,000}$

Total indemnity = $180,000

Example 6   **Coinsurance Not Met**

Centex Distributor's building (Example 4) is valued at $750,000 but is insured for only $500,000 under an 80% coinsurance requirement. Assume now that this $500,000 coverage is divided between Company A ($300,000) and Company B ($200,000). (a) What part of any loss is covered? (b) How much reimbursement will Centex receive after a $120,000 fire? (c) What is each insurance company's payment after the $120,000 loss?

(a) As in Example 4, the required coverage is 80% × $750,000, or $600,000. Thus, the part of any loss that is covered (up to the face value of the policy) as before is

$$\frac{\text{Insurance carried}}{\text{Insurance required}} = \frac{\$500,000}{\$600,000} = \frac{5}{6}$$

(b) Again, following a $120,000 fire,

$\frac{5}{6} \times \$120,000 = \$100,000$ Total indemnity (to be shared by all insurance carriers)

(c) Each company's share of this $100,000 indemnity (due following the $120,000 loss) is the ratio of its policy to the total insurance carried ($500,000):

Company A: $\frac{\$300,000}{\$500,000}$ or $\frac{3}{5} \times \$100,000 = \$ 60,000$

Company B: $\frac{\$200,000}{\$500,000}$ or $\frac{2}{5} \times \$100,000 = \underline{\quad 40,000}$

Total indemnity = $100,000

Thus, with multiple carriers, the $100,000 will be reimbursed as $60,000 from Company A and $40,000 from Company B. Centex assumes the remaining $20,000 loss itself.

Several other types of insurance are carried by most businesses. The following paragraphs give a brief description of this additional insurance.

BUSINESS INTERRUPTION

Business interruption insurance is purchased as a supplement to fire and property insurance. Whereas fire and property insurance provides coverage for the actual physical damage resulting from a fire or natural disaster (earthquake, flood, hurricane, tornado, and so on), business interruption insurance provides protection against loss of income until the physical damage can be repaired and normal business resumed. This insurance provides money to meet business expenses that continue despite the lapse in operations: interest, loan payments, mortgage or rent payments, taxes, utilities, insurance, and advertising, as well as the salaries of key members of the firm and ordinarily even the anticipated net profit.

LIABILITY

Businesses may be held financially liable, or responsible, for many different occurrences; general **liability** insurance is a business essential in order to provide this protection. An accident causing personal injury to a customer often results in the customer's suing the business. When firms are declared negligent of their responsibility for the customer's safety, judgments of large sums of money are often granted by the courts. Hence, no well-managed company would operate without *public liability* insurance.

Another area in which court judgments have been quite large in recent years is bodily injury resulting from defective or dangerous products. Companies that manufacture mechanical products, food products, or pharmaceutical products that could conceivably be harmful can be insured by *product liability* policies.

Liability for the safety of employees is covered under **worker's compensation** policies. This insurance provides payments to an employee who loses work time because of sickness or accidental injury resulting directly from work responsibilities. State laws determine the amount of benefits paid under different situations. Disability income is paid to workers who are totally disabled temporarily, or is paid permanently to those who are partially disabled on a permanent basis. The national average is about 75% of the worker's take-home pay. Some states set no limits on length of payments, whereas other states set a limit in years or a maximum dollar amount paid to the worker. Most worker's compensation policies also provide death benefits in case of a fatality.

The labor statutes in 47 states require employers to carry this insurance; some states have a state fund for worker's compensation, although in most cases it is purchased from commercial insurance companies. As would be expected, worker's compensation rates vary greatly according to the occupational hazards of different jobs. These rates are usually quoted as an amount per $100 in wages. The premiums are usually based only on regular wages; that is, premiums are not paid on overtime wages.

A business firm's liability for the operation of motor vehicles is similar to that of the general public; both are included in the following section of this chapter. Business policies, however, often contain special provisions covering fleets of cars, hired cars, or the cars of employees being used in company business.

Money and Securities

The theft of money and securities (such as checks, stocks, titles of ownership, and so forth) is a constant hazard in many businesses. A number of different policies cover the various conditions under which money and valuable securities may be stolen, such as burglary from an office safe or robbery of a messenger carrying money or securities to or from the business.

General theft policies do not cover the dishonesty of employees, which must be provided for by separate policies. *Fidelity bonds* protect against embezzlement or other loss by employees who handle large sums of money. These bonds are essential in banks, retail stores, and other offices where large amounts of cash or securities are involved.

The worst financial loss experienced by most retail stores results from shoplifting (by both customers and employees), a problem of major national proportion that is increasing annually. There is presently no insurance available that adequately protects against shoplifting losses; thus, the business's only means of recovery is to pass the loss along to consumers in the form of higher prices.

Life Insurance

Particularly in small businesses, the death of a partner or a key person in the firm may jeopardize the entire business operation. Many firms, therefore, purchase life insurance on critical personnel to protect against such financial loss. Such policies also enable the firm to buy from the estate the business interest of a deceased partner. Life insurance is discussed in more detail in a later section.

Group Insurance

Group insurance available at many companies provides life insurance and/or health and accident coverage for employees and their families. This coverage, unlike worker's compensation, provides coverage for illnessses or accidents not related to occupational duties. Most group life insurance plans offer term life insurance similar to the policies described later in this chapter. These group life policies are also available to many individuals through the official associations of their business or profession, alumni associations, or other special interest groups.

Typical group hospitalization plans provide basic benefits for either accidental injuries or medical/surgical care. Each policy contains a long list of covered hospital services and surgical procedures, each specifying the maximum amount that will be paid to the hospital and surgeon. Maternity benefits are generally included in group policies, although a specified waiting period is often required for this coverage, as well as for coverage of certain preexisting medical conditions.

Some group policies also include a major medical supplement, which provides for certain expenses not covered by the basic benefits. For example, major medical coverage will pay a specified percent of the costs of physicians' fees, prescriptions, and many services or procedures not otherwise covered, as well as excess expenses above those reimbursed through the basic policy. All major medical plans and some basic policies specify a deductible amount that the patient must first pay before insurance coverage becomes effective.

The employer often pays half or more of the group insurance premiums as one of the organization's fringe benefits. Few companies pay the full amount of employees' premiums anymore, however. Even when the employee pays the entire cost, group insurance is cheaper than similar coverage purchased privately. Also, group policies may usually be purchased without a physical examination, which permits some people to receive insurance protection when they would not ordinarily qualify. Coverage usually terminates immediately when a worker leaves the firm, although legally the worker can remain on the group policy for a limited time by assuming the full cost. As another alternative, a group policy may often be converted to a private policy (at an increased rate, of course).

In many cases, group insurance provides standard coverage for all employees. However, certain benefits may sometimes be increased according to the employee's salary or years of service to the business. A recent trend at many businesses, called "cafeteria coverage," lets employees choose from the various policies available, in order to meet the particular needs of themselves or their individual families.

SECTION 1   PROBLEMS

Compute the fire insurance premiums required for the following homes, using the premiums listed in Table 6-1.

		STRUCTURE VALUE	CONTENTS VALUE	CLASS	TERRITORY	TOTAL PREMIUM
1.	a.	$110,000	$40,000	C	2	
	b.	220,000	50,000	D	3	
2.	a.	$350,000	$64,000	B	1	
	b.	580,000	78,000	A	2	

Compute the premium and the refund due (if applicable) for the following amounts of insurance for the terms indicated. Assume an annual rate of $0.78 per $100. Use the short-term rates from Table 6-2 when applicable.

		AMOUNT OF INSURANCE	TERM	CANCELED BY	PREMIUM	REFUND DUE
3.	a.	$ 365,000	1 year	X		X
	b.	670,000	5 months	X		X
	c.	820,000	8 months	Insured		
	d.	1,000,000	9 months	Carrier		
4.	a.	$ 290,000	1 year	X		X
	b.	770,000	6 months	X		X
	c.	830,000	9 months	Insured		
	d.	1,500,000	4 months	Carrier		

Compute the compensation (indemnity) that will be paid under each of the following conditions.

		PROPERTY VALUE	COINSURANCE CLAUSE	INSURANCE REQUIRED	INSURANCE CARRIED	AMOUNT OF LOSS	INDEMNITY
5.	a.	$ 400,000	80%		$200,000	$140,000	
	b.	680,000	80		544,000	100,000	
	c.	300,000	90		210,000	180,000	
	d.	800,000	80		560,000	800,000	
	e.	700,000	70		420,000	630,000	
6.	a.	$ 500,000	80%		$200,000	$250,000	
	b.	440,000	80		352,000	90,000	
	c.	600,000	90		450,000	420,000	
	d.	900,000	80		600,000	780,000	
	e.	1,000,000	80		700,000	950,000	

Determine the compensation to be paid by each carrier below (assume that coinsurance clause requirements are met).

		COMPANY	AMOUNT OF POLICY	RATIO OF COVERAGE	AMOUNT OF LOSS	COMPENSATION
7.	a.	A	$360,000		$350,000	
		T	480,000			
	b.	R	$25,000		$80,000	
		I	40,000			
		S	15,000			
		K	20,000			

	COMPANY	AMOUNT OF POLICY	RATIO OF COVERAGE	AMOUNT OF LOSS	COMPENSATION
8. a.	N	$440,000		$477,000	
	O	620,000			
b.	I	$750,000		$840,000	
	N	400,000			
	S	250,000			

Calculate the settlement due from each carrier below under the given coinsurance terms.

	PROP. VALUE	COINS.	INSURANCE REQUIRED	INSURANCE CARRIED	FIRE LOSS	TOTAL INDEM- NITY	COM- PANY	POLICY VALUE	CO. RATIO	CO. PAY- MENT
9. a.	$500,000	80%			$300,000		E	$100,000		
							F	260,000		
b.	700,000	90			90,000		G	224,000		
							H	280,000		
10. a.	$600,000	80%			$450,000		I	$200,000		
							J	120,000		
b.	800,000	70			770,000		K	280,000		
							L	200,000		

11. A boat manufacturer purchased a 6-month fire policy to protect its boats during the fall-winter months. The premium was based on an annual rate of $1.10 per $100. Find the short-term premium on its $250,000 inventory.

12. The Southern Co. insures its merchandise that is warehoused for a 3-month period. The annual rate for this fire insurance policy is $0.60 per $100. How much does this short-term policy cost if the merchandise is insured for $180,000?

13. A fire policy covering a $750,000 warehouse cost $1.36 per $100. Determine the premium charged and the refund due if the policy is canceled in 9 months
   a. By the insured
   b. By the carrier

14. An annual rate of $0.90 per $100 was charged on a fire insurance policy covering a $630,000 building. Determine the premium charged and the refund due if the policy is canceled after 5 months
   a. By the insured
   b. By the insurer

15. A 90% coinsurance clause is written into the fire insurance policy covering an $800,000 plant. If the plant is insured for $750,000, how much will the policy pay after
   a. A $24,000 fire
   b. A $420,000 fire
   c. A $700,000 fire

16. A fire insurance policy containing an 80% coinsurance clause is written for $600,000. The building covered by this policy is valued at $680,000. How much indemnity would be paid on

    a. An $80,000 loss

    b. A $500,000 loss

    c. A $600,000 loss

17. A supermarket valued at $1,000,000 is insured for $700,000. If the fire insurance policy includes an 80% coinsurance clause, determine the compensation due when a fire resulted in

    a. A $180,000 loss

    b. A $720,000 loss

    c. A total loss

18. An office park valued at $5,000,000 was insured for $3,500,000. The fire insurance policy contained a 90% coinsurance clause. What settlement would be due following

    a. A $36,000 loss

    b. A $2,700,000 loss

    c. A total loss

19. A print shop valued at $850,000 was insured for $612,000. If the fire insurance policy included a 90% coinsurance clause, determine the compensation due when a fire resulted in

    a. $50,000 damages

    b. $450,000 damages

    c. $790,000 damages

20. The owner of a small business housed in a $300,000 building purchased a $240,000 fire insurance policy to cover the building. The policy included a 90% coinsurance clause. How much indemnity would be paid when a fire caused

    a. $81,000 damages

    b. $189,000 damages

    c. $288,000 damages

21. Atlantic Sports Center, Inc. is insured for $260,000 with Company S, $125,000 with Company P, and $115,000 with Company T. The policies meet coinsurance requirements. Determine each company's settlement if there is

    a. A $60,000 fire loss

    b. A $408,000 fire loss

    c. A $500,000 fire loss

22. A hardware store is insured for $240,000 with Company A, $340,000 with Company B, and $420,000 with Company C. The policies meet all coinsurance requirements. What settlement would each insurer make as a result of

    a. A $50,000 fire

    b. A $900,000 fire

    c. A $1,000,000 fire

23. A paint store carries the following policies, which meet all coinsurance clauses: $200,000 with Company E, $100,000 with Company F, $180,000 with Company G, and $220,000 with Company H. What is each company's share of

    a. A $63,000 loss

    b. A $210,000 loss

    c. A $700,000 loss

24. A meat-processing plant has fire protection with four underwriters: $300,000 with AA Company, $350,000 with BB Company, $250,000 with CC Company, and $100,000 with DD Company. Each policy meets coinsurance requirements. What is each company's share of

a. A $72,000 loss

b. A $600,000 loss

c. A $1,000,000 loss

25. The Sullivan Company is insured by Company X for $400,000 and Company Y for $400,000. Both policies require 90% coinsurance. If there is a fire loss of $918,000 on its $1,000,000 building,

a. What total amount will Sullivan Company receive?

b. How much will each insurance company pay in settlement?

26. A jewelery store is insured by American Insurance Co. for $700,000 and by Liberty Insurance Co. for $800,000. Its merchandise and building are valued at $2,500,000. If both companies have 80% coinsurance requirements,

a. What total amount would the store receive on fire damages of $2,200,000?

b. How much will each insurance company pay in indemnities?

27. Swartz Hospital Supply Co. is valued at $4,800,000. The fire insurance policies are with Company C for $600,000, Company D for $1,500,000, and Company E for $900,000. Each policy contains a 90% coinsurance clause.

a. What is the total compensation that would be due after a $3,960,000 fire loss?

b. How much would each company pay toward the damages?

28. Mike's Computer Supply Co. is valued at $6,000,000. The store has fire insurance policies with Dickson Insurance Agency for $1,200,000, Eagle Insurance Company for $1,800,000, and Franklin Insurance Co. for $1,000,000. Each policy contains an 80% coinsurance clause. Determine:

a. The total indemnity paid for a fire loss of $4,500,000

b. How much each company would pay toward this loss

SECTION 2

MOTOR VEHICLE INSURANCE

Automobile and truck insurance is carried by responsible owners (private or industrial) of most vehicles. Insurance policies on motor vehicles are usually written for a maximum term of 1 year. The primary reason for the 1-year policy is the alarmingly high nationwide accident rate—each year, there are more than 25 million auto accidents in the United States. These accidents constantly increase rates for automobile insurance, because repair costs continue to spiral. Since insurance rates always reflect the risk taken by the insurance company, the driving record (and age) of the operator is another major factor in determining the cost of vehicle insurance. This further accounts for the 1-year policies, since a driver's record may fluctuate over a period of years.

Insurance on motor vehicles is divided into two basic categories: (1) liability coverage and (2) comprehensive/collision coverage.

LIABILITY INSURANCE

Liability insurance provides financial protection for damage inflicted by the **policyholder** to the person or property of other people. As a protection to its citizens, virtually all states require the owners of all vehicles registered in the state to purchase minimum amounts of liability insurance. Liability insurance covers **bodily injury liability** (liability for physical injury to others resulting from the policyholder's negligence, such as a pedestrian or occupants of another car being struck by the insured) and **property damage liability** (liability for damage caused by the policyholder to another vehicle or other property).

Also available on an optional basis is **medical payment** insurance (commonly called "medical pay"). Medical payment insurance pays for necessary medical and/or funeral services resulting from an auto accident. This insurance supplements other health and accident insurance the policyholder may carry, and it will begin paying medical costs immediately, without waiting for the courts to rule on a liability suit. Under liability coverage, the insurance company promises to pay for bodily injury only to the extent that the insured is legally responsible for having caused the injury. Medical pay covers the insured driver, family members when they are either passengers or pedestrians struck by another motor vehicle, and any other person occupying the covered auto. Occupants of other cars struck by the insured or by a family member are not covered by this part of the policy.

Each state determines the minimum liability coverage for its drivers. For example, the state of Virginia has set "25/50/20" as the minimum coverage. This means that the policy will pay up to $25,000 for the bodily injury caused to a single person or, when more than one person is involved in the accident, a maximum total of $50,000 for the injuries inflicted to all victims. Also, a maximum of $20,000 will be paid for the property damage resulting from a single accident. On a nationwide basis, 25/50 for bodily injury and 10 to 30 for property damage is the most common coverage. Much higher liability coverage is available at rates only slightly higher than those for minimum coverage. Thus, many drivers purchase bodily injury coverage up to $300,000/ $300,000 and property damage coverage up to $100,000.

If a court suit should result in a victim's being awarded a settlement that exceeds the policyholder's coverage, the additional sum would have to be paid by the insured. For example, suppose that a driver carried 15/30/10 liability insurance and a court awards $18,000 to one victim and $2,000 to each of three other victims. Even though the total awarded to all victims ($24,000) is less than the maximum total coverage ($30,000), the insurance company will not pay the total award: Since the maximum liability for a single victim is $15,000, the insurance company is responsible only for $21,000 ($15,000 to the first victim and $2,000 each to the other three victims). The remaining $3,000 of the $18,000 judgment must be paid by the policyholder.

Two factors determine the standard rates charged for liability insurance: the classification of the vehicle (according to how much it is driven, whether used for business, and the age of the driver) and the territory in which the vehicle is operated. As would be expected, rates are higher in populous areas, since traffic congestion results in more accidents. If a driver has a record of accidents and/or speeding tickets, insurance will be available only at greatly increased rates through an *assigned risk plan*, a joint underwriting association, or a reinsurance facility.

Table 6-3 contains excerpts of the driver classifications used to determine

TABLE 6-3 DRIVER CLASSIFICATIONS

Multiples of Base Annual Automobile Insurance Premiums

			PLEASURE; LESS THAN 3 MILES TO WORK EACH WAY	DRIVES TO WORK 3 TO 9 MILES EACH WAY	DRIVES TO WORK, 10 MILES OR MORE EACH WAY	USED IN BUSINESS
No young operators	Only operator is female, age 30–64		0.90	1.00	1.30	1.40
	One or more operators age 65 or over		1.00	1.10	1.40	1.50
	All others		1.00	1.10	1.40	1.50
Young females	Age 16	DT[a]	1.40	1.50	1.80	1.90
		No DT	1.55	1.65	1.95	2.05
	Age 20	DT	1.05	1.15	1.45	1.55
		No DT	1.10	1.20	1.50	1.60
Young males (married)	Age 16	DT	1.60	1.70	2.00	2.10
		No DT	1.80	1.90	2.20	2.30
	Age 20	DT	1.45	1.55	1.85	1.95
		No DT	1.50	1.60	1.90	2.00
	Age 21		1.40	1.50	1.80	1.90
	Age 24		1.10	1.20	1.50	1.60
Young unmarried males (not principal operator)	Age 16	DT	2.05	2.15	2.45	2.55
		No DT	2.30	2.40	2.70	2.80
	Age 20	DT	1.60	1.70	2.00	2.10
		No DT	1.70	1.80	2.10	2.20
	Age 21		1.55	1.65	1.95	2.05
	Age 24		1.10	1.20	1.50	1.60
Young unmarried males (owner or principal operator)	Age 16	DT	2.70	2.80	3.10	3.20
		No DT	3.30	3.40	3.70	3.80
	Age 20	DT	2.55	2.65	2.95	3.05
		No DT	2.70	2.80	3.10	3.20
	Age 21		2.50	2.60	2.90	3.00
	Age 24		1.90	2.00	2.30	2.40
	Age 26		1.50	1.60	1.90	2.00
	Age 29		1.10	1.20	1.50	1.60

[a] "DT" indicates completion of a certified driver training course.

TABLE 6-4 AUTOMOBILE LIABILITY AND MEDICAL PAYMENT INSURANCE

Base Annual Premiums

	BODILY INJURY				PROPERTY DAMAGE		
COVERAGE	TERRITORY 1	TERRITORY 2	TERRITORY 3	COVERAGE	TERRITORY 1	TERRITORY 2	TERRITORY 3
15/30	$ 81	$ 91	$112	$ 5,000	$83	$ 95	$100
25/25	83	94	115	10,000	85	97	103
25/50	86	97	120	25,000	86	99	104
50/50	88	101	125	50,000	87	101	107
50/100	90	103	129	100,000	90	103	108
100/100	91	104	131		MEDICAL PAYMENT		
100/200	94	108	136				
100/300	95	110	139	$ 1,000	$62	$ 63	$ 64
200/300	98	112	141	2,500	65	66	67
300/300	100	115	144	5,000	67	68	69
				10,000	70	72	74

liability insurance premiums. As you can see from the table, drivers under 21 are intricately classified according to sex, age, and whether or not they have taken a driver training course. Males under 25 are classified according to marital status and whether they are the owners or principal operators of the automobiles. Unmarried males may continue to be so classified until they reach age 30. Most mature drivers would be classified "all others." (An automobile is classified according to the status of the youngest person who operates the car or according to the person whose driver classification requires the highest multiple.)

There is a positive correlation between scholastic records and driving records of full-time students. Full-time high school and college students who have a B average or better or are in the upper 20% of their class may often qualify for a *good student discount*. This discount is usually a 15% to 20% reduction in the annual premiums paid to the insurance company.

Table 6-4 presents some typical base annual rates for automobile liability insurance (bodily injury and property damage) and medical payment insurance in three different territories. The total premium cost for two or more vehicles would be reduced by a certain percent when the vehicles are covered under one insurance policy.

Example 1 Edwin Carter is 32 years old and drives his car to work each day, a one-way distance of 6 miles. The area in which he lives is classified Territory 1. (a) Determine the cost of 15/30/10 liability coverage on his automobile and $2,500 medical payment insurance. (b) What would be Carter's premium if he increased his insurance to 50/100 bodily injury and $25,000 property damage coverage (with the same $2,500 medical pay coverage)?

(a)   Table 6-4 is used to compute the base annual premium of Carter's liability ($15,000 single and $30,000 total bodily injury; $10,000 property damage) and medical payment ($2,500) coverage in Territory 1:

$$
\begin{array}{rl}
\$\ 81 & \text{Bodily injury (15/30)} \\
85 & \text{Property damage ($10,000)} \\
+\ \underline{\ 65} & \text{Medical payment ($2,500)} \\
\$231 & \text{Total base premium}
\end{array}
$$

The base annual premium in Territory 1 is $231 for 15/30/10 liability and $2,500 medical pay coverage. Because Carter is 32 years old, he falls into the category "all others" in the driver classification table. Because he drives 6 miles to work, Table 6-3 indicates that the multiple 1.10 should be used to compute Carter's annual premium:

$$
\begin{array}{rl}
\$\quad 231 & \text{Base annual premium} \\
\times\ \underline{\quad 1.10} & \\
\$254.10 & \text{Actual annual premium}
\end{array}
$$

(b)   Increased coverage would cost as follows:

$$
\begin{array}{rl}
\$\quad 90 & \text{50/100 Bodily injury} \\
86 & \text{$25,000 Property damage} \\
+\ \underline{\quad 65} & \text{$2,500 Medical payment} \\
\$\quad 241 & \text{Base annual premium} \\
\times\ \underline{\quad 1.10} & \\
\$265.10 & \text{Total annual premium}
\end{array}
$$

Note.   In an actual insurance office, the agent is not required to use the driver multiple (like 1.10 above) because extensive tables are available that show the rates in every category, with the appropriate multiple already applied. It is beneficial for students to compute rates using the multiples, however, since it enables you to see clearly how the significantly different premium rates for various ages and categories are obtained.

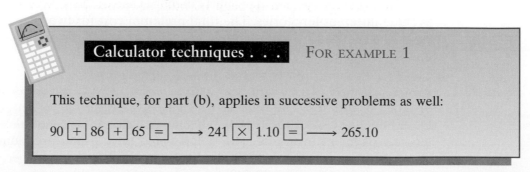

Calculator techniques . . .   FOR EXAMPLE 1

This technique, for part (b), applies in successive problems as well:

90 [+] 86 [+] 65 [=] ⟶ 241 [×] 1.10 [=] ⟶ 265.10

Example 2   Bill Hold of Hold Construction Co. was driving a company truck when it struck an automobile, injuring a mother and her son. Hold Construction was subsequently sued for $120,000 personal injuries and $4,500 property damage. The court awarded the victims a $60,000 personal injury judgment and the entire property damage suit. Hold Construction carries 25/50/10 liability coverage. (a) How much will the insurance company pay? (b) How much of the award will Hold Construction have to pay?

(a)   The insurance company will pay the $50,000 maximum total bodily injury coverage ($25,000 maximum per person) and the total award for property damage ($4,500). Thus, the insurance company will pay a total of $54,500.

(b)   Since the court awarded the claimants more than Hold Construction's bodily injury coverage, the firm will be responsible for the excess:

$60,000	Bodily injury awarded
− 50,000	Liability coverage for bodily injury
$10,000	Amount to be paid by Hold Construction

NO-FAULT INSURANCE

No-fault insurance has been tried, with varying degrees of success, in nearly 20 states. At one time Congress considered making a standard no-fault policy mandatory nationwide, but the proposed legislation failed to pass.

Under **no-fault insurance,** each motorist collects for *bodily injuries* from his or her own insurance company after an accident, regardless of who is at fault. Reimbursement is made for medical expenses and the value of lost wages and/or services (such as the cost of a housekeeper for an injured mother). The victim then forfeits any right to sue, unless the accident was so severe that the damages exceed a certain specified amount.

No-fault insurance was proposed as a means whereby motorists could save money on insurance but still be protected financially in case of accident. However, the savings derived through no-fault insurance pertain only to reduced bodily injury premiums, since small suits are disallowed. Premiums for property damage liability, as well as comprehensive and collision insurance on the driver's own vehicle (described below), are not reduced by the adoption of no-fault insurance. Furthermore, considerably more than half of a motorist's premiums apply toward these latter types of insurance. Thus, rising automobile repair costs, in particular, plus the tendency of victims to sue for large amounts, as well as the increased operating costs of the insurance companies themselves, have combined to produce continually increasing premiums in most areas, despite the adoption of no-fault insurance. Since the originally high expectations for no-fault insurance have not been achieved, enthusiasm for its adoption has also lessened.

COMPREHENSIVE AND COLLISION INSURANCE

Comprehensive and collision insurance protect against damage to the policyholder's own automobile. **Comprehensive insurance** covers damage to the car resulting from fire, theft, vandalism, acts of nature, falling objects, and so forth. Items such as coats left in a car are not covered. Equipment not built into the car, such as a CB radio, generally is not covered. Extra insurance can be purchased on these items.

Protection against collision or upset damage may be obtained through **collision insurance.** (That is, collision insurance also covers damage resulting from one-car accidents—including damage to runaway, driverless vehicles—where no collision of two vehicles occurred.) Collision insurance pays for repairs to the vehicle of the insured when the policyholder is responsible for an accident, when the insured's car was damaged by a hit-and-run driver, or when another driver was responsible for the collision but did not have liability insurance and was unable to pay for the property damaged caused. Collision insurance does not pay for loss or damage to other vehicles or property. Because collision insurance (as well as comprehensive) will pay for damages regardless of who is at fault, it is in this respect already a type of no-fault insurance and thus is not affected whenever no-fault insurance is adopted in a state.

Also available is **uninsured motorists insurance,** which offers financial protection for the policyholder's own bodily injuries or those of a family member when hit by a driver who did not carry bodily injury liability insurance. It does not cover the uninsured motorist.

Since comprehensive and collision insurance provides compensation only to the insured, it is not required by state law. Under certain circumstances, however, this coverage is mandatory. If an automobile is purchased by monthly installment payments, the institution financing the purchase retains legal title to the automobile until all installments are paid. To protect their investment, these financial agencies ordinarily require the purchaser to carry comprehensive and collision insurance.

A vehicle's owner may purchase collision insurance which pays the entire cost of repairing damage; however, such insurance is quite expensive. Most collision insurance is sold with a "deductible" clause, which means that the policyholder pays a specified part of any repair cost (an amount specified in the deductible clause). For instance, if "$250 deductible" collision insurance is purchased, the insured must pay the first $250 in damage resulting from any one collision, and the carrier will pay the remaining cost, up to the value of the vehicle. (If damage is less than $250, the owner must pay the entire cost.)

TABLE 6-5 COMPREHENSIVE AND COLLISION INSURANCE

Base Annual Premiums

MODEL CLASS	AGE GROUP	TERRITORY 1			TERRITORY 2			TERRITORY 3		
		COMPRE-HENSIVE	$250-DEDUCTIBLE COLLISION	$500-DEDUCTIBLE COLLISION	COMPRE-HENSIVE	$250-DEDUCTIBLE COLLISION	$500-DEDUCTIBLE COLLISION	COMPRE-HENSIVE	$250-DEDUCTIBLE COLLISION	$500-DEDUCTIBLE COLLISION
(1) A-G	1	$55	$ 82	$ 76	$59	$ 92	$ 80	$73	$100	$ 91
	2,3	52	77	73	56	86	76	58	94	85
	4	49	71	67	51	79	70	54	85	78
(3) J-K	1	63	111	101	69	128	108	75	141	127
	2,3	59	103	95	64	118	101	68	131	118
	4	54	93	86	57	106	91	61	116	105
(4) L-M	1	68	123	112	76	143	120	83	169	142
	2,3	64	125	104	70	133	112	75	138	132
	4	57	102	94	62	117	100	66	129	117
(5) N-O	1	77	140	126	86	164	136	95	183	162
	2,3	70	130	117	77	151	126	85	168	150
	4	62	115	105	68	133	112	73	147	132

Collision policies are available with various deductible amounts specified; $250-deductible policies are most common, but deductibles of higher amounts are becoming more frequent. Whereas comprehensive policies formerly were usually written without deductible amounts, such policies now often have a deductible. Both collision and comprehensive coverage become considerably less expensive as the policyholder pays a higher amount of each repair cost.

Recall that standard liability rates are determined by the classification of vehicle use and the territory in which the vehicle is operated. In addition to the previous considerations, comprehensive/collision rates also depend on the make, model, and age of the vehicle (example: "Chevrolet, Camaro, 2 years old"). Each automobile model is assigned an identification letter from A to Z. (Less expensive models are identified by letters near the beginning of the alphabet, and successive letters identify increasingly expensive models.)

The age of the automobile influences comprehensive and collision costs, since newer automobiles are more expensive to repair. Automobiles of the current model year are classified as "Age Group 1"; cars in Group 2 (the first preceding model year) and Group 3 (the second preceding model year) both pay the same rate; and models 3 or more years old are classified in Group 4. Table 6-5 lists excerpts of base annual comprehensive and collision insurance rates in three territories for vehicles of various ages. As in the case of liability insurance, the base annual comprehensive/collision premiums must be multiplied by a multiple reflecting the driver's classification.

Example 3   Juan Lopez, who is 26 and single, uses his car in business in Territory 3. His car is a Model K, less than 1 year old. Find his annual premium for the following insurance: 50/50/10 liability insurance, $1,000 medical payment, full comprehensive, and $500-deductible collision.

Table 6-4 is used to determine the base annual liability (bodily injury and property damage) and medical pay premiums, as before. The comprehensive and collision base premiums are shown in Table 6-5. The total of all these base annual premiums must then be multiplied by the driver classification multiple from Table 6-3 to obtain the total annual premium.

$ 125	50/50 Bodily injury
103	$10,000 Property damage
64	$1,000 Medical payment
75	Comprehensive (Model K, Age Group 1, Territory 3)
+ 127	$500-deductible collision (classified as above)
$494	Base annual premium
× 2.00	Driver classification (unmarried male, age 26, business use)
$988	Total annual premium

Example 4   Harold Granger carries 10/20/5 liability insurance and $250-deductible collision insurance. Granger was at fault in an accident which caused $900 damage to his own car and $1,200 damage to the other vehicle. (a) How much of this property damage will Granger's insurance company pay? (b) Suppose that a court suit results in a $12,000 award for personal injuries to the other driver. How much would the insurance company pay and how much is Granger's responsibility?

(a)   Property damage to the other car (under Granger's
      $5,000 property damage policy) $1,200
      Property damage to Granger's car ($900 − $250
        deductible)   +   650
            Total property-damage settlement $1,850

(b)   Under Granger's $10,000 single and $20,000 total bodily injury policy, the maximum amount the insurance company will pay to any one victim is $10,000. Thus, Granger is personally liable for the remaining $2,000 of the $12,000 settlement (making his total obligation $2,250, including the $250 deductible).

SECTION 2   PROBLEMS

Compute the following motor vehicle insurance problems using the tables in this section. Compute the total of liability and medical payment premiums for each of the following.

	LIABILITY COVERAGE	MEDICAL PAYMENT	TERRITORY	DRIVER CLASSIFICATION	TOTAL LIABILITY AND MEDICAL PYMT. PREMIUM
1. a.	25/50/10	$ 1,000	1	Male, 47, drives 8 miles to work	
b.	50/50/5	2,500	2	Female, 32, business use	
c.	50/100/25	10,000	3	Unmarried male, not principal operator, 16, driver's training, drives 3 miles to work	
2. a.	15/30/10	$ 1,000	2	Female, 20, no driver's training, drives 6 miles to work	
b.	50/50/25	5,000	1	Female, 30, drives 12 miles to work	
c.	100/200/50	10,000	3	Unmarried male, 21, not principal operator, drives 5 miles to work	

Determine the total comprehensive/collision premiums for the following.

	MODEL	AGE GROUP	TERRITORY	DEDUCTIBLE ON COLLISION	DRIVER CLASSIFICATION	TOTAL COMPREHENSIVE/ COLLISION PREMIUM
3. a.	G	2	2	$250	Female, 54, drives 1 mile to work	
b.	L	4	1	500	Male, 35, drives 11 miles to work	
c.	N	1	3	500	Unmarried male, owner, 24, drives 7 miles to work	
4. a.	A	3	1	$250	Female, 16, driver's training, pleasure driving	
b.	J	1	2	500	Married male, 20, no driver's training, drives 12 miles to work	
c.	L	2	2	500	Female, 65, drives 15 miles to work	

5. C. C. Netherton, age 39, lives in Territory 3. Each day he drives 5 miles each way to the college where he teaches. His liability insurance includes $50,000 for single bodily injury, $100,000 for total bodily injury, $10,000 for property damage, and $2,500 for medical payment. Determine his annual payment.

6. Jim Chang, age 20 and unmarried, lives in Territory 2 and drives 6 miles each way to a part-time job. He has had driver's training and is not the principal operator. How much will he pay for liability coverage that includes $25,000 for single bodily injuries, $50,000 for total bodily injuries, $10,000 for property damages, and $1,000 for medical payment coverage?

7. Kristin Dixon, age 16, has completed a driver training course. She lives in Territory 2 and drives a Model L, Age Group 2 car for pleasure. Determine her annual premium for comprehensive and $500-deductible collision coverage.

8. Becky Karolyi, age 27, lives in Territory 1 where she uses her car as a real estate agent. She drives a Model N, Age Group 1 car. Her bank requires that she carry comprehensive and $250-deductible collision insurance until her car loan is paid in full. What will be the annual cost of this insurance coverage?

9. George North, age 26, unmarried and the owner of his car, bought a new Model D car. Each day he commutes 15 miles each way to work from his home in Territory 1. His auto insurance includes 25/50/10 liability coverage and $5,000 medical payment. He also has comprehensive and $250-deductible collision coverage. How much is his annual premium?

10. Bob Burton, age 48, lives in Territory 3. He drives 2½ miles each way daily to his computer analyst job in his Model B car that is 2 years old. His insurance coverage includes 100/100/50 liability and $10,000 medical payment. He also carries comprehensive and $500-deductible collision insurance. Determine his annual car insurance cost.

11. Jack Gleason, age 21 and married, daily drives 4 miles each way to work in Territory 2. He drives a Model A car that is 2 years old. His insurance includes 25/25/5 liability, $1,000 medical pay, comprehensive, and $500-deductible collision. What is his annual premium?

12. Erika Chavez, age 22, drives 11 miles each way to her position as a word processor at an insurance agency. She lives in Territory 2 and drives her Model K car that is 4 years old. Chavez's insurance includes 50/50/10 liability coverage, $2,500 medical pay, comprehensive, and $500-deductible collision protection. Find her annual cost.

13. Joyce Watson's car struck another car and injured two people inside. The other car received $11,300 damages. The court awarded $40,000 to one of the people and $18,000 to the other for medical expenses. Miss Watson carries 25/50/10 liability insurance.

   a. How much of this court settlement will the insurance company pay?

   b. How much is Miss Watson's personal responsibility?

14. Peter Conroy was liable for an accident in which two people were injured. His insurance included 50/100/10 liability protection. The court awarded one person $30,000 and the other $65,000 for personal bodily injuries. An award of $25,000 for damages to the other car was also part of the court settlement.

   a. How much of the court settlement will the insurance company pay?

   b. How much of the expense must Conroy pay personally?

15. During a thunderstorm, Tony Purzer's car was struck by a falling tree branch, which caused $800 in damages. Purzer carries comprehensive and $250-deductible collision insurance.

   a. How much of the damages will his insurance cover?

   b. How much of the repair bill will Purzer pay?

16. Janet Bruce's car was stolen while parked outside her home. The car had a market value of $20,000. Bruce carries comprehensive and $500-deductible collision insurance.

   a. How much indemnity will the insurance company pay?

   b. How much out-of-pocket expense will Bruce incur to replace the car, assuming she can buy another one for $20,000?

17. Bea Morgan sideswiped Ann Kinard's car at a shopping center, causing $7,200 in damages to Kinard's car. Ms. Kinard suffered internal injuries requiring medical expenses of $35,000. Ms. Morgan's car repairs totaled $1,500. Ms. Morgan's insurance policy includes 25/50/5 liability, comprehensive, $500-deductible collision, and $2,500 medical payment.

   a. What amount will her insurance company pay toward these costs?

   b. How much will Ms. Morgan have to pay?

18. A company van owned by Gault Service Corp. was responsible for an accident with a passenger car. The driver and an occupant of the car suffered bodily injuries with medical costs totaling $48,000 and $60,000, respectively. Damage to the van totaled $7,000 and to the car, $12,000. Gault Service Corp. carries 15/30/5 liability, comprehensive, and $500-deductible collision.

   a. What indemnity will the insurance company pay?

   b. How much of the total expense will Gault Service Corp. have to pay?

19. The brakes failed on Roger Steven's car on a rainy day, and his car crashed into another car, killing the passenger of the other car and seriously injuring the other driver. The court awarded $500,000 to the widow of the deceased and $120,000 to the injured driver. Mr. Stevens' hospital and doctor bills totaled $22,000. Property damages were $15,000 to the other car and $4,000 to the defendant's car. Mr. Stevens carries 200/300/10 liability, $10,000 medical payment, comprehensive, and $500-deductible collision insurance.

   a. Determine the total amount the insurance company should pay.

   b. For what amount is Mr. Stevens responsible?

20. A truck driven by Tom Edmunds crashed into a car driven by Tim Gemini, killing the passenger of the car and injuring Mr. Gemini. After a court suit, the widow of the passenger was awarded $400,000 on the death of her husband, and $110,000 was awarded to Mr. Gemini. Mr. Edmunds suffered $3,000 of personal injuries. Property damages were $15,000 to the car and $2,000 to the truck. Mr. Edmunds carries 100/200/25 liability, comprehensive, $250-deductible collision, and medical payment of $1,000.

   a. How much of the total cost will the insurance company pay?

   b. How much will Mr. Edmunds have to pay?

SECTION 3
LIFE INSURANCE

The basic purpose of life insurance is to provide compensation to survivors following the death of the insured. Whereas other types of insurance pay damages only up to the actual value of the insured property, life insurance companies make no attempt to assign a specific value to any life. A person (in good health and with normal expectancy of survival) may purchase any amount of insurance corresponding to the policyholder's income and standard of living; and, when death occurs, the full value of the life insurance coverage will be paid.

The responsibility that most heads of households feel for the financial security of their families induces them to purchase life insurance. Life insurance is owned by 78% of American households. The Life Insurance Marketing and Research Association reports that the average amount of life insurance is $143,100 per insured household in the United States (compared with only $111,600 per household when all households are considered). Some 71% of adult men have life insurance, whereas 65% of adult women are insured. The average (mean) amount of total life insurance on an adult man is $105,500, which is twice the average amount held by an insured adult woman ($52,700). Group life insurance is steadily increasing, having become a standard employee benefit. Group insurance amounts to 40.1% of the life insurance in force.

Many businesses purchase life insurance on key members of the firm whose death would cause a severe loss to the business. Partners sometimes carry insurance on each other so that, in event of a death, the surviving partners can purchase the deceased partner's share of the business from the estate.

TERM

The **term** policy is so named because it is issued for a specific period of time. The policyholder is insured during this time and, if the insured is still alive when the term expires, the coverage then ceases. Term insurance is considerably less expensive than the other types of policies. The principal reason for the low cost is that term policies do not build investment value for the policyholder. (Investment values are discussed in the section "Nonforfeiture Options.") Term insurance accounts for 22% of all new policies written today.

Term policies are commonly issued for terms of 5, 10, or 15 years. Many heads of families feel that their need for insurance is greatest while they are younger and have children to support and educate. Term policies can provide maximum insurance protection during these years at a minimum cost. Another typical use of term policies is the life insurance purchased for a plane or boat trip; the term in this case is the duration of the trip.

Term policies may include the provision that, at the end of the term, the policy may be renewed for another term or converted to another form of insurance (both at increased premium rates). A variation of the ordinary term policy is *decreasing term* insurance; under this policy, the face value is largest when issued and decreases each year until it reaches zero at the end of the term. A typical example of decreasing term insurance is mortgage insurance, which decreases as the home loan decreases.

WHOLE LIFE

The **whole life** (or **straight life**) policy provides protection throughout the insured's lifetime. Payment of the face amount is paid upon death of the insured regardless of when death occurs. The whole life policy is by far the life insurance coverage most often purchased in the United States. Approximately 58% of all policies purchased are whole life. In order to keep the original policy in force, premiums must be paid for as long as the insured lives. The premium cost depends upon the age of the policyholder when the insurance is first purchased, and this premium remains the same each year thereafter. In effect, the insurance cost is averaged over the policyholder's lifetime. An important feature of whole life insurance is that it accumulates a cash value, similar to a savings account.

LIMITED PAYMENT LIFE

The **limited payment life** policy is a variation of the whole life policy. With limited payment insurance, the policyholder makes payments for only a limited number of years. If the policyholder survives beyond this time, no further payments are required; however, the same amount of insurance coverage remains in force for the remainder of the policyholder's life. This coverage is very similar to whole life insurance, except that the payments are sufficiently larger so that the policyholder will pay, during the limited payment period, an amount approximately equivalent to the total that would be paid during an average lifetime for whole life insurance.

Limited payment policies are frequently sold with payment periods of 20 or 30 years. Such policies are called "20-payment life" or "30-payment life" policies. The limited payment policy is also commonly sold with a payment period so that the policy is "paid up" at age 65.

UNIVERSAL AND VARIABLE LIFE

The **universal life** policy is a flexible-premium, adjustable death benefit policy. Offered as an inducement to investors, universal life provides low-cost term insurance, with the balance of each premium going into a cash account that earns interest. A minimum rate is guaranteed (about 4% or 5%) but may go much higher depending on how successful the company's own investments are. The actual rate of growth will vary from year to year.

The universal life policy is highly flexible, allowing the policyholder to raise or lower the face amount as circumstances change. The premium payment may also vary, or, if the policyholder fails to make a payment, the premium is simply deducted from the cash account. (Regular payments would be necessary, however, to keep coverage in force for the full term and to build investment value.)

The **variable life** policy is similar to universal life, but was developed to offer investors more options, with the potential for higher rates of return. A set premium is paid, and the investment portion is distributed among any of several different options selected by the policyholder—typically, stock funds, bond funds, and money-market accounts. These funds are all subject to the interest-rate fluctuations of the national economic climate.

Certain minimums are guaranteed, but in general both the interest rate and the death benefit may rise or fall, depending on the performance of the policyholder's investment choices. Thus, with a variable life policy, the customer assumes more responsibility for the investment risk, whereas the return on a universal life policy depends on the success of the company's investments overall. Universal and variable life policies together account for about 20% of policies written today.

Since many people are uncomfortable with an uncertain death benefit, the insurance industry recently developed a policy intended to offer the best of both worlds. Called **variable universal life,** this policy provides a guaranteed death benefit while still allowing the policyholder to select from various types of funds for the cash investment. As with universal life, there is much flexibility regarding when premiums are paid and their amount, but if interest rates should fall, premiums must be paid throughout the term of this policy, to keep it in force.

Note.   This policy may also be called "flexible-premium variable life."

There are arguments pro and con regarding the advisability of using insurance policies as an investment. The insurance industry therefore attempts to offer something for everyone—from pure insurance with no investment (term policies) to policies that are primarily investments (such as single-premium variable life policies that are fully paid upon purchase and are bought simply to build value that will provide monthly annuity payments after retirement).

PARTICIPATING AND NONPARTICIPATING

Insurance policies are further classified as participating or nonparticipating. **Participating** policies are so named because the policyholders participate in the profits earned on these policies. The premium rates charged for participating policies are usually somewhat higher, because the insurance companies charge more than they expect to need in order to pay claims and operating expenses. However, the excess funds at the end of the year are then refunded to policyholders in the form of *dividends*.

At the time a participating policy is purchased, the policyholder selects the form in which dividends will be paid. The usual choices available to the policyholder are that the dividends may be (1) paid in cash, (2) used to help pay the premiums, (3) used to purchase additional paid-up insurance (paid-up insurance is discussed in the section "Nonforfeiture Options"), or (4) left invested with the company to accumulate with interest. A fifth option sometimes available is that the dividends may be (5) used to purchase a 1-year term insurance policy.

Nonparticipating policies, on the other hand, generally offer lower premium rates than participating policies; however, no dividends are paid on nonparticipating policies. Thus, the actual annual cost of the life insurance is usually less for participating policyholders than for the owners of nonparticipating policies.

Most participating policies are sold by mutual companies. **Mutual life insurance companies** do not have stockholders, but are owned by the policyholders themselves. The companies are governed by a board of directors elected by the policyholders. Life insurance companies may also be *stock* *companies,* which are corporations operated for the purpose of earning profits and which are owned by stockholders. As is the usual case for corporations, the board of directors of a stock life insurance company is elected by the stockholders, and the profits earned by the company are paid to the stockholders. Thus, most stock companies issue nonparticipating policies, although a few do issue some participating policies.

PREMIUMS

The risk assumed by insurance companies for other types of policies (fire or automobile, for example) is the probability that a loss will occur times the expected size of the loss (it may be only a partial loss). With life insurance, however, the risk is not "whether" a loss will occur but "when" the loss will occur, for death will inevitably come to all policyholders, and the full value of each effective policy will always be paid after a death.

Premium rates for life insurance are computed by an *actuary,* a highly skilled person trained in mathematical probability as well as business administration. To help calculate premiums and benefits, the actuary uses a *mortality table;* this is an extensive table showing how many people die at each age and the death rate (per 100,000 people) for each age group. Table 6-6 shows an excerpt from a table published by the U.S. Department of Health and Human Services. (Collecting and tabulating these statistics requires a delay of two or more years between the date of the data and their publication.)

If a group of people, all of the same age, take out insurance during the same year, the mortality table enables the actuary to know how many claims to expect during each succeeding year. Premium rates must be set so that the total amount paid by all policyholders is sufficient to pay all these claims, as well as to finance business expenses.

Almost all life insurance sold today is issued on the *level premium system.* This means that the policyholder pays the same premium each year for life insurance protection. Thus, the policyholders pay more than the expected claims during the early years of their policies, and during the later years they pay less than the cost of the claims. The excess paid during the early years is invested by the insurance company, and interest earned on these investments pays part of the cost of the insurance. Thus, the total premium cost over many years is less than if the policyholders paid each year only the amount which the company expected to need.

A physical examination is often required before an individual life insurance policy will be issued. (Group policies, on the other hand, are generally issued without physical examinations.) Annual premiums on new policies are naturally less expensive for younger people, since their life expectancy is longer and the insurance company expects them to pay for more years than older persons would. (This emphasizes the advisability of setting up an adequate insurance program while one is still young. The high cost of insurance taken out at an older age makes new coverage too expensive for many older persons to afford.)

| TABLE 6-6 | DEATHS AND DEATH RATES, BY AGE, RACE, AND SEX: UNITED STATES, 1992 |

[Rates per 100,000 Population in Specified Group]

| | All races | | |
Age	Both sexes	Male	Female
All ages	2,175,613	1,122,336	1,053,277
Under 1 year.	34,628	19,545	15,083
1–4 years.	6,764	3,809	2,955
5–9 years.	3,739	2,231	1,508
10–14 years	4,454	2,849	1,605
15–19 years	14,411	10,747	3,664
20–24 years	20,137	15,460	4,677
25–29 years	24,314	18,032	6,282
30–34 years	34,167	24,863	9,304
35–39 years	42,089	29,641	12,448
40–44 years	49,201	33,354	15,847
45–49 years	56,533	36,622	19,911
50–54 years	68,497	42,649	25,848
55–59 years	94,582	58,083	36,499
60–64 years	146,409	88,797	57,612
65–69 years	211,071	124,228	86,843
70–74 years	266,845	149,937	116,908
75–79 years	301,736	158,257	143,479
80–84 years	308,116	141,640	166,476
85 years and over	487,446	161,236	326,210
Not stated	474	356	118
All ages[1]	852.9	901.6	806.5
Under 1 year[2]	865.7	956.6	770.8
1–4 years.	43.6	48.0	39.0
5–9 years.	20.4	23.7	16.8
10–14 years	24.6	30.7	18.2
15–19 years	84.3	122.4	44.0
20–24 years	105.7	159.4	50.1
25–29 years	120.5	178.0	62.5
30–34 years	153.5	224.0	83.3
35–39 years	199.5	282.8	117.2
40–44 years	261.6	359.1	166.5
45–49 years	368.0	485.7	254.6
50–54 years	568.2	728.1	417.1
55–59 years	902.1	1,156.5	668.2
60–64 years	1,402.2	1,815.2	1,038.2
65–69 years	2,114.8	2,775.4	1,577.7
70–74 years	3,146.8	4,109.3	2,419.9
75–79 years	4,705.9	6,202.4	3,716.8
80–84 years	7,429.1	9,726.0	6,186.1
85 years and over	14,972.9	17,740.4	13,901.0

[1] Figures for age not stated are included in All ages but not distributed among age groups.
[2] Death rates under 1 year (based on population estimates) differ from infant mortality rates (based on live births).
Source: Kochanek, K. D., and Hudson, B. L. Advance report of final mortality statistics, 1992. Monthly vital statistics report; vol. 43 no. 6, suppl. Hyattsville, MD: National Center for Health Statistics, 1995.

Life insurance taken out at any given age is usually less expensive for women than for men, because women have a longer life expectancy.

Life insurance premiums on new policies are determined by the applicant's age to the nearest birthday. Thus, applicants who are $26\frac{1}{2}$ would pay premiums as if they were 27 at the time when the policy was issued.

Life insurance policies are written with face values in multiples of $1,000. The rate tables list the cost for a $1,000 policy, and this rate must be multiplied by the number of thousands of dollars in coverage that is being purchased. Table 6-7 gives typical annual premiums per $1,000 of face value for each type of life insurance policy, when taken out at various ages. (The policyholder would pay a slightly higher annual total if premiums were paid semiannually, quarterly, or monthly. Each semiannual premium is approximately 51.5% of the annual premium; quarterly premiums are approximately 26.3% of the annual; and each monthly premium would be approximately 8.9% of the annual.)

The premiums in Table 6-7 represent an approximate average between participating and nonparticipating rates. The premiums for variable universal policies can vary widely, depending on the policyholder's investment goal, but they probably average about the same as whole life policies. These rates are for example purposes only.

Since life expectancy is longer for women than for men, different premium rates are used. There would be a separate premium table for women, or an

TABLE 6-7 ANNUAL LIFE INSURANCE PREMIUMS

Per $1,000 of Face Value for Male Applicants[1]

AGE ISSUED (YEARS)	TERM 10-YEAR	WHOLE LIFE	VARIABLE UNIVERSAL	LIMITED PAYMENT 20-YEAR
18	$ 6.81	$14.77	$18.36	$24.69
20	6.88	15.46	19.81	25.59
22	6.95	16.12	20.28	26.53
24	7.05	16.82	21.33	27.53
25	7.10	17.22	22.17	28.06
26	7.18	17.67	23.43	28.63
28	7.35	18.64	24.79	29.84
30	7.59	19.73	26.05	31.12
35	8.68	23.99	30.87	35.80
40	10.64	28.26	35.55	40.34
45	14.52	33.79	41.24	46.01
50	22.18	40.77	48.11	53.24
55	32.93	51.38	57.62	64.77
60	—	59.32	—	70.86

[1] Because of women's longer life expectancy, premiums for women approximately equal those of men who are 5 years younger.

adjustment of 3 to 5 years can be made in the men's table. An adjustment or setback of 5 years in Table 6-7 will be used in this text for examples and problems involving women.

Example 1   Paul Kirkland is 30 and wishes to purchase $25,000 of life insurance. Determine the annual cost of (a) a whole life policy and (b) a 20-payment life policy.

In each case, the insurance coverage is to be $25,000. Thus, each premium rate from Table 6-7 must be multiplied by 25.

(a) _Whole Life_		(b) _20-Payment Life_
$ 19.73	← Premium per $1,000 →	$ 31.12
× 25		× 25
$493.25	← Premium on $25,000 →	$778.00

Kirkland would pay an annual premium of (a) $493.25 for a whole life policy or (b) $778 for similar coverage under a 20-payment life policy.

Example 2   Suppose that Paul Kirkland (Example 1) lived for 38 years after purchasing his life insurance. (a) How much would he pay altogether in premiums for each policy? (b) Which policy's premiums would cost more, and how much more?

(a) _Whole Life_		_20-Payment Life_
$ 493.25	Annual premiums	$ 778.00
× 38	Number of premiums	× 20
$18,743.50	Total premium cost	$15,560.00

The total premium cost would be $18,743.50 for a $25,000 whole life policy and only $15,560 for an equivalent 20-payment life policy. Thus,

(b)	$18,743.50	Total premium cost of whole life policy
	− 15,560.00	Total premium cost of 20-payment life policy
	$ 3,183.50	Extra premium cost of whole life policy

The whole life policy would cost Kirkland $3,183.50 more if he lived to be 68.

Example 3   Donna Pitman, age 30, is considering the purchase of either a whole life policy or a variable universal policy with a face value of $50,000. Determine the annual cost of (a) a whole life policy and (b) a variable universal policy.

Using Table 6-7, the premium rates for Ms. Pitman will be "set back" 5 years from the male applicant rates, arriving at the age 25 rate.

(a) *Whole Life*   (b) *Variable Universal*

$ 17.22	← Premium per $1,000 →	$ 22.17
× ____50		× ____50
$860.00	← Premium on $50,000 →	$1,108.05

Actually, premiums can vary greatly from company to company on all policies, as do the nonforfeiture values, which will be discussed later. Besides the basic life insurance coverage, several types of additional protection may be purchased at a slight increase in premiums. Among the additional features most frequently purchased are the **waiver of premium benefit** (the company agrees to assume the cost of the policyholder's insurance in case of total and permanent disability) and the **double indemnity benefit** (the insurance company will pay twice the face value of the policy if death results from accidental causes rather than illness).

Since many different provisions are available on different policies, buyers trying to find the best value found themselves trying "to compare apples and oranges." To avoid this confusion, insurance companies are now required to disclose the **interest-adjusted net cost** of each policy—the total cost strictly for life insurance coverage, disregarding the investment values. This cost is also expressed as the **net cost index**—an annual cost equivalent to the total interest-adjusted net cost. Therefore, if a buyer is considering two similar insurance policies from equally reliable companies, the better choice may well be the policy with the lower interest-adjusted net cost.

NONFORFEITURE OPTIONS

The most obvious benefit from an insurance policy is the death benefit: The policyholder designates a person, called the **beneficiary,** to receive the face value of the policy following the death of the insured. However, there are a number of alternative benefits available to the policyholders themselves which should be taken into consideration when selecting a policy.

Whereas the premiums paid for other types of insurance policies purchase only protection, the premiums paid for life insurance policies (except term policies) also build an investment for the policyholder. This investment accumulates because the level premium system has caused excess funds to be paid into the company. Also, on the newer types of policies, the policyholder intentionally contributes extra funds for investment purposes.

Because of selling and bookkeeping costs, a policyholder is not usually considered to have accumulated any investment until 2 or 3 years after the policy was written. Any time thereafter, however, the policy possesses certain values, called **nonforfeiture values** (or **nonforfeiture options**) to which the policyholder is entitled. These values may often be claimed only if the insured stops paying premiums for some reason, or if the policyholder turns in the policy (and is thus no longer insured). The principal nonforfeiture options (cash value, paid-up insurance, and extended term insurance) are outlined briefly in the following paragraphs.

Included in each life insurance policy (excluding term policies) is a table showing the **cash value** of the policyholder's investment after each year. As discussed earlier, even the universal and variable life policies are guaranteed a minimum cash value, approximately equal to that of whole life policies with the same face value.

If the policyholder wishes to terminate the insurance coverage, (s)he may surrender the policy and receive this cash value. If the policyholder needs some money but prefers to maintain the insurance protection, (s)he may borrow up to the cash value of the policy. The loan must be repaid with interest at a moderate rate, and the full amount of insurance coverage will remain in force during this time.

If the insured wishes to stop paying premiums and surrender the policy, the cash value of the policy may be used to purchase a reduced level of **paid-up insurance.** This means that the company will issue a policy which has a smaller face value but which is completely paid for. That is, without paying any further premiums, the policyholder will have a reduced amount of insurance that will remain in effect for the remainder of his or her life. Life insurance policies (except term policies) also contain a table showing the amount of paid-up insurance that the cash value would purchase after each year. (The newer types of policies do not build paid-up insurance values.)

When a policyholder simply stops paying premiums without notifying the company of any intent and without selecting a nonforfeiture option, the company usually automatically applies the third nonforfeiture option: **extended term** insurance. Under this plan the policy remains in effect, at its full face value, for a limited period of time. Stated another way, the cash value of the policy is used to purchase a term policy with the same face value and for the

| **TABLE 6-8** | NONFORFEITURE OPTIONS[a] ON TYPICAL LIFE INSURANCE POLICIES |

Issued at Age 25

| YEARS IN FORCE | WHOLE LIFE | | | | 20-PAYMENT LIFE | | | | VARIABLE UNIVERSAL |
| | CASH VALUE | PAID-UP INSURANCE | EXT. TERM | | CASH VALUE | PAID-UP INSURANCE | EXT. TERM | | CASH VALUE |
			YEARS	DAYS			YEARS	DAYS	
3	$ 9	$ 19	1	190	$ 32	$ 94	10	84	$ 44
5	31	87	9	200	74	218	19	184	95
10	93	248	18	91	187	507	28	186	321
15	162	387	20	300	319	768	32	164	595
20	251	535	22	137	470	1,000	Life		1,030
40	576	827	—	—	701	—	—	—	2,979

[a] The "cash value" and "paid-up insurance" nonforfeiture values are per $1,000 of life insurance coverage. The time period for "extended term insurance" applies as shown to all policies, regardless of face value. The cash value for variable universal life represents a reasonable estimate of the increase in value rather than a guaranteed amount.

maximum time period that the cash value will finance. Each policy also includes a table indicating after each year the length of time that an extended term policy would remain in force. This provision does not apply to term policies, since they do not build cash value. Also, if the policyholder of a universal or variable life policy fails to make payments, the company will simply deduct them from the cash account (rather than converting to extended term or paid-up insurance).

Note. Most insurance companies allow a *grace period* (usually 31 days following the date a premium is due) during which time the overdue premium may be paid without penalty. The policy remains in effect during this time.

The nonforfeiture options shown in Table 6-8 are typical provisions of life insurance policies. As stated previously, expanded tables are provided in individual policies. Example 4 illustrates the use of Table 6-8 in determining the three principal nonforfeiture options.

Example 4   Carl Best purchased a $50,000 20-payment life insurance policy at age 25. Determine the following values for his policy after it had been in force for 10 years: (a) cash value, (b) the amount of paid-up insurance he could claim, and (c) the time period for which extended term insurance would remain in effect.

(a)   As shown in Table 6-8, the cash value per $1,000 of face value after 10 years on a 20-payment life policy issued at age 25 is $187. Therefore, the cash value of a $50,000 policy is:

$$
\begin{array}{rl}
\$\ 187 & \text{Cash value per \$1,000} \\
\times\ \underline{\quad 50} & \\
\$9,350 & \text{Cash value of a \$50,000 policy}
\end{array}
$$

The cash value of Best's $50,000 policy would be $9,350 after the policy has been in effect for 10 years. Best could receive this amount by turning in his policy and forgoing his insurance coverage, or he could borrow this amount and pay it back with interest without losing his insurance protection.

(b)   The amount of paid-up insurance available is:

$$
\begin{array}{rl}
\$\ \ \ 507 & \text{Paid-up insurance per \$1,000} \\
\times\ \underline{\quad 50} & \\
\$25,350 & \text{Paid-up insurance on a \$50,000 policy}
\end{array}
$$

If Best surrenders his policy, he could receive a policy of $25,350, which would remain in force until his death without any premiums being required.

(c)   From Table 6-8, we see that if Best stops paying premiums, for whatever reason, extended term insurance would allow his full $50,000 coverage to remain in effect for 28 years and 186 days.

SETTLEMENT OPTIONS

When a life insurance policyholder dies, there are several ways in which the death benefits may be paid, called **settlement options.** According to the American Council of Life Insurance, life insurance companies paid out $100.0 billion to beneficiaries in 1993. Whole life policies accounted for 56% of the payments. The beneficiary may receive the face value in one lump-sum payment, or the benefits may be left invested with the company in order to earn interest.

An alternative method of receiving benefits, which deserves consideration, is to receive the benefits in the form of an annuity. An *annuity* is a series of payments, equal in amount and paid at equal intervals of time. (We shall consider here only annuities paid monthly.)

There are several options available to a beneficiary or an *annuitant* (person receiving an annuity) concerning the type of annuity (or installment) to choose. These choices are described briefly here.

1. The beneficiary may choose a monthly installment of a *fixed amount.* In this case, the specified amount will be paid each month for as long as the insurance money (plus interest earned on it) lasts. For example, a beneficiary may decide to receive a payment of $150 per month; this will continue until the account is depleted. The amount of income is the primary consideration when choosing this option.

2. The beneficiary may prefer the security of a monthly payment for a *fixed number of years.* For example, the beneficiary may want to receive monthly payments for 15 years. The insurance agent will then determine how much the beneficiary may receive monthly in order for the funds to last the specified number of years.

3. A third choice is to take the benefits in the form of an *annuity for life.* That is, depending on the age and sex of the beneficiary, the insurance company will agree to pay a monthly income to the beneficiary for as long as he or she lives. No further payments are made to anyone after the primary beneficiary dies.

4. The remaining possibility is to receive a *life annuity, guaranteed for a certain number of years.* For example, if the beneficiary chooses a life annuity guaranteed for 15 years, a monthly annuity will be paid for a minimum of 15 years or as long thereafter as the beneficiary lives. The monthly income from this annuity is somewhat lower than from the life annuity in plan 3, but a guaranteed annuity is selected much more often than an ordinary life annuity. The reason for this is that the unqualified life annuity is payable only to the beneficiary; when the beneficiary dies, even if only one annuity payment has been made, no further benefits will be paid. With the guaranteed annuity, however, installments are payable for as long as the primary beneficiary lives, or, if the primary beneficiary dies before a predetermined number of years, the company will continue the installments to a secondary beneficiary until the period ends. Thus, most people with dependents naturally choose the guaranteed

TABLE 6-9	SETTLEMENT OPTIONS

Monthly Installments per $1,000 of Face Value

OPTIONS 1 AND 2: FIXED AMOUNT OR FIXED NUMBER OF YEARS		OPTIONS 3 AND 4: INCOME FOR LIFE				
		AGE WHEN ANNUITY BEGINS			LIFE WITH 10 YEARS	LIFE WITH 20 YEARS
YEARS	AMOUNT	MALE	FEMALE	LIFE ANNUITY	CERTAIN	CERTAIN
10	$9.60	40	45	$4.60	$4.56	$4.44
12	8.52	45	50	5.16	5.07	4.80
14	7.71	50	55	5.30	5.28	5.00
15	6.89	55	60	6.13	6.00	5.46
16	6.43	60	65	6.56	6.31	5.65
18	6.08	65	70	7.22	6.70	5.78
20	5.66					

annuity. Even persons without dependents often select the guaranteed annuity, feeling that they would rather have someone receive the payments if they should die, instead of risking that most of the face value of the policy might be lost.

It should be noted that these settlement options are also available to policyholders themselves, if they surrender the policy. Older people often feel that they no longer need the amount of life insurance coverage carried since they were younger and had children to support. Therefore, many retired persons convert some of their life insurance policies to annuities in order to supplement other retirement income. When whole life or limited payment life policies are converted to annuities, the monthly installment depends on the *cash value* of the policy rather than upon its face value.

Table 6-9 lists the monthly income per $1,000 of face value available under the various settlement options. Life insurance policies also include tables similar to Table 6-9.

Example 5   Martha Donaldson, age 60 and the beneficiary of a $30,000 life insurance policy, has decided to receive the money as monthly payments rather than one lump sum. (a) Find the monthly payment if she chooses to receive payments for 15 years. (b) If Mrs. Donaldson arranges a monthly income of approximately $255, how many years will the payments continue?

(a)   Table 6-9 shows that for a $1,000 policy, the monthly installment for the fixed number of years (15) would be $6.89. Thus, Mrs. Donaldson could receive $206.70 monthly for 15 years, as follows:

$\begin{array}{r} \$\ 6.89 \\ \times\ \underline{\quad 30} \\ \$206.70 \end{array}$ $\begin{array}{l} \text{Monthly installment per } \$1,000 \\ \\ \text{Monthly installment from } \$30,000 \text{ policy (for 15 years)} \end{array}$

(b) Mrs. Donaldson will receive 30 times any amount shown in Table 6-9. Thus, 30 × $8.52 will provide approximately $255 per month. The table indicates that she could receive $255.60 monthly for 12 years before the money (and interest paid by the insurance company) has been depleted.

Note. Notice that Mrs. Donaldson's age had no bearing on this problem, because she was not considering an annuity that would pay for the remainder of her life.

Example 6 Walter Carson became the beneficiary of a $25,000 life insurance policy at age 55. (a) What would be the monthly income from a life annuity? (b) From an annuity guaranteed for 20 years?

Carson's age affects both these options because either annuity pays until his death. (Notice that an older person would receive more per month because payments are not expected to be made for as many years.) From Table 6-9 for a male, age 55, the two annuities would pay as follows:

(a) *Life Annuity*		(b) *Life, 20-Year Certain*
$ 6.13	← Monthly per $1,000 →	$ 5.46
× 25		× 25
$153.25	← Monthly for $25,000 policy →	$136.50

Option (a) indicates that Carson could receive $153.25 monthly for life. These payments would stop immediately upon his death, however, regardless of how few monthly installments might have been paid. Option (b) will pay $136.50 monthly for as long as Carson lives (even if this is more than 20 years). If he dies in less than 20 years, the payments will continue to his heirs until the guaranteed 20 years have passed.

Example 7 Mary Reynolds was widowed at age 60, inheriting a $40,000 life insurance policy. She chose to receive all the benefits over a 10-year period. (a) What was her monthly installment? (b) How much would she have received monthly by choosing a life annuity guaranteed for 10 years? (c) Suppose that Mrs. Reynolds actually lived another 13 years. From which option above would she have received more total income, and how much more?

(a) *10-Year Annuity*		(b) *Life, 10-Year Certain*
$ 9.60	← Monthly per $1,000 →	$ 6.00
× 40		× 40
$384.00	← Monthly for $40,000 policy →	$240.00

Option (a) paid Mrs. Reynolds $384 per month for 10 years, although she lived longer. Had she chosen option (b), she would have received $240.00 monthly for the entire 13 years she lived.

(c) The total payments received under each settlement would have been as follows:

10-Year Annuity		Life Annuity Guaranteed 10 Years
12	Months per year	12
× 10	Years	× 13
120	Payments	156
$ 384	Monthly annuity	$ 240
× 120	Payments	× 156
$46,080	Total received	$37,440

Mrs. Reynolds received $46,080 − $37,440 = $8,640 more income by selecting the fixed 10-year annuity, despite the fact that the guaranteed life annuity would have paid for 3 years longer. Observe that no further payments would be made to her heirs, because the guaranteed period is past for both annuities.

SECTION 3 PROBLEMS

Unless other rates are given, use the tables in this chapter to compute the following life insurance problems. Assume that each age given is the "insurable age" (age to the nearest birthday) of the applicant. For female applicants, use a 5-year "setback" in age when obtaining rates from a table for males.

Using the information given and Table 6-7, compute the annual premium for each life insurance policy.

APPLICANT, AGE	TYPE OF POLICY	FACE VALUE	ANNUAL PREMIUM
1. a. Male, 18	10-year term	$ 10,000	
b. Male, 24	20-payment life	35,000	
c. Female, 30	Whole life	50,000	
d. Female, 40	Variable universal	25,000	
2. a. Male, 20	10-year term	$ 15,000	
b. Male, 45	Whole life	50,000	
c. Female, 25	20-payment life	20,000	
d. Female, 30	Variable universal	100,000	

Determine the nonforfeiture values of the following policies, based on the rates in Table 6-8 (for policies issued at age 25).

		YEARS IN FORCE	TYPE OF POLICY	FACE VALUE	NONFORFEITURE OPTIONS	NONFORFEITURE VALUE
3.	a.	10	Whole life	$ 25,000	Cash value	
	b.	15	20-payment life	40,000	Paid-up insurance	
	c.	3	20-payment life	50,000	Extended term	
	d.	15	Variable universal	70,000	Cash value	
4.	a.	20	Whole life	$100,000	Cash value	
	b.	5	20-payment life	50,000	Paid-up insurance	
	c.	10	Whole life	125,000	Extended term	
	d.	3	Variable universal	150,000	Cash value	

Use Table 6-9 to find the monthly annuity (or total years of the annuity) that the beneficiary may select in settlement of the following policies.

		BENEFICIARY		FACE VALUE	SETTLEMENT OPTION CHOSEN	MONTHLY ANNUITY	
		SEX	AGE			YEARS	AMOUNT
5.	a.	M	58	$ 20,000	Fixed number of years	15	
	b.	F	50	45,000	Fixed amount per month		$310
	c.	F	55	30,000	Life annuity	—	
	d.	F	60	40,000	Guaranteed annuity	10	
6.	a.	M	60	$ 55,000	Fixed number of years	16	
	b.	F	50	75,000	Fixed amount per month		$425
	c.	F	65	100,000	Life annuity	—	
	d.	F	45	150,000	Guaranteed annuity	20	

7. Sam Pruitt was 30 years old when he purchased his 20-payment life policy with a face value of $70,000.

a. What was the annual premium for the policy?

b. If Pruitt purchased a whole life policy for the same face value, how much would be the annual premium?

Assume that Pruitt lives for 27 years after purchasing the policy. How much would he pay in total premiums for

c. The 20-payment life policy?

d. The whole life policy?

8. Charles Simon purchased a 20-payment life policy with a face value of $85,000 when he was 25.

a. What was the annual premium for the policy?

b. If Simon purchased a whole life policy for the same face value, how much would be the annual premium?

Assume that Simon lives 30 years after purchasing the policy. What would be his total premiums for

   c. The 20-payment life policy?

   d. The whole life policy?

9. Susan Williams purchased a $25,000, 10-year term life policy at age 25.

   a. What was the annual premium?

   b. How much will she pay in total premiums?

   c. If Williams had selected a whole life policy, how much would premiums have cost during the same 10 years?

   d. How much would be saved in premiums by purchasing the 10-year term policy?

   e. If Williams died in the following year after the term policy expired, how much life insurance would her beneficiary receive?

10. Christine Rubella purchased a $50,000, 10-year term policy at age 30.

   a. What was the annual premium?

   b. How much in total premiums will she pay over 10 years?

   c. What would be the equivalent whole life cost during the same 10-year period?

   d. How much would she save in premium cost by purchasing the term policy?

   e. If Christine died the year after the 10-year term, how much life insurance would her beneficiary receive?

11. When Martin Andrews was 40, he bought a $40,000, variable universal policy.

   a. What would have been the annual savings on premiums if Andrews had taken out the policy at age 20?

   b. How much will the premiums cost to age 60?

   c. What would the total premiums have been on a policy taken out at age 20 assuming he lives to age 60?

   d. What other factors should be considered?

12. At age 30, James Godwin purchased a $75,000 whole life policy.

   a. How much less would he pay annuallly on premiums if he had started this policy at age 20?

   b. What total amount will he pay for the premiums to age 65?

   c. What would the premiums have cost to age 65 if Godwin had started the policy at the age of 20?

   d. What other factors should be considered?

13. Chuck Stearns purchased a $50,000 whole life policy when he was 25.

   a. What was the annual premium?

   b. After 5 years, Stearns decided to terminate the coverage. What is the cash value of the policy at this point?

   c. How much insurance paid up for life would this policy provide after 5 years?

   d. After 5 years in force, for how long would the same $50,000 coverage be continued under extended term insurance?

14. Roland Burke, age 25, purchased a $100,000, 20-payment life policy.

   a. What was his annual premium?

   b. After 10 years, Burke surrendered his policy. What was the cash value of the policy at that time?

   c. How much paid-up insurance for life would this policy provide after 10 years in force?

   d. For how long would the same coverage remain in force if he discontinued paying premiums?

15. Edwin Valburn purchased a whole life insurance policy 20 years ago when he was 25. He is now interested in borrowing on his $80,000 policy without surrendering his insurance coverage. How much could he now borrow against the policy?

16. At age 25, Tanya Dean purchased a $40,000, 20-payment life policy. It has been in force for 15 years. She is now faced with college tuition payments for her daughter. How much can she borrow on her $40,000 policy without surrendering her coverage?

17. Bob Rand purchased a $20,000 variable universal life policy when he was 25.

    a. How much cash value can he expect after 10 years?

    b. How much would the cash value be after 10 years on a 20-payment life policy of the same amount?

18. Marlow Jones purchased a $150,000 variable universal policy at age 25.

    a. What is the cash value after 20 years?

    b. What would be the cash value after 20 years on a whole life policy of the same amount?

19. Margaret Dupree, who is age 55, is the beneficiary of her husband's $25,000 life insurance policy.

    a. If she decides to have the benefits paid over a 10-year period, what annuity will Mrs. Dupree receive each month during that time?

    b. If she wants to receive a monthly benefit of about $215, approximately how many years will the benefits last?

20. At age 52, Hilda Strothers became the beneficiary of an $80,000 life insurance policy.

    a. If she decides to have the benefits paid over a 14-year period, what monthly annuity will she receive?

    b. If she decides to receive a monthly annuity of about $770, for approximately how many years will she receive the benefits?

21. Dan Kingston inherited a $30,000 life insurance policy when he was 60.

    a. Determine the monthly payment Kingston will receive if he chooses the settlement option of a life annuity.

    b. If he chooses a life annuity guaranteed for 10 years, how much will he receive each month?

22. Amilla Pashayev has become the beneficiary of a $125,000 life insurance policy at the age of 45.

    a. Determine the monthly payment she will receive if she selects a life annuity.

    b. If she selects a life annuity guaranteed for 20 years, how much will she receive each month?

23. At age 50, Sid Berlin decided to convert the $60,000 cash value of his life insurance to an annuity. Mr. Berlin decided to pick a 20-year annuity. He lived to age 74.

    a. Determine the total amount he received from the annuity.

    b. If he had chosen a life annuity with 20 years certain, how much would he have received?

    c. Did he gain or lose by his annuity choice, and how much?

    Hint. Recall that an annuity is based on the cash value (rather than the face value) when the policyholder converts a policy.

24. Sandra Tolson decided to convert the $50,000 cash value of her life insurance policy to a 10-year annuity when she reached age 55. She lived to age 68.

    a. Determine the total amount she received from the 10-year annuity.

    b. If she had chosen a life annuity with 10 years certain, how much would she have received?

    c. Did she gain or lose by her annuity choice and how much?

    Hint. An annuity is based on the cash value rather than the face value when the policyholder converts a policy.

25. Sarah Field, at age 60, became the beneficiary of a $40,000 life insurance policy. Mrs. Field selected as a settlement option a monthly annuity for life.

    a. How much did she receive each month?

    b. Mrs. Field lived to be 78. What total amount did the annuity pay?

c. Suppose Mrs. Field had chosen a life annuity guaranteed for 20 years. What annuity would she have received each month?

d. Did she gain or lose by her choice of settlement options, and how much?

26. At age 50, Jennifer Holt became the beneficiary of a $90,000 life insurance policy. Ms. Holt chose a monthly annuity for life as her settlement option.

a. What was her monthly income from the annuity?

b. If Ms. Holt lived to age 67, what total amount did she receive?

c. If she had selected a life annuity guaranteed for 20 years, what montly annuity would she have received?

d. Did she gain or lose by her choice of settlements, and how much?

CHAPTER 6 GLOSSARY

Annuity. The payment of a set amount at a regular interval. (Often used in settlement of a life insurance claim, either for a fixed number of years or for life—sometimes for life with 10 or 20 years certain.)

Beneficiary. The person designated to receive the face value of a life insurance policy after death of the insured.

Bodily injury liability. The portion of standard motor vehicle insurance that pays for physical injuries to other persons when the insured is at fault in an accident.

Business interruption insurance. Insurance that provides income after a fire to meet continuing business expenses and net income until business operations resume.

Carrier. An insurance company (also called an "underwriter" or an "insurer").

Cash value. An investment value earned by all life insurance policies except term, which may be borrowed against or may be paid in cash to a policyholder who terminates the policy.

Coinsurance clause. A clause in many fire policies requiring coverage of a stipulated percent (often 80%) of value at the time of a fire in order to qualify for full reimbursement for damages.

Collision (and upset) insurance. Insurance that pays for damages to the policyholder's own vehicle (1) if the insured is at fault in an accident or (2) if the insured's vehicle is damaged when there is no other responsible driver whose liability insurance will pay.

Comprehensive insurance. Insurance that pays for damages of any kind to the policyholder's own vehicle, except those caused by collision or one-vehicle upset.

Double indemnity. An optional provision of many life insurance policies that pays twice the face value in case of accidental death.

Extended coverage. A supplement to standard fire policies that insures against related damages such as smoke and water damage.

Extended term. A nonforfeiture option earned by life insurance policies (except term or variable universal) whereby the same level of coverage may be maintained for a specified time after premiums are terminated.

Face value. The full (maximum) amount of insurance specified by an insurance policy.

Group insurance. Life and/or health/accident insurance available at a reduced cost to employees of a business or to members of a business/professional organization.

Indemnity. The payment made by an insurance company after an insured loss.

Insured. A person or business that purchases insurance coverage (also called a "policyholder"), or a person whose life is covered by a life insurance policy.

Insurer. An insurance company (also called a "carrier" or an "underwriter").

Interest-adjusted net cost. The method now used to determine the cost of life insurance protection only, not including the portion of premiums used to build investment value.

Interest-adjusted net cost index. The annual cost index of a life insurance policy: Total interest-adjusted net cost ÷ Years.

Liability insurance. The portion of standard motor vehicle insurance and general business insurance that provides financial protection when the person or business is held legally responsible for personal injuries and/or physical damages.

Limited payment life insurance. Insurance that provides life insurance and, within the specified payment period, builds paid-up insurance of the full face value of the policy.

Medical payment insurance. The optional portion of standard motor vehicle insurance that pays medical costs (supplementing other policies) for injuries to the policyholder's own self, family, and passengers, regardless of who is at fault in an accident.

Mutual life insurance company. A life insurance company owned by policyholders rather than stockholders; sells participating policies, which pay dividends to the policyholder.

Net cost index. An annual cost equivalent to thc total interest-adjusted net cost. Also called the "interest-adjusted net cost index."

No-fault insurance. Vehicle insurance in which each motorist collects for bodily injuries from his or her insurance company after an accident, regardless of who is to blame.

Nonforfeiture value (or option). The investment value (cash value, paid-up

insurance, or extended term) available when a life insurance policy is terminated (except term; variable/universal has cash value only).

Nonparticipating policy. A life insurance policy that pays no dividends and usually has lower premiums than participating policies; the type of policy usually sold by stock life insurance companies.

Paid-up insurance. A nonforfeiture value earned by life insurance policies (except term or variable/universal policies) whereby a specified level of insurance may be maintained after premiums are terminated.

Participating policy. A life insurance policy that pays dividends and usually has higher premiums than nonparticipating policies; the type of policy usually sold by mutual life insurance companies.

Policy. An insurance contract that specifies in detail the provisions and limitations of the coverage.

Policyholder. A person or business that purchases insurance (also called the "insured").

Premium. The payment made to purchase insurance coverage.

Property damage liability. The portion of standard motor vehicle insurance that pays for damages caused by the policyholder to another vehicle or other property.

Settlement option. Any of several types of annuities available to the beneficiary of a life insurance policy instead of a lump-sum payment.

Short-term policy. An insurance policy in force for less than 1 year (usually referring to a fire insurance policy).

Straight life insurance. (See "Whole life insurance.")

Term (of a policy). The time period for which a policy will remain in effect.

Term life insurance. Provides life insurance only and builds no nonforfeiture values for the policyholder; the least expensive type of life insurance policy.

Underwriter.   An insurance company (also called the "carrier" or "insurer").

Uninsured motorists insurance.   An optional portion of standard motor vehicle insurance that provides financial protection for the policyholder's own bodily injuries when hit by a driver without bodily injury liability insurance.

Universal life insurance.   A policy providing term life coverage, with the remainder of each premium going into a cash account that earns interest at a rate reflecting the company's overall profit. Allows flexibility in paying premiums and changing policy provisions.

Variable life insurance.   A policy providing term life coverage, with the remainder of each set premium going into stocks, bonds, money market accounts, or other funds the policyholder selects. The death benefit may fluctuate along with the value of the investment accounts.

Variable universal life insurance.   Insurance that combines aspects of the variable life and universal life policies. Provides a set death benefit, invests cash into various financial funds, and allows flexibility in premium payments. (Also called flexible-premium variable life.)

Waiver of premium benefit.   An optional provision of many life insurance policies whereby the company bears the cost of the premiums if the policyholder is totally and permanently disabled.

Whole (ordinary, or straight) life insurance.   The most common type of policy, in which premiums are paid for as long as the insured lives in order to maintain coverage of full face value.

Worker's compensation.   A form of business liability insurance that pays employees for sickness or accidental injury resulting directly from work responsibilities.

7

CHECKBOOK AND CASH RECORDS

OBJECTIVES

Upon completion of Chapter 7, you will be able to:

1. Define and use correctly the terminology associated with each topic.
2. Complete check stub balances using the information given (Section 1: Problems 1, 2).
3. Prepare a bank reconciliation (Section 1: Example 1):
   a. Using given information (Problems 3–8)
   b. Using a check register and bank statement (Problems 9–12).
4. Complete the following cash forms (Section 2):
   a. Daily cash record (Example 1; Problems 1–4)
   b. Over-and-short summary (Example 2; Problems 5, 6).

Practically all business obligations today are paid by check. Checks are not only more convenient and safer than cash, but the canceled check also constitutes a legal receipt. Nevertheless, retail businesses do handle a large amount of cash received on customer purchases. Both checking and cash transactions require certain records, which are the subject of this unit.

SECTION 1

CHECKBOOK RECORDS

A check, like a bank note, is a **negotiable instrument,** which means that it may be transferred from one party to another for its equivalent value; the second party then holds full title to the check. Virtually all checks, from both businesses and individuals, are written on preprinted forms obtained from the individual bank. Although checks may vary in appearance somewhat, the following requirements must be met: (1) It must be dated; (2) the bank (or **drawee**) must be positively identified (with address, if necessary); (3) the check must name the **payee** (the person, firm, "cash," or to whomever it is payable); (4) the amount must be shown; and (5) the check must be signed by the **drawer** or **maker** (the person writing the check).

Figure 7-1 illustrates the parts of a check. The amount of the check is written twice, once in figures and then in words. The written numbers are considered the legal amount of the check; that is, if there is a discrepancy between the figures ($50) and the written words (Fifty and no/100 dollars), the written words are considered to be the correct and legal amount of the check.

The magnetic ink figures at the bottom of the check are used for identifying the check and for sorting the check electronically during the routing process from the payee's (Davis Supply Co.) bank through the Federal Reserve District Bank back to the drawer's (James C. Morrison and Associates) bank.

The check stub balance should always be kept current. All checks and

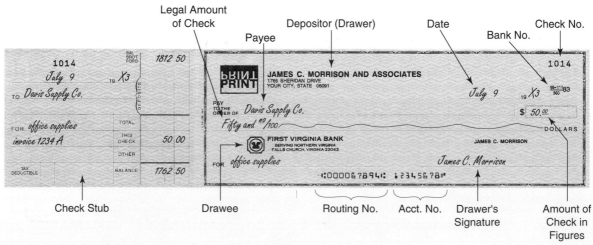

Courtesy of First Virginia Bank

FIGURE 7-1   Parts of Check

deposits should be recorded immediately and the new balance computed. If a check is not used due to errors in writing the check or for other reasons, it should be voided. The word "void" should be written on the check stub and on the check itself to provide a record for that particular check. The check stub also provides space to record pertinent information about the check: what the check was for, how much sales tax (or interest) was included, how much cash discount was given, and so forth.

The **voucher check** (Figure 7-2) is often used by businesses. The "voucher" attached to the check itself explains which invoice the check covers, itemizes any deductions, or otherwise accounts for the amount shown on the check. The voucher is not negotiable and is torn off at the perforation before the check is cashed or deposited. The company issuing the check retains one or more carbon copies of the voucher for its records. In larger firms, these voucher checks are typically generated by a computer, which also updates the account for their bank balance at the same time.

Smaller firms using voucher checks ordinarily keep a **check register,** which is a concise record of checks and deposits. Figure 7-3 illustrates a check register.

ORIGINAL

PAYEE SHOULD DETACH TOP
AND DEPOSIT VOUCHER AT ONCE

STATE BOARD OF EDUCATION   VOUCHER NO / 1,308

DEPARTMENT OF COMMUNITY COLLEGES

PURCHASE ORDER NO	DATE OF INVOICE	BUDGET POS NO	PARTICULARS	CODE	GROSS AMOUNT	DEDUCTIONS	NET AMOUNT
819	Sept. 23	14	30 Lab Stools	16–D	$345.00	$ 7.90	$337.10
834	Sept. 26	21	60 #243A Desks	9–A	900.00	18.00	882.00
							$1,219.10

IBAA 35

DRAWN BY (NAME AND ADDRESS OF INSTITUTION)

Blue Ridge Community College
Asheville, North Carolina

STATE BOARD OF EDUCATION · DEPARTMENT OF COMMUNITY COLLEGES

PRESENT
**TO STATE TREASURER—
STATE OF NORTH CAROLINA**
RALEIGH, N. C.

66-1059
512

VOUCHER NO. 1,308

PAYABLE AT PAR THROUGH THE FEDERAL RESERVE SYSTEM

DATE November 9 ____ 19____

THIS CHECK VOID AFTER 60 DAYS FROM DATE

PAY Twelve hundred nineteen and 10/100 ------------------------- DOLLARS $ 1,219.10 ____

	Whitman Technical Furniture, Inc.	Specimen	G-185
TO THE ORDER OF	Box 823 High Point, North Carolina	Priscilla K. Stokes AUTHORIZED SIGNATURE	
		William R. Marten COUNTERSIGNED	

⑈0512⑈1059⑈ 4⑈000⑈574⑈

FIGURE 7-2   Voucher Check

DATE	CHECK NUMBER	ORDER OF	FOR	AMOUNT CHECK	AMOUNT DEPOSIT	BALANCE
19___						$ 919.81
July 16	362	The May Co.	Brass fixtures	$ 86.15		833.66
18	363	Warren's, Inc.	Office supplies	15.83		817.83
18		Deposit			$840.00	1,657.83
19	364	Payroll	Weekly payroll	789.25		868.58
21	365	Carson's, Inc.	Equipment rental	27.44		841.14

FIGURE 7-3   Check Register

DEPOSITS

Before a company can deposit or cash the checks it has received, the checks must be endorsed by the payee. The endorsement should be on the back left-hand side of the check in a designated, boxed area that is within $1\frac{1}{2}$ inches from the edge of the check. The space below this boxed area is used by the Federal Reserve.

There are three forms of **endorsement** in common use: the blank endorsement, the restrictive endorsement, and the special endorsement.

The **blank** endorsement is just the name of the person or the firm (and often the individual representing that firm) to which the check is made payable. This endorsement makes the check negotiable by anyone who has possession of it. That is, such a check could be cashed by anyone if it should become lost or stolen.

The other two endorsements are somewhat similar. The **restrictive** endorsement is usually applied to indicate that the check must be deposited in the firm's bank account. Many firms endorse all their checks in this manner as soon as they are received, as a precautionary measure against theft. The **special**

ENDORSE HERE
x James C. Morrison

DO NOT WRITE, STAMP OR SIGN BELOW THIS LINE
RESERVED FOR FINANCIAL INSTITUTION USE.

BLANK

ENDORSE HERE
x For Deposit at
First Virginia Bank
Acct. # 12345678

DO NOT WRITE, STAMP OR SIGN BELOW THIS LINE
RESERVED FOR FINANCIAL INSTITUTION USE.

RESTRICTIVE

ENDORSE HERE
x Pay to the order of
Union Fund
James C. Morrison

DO NOT WRITE, STAMP OR SIGN BELOW THIS LINE
RESERVED FOR FINANCIAL INSTITUTION USE.

SPECIAL

(or **full**) endorsement indicates a check is to be passed on to another person, firm, or organization. The special endorsement's great advantage over the blank endorsement is that, should the check be lost or stolen, it is negotiable only by the party named in the special endorsement.

After being endorsed, the checks are listed individually on a bank deposit slip. Checks are often listed according to the name of the bank on which they were written, or by the name of the person or firm who wrote them; however, many banks prefer that checks be listed by *bank number*. Most printed checks contain the complete identification number of the bank on which the check is written. For example,

$$\frac{66\text{-}1059}{512}$$

The "512" represents the Federal Reserve District within which the bank is located; the "66" identifies the location (state, territory, or city) of the bank; and the "1059" indicates the particular bank in that location. Notice in Figure 7-2 that the bank number "0512···1059" is coded in magnetic ink at the bottom of the check. Some checks contain only the numbers indicating location and official bank number (for example, 66-1059). It is sufficient to use this hyphenated number when listing checks on the deposit slip. A completed deposit slip is shown in Figure 7-4.

BANK SERVICES AND STATEMENTS

Banks today offer numerous services. Charges are made for these services and deducted by the bank from the customer's (or drawer's) account. The customer many times does not know of these charges until the bank statement is received,

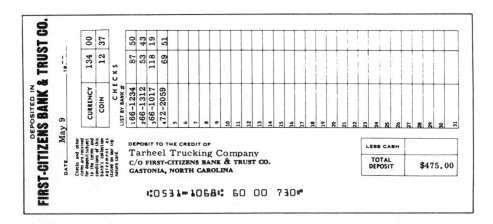

FIGURE 7-4   Deposit Slip

and thus has not recorded them on the check stub or in the check register. Typical bank charges are **service charges** (SC) for checks written, **debit memos (DM)** (charges for printing new checks or other services), corrections in deposits, **returned checks** (RT), **overdrafts** (OD), and **insufficient funds** (ISF). A charge for a cash withdrawal from an automatic teller machine (**ATM**) is another example. A returned check is a check previously *deposited* into the checking account when the maker's account did not contain enough money to cover the check. Because the depositor's bank could not collect the funds, the check amount is deducted from the depositor's account—the RT (return) charge—and the check is returned to the depositor. The check may be redeposited after the maker has had time to replenish the funds in his or her account.

When the owner of the checking account (the maker) writes a check that would overdraw the account, the bank charges a fee that is deducted from the maker's account—an OD (overdraft) or ISF (insufficient funds) charge. Most banks offer **overdraft protection.** Some banks automatically transfer money from savings accounts to cover the insufficient funds. Other banks use the customer's credit card as backing. A third type of overdraft protection is the **line of credit.** A line of credit is the limit to the amount of credit that is given to a customer. In this case, a line of credit is assigned to each individual account. If the maker or drawer overdraws the account balance, the bank will honor checks up to this line of credit. The owner of the account must repay the bank for any amounts used under the provisions of overdraft protection, plus interest.

Electronic funds transfer is a computerized system for depositing and/or paying funds out of the customer's account. Many employees have their employer transfer their wages directly from the company's bank account into their personal accounts. Another variation of electronic banking is the use of point-of-sale terminals, which are located at the merchant's business and connected to a bank computer. When the customer makes a purchase using a **debit card,** money from the customer's bank account is automatically and instantly transferred to the merchant's account. Another variation of electronic banking is the use of personal computers that are linked by telephone to the bank computer. In response to instructions from the customer's personal computer, the bank automatically transfers money from the customer's account to the merchant's account. Wide acceptance of electronic funds transfer could effectively make us a cashless society.

If the customer forgets to record any of these deductions on the check stub or in the check register, the discrepancy will appear when the bank statement (Figure 7-5) and check register are reconciled.

In some instances, the bank statement contains a credit that will increase the checkbook balance—typically interest. Many financial institutions are presently offering **negotiable order of withdrawal (NOW) accounts,** a combination checking-savings account that earns interest on its average daily balance, provided the average meets or exceeds a specified minimum. The interest on this

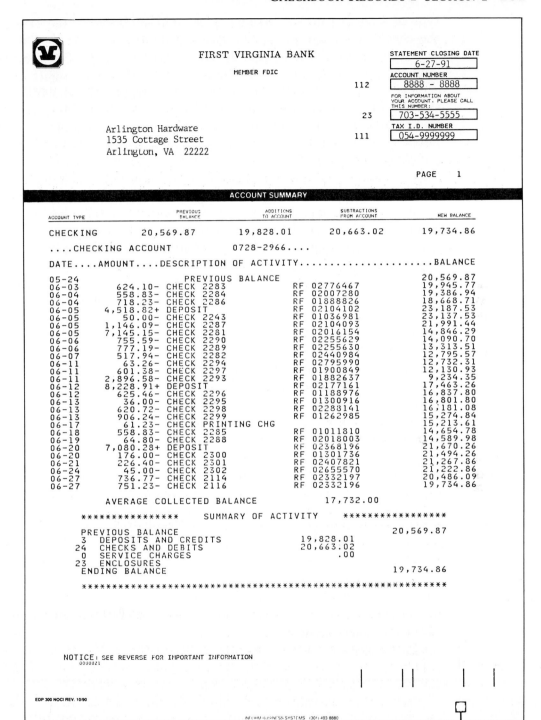

Courtesy of First Virginia Bank

FIGURE 7-5   Sample Bank Statement

type of account is treated like a deposit, and the bookkeeper should add it to the checkbook balance. (Similar types of interest-earning checking accounts at a savings and loan association are sometimes called "money-market accounts.")

Frequently, some checks and possibly a deposit (particularly those near the end of the month) will not have been processed in time to be recorded and returned with the bank statement. Such checks are known as **outstanding checks.** The bookkeeper should compare the bank statement against the check stubs to determine which checks (and deposits) are outstanding. The bookkeeper must determine the total of the outstanding checks and deduct this amount from the bank statement balance to obtain the adjusted bank balance. (A deposit not included on the bank statement must be added to the bank statement balance to determine the adjusted bank balance.)

After these adjustments have been made, the adjusted bank balance and the adjusted checkbook balance will be the same, provided that all transactions were recorded correctly and that there has been no mathematical mistake. This adjusted balance represents the true balance available to spend.

Example 1   A florist's checkbook shows a balance if $935.50, whereas the balance on the bank statement is $1,520.36. Outstanding checks amount to $707.90, and there is a deposit of $244.30 not recorded by the bank. The bank statement shows the following: a service charge of $2.50; interest on the NOW account of $4.76; an electronic funds deposit of $210; and a returned check for $100.00. The bookkeeper made a subtraction error of $9.00 in recording check number 125. Reconcile the bank statement.

BANK RECONCILIATION

Bank Statement Balance	$1,520.36	Checkbook Balance		$ 935.50
Add: Outstanding deposit	+ 244.30	Add: Interest	$ 4.76	
	$1,764.66	Electronic deposit	210.00	
		Recording error	9.00	+ 223.76
				$1,159.26
		Less: Service charge	$ 2.50	
		Returned check	100.00	− 102.50
Less: Outstanding checks	− 707.90			
Adjusted Balance	$1,056.76	Adjusted Balance		$1,056.76

SECTION 1   PROBLEMS

Complete the check stubs below with the information given and update the account balance after each check is written:

1. Balance brought forward: $2,456.21

Check No. 4600
Date: Oct. 8, 19xx
Amount: $756.00
To: Modern Furniture Co.
For: Sofa and 2 chairs, $720
           plus sales tax, $36
Deposit: 0.

Check No. 4601
Date: Oct. 9, 19xx
Amount: $312.00
To: All-State Distributors
For: Merchandise
Deposit: $650.23

Check No. 4602
Date: Oct. 9, 19xx
Amount: $430.00
To: Hurd Insurance Co.
For: Quarterly payment
           of liability premium
Deposit: 0.

Check No. 4603
Date: Oct. 10, 19xx
Amount: $150.00
To: B. Vaccarella, petty
           cashier
For: Petty cash
           replenishment
Deposit: 0.

#	Balance Brought Forward	_____
_____ 19 ____		
To _____		
For _____		
	Amount Check	_____
	Deposit	_____
	Balance	_____

#	Balance Brought Forward	_____
_____ 19 ____		
To _____		
For _____		
	Amount Check	_____
	Deposit	_____
	Balance	_____

#	Balance Brought Forward	_____
_____ 19 ____		
To _____		
For _____		
	Amount Check	_____
	Deposit	_____
	Balance	_____

#	Balance Brought Forward	_____
_____ 19 ____		
To _____		
For _____		
	Amount Check	_____
	Deposit	_____
	Balance	_____

2. Balance brought forward: $3,645.26

Check No. 360	Check No. 361
Date: Nov. 2, 19xx	Date: Nov. 3, 19xx
Amount: $183.92	Amount: $982.30
To: Gold Supplies	To: Baltimeier Sales
For: Office supplies, $176 plus sales tax, $7.92	For: Computer, $940 plus sales tax, $42.30
Deposit: 0.	Deposit: 0.

Check No. 362	Check No. 363
Date: Nov. 3, 19xx	Date: Nov. 4, 19xx
Amount: $1,250.00	Amount: $64.00
To: Garman, Inc.	To: US Postal Service
For: Merchandise	For: Postage stamps
Deposit: $374.46	Deposit: 0.

```
#                    Balance
                     Brought
                     Forward     _____
_____ 19 ___
To _____
   _____
For _____
   _____
                     Amount
                     Check      _____
                     Deposit    _____
                     Balance    _____
```

```
#                    Balance
                     Brought
                     Forward     _____
_____ 19 ___
To _____
   _____
For _____
   _____
                     Amount
                     Check      _____
                     Deposit    _____
                     Balance    _____
```

```
#                    Balance
                     Brought
                     Forward     _____
_____ 19 ___
To _____
   _____
For _____
   _____
                     Amount
                     Check      _____
                     Deposit    _____
                     Balance    _____
```

```
#                    Balance
                     Brought
                     Forward     _____
_____ 19 ___
To _____
   _____
For _____
   _____
                     Amount
                     Check      _____
                     Deposit    _____
                     Balance    _____
```

Prepare a bank reconciliation for each of the following.

3. The following checks were outstanding: $68.87, $129.95, $98.33, and $253.43. A deposit in transit for $300 was not recorded on the bank statement, but a service charge of $9.40 and a returned check for $350 were listed. The checkbook balance was $745.21, and the balance on the bank statement was $636.39.

4. The checkbook balance was $142.12, and the bank statement balance was $605.75. Outstanding checks were in the amounts of $86.20, $122.48, $250.39, and $12.56. A service charge of $8 had been deducted on the bank statement.

5. The bank statement balance was $942.40, while the check register balance was $1,207.66. There was a returned check for $45 and a debit memo of $10 listed on the statement. Checks outstanding were $25.64, $148.80, and $67.30. There were also two outstanding deposits of $300 and $152.

6. The bank statement balance was $1,814.41, and the checkbook balance was $1,077.97. A deposit of $334.82 had not been entered on the bank statement. A service charge of $10.45 and a check printing charge of $9.20 were listed on the statement. There was a returned check of $50. Interest on the NOW account was $5.15. Outstanding checks were for $176, $440.60, and $519.16.

7. The check register balance was $1,855.67, while the bank statement shows a balance of $2,340.50. The statement shows a deposit by electronic transfer of $230 that was not recorded in the check register. Interest on the NOW account of $6.42 was not recorded in the check register either. The bookkeeper subtracted $5 too much in the check register while recording a check. Outstanding checks amounted to $976.41. There was an ATM withdrawal of $50, plus a charge of $1, and a debit memo for printing new checks of $22 listed on the bank statement. A deposit of $660 was not included on the statement.

8. The checkbook showed a balance of $1,348.29 whereas the bank statement had a balance of $845.32. An outstanding deposit totaled $800. Outstanding checks amounted to $477.75. The statement showed a charge of $2.00 to correct a deposit. The service charge was $6.55. A returned check amounted to $165.65. There was a debit of $15 for printing new checks on the statement. The account earned interest of $8.48.

> *Four check registers and bank statements are given in Problems 9–12 on the following pages. Determine which checks and deposits have been processed, which are still outstanding, and what bank charges have been made. (Use the "√" column of the register to check off items that appear on the bank statement. Some have already been checked the preceding month.) Each check register shows the adjustment made last month to balance the register. Based on the information found, prepare a bank reconciliation for each problem, and determine the adjusted balance.*

9.

CHECK REGISTER

CHECK NUMBER	DATE	CHECK ISSUED TO	AMOUNT OF CHECK	√	AMOUNT OF DEPOSIT	BALANCE 1,282.22
341	6/27	Morgan Supply	133.18			1,149.04
342	6/28	Packard Co.	98.61			1,050.43
343	6/28	Bo's Auto Co.	400.50			649.93
	6/28	DEPOSIT			1,994.60	2,644.53
344	6/28	Payroll	1,540.88			1,103.65
345	7/1	Adams Sales	63.77			1,039.88
	7/4	DEPOSIT			655.23	1,695.11
346	7/8	Lopez & Sons	75.34			1,619.77
347	7/8	AB Insurance	108.55			1,511.22
348	7/10	Cook Roofing	425.40			1,085.82
349	7/12	Fast Freight	300.00			785.82
	7/12	Adj. for debit memo	7.50	√		778.32
350	7/12	Colonial Garage	54.41			723.91
	7/14	DEPOSIT			1,680.54	2,404.45
351	7/15	Payroll	1,432.10			972.35
352	7/16	Reed Printers	225.22			747.13
	7/22	DEPOSIT			842.24	1,589.37
353	7/24	Tuss Sales & Svc.	206.64			1,382.73
354	7/25	ComputerPlus	887.75			494.98
355	7/29	Petty Cashier	32.26			462.72
	7/30	DEPOSIT			411.12	873.84

BANK STATEMENT

CHECK NUMBER	CHECKS AND OTHER DEBITS	DEPOSITS	DATE	BALANCE
			7/1	1,274.72
		1,994.60	7/1	3,269.32
	50.30 RT		7/2	3,219.02
343	400.50		7/2	2,818.52
344	1,540.88		7/3	1,277.64
341	133.18		7/8	1,144.46
		655.23	7/10	1,799.69
346	75.34		7/11	1,724.35
342	98.61		7/15	1,625.74
		36.18 IN	7/15	1,661.92
349	300.00		7/15	1,361.92
		1,680.54	7/16	3,042.46
351	1,432.10		7/17	1,610.36
347	108.55		7/23	1,501.81
352	225.22		7/24	1,276.59
	10.00 SC		7/31	1,266.59

10.

CHECK REGISTER

Check Number	Date	Check Issued To	Amount of Check	√	Amount of Deposit	Balance 2,648.21
220	2/14	McDonald Co.	85.26	√		2,562.95
221	2/15	Payroll	1,808.66			754.29
	2/15	DEPOSIT			2,004.60	2,758.89
222	2/18	Hardison Co.	264.34			2,494.55
223	2/19	VHG, Inc.	91.75			2,402.80
224	2/19	West Ins. Co.	376.28			2,026.52
	2/20	Adj. for bank charges	3.27	√		2,023.25
225	2/20	Dean Gas Co.	559.42			1,463.83
226	2/21	Chavez Assoc.	130.90			1,332.93
227	2/21	HIH Services	228.48			1,104.45
228	2/21	Cain & Dole	240.30			864.15
229	2/22	Bradshaw Sales	54.62			809.53
	2/22	DEPOSIT			699.49	1,509.02
230	2/25	Powers & Co.	150.96			1,358.06
231	2/25	Smith Couriers	62.25			1,295.81
	2/26	DEPOSIT			1,359.00	2,654.81
232	2/27	Chairs, Inc.	282.44			2,372.37
233	2/27	Payroll	1,856.43			515.94
	2/27	DEPOSIT			836.24	1,352.18

BANK STATEMENT

Check Number	Checks and Other Debits	Deposits	Date	Balance
			2/14	2,559.68
		2,004.60	2/15	4,564.28
226	130.90		2/23	4,433.38
221	1,808.66	699.49	2/23	3,324.21
222	264.34		2/24	3,059.87
225	559.42		2/26	2,500.45
228	240.30		2/26	2,260.15
223	91.75	1,359.00	2/27	3,527.40
	400.00 RT		2/27	3,127.40
227	228.48		2/28	2,898.92
230	150.96		2/28	2,747.96
231	62.25		2/29	2,685.71
	10.00 SC		2/29	2,675.71

11.

CHECK REGISTER

CHECK NUMBER	DATE	CHECK ISSUED TO	AMOUNT OF CHECK	√	AMOUNT OF DEPOSIT	BALANCE 1,148.16
223	10/27	Allied Freight	18.53	√		1,129.63
224	10/28	IRS	57.49			1,072.14
225	10/28	Great Lakes Lab.	24.21			1,047.93
	10/30	DEPOSIT			1,476.00	2,523.93
226	11/3	Heavenly Beauty Sup.	35.89			2,488.04
227	11/4	Payroll	442.12			2,045.92
	11/5	Adj. for bank charges	3.59	√		2,042.33
228	11/10	Midwest Realty	315.00			1,727.33
229	11/10	Barton Drug Supply	207.65			1,519.68
230	11/10	Commercial Sales Inc.	45.72			1,473.96
231	11/10	Protective Ins. Agcy.	78.33			1,395.63
232	11/10	Midwestern Bell Tel.	21.41			1,374.22
233	11/10	City of Chicago	86.64			1,287.58
234	11/10	Health Products Inc.	71.96			1,215.62
235	11/11	Payroll	407.78			807.84
	11/12	DEPOSIT			891.43	1,699.27
236	11/15	Great Lakes Lab.	169.35			1,529.92
237	11/17	Nationwide Sales Inc.	262.13			1,267.79
	11/18	DEPOSIT			455.52	1,723.31
238	11/18	Payroll	543.29			1,180.02
239	11/23	Allied Freight	4.34			1,175.68
	11/24	DEPOSIT			509.60	1,685.28
240	11/25	Payroll	428.80			1,256.48
241	11/27	Milford Laboratory	50.09			1,206.39
	11/28	DEPOSIT			234.03	1,440.42
242	11/29	Grange Drug Supply	361.15			1,079.27

11. (*cont.*)

BANK STATEMENT

CHECK NUMBER	CHECKS AND OTHER DEBITS	DEPOSITS	DATE	BALANCE
			10/31	1,126.04
225	24.21	1,476.00	11/5	2,577.83
227	442.12		11/8	2,135.71
	18.45 RT		11/8	2,117.26
234	71.96		11/12	2,045.30
226	35.89		11/12	2,009.41
233	86.64		11/15	1,922.77
228	315.00		11/15	1,607.77
232	21.41		11/15	1,586.36
		891.43	11/16	2,477.79
235	407.78		11/17	2,070.01
230	45.72		11/17	2,024.29
236	169.35	455.52	11/20	2,310.46
238	543.29		11/21	1,767.17
229	207.65		11/21	1,559.52
		509.60	11/26	2,069.12
240	428.80		11/28	1,640.32
224	57.49		11/28	1,582.83
241	50.09		11/30	1,532.74
	7.40 SC		11/30	1,525.34

12.

CHECK REGISTER

CHECK NUMBER	DATE	CHECK ISSUED TO	AMOUNT OF CHECK	√	AMOUNT OF DEPOSIT	BALANCE 2,941.38
824	5/15	Simpson Supply	124.26	√		2,817.12
825	5/15	Carter Repair Shop	260.40			2,556.72
826	5/16	Payroll	950.18			1,606.54
	5/19	DEPOSIT			1,455.60	3,062.14
827	5/19	Link Insurance	360.00			2,702.14
828	5/19	Utah Savings Bank	210.55			2,491.59
829	5/20	Cobb Security Co.	423.46			2,068.13
830	5/20	Petty Cashier	51.98			2,016.15
	5/21	Debit memo	20.00	√		1,996.15
831	5/21	LaSasso Interiors	84.42			1,911.73
832	5/21	Biles Service Corp.	163.35			1,748.38
	5/21	DEPOSIT			1,643.52	3,391.90
833	5/23	Payroll	1,050.45			2,341.45
834	5/23	Rupps Freight Co.	96.33			2,245.12
835	5/23	Gray Accounting Co.	200.00			2,045.12
	5/26	DEPOSIT			634.12	2,679.24
836	5/27	Bruce Transit Co.	105.00			2,574.24
837	5/27	Brown Ofc. Mach.	82.42			2,491.82
838	5/28	Petty Cashier	50.49			2,441.33
839	5/30	Payroll	1,008.96			1,432.37
840	5/30	Salt Lake Power Co.	130.50			1,301.87
841	5/30	Parker Printing Co.	50.40			1,251.47

12. (*cont.*)

BANK STATEMENT

CHECK NUMBER	CHECKS AND OTHER DEBITS	DEPOSITS	DATE	BALANCE
			5/15	2,797.12
826	950.18	9.00 CM	5/16	1,855.94
		1,455.60	5/19	3,311.54
828	210.55		5/20	3,100.99
825	260.40		5/20	2,840.59
830	51.98		5/20	2,788.61
827	360.00		5/23	2,428.61
831	84.42		5/23	2,344.19
832	163.35	1,643.52	5/23	3,824.36
833	1,050.45		5/26	2,773.91
838	50.49		5/28	2,723.42
837	82.42		5/29	2,641.00
834	96.33		5/30	2,544.67
836	105.00		5/30	2,439.67
	8.50 SC		5/30	2,431.17

SECTION 2

CASH RECORDS

There is some variation, of course, but one good indication of business stability is the consistency of sales. These sales are recorded daily on the cash register tapes of a business and are given careful attention by officials of the company.

At the beginning of the business day, a stipulated amount is placed in each cash register drawer; this is the **change fund,** which must be available so that early customers may receive correct change. The fund ($20 to $200) will contain a certain number of the various denominations of money.

At the close of the business day, the clerk completes a **daily cash record.** (S)he counts each denomination of money separately, lists each check separately, and finds the total. The original change fund is then deducted from this total and the remaining sum (the **net cash receipts**) should equal the day's total cash sales shown on the cash register tape.

Human error being what it is, however, it frequently occurs that the cash receipts and the cash register total are not equal. If the cash receipts are less than the total sales, the cash is **short.** If there is more cash than the total of sales, the cash is **over.**

A summary of these daily cash records, commonly called an **over-and-short summary,** may be kept by the week or by the month. Over a period of time, the totals of the amounts in the "cash over" and in the "cash short" tend nearly to cancel each other.

Example 1

THE PERFECT GIFT SHOP
Daily Cash Record

Date: February 8, 19X1
Register: 1-B
Clerk: Sharie Holmes

Pennies	. . . .	$ 0.73
Nickels	. . . .	1.85
Dimes	. . . .	2.60
Quarters	. . . .	6.75
Halves	. . . .	2.50
Ones	. . . .	21.00
Fives	. . . .	50.00
Tens	. . . .	40.00
Twenties	. . . .	80.00
Other currency	. . . .	

Checks (list separately): $ 5.15
　　　　　　　　　　　　12.67
　　　　　　　　　　　　23.80
　　　　　　　　　　　　15.00
　　　　　　　　　　　　32.50　　　　　　　　　　89.12

	Total	$294.55
Less:	Change fund	− 30.00
	Net cash receipts	$264.55
	Cash over (subtract)	
	Cash short (add)	+ 0.05
	Cash register total	$264.60

Example 2

THE PERFECT GIFT SHOP
Over-and-Short Summary

DATE	CASH REGISTER TOTALS	NET CASH RECEIPTS	CASH OVER	CASH SHORT
Feb. 8, 19X1	$ 552.59	$ 552.23		$0.46
9	615.87	616.12	$0.25	
10	581.69	580.69		1.00
11	643.81	643.08		0.73
12	577.16	578.25	1.09	
13	601.12	601.62	0.50	
Totals	$3,572.34	$3,571.99	$1.84	$2.19

Total cash receipts	$3,571.99
Total cash short (add)	+ 2.19
	$3,574.18
Total cash over (subtract)	− 1.84
Total cash register readings	$3,572.34

SECTION 2   PROBLEMS

1. Prepare a daily cash report for clerk Mariana Adams on register 1 of Atwater Grocery Store using today's date. The number of each denomination of money is as follows: pennies, 38; nickels, 51; dimes, 80; quarters, 45; halves, 3; ones, 24; fives, 18; tens, 8; twenties, 6; other currency, one $50. Checks were for $25.45, $81.80, $65.14, $38.25, $44.44, $56.62, $28.75, and $52.63. The cash register total for the day is $730.51. There is a $50 change fund.

2. Prepare a daily cash report for clerk Tom Bruno on register 9 of Kitchen, Inc. using today's date. The number of each denomination of money in the cash drawer is as follows: pennies, 52; nickels, 68; dimes, 30; quarters, 74; halves, 2; ones, 34; fives, 12; tens, 15; twenties, 10; other currency, one $100. Checks were for $50.25, $36.36, $15.98, $48.00, and $67.57. The cash register total for the day is $689.48. There is a change fund of $100.

3. The following denominations were found in the cash drawer at the end of the day, December 3, by Susan Stanberry, the clerk responsible for this cash drawer on register 2 at Ellis Electronics Co.: pennies, 54; nickels, 15; dimes, 25; quarters, 31; halves, 10; ones, 14; fives, 21; tens, 16; twenties, 19. Checks in the drawer were for $16.14, $39.95, $57.82, and $66.18. The change fund contained $75. The cash register tape reveals $781.82 for the day's cash sales. Prepare a daily cash report.

4. Prepare a daily cash report for the cash drawer at Hair, Inc. using today's date. Several people collect money and make change from the drawer, so no one person is responsible for the register. The following denominations are in the drawer: pennies, 22; nickels, 74; dimes, 53; quarters, 15; halves, 1; ones, 36; fives, 18; tens, 12; twenties, 5; other currency, one $50. The checks in the drawer were: $25, $30, $42, $30, and $35. The cash register total for the day was $496.22. There was a change fund of $75.

5. Compute a weekly over-and-short summary for Auto Parts, Inc. Each day's cash register totals and net cash receipts are as follows:

DATE	TOTAL SALES	NET CASH RECEIPTS
March 5	$1,252.39	$1,251.39
March 6	1,334.01	1,334.48
March 7	1,298.60	1,299.86
March 8	1,276.50	1,276.46
March 9	1,239.84	1,240.44

6. Compute a weekly over-and-short summary for USA Chips, Inc. Each day's cash register totals and net cash receipts are as follows:

DATE	TOTAL SALES	NET CASH RECEIPTS
June 3	$1,045.68	$1,045.70
June 4	1,172.49	1,172.49
June 5	1,099.46	1,100.00
June 6	1,153.33	1,153.25
June 7	1,155.21	1,155.16

CHAPTER 7 GLOSSARY

Automatic teller machine. A terminal used to perform simple banking transactions without the aid of a teller.

Blank endorsement. Name (signature) of person or firm to whom a check was payable and which transfers ownership of the check to anyone in possession of it.

Change fund. A set amount of money placed in a cash register at the start of each day, in order to provide change for early customers.

Check register. A record of checks and deposits made by the owner of a checking account.

Daily cash record. A form that compares the net cash receipts with the cash register's total cash sales, in order to determine whether the cash drawer is over or short.

Debit card. A card that decreases a person's bank account automatically when used to make purchases.

Debit memo. A fee that the bank charges for service to an account, such as the charge for printing checks.

Drawee. The bank from which funds of a check are drawn (paid).

Drawer (or maker). The person (or firm) who draws funds (pays) by check.

Electronic funds transfer. A computerized system for depositing funds and/or decreasing a bank account balance.

Endorsement. The inscription on the back of a check (or negotiable instrument) by which ownership is transferred to another party.

Insufficient funds. A situation in which a checking account does not contain enough money for the bank to honor a check written on the account.

Line of credit. The limit to the amount of credit allowed a customer.

Maker (or drawer). The writer of a check.

Negotiable instrument. A document (check or bank note) that may be transferred from one party to another for its equivalent value.

Net cash receipts. A clerk's total (of money and checks) remaining after the change fund has been deducted; equals the total cash sales, if cash is neither over nor short.

NOW (negotiable order of withdrawal) account. A combination checking-savings account that earns interest on its average daily balance, provided a certain minimum balance is maintained.

Outstanding check. A check given in payment but not yet deducted from the writer's bank account.

Over. Term meaning that net cash receipts exceeded the cash register total.

Over-and-short summary. A weekly or monthly list of the total daily cash records of a business.

Overdraft. Term meaning that a customer has written a check for which there are insufficient funds in the account.

Payee. The party named to receive the funds of a check.

Reconciliation. Adjustments to the checkbook and the bank statement whereby the owner finds the actual spendable balance and determines that both records agree.

Restrictive endorsement. An endorsement restricting the use of a check; usually indicates that the check must be deposited to the endorser's account.

Returned check. A check returned to the party who deposited it because the check writer's account did not contain sufficient funds.

Service charge. A fee paid to the bank by the owner of a bank account for checking account services.

Short.   Term meaning that net cash receipts were less than the cash register total.

Special (or full) endorsement.   An endorsement naming the next party to whom a check is to be transferred.

Voucher check.   A check with a perforated attachment which explains or otherwise accounts for the amount shown on the check.

WAGES AND PAYROLLS

OBJECTIVES

Upon completion of Chapter 8, you will be able to:

1. Define and use correctly the terminology associated with each topic.

2. Compute gross wages and payrolls based on:
   a. Salary (Section 1: Example 1; Problems 1, 2)
   b. Commission (including quota and override) (Section 2: Examples 1–4; Problems 3–20)
   c. Hourly rates (Section 3: Example 1; Problems 1, 2)
   d. Production (including incentives and dockings) (Section 4: Examples 1, 3, 4; Problems 1–6, 9, 10).

3. Complete the account sales and account purchase (Section 2: Examples 5, 6; Problems 21–24).

4. For overtime work, compute gross wages and payrolls using:
   a. Hourly rates (at standard overtime or overtime excess) (Section 3: Examples 2, 3, 5; Problems 3–12)
   b. Salary (Section 3: Example 4; Problems 7, 8)
   c. Production (Section 4: Example 2; Problems 7, 8).

5. Complete net weekly payrolls, including standard deductions for (Section 5: Examples 1–3; Problems 1–12):
   a. Social Security
   b. Feceral income tax
   c. Other common deductions, as given.

6. a. Complete the basic portion of the Employer's Quarterly Federal Tax Return (Section 6: Examples 1, 3; Problems 1–6, 13–16)
   b. Determine employees' social security taxable wages and Medicare taxable wages (Example 2; Problems 7–12, 21, 22).

7. a. Determine employees' taxable wages for unemployment (Section 6: Example 4; Problems 17–20)
   b. Compute employers' state and federal unemployment taxes (Example 4: Problems 17–22).

In any business math course, many topics apply primarily to certain types of businesses. The payroll, however, is one aspect of business math that applies to every conceivable type of business, no matter how small or how large.

There is probably no other single factor so important to company–employee relations as the negotiation of wages and completion of the payroll. Wage negotiation within a company—whether by a union or on an individual basis—is not a responsibility of the payroll clerk, however. We shall thus be concerned only with computation of the payroll, which is also a very important responsibility. To find a mistake in their pay is very damaging to employees' morale. It is imperative, therefore, that everyone connected with the payroll exercise extreme care to ensure absolute accuracy.

There are four basic methods for determining **gross wages** (that is, wages before deductions): *salary, commission, hourly rate basis,* and *production basis.* We shall make a brief study of each.

SECTION *1*
COMPUTING WAGES—SALARY

Employees on a salary receive the same wages each pay period, whether this be weekly, biweekly, semimonthly, or monthly. Most executives, office personnel, and many professional people receive salaries.

If a salaried person misses some work time because of illness or for personal reasons, (s)he will usually still receive the same salary on payday. However, if business conditions require a salaried person to work extra hours, the payment of additional compensation may vary according to the person's position in the firm. General office personnel who are subject to the Fair Labor Standards Act (a topic in Section 3) must receive extra compensation as though they were normally paid on an hourly rate basis. However, highly paid executive and administrative personnel, whose positions are at the supervisory and decision-making levels of the company, normally receive no additional compensation for extra work. Professional people (members of the medical professions, lawyers, accountants, educators, and so on) are similarly exempt from extra-pay provisions.

The employer will quote an annual salary to a prospective employee, but this amount normally is not paid annually or once a year. The annual salary will be paid weekly, biweekly, semimonthly, or monthly. The employee will receive approximately the same amount each pay period with the total of each period's wages equaling the annual salary.

Example 1   Jan Jones will receive an annual salary of $36,000. Determine her salary per pay period assuming wages are paid (a) monthly, (b) semimonthly, (c) weekly, and (d) biweekly.

(a)   There will be 12 pay periods if Ms. Jones receives her salary monthly.

$$\frac{\$36,000}{12} = \$3,000 \text{ monthly}$$

(b)   There will be 24 pay periods if she is paid semimonthly.

$$\frac{\$36,000}{24} = \$1,500 \text{ semimonthly}$$

(c)   If she receives her salary weekly, she will receive 52 payments.

$$\frac{\$36,000}{52} = \$692.31 \text{ weekly}$$

(d)   If she is paid biweekly, she will be paid every 2 weeks or a total of 26 payments.

$$\frac{\$36,000}{26} = \$1,384.62 \text{ biweekly}$$

SECTION 2
COMPUTING WAGES—COMMISSION

Business people whose jobs consist of buying or selling merchandise are often compensated for their work by means of a commission. The true (or **straight**) **commission** is determined by finding some percent of the individual's net sales. Thus, we have another business formula based on the basic percent form "____% of ____ = ____":

____% of Net sales = Commission

or simply

$$\% \cdot S = C$$

Certain business people receive commissions without actually calling them commissions. For instance, a stockbroker's fee for buying stock is a certain percentage of the purchase price of that stock; however, this fee is called a **brokerage.** A lawyer's fee for handling the legal aspects of a real estate sale is generally some percentage of the value of the property; this constitutes a commission, although it probably will not be called one.

STRAIGHT COMMISSION

Example 1   (a) Edith Reilly receives an 8% commission on her net sales. Find her commission if her gross (original) sales were $5,680 and sales returns and allowances were $180.

Note. "Sales returns and allowances" are first deducted from gross sales, because the customer either returned the merchandise or was later allowed some reduction in sales price.

Gross sales	$5,680		$\%S = C$
Less: Sales returns			$8\% = \$5,500 = C$
and allowances	− 180		
Net sales	$5,500		$\$440 = C$

(b) A stockbroker received a $240 commission on a stock sale. What was the value of the stock, if the broker charged 2%?

$$\%S = C$$

$$2\%S = \$240$$

$$\frac{\cancel{0.02}\,S}{\cancel{0.02}} = \frac{240.00}{0.02}$$

$$S = \$12,000$$

Travel, meals, entertainment, and similar expenses of salespeople often remain approximately the same regardless of their sales. Thus, once a salesperson has made sufficient sales to cover expenses, the company may be willing to pay a higher commission on additional sales. A commission rate that increases as sales increase is known as a **sliding-scale commission.**

Example 1   (c) Determine George Butler's commission on his net sales of $14,000.
(cont.)   He receives 5% on his first $5,000 sales, 6% on the next $7,000, and 7.5% on all net sales over $12,000.

$$\%S = C$$

$$0.05 \times \$\ 5,000 = \$250$$

$$0.06 \times \quad 7,000 = \quad 420$$

$$0.075 \times \quad \underline{2,000} = \quad \underline{150}$$

Net sales	$14,000	$820 Commission

COMMISSION VARIATIONS

Salary.   Some salespeople are guaranteed a minimum income in the form of a salary. They are often then paid commissions at a lower rate than is usually earned by those on straight commission.

Example 2

(a)   Suppose that Edith Reilly [Example 1(a)] also receives a $100 salary. Her total gross wages would then be:

Salary	$100
Commission	+ 440
Gross wages	$540

Drawing Account.   A salesperson on commission may be allowed a **drawing account** from which funds may be withdrawn to defray business or personal expenses. This is an advance on earnings, and the drawings are later deducted from the salesperson's commission. Like a salary, a drawing account assures a minimum income. However, the earned commission must consistently either equal or exceed the drawings, or else the employee would be released from the company.

Example 2
(cont.)

(b)   Suppose that Edith Reilly [Example 1(a)] is entitled to a drawing account of $300 and that she has withdrawn the $300 before pay day.

Commission	$440
Less: Drawings	− 300
Amount due	$140

Quota.   Many persons on commission, particularly those who also receive a salary, are required to sell a certain amount (or **quota**) before the commission starts. This plan is designed to ensure the company a certain level of performance for the salary it pays.

Example 2
(cont.)

(c)   Bob Hendrix has net sales of $12,000 and must meet a $4,000 quota. His 5% commission is then computed as follows:

Total sales	$12,000	$\%S = C$
Less: Quota	− 4,000	$5\% \times \$8,000 = C$
Sales subject to commission	$ 8,000	$\$400 = C$

Override.   Department heads in retail stores are often allowed a special commission on the net sales of all other clerks in their departments. This commission, called an **override,** is given to offset the time a department head must devote to nonselling responsibilities (such as pricing and stocking merchandise, taking inventory, or placing orders). An override is computed exactly like an ordinary commission, but at a lower rate than the commission rate for one person's sales. The override is paid in addition to the department head's own commission and/or salary.

Example 3   Dana Harvey is a department head in a retail store. From the following information, determine her week's wages.

Personal sales	$ 4,734	Salary	$225
Returns and allowances	54	Quota	$1,800
Department sales	16,455	Commission rate	5%
Dept. returns and allowances	375	Override rate	$\frac{1}{2}$%

Personal sales	$4,734	Department sales	$16,455
Less: Sales returns	$-$ 54	Less: Sales returns	$-$ 375
	$4,680	Sales subject to	$16,080
Less: Quota	$-$ 1,800	override	
Sales subject to commission	$2,880		

$$\%S = C \qquad\qquad \%S = OR$$

$$0.05 \times \$2,880 = C \qquad 0.005 \times \$16,080 = OR$$

$$\$144 = C \qquad\qquad \$80.40 = OR$$

Salary	$225.00
Commission	144.00
Override	+ 80.40
Total wages	$449.40

Example 4   The following are examples of commission payrolls. The second example includes the calculation of net sales, as well as quotas and salaries.

(a)   **TOP DRAWER FASHIONS**
Payroll for Week Ending September 6, 19XX

| | NET SALES | | | | | | COMM. | GROSS |
NAME	M	T	W	T	F	TOTAL	RATE	COMM.
Albright, K.	$ 950	$1,075	$ 920	$1,070	$1,085	$5,100	5%	$ 255.00
Carver, D.	1,100	1,225	1,380	1,320	1,275	6,300	4	252.00
Foyt, B.	930	—	1,060	1,175	1,350	4,515	6	270.90
Gantt, M.	1,040	975	975	1,050	1,220	5,260	5	263.00
							Total	$1040.90

(b)   **VALLEY SALES DISTRIBUTORS, INC.**

Payroll for Week Ending June 13, 19XX

| NAME | SALES | | | QUOTA | COMM. SALES | COMM. RATE | GROSS COMM. | SALARY | GROSS WAGES |
	Gross	R&A	NET						
Barton, H.	$3,518	$18	$3,500	$1,200	$2,300	4%	$ 92.00	$ 275.00	$367.00
Dixon, R.	5,432	52	5,380	—	5,380	3	161.40	250.00	411.40
Hill, C.	4,776	86	4,690	2,000	2,690	4	107.60	280.00	387.60
Landers, J.	3,084	34	3,050	—	3,050	6	183.00	200.00	383.00
						Totals	$544.00	$1,005.00	$1,549.00

Note.   On the bottom line, the total Gross Commissions plus total Salaries must equal the total Gross Wages ($544 + $1,005 = $1,549).

Calculator techniques . . .   FOR EXAMPLE 4

For payrolls, work horizontally across each line, using $\boxed{\text{M+}}$ for each worker's total; then press $\boxed{\text{MR}}$ for the column total. This illustrates calculations for the first employee in part (a):

950 $\boxed{+}$ 1,075 $\boxed{+}$ 920 $\boxed{+}$ 1,070 $\boxed{+}$ 1,085 $\boxed{=}$ ⟶ 5,100

(5,100) $\boxed{\times}$ 5 $\boxed{\%}$ ⟶ 255 $\boxed{\text{M+}}$

. . . After all workers are processed, $\boxed{\text{MR}}$ ⟶ 1,040.9

Payroll (b) deducts returns and quotas before computing the commission, then adds the salary and uses $\boxed{\text{M+}}$ to insert each worker's total:

3,518 $\boxed{-}$ 18 $\boxed{=}$ ⟶ 3,500 $\boxed{-}$ 1,200 $\boxed{=}$ ⟶ 2,300

(2,300) $\boxed{\times}$ 4 $\boxed{\%}$ ⟶ 92 $\boxed{+}$ 275 $\boxed{\text{M+}}$ ⟶ 367

. . . After every employee is calculated, $\boxed{\text{MR}}$ ⟶ 1,549

Note.   Parentheses around the first number on the second line of each example mean to use that displayed value without reentering it.

Commission Agents

The entire operation of some agencies consists of buying and/or selling products for others. Agents employed for this purpose are known as **commission agents, commission merchants, brokers,** or **factors.** The distinguishing characteristic of such agents is that whereas they are authorized to act for another individual or firm, at no time do they become the legal owners of the property involved.

Many items—particularly foodstuffs—are produced in areas which are some distance from the market. The producer (or **consignor**) ships the goods "on consignment" to a commission agent (or **consignee**) who sells them at the best possible price. The price obtained by the agent is the **gross proceeds.** From these gross proceeds the agent deducts a commission, as well as any expenses incurred (such as storage, freight, insurance, and so forth). The remaining sum, called the **net proceeds,** is then sent to the producer. Accompanying the net proceeds is an **account sales,** which is a detailed account of the sale of the merchandise and of the expenses involved. (Gross proceeds − Total charges = Net proceeds.)

Example 5   The following example shows an account sales.

		ACCOUNT SALES		
		HARCOURT & WYMAN		
		2519 South Shore Blvd.		
		Chicago, Illinois 60649		
REC'D: Jan. 20, 19XX			NO: B27116	
VIA: Air freight			DATE: Feb. 1, 19XX	
	Aloha Floral Distributors			
	7401 Kamani Blvd.			
	Honolulu, Hawaii 96813			

19- -		Sales:		
Jan.	21	50 boxes — orchids @ $36	$1,800.00	
	22	25 boxes — bird of paradise @ $60	1,500.00	
	22	15 boxes — ti @ $40	600.00	
		Gross proceeds		$3,900.00
		Charges:		
		Commission, 6% of $3,900	$ 234.00	
		Refrigerator storage	24.75	
		Delivery charges	36.50	
				− 295.25
		Net proceeds		$3,604.75

Calculator Tip.   Add the Sales and press M+ ; add the Charges and press M− ; then, pressing MR displays the Net proceeds.

Other commission agents act as buyers for their clients. When a store wishes to purchase an item not regularly carried in its inventory and not obtainable through regular wholesale distributors, the store may commission an agent to secure the items. Typical examples of such items might be art objects, specialized furniture, carpets, exclusive dry goods, exotic foods, and so on, which are specialties of a particular locality or country. Much international as well as domestic purchasing is accomplished in this manner.

In general, a commission agent is employed whenever it is inconvenient to send a company buyer or when a special knowledge of the product or the market conditions is required. The agent submits an itemized account of the purchase, called an **account purchase.** The account purchase includes the actual price paid by the agent (**prime cost**) plus the agent's commission on the prime cost, as well as any expenses incurred during the purchase. This total (or **gross cost**) is the actual cost of the merchandise to the company and the amount to be remitted to the commission agent. (Prime cost + Total charges = Gross cost.)

Example 6 The following example shows an account purchase.

ACCOUNT PURCHASE				
IMPORT SPECIALTIES, INC.				
32714 Santa Monica Blvd.				
Los Angeles, California 90037				

BOUGHT FOR THE ACCOUNT OF:

   The Import House
   7709 Granville Rd.
   St. Paul, Minnesota 55101

YOUR ORDER: #5415   NO: G42081
          DATE: April 9, 19XX   DATE: May 1, 19XX

19XX				
April	18	Purchases:		
		15 — Philippine wicker fan chairs @ $96	$1,440.00	
	20	12 — Philippine wicker hooded		
		swings @ $75	900.00	
	21	20 — Wicker chests @ $25	500.00	
		Prime cost		$2,840.00
		Charges:		
		Shipping	$ 78.85	
		Insurance	15.95	
		Import duty	56.80	
		Commission, 5% of $2,840	142.00	
				+ 293.60
		Gross cost		$3,133.60

SECTIONS 1 AND 2   PROBLEMS

Determine the salary.

	ANNUAL	MONTHLY	SEMIMONTHLY	WEEKLY	BIWEEKLY
1. a.	$18,000				
b.		$5,000			
c.			$2,250		
2. a.	$48,000				
b.		$6,000			
c.			$1,000		

Complete the following.

	SALARY	QUOTA	RATE	NET SALES	COMMISSION	GROSS WAGES
3. a.	X	X	8%	$ 7,400		X
b.	X	X	5		$450	X
c.	X	X		8,300	332	X
d.	$150	X	6	4,800		
e.	75	X	7			$915
f.	X	$ 500	5	6,600		X
g.	100	1,000	6	10,000		
h.	200	2,000	4			800
i.		1,000	8	12,000		960
4. a.	X	X	6%	$ 9,500		X
b.	X	X	$5\frac{1}{2}$		$396	X
c.	X	X		6,100	244	X
d.	$200	X	5	4,400		
e.	110	X	7			$530
f.	X	$ 500	$4\frac{1}{2}$	9,900		X
g.	175	200	8	5,700		
h.	80	1,000	6			404
i.		1,400	5	13,800		820

5. Dave Pape receives a 7% commission on his net sales. If his sales totaled $6,700 but customers returned $200 of merchandise, what were his gross wages for the week?

6. Gail Johnson had sales totaling $7,500, with sales returns of $350. If she receives a 6% commission on net sales, what were her gross wages?

7. Jessie Gray's gross wages were $504. If her commission rate was 7%, how much were her net sales?

8. A real estate agent received $10,500 as a commission on a house sale. If his commission rate is 7%, what was the amount of the house sale?

9. Mary Klinko earns a sliding-scale commission of $2\frac{1}{2}\%$ on the first $4,000 of net sales, 3% on the next $3,000 of net sales, and 6% on all net sales over $7,000. She is allowed a monthly drawing account of $200. Ms. Klinko's net sales for the month were $15,000.

   a. What was her total commission?

   b. Ms. Klinko withdrew the $200 for expenses. What amount is still due her at the end of the month?

10. Mike Perry receives a sliding-scale commission of $3\frac{1}{2}\%$ on the first $2,000 of net sales, 5% on the next $4,000 of net sales, and 6% on all net sales over $6,000. He is allowed a drawing account of $350. Mr. Perry's net sales for the week were $13,600.

    a. What was his total commission?

    b. If Mr. Perry withdrew the $350 during the week, what amount is due him at the end of the week?

11. Carol Morgan earned a gross wage of $720, which included a salary of $90 and a 6% commission. How much were her net sales?

12. Lisa Wilson earned a gross wage of $300, which included a $120 salary and a 5% commission. How much were her net sales last week?

13. Jerry Gale earns a weekly salary of $150 plus a 6% commission on sales in excess of $350. Determine his gross wages for a week in which he had sales of $5,440 and sales returns of $60.

14. Robin Alterman earns a weekly salary of $175 plus a 4% commission on net sales in excess of $200. What were her gross wages for a week in which she had sales of $3,000 and sales returns of $50?

15. Marvin Palguta, a manager of a men's clothing department, receives a salary of $60 each week, plus an 8% commission on his net sales in excess of $400 and a 1% override on the net sales of the sales associates working under him. His sales last week were $5,000. The other employees had sales of $24,350 with sales returns of $350. What were his gross wages?

16. Rebecca Marshall manages the women's sportswear department of Woodwind Co. She receives a salary of $160; in addition, she receives a 6% commision on her net sales in excess of $500 and a $1\frac{1}{2}\%$ override on the net sales of the sales associates in her department. Her sales totaled $2,200 last week. Other employee sales totaled $10,100 with sales returns of $50. Determine her weekly gross wages.

Calculate the commission payrolls in Problems 17–20.

17.
FORD SPORTING GOODS CO.
Payroll for Week Ending May 15, 19XX

NAME	NET SALES							COMM. RATE	GROSS COMM.
	M	T	W	T	F	S	TOTAL		
Bazzle, C.	$325	$342	$457	$396	$350	$558		5%	
Holt, B.	256	345	384	390	400	450		6	
Laski, P.	560	582	578	543	550	581		6	
Noor, A.	366	384	390	425	400	305		7	
Sovine, K.	465	505	512	496	480	500		8	
								Total	

18.

OZARK OFFICE SUPPLY
Payroll for Week Ending June 11, 19XX

| NAME | NET SALES | | | | | | TOTAL | COMM. RATE | GROSS COMM. |
	M	T	W	T	F	S			
Dempsey, J.	$396	$375	$280	$195	$304	$400		6%	
Gold, K.	266	483	501	442	455	423		5	
Keller, E.	322	388	404	425	421	500		6	
Miller, M.	329	284	292	275	360	410		5	
Wright, C.	340	465	488	492	502	613		7	
								Total	

19.

CHATMAN ENTERPRISES, INC.
Payroll for Week Ending November 4, 19XX

| NAME | SALES | | | QUOTA | COMM. SALES | COMM. RATE | GROSS COMM. | SALARY | GROSS WAGES |
	Gross	R&A	Net						
Adams, D.	$5,456	$36		—		4%		$110	
Chong, Y.	5,680	80		$1,000		7		—	
Dudley, W.	6,744	64		—		5		200	
Stevens, S.	6,200	—		1,200		6		150	
Woods, J.	7,756	26		730		8		—	
							Total		

20.

ROZELLA COMPUTER CO.
Payroll for Week Ending October 4, 19XX

| NAME | SALES | | | QUOTA | COMM. SALES | COMM. RATE | GROSS COMM. | SALARY | GROSS WAGES |
	Gross	R&A	Net						
Bellis, E.	$6,340	$40		—		4%		$ 80	
Chambers, D.	5,820	70		$ 800		6		—	
Duggan, C.	4,690	50		—		5		100	
Meyer, C.	6,500	—		1,000		6		120	
Quader, E.	6,332	82		500		8		—	
							Total		

21. Complete the following account sales.

		ACCOUNT SALES		
		GATES & ASSOCIATES		
		Commission Merchants		
		Alpharetta, Georgia		
		Sold for Account of:		
		E. Myer Fruit Co.		
		Florence, South Carolina		

19xx				
July	8	450 lbs. peaches @ $0.60		
	8	700 lbs. tomatoes @ $0.42		
	15	150 bu. white corn @ $5.00		
	22	200 bu. bush beans @ $6.40		
		GROSS PROCEEDS		
		Charges:		
		Freight	$120.50	
		Insurance	32.50	
		Commission, 5%		
		NET PROCEEDS		

22. Prepare an account sales for the following transactions: Rivera Farms, Homestead, Florida, consigned a shipment to Thompson Produce, Nashville, Tennessee. The shipment contained 220 lbs. tomatoes sold on March 5 @ $0.65; 175 lbs. carrots sold on March 5 @ $0.52; 460 lbs. red potatoes sold on March 10 @ $0.48; and 200 bu. lima beans sold on March 12 @ $6.30. Freight charges totaled $98.40 and insurance charges totaled $100. The commission agents' charge was 6%.

23. Complete the following account purchase.

```
┌──────────────────────────────────────────────────────────────────┐
│                        ACCOUNT PURCHASE                            │
│                                                                    │
│                        BRACY & SONS                                │
│                      Commission Merchants                          │
│                      Alexandria, VA    22302                       │
│                                                                    │
│    BOUGHT FOR THE ACCOUNT OF:                                      │
│                                                                    │
│    ┌                     ┐                                         │
│      Northern Virginia Imports                                     │
│      Alexandria, VA  22304                                         │
│    └                     ┘                                         │
├──────────────────────────────────────────────────────────────────┤
│  19xx │                                                            │
│       │                                                            │
│  Feb. │ 15  20 ginger jar lamps @ $88.50                          │
│       │ 15  36 Ming vases @ $62.00                                │
│       │ 19  12 4' x 8' silksreens @ $45                           │
│       │ 25  5 5' x 7' Oriental carpets @ $190                     │
│       │                                                            │
│       │                            PRIME COST                      │
│       │                                                            │
│       │   Charges                                                  │
│       │                                                            │
│       │   Shipping                         $110.80                 │
│       │   Insurance                          54.20                 │
│       │   Commission, 4%                                           │
│       │                                                            │
│       │                            GROSS COST                      │
└──────────────────────────────────────────────────────────────────┘
```

24. Prepare an account purchase for the following purchases: The China Closet in Jackson, Mississippi, commissioned Taylor Imports in Baltimore, Maryland, to purchase the following crystal pieces from Ireland: On October 1, 5 cake plates at $70 and 10 goblets at $36; on October 10, 3 7″ vases at $61; on October 18, 2 8″ × 10″ picture frames at $57 and 4 pairs of 6″ candlesticks at $45. Shipping charges were $96 and insurance totaled $55. Taylor Imports charges a $5\frac{1}{2}\%$ commission.

SECTION 3
COMPUTING WAGES—HOURLY RATE BASIS

The hourly rate basis for determining wages is the method under which wages are computed for the majority of the American working force. Under this plan, the employee receives a stipulated amount for each hour's work. One or more of several factors may determine the hourly rate: the individual's ability, training, and experience, as well as federal minimum-wage laws or union contracts. Once the hourly rate has been established, however, the employees' quality of performance or amount of production in no way influences their wages.

The **Fair Labor Standards Act** governs the employment of more than 80 million workers. Previously affecting primarily workers engaged in some form of interstate, international, or other large-scale commercial enterprises, the Act has now been extended to include a wide range of occupations. Among covered workers are nonprofessional employees of hospitals and nursing homes, public or private schools, cleaners and laundries, transportation services, construction industries, domestic workers, and all levels of federal government. Beginning in 1986, the Act also covers workers in city, state, and county governments. It is expected that the Act will eventually be extended to cover essentially all full-time workers.

As of 1995, the Act prescribed a 40-hour workweek and required a minimum wage for all covered employees of $4.25 per hour. Many states have also established minimum wages for employees not subject to the federal regulations. The federal act applies equally to men and women. It should be emphasized, however, that certain employees are specifically excluded from the provisions of the Act—executive and administrative personnel in businesses, professional people (such as medical personnel, educators, lawyers, and so forth), as well as outside salesworkers, employees of very small retail or service establishments, and many seasonal and small-farm workers.

OVERTIME

In accordance with the federal law covering the 40-hour workweek, most employees' regular week consists of five 8-hour working days. Under the federal regulation, **overtime** must be paid to covered employees for work in excess of 40 hours. Many employers, even though not subject to the federal act, follow this policy voluntarily.

The standard overtime rate is $1\frac{1}{2}$ times the worker's regular hourly rate ("time-and-a-half"). "Double time" is often paid for work on Sundays or on holidays. It should be noted, however, that the Fair Labor Standards Act does *not* require extra pay for Saturdays, Sundays, or holidays (except for hours in excess of 40); neither does it require pay for time off during vacations or holidays. Whether premium rates are paid or time off is granted depends entirely on the employment (or union) agreement.

In some instances, overtime is paid when an employee works more than 8 hours on any one day. (Some union contracts contain this provision.) As a general practice, however, a worker is required to have worked 40 hours before receiving any overtime compensation; otherwise, the worker might be paid some overtime without having completed the prescribed 40-hour week. If a company nonetheless pays overtime on a daily basis, the payroll clerk must keep a careful record of each employee's hours to ensure that the same time is not counted twice; that is, hours which qualify as daily overtime must not be applied toward the regular 40-hour week, after which all work would be at overtime rates.

Union contracts often contain overtime provisions regarding any number of other situations, such as working two complete shifts in 1 day or working 6 or 7 consecutive days.

Large firms usually have timeclocks that record entering and leaving times on each employee's timecard. Total work hours are then calculated from this record. Just as employees are compensated for overtime work, they are likewise penalized for tardiness. In smaller firms, a supervisor may record the whole crew's hours; or, in some instances, employees report their own hours. Total hours worked are usually kept to the nearest quarter-hour each day.

Example 1   Compute the gross wages of each employee.

B & J PAINT CO.
Payroll for Week Ending May 9, 19XX

NAME	M	T	W	T	F	S	TOTAL HOURS	RATE PER HOUR	GROSS WAGES
Anderson, James	9	7	8	8	8	0	40	$7.50	$ 300.00
Bailey, Carl	8	8	10	0	10	4	40	8.75	350.00
Carson, W. J.	8	6	0	5	8	4	31	6.80	210.80
Davis, Willard	9	8	8	8	6	0	39	8.00	312.00
								Total	$1,172.80

Example 2   Compute the gross wages of each employee of Wilkins-Casey, Inc. Overtime is paid at $1\frac{1}{2}$ times the regular rate on hours in excess of 40 per week.

WILKINS-CASEY, INC.
Payroll for Week Ending December 12, 19XX

NAME	M	T	W	T	F	S	TOTAL HOURS	REG-ULAR HOURS	HOURLY RATE	OVER-TIME HOURS	OVER-TIME RATE	REG-ULAR WAGES	OVER-TIME WAGES	TOTAL GROSS WAGES
Elliot, D. R.	8	8	10	9	10	4	49	40	$5.00	9	$ 7.50	$ 200.00	$ 67.50	$ 267.50
Farmer, Ray	10	8	9	7	7	4	45	40	6.60	5	9.90	264.00	49.50	313.50
Givens, Troy	8	8	8	8	8	0	40	40	6.75	—	10.13	270.00	—	270.00
Harper, Tony	5	8	6	7	8	4	38	38	6.50	—	9.75	247.00	—	247.00
Ivey, Marion	9	8	6	8	9	4	44	40	5.80	4	8.70	232.00	34.80	266.80
											Total	$1,213.00	$151.80	$1,364.80

Note.   The Regular Wages total plus the Overtime Wages total must equal the total Gross Wages ($1,213 + $151.80 = $1,364.80).

Calculator techniques . . .   FOR EXAMPLE 2

First, total the hours of each worker and determine each person's overtime rate per hour. Where multiple earnings or deductions columns must be calculated (here, regular and overtime wages), it is convenient to compute the first column for all workers, then the next column for everyone, and so on. If your calculator has a constant key, this method allows the 40 hours and the 1.5 overtime factor to be used as constants. (Parentheses on the last line enclose a number to be used without reentering it.)

$$40 \; \boxed{\times} \; 5 \; \boxed{M+} \longrightarrow 200; \quad 40 \; \boxed{\times} \; 6.6 \; \boxed{M+} \longrightarrow 264;$$

$$\ldots \; 40 \; \boxed{\times} \; 5.8 \; \boxed{M+} \longrightarrow 232; \quad \boxed{MR} \longrightarrow 1{,}213$$

Clear memory.

$$9 \; \boxed{\times} \; 7.5 \; \boxed{M+} \longrightarrow 67.5; \quad 5 \; \boxed{\times} \; 9.9 \; \boxed{M+} \; 49.5;$$

$$4 \; \boxed{\times} \; 8.7 \; \boxed{M+} \longrightarrow 34.8; \quad MR \longrightarrow 151.8$$

$$(151.8) \; \boxed{+} \; 1{,}213 \; \boxed{=} \longrightarrow 1{,}364.8$$

Then, after clearing memory, enter the total wages for each worker into memory with $\boxed{M+}$ (for instance, $200 + 67.5 \; \boxed{M+} \longrightarrow 267.5$), and finally press $\boxed{MR}$ to verify that the total Gross Wages equals the same 1,364.80 as before.

Example 3   Compute David Pressley's gross wages based on the following: Overtime at time-and-a-half is paid daily on work in excess of 8 hours and weekly on work in excess of 40 regular hours; double time is paid for all Sunday work. His hours each day were as follows:

	M	T	W	T	F	S	S
Pressley, David	8	9	7	10	7	5	5

		M	T	W	T	F	S	S	TOTAL HOURS	RATE PER HOUR	BASE WAGES	TOTAL GROSS WAGES
Pressley, D.	RT[a]	8	8	7	8	7	2		40	$ 6.80	$272.00	
	$1\frac{1}{2}$		1		2		3		6	10.20	61.20	
	DT							5	5	13.60	68.00	$401.20

[a] RT indicates "Regular time"; $1\frac{1}{2}$, "Time-and-a-half"; and DT, "Double time."

Calculator Tip. For this problem and also in the succeeding examples, place the regular and overtime components into memory and use $\boxed{\text{MR}}$ to determine each person's total wages: 40 $\boxed{\times}$ 6.8 $\boxed{\text{M+}}$ → 272; 6 $\boxed{\times}$ 10.2 $\boxed{\text{M+}}$ → 61.2; 5 $\boxed{\times}$ 13.6 $\boxed{\text{M+}}$ → 68; $\boxed{\text{MR}}$ → 401.2.

OVERTIME FOR SALARIED PERSONNEL

The following example illustrates the computation of gross wages for salaried personnel who must be paid overtime wages. This procedure is required under the Fair Labor Standards Act for salaried personnel subject to overtime provisions. (Refer also to Section 1, "Computing Wages—Salary.")

Example 4
(a) A file clerk receives a $180 salary for her regular 40-hour week. What should be her gross wages for a week when she works 44 hours?

The clerk's regular wages are equivalent to $\dfrac{\$180}{40 \text{ hours}}$ or $4.50 per hour. She must be paid overtime at $1\frac{1}{2}$ times this rate, or $6.75 per hour. Her wages for a 44-hour week would thus be $207:

40 hours @ $4.50 = $180	Regular salary
4 hours @ $6.75 = 27	Overtime
44-hour week = $207	Gross wages

(b) An administrative assistant receives a $217 weekly salary. If a regular workweek contains 35 hours, how much compensation would she receive when she works 43 hours?

$$\frac{\$217}{35 \text{ hours}} = \$6.20 \text{ per hour}$$

$$\text{Overtime rate} = 1\tfrac{1}{2} \times \$6.20 = \$9.30 \text{ per hour}$$

40 hours @ $6.20 = $248.00	
3 hours @ $9.30 = 27.90	
43-hour week = $275.90	Gross wages

(c) A service manager's salary is $324, although his hours vary each week. What are his gross wages for a week when he works 45 hours?

$$\frac{\$324}{45 \text{ hours}} = \$7.20 \text{ per hour}$$

Overtime rate = $1\frac{1}{2} \times \$7.20 = \$10.80$ per hour

40 hours @ $ 7.20 = $288.00
 5 hours @ $10.80 =   54.00
45-hour week   = $342.00   Gross wages

Observe that, for any employee whose hours vary each week as do this man's hours, the employee's regular rate per hour will also vary each week. That is, his regular rate is $7.20 per hour only during a week when he works 45 hours. If he works 54 hours the following week, then his regular rate that week will be

$$\frac{\$324}{54} = \$6.00 \text{ per hour on the 40 hours}$$

plus overtime for 14 hours at $9.00 per hour.

Overtime Excess

All the examples above show regular rates for 40 hours and an overtime rate of $1\frac{1}{2}$ times the regular rate. An alternative method will produce the same result: Calculate the total hours at the regular rate plus the overtime hours at one-half the regular rate. The additional compensation calculated by this method (overtime hours times one-half the regular rate) is called the **overtime excess.**

Example 5   A worker who earns $5 per hour works 46 hours. Compare these earnings (a) by standard overtime and (b) by overtime excess.

(a)   40 hours @ $5.00 = $200.00   Straight-time wages
       6 hours @ $7.50 =   45.00   Overtime wages
      46 hours   = $245.00   Total gross wages

(b)   One-half the regular rate is $\frac{1}{2}$ of $5.00 = $2.50 per hour for the excess hours over 40:

      46 hours @ $5.00 = $230.00   Regular-rate wages
       6 hours @ $2.50 =   15.00   Overtime excess
      46 hours   = $245.00   Total gross wages

Some firms prefer the overtime excess method for their own cost accounting, since it emphasizes the extra expense (and lost profit) resulting from overtime work. (Notice above that if the extra 6 hours' work could have been delayed until the following week and all done at the regular rate, the company's cost would have been only $230, rather than $245.)

Use of the overtime excess method is becoming more widespread. Although either overtime method is permitted under the Fair Labor Standards Act, the Wage and Hour Division (which enforces the act) uses the overtime excess method. A major reason for using overtime excess is that workers' compensation insurance premiums are based on a worker's regular-rate wages (that is, the $230 above).

Section 3 Problems

1. Complete the following gross payroll.

THE KELLER MUSIC CO.
Payroll for Week Ending June 2, 19XX

Name	M	T	W	T	F	S	Total Hours	Rate Per Hour	Gross Wages
Corbett, B.	8	7	7	8	8	0		$8.00	
Eyer, P.	6	8	8	8	8	2		8.50	
Gwaltney, M.	9	5	8	7	8	0		7.80	
Palumbo, L.	8	8	6	7	7	4		9.50	
Sichenze, C.	5	5	6	9	9	3		7.25	
Sass, S.	9	8	7	8	8	0		9.40	
Wigle, E.	8	8	9	7	4	0		8.15	
								Total	

2. Complete the following gross payroll.

CRISTO CLEANING SERVICE
Payroll for Week Ending May 5, 19XX

Name	M	T	W	T	F	S	Total Hours	Rate Per Hour	Gross Wages
Becker, P.	8	8	7	7	6	0		$6.50	
Doyle, D.	8	7	8	9	8	0		6.75	
Magro, D.	5	8	7	8	8	3		5.80	
Murray, A.	8	7	7	7	7	2		6.25	
Richards, B.	6	6	6	6	6	6		6.00	
Stark, R.	8	9	9	8	6	0		7.10	
Wu, Y.	9	8	8	7	5	0		7.00	
								Total	

3. The employees of McDaniel Electronics, Inc. are required to work a regular 40-hour week, after which time they receive standard overtime for any additional hours. Complete the payroll record below.

MCDANIEL ELECTRONICS, INC.
Payroll for Week Ending July 24, 19XX

NAME	M	T	W	T	F	S	TOTAL HOURS	REG-ULAR HOURS	HOURLY RATE	OVER-TIME HOURS	OVER-TIME RATE	REG. WAGES	OVER-TIME WAGES	TOTAL GROSS WAGES
Ford	9	8	8	7	8	0			$6.75					
Gale	8	9	9	8	8	0			9.00					
Gray	6	7	8	8	6	3			8.50					
Myer	9	9	8	9	6	5			7.20					
Scott	8	8	7	9	8	4			9.70					
											Totals			

4. The employees of Saunders, Inc. are required to work a regular 40-hour week. Standard overtime is paid for any additional hours. Complete the following gross payroll.

SAUNDERS, INC.
Payroll for Week Ending December 6, 19XX

NAME	M	T	W	T	F	S	TOTAL HOURS	REG-ULAR HOURS	HOURLY RATE	OVER-TIME HOURS	OVER-TIME RATE	REG. WAGES	OVER-TIME WAGES	TOTAL GROSS WAGES
Agnew	8	8	8	8	8	0			$7.50					
Barski	8	7	9	8	7	3			6.50					
Clancy	6	8	7	8	8	0			6.00					
Ibar	7	8	8	9	8	5			7.20					
Nozek	8	8	6	9	8	4			7.00					
											Totals			

5. Daily overtime (in excess of 8 hours) and weekly overtime (after 40 regular hours) are paid to the employees at Avery Office Systems, Inc. They also receive double time for work done on Sundays or national holidays. Monday of the week shown was a national holiday. Determine the number of hours in each category and complete the gross payroll.

	M	T	W	T	F	S	S
Bohling, N.	5	8	9	8	9	8	0
Chatman, B.	8	8	10	8	9	6	0
Otey, W.	0	10	8	8	8	5	5

NAME		M	T	W	T	F	S	S	TOTAL HOURS	RATE PER HOUR	BASE WAGES	TOTAL GROSS WAGES
Bohling, N.	RT $1\frac{1}{2}$ DT									$8.50		
Chatman, B.	RT $1\frac{1}{2}$ DT									9.00		
Otey, W.	RT $1\frac{1}{2}$ DT									9.40		
											Total	

6. Daily overtime (in excess of 8 hours) and weekly overtime (after 40 regular hours) are paid to the employees listed below. They also receive double time for Sunday work. From the total daily hours given, determine the number of hours in each category and compute the gross wages.

	M	T	W	T	F	S	S
Coffey, D.	6	8	7	9	7	0	4
Hanson, C.	7	8	10	8	7	8	0
Yoder, Z.	9	5	8	9	8	4	3

NAME		M	T	W	T	F	S	S	TOTAL HOURS	RATE PER HOUR	BASE WAGES	TOTAL GROSS WAGES
Coffey, D.	RT $1\frac{1}{2}$ DT									$8.00		
Hanson, C.	RT $1\frac{1}{2}$ DT									8.50		
Yoder, Z.	RT $1\frac{1}{2}$ DT									8.60		
											Total	

Compute the gross weekly wages for the salaried employees shown who are all subject to the Fair Labor Standards Act.

	POSITION	REGULAR HOURS PER WEEK	REGULAR SALARY	HOURS WORKED	GROSS WAGES
7. a.	Telemarketer	40	$240	43	
b.	Secretary	35	315	46	
c.	Supervisor	(Varies)	312	48	
8. a.	Sales manager	40	$336	45	
b.	Office manager	38	304	44	
c.	Plant manager	(Varies)	437	46	

For each of the employees listed below, compute gross earnings (1) with overtime at time-and-a-half and (2) by using overtime excess.

	EMPLOYEE	HOURS	RATE PER HOUR	GROSS EARNINGS
9. a.	D	42	$10.00	
b.	E	46	8.30	
c.	F	49	9.90	
10. a.	G	43	$5.60	
b.	H	45	7.70	
c.	I	47	8.50	

11. Complete the following payroll register, where the overtime excess method is used for work above 40 hours per week.

WURZER MOTORCYCLE SALES AND SERVICE
Payroll for Week Ending December 23, 19XX

NAME	M	T	W	T	F	S	TOTAL HOURS	RATE PER HOUR	OVER-TIME HOURS	OVER-TIME EXCESS RATE	REG-ULAR RATE WAGES	OVER-TIME EXCESS	TOTAL GROSS WAGES
Byrd, J.	8	8	9	9	8	0		$ 6.20					
Evans, G.	7	8	8	8	8	6		7.50					
Gross, E.	8	8	8	8	8	4		8.40					
Smith, V.	9	8	7	8	8	8		10.50					
Zais, L.	8	8	5	8	6	11		9.60					
										Totals			

12. Use the overtime excess method in computing the following gross payroll.

KING APPLIANCES, INC.
Payroll for Week Ending February 15, 19XX

NAME	M	T	W	T	F	S	TOTAL HOURS	RATE PER HOUR	OVER-TIME HOURS	OVER-TIME EXCESS RATE	REG-ULAR RATE WAGES	OVER-TIME EXCESS	TOTAL GROSS WAGES
Dinh, T.	8	8	10	8	9	5		$5.50					
Horn, S.	8	9	8	9	9	6		6.40					
Rook, E.	7	8	10	10	8	4		7.30					
Soka, V.	9	9	8	8	8	0		7.90					
Vern, H.	8	9	9	6	10	8		7.60					
										Totals			

SECTION 4
COMPUTING WAGES— PRODUCTION BASIS

In many factories, each employee's work consists of repeating the same task— either some step in the construction of a product or inspecting or packing finished products. The employee's value to the company can thus be measured by how often this task is completed. Therefore, many plants compensate employees at a specified rate per unit completed; this constitutes the **production** basis of computing wages. This type of wage computation is also known as the **piecework** or **piecerate** plan.

Naturally, some steps in the production of a product require more time than others. Thus, the rate per unit for that step would be more than the rate on another task which could be completed much more quickly. In order to keep the entire plant production moving at an even pace, more employees are hired to perform time-consuming tasks and fewer workers are employed to perform easier tasks.

On the straight production plan, the employee receives the same rate for each piece; thus,

192 pieces at $1.55 = $297.60 wages

Example 1   The following example shows a payroll based on straight production.

TEXAS BOOT CO.

Payroll for Week Ending April 5, 19XX

NAME	NET PIECES PRODUCED						TOTAL NET PIECES	RATE PER PIECE	GROSS WAGES EARNED
	M	T	W	T	F	S			
Cruz, Ramon	17	14	18	19	16	8	92	$3.25	$ 299.00
Gelernter, Art	39	46	39	37	34	19	214	1.45	310.30
Kearns, D. M.	125	138	140	131	129	65	728	0.50	364.00
Lotus, Kim	71	67	74	79	70	34	395	0.72	284.40
Steiner, Lois	76	82	81	77	80	40	436	0.72	313.92
								Total	$1,571.62

INCENTIVES (PREMIUM RATES)

At first glance, the production plan would seem to be quite acceptable to employees, since compensation is determined entirely by individual performance. Such has not been the case, however. Most companies that use the piecework plan have established quotas or minimum production standards for each job. As an extra incentive to production, workers receive a higher rate per unit on all units they complete in excess of this minimum. It is these quotas which historically have caused such dissatisfaction with the production plan. In the past, many companies set "production" (quotas) at such high levels that only a few of the most skillful and experienced employees could attain them. Furthermore, employees complained that when they finally did master the task well enough to "make production," the company would reassign them to a new, unfamiliar task.

As a result of the general dissatisfaction over piecework plans, many union contracts now contain provisions that members must be paid on the hourly rate basis. Those plants which do still use the production basis have tended to reestablish quotas that are more in line with the production that average employees can attain.

OVERTIME

Piecework employees who are subject to the Fair Labor Standards Act must likewise be paid time-and-a-half for work in excess of 40 hours per week. This overtime may be computed similarly to that of salaried workers (by finding the hourly rate to which their regular-time earnings are equivalent and paying $1\frac{1}{2}$ this hourly rate for overtime hours). However, the overtime wages of a production worker may be computed at $1\frac{1}{2}$ times the regular rate per piece, provided that certain standards are met to ensure that workers are indeed being paid the required minimum wage for regular-time hours, as well as time-and-a-half for overtime hours. This method is preferred because it is simpler to compute and will therefore be used for our problems. We shall not, however, attempt to compute overtime where premium production is also involved.

Example 2   In the following production payroll, all work performed on Saturday is overtime and is paid at $1\frac{1}{2}$ times the regular piece rate.

PRECISION INSTRUMENTS, INC.
Payroll for Week Ending August 30, 19XX

NAME	NET PIECES PRODUCED						REG. TIME PROD.	RATE PER PIECE	OVER-TIME PROD.	OVER-TIME RATE	REG. WAGES	OVER-TIME WAGES	TOTAL GROSS WAGES
	M	T	W	T	F	S							
Ball	214	220	218	216	219	112	1,087	$0.25	112	$0.375	$ 271.75	$ 42.00	$ 313.75
Grange	94	90	97	99	101	51	481	0.50	51	0.75	240.50	38.25	278.75
Mills	68	70	72	69	71	30	350	0.65	30	0.975	227.50	29.25	256.75
Trent	151	155	153	150	152	75	761	0.36	75	0.54	273.96	40.50	314.46
										Totals	$1,013.71	$150.00	$1,163.71

Calculator Tip.   Use the same technique as for Example 2 of Section 3. That is, find (a) the total production for each worker and (b) each person's overtime rate. Then, compute (c) the Regular Wages column and (d) the Overtime Wages column, and find the total for the two columns. Finally, (e) total each person's wages with M+ and use MR to verify that the Total Gross Wages column shows the same total you obtained after step (d).

DOCKINGS (SPOILAGE CHARGEBACKS)

Although companies are willing to pay premium rates as incentives to high production, they also wish to discourage haste resulting in carelessness and mistakes. An employee who makes an uncorrectable mistake has not only destroyed the raw materials involved but has also wasted the work done previously by other employees. Thus, an unreasonably high amount of spoilage would be quite expensive to a company.

To discourage carelessness, many plants require employees to forfeit a certain amount for each piece they spoil. These penalty **dockings** (or **charge-backs**) are generally at a lower rate than the employee ordinarily receives for each task, because a small amount of spoilage is to be expected. The worker is actually penalized twice for the spoilage: First, the number of spoiled units is subtracted from the worker's gross units to obtain the net units earning at the base (or quota) rate; later, the spoiled units are multiplied by the chargeback rate and this penalty subtracted from earnings. In general,

Gross wages = Base production + Premium − Chargeback

"Base production" is computed on the number of pieces (gross production − spoilage) that apply toward meeting quota. "Premium" is computed on any excess pieces (those still above quota). "Chargeback" is computed on the spoiled pieces. (See Example 3 for an illustration.)

Example 3   Cynthia Simmons sews zippers into dresses for a garment manufacturer. Her base rate is $1.25 per dress, and a premium rate of $1.52 per dress is paid after she exceeds the weekly production quota of 240 zippers. She is docked 75¢ for each zipper that does not pass inspection. Compute her gross earnings for the week:

	M	T	W	T	F	Total
Gross production	54	51	55	53	52	265
Less: Spoilage	2	0	0	3	1	6
Net production	52	51	55	50	51	259

Cynthia will receive the premium rate on 259 − 240 = 19 zippers. Her Gross wages = Base production + Premium − Chargeback is:

	TOTAL NET PRODUCTION	×	RATE	=	GROSS WAGES
Base production	240	@	$1.25		$300.00
Premium production	19	@	1.52		+ 28.88
					$328.88[a]
Spoilage chargeback	6	@	0.75		− 4.50
					$324.38

[a] Wage computation would end at this point for any production tasks not subject to chargeback.

Example 4   Complete the following production payroll, which includes incentive and penalty chargeback rates:

HANOVER GAME CO.
Payroll for Week Ending October 17, 19XX

NAME		DAILY PRODUCTION M	T	W	T	F	TOTAL PRODUCTION	NET QUOTA	PRODUCTION	RATE (CENTS)	BASE WAGES	TOTAL GROSS WAGES
Eaton, C. C.	GP[a]	159	162	152	155	158	786		Base: 750	45¢	$337.50	
	S	7	3	0	2	2	14		Premium: 22	56	12.32	$344.92
	NP	152	159	152	153	156	772	750	Chargeback: 14	35	4.90	
Gray, N. S.	GP	126	128	131	125	129	639		Base: 632	47¢	$297.04	
	S	2	1	3	0	1	7		Premium: —	59	—	294.38
	NP	124	127	128	125	128	632	650	Chargeback: 7	38	2.66	
Spann, B. R.	GP	143	141	145	142	144	715		Base: 700	46¢	$322.00	
	S	2	0	0	1	2	5		Premium: 10	58	5.80	325.95
	NP	141	141	145	141	142	710	700	Chargeback: 5	37	1.85	
											Total	$965.25

[a] GP indicates "gross production"; S, "spoilage"; and NP, "net production."

SECTION 4   PROBLEMS

1. Elena Thomas cleans crabs at a fish processing plant. Her base rate is 88¢ per pound of crabmeat. She receives a premium rate of $1.00 per pound when she cleans more than 200 pounds per week. Determine her wages for the following week:

	M	T	W	T	F	TOTAL
Thomas, Elena	60	62	68	66	70	

2. Tom Stone operates a machine that bolts two pieces of an automobile engine together. His base rate is 52¢ per bolt, with a premium rate of 70¢ per bolt in excess of 500 bolts per week. Determine his gross wages for the following week:

	M	T	W	T	F	TOTAL
Stone, Tom	128	140	155	158	167	

3. Sarah Hubbard assembles parts in microcomputers. Her base rate is $1.75 per assembly. She receives an incentive rate of $2.50 per assembly in excess of 100 net assemblies. She is docked 50¢ for each assembly that does not pass inspection. Calculate her wages for the week when she had the following production:

	M	T	W	T	F	TOTAL
Gross production	26	34	28	34	29	
Less: Spoilage	1	2	0	3	0	
Net production						

4. Alice Pollen, working in a paper plant, operates a machine that attaches glue to assembled pads of paper. Her base rate is 15¢ per pad, and a premium rate of 24¢ is paid for net production over 1,000 pads. She is docked 8¢ for each pad that does not pass inspection. Calculate her wages for the following week:

	M	T	W	T	F	TOTAL
Gross production	325	344	346	350	342	
Less: Spoilage	5	6	2	0	1	
Net production						

5. Complete the following payroll based on straight production:

BRISTOL FABRICS
Payroll for Week Ending February 4, 19XX

NAME	NET PIECES PRODUCED					TOTAL NET PIECES	RATE PER PIECE	GROSS WAGES EARNED
	M	T	W	T	F			
Cobb, S.	89	87	99	96	93		50¢	
Ford, B.	95	100	98	91	96		70	
Hurl, M.	102	110	99	105	104		90	
Less, A.	80	79	84	80	77		57	
Pia, D.	112	126	98	104	120		60	
Will, D.	100	132	96	112	110		96	
Wren, E.	69	71	74	86	90		80	
							Total	

6. Complete the following production payroll:

PACIFIC TOOL CO.
Payroll for Week Ending April 5, 19XX

NAME	NET PIECES PRODUCED					TOTAL NET PIECES	RATE PER PIECE	GROSS WAGES EARNED
	M	T	W	T	F			
Abbott, A.	64	78	76	84	90		$1.00	
Breck, J.	71	77	78	79	86		1.03	
Dunn, M.	56	58	62	65	67		1.01	
Jones, B.	80	89	88	85	97		1.04	
Simon, C.	75	75	82	86	89		1.02	
Woods, V.	84	85	89	86	91		1.05	
Yoe, T.	88	85	88	90	92		1.06	
							Total	

7. Compute the following production payroll. Work performed on Saturday is overtime and should be reimbursed at $1\frac{1}{2}$ times the standard rate per piece.

NAME	NET PIECES PRODUCED						REG. TIME PROD.	RATE PER PIECE	OVER-TIME PROD.	OVER-TIME RATE	REG. WAGES	OVER-TIME WAGES	TOTAL GROSS WAGES
	M	T	W	T	F	S							
Blue, K.	60	66	71	78	85	45		$0.55					
Downs, J.	92	99	96	97	91	61		0.60					
Kent, H.	74	76	73	80	62	32		0.80					
Hamm, I.	78	80	85	76	81	36		0.90					
Wilks, O.	65	68	70	89	83	44		0.80					
										Totals			

8. Complete the following production payroll. Overtime will be paid for Saturday work at $1\frac{1}{2}$ times the standard rate per piece.

STROZIER MANUFACTURING CO.
Payroll for Week Ending January 18, 19XX

NAME	NET PIECES PRODUCED						REG. TIME PROD.	RATE PER PIECE	OVER- TIME PROD.	OVER- TIME RATE	REG. WAGES	OVER- TIME WAGES	TOTAL GROSS WAGES
	M	T	W	T	F	S							
Chang, A.	56	58	60	62	66	50		$0.60					
Evans, R.	60	61	64	67	70	45		0.68					
Hall, E.	72	75	75	79	78	70		0.70					
Long, B.	65	70	76	71	75	63		0.74					
Thomas, A.	61	71	73	76	74	68		0.86					
										Totals			

9. Find the gross wages for the following payroll:

HOMER TEXTILES, INC.
Payroll for Week Ending May 6, 19XX

NAME		DAILY PRODUCTION					TOTAL PRODUC- TION	NET QUOTA	PRODUCTION	RATE (CENTS)	BASE WAGES	TOTAL GROSS WAGES
		M	T	W	T	F						
Adam	GP	80	94	82	70	76			Base:	50¢		
	S	1	0	2	2	2			Premium:	60		
	NP							300	Chargeback:	30		
Boyd	GP	65	72	85	92	56			Base:	64¢		
	S	0	1	2	1	1			Premium:	70		
	NP							200	Chargeback:	20		
Cole	GP	96	99	94	90	82			Base:	80¢		
	S	0	0	2	4	5			Premium:	90		
	NP							360	Chargeback:	40		
Hall	GP	76	75	71	70	64			Base:	56¢		
	S	1	0	3	2	0			Premium:	64		
	NP							250	Chargeback:	40		
Wills	GP	60	59	65	68	71			Base:	68¢		
	S	0	1	2	1	1			Premium:	75		
	NP							300	Chargeback:	30		
											Total	

10. Find the gross wages for the following payroll:

SCHNELL ATHLETIC EQUIPMENT
Payroll for Week Ending August 12, 19XX

NAME		DAILY PRODUCTION					TOTAL PRODUC- TION	NET QUOTA	PRODUCTION	RATE (CENTS)	BASE WAGES	TOTAL GROSS WAGES
		M	T	W	T	F						
Allen	GP S NP	39 1	48 2	50 0	51 1	46 2		200	Base: Premium: Chargeback:	88¢ 95 75		
Dill	GP S NP	36 0	41 2	49 2	54 0	50 0		150	Base: Premium: Chargeback:	92¢ 99 60		
Edward	GP S NP	54 3	58 1	64 1	60 0	63 0		250	Base: Premium: Chargeback:	70¢ 80 50		
Jenks	GP S NP	48 1	49 0	53 2	59 1	66 0		300	Base: Premium: Chargeback:	85¢ 93 40		
Strong	GP S NP	65 3	68 1	73 1	75 2	79 1		275	Base: Premium: Chargeback:	86¢ 96 45		
										Total		

SECTION 5
NET WEEKLY PAYROLLS

Employers are required by law to make certain deductions from the gross earnings of their employees. The amount remaining after deductions have been withheld is known as **net wages.** The same methods are used to determine net wages from gross wages regardless of which payroll plan has been used to compute gross wages.

Two compulsory federal deductions are made in all regular payrolls: FICA and personal income tax. In many states, the employer is also required to withhold state income tax. Certain other deductions, either compulsory or voluntary, may also be made. The following paragraphs discuss important aspects of the various deductions.

FEDERAL INSURANCE CONTRIBUTIONS ACT (FICA)

The original **Federal Insurance Contributions Act** was one of the emergency measures passed during the Great Depression of the 1930s; it has been amended and extended many times and now draws payments from over 117 million workers. **FICA** includes both **Social Security** and **Medicare.** Social Security is not a tax in the normal sense; it could be described more accurately as a compulsory savings or insurance program. Social Security funds are used to pay monthly benefits to retired or disabled workers and to pay benefits to the surviving spouse and minor children of a deceased worker. Social Security accounted for 40% of the total income of senior citizens and kept 1.1 million children out of poverty in 1993. Currently, 46 million people receive payments from Social Security.

The Social Security program has been the subject of much concern in recent years, because paying benefits to retirees and other qualified recipients has required spending all the funds that current workers pay in—and the number of recipients continues to increase each year. Although the program was originally intended only to supplement individual savings and retirement plans, their FICA benefits have in reality become the sole source of income for many older persons. Moreover, as the number of older people in the population increases, a smaller proportion of younger workers will be paying into the Social Security fund. Thus, the prospect of a bankrupt Social Security system has serious national impact. So far, the situation has primarily been handled by frequently raising both the FICA percentage rate and the base amount of wages on which FICA is computed. Other major changes can also be expected in the near future to ensure that the Social Security program survives.

The Medicare program was enacted by Congress in 1965. Medicare is a two-part program of health insurance available to persons 65 or older who are also eligible for Social Security benefits or Railroad Retirement benefits. There are exceptions for certain disabled workers and disabled widows and widowers under age 65. Part A pays for hospital, hospice, and skilled nursing facility services. Part B is optional and pays for most doctor bills, medical equipment, and certain outpatient services. There is a monthly premium for Part B.

Prior to 1991, there was one tax for both Social Security and Medicare levied on earnings up to a maximum taxable wage. A 1990 tax law provided for two separate tax rates and separate maximum taxable wages, one for Social Security and one for Medicare. The maximum taxable wage for Medicare was dropped in 1994.

This text uses the 1995 withholding rate for Social Security, 6.2% on the first $61,200; and for Medicare, 1.45% on all gross earnings. All FICA deductions must be matched by an equal contribution from the employer (making the total rate 15.3%). Self-employed persons pay 12.4% on the first $61,200 for Social Security and 2.9% on gross earnings for Medicare.

Although the two taxes are separate, Table C-11 in Appendix C will combine them into one rate of 7.65% for convenience. In an office, however, careful attention must be given to an employee's total gross earnings reaching the $61,200 maximum taxable wage for Social Security. Gross wages over $61,200 are still subject to the Medicare tax of 1.45%.

The total FICA deduction to be withheld may be computed by finding

7.65% of the employee's gross wages (for wages up to $61,200). However, many payroll clerks use FICA tax tables furnished by the government. These tables list wage intervals and the deduction to be made within each interval for wages under $100. This deduction must then be added to the deduction due on the even hundred dollars (for instance, $100 or $200) to obtain the total FICA deduction. For convenience, we will use an expanded FICA tax table (Table C-11, Appendix C) containing entries for higher intervals as well.

Today, most payroll systems are computerized, but an understanding of the calculations is essential for employer and employee. Most computerized payroll systems are programmed to cease Social Security deductions when $61,200 in wages is reached.

Example 1 (a) Ewald Straiton earns a gross salary of $380 each week. What FICA deduction must be made?

By percentage,

$$7.65\% \text{ of } \$380 = \$29.07$$

By the FICA tax table: Under the Wages heading, consult the interval where gross earnings are "at least $379.94, but less than $380.07." The next column shows that the "tax to be withheld" is $29.07.

(b) Mary Gibson earned gross wages of $250. What is her FICA tax deduction?

By the FICA tax table, $250.00 appears on two lines. On the first line, however, Gibson's wages appear under the heading "but less than $250.00." This does not apply because Gibson did not earn "less than $250.00." We must use the second line, where the wages earned were "at least $250.00." The "tax to be withheld" column indicates that Gibson's deduction will be $19.13.

Although the $61,200 base for Social Security means that most workers make contributions from their entire year's earnings, many executives and some highly paid workers have earnings above $61,200 that are not subject to Social Security. The payroll clerk should thus be alert so that no Social Security deductions are made on earnings in excess of $61,200. Medicare deductions (1.45%) will continue on earnings over $61,200, since there is not a cumulative maximum earnings figure for Medicare.

FEDERAL INCOME TAX

Employers are required to make deductions from each worker's wages for federal **income tax,** and most businesses use federal income tax **withholding**

Form **W-4** Department of the Treasury Internal Revenue Service

Employee's Withholding Allowance Certificate

▶ **For Privacy Act and Paperwork Reduction Act Notice, see reverse.**

OMB No. 1545-0010 **1995**

1 Type or print your first name and middle initial: Sarah B. Last name: Bingham

2 Your social security number: 123 45 6789

Home address (number and street or rural route): 4120 Ft. Worth Place

3 ☒ Single ☐ Married ☐ Married, but withhold at higher Single rate.
Note: *If married, but legally separated, or spouse is a nonresident alien, check the Single box.*

City or town, state, and ZIP code: Birmingham, AL 35209

4 If your last name differs from that on your social security card, check here and call 1-800-772-1213 for a new card . . . . . ▶ ☐

5 Total number of allowances you are claiming (from line G above or from the worksheets on page 2 if they apply) . **5** 1

6 Additional amount, if any, you want withheld from each paycheck . . . . **6** $

7 I claim exemption from withholding for 1995 and I certify that I meet **BOTH** of the following conditions for exemption:
- Last year I had a right to a refund of **ALL** Federal income tax withheld because I had **NO** tax liability; **AND**
- This year I expect a refund of **ALL** Federal income tax withheld because I expect to have **NO** tax liability.

If you meet both conditions, enter "EXEMPT" here . . . . ▶ **7**

Under penalties of perjury, I certify that I am entitled to the number of withholding allowances claimed on this certificate or entitled to claim exempt status.

Employee's signature ▶ Sarah B. Bingham **Date** ▶ August 21, 1995

8 Employer's name and address (Employer: Complete 8 and 10 only if sending to the IRS)
MacArthur Supply Co.
2021 First Avenue, South
Birmingham, AL 35211

9 Office code (optional)

10 Employer identification number: 98 765432

FIGURE 8-1 Employee's Withholding Allowance Certificate, Form W-4

tables to determine these deductions. Beginning in 1984, the wage intervals have been adjusted annually for inflation—so that a worker's earnings could increase with inflation without the person having to pay tax at a higher percentage rate. This text uses the income tax withholding tables effective in the fourth quarter of 1995.

Married and single persons are subject to different income tax rates, the rates being higher for single persons. The income tax withholding tables (Table C-12, Appendix C) therefore contain separate listings for married and single persons. The tax to be withheld is dependent upon three factors: (1) the worker's gross earnings, (2) whether married or single, and (3) the number of allowances claimed.

Immediately after a new employee is hired, the payroll clerk should ask the employee to complete Form W-4, the Employee's Withholding Allowance Certificate (Figure 8-1). This form indicates the employee's marital status, Social Security number, and the total number of withholding allowances claimed. One allowance may be taken for each member of the family. This total number may include a grandparent or other persons who live with the family and/or qualify as dependents.* These allowances are the same ones that are claimed on the personal income tax return, Form 1040.

*To qualify, the persons (1) must either be specified relatives or reside with the taxpayer, (2) must have received more than one-half their support from the taxpayer, (3) must have earned less than a certain gross income of their own (except youths under 19 or qualified students), (4) must not have filed a joint return with a spouse, and (5) must be a citizen of the United States or of certain other specified places.

Care should be taken to ensure the accuracy of this form as it controls the amount of federal and in many cases state income tax withheld from the employee's gross pay. Civil and criminal fines can be imposed for the claiming of too many withholding allowances, thereby reducing the amount of income taxes withheld.

Generally, the number of withholding allowances permitted is governed by the number of taxpayers and qualifying dependents to be claimed on the employee's income tax return. The employee can claim fewer withholding allowances than (s)he is entitled to if (s)he wishes to increase the amount of withholding. Although the withholding laws are complex, some general guidelines are given.

1. Husband and wife who will file a joint tax return may split the withholding allowances as they choose, but should be careful not to duplicate the allowances. Married persons who will file separate income tax returns should realize that only the amounts withheld from their wages can be credited on their income tax return, and that the rules for claiming taxpayer and dependency exemptions on their income tax return are less flexible.

2. Changes in the claimed withholding allowances can be made as new situations occur.

3. In addition to the withholding allowances for the taxpayer and qualifying dependents, additional allowances may be claimed in certain situations to avoid overwithholding of taxes:
   a. A single taxpayer or a married taxpayer with a nonworking spouse may claim an additional allowance.
   b. Additional allowances are permitted when the employee has large itemized deductions, alimony payments, business losses, capital losses, and certain other deductions.

4. Some employees may be eligible for exemption from income tax withholding by filing a special form W-4E.

Ultimately the employee is responsible for the withholding allowances claimed. The claiming of too many allowances can result in fines and penalties for the underpayment of taxes. Conversely, the claiming of too few can result in the overwithholding of taxes and the interest-free use of the employee's money by the government.

Example 2   Sharie Chiles and Sue Watson both earn salaries of $350 per week. Each claims only herself as a dependent; however, Chiles is married and Watson is single. How much income tax will be withheld from each salary?

Since Sharie Chiles is married and claims only herself, the allowance indicated on her payroll sheet is "M-1." Using the income tax withholding table for "Married Persons" paid on a "Weekly Payroll Period," we find the gross wage interval containing wages that are "at least $350 but less than $360." The columns at the right have headings indicating the number of allowances claimed. For Chiles, we use the column which shows "1 allowance claimed" and see that the tax to be withheld is $28. If Chiles's salary does not change, she will have income tax of $28 withheld each week during the entire year.

Sue Watson's payroll sheet indicates that her allowance is "S-1." Using the income tax withholding table for "Single Persons" paid on a "Weekly" basis, we find the same gross wage interval of "at least $350 but less than $360." In the column for "1 allowance claimed," we see that income tax of $39 will be deducted from Watson's salary each week. Thus, Watson, because she is single, pays $39 − $28, or $11 more income tax each week than Chiles pays.

STATE DEDUCTIONS

In many states, state income tax withholding is also a compulsory deduction. Withholding tables are usually available for the state tax, which is typically at a considerably lower rate than the federal income tax.

The Social Security (FICA) Act contains provisions for an **unemployment insurance tax.** The funds from this tax provide compensation to able-bodied workers who are willing to work but who have been "laid off" by their employers because there is no work available. Under the federal act, this unemployment tax is paid by the *employer only,* with most of the funds being collected by the individual states. (That is, no deduction is made from the worker's pay for federal unemployment tax.) In a few states, however, the state has levied its own unemployment tax in addition to the federal tax. In these states, a compulsory unemployment tax may be deducted from the worker's wages.

OTHER DEDUCTIONS

There are any number of other deductions that may be withheld from the worker's wages. One of the most common is group insurance. Many firms sponsor a group policy whereby employees may purchase health, accident, hospitalization, and sometimes even life insurance at rates substantially lower than they could obtain individually.

Union dues or dues to professional organizations may be deducted. Employees often contribute to a company or state retirement plan which will supplement their Social Security retirement benefits. Contributions to charitable causes (such as the United Way Fund) may be withheld. Some businesses participate in a payroll savings plan whereby deductions are made to purchase U.S. Savings Bonds or to buy shares of the corporation's stock (often at a reduced price). Numerous other examples could be cited.

Example 3 Below is a net payroll of Everett Paint Co. In addition to the standard deductions, 1.5% of each employee's gross wages is withheld for a company retirement plan.

EVERETT PAINT CO.
Payroll for Week Ending June 3, 19XX

Employee	Allow-ances	Gross Wages	FICA	Federal Income Tax	Other (Ret.)	Other	Other	Total Deduc-tions	Net Wages Due
Camden, S.	M-3	$ 345.00	$ 26.39	$ 12.00	$ 5.18			$ 43.57	$ 301.43
Duncan, E.	M-2	345.00	26.39	19.00	5.18			50.57	294.43
Hager, W.	S-2	345.00	26.39	30.00	5.18			61.57	283.43
Ingles, C.	M-1	278.00	21.27	16.00	4.17			41.44	236.56
Mott, C.	M-3	367.00	28.08	15.00	5.51			48.59	318.41
Reid, M.	S-1	290.00	22.19	30.00	4.35			56.54	233.46
Trent, B.	M-4	380.00	29.07	10.00	5.70			44.77	335.23
Totals		$2,350.00	$179.78	$132.00	$35.27			$347.05	$2,002.95

Note. The totals of the individual deducations, when added together, must equal the total in the Total Deductions column ($179.78 + $132.00 + $35.27 = $347.05). The total of Gross Wages minus the Total Deductions total must give the total of Net Wages Due ($2,350 − $347.05 = $2,002.95).

Hint. A net payroll can be completed more efficiently if all deductions of the same type are done at one time. That is, list FICA for all employees, then determine federal income tax for all employees, and so forth.

Calculator techniques . . .   FOR EXAMPLE 3

As the Hint suggests, each deductions column should be completed before you begin the next column. You can multiply by 7.65% to find each FICA value, if you wish, rather than using the table; then round each result to the nearest cent before writing it in the payroll. To compute the Retirement column, 0.015 can similarly be used for repetitive multiplication. (We don't accumulate into memory here, because the rounded values produce a different total. Remember to clear the memory between steps.)

0.0765 $\boxed{M+}$;   (0.0765) $\boxed{\times}$ 345 $\boxed{=}$ ⟶ 26.3925;

$\boxed{MR}$ $\boxed{\times}$ 278 $\boxed{=}$ ⟶ 21.267; $\boxed{MR}$ $\boxed{\times}$ 367 $\boxed{=}$ ⟶ 28.0755;

$\boxed{MR}$ $\boxed{\times}$ 290 $\boxed{=}$ ⟶ 22.185; $\boxed{MR}$ $\boxed{\times}$ 380 $\boxed{=}$ ⟶ 29.07

After filling in each deduction, vertically add the FICA, Income Tax, and Retirement columns, and find their combined total. Then, use memory to complete the Total Deductions column, as follows:

26.39 $\boxed{+}$ 12 $\boxed{+}$ 5.18 $\boxed{M+}$ ⟶ 43.57;

26.39 $\boxed{+}$ 19 $\boxed{+}$ 5.18 $\boxed{M+}$ ⟶ 50.57;

. . . 29.07 $\boxed{+}$ 10 $\boxed{+}$ 5.7 $\boxed{M+}$ ⟶ 44.77; $\boxed{MR}$ ⟶ 347.05

Verify that this column total is the same Total Deductions value you obtained before. Then, again use memory to find the Net Wages Due for each worker:

345 $\boxed{-}$ 43.57 $\boxed{M+}$ ⟶ 301.43;   345 $\boxed{-}$ 50.57 $\boxed{M+}$ ⟶ 294.43;

. . . 380 $\boxed{-}$ 44.77 $\boxed{M+}$ ⟶ 335.23; $\boxed{MR}$ ⟶ 2,002.95

Verify this result for Net Wages Due by subtracting Gross Wages − Total Deductions ($2,350.00 − 347.05 = $2,002.95).

SECTION 5   PROBLEMS

1. Compute each person's net wages due. Deduct FICA and federal income tax. All employees also pay $7.50 for their own group insurance plus $3.00 for each additional dependent. (The number of additional dependents for all employees is shown in parentheses beside their names.)

RUDOLPH PRINTING CO.
Payroll for Week Ending February 3, 19XX

EMPLOYEE (ADDL. DEPS.)	ALLOW-ANCES	GROSS WAGES	DEDUCTIONS					NET WAGES DUE
			FICA	FEDERAL INCOME TAX	OTHER (INS.)	OTHER	TOTAL DEDUC-TIONS	
Bosin, J. (0)	M-2	$345.00						
Dunhan, B. (1)	M-2	237.00						
Free, A. (0)	S-1	221.75						
Good, L. (2)	M-3	400.50						
Stuart, S. (3)	M-4	390.40						
	Totals							

2. Complete the following salary payroll. Withhold FICA and federal income tax. All workers pay $8.00 for their own health insurance plus $4 for each additional dependent. (The number of additional dependents for all employees is shown in parentheses beside their names.)

HERNDON GARDEN SHOP
Payroll for Week Ending April 26, 19XX

EMPLOYEE (ADDL. DEPS.)	ALLOW-ANCES	GROSS WAGES	DEDUCTIONS					NET WAGES DUE
			FICA	FEDERAL INCOME TAX	OTHER (INS.)	OTHER	TOTAL DEDUC-TIONS	
Brady, J. (0)	M-2	$367.00						
Hope, H. (2)	M-3	322.80						
LaDow, O. (0)	S-1	261.45						
Marsh, M. (1)	S-2	300.50						
Widden, S. (3)	M-4	401.00						
	Totals							

3. Compute the net payroll for Johnson Sales, Inc., whose employees are paid on an hourly basis. Deductions are made for FICA, federal income tax, and union dues. These dues are deducted according to the employee's hourly rate as follows:

RATE PER HOUR	DUES	RATE PER HOUR	DUES
$4.00–4.49	$7.50	$5.50–6.99	$ 9.00
4.50–5.49	8.00	7.00–8.49	10.00

JOHNSON SALES, INC.
Payroll for Week Ending March 22, 19XX

| EMPL. NO. | ALLOW-ANCES | TOTAL HOURS | RATE PER HOUR | GROSS WAGES | DEDUCTIONS | | | | | NET WAGES DUE |
					FICA	FED. INC. TAX	OTHER (DUES)	OTHER	TOTAL DEDS.	
195	M-1	40	$5.60							
196	M-2	40	6.80							
197	S-1	36	6.45							
198	S-2	40	7.70							
199	M-3	38	8.25							
			Totals							

4. Complete the weekly payroll for Graystone Hardware Co. The employees are paid on an hourly basis, and deductions are withheld for FICA, federal income tax, and union dues. Their union dues are deducted according to the chart provided in Problem 3.

GRAYSTONE HARDWARE CO.
Payroll for Week Ending June 14, 19XX

| EMPL. NO. | ALLOW-ANCES | TOTAL HOURS | RATE PER HOUR | GROSS WAGES | DEDUCTIONS | | | | | NET WAGES DUE |
					FICA	FED. INC. TAX	OTHER (DUES)	OTHER	TOTAL DEDS.	
201	S-1	40	$6.80							
202	M-1	36	6.00							
203	M-2	40	7.50							
204	S-2	40	7.75							
205	S-1	40	5.25							
			Totals							

5. The following commission payroll has standard deductions for FICA and federal income tax. The sales representatives have each pledged 10% of their gross wages for the week as contributions to the United Way Campaign. Determine the net payroll for the week.

NEWBERRY DISTRIBUTORS
Payroll for Week Ending December 4, 19XX

EMPL. NO.	ALLOW-ANCES	NET SALES	COMM. RATE	GROSS WAGES	DEDUCTIONS					NET WAGES DUE
					FICA	FED. INC. TAX	OTHER (CONTR.)	OTHER	TOTAL DEDS.	
A-3	M-1	$9,500	4.0%							
A-5	M-2	9,800	3.5							
B-1	S-2	8,500	4.0							
B-4	S-1	7,600	5.0							
C-2	M-4	8,200	4.5							
			Totals							

6. Employees at Swarts Sales Corp. are paid on a commission basis. Standard deductions are made. Each employee has pledged 5% of gross wages for the week as contributions to a company charity fund. Determine the net payroll for the week.

SWARTS SALES CORP.
Payroll for Week Ending February 17, 19XX

EMPL. NO.	ALLOW-ANCES	NET SALES	COMM. RATE	GROSS WAGES	DEDUCTIONS					NET WAGES DUE
					FICA	FED. INC. TAX	OTHER (CONTR.)	OTHER	TOTAL DEDS.	
G-4	S-0	$3,500	6.5%							
G-5	M-2	4,900	5							
G-6	M-1	5,200	4.5							
G-7	M-3	5,500	6							
G-8	S-1	6,400	5							
			Totals							

7. Burkehalter Seafood Corp. pays its employees on the production basis. Complete the company's net payroll, which includes the standard deductions as well as group insurance deducted at the rate of $6 for each worker and $2 for each additional dependent. (The total number of additional dependents is listed in parentheses beside each employee number.)

BURKEHALTER SEAFOOD CORP.
Payroll for Week Ending October 8, 19XX

| EMPL. NO. | ALLOW-ANCES | NET PRODUC-TION | RATE PER PIECE | GROSS WAGES | DEDUCTIONS | | | | | NET WAGES DUE |
					FICA	FED. INC. TAX	OTHER (INS.)	OTHER	TOTAL DEDS.	
22 (2)	M-3	2,200	17¢							
23 (1)	S-1	1,850	18							
24 (0)	S-1	1,900	16							
25 (3)	M-4	1,920	20							
26 (0)	M-2	1,500	14							
				Totals						

8. The production plan is used by King Associates, Inc. to pay its employees. Deductions are made for FICA and federal income tax plus health insurance at the rate of $7 for each employee and $2 for each additional dependent. (The number of additional dependents is listed in parentheses beside each employee number.)

KING ASSOCIATES, INC.
Payroll for Week Ending June 11, 19XX

| EMPL. NO. | ALLOW-ANCES | NET PRODUC-TION | RATE PER PIECE | GROSS WAGES | DEDUCTIONS | | | | | NET WAGES DUE |
					FICA	FED. INC. TAX	OTHER (INS.)	OTHER	TOTAL DEDS.	
54 (2)	S-2	3,500	10¢							
55 (1)	M-2	2,800	12							
56 (2)	M-3	2,600	15							
57 (0)	S-1	3,000	13							
58 (1)	S-1	3,700	16							
				Totals						

9. Leonard Industries pays employees on an hourly basis. Standard overtime is paid for work after 40 hours per week. Complete the net payroll, including a 1% deduction for company retirement.

LEONARD INDUSTRIES
Payroll for Week Ending September 8, 19XX

EMPL.	ALLOW-ANCES	HOURS WORKED	REG. WAGES	OVER-TIME WAGES	TOTAL GROSS WAGES	FICA	FED. INC. TAX	OTHER (RET.)	TOTAL DEDS.	NET WAGES DUE
A	M-1	44	$200							
B	M-2	46	320							
C	M-4	42	240							
D	S-1	48	260							
		Totals								

10. McVeigh Industries pays standard overtime for work over 40 hours per week. Group insurance of $10 is withheld for each employee. Determine the company's net payroll for the week.

MCVEIGH INDUSTRIES
Payroll for Week Ending December 20, 19XX

EMPL.	ALLOW-ANCES	HOURS WORKED	REG. WAGES	OVER-TIME WAGES	TOTAL GROSS WAGES	FICA	FED. INC. TAX	OTHER (INS.)	TOTAL DEDS.	NET WAGES DUE
E	M-3	41	$280							
F	M-2	45	300							
G	S-0	48	320							
H	S-1	43	264							
		Totals								

11. Compute the net payroll of Durham Enterprises, Inc. Standard overtime is paid at 1½ times the regular piece rate. Deduct 3% for a company retirement plan.

DURHAM ENTERPRISES, INC.

Payroll for Week Ending May 15, 19XX

									DEDUCTIONS				
EMPL.	ALLOW-ANCES	NET REG. PRODUC-TION	OVER-TIME PRODUC-TION	REG. PIECE RATE	REG. WAGES	OVER-TIME WAGES	TOTAL GROSS WAGES	FICA	FED. INC. TAX	OTHER (RET.)	TOTAL DEDS.	NET WAGES DUE	
#10	M-1	350	30	70¢									
11	S-1	275	55	68									
12	M-3	400	24	60									
13	M-5	360	20	74									
				Totals									

12. Complete the net payroll for Dawes Equipment, Inc. Overtime is paid at 1½ times the regular piece rate. In addition to standard deductions, deduct 4% for the company retirement plan.

DAWES EQUIPMENT, INC.

Payroll for Week Ending January 5, 19XX

									DEDUCTIONS				
EMPL.	ALLOW-ANCES	NET REG. PRODUC-TION	OVER-TIME PRODUC-TION	REG. PIECE RATE	REG. WAGES	OVER-TIME WAGES	TOTAL GROSS WAGES	FICA	FED. INC. TAX	OTHER (RET.)	TOTAL DEDS.	NET WAGES DUE	
#20	M-2	400	25	60¢									
21	M-0	390	46	64									
22	S-2	416	52	72									
23	M-1	375	30	88									
				Totals									

SECTION 6
QUARTERLY EARNINGS REPORT AND RETURNS

EMPLOYEE'S QUARTERLY EARNINGS REPORT

Employers must keep an individual payroll record showing each employee's earnings, deductions, and other pertinent information. Figure 8-2 presents an **employee's quarterly earnings report.** The report shows quarterly and weekly, biweekly, semimonthly, or monthly totals, which are used in computing the

UNLIMITED HORIZONS
Quarterly Earnings Report
For the Period From Jan 1, 1995 to Mar 31, 1995

Employee ID Employee SS No	Date Reference	Amount	Gross Fed_Income FUTA_ER	Soc_Sec State_Incom SUI_ER	Medicare Soc_Sec_ER	Health Ins Medicare_ER
Beginning Balance for JOHN HORTON						
HORTON JOHN HORTON 198-78-2498	1/2/95 1264	343.61	414.00 -29.22 -3.31	-25.67 -10.35	-6.00 -25.67	-9.50 -6.00
HORTON JOHN HORTON 198-78-2498	1/9/95 1400	385.37	468.00 -37.32 -3.74	-29.02 -11.70	-6.79 -29.02	-9.50 -6.79
HORTON JOHN HORTON 198-78-2498	1/16/95 1425	354.05	427.50 -31.24 -3.42	-26.51 -10.69	-6.20 -26.51	-9.50 -6.20
HORTON JOHN HORTON 198-78-2498	1/23/95 1456	301.84	360.00 -21.12 -2.88	-22.32 -9.00	-5.22 -22.32	-9.50 -5.22
HORTON JOHN HORTON 198-78-2498	1/30/95 1489	395.83	481.50 -39.34 -3.85	-29.85 -12.04	-6.98 -29.85	-9.50 -6.98
HORTON JOHN HORTON 198-78-2498	2/6/95 1565	298.37	355.50 -20.44 -2.84	-22.04 -8.89	-5.15 -22.04	-9.50 -5.15
HORTON JOHN HORTON 198-78-2498	2/13/95 1595	322.73	387.00 -25.17 -3.10	-23.99 -9.68	-5.61 -23.99	-9.50 -5.61
HORTON JOHN HORTON 198-78-2498	2/20/95 1645	298.37	355.50 -20.44 -2.84	-22.04 -8.89	-5.15 -22.04	-9.50 -5.15
HORTON JOHN HORTON 198-78-2498	2/27/95 1685	374.94	454.50 -35.29 -3.64	-28.18 -11.36	-6.59 -28.18	-9.50 -6.59
HORTON JOHN HORTON 198-78-2498	3/6/95 1789	333.17	400.50 -27.19 -3.20	-24.83 -10.01	-5.81 -24.83	-9.50 -5.81
HORTON JOHN HORTON 198-78-2498	3/13/95 1815	280.95	333.00 -17.07 -2.66	-20.65 -8.32	-4.83 -20.65	-9.50 -4.83
HORTON JOHN HORTON 198-78-2498	3/20/95 1864	427.15	522.00 -45.42 -4.18	-32.36 -13.05	-7.57 -32.36	-9.50 -7.57
HORTON JOHN HORTON 198-78-2498	3/27/95 1987	364.50	441.00 -33.27 -3.53	-27.34 -11.03	-6.39 -27.34	-9.50 -6.39
Total 1/1/95 thru 3/31/95		4,480.88	5,400.00 -382.53 -43.19	-334.80 -135.01	-78.29 -334.80	-123.50 -78.29
Total for JOHN HORTON		4,480.88	5,400.00 -382.53 -43.19	-334.80 -135.01	-78.29 -334.80	-123.50 -78.29

FIGURE 8-2   Employee's Earnings Report

state and federal quarterly payroll reports that every company must file. The format of the report will vary depending on the software used if the payroll is computerized. Some explanation is needed for this particular report. In the heading, "Reference" refers to the check number issued. "Gross" is the gross weekly wages, and "Fed–Income" is the federal income tax deduction. "FUTA–ER" is the employer's federal unemployment tax liability. "Soc–Sec" refers to the employee's Social Security deduction, whereas "Soc–Sec–ER" is the employer's contribution of the Social Security tax. "SUI–ER" is the employer's liability for state unemployment insurance. "Medicare–ER" is the employer's share of the Medicare tax; the employee's deduction is shown under "Medicare."

The computer program used for this quarterly earnings report utilizes formulas, so the income tax figures are slightly different from the tax table figures in Appendix C (Table C-12). The appendix tax tables are rounded to whole dollars. In addition, the quarterly earnings report will show the cutoff points for Social Security and unemployment taxes when cumulative earnings hit those maximum amounts. It also contains the quarterly and yearly totals of earnings and deductions required for personal income tax purposes, which are distributed to the employee and the Internal Revenue Service on Form W-2, the Wage and Tax Statement (Figure 8-3).

1 Control number	**22222**	For Official Use Only ▶ OMB No. 1545-0008								
2 Employer's name, address, and ZIP code			6 Statutory employee ☐	Deceased ☐	Pension plan ☐	Legal rep. ☐	942 emp. ☐	Subtotal ☐	Deferred compensation ☐	Void ☐
Maxwell Restaurant 456 Peachtree Dr. Atlanta, GA 30306			7 Allocated tips				8 Advance EIC payment			
			9 Federal income tax withheld **1,860.00**				10 Wages, tips, other compensation **16,000.00**			
3 Employer's identification number **12–3456789**	4 Employer's state I.D. number		11 Social security tax withheld **992.00**				12 Social security wages **10,000.00**			
5 Employee's social security number **111–22–3333**			13 Social security tips **6,000.00**				14 Medicare wages and tips **16,000.00**			
19a Employee's name (first, middle, last)			15 Medicare tax withheld **232.00**				16 Nonqualified plans			
William O. Little 128 Green Street Atlanta, GA 30305 19b Employee's address and ZIP code			17 See Instrs. for Form W-2				18 Other			
20		21		22 Dependent care benefits				23 Benefits included in Box 10		
24 State income tax **372.00**	25 State wages, tips, etc. **16,000.00**	26 Name of state **GA**	27 Local income tax		28 Local wages, tips, etc.		29 Name of locality			

Copy A For Social Security Administration　　　　　　　　　　　Department of the Treasury—Internal Revenue Service

FIGURE 8-3   Wage and Tax Statement, Form W-2

QUARTERLY PAYROLL RETURNS

There are several payroll reports which must be completed at the end of each quarter. The exact reports vary somewhat from state to state, but there are two types which must be completed in every state: the *Employer's Quarterly Federal Tax Return* (Form 941) and the state *unemployment returns* required under the Federal Insurance Contributions Act.

Employer's Quarterly Federal Tax Return. This form reports the amount of Social Security paid (by both the employer and employee) and the income tax that has been withheld by the company. Figure 8-4 shows a complete Form 941, **Employer's Quarterly Federal Tax Return.**

Notice that lines 2–5 of Form 941 report the total (gross) wages and tips earned during the quarter and the income tax that has been deducted on them. Taxable Social Security wages (line 6a), taxable Social Security tips (line 6b), and taxable Medicare wages and tips (line 7) are listed separately. The amounts on lines 6a and 6b are multiplied times 12.4% to obtain the total Social Security obligation for both the employees and the employer. Likewise, the amount on line 7 is multiplied times 2.9% to obtain the total Medicare obligation. For convenience, the Internal Revenue Service allows taxpayers to round amounts to whole dollars.

Backup withholding (lines 11–13) applies to financial or other institutions that pay interest or dividends to clients. For any clients who fail to provide their taxpayer identification numbers (usually the Social Security numbers) for earnings reports to the Internal Revenue Service, 20% of their interest or dividend must be withheld and submitted with the institution's quarterly report. Then (lines 14–16) the quarter's total taxes are the sum of the total income tax (line 5), plus total FICA (line 10), plus total backup withholding (line 13), less earned income credit, if any (line 15).

The lower portion of Form 941 divides the quarter into its three months. The employer must indicate, for each month, when each payday occurred and the total tax liability (income tax plus FICA, plus any backup withholding) associated with each payroll. The 3-month total on line IV must equal the net taxes computed on line 16 above. The amount previously deposited (as subsequently explained) is entered on line 17. This total should normally equal the total taxes for the quarter (shown on line 16, assuming no adjustments were required). Any taxes still due are computed on line 18 and may be submitted with the quarterly return, if less than $500.

Whenever the Social Security (of both the employee and the employer) plus income tax withheld (plus any backup withholding) during one month exceeds $500, the employer is required to deposit these funds in an authorized commercial or Federal Reserve bank on a monthly basis, rather than waiting until the quarter ends. (Similarly, the taxes must be deposited when the cumulative total for two months exceeds $500.) Most firms with only a few employees will have at least $500 in taxes for any one month and will thus have already deposited all taxes before the 941 Quarterly Return is completed. Large firms whose total taxes exceed $3,000 in less than one month must deposit their

Form 941
(Rev. January 1992)
Department of the Treasury
Internal Revenue Service

4141

Employer's Quarterly Federal Tax Return

► **See Circular E for more information concerning employment tax returns.**

Please type or print.

Your name, address, employer identification number, and calendar quarter of return. (If not correct, please change.)

If address is different from prior return, check here ►

Name (as distinguished from trade name)	Date quarter ended
George R. Brown	Sept, 19XX
Trade name, if any	Employer identification number
Brown's Diner	98-7654321
Address (number and street)	City, state, and ZIP code
4565 Allison Street	
Lexington, KY 40502	

OMB No. 1545-0029
Expires 5-31-93

T	
FF	
FD	
FP	
I	
T	

IRS Use

1 1 1 1 1 1 1 1   2   3 3 3 3 3   4 4 4
5 5 5   6   7   8 8 8 8 8   9 9 9   10 10 10 10 10 10 10 10 10

If you do not have to file returns in the future, check here . ► ☐   Date final wages paid . . ►
If you are a seasonal employer, see **Seasonal employers** on page 2 and check here . ► ☐

1	Number of employees (except household) employed in the pay period that includes March 12th ►	1	7	
2	Total wages and tips subject to withholding, plus other compensation . . . . . ►	2	23,200	00
3	Total income tax withheld from wages, tips, pensions, annuities, sick pay, gambling, etc. ►	3	3,528	00
4	Adjustment of withheld income tax for preceding quarters of calendar year (see instructions) ►	4		
5	Adjusted total of income tax withheld (line 3 as adjusted by line 4—see instructions) . .	5	3,528	00
6a	Taxable social security wages (Complete line 7) $ 22,000 00 . 12.4% (.124) =	6a	2,728	00
b	Taxable social security tips $ 1,200 00 . 12.4% (.124) =	6b	149	00
7	Taxable Medicare wages and tips . . . $ 23,200 00 . 2.9% (.029) =	7	673	00
8	Total social security and Medicare taxes (add lines 6a, 6b, and 7)	8	3,550	00
9	Adjustment of social security and Medicare taxes (see instructions for required explanation) .	9		
10	Adjusted total of social security and Medicare taxes (line 8 as adjusted by line 9—see instructions) . ►	10	3,550	00
11	Backup withholding (see instructions)	11		
12	Adjustment of backup withholding tax for preceding quarters of calendar year	12		
13	Adjusted total of backup withholding (line 11 as adjusted by line 12)	13		
14	**Total taxes** (add lines 5, 10, and 13)	14	7,078	00
15	Advance earned income credit (EIC) payments made to employees, if any . . . . . ►	15		
16	Net taxes (subtract line 15 from line 14). **This should equal line IV below** (plus line IV of Schedule A (Form 941) if you have treated backup withholding as a separate liability) . . .	16	7,078	00
17	**Total deposits for quarter**, including overpayment applied from a prior quarter. from your records . ►	17	7,078	00
18	**Balance due** (subtract line 17 from line 16). This should be less than $500. Pay to Internal Revenue Service . ►	18	-0-	
19	**Overpayment**, if line 17 is more than line 16. enter excess here ► $ _____ and check if to be:			

☐ Applied to next return **OR** ☐ Refunded.

Record of Federal Tax Liability (You must complete if line 16 is **$500** or more and Schedule B is not attached.) See instructions before checking these boxes.
If you made deposits using the 95% rule, check here ► ☐   If you are a first time 3-banking-day depositor, check here . ► ☐

Date wages paid	Show tax liability here. **not deposits.** The IRS gets deposit data from FTD coupons.					
	First month of quarter		Second month of quarter		Third month of quarter	
1st through 3rd	A		I		Q	
4th through 7th	B		J		R	
8th through 11th	C		K		S	
12th through 15th	D	600.00	L	863.00	T	1,250.00
16th through 19th	E		M		U	
20th through 22nd	F		N		V	
23rd through 25th	G		O		W	
26th through the last	H	1,400.00	P	1,500.00	X	1,465.00
Total liability for month	I	2,000.00	II	2,363.00	III	2,715.00

DO NOT Show Federal Tax Deposits Here

► IV Total for quarter (add lines I, II, and III). This should equal line 16 above . . . . . ► | 7,078.00

Sign Here

Under penalties of perjury. I declare that I have examined this return. including accompanying schedules and statements. and to the best of my knowledge and belief. it is true. correct. and complete.

Signature ► *George Brown*   Print Your Name and Title ► President   Date ► Sept. 30, 19XX

Social Security Income Tax

Total Tax

FIGURE 8-4   Employer's Quarterly Federal Tax Return, Form 941

taxes immediately after the $3,000 level is reached, which may occur every few days.* Companies so equipped are also encouraged to use magnetic computer-tape reporting rather than filing a paper Form 941.

Most employee earnings reports include (after the end of each month and each quarter) a total of the worker's cumulative earnings and deductions for the year (see Figure 8-2). Most of the figures required for the quarterly return can thus be obtained simply by computing company totals from these earnings reports—the total gross wages (and tips, if applicable) for line 2, and the total income tax withheld for line 3. During the early quarters, all wages earned are FICA taxable; thus, line 6a will be the same as the company's gross wages.

Example 1   Complete lines 2–18 of the Employer's Quarterly Federal Tax Return for Taylor Appliance Sales, based on the following information: Total wages (all FICA taxable) for the quarter total $12,000, and income tax of $1,872 was withheld. Verify that all taxes have been paid, if monthly deposits of $1,295, $1,218, and $1,195 have been made. (No tips are earned by these employees.)

The completed upper portion (lines 2–18) of Form 941 as shown here indicates that the total taxes due are $3,708. The total monthly deposits for the quarter are $1,295 + $1,218 + $1,195 = $3,708. Thus, all taxes have been deposited before Form 941 is filed.

#	Description	Line	Amount	
2	Total wages and tips subject to withholding, plus other compensation ▸	2	12,000	00
3	Total income tax withheld from wages, tips, pensions, annuities, sick pay, gambling, etc. ▸	3	1,872	00
4	Adjustment of withheld income tax for preceding quarters of calendar year (see instructions) ▸	4		
5	Adjusted total of income tax withheld (line 3 as adjusted by line 4—see instructions)	5	1,872	00
6a	Taxable social security wages (Complete line 7) $ 12,000 00 × 12.4% (.124) =	6a	1,488	00
b	Taxable social security tips $ × 12.4% (.124) =	6b		
7	Taxable Medicare wages and tips $ 12,000 00 × 2.9% (.029) =	7	348	00
8	Total social security and Medicare taxes (add lines 6a, 6b, and 7)	8	1,836	00
9	Adjustment of social security and Medicare taxes (see instructions for required explanation)	9		
10	Adjusted total of social security and Medicare taxes (line 8 as adjusted by line 9—see instructions) ▸	10	1,836	00
11	Backup withholding (see instructions)	11		
12	Adjustment of backup withholding tax for preceding quarters of calendar year	12		
13	Adjusted total of backup withholding (line 11 as adjusted by line 12)	13		
14	**Total taxes** (add lines 5, 10, and 13)	14	3,708	00
15	Advance earned income credit (EIC) payments made to employees, if any ▸	15		
16	Net taxes (subtract line 15 from line 14). **This should equal line IV below** (plus line IV of Schedule A (Form 941) if you have treated backup withholding as a separate liability)	16	3,708	00
17	**Total deposits for quarter**, including overpayment applied from a prior quarter, from your records ▸	17	3,708	00
18	**Balance due** (subtract line 17 from line 16). This should be less than $500. Pay to Internal Revenue Service. ▸	18	0	

Particular attention must be paid, however, to the "taxable Social Security wages" (line 6a of Form 941). Recall that Social Security applies only to the first $61,200 earned; thus, many executives and some highly paid workers will earn wages during the latter quarter(s) that are exempt from Social Security

*A proposal to require these tax deposits no more often than once a month, in order to reduce the accounting expenses imposed on large businesses, has been pending in Congress for years.

deductions. When that occurs, the "taxable Social Security wages" on line 6a will be less than the "total wages and tips" on line 2. All employees' earnings will continue to have Medicare taxes withheld. (All earnings, of course, are subject to income tax deductions.)

For instance, if an executive had earned $50,700 during the first three quarters, then his Social Security wages during the fourth quarter would be computed as follows:

	Maximum Social Security wages	Previous − total wages =	Taxable Social Security wages for quarter
	$61,200	− $50,700 =	$10,500

Thus, only $10,500 of the executive's fourth quarter wages would be included in the firm's taxable Social Security wages on line 6a.

Assume that the executive earns $15,000 during the fourth quarter. Then his Medicare wages for the fourth quarter, $15,000, would be included on line 7.

Example 2 Chris Dean, a sales representative for computer software, earned quarterly wages of $16,000, $16,200, $16,500, and $16,600. How much of her 4th quarter wages were taxable?

The quarterly totals of Dean's earnings were:

$16,000 + $16,200 + $16,500   = $48,700   End of 3rd quarter

$16,000 + $16,200 + $16,500 + $16,600 = $65,300   End of 4th quarter

Since Dean passed the $61,200 base amount for Social Security during the 4th quarter, her taxable FICA earnings were:

$61,200 − $48,700 = $12,500   FICA Taxable wages
for 4th quarter

Therefore, Social Security tax was deducted on $12,500 of Dean's $16,600 4th quarter earnings, and $12,500 is the amount included in the company's Social Security taxable wages on line 6a. However, $16,600 is included in line 7 for Medicare taxable wages, and income tax was also deducted on her entire $16,600, which is included on line 2 for "Total wages subject to (income tax) withholding."

Example 3 Complete lines 2–18 of the employer's quarterly tax return for the small computer firm above. Assume that the *other* employees of the company earned gross wages (all FICA taxable) of $28,200. (See Example 2 for Chris Dean's wages.) Income tax (for everyone) of $6,200 was withheld. Also, determine whether all required taxes have been paid, if deposits were made in the amounts of $4,086, $4,201, and $4,259.

Line 2 of Form 941 contains the "Total (gross) wages subject to (income tax) withholding" for the other employees and Dean ($28,200 + $16,600 = $44,800). Line 6a computes 12.4% of "Taxable Social Security wages" of $28,200 + $12,500 = $40,700. Line 7 computes 2.9% of "Taxable Medicare wages" of $28,200 + $16,600 = $44,800. Since the previous deposits total $4,086 + $4,201 + $4,259 = $12,546, the entire amount due for this quarter has already been paid.

2	Total wages and tips subject to withholding, plus other compensation . . . . ▶			**2**	44,800	00
3	Total income tax withheld from wages, tips, pensions, annuities, sick pay, gambling, etc. . . ▶			**3**	6,200	00
4	Adjustment of withheld income tax for preceding quarters of calendar year (see instructions) . . ▶			**4**		
5	Adjusted total of income tax withheld (line 3 as adjusted by line 4—see instructions)			**5**	6,200	00
6a	Taxable social security wages ⎛**Complete**⎞ $ 40,700	× 12.4% (.124) =	**6a**	5,047	00
b	Taxable social security tips ⎝ **line 7** ⎠ $	× 12.4% (.124) =	**6b**		
7	Taxable Medicare wages and tips . . . . . $ 44,800	00	× 2.9% (.029) =	**7**	1,299	00
8	Total social security and Medicare taxes (add lines 6a, 6b, and 7)			**8**	6,346	00
9	Adjustment of social security and Medicare taxes (see instructions for required explanation) . . .			**9**		
10	Adjusted total of social security and Medicare taxes (line 8 as adjusted by line 9—see instructions) ▶			**10**	6,346	00
11	Backup withholding (see instructions)			**11**		
12	Adjustment of backup withholding tax for preceding quarters of calendar year. . . . . . . ▶			**12**		
13	Adjusted total of backup withholding (line 11 as adjusted by line 12)			**13**		
14	**Total taxes** (add lines 5, 10, and 13)			**14**	12,546	00
15	Advance earned income credit (EIC) payments made to employees, if any . . . . ▶			**15**		
16	Net taxes (subtract line 15 from line 14). **This should equal line IV below** (plus line IV of Schedule A (Form 941) if you have treated backup withholding as a separate liability)			**16**	12,546	00
17	**Total deposits for quarter,** including overpayment applied from a prior quarter, from your records. ▶			**17**	12,546	00
18	**Balance due** (subtract line 17 from line 16). This should be less than $500. Pay to IRS . . . . ▶			**18**	0	

Unemployment Tax Returns. As mentioned earlier, the employer's **federal unemployment tax** (FUTA) was passed as part of the FICA Act. The funds collected under this tax are used to provide limited compensation, for a limited period of time, to workers who are unemployed because there is no work available to them. The federal tax is levied only against employers, with most of the funds being collected by the states; the federal unemployment tax is *not* deducted from workers' earnings. (Only a few states have levied additional unemployment taxes of their own, which may be deducted from the workers' gross earnings.) The portion of unemployment taxes paid to the state must be reported quarterly; the federal return, Form 940, is submitted each January for the preceding calendar year, although deposits may be required quarterly. If the tax liability for any quarter is over $100, deposits must be made to an authorized depository. Since unemployment forms are not uniform from state to state, however, we shall omit any forms and study only the computation of these taxes.

Employers are subject to federal unemployment tax if they employ one or more persons during any 20 weeks of the year or if they paid $1,500 in salaries during any quarter. Employers of certain farmworkers and household workers must also pay. The unemployment tax is similar to the Social Security tax in that it applies only to wages up to a specified maximum—currently the first $7,000 earned by each employee. The FUTA rate for 1995 was as follows:

$$5.4\% \quad \text{Paid to the state}$$
$$\underline{0.8\%} \quad \text{Paid to the federal government}$$
$$6.2\% \quad \text{Total federal unemployment tax}$$

Employers who have a good history of stable employment may have their state rates substantially reduced, while still paying only 0.8% to the federal government, so that their actual rate is well below 6.2%. Conversely, employers with a history of high unemployment may be required to pay substantially more than 5.4% for the state portion of their unemployment insurance. However, 5.4% is the maximum that could be claimed against the 6.2% federal unemployment rate; that is, all employers must pay at least 0.8% to the federal government. Employers in some states must pay more than 0.8% (that is, they cannot claim the full 5.4% to the state) because their state has borrowed from the federal fund and has kept an outstanding balance for at least two years. The following example illustrates the computation of the federal unemployment tax (FUTA).

Example 4   Angela Klein is a salaried part-time employee who earned $2,500 each quarter.

(a)   Determine the amount of Ms. Klein's wages that are subject to unemployment tax each quarter.

The cumulative totals of her earnings after each quarter are:

$$1 \times \$2,500 = \$\ 2,500 \quad \text{1st quarter}$$

$$2 \times \$2,500 = \$\ 5,000 \quad \text{2nd quarter}$$

$$3 \times \$2,500 = \$\ 7,500 \quad \text{3rd quarter}$$

$$4 \times \$2,500 = \$10,000 \quad \text{4th quarter}$$

She had earned $5,000 at the end of the 2nd quarter; thus, her wages passed the $7,000 cutoff point for unemployment tax during the 3rd quarter. In general, for the quarter when the maximum is reached,

$$\frac{\text{Maximum}}{\text{unemployment wages}} - \frac{\text{Previous}}{\text{total wages}} = \frac{\text{Taxable wages}}{\text{for quarter}}$$

$$\$7,000 \quad - \quad \$5,000 \quad = \quad \$2,000$$

Thus, Ms. Klein's employer must pay unemployment tax on $2,000 of the wages she earns during the 3rd quarter. No tax will be due on her 4th quarter earnings.

(b)   Assume that the employer in part (a) has a reduced state unemployment tax rate of 4.0%. Compute the company's state unemployment taxes for each quarter, based on the following wages:

	TAXABLE WAGES FOR UNEMPLOYMENT			
	1ST QUARTER	2ND QUARTER	3RD QUARTER	4TH QUARTER
Other employees	$6,000	$5,300	$3,800	0
Ms. Klein	2,500	2,500	2,000	0
Totals	$8,500	$7,800	$5,800	0

$$\$8,500 \times 4.0\% = \$340 \quad \text{1st quarter}$$

$$7,800 \times 4.0\% {}^- 312 \quad \text{2nd quarter}$$

$$5,800 \times 4.0\% = 232 \quad \text{3rd quarter}$$

No unemployment tax will be owed to the state during the fourth quarter.

(c)   Find the federal unemployment tax due each quarter, assuming the standard federal rate.

$$\$8,500 \times 0.8\% = \$68.00 \quad \text{1st quarter}$$

$$7,800 \times 0.8\% = \$62.40 \quad \text{2nd quarter}$$

$$5,800 \times 0.8\% = \$46.40 \quad \text{3rd quarter}$$

No federal unemployment tax will be due in the 4th quarter.

Note.   The company is not required to make a deposit until the second quarter, when total federal unemployment tax exceeds $100.

SECTION 6   PROBLEMS

Complete lines 2–18 of the Employer's Quarterly Federal Tax Return (Form 941) for Problems 1–6, based on the informaton given. (Assume that only taxable FICA wages are earned, except where noted.) Determine whether the full tax obligation has already been deposited.

2	Total wages and tips subject to withholding, plus other compensation . . . . ▶				**2**	
3	Total income tax withheld from wages, tips, pensions, annuities, sick pay, gambling, etc. . . ▶				**3**	
4	Adjustment of withheld income tax for preceding quarters of calendar year (see instructions) . . ▶				**4**	
5	Adjusted total of income tax withheld (line 3 as adjusted by line 4—see instructions)				**5**	
6a	Taxable social security wages (Complete	$		× 12.4% (.124) =	**6a**	
b	Taxable social security tips line 7)	$		× 12.4% (.124) =	**6b**	
7	Taxable Medicare wages and tips	$		× 2.9% (.029) =	**7**	
8	Total social security and Medicare taxes (add lines 6a, 6b, and 7)				**8**	
9	Adjustment of social security and Medicare taxes (see instructions for required explanation) . . .				**9**	
10	Adjusted total of social security and Medicare taxes (line 8 as adjusted by line 9—see instructions) ▶				**10**	
11	Backup withholding (see instructions)				**11**	
12	Adjustment of backup withholding tax for preceding quarters of calendar year. . . . . . . ▶				**12**	
13	Adjusted total of backup withholding (line 11 as adjusted by line 12)				**13**	
14	**Total taxes** (add lines 5, 10, and 13)				**14**	
15	Advance earned income credit (EIC) payments made to employees, if any . . . . . . ▶				**15**	
16	Net taxes (subtract line 15 from line 14). **This should equal line IV below** (plus line IV of Schedule A (Form 941) if you have treated backup withholding as a separate liability) . .				**16**	
17	**Total deposits for quarter,** including overpayment applied from a prior quarter, from your records. ▶				**17**	
18	**Balance due** (subtract line 17 from line 16). This should be less than $500. Pay to IRS . . . . ▶				**18**	

	TAXABLE FICA WAGES	TAXABLE TIPS	INCOME TAX WITHHELD	MONTHLY DEPOSITS		
				OCT.	NOV.	DEC.
1.	$20,000	—	$5,333	$2,680	$3,563	$2,150
2.	15,000	—	2,250	1,505	1,615	1,425
3.	32,000	$4,000	5,400	4,100	3,250	3,558
4.	44,000	6,000	7,500	4,050	5,500	5,600
5.	15,000[a]	2,000	2,850	1,400	2,100	2,009
6.	26,000[a]	1,500	4,200	2,596	2,800	3,070

[a]Assume that income tax was deducted on an additional $2,000 in earnings that were not Social Security taxable but were Medicare taxable.

Quarterly earnings for three executives are listed in each of the following problems. Assuming that $61,200 is the maximum Social Security taxable wages, determine each person's taxable earnings for each quarter.

		GROSS WAGES EARNED			
	EMPLOYEE	1ST QUARTER	2ND QUARTER	3RD QUARTER	4TH QUARTER
7.	D	$ 6,000	$16,000	$18,000	$18,000
	E	17,500	17,400	17,000	17,500
	F	26,000	27,500	27,000	27,000
8.	G	$19,500	$19,000	$20,000	$20,000
	H	21,000	20,500	18,800	19,500
	I	24,400	26,500	25,600	26,000

The cumulative earnings for the first 3 quarters are listed below. Determine each person's earnings during the 4th quarter that will be subject to Social Security tax deductions and to Medicare deductions.

	EMPLOYEE	CUMULATIVE EARNINGS	4TH QTR. EARNINGS	SOCIAL SECURITY EARNINGS	MEDICARE EARNINGS
9.	4	$ 48,000	$16,000		
	5	60,000	20,000		
	6	120,000	24,000		
10.	7	$ 56,000	$18,000		
	8	45,000	17,000		
	9	80,000	25,000		

Listed in each problem below are all the employees of a small business, along with their total gross earnings and income tax deductions for each quarter.

(a) Determine each employee's FICA taxable wages for each quarter.

(b) Find the firm's total FICA taxable wages for each quarter.

	EMPLOYEE	1ST QUARTER		2ND QUARTER		3RD QUARTER		4TH QUARTER	
		GROSS WAGES	INCOME TAX	GROSS WAGES	INCOME TAX	GROSS WAGES	INCOME TAX	GROSS WAGES	INCOME TAX
11.	L	$ 7,000	$ 980	$ 8,500	$1,190	$ 8,800	$1,232	$ 9,000	$1,260
	M	12,500	1,750	14,000	1,960	14,300	2,002	13,500	1,890
	N	18,400	2,500	18,600	2,600	21,000	2,940	20,000	2,800
	O	20,500	2,870	26,000	3,640	25,000	3,500	25,000	3,500
	Totals								
12.	F	$ 6,800	$ 952	$ 7,400	$1,036	$ 7,500	$1,050	$ 7,900	$1,106
	I	14,000	1,960	14,800	2,072	15,000	2,100	15,300	2,142
	C	17,800	2,492	18,600	2,604	18,700	2,618	18,500	2,590
	A	29,000	4,060	28,500	3,990	28,900	4,046	28,800	4,032
	Totals								

Compute lines 2–18 of the Employer's Quarterly Federal Tax Return (Form 941) for Problems 13–16, given the monthly deposits of withheld income tax and FICA tax.

2	Total wages and tips subject to withholding, plus other compensation . . . . ▶	2
3	Total income tax withheld from wages, tips, pensions, annuities, sick pay, gambling, etc. . . ▶	3
4	Adjustment of withheld income tax for preceding quarters of calendar year (see instructions) . . ▶	4
5	Adjusted total of income tax withheld (line 3 as adjusted by line 4—see instructions)	5
6a	Taxable social security wages **(Complete** $ × 12.4% (.124) =	6a
b	Taxable social security tips **line 7)** $ × 12.4% (.124) =	6b
7	Taxable Medicare wages and tips . . . . . $ × 2.9% (.029) =	7
8	Total social security and Medicare taxes (add lines 6a, 6b, and 7)	8
9	Adjustment of social security and Medicare taxes (see instructions for required explanation) . . .	9
10	Adjusted total of social security and Medicare taxes (line 8 as adjusted by line 9—see instructions) ▶	10
11	Backup withholding (see instructions)	11
12	Adjustment of backup withholding tax for preceding quarters of calendar year. . . . . . . ▶	12
13	Adjusted total of backup withholding (line 11 as adjusted by line 12)	13
14	**Total taxes** (add lines 5, 10, and 13)	14
15	Advance earned income credit (EIC) payments made to employees, if any . . . . ▶	15
16	Net taxes (subtract line 15 from line 14). **This should equal line IV below** (plus line IV of Schedule A (Form 941) if you have treated backup withholding as a separate liability)	16
17	**Total deposits for quarter,** including overpayment applied from a prior quarter, from your records. ▶	17
18	**Balance due** (subtract line 17 from line 16). This should be less than $500. Pay to IRS. . . . ▶	18

			MONTHLY DEPOSITS		
	REFER TO PROBLEM #	QUARTER TO COMPUTE	1ST MO.	2ND MO.	3RD MO.
13.	11	3rd	$6,750	$6,810	$5,409
14.	12	3rd	6,820	5,180	5,415
15.	11	4th	4,850	4,900	4,845
16.	12	4th	5,600	5,000	4,948

17. Refer to the employees in Problem 7.

   a. Determine the taxable wages for unemployment of each worker for each quarter.

   b. Find the total taxable wages for unemployment of the company for each quarter, assuming that these are the only employees.

   c. How much unemployment tax will the company pay to its state each quarter? What are the annual taxes due to the federal government? Assume that the firm has a state rate of 3.6%. (The federal rate is never reduced.)

18. Refer to the employees in Problem 8.

   a. Determine the taxable wages for unemployment of each worker for each quarter.

   b. Determine the total taxable wages for unemployment of the company for each quarter, assuming that these are the only employees.

   c. How much unemployment tax will the company pay to its state each quarter? What are the annual taxes due to the federal government? Assume that the firm has a state rate of 3.6%. (The federal rate is never reduced.)

19. Repeat Problem 17, this time based on the employees in Problem 11. Assume that the company has a reduced state unemployment tax rate of 2.9%.

20. Repeat Problem 18, this time basing your answers on the employees in Problem 12. Assume that the business must pay state unemployment tax at the rate of 3.5%.

21. A professional model has gross earnings of $150,000 for the year. Determine
    a. Her total Social Security tax deduction for the year
    b. Her total Medicare tax deduction for the year
    c. The amount of federal unemployment tax her employer must pay
    d. The amount of state unemployment tax her employer must pay, assuming a state rate of 5.4%.

22. The chief executive officer of a corporation earned $400,000 for the year. Determine
    a. His total Social Security tax deduction for the year
    b. His total Medicare tax deduction for the year
    c. The amount of federal unemployment tax the corporation must pay
    d. The amount of state unemployment tax the corporation must pay, assuming a state rate of 5.4%.

CHAPTER 8 GLOSSARY

Account purchase.   Itemized statement (Prime cost + Charges = Gross cost) for merchandise bought by a commission agent.

Account sales.   Itemized statement (Gross proceeds − Charges = Net proceeds) for the merchandise sold by a commission agent.

Broker.   (See "Commission agent.")

Brokerage.   (See "Commission.")

Chargeback (or docking).   A deduction made from base production or base earnings, because of production work that did not pass inspection.

Commission.   Wages computed as a percent of the value of items sold or purchased; this is also called "straight commission" or "brokerage."

Commission agent.   One who buys or sells merchandise for another party without becoming legal owner of the property; also called a "commission merchant," "broker," or "factor."

Consignee.   Agent to whom a shipment is sent in order to be sold.

Consignor.   Party who sends a shipment to an agent who will sell it.

Docking.   (See "Chargeback.")

Drawing account.   Fund from which a salesperson may receive advance wages to defray business or personal expenses.

Employee's Quarterly Earnings Report.   The summary of an employee's gross wages, itemized deductions, and net pay for each pay period during a quarter.

Employee's Withholding Allowance Certificate (Form W-4).   Federal form on which an employee lists marital status and computes the number of qualified allowances for income tax withholding.

Employer's Quarterly Federal Tax Return (Form 941).   Quarterly form on which a company reports wages and tips earned (taxable and total) and the amount of Social Security, Medicare, and income tax that has been withheld (and matched, for Social Security and Medicare).

Factor.   (See "Commission agent.")

Fair Labor Standards Act.   Federal law requiring the payment of a minimum hourly wage and overtime (at "time-and-a-half" for hours worked in excess of 40 per week) to all employees covered.

Federal Insurance Contributions Act (FICA).   Deduction of a specified percent made on a base amount of each worker's yearly wages (and matched by the employer) to fund benefits for Social Security and Medicare. (See also "Social Security" and "Medicare.")

Federal Unemployment (FUTA) Tax.   Federal tax (paid partially to the state) to provide some compensation to workers who are unemployed because there is no work available; paid by the employer only.

Gross cost.   Total cost, after commission and other charges are added, of merchandise purchased by a commisison agent.

Gross proceeds.   Initial sales value, before commission and other charges are deducted, of merchandise sold by a commission agent.

Gross wages.   Wages earned by an employee before FICA, income tax, or other deductions.

Income tax (federal).   Tax withheld from each employee's pay, as determined by the person's gross earnings, marital status, and number of allowances.

Medicare.   Percentage deduction based on an employee's gross earnings (and matched by the employer) to fund health insurance for persons age 65 or older and eligible for Social Security or Railroad Retirement benefits.

Net proceeds.   Remaining sales value, after commission and other charges are deducted, of merchandise sold by a commission agent.

Net wages.   The actual wages (after all deductions) paid to a worker.

Override.   A small commission on the net sales of all other salespersons, paid to reimburse a supervisor for nonselling duties.

Overtime.   (See "Fair Labor Standards Act.")

Overtime excess.   The extra amount (at one-half the regular rate) that must be paid for work done during overtime hours.

Piecework or piecerate.   (See "Production method.")

Premium rates.   Higher-than-regular rates, paid for overtime work or work that exceeds a quota.

Prime cost.   Initial cost, before commission and other charges are added, of merchandise purchased by a commission agent.

Production method.   Wage computation determined by the number of times a task is completed; also called the "piecework" or "piecerate" method.

Quarterly return.   (See "Employer's Quarterly Federal Tax Return.")

Quota.   Minimum sales or production level required in many jobs (usually in order to qualify for pay at a higher rate).

Sliding-scale commission.   A commission rate that increases, once sales exceed the specified level(s), on all sales above the set level(s).

Social Security.   Percentage deduction based on an employee's gross earnings up to a maximum amount (and matched by the employer) to fund benefits for retired or disabled workers, and for surviving spouses and children.

Straight commission.   (See "Commission.")

Unemployment insurance tax.   (See "Federal Unemployment [FUTA] Tax.")

Wage and Tax Statement (Form W-2).   The form completed by the employer to record the employee's yearly totals of earnings and deductions. The W-2 form is distributed to the employee and to the Internal Revenue Service; it must also be included with the employee's personal income tax return.

Withholding.   Deductions made by employers from the gross earnings of their employees—including required FICA and income tax deductions. (See also "Employee's Withholding Allowance Certificate.")

9

DEPRECIATION AND OVERHEAD

OBJECTIVES

Upon completion of Chapter 9, you will be able to:

1. Define and use correctly the terminology associated with each topic.
2. Compute cost recovery or depreciation (including schedules) by the following five methods (Section 1):
   a. Accelerated cost recovery/MACRS (Examples 1, 2; Problems 1–8)
   b. Straight-line (Example 3; Problems 1, 2, 9–14)
   c. Declining-balance (Example 4; Problems 1, 2, 15–20)
   d. Sum-of-the-digits (Example 5; Problems 21, 22)
   e. Units-of-production (Example 6; Problems 23, 24).
3. Prorate depreciation for partial years by the following methods (Section 2):
   a. Straight-line (Example 1; Problems 1–4)
   b. Declining-balance (Example 2; Problems 1, 2, 5, 6)
   c. Sum-of-the-digits (Example 3; Problems 1, 2, 7, 8).
4. Compute overhead as a ratio of (Section 3):
   a. Floor space per department (Example 1; Problems 1, 2, 7, 8)
   b. Net sales (Example 2; Problems 3, 4, 9, 10)
   c. Employees per department (Example 3; Problems 5, 6, 11, 12).

When expenses are calculated for a business, two of the most important items are depreciation and overhead. To determine these expenses, you will have to learn certain mathematical techniques. The procedures you will need to know are explained in the following sections.

SECTION *1*
DEPRECIATION

The **plant assets** of a business are its buildings, machinery, equipment, land, and similar tangible properties that will be used for more than one year. An older term for plant asset is **fixed asset**. All plant assets except land should be depreciated. In business, **depreciation** is the systematic allocation of the asset's cost over its period of economic utility (its usefulness). The allocation of cost is necessary because the asset loses its economic utility due to such things as technological obsolescence, functional inadequacy, and physical deterioration.

Because a plant asset will be used for several years, the Internal Revenue Service does not allow a business to list the entire business cost of the asset as an expense during the year in which it was obtained. However, a business is permitted to recover the cost of a plant asset over a prescribed period of years. Thus, the cost recovery (or depreciation) currently claimed does not normally represent money actually spent during the present year; rather, it represents this year's share of a larger expenditure that occurred during some previous year. However, this annual calculation is considered an operating expense (just as are salaries, rent, insurance, and so forth) and is deducted from business profits when determining the year's taxable income. Because cost recovery (or depreciation) is a tax-deductible item on business income tax returns, the Internal Revenue Service regulates carefully the conditions under which it may be computed.

The Economic Recovery Tax Act of 1981 made enormous changes in both the concept of depreciation and the methods of calculation. Previously, the amount of depreciation charged to an asset each year would approximately equal the amount by which its value decreased. The concept of depreciation thus reflected a "using up" of the asset. Moreover, at no time was a business allowed to depreciate an asset below a reasonable estimate of its actual market value.

Under the Economic Recovery Act, "depreciation" was replaced by the concept of **cost recovery:** Assets were placed into designated categories, and each year a business was entitled to recover a certain percentage of the original cost. Eventually, the entire cost would be claimed as an operating expense. Cost recovery thus has little connection with the actual value that an asset may have or the actual length of its useful life. Also, no distinction is made between new or used property.

The 1981 system, called the **accelerated cost recovery system** (ACRS), greatly simplified previous methods of tax depreciation. All property was placed into a cost-recovery class of either 3, 5, 10, or 15 years (plus later adjustments to 19 years for real estate). All property of the same class acquired during one year was grouped together, and the ACRS tables stipulated the percent to be multiplied times the total cost of the class during each year of the recovery period. This ACRS method remains in effect for assets purchased and placed in service during the years 1981 through 1986.

MODIFIED ACRS METHOD

The Tax Reform Act of 1986 modified the ACRS method considerably for assets placed into service in 1987 or after. Additional cost-recovery classes were created, and percentage calculations were replaced by other prescribed methods of cost recovery. As before, there is no distinction between the purchase of new versus used assets, for cost-recovery purposes.

The Act adds a 7-year class for personal (tangible) property, as well as three classes of 20 years or more for real estate and utilities. A primary difference is that most assets are shifted from their previous class to the next longer class; thus, more years are now required for a business to recover its investment cost. The 3-year class is limited primarily to specialized tools for industry. Cars, trucks (light or heavy), computers, research and experimental equipment, construction assets, and certain manufacturing assets compose the 5-year class. The new 7-year class contains office furniture, fixtures, and equipment, as well as most manufacturing equipment; this category also covers *any assets not specifically assigned elsewhere*. The longer 10-, 15-, and 20-year classes include heavy manufacturing and utility properties, as well as certain other specified assets. Residential rental property composes a 27.5-year class; all other real property falls into a 31.5-year class.

The 3-, 5-, 7- and 10-year classes are depreciated using a 200% declining-balance method, switching to the straight-line method about halfway through the recovery period. The 15- and 20-year classes use a 150% declining-balance method, with a later switch to straight-line. Both the 27.5- and 31.5-year classes for real estate use straight-line depreciation.

The straight-line method and the declining-balance method of depreciation are both long-established accounting procedures. Each of these methods is described in detail later in this section. However, since their application in the accelerated cost recovery systems (ACRS and MACRS) requires advanced knowledge, an asset's cost recovery is often computed using a table similar to Table C-13 in Appendix C and Table 9-1, which contain multiplicative factors for each year during the cost-recovery period. Although the 1986 Tax Reform Act does not specifically include a table, most accounting firms use similar tables for convenience.

The 1986 Act maintains many of the previous characteristics of the ACRS method: All property of the same class acquired during the year is grouped together. You can look under the corresponding class of the MACRS table to obtain an appropriate factor to multiply times the original cost (see Table 9-1). The smaller first-year factors assume that the various assets were acquired

TABLE 9-1 MACRS COST RECOVERY FACTORS

For Property Placed into Service in 1987 and Thereafter

RECOVERY YEAR	3-YR. CLASS	5-YR. CLASS	7-YR. CLASS	10-YR. CLASS
1	0.333333	0.200000	0.142857	0.100000
2	0.444444	0.320000	0.244898	0.180000
3	0.148148	0.192000	0.174927	0.144000
4	0.074074	0.115200	0.124948	0.115200
5	—	0.115200	0.089249	0.092160
6	—	0.057600	0.089249	0.073728
7	—	—	0.089249	0.065536
8	—	—	0.044624	0.065536
9	—	—	—	0.065536
10	—	—	—	0.065536
11	—	—	—	0.032768

RECOVERY YEAR(S)	15-YR. CLASS	20-YR. CLASS	27.5-YR. REAL ESTATE	31.5-YR. REAL ESTATE
1	0.050000	0.037500	0.034848	0.030423
2	0.095000	0.072188	0.036364	0.031746
3	0.085500	0.066773	0.036364	0.031746
4	0.076950	0.061765	0.036364	0.031746
5	0.069255	0.057133	0.036364	0.031746
6	0.062330	0.052848	0.036364	0.031746
7	0.059049	0.048884	0.036364	0.031746
8	0.059049	0.045218	0.036364	0.031746
9–15	0.059049	0.044615	0.036364	0.031746
16	0.029525	0.044615	0.036364	0.031746
17–20	—	0.044615	0.036364	0.031746
21	—	0.022308	0.036364	0.031746
22–27	—	—	0.036364	0.031746
28	—	—	0.019697	0.031746
29–31	—	—	—	0.031746
32	—	—	—	0.017196

throughout the year and thus average a half-year of service. (In fact, if 40% or more of all assets are acquired during the last quarter of the business year, the firm must use a different method that depreciates them according to the quarter in which they were acquired.) Over all the years in each class, 100% of the purchase cost is recovered.

Note. Cost recovery or depreciation may never be claimed on land and may not ordinarily be claimed by an individual on personal property. However, some plant assets, particularly automobiles, are frequently used for both business and pleasure. Careful records must be kept of the amount of business use, and a corresponding portion of standard cost recovery may then be claimed by the owner (unless IRS-approved reimbursement has been paid for the business use).

Example 1   New processing equipment (7-year class) cost Nu-way Cleaners $14,000 during 1994. Determine the amount of cost recovery the cleaners can claim each year.

Using the MACRS table, the years of use appear in the left-hand column, and the corresponding factor for each year is found on the same line in the column entitled "7-yr. class." Each year's cost recovery amount is then found by multiplication.

Year	Cost × Factor =	Cost Recovery
1	$14,000 × 0.142857 =	$ 2,000.00
2	14,000 × 0.244898 =	3,428.57
3	14,000 × 0.174927 =	2,448.98
4	14,000 × 0.124948 =	1,749.27
5	14,000 × 0.089249 =	1,249.49
6	14,000 × 0.089249 =	1,249.49
7	14,000 × 0.089249 =	1,249.49
8	14,000 × 0.044624 =	624.74
	Total	$14,000.03

Thus, the total $14,000 cost is recovered over 7 years although the equipment will probably be used for much longer. The MACRS method is called "accelerated" because owners can recover their investment cost in fewer years than the asset's expected useful life.

A *cost-recovery schedule* (or *depreciation schedule*) is usually kept for every asset class acquired each year. This is a record showing the amount of cost recovered each year, the total amount recovered to date, and the amount as yet unrecovered. This remaining portion of the cost may also be referred to as the **book value**. Book value is used in calculating the value of a company's capital assets for the year. (Under the MACRS method, however, this book value is considerably less than the actual market value of the assets.)

Example 2   Nu-way Cleaners wishes to prepare a cost-recovery schedule for its 7-year property shown in Example 1.

The cost-recovery schedule is completed as follows:

1. A year "0" is marked down to indicate when the property was new, and the original cost is entered under "book value." Since no cost recovery has yet occurred when the property is new, no other entries are made on line 0.

2. Each year the "annual recovery" is entered.

3. The "accumulated recovery" is found each year by adding the current year's recovery amount to the previous year's accumulated recovery.

4. The "book value" is determined by subtracting the current year's (annual) recovery from the previous year's book value.

For practical purposes, most accountants round dollar amounts to whole dollars because depreciation and cost recovery are only estimations. We will follow this practice in the examples and problems in this chapter.

COST-RECOVERY SCHEDULE
MACRS Method

YEAR	BOOK VALUE (END OF YEAR)	ANNUAL RECOVERY	ACCUMULATED COST RECOVERY
0	$14,000	—	—
1	12,000	$2,000	$ 2,000
2	8,571	3,429	5,429
3	6,122	2,449	7,878
4	4,373	1,749	9,627
5	3,124	1,249	10,876
6	1,875	1,249	12,125
7	626	1,249	13,374
8	0	626	14,000

Note. Observe that the final book value should be zero and that the final accumulated cost recovery should equal the cost of the property. That is the reason the annual cost recovery in year 8 is $626 instead of the $625 as shown in Example 1. An adjustment was made in that year so that the annual cost recovery equals the book value at the beginning of that year. The $1 difference is a rounding error. Also observe that a mistake at any point in a recovery schedule will cause the succeeding entries to be incorrect, so you should also use the following checks:

(a) For any year, the sum of all the annual amounts to that point should equal that year's accumulated cost recovery. Example:

Year 3: $2,000 + $3,429 + $2,449 = $7,878

(b) During any year, the book value plus the accumulated cost recovery must equal the original cost. Example:

Year 4: $4,373 + $9,627 = $14,000

Calculator techniques . . .   FOR EXAMPLE 2

For the MACRS method, you should calculate the entire Annual Recovery column first, using the factors as shown in Example 1. Then complete the Accumulated Cost Recovery column by adding the annual amounts. As you press $\boxed{+}$ preceding the third value (and for every year thereafter), notice that the current subtotal briefly appears; this provides the corresponding year's accumulated value, which you can fill in before proceeding:

$$2,000 \boxed{+} 3,429 \boxed{+} \qquad [\longrightarrow 5,429]$$
$$2,449 \boxed{+} \qquad [\longrightarrow 7,878]$$
$$1,749 \boxed{+} \qquad [\longrightarrow 9,627] \text{ and so on.}$$

Similarly, the Book Value column can be completed by successively subtracting the annual recovery amounts from the cost. The current subtotal appears as you press $\boxed{-}$, prior to the second and all subsequent annual recovery amounts:

$$14,000 \boxed{-} 2,000 \boxed{-} \qquad [\longrightarrow 12,000]$$
$$3,429 \boxed{-} \qquad [\longrightarrow 8,571]$$
$$2,449 \boxed{-} \qquad [\longrightarrow 6,122] \text{ and so on.}$$

Also check that the book value plus accumulated cost recovery for any year equals the original cost, as illustrated in part (b) of the Note preceding this Calculator Technique. The calculator procedures described here for an MACRS cost-recovery schedule also apply for depreciation schedules under the straight-line method, described next.

The ACRS and MACRS methods of cost recovery were established by Congress to provide incentives to businesses to make capital investments. With 100% of asset cost being recoverable—and in less time than previously allowed—the extra depreciation claimed each year means that companies have higher tax deductions and thus more after-tax funds. This was seen as a way to stimulate widespread sales and production and thereby to improve the country's economic status.

Because depreciation accounting has been a standard business procedure for so long, calculations under both ACRS and MACRS systems of cost recovery continue to be referred to as "depreciation" in common practice. It must be emphasized that in most cases the ACRS and MACRS methods apply *only* for tax accounting purposes, and a business may use any of the single, traditional depreciation methods for its own internal accounting. Thus, presentations follow for the traditional depreciation methods used: *straight-line method, declining-balance method, sum-of-the-digits method*, and *units-of-production method*. As previously noted, the ACRS and MACRS method are, in fact, a combination of the straight-line and declining-balance methods. Furthermore, the straight-line method is acceptable for tax purposes in many cases as an alternative to the ACRS and MACRS methods, and it used entirely for real estate (in a variation which assumes that usage began at midmonth.)

QUICK PRACTICE

Compute the annual cost recovery for each of the following using the MACRS method and the data given. Construct a cost-recovery schedule for the (c) parts only.

	COST	CLASS/YEARS			COST	CLASS/YEARS
1. a.	$ 850	3		**2.** a.	$3,000	3
b.	2,400	5		b.	5,800	7
c.	9,200	10		c.	4,500	5

STRAIGHT-LINE METHOD

The **straight-line method** is the simplest method for computing depreciation. By this method, the total allowable depreciation is divided evenly among all the years. That is, the same amount of depreciation is charged each full year.

The straight-line method of depreciation is used during the latter years for MACRS calculations on personal property and during all years for the real estate classes. Straight-line depreciation is also permitted as an alternative to the MACRS cost-recovery method, using the regular recovery period or optional longer recovery periods. However, most businesses select the MACRS method for tax purposes, since their asset costs are deducted from taxable income more quickly using MACRS.

Use of the straight-line method requires determining a depreciable value—the total amount to be depreciated. Recall that under the MACRS method, the entire cost may be recovered. For internal cost accounting purposes, however, a business may use the historical method, which first requires the estimation of the asset's **residual value.** The residual value is the expected value of the asset at the end of its useful life. The residual value is often called the asset's **trade-in value** (the expected amount to be received if the asset is traded), **resale value** (the expected amount to be received if the asset is sold), or **scrap** or **salvage value** (the expected amount if the asset is junked). The asset's residual value is subtracted from the asset's cost basis to determine the **depreciable value** (sometimes called the **wearing value**). The depreciable value is the maximum total depreciation expense that can be written off on the asset over its useful life.

Example 3   A machine that manufactures keys is purchased for $5,100. It is expected to last for 5 years and have a trade-in value of $900. Prepare a depreciation schedule by the straight-line method.

First, the depreciable value must be computed and divided by the years of use to determine each year's depreciation charge:

Original cost	$5,100
Trade-in value	− 900
Depreciable value	$4,200

$$\frac{\text{Depreciable value}}{\text{Years}} = \frac{\$4,200}{5}$$

$$= \$840 \quad \text{Annual depreciation}$$

A depreciation of $840 would thus be claimed each year for 5 years, by the traditional straight-line method. A schedule for the foregoing depreciation (or cost recovery) is completed below as described in Example 2.

DEPRECIATION SCHEDULE
Straight-Line Method

YEAR	BOOK VALUE (END OF YEAR)	ANNUAL DEPRECIATION	ACCUMULATED DEPRECIATION
0	$5,100	—	—
1	4,260	$840	$ 840
2	3,420	840	1,680
3	2,580	840	2,520
4	1,740	840	3,360
5	900	840	4,200

Residual value   Depreciable value

If the straight-line depreciation schedule is done correctly, a check for its accuracy can be done visually very quickly. The last year's book value will equal the residual value, and the last year's total accumulated depreciation will equal the depreciable value.

QUICK PRACTICE

Compute the annual depreciation by the straight-line method using the data given. Construct depreciation schedules for the (c) parts only.

		COST	RESIDUAL VALUE	USEFUL LIFE (YEARS)
1.	a.	$80,000	$5,000	30
	b.	5,300	400	7
	c.	2,200	700	5

		COST	RESIDUAL VALUE	USEFUL LIFE (YEARS)
2.	a.	$7,500	$300	6
	b.	5,800	400	9
	c.	2,900	700	4

DECLINING-BALANCE METHOD

Another method frequently used for computing depreciation is the **declining-balance method** (sometimes called the *constant percent method*). This method also offers the advantage of greatest annual depreciation during the early years, progressively declining as time passes. This is the method used for MACRS calculations (for the non–real estate classes) during the early years of cost recovery.

This method is often referred to as "double declining balance" because the rate most often used is twice the straight-line rate. For example, the straight-line rate for an asset used 6 years is $\frac{1}{6}$ each year. The typical declining-balance rate (percent) for the same 6 years is thus

$$2 \times \frac{1}{6} = \frac{2}{6} = \frac{1}{3} \qquad \text{or} \qquad 33\frac{1}{3}\% \text{ annually}$$

Stated differently, the usual declining-balance rate is twice the reciprocal of the number of years. (This is normally the maximum rate; a lower rate may be used for internal accounting.) Observe that this is the method initially used for MACRS calculations. Because the method uses a factor of 2, it is also often called the "200% declining-balance" method.

When the declining-balance method is used, depreciation is computed using the same rate (percent) each year. For the *first year*, this *constant rate* is multipled times the *cost* (not the depreciable value) to determine the first year's depreciation. This amount is subtracted from the cost to yield the book value at the end of the first year. For each succeeding year, the book value at the end of the previous year is multiplied by the constant rate. This method gets its name from the procedure of applying a constant rate to a decreasing book value (declining balance) each year.

If there is a residual value, caution must be used to ensure that the asset is not depreciated beyond its residual value. This often requires adjusting the final annual depreciation amount so that the ending book value equals the residual value.

Example 4   (a) A new vending machine that cost $4,800 has an expected residual value of $600 and a life of 5 years. Prepare a depreciation schedule by the declining-balance method.

Because the machine will be used for 5 years, it can be depreciated at a rate of

$$2 \times \frac{1}{5} = \frac{2}{5} = 40\% \text{ annually}$$

Each succeeding book value will be multiplied by 40% (or 0.4) to find the next annual depreciation. The following depreciation schedule shows dollar amounts rounded to whole dollars.

DEPRECIATION SCHEDULE
Declining-Balance Method

YEAR	BOOK VALUE (END OF YEAR)	ANNUAL DEPRECIATION	ACCUMULATED DEPRECIATION
0	$4,800	—	—
1	2,880	$1,920	$1,920
2	1,728	1,152	3,072
3	1,037	691	3,763
4	622	415	4,178
5	600	22[a]	4,200

[a] This is an adjusted final depreciation value, not a declining-balance calculation.

Notice that an adjustment was made to the annual depreciation in the fifth year so that the end-of-year book value would equal the residual value, $600. An adjustment may often be necessary before the final year. As the estimated residual value increases in amount, the probability of the adjustment to annual depreciation before the final year increases. For example, assume the residual value is $800. The book value at the end of the third year is $1,037, so the fourth-year annual depreciation would be $237 (the difference between the $1,037 and $800), and the book value at the end of that year would be $800. No further depreciation would be taken. This is illustrated in the following example.

Example 4 (cont.)   (b) Cost = $4,800
Residual value = $800
Life = 5 years

YEAR	BOOK VALUE (END OF YEAR)	ANNUAL DEPRECIATION	ACCUMULATED DEPRECIATION
0	$4,800	—	—
1	2,880	$1,920	$1,920
2	1,728	1,152	3,072
3	1,037	691	3,763
4	800	237	4,000
5	—	—	—

If the same business assumed a residual value of zero, then the adjustment would be made in the fifth year as shown in Example 4(c).

Example 4
(cont.)

(c) Cost = $4,800
Residual value = 0
Life = 5 years

Year	Book value (end of year)	Annual depreciation	Accumulated depreciation
0	$4,800	—	—
1	2,880	$1,920	$1,920
2	1,728	1,152	3,072
3	1,037	691	3,763
4	622	415	4,178
5	—	622	4,800

Calculator techniques . . . FOR EXAMPLE 4

For declining-balance depreciation, each year's Annual Depreciation is based on the preceeding year's Book Value; thus, those two columns must be computed simultaneously. Keep in mind that parentheses here denote displayed values that are used without reentering them.

Note. If your calculator has a decimal control key, place the control on 0 so that the amounts will be rounded to whole dollars. If your calculator does not have this feature, you must round the number in your head, record the rounded value in the annual column, and (without pressing any operation key) also enter the rounded value into the computer display; then press ⌷M−⌷ to deduct the rounded value. Observe that the book value at the end of the fourth year is $622.08 or $622, and also that the desired residual value is $600, so we stop the calculations at this point. (If the annual value computes to even dollars, you press ⌷M−⌷ directly, without entering the annual value.)

4,800 ⌷M+⌷

(4,800) ⌷×⌷ 0.4 ⌷=⌷ ⟶ 1,920 ⌷M−⌷;   ⌷MR⌷ ⟶ 2,880

(2,880) ⌷×⌷ 0.4 ⌷=⌷ ⟶ 1,152 ⌷M−⌷;   ⌷MR⌷ ⟶ 1,728

(1,728) ⌷×⌷ 0.4 ⌷=⌷ ⟶ 691.2; 691 ⌷M−⌷;   ⌷MR⌷ ⟶ 1,037

(1,037) ⌷×⌷ 0.4 ⌷=⌷ ⟶ 414.8; 415 ⌷M−⌷;   ⌷MR⌷ ⟶ 622

The Accumulated Depreciation column is then computed as described previously.

From a commercial standpoint, most machinery actually undergoes greater depreciation during its early years and less depreciation during later years.

The declining-balance method of depreciation is thus more "realistic" than the straight-line method, even for internal accounting purposes. However, when listing the net worth of assets on its balance sheet, a business may want them to show the greatest possible value. In that case, the straight-line method would be preferred, to depreciate them more slowly.

A characteristic of declining-balance depreciation is that the book value never reaches zero in the natural course of events. It must be forced to do so by the accountant. Otherwise, the process can go on for many years before it even approaches zero. By contrast, companies using the MACRS method for tax purposes want to recover their full cost as soon as possible. However, it works out mathematically that, about halfway through the recovery period, if the remaining book value is depreciated *evenly* among all the remaining years, then the annual depreciation thereafter will actually be larger than if the company had continued using the declining-balance method. This is the basis for the MACRS method, which uses 200% declining-balance for the first several years, switching to straight-line at a point that maximizes recovery.

QUICK PRACTICE

Compute the annual depreciation and construct a depreciation schedule by the declining-balance method using the data given for each part.

	COST	RESIDUAL VALUE	USEFUL LIFE (YEARS)		COST	RESIDUAL VALUE	USEFUL LIFE (YEARS)
1. a.	$2,250	$ 0	5	2. a.	$1,900	$ 0	5
b.	3,000	0	6	b.	800	0	4
c.	1,200	50	3	c.	1,800	150	6

SUM-OF-THE-DIGITS METHOD

Another traditional depreciation method is the **sum-of-the-digits method.** This method, like the MACRS and declining-balance methods, allows more depreciation to be claimed during the earlier years of the schedule.

Example 5   Prepare a depreciation schedule for Example 3 using the sum-of-the-digits method. (A $5,100 machine that molds keys was expected to be worth $900 after 5 years.)

The sum-of-the-digits method derives its name from step 1 in the following procedure:

1. The digits comprising the life expectancy of the asset are added. Thus, since the life expectancy is 5 years,

$$1 + 2 + 3 + 4 + 5 = 15*$$

2. This sum becomes the denominator of a series of fractions. The years' digits, taken in reverse order, are the numerators of the fractions.

3. The fractions derived in step 2 are then multiplied times the depreciable value to compute each year's depreciation:

Depreciable value = $5,100 − $900 = $4,200

YEAR	ANNUAL DEPRECIATION
1	$\frac{5}{15} \times \$4,200 = \$1,400$
2	$\frac{4}{15} \times 4,200 = 1,120$
3	$\frac{3}{15} \times 4,200 = 840$
4	$\frac{2}{15} \times 4,200 = 560$
5	$\frac{1}{15} \times 4,200 = 280$

A depreciation schedule could then be completed as previously illustrated.

QUICK PRACTICE

Compute annual depreciation by the sum-of-the-digits method using the data given. Construct depreciation schedules for the (c) parts only.

	COST	RESIDUAL VALUE	USEFUL LIFE (YEARS)		COST	RESIDUAL VALUE	USEFUL LIFE (YEARS)
1. a.	$1,400	$200	4	**2.** a.	$2,000	$320	6
b.	3,120	600	8	b.	1,700	200	3
c.	5,000	500	9	c.	2,100	700	7

*A short-cut method for determining the sum of the digits is to use the formula $\frac{n(n + 1)}{2}$, where the symbol n represents the number of years. In Example 5, substituting 5 for n yields 15 as before:

$$\frac{5(5 + 1)}{2} = 15$$

UNITS-OF-PRODUCTION METHOD

Another historical depreciation method, the **units-of-production method,** is applicable for machinery used in manufacturing distinct products. By this method, the total depreciable value of the machine is first divided by the total number of products the machine is expected to produce during its useful life, in order to obtain a per-unit depreciation charge. The annual depreciation is then computed as each year's total number of units times the per-unit depreciation rate. A depreciation schedule would then be completed as for other methods.

Example 6 Use the units-of-production method to determine annual depreciation for the key machine in Example 3. (A $5,100 machine was expected to be worth $900 after 5 years.) The number of units (keys) produced annually is given below. The machine is expected to produce 560,000 keys during its lifetime.

The depreciable value ($5,100 − $900 = $4,200) is divided by the expected 560,000 units to obtain the per-unit depreciation:

$$\frac{\text{Depreciable value}}{\text{Total units}} = \frac{\$4{,}200}{560{,}000} = \$0.0075 \text{ per-unit depreciation}$$

YEAR	ANNUAL UNITS	×	UNIT RATE	=	ANNUAL DEPRECIATION
1	120,000	×	$0.0075	=	$ 900
2	160,000	×	0.0075	=	1,200
3	40,000	×	0.0075	=	300
4	80,000	×	0.0075	=	600
5	160,000	×	0.0075	=	1,200
Totals	560,000				$4,200

Although the units-of-production method is convenient mathematically, it has not been recommended by accountants. It is difficult to estimate in advance how many units a machine will produce, thus the depreciated value may be too high or too low. If the machine becomes obsolete and is taken out of service, it would then appear not to be depreciating—an extremely unrealistic situation. By contrast, the units-of-production method can cause more depreciation in the latter years (as shown above), which reverses the pattern of actual economic depreciation.

QUICK PRACTICE

Compute the annual depreciation by the units-of-production method using the data given.

	COST	RESIDUAL VALUE	USEFUL LIFE (UNITS)	ANNUAL PRODUCTION
1. a.	$3,100	$600	50,000	5,000; 10,000; 12,000; 19,000; 4,000
b.	8,800	500	100,000	20,000; 40,000; 30,000; 10,000
c.	4,900	900	25,000	3,000; 8,000; 5,000; 9,000
2. a.	$7,500	$300	90,000	18,000; 20,000; 25,000; 19,000; 8,000
b.	8,100	600	30,000	10,000; 8,000; 7,000; 5,000
c.	2,600	200	200,000	35,000; 44,000; 40,000; 33,000; 28,000; 20,000

REVIEW

Stated simply, every depreciation method involves multiplying a depreciation *rate times a base value* to find an annual amount of depreciation. The methods differ in how to determine the depreciation rate (most use the same rate every year; one doesn't) and in how to determine the base value to which the rate will be applied (here again, some methods use the same base value every year; others don't). In review, let's look at the various depreciation formulas and see how the methods differ. The following review assumes full years.

Straight-Line Method

Under the straight-line method, a constant rate is multiplied times the depreciable value each year. The depreciable value is the difference between the asset's cost and its residual value. The annual depreciation will be the same amount each year.

$$\frac{\text{Cost} - \text{Residual value}}{\text{Years}} = \text{Annual depreciation}$$

Note from the preceding formula that dividing by years is the same as multiplying the depreciable value by $\frac{1}{\text{Years}}$.

Declining-Balance Method

The declining-balance method also uses a constant rate each year—typically twice the straight-line rate. This rate is multiplied times the full cost the first year (not times the depreciable value). For each succeeding year, the constant rate is multiplied times the preceding year's book value.

1. $2 \times$ (straight-line rate)   = Rate
2. Rate $\times$ cost   = First year's depreciation
3. Cost − depreciation   = Book value
   (from previous step)

4.   Rate × book value   = Subsequent year's depreciation

5.   Book value − depreciation = New book value

6.   Continue steps 4 and 5.

Sum-of-the-Digits Method

For the sum-of-the-digits method, first determine the denominator of the fraction (rate) either by adding the consecutive digits for the number of years or by using a short-cut formula. The numerators of the fractions will be the years' digits in reverse. Multiply each year's fraction times the depreciable value to determine the annual depreciation.

1. $\dfrac{n(n + 1)}{2}$ = Denominator

2.   Reverse the years' digits for numerators

3.   Each year's fraction × Depreciable value = Annual depreciation

Units-of-Production Method

The units-of-production method is very similar to the straight-line method. First determine the depreciable value, and then divide by the total number of units to be produced. This yields a rate per unit. The rate per unit is multiplied times each year's production to give the annual depreciation each year.

1. $\dfrac{\text{Cost} - \text{Residual value}}{\text{Total number of units}}$ = Rate

2.   Rate × each year's production = Annual depreciation

Note.   Although many of the problems in this section describe a single asset, remember as you work these problems that the MACRS method requires all assets of the same class that are purchased during one year to be computed as a group.

SECTION 1   PROBLEMS

Compute the annual cost recovery (depreciation) and complete a schedule for each of the following, using the data and method indicated. Round answers to whole dollars.

	COST	SCRAP VALUE	CLASS/ YEARS	COST-RECOVERY OR DEPRECIATION METHOD
1. a.	$8,100	$ 0	5	MACRS
b.	7,800	0	7	MACRS
c.	5,300	500	8	Straight-line
d.	9,900	800	6	Declining-balance
2. a.	$1,400	$ 0	3	MACRS
b.	3,600	0	5	MACRS
c.	6,000	400	7	Straight-line
d.	4,000	200	4	Declining-balance

3. In 1996, Creative Construction Co. purchased $8,400 of equipment. Prepare a cost-recovery schedule by the MACRS method for this 5-year-class equipment.

4. HHI Indutries purchased $6,200 of equipment in 1995. Prepare a cost-recovery schedule by the MACRS method for this 5-year-class equipment.

5. Ashley Realty purchased an apartment complex for $1,200,000 in 1996. Using the 27.5-year class and the MACRS method, determine the cost recovery for the first 3 years only.

6. Afshar Investments purchased an office building for $800,000. Using the 31.5-year class, determine the cost recovery for the first 3 years only by the MACRS method.

7. McLeod Realty purchased a warehouse for $500,000 and an apartment building for $900,000 in 1996. The warehouse is classified as 31.5-year property, and the apartment building is classified as 27.5-year property under MACRS. How much cost recovery will their tax return include each year during the first 3 years? Hint: Compute each class separately and sum the amounts for each year.

8. Charleston Manufacturing Co. purchased a warehouse for $600,000 in the 27.5-year class and shop equipment for $25,000 in the 7-year class under MACRS. Determine the cost recovery for tax purposes each year during the first 3 years.

9. Chambers Advertising Co. purchased a computer for $1,450. The company plans to use the computer for 4 years. Construct a depreciation schedule using the straight-line method, assuming that the trade-in value will be $350.

10. DDG purchased office equipment costing $6,800. The useful life will be 7 years. The residual value is $500. Construct a depreciation schedule using the straight-line method.

11. Determine the annual depreciation by the straight-line method for office desks and chairs costing $5,200, with a salvage value of $500 and a useful life of 10 years.

12. State Supply Store purchased equipment for $1,200. The salvage value was estimated to be $300, and the useful life was 6 years. Determine the annual depreciation by the straight-line method.

13. The Children's Clinic paid $6,300 for its waiting room furnishings. The furniture will be worth $800 after 5 years of use. Using the straight-line method, determine the annual depreciation for these furnishings.

14. The Mason-Dixon Restaurant paid $38,000 for new kitchen equipment. The equipment will be worth $6,000 after 8 years of use. What is the annual depreciation expense by the straight-line method?

15. New personal computers cost an accounting firm $10,000. For financial reporting, the firm has chosen the declining-balance method for depreciating these computers. The life is estimated to be 5 years, and the trade-in value at that time will be $500. Construct a depreciation schedule.

16. Using the declining-balance method, construct a depreciation schedule for office furniture costing $12,000. After the estimated 6-year life, the furniture should have a residual value of $800.

17. TV station XYZ paid $7,290 for a new video camera, which should remain serviceable for 6 years, at which time the residual value should be $400. Make a depreciation schedule by the declining-balance method.

18. Factory equipment that cost $6,000 will be obsolete after 4 years of use. The scrap value should be zero. Determine the annual depreciation by the declining-balance method and construct a depreciation schedule.

19. A photo-processing machine cost Color Corp. $5,000. The machine will be in use for 3 years and have a residual value of $200. Make a depreciation schedule using the declining-balance method at twice the straight-line rate.

20. The Aerials Gymnastics Center spent $800 on a new balance beam. The useful life of the beam is 8 years, at which time the Center wants to sell it for $100. Construct a depreciation schedule using the declining-balance method.

Use the sum-of-the-digits method to calculate the annual depreciation for each asset.

	COST	SCRAP VALUE	YEARS			COST	SCRAP VALUE	YEARS
21. a.	$12,000	$3,000	5	**22.** a.		$2,000	$ 320	6
b.	3,200	400	7	b.		4,900	400	5
c.	18,000	3,600	4	c.		5,600	200	9
d.	8,300	1,100	8	d.		4,500	580	7
e.	3,300	150	9	e.		5,200	250	10

Use the units-of-production method to compute the unit rate and the annual depreciation for each machine below. In each case, assume that the machine is expected to produce 300,000 units during its useful life.

	COST	SCRAP VALUE	USEFUL LIFE (YEARS)	ANNUAL PRODUCTION
23. a.	$9,300	$900	5	25,000; 50,000; 70,000; 95,000; 60,000
b.	3,150	750	4	40,000; 50,000; 115,000; 95,000
c.	5,400	600	6	25,000; 40,000; 50,000; 55,000; 60,000; 70,000
24. a.	$3,000	$300	4	75,000; 82,000; 76,000; 67,000
b.	5,700	300	4	70,000; 80,000; 88,000; 62,000
c.	6,800	500	6	56,000; 58,000; 60,000; 55,000; 40,000; 31,000

SECTION 2
PARTIAL-YEAR DEPRECIATION

The discussion so far has described plant assets depreciated for full years. Not all assets, however, are purchased and placed into service on January 1 nor sold or discarded on December 31. How does a business prorate the depreciation for a partial year of use? With the MACRS method, this is not a concern because its table uses the *half-year convention*. The half-year convention automatically allows six months of depreciation during the first year, regardless of when the asset is actually placed into service. (Remember the 40% exception mentioned on p. 236.) Under MACRS, a plant asset acquired April 1 would have the same cost recovery in the first year as if it had been acquired October 1.

Under the traditional depreciation methods (straight-line, declining-balance, and sum-of-the-digits), the annual depreciation should be prorated for partial years of service, normally by using the *mid-month convention*. That is, if an asset is placed into service on or before the 15th of the month, it is depreciated as if it were placed into service on the first of the month. Conversely, if the asset is placed into service after the 15th of the month, depreciation does not start until the first day of the next month.

Example 1   A machine costing $6,800 is to be depreciated by the straight-line method. It was purchased on April 1. The residual value is to be $800 and the life is estimated to be 5 years.

$$\text{April 1 to December 31} = 9 \text{ months} = \frac{9}{12} \text{ year}$$

$$\frac{\$6,800 - \$800}{5} = \frac{\$6,000}{5} = \$1,200 \quad \text{Annual depreciation}$$

$$\frac{9}{12}(\$1,200) = \$900 \quad \text{First year's depreciation}$$

The first year's depreciation will be $900, and $1,200 will be the depreciation for each of the remaining full years. However, the business is entitled to 5 full years of depreciation, so the depreciation schedule will cover 6 calendar years as shown subsequently. The first year's depreciation is for 9 months, the depreciation for the 2nd through 5th years represent 12 months each, and the 6th year's depreciation represents 3 months.

YEAR	BOOK VALUE (END OF YEAR)	ANNUAL DEPRECIATION	ACCUMULATED DEPRECIATION
0	$6,800	—	—
1	5,900	$ 900	$ 900
2	4,700	1,200	2,100
3	3,500	1,200	3,300
4	2,300	1,200	4,500
5	1,100	1,200	5,700
6	800	300	6,000

Example 2   Assume that the business in Example 1 decides to depreciate the asset by the declining-balance method. How much annual depreciation will be taken each year?

$$\text{Rate} = 2\left(\frac{1}{5}\right) = \frac{2}{5} = 40\%$$

In the following yearly calculations, the depreciable value for each successive year is obtained by first computing a new book value in the subsequent depreciation schedule.

Year

1    $\frac{9}{12} \times 0.40 \times \$6,800 = \$2,040$

2    $0.40 \times \$4,760 = 1,904$

3    $0.40 \times \$2,856 = 1,142$

4    $0.40 \times \$1,714 = 686$

5    $0.40 \times \$1,028 = 411$ (adjusted to \$228)

Year	Book value (end of year)	Annual depreciation	Accumulated depreciation
0	\$6,800	—	—
1	4,760	\$2,040	\$2,040
2	2,856	1,904	3,944
3	1,714	1,142	5,086
4	1,028	686	5,772
5	800	228	6,000

Prorating the annual depreciation for partial years under the sum-of-the-digits method is more complicated. Because this method accelerates the annual depreciation in the early years, it is to the firm's advantage to use the highest rates possible each year. In the following example, the business is entitled to use a rate of $\frac{5}{15}$ for a full 12 months. Because it only got to use 9 months of the $\frac{5}{15}$ rate in year 1, it is entitled to carry over 3 months of the $\frac{5}{15}$ rate in year 2 and then must drop to a rate of $\frac{4}{15}$ for the last 9 months of year 2. Proration continues throughout the schedule as shown in the following illustration.

Example 3   Now assume that the business in Example 1 uses the sum-of-the-digits method. How much depreciation can be taken each year?

$$\frac{n(n+1)}{2} = \frac{5(5+1)}{2} = 15$$

$$\$6,800 - \$800 = \$6,000 \text{ Depreciable value}$$

Year

1    $\frac{5}{15} (\$6,000) \times \frac{9}{12} = \$1,500$

2    $\frac{5}{15} (\$6,000) \times \frac{3}{12} = \$\ 500$
     $\frac{4}{15} (\$6,000) \times \frac{9}{12} = \underline{1,200}$
     $\$1,700$

3     $\frac{4}{15}$ ($6,000) $\times \frac{3}{12}$ = $\ $ 400
$\frac{3}{15}$ ($6,000) $\times \frac{9}{12}$ = $\underline{900}$
$1,300

4     $\frac{3}{15}$ ($6,000) $\times \frac{3}{12}$ = $\ $ 300
$\frac{2}{15}$ ($6,000) $\times \frac{9}{12}$ = $\underline{600}$
$\ $ 900

5     $\frac{2}{15}$ ($6,000) $\times \frac{3}{12}$ = $\ $ 200
$\frac{1}{15}$ ($6,000) $\times \frac{9}{12}$ = $\underline{300}$
$\ $ 500

6     $\frac{1}{15}$ ($6,000) $\times \frac{3}{12}$ = $\ $ 100

Proration does not apply to the units-of-production method, because that method is based on actual usage regardless of the time that the asset is held.

SECTION 2   PROBLEMS

Compute the annual depreciation and complete a cost recovery or depreciation schedule for each of the following assets, using the data and method indicated. Use the mid-month convention where applicable to prorate depreciation according to the date of purchase. Assume that the asset was placed in service on the purchase date.

	COST	RESIDUAL VALUE	USEFUL LIFE (YEARS)	PURCHASE DATE	METHOD
1. a.	$4,200	$500	3	5/25	MACRS
b.	4,000	400	6	2/24	Straight-line
c.	3,000	0	5	7/1	Declining-balance
d.	1,800	600	3	8/1	Sum-of-the-digits
2. a.	$9,000	$ 0	5	10/21	MACRS
b.	4,900	400	5	9/10	Straight-line
c.	2,600	0	4	4/4	Declining-balance
d.	2,500	100	3	3/12	Sum-of-the-digits

3. On July 1, Superior Plumbing Service paid $35,000 for a service truck. The company expects to use the truck for 8 years. It estimates the trade-in value to be $1,000. Construct a depreciation schedule using the straight-line method.

4. Determine the annual depreciation by the straight-line method on equipment costing $4,600, if the trade-in value is estimated to be $400 and the useful life will be 7 years. Construct a depreciation schedule for this equipment that was purchased on November 1.

5. How much depreciation would be claimed by the declining-balance method on factory machinery costing a business $27,000, if the trade-in value is estimated to be $3,000 and the useful life will be 6 years? The business purchased the machinery on September 1. Construct a depreciation schedule.

6. On August 5, a business purchased a printer costing $900. The estimated life is 3 years, and the residual value is expected to be $50. Use the declining-balance method, and construct a depreciation schedule.

7. Determine the annual depreciation by the sum-of-the-digits method for equipment purchased on July 1 for $9,000. The estimated salvage value is $1,500, and the life is 5 years. Construct a depreciation schedule.

8. On April 6, a shopping mall purchased cleaning equipment costing $3,000. The trade-in value is estimated to be $200, and the useful life is 4 years. Construct a depreciation schedule using the sum-of-the-digits method.

SECTION 3
OVERHEAD

In addition to the cost of materials or merchandise, other expenses are incurred in the operation of any business. Examples of such *operating expenses* are salaries, rent, utilities, office supplies, taxes, depreciation, insurance, and so on. These additional expenses are known collectively as **overhead.**

   Overhead contributes indirectly to the actual total cost of the merchandise being manufactured or sold by the concern. In order to determine the efficiency of various departments, to determine the total cost of manufacturing various items, or to determine whether certain items are being sold profitably, a portion of the total plant overhead is assigned to each department. From the many methods of distributing overhead, we shall consider three: according to total floor space, according to total net sales, and according to the number of employees in each department.

Example 1   The mean monthly overhead of $96,000 at Johnson Manufacturing Corp. is distributed *according to the total floor space* of each department. The floor space of each department is as follows:

DEPARTMENT	FLOOR SPACE (SQ. FT)
A—Receiving and raw materials	4,000
B—Assembling	9,000
C—Inspecting and shipping	8,000
D—Administration	+ 3,000
Total	24,000

   Each department therefore has the following ratio of floor space and is assigned the following amount of overhead expense:

DEPARTMENT	RATIO OF FLOOR SPACE	OVERHEAD CHARGE	
A	$\dfrac{4,000}{24,000} = \dfrac{1}{6}$	$\dfrac{1}{6} \times \$96,000 =$	$16,000
B	$\dfrac{9,000}{24,000} = \dfrac{3}{8}$	$\dfrac{3}{8} \times \$96,000 =$	36,000
C	$\dfrac{8,000}{24,000} = \dfrac{1}{3}$	$\dfrac{1}{3} \times \$96,000 =$	32,000
D	$\dfrac{3,000}{24,000} = \dfrac{1}{8}$	$\dfrac{1}{8} \times \$96,000 =$	+ 12,000
		Total overhead =	$96,000

Calculator techniques . . .   FOR EXAMPLE 1

Set up the problem as shown in the example, and reduce each fraction representing the ratio of floor space. Then, the overhead charge is computed as follows:

96,000 ÷ 6 M+ ⟶ 16,000;

3 × 96,000 ÷ 8 M+ ⟶ 36,000;

96,000 ÷ 3 M+ ⟶ 32,000;

96,000 ÷ 8 M+ ⟶ 12,000;   MR ⟶ 96,000

Other methods of overhead distribution use similar calculator techniques.

Example 2   Avery Hardware wishes to distribute its $20,000 July overhead *according to the total net sales* of each department. These sales and the overhead distribution are as follows:

DEPARTMENT	NET SALES	RATIO OF SALES	OVERHEAD CHARGE	
Lumber/glass	$24,000	$\dfrac{\$24,000}{\$80,000} = \dfrac{3}{10}$	$\dfrac{3}{10} \times \$20,000 =$	$ 6,000
Tools/electrical	16,000	$\dfrac{\$16,000}{\$80,000} = \dfrac{1}{5}$	$\dfrac{1}{5} \times 20,000 =$	4,000
Building supplies	20,000	$\dfrac{\$20,000}{\$80,000} = \dfrac{1}{4}$	$\dfrac{1}{4} \times 20,000 =$	5,000
Housewares	8,000	$\dfrac{\$8,000}{\$80,000} = \dfrac{1}{10}$	$\dfrac{1}{10} \times 20,000 =$	2,000
Sports equipment	$12,000	$\dfrac{\$12,000}{\$80,000} = \dfrac{3}{20}$	$\dfrac{3}{20} \times 20,000 =$	+ 3,000
Total net sales =	$80,000		Total overhead =	$20,000

Example 3   The $15,000 monthly office overhead at White Plumbing & Heating is charged to the various divisions *according to the number of employees* in the office who work directly with each division. This results in the following overhead charges:

Department	Number of Employees	Ratio of Employees	Overhead Charge
Plumbing	3	$\dfrac{3}{10}$	$\dfrac{3}{10} \times \$15,000 =$ $\$ 4,500$
Heating	1	$\dfrac{1}{10}$	$\dfrac{1}{10} \times \$15,000 =$ $1,500$
Air conditioning	4	$\dfrac{4}{10} = \dfrac{2}{5}$	$\dfrac{2}{5} \times \$15,000 =$ $6,000$
Appliances	$+\ 2$	$\dfrac{2}{10} = \dfrac{1}{5}$	$\dfrac{1}{5} \times \$15,000 = +\ \ 3,000$
	Total employees $= 10$		Total overhead $=$ $\$15,000$

SECTION 3   PROBLEMS

In Problems 1–6, distribute overhead according to floor space, net sales, or number of employees, depending on the information given.

	Department	Floor Space (Sq. Ft.)	Total Overhead		Department	Floor Space (Sq. Ft.)	Total Overhead
1. a.	E	3,000	$42,600	**2. a.**	#1	800	$13,500
	F	2,000			2	1,000	
	G	8,000			3	1,200	
	H	7,000			4	1,500	
b.	I	1,000	$16,000	b.	#16	500	$32,000
	J	900			17	300	
	K	600			18	400	
	L	1,500			19	800	

	DEPARTMENT	NET SALES	TOTAL OVERHEAD			DEPARTMENT	NET SALES	TOTAL OVERHEAD
3. a.	N	$25,000	$64,000	4. a.		W	$30,000	$56,000
	O	18,000				X	15,000	
	P	30,000				Y	14,000	
	Q	27,000				Z	11,000	
b.	S	$18,000	$45,500	b.		AA	$20,000	$36,000
	T	24,000				BB	22,000	
	U	11,000				CC	14,000	
	V	12,000				DD	16,000	

	DEPARTMENT	NUMBER OF EMPLOYEES	TOTAL OVERHEAD			DEPARTMENT	NUMBER OF EMPLOYEES	TOTAL OVERHEAD
5. a.	R	6	$90,000	6. a.		#22	40	$45,000
	S	10				33	10	
	T	20				44	30	
	U	9				55	20	
b.	W	4	$96,000	b.		#10	2	$30,000
	X	3				11	4	
	Y	5				12	3	
	Z	8				13	6	

7. A grocery store has total overhead of $150,000. The store apportions its operating expenses according to the floor space of each department. Find the overhead for each department.

DEPARTMENT	FLOOR SPACE (SQ. FT.)	OVERHEAD
Produce	1,000	
Meats	500	
Frozen foods	1,200	
Bakery	800	
Household products	1,500	

8. A department store allocates operating expenses of $60,000 according to the total floor space occupied by each department. Find the overhead for each department.

DEPARTMENT	FLOOR SPACE (SQ. FT.)	OVERHEAD
Accounting	200	
General office	300	
Marketing	400	
Service/repairs	600	

9. Crosby's Department Store distributes its overhead according to each department's ratio of total net sales. If the August operating expenses were $180,000, what was each department's share of the overhead?

DEPARTMENT	NET SALES	OVERHEAD
Shoes	$26,000	
Men's clothing	70,000	
Cosmetics	28,000	
Women's clothing	86,000	
Children's clothing	30,000	

10. William's Book Store distributes its overhead according to each department's ratio of total net sales. Allocate operating expenses of $40,000.

DEPARTMENT	NET SALES	OVERHEAD
Biographies	$60,000	
Adventure/mysteries	44,000	
Computers/software	42,000	
Romance	38,000	
Sports	16,000	

11. The overhead expenses at George's Hardware are apportioned among its departments according to the number of employees in each department. If the total overhead during February was $90,000, determine how much should be charged to each department.

DEPARTMENT	NUMBER OF EMPLOYEES	OVERHEAD
Lawn/garden	7	
Appliances	8	
Automotive	5	
Building supplies	6	
Paint	4	

12. Allocate the overhead expenses of $50,000 among the branch offices of Pickford Associates according to the number of employees.

BRANCH	NUMBER OF EMPLOYEES	OVERHEAD
A	5	
B	9	
C	8	
D	6	
E	12	

CHAPTER 9 GLOSSARY

ACRS and MACRS. "Accelerated cost recovery system" and "Modified accelerated cost recovery system" for determining the amount of asset cost to recover as annual tax deductions. As substantially revised by the Tax Reform Act of 1986, the MACRS method prescribes recovery categories (five for personal property and two for real estate) that enable 100% of the purchase cost to be deducted over the entire recovery period. This method employs a combination of the declining-balance method and the straight-line method of depreciation to accelerate cost recovery for tax purposes. (See also "Cost recovery.")

Book value. The calculated value of a plant asset after depreciation or cost recovery has been deducted.

Cost recovery. Reduction in value of a plant asset, based on set annual portions of the original cost; a tax-deductible business expense. The amount of cost recovery should not be confused with the decrease in value that an asset undergoes from the standpoint of its actual market value.

Cost recovery (or depreciation) schedule. A record showing annual cost recovery (or depreciation), current book value, and accumulated recovery (or depreciation) of a plant asset.

Declining-balance method. The depreciation method in which a constant percent is annually multiplied times book value (starting with cost). Formerly called "double (or 200%) declining-balance." The double declining-balance method employs twice the straight-line rate.

Depreciable value. The total amount of depreciation that is computed over the life of an asset. Depreciable value = Total cost of asset − Residual value.

Depreciation. The systematic allocation of the asset's cost over its period of economic utility (its usefulness).

Fixed asset. An older term for plant asset. (See "Plant asset.")

Overhead. Business expenses other than the cost of materials or merchandise. (Examples: Salaries, rent, utilities, supplies, advertising.)

Plant asset. A business possession that will be used for more than 1 year. (Examples: Buildings, vehicles, machinery, equipment.)

Resale value. The expected value to be received if the asset is sold.

Residual value. The expected value of the asset at the end of its useful life.

Salvage (scrap) value. The value of the materials in a plant asset at the end of its useful life.

Straight-line method. The depreciation method in which the depreciable value of an asset is evenly divided among the years.

Legal for tax purposes as an alternative to the MACRS method.

Straight-line rate.   The depreciation rate that is the reciprocal of the number of years of useful life. (Example: For an asset used 5 years, the annual straight-line rate is $\frac{1}{5}$.)

Sum-of-the-digits method.   An accelerated depreciation method whereby depreciable value is multiplied by a series of fractions whose denominators are the total of the years' digits and whose numerators are the years' digits taken in reverse order.

Trade-in value.   The expected residual value to be received if the asset is traded.

Units-of-production method.   The depreciation method in which depreciable value is divided by total expected production to obtain a per-unit depreciation rate, which is then multiplied times the number of units produced each year, in order to determine annual depreciation.

Useful life.   The number of years that a plant asset is efficiently used and depreciated in a business. (Useful life differs from the recovery period used in the MACRS method.)

Wearing value.   An older term for depreciable value. (See "Depreciable value.")

10

FINANCIAL STATEMENTS AND RATIOS

OBJECTIVES

Upon completion of Chapter 10, you will be able to:

1. Define and use correctly the terminology associated with each topic.

2. On an income statement with the individual amounts given (Section 1: Examples 1, 2; Problems 1, 2):

   a. Compute the net profit.

   b. Find the percent of net sales that each item represents.

3. On a balance sheet with individual amounts given, compute (Section 2: Example 1; Problems 3, 4):

   a. Total assets

   c. Net worth

   b. Total liabilities

   d. Percent each item represents.

4. On comparative income statements and comparative balance sheets, compute (Section 1: Example 3; Problems 5, 6; Section 2: Example 2; Problems 7, 8):

   a. Subtotals and totals

   b. Increase or decrease between years (amount and percent)

   c. Percent of net sales for each year (on income statements)

   d. Percent of total assets for each year (on balance sheets).

5. Select from income statements and balance sheets the amounts required to compute common business ratios (Section 2: Example 3; Problems 9–12).

6. Compute (both at cost and at retail) (Section 3):

   a. Average inventory (Example 1; Problems 1–4, 7–12)

   b. Inventory turnover (Examples 2, 3; Problems 5–12).

7. **a.** Evaluate inventory by the methods of (Section 4: Example 1; Problems 1–8):
   (1) Weighted average, (2) FIFO, and (3) LIFO.

   b. On an income statement, compare gross profit (amount and percent) under each inventory method.

At least once each year (sometimes more often), the accounting records of every business are audited to determine the financial condition of the business: what the volume of business was, how much profit (or loss) was made, how much these things have changed in the past year, how much the business is actually worth, and so on. As a means of presenting this important information, two financial statements are normally prepared and analyzed: the *income statement* and the *balance sheet.*

SECTION 1
INCOME STATEMENT

The **income statement** shows a business's sales, expenses, and profit (or loss) during a certain period of time. (The time might be a month, a quarter, 6 months, or a year.) The basic calculations of an income statement are these:

$$
\begin{array}{l}
\text{Net sales} \\
-\,\underline{\text{Cost of goods sold}} \\
\text{Gross profit} \\
-\,\underline{\text{Operating expenses}} \\
\text{Net profit (or net loss)}
\end{array}
$$

Notice that **gross profit** (also known as **margin** or **markup**) is the amount that would remain after the merchandise has been paid for. Out of this gross profit must be paid all of the **operating expenses** (**overhead,** such as salaries, rent, utilities, supplies, insurance, and so forth). The amount that remains is the clear or spendable profit—the **net income** or **net profit.**

In order to effectively analyze expenses or to compare an income statement with preceding ones, it is customary to convert the dollar amounts to percents of the total net sales. These percents are computed as follows:

$$\_\_\_\ \% \text{ of Net sales} = \text{Each amount}$$

$$\frac{\_\_\_\ \% \times \cancel{\text{Net sales}}}{\cancel{\text{Net sales}}} = \frac{\text{Each amount}}{\text{Net sales}}$$

$$\_\_\_\ \% = \frac{\text{Each amount}}{\text{Net sales}}$$

Example 1 During April, Williams Pharmacy had net sales of $60,000. The drugs and merchandise cost $45,000 wholesale, and operating expenses totaled $12,000. How much net profit was made during April, and what was the percent of net sales for each item?

WILLIAMS PHARMACY

Income Statement for Month Ending April 30, 19XX

Net sales	$60,000	100%
Cost of goods sold	− 45,000	− 75%
Gross profit	$15,000	25%
Operating expenses	− 12,000	− 20%
Net profit	$ 3,000	5%

The percents above were obtained as follows:

$$\underline{\quad}\% = \frac{\text{Each amount}}{\text{Net sales}}$$

(a) $\dfrac{\text{Cost of goods}}{\text{Net sales}} = \dfrac{\$45,000}{\$60,000} = 75\%$

(b) $\dfrac{\text{Gross profit}}{\text{Net sales}} = \dfrac{\$15,000}{\$60,000} = 25\%$

(c) $\dfrac{\text{Operating expenses}}{\text{Net sales}} = \dfrac{\$12,000}{\$60,000} = 20\%$

(d) $\dfrac{\text{Net profit}}{\text{Net sales}} = \dfrac{\$ 3,000}{\$60,000} = 5\%$

Thus, we find that 75% of the income from sales was used to pay for the drugs and merchandise itself, leaving a gross profit ($15,000) of 25% of sales. Another 20% of the income from sales was used to pay the $12,000 operating expenses. This leaves a clear, net profit of $3,000, which is 5% of the sales income.

Note.   Observe that the cost of goods, the overhead, and the net profit together account for all the income from sales. Also, the overhead (operating expenses) plus the net profit comprise the margin (gross profit). In equation form,

$$\text{Cost} + \text{Overhead} + \text{Net profit} = \text{Sales}$$

$$\$45,000 + \underbrace{\$12,000 + \$3,000} = \$60,000$$

$$\text{Cost} + \qquad \text{Margin} \qquad = \text{Sales}$$

$$\$45,000 + \qquad \$15,000 \qquad = \$60,000$$

This fundamental relationship is the basis for the equation $C + M = S$, which is applied by merchants to determine the selling price of their merchandise. (This is the subject of Chapter 13.)

These relationships, which are true for the amounts of money shown on the income statement, are also true of the percents for each item:

$$\text{Cost} + \text{Overhead} + \text{Net profit} = \text{Sales}$$

$$75\% + \underbrace{20\% + 5\%} = 100\%$$

$$\text{Cost} + \text{Margin} = \text{Sales}$$

$$75\% + 25\% = 100\%$$

The foregoing income statement was a highly simplified example. In actual practice, the basic topics used to calculate profit (net sales, cost of goods sold, and operating expenses) are broken down in detail, showing the contributing factors of each. The next example presents a more complete income statement.

Of the various topic breakdowns, only cost of goods sold needs particular mention. To the value of the inventory on hand at the beginning of the report period is added the total of all merchandise bought during the period (less returned merchandise and plus freight charges paid separately). This gives the total value of all merchandise that the company had available for sale during the period. Since some of this merchandise is not sold, however, the value of the inventory on hand at the end of the period must be deducted, in order to obtain the value of the merchandise that was actually sold.

Notice that, just as the individual amounts are added or subtracted to obtain subtotals, so may their corresponding percents be added or subtracted to obtain the percents of the subtotals. This is possible because each amount is represented as a percent of the same number (net sales).

The percents indicated in the right-hand column are the ones of particular interest to management and potential investors: (1) What percent of sales was the cost of goods? (2) What percent of sales was the gross profit (margin)? (3) What percent of sales was the operating expense (overhead)? But most important by far to the owners and possible investors is the final percent: (4) What percent of sales was the net profit? This is usually the first question asked by anyone considering investing money in a business.

Example 2   **RANIER ELECTRONICS, INC.**
                    Income Statement for Year Ending December 31, 19XX

Income from sales:				
Total sales		$123,000		102.5%
Less: Sales returns and				
allowances		− 3,000		− 2.5
Net sales			$120,000	100.0%
Cost of goods sold:				
Inventory, January 1		$ 31,300		
Purchases	$71,600			
Less: Returns and				
allowances	− 1,300			
Net purchases	$70,300			
Add: Freight in	+ 800			
Net purchase cost		+ 71,100		
Goods available for sale		$102,400		
Inventory, December 31		− 27,400		
Cost of goods sold			− 75,000	− 62.5
Gross profit on sales			$45,000	37.5%
Operating expenses:				
Office salaries		$ 24,000		20.0%
Rent and utilities		4,800		4.0
Office supplies		1,200		1.0
Insurance		720		0.6
Advertising		1,800		1.5
Depreciation		3,000		2.5
Miscellaneous		+ 480		+ 0.4
Total operating expenses			− 36,000	− 30.0
Net income from operations			$ 9,000	7.5%

Calculator Tip.   Store the net sales, $120,000, into memory with $\boxed{M+}$. Then, each financial item $\boxed{÷}$ $\boxed{MR}$ $\boxed{\%}$ → its corresponding percent. For instance, 9,000 $\boxed{÷}$ $\boxed{MR}$ $\boxed{\%}$ → 7.5. Recall that the $\boxed{\%}$ key includes the "=" process as it converts the result to its equivalent percent.

The percents derived in Examples 1 and 2 represent **vertical analysis.** That is, each separate item is represented as a percent of the total during a *single* time period. In order to analyze business progress and potential more effectively, however, recent figures should be compared with corresponding figures from the previous period (or periods). When comparisons are made between corresponding items during a *series* of two or more time periods, this is known as **horizontal analysis.**

The following example illustrates horizontal analysis. On this comparative income statement, both the dollar increase (or decrease) and the percent of change are shown. Figures for the earlier year are used as the basis for finding the percent of change, using the formula "_____ % of original = change?"

Example 3

GOLDSTEIN'S INC.

Comparative Income Statement for Fiscal Years Ending June 30, 19X2 and 19X1

	19X2	19X1	INCREASE OR (DECREASE) Amount	Percent	PERCENT OF NET SALES 19X2	19X1
Income:						
Net sales	$275,000	$250,000	$25,000	10.0%	100.0%	100.0%
Cost of goods sold:						
Inventory, July 1	$ 60,000	$ 70,000	($10,000)	(14.3%)	21.8%	28.0%
Purchases	205,000	165,000	40,000	24.2	74.5	66.0
Goods available for sale	$265,000	$235,000	$30,000	12.8%	96.4%	94.0%
Inventory, June 30	75,000	60,000	15,000	25.0	27.3	24.0
Cost of goods sold	190,000	175,000	15,000	8.6	69.1	70.0
Gross margin	$ 85,000	$ 75,000	$10,000	13.3%	30.9%	30.0%
Expenses:						
Administration	$ 50,000	$ 45,000	$ 5,000	11.1%	18.2%	18.0%
Occupancy	7,500	5,000	2,500	50.0	2.7	2.0
Sales	16,000	10,000	6,000	60.0	5.8	4.0
Miscellaneous	1,500	2,500	(1,000)	(40.0)	0.5	1.0
Total expenses	75,000	62,500	12,500	20.0	27.3	25.0
Net income (before taxes)	$ 10,000	$ 12,500	($ 2,500)	(20.0%)	3.6%	5.0%

The following example (for net sales) illustrates how the preceding percents of change were computed. Net sales increased from $250,000 to $275,000, a "change" of $25,000. The "original" is the first year, 19X1. Thus,

$$\text{\_\_\_\_ \% of Original} = \text{Change}$$

$$\text{\_\_\_\_ \% of \$250,000} = \$25,000$$

$$\frac{250,000 \, x}{250,000} = \frac{25,000}{250,000}$$

$$x = 10\%$$

Notice that, whereas the "amounts" of increase or decrease may be added or subtracted to obtain the amounts of the subtotals, the "percents" of increase or decrease *cannot* be added or subtracted to obtain the percents of change in the subtotals. This is because each percent of change is based on a different number (the 19X1 amount of that particular item).

Calculator techniques . . .   FOR EXAMPLE 3

After the 19X2 and 19X1 columns are totaled, you work *horizontally* to calculate the Amount and Percent of increase or decrease. To illustrate, this computation determines Inventory (July 1) and Purchases. Recall that parentheses indicate you should use a displayed number without reentering it. Although the $\boxed{\%}$ includes "=," you must round each resulting percent to the nearest tenth.

$60,000 \boxed{-} 70,000 \boxed{=} \longrightarrow -10,000$

$\quad\quad (-10,000) \boxed{\div} 70,000 \boxed{\%} \longrightarrow -14.285714;$

$205,000 \boxed{-} 165,000 \boxed{=} \longrightarrow 40,000$

$\quad (40,000) \boxed{\div} 165,000 \boxed{\%} \longrightarrow 24.242424; \quad$ and so on.

Later, the two columns for Percent of Net Sales are calculated *vertically,* using the year's Total Sales from memory. For 19X2, this technique finds the percent of net sales represented by Inventory and Purchases:

$275,000 \boxed{M+}$

$60,000 \boxed{\div} \boxed{MR} \boxed{\%} \longrightarrow 21.818181;$

$205,000 \boxed{\div} \boxed{MR} \boxed{\%} \longrightarrow 74.545454; \quad$ and so on.

The 19X1 column is computed similarly, with net sales of $250,000 in memory. This technique also applies for comparative balance sheets, presented in the next section.

SECTION 2
BALANCE SHEET

The purpose of a **balance sheet** is to give an overall picture of what a business is worth at a specific time: how much it owns, how much it owes, and how much the owners' investment is worth. On a balance sheet are listed all of a business's **assets** (resources owned and money owed to the business) and all its **liabilities** (money that the business owes to someone else). The balance remaining after the liabilities are subtracted from the assets is the **net worth** of the business. Thus, the assets must balance the liabilities and net worth, or

$$Assets = Liabilities + Net\ worth$$

The general heading of assets on a balance sheet is ordinarily broken down into current and plant categories, as follows:

CURRENT ASSETS	PLANT ASSETS
Cash	Buildings
Accounts receivable	Furnishings
Notes receivable	Machinery
Office supplies	Equipment
Merchandise inventory	Land

Current assets are the resources that are owned by the business that will be converted to cash or consumed within one operating cycle or one year of the balance sheet date. Cash includes the cash on hand and cash in checking and savings accounts in the bank. **Accounts receivable** are accounts owed to the business by customers who have bought merchandise or services "on credit." Notes receivable indicate money owed by customers who have signed a written promise to pay by a certain date.

Plant assets are the long-term, tangible assets that are used repeatedly by the business, such as equipment. Equipment and other plant assets have a long life and can be used for several years. Because these assets are used for more than one year, their cost is spread over the useful life of the assets; that is, the plant assets are depreciated and are shown on the balance sheet at their current book value. The only exception is for land, which is not depreciated.

The liabilities of a business are separated according to current liabilities and long-term liabilities, as indicated:

CURRENT LIABILITIES	LONG-TERM LIABILITIES
Accounts payable	Mortgages
Notes payable	Long-term notes payable
Interest payable	Bonds
Taxes payable	

Current liabilities are debts that will be paid within one operating cycle or one year of the balance sheet date. They include **accounts payable,** which are accounts owed to others for goods or services bought on credit by the business. The notes payable, interest, and taxes are also items that are due within a short time.

Long-term liabilities are the debts that do not mature or come due until more than one year after the balance sheet date. A good example is a 25-year mortgage on a building. The long-term notes payable are those promissory notes that will be due after one year from the balance sheet date.

Other common terms used instead of net worth are **net ownership, net investment, proprietorship** (for an individually owned business), **owner's equity,** and **stockholders' equity** (for a corporation). Stockholders' equity is composed of two parts: **Contributed capital** represents the investments of owners in the business, and **retained earnings** represents the part of net profits that is kept for use in the business. Cash dividends are paid from a corporation's retained earnings.

The following balance sheet again illustrates vertical analysis. That is, each asset is represented as a percent of the total assets; and each liability, as well as stockholders' equity (net worth), is presented as a percent of the total liabilities plus stockholders' equity (which is the same as the total assets).

Example 1

COLLIER-WHITWORTH CORP.

Balance Sheet, December 31, 19X1

Assets			*Percent*	
Current assets:				
Cash on hand		$ 500	0.2%	
Cash in bank		9,000	3.6	
Accounts receivable		42,000	16.8	
Inventory		23,500	9.4	
Total current assets		$ 75,000		30.0%
Plant assets:				
Plant site		$25,000	10.0%	
Building	$150,000			
Less: Depreciation	− 55,000			
		95,000	38.0	
Equipment	$ 75,000			
Less: Depreciation	− 20,000			
		55,000	22.0	
Total plant assets		175,000		70.0
Total assets		$250,000		100.0%
Liabilities and Stockholders' Equity				
Current liabilities:				
Accounts payable	$37,500		15.0%	
Notes payable	10,000		4.0	
Total current liabilities		$ 47,500		19.0%
Long-term liabilities:				
Mortgage		102,500		41.0
Total liabilities		$150,000		60.0%
Stockholders' equity		100,000		40.0
Total liabilities and stockholders' equity		$250,000		100.0%

The following illustrations show how the preceding percents were computed. The percents are found by dividing each amount by total assets:

(a) Total current assets are what percent of total assets?

$$\underline{\quad\quad}\% \text{ of Total assets} = \text{Current assets}$$

$$\underline{\quad\quad}\% \text{ of } \$250,000 = \$75,000$$

$$250,000x = 75,000$$

$$x = \frac{75,000}{250,000}$$

$$x = 30\%$$

(b) The plant site is what percent of total assets?

$$\underline{\quad\quad}\% \text{ of Total assets} = \text{Plant site}$$

$$\underline{\quad\quad}\% \text{ of } \$250,000 = \$25,000$$

$$250,000x = 25,000$$

$$x = \frac{25,000}{250,000}$$

$$x = 10\%$$

Just as horizontal analysis was helpful in studying the income statement, so is it useful in determining the significance of a balance sheet. The next example presents such a balance sheet.

Example 2

THE DANVILLE CORP.
Comparative Balance Sheet, January 31, 19X2 and 19X1

	19X2	19X1	INCREASE OR (DECREASE) AMOUNT	PERCENT	PERCENT OF TOTAL ASSETS 19X2	19X1
Assets						
Cash	$ 11,000	$ 14,000	($ 3,000)	(21.4%)	4.6%	7.0%
Accounts receivable	32,000	28,000	4,000	14.3	13.3	14.0
Inventory	57,000	38,000	19,000	50.0	23.8	19.0
Current assets	$100,000	$ 80,000	$20,000	25.0%	41.7%	40.0%
Plant assets (net)	140,000	120,000	20,000	16.7	58.3	60.0
Total assets	$240,000	$200,000	$40,000	20.0%	100.0%	100.0%
Liabilities and Stockholders' Equity						
Current liabilities	$ 45,000	$ 36,000	$ 9,000	25.0%	18.7%	18.0%
Long-term liabilities	90,000	74,000	16,000	21.6	37.5	37.0
Total liabilities	$135,000	$110,000	$25,000	22.7%	56.2%	55.0%
Common stock	$ 75,000	$ 65,000	$10,000	15.4%	31.3%	32.5%
Retained earnings	30,000	25,000	5,000	20.0	12.5	12.5
Total stockholders' equity	105,000	90,000	15,000	16.7	43.8	45.0
Total liabilities and stockholders' equity	$240,000	$200,000	$40,000	20.0%	100.0%	100.0%

In the preceding balance sheet, a horizontal analysis was done to find the "increase or decrease" amounts and percents. After the amount of increase or decrease is determined, the "percent of change" formula is used to find the percent of increase or decrease.

(a)   What is the change in cash from 19X1 to 19X2?

$$\begin{array}{ll} \$14,000 & \text{19X1} \\ -\ \underline{11,000} & \text{19X2} \\ (\$\ 3,000) & \text{Decrease in cash} \end{array}$$

(b)   What is the percent of change from 19X1 to 19X2?

_____ % of Original year = Change

_____ % of $14,000   = ($3,000)

$$14,000x = (3,000)$$

$$x = \frac{(3,000)}{14,000}$$

$$x = (21.4\%)$$

A vertical analysis is done for each year separately to find the "Percent of Total Assets." The percents in the columns are found by dividing each item in a particular year by the total assets for that same year.

(c) Cash in 19X2 is what percent of total assets?

$$\underline{\qquad}\ \%\ \text{of Total assets} = \text{Cash}$$

$$\underline{\qquad}\ \%\ \text{of}\ \$240,000 = \$11,000$$

$$240,000x = 11,000$$

$$x = \frac{11,000}{240,000}$$

$$x = 4.6\%$$

(d) Cash in 19X1 is what percent of total assets?

$$\underline{\qquad}\ \%\ \text{of Total assets} = \text{Cash}$$

$$\underline{\qquad}\ \%\ \text{of}\ \$200,000 = \$14,000$$

$$200,000x = 14,000$$

$$x = \frac{14,000}{200,000}$$

$$x = 7\%$$

Business Ratios

In addition to the analysis possible from single or comparative financial statements, a number of other business ratios are commonly used to provide a more thorough indication of a business's financial condition. These ratios are computed using figures readily available from the income statement and the balance sheet. The next example contains an explanation and illustration of some of the more common ratios.

Example 3   The six ratios given below are based on the balance sheet of The Danville Corp. in Example 2. Its income statement for 19X2 was as follows:

THE DANVILLE CORP.
Condensed Income Statement for Fiscal Year Ending January 31, 19X2

Net sales	$300,000	100.0%
Cost of goods sold	− 195,000	− 65.0
Gross margin	$105,000	35.0%
Operating expenses	− 90,000	− 30.0
Net income (before taxes)	$ 15,000	5.0%
Income taxes	− 5,000	− 1.7
Net income (after taxes)	$ 10,000	3.3%

Working Capital Ratio

The **working capital ratio** is often called the **current ratio,** because it is the ratio of current assets to current liabilities. This ratio indicates the ability of the business to meet its current obligations. The Danville Corp.'s 19X2 current ratio is computed as follows:

$$\frac{\text{Current assets}}{\text{Current liabilities}} = \frac{\$100,000}{\$45,000}$$

$$= \frac{2.2}{1} \quad \text{or} \quad 2.2:1$$

The Danville Corp.'s current ratio is 2.2 : 1. A current ratio of 2 : 1 is generally considered the minimum acceptable ratio; bankers would probably hesitate to lend money to a firm whose working capital ratio was lower than 2 : 1.

Acid-Test Ratio

The current ratio included all current assets. In order to provide some measure of a firm's ability to obtain funds quickly, the **acid-test** or **quick ratio** is computed. This is a ratio of the quick assets to the current liabilities. **Quick assets** (sometimes called **liquid assets**) are those which can be easily and quickly converted to cash, for example, accounts receivable, notes receivable and marketable securities, as well as cash itself. (The inventory and supplies are not included in the acid-test ratio, because it often proves difficult to convert both of these assets to cash quickly.)

$$\frac{\text{Quick assets}}{\text{Current liabilities}} = \frac{\$43,000}{\$45,000}$$

$$= \frac{0.96}{1} \quad \text{or} \quad 1.0:1$$

As an indication of a firm's ability to meet its obligations, the acid-test ratio should be at least $1:1$. The Danville Corp. barely meets this minimum requirement in 19X2. This may be only a temporary drop, however (from $1.2:1$ the previous year). Observe that during 19X2 Danville greatly increased its inventory, with a corresponding increase in current liabilities and a drop in cash. If the increased inventory results in higher net sales and an improved cash position, the acid-test ratio may also soon improve.

Net Income (after Taxes) to Average Net Worth

Since businesses are organized to make profits, of primary concern to the owners is the return on their investment. The **ratio of net income (after taxes) to average net worth** indicates the rate of return on the average owners' equity. Using the 19X1 and 19X2 figures for total stockholders' equity of The Danville Corp., we obtain

$$\text{Average net worth} = \frac{\$90,000 + \$105,000}{2}$$

$$= \frac{\$195,000}{2}$$

$$= \$97,500$$

$$\frac{\text{Net income (after taxes)}}{\text{Average net worth}} = \frac{\$10,000}{\$97,500}$$

$$= 10.3\%$$

Net Sales to Average Total Assets

The **ratio of net sales to average total assets** gives some indication of whether a business is using its assets to best advantage—that is, of whether its sales volume is in line with what the company's capital warrants. This ratio (sometimes called the **total capital turnover**) would indicate inefficiency if it is too low.

Note. The "total assets" used here should exclude any assets held by the company which do not contribute to the business operation.

$$\text{Average total assets} = \frac{\$200,000 + \$240,000}{2}$$

$$= \frac{\$440,000}{2}$$

$$= \$220,000$$

$$\frac{\text{Net sales}}{\text{Average total assets}} = \frac{\$300,000}{\$220,000}$$

$$= \frac{1.4}{1} \quad \text{or} \quad 1.4:1$$

The ratio of net sales to average total assets at The Danville Corp. is 1.4:1; or, the turnover of total capital was 1.4 times during 19X2.

Accounts Receivable Turnover

Accounts receivable turnover is the ratio of net sales to average accounts receivable. (Notes receivable, if they are held, should also be included with the accounts receivable.) This is a measure of how fast a business converts its accounts receivable into cash.

$$\text{Average accounts receivable} = \frac{\$28,000 + \$32,000}{2}$$

$$= \frac{\$60,000}{2}$$

$$= \$30,000$$

$$\frac{\text{Net sales}}{\text{Average accounts receivable}} = \frac{\$300,000}{\$30,000}$$

$$= 10 \text{ times}$$

Accounts receivable are not actually an asset to a business until they are collected. Prompt collection of the accounts receivable makes funds available to the business to meet its own obligations and to take advantage of cash discounts and special offers. Furthermore, the older an account becomes, the less the likelihood that it will be collected. Thus, a minimum of the firm's assets should be tied up in receivables, and they should turn over frequently.

Average Age of Accounts Receivable

This ratio, the **average age of accounts receivable,** in conjunction with the preceding one, indicates whether a business is keeping its accounts receivable up to date. The average age in days is found by dividing 365 days by the turnover in accounts receivable.

$$\frac{365}{\text{Accounts receivable turnover}} = \frac{365}{10}$$

$$= 36.5 \text{ days}$$

The average age of the accounts receivable is 36.5 days. If The Danville Corp. has an allowable credit period of 30 days, this average age of 36.5 days indicates that it is not keeping up with collections.

There are quite a few other ratios of interest to owners and management. One of these, the *equity ratio*, is already indicated on the balance sheet; the equity ratio is the ratio of owners' equity to total assets (43.8%). The *ratio of current liabilities to net worth* and the *ratio of total liabilities to net worth* are ratios based on balance sheet figures. Other common ratios relating to net sales are the *ratio of net sales to net worth*, the *total asset turnover* (the ratio of net sales to average total assets), and the *ratio of net sales to net working capital*:

$$\frac{\text{Net sales}}{\text{Current assets} - \text{Current liabilities}}$$

Also, the *inventory turnover* (the ratio of net sales to average inventory) is discussed in Section 3.

Additional ratios involving net income are the *ratio of net income to total assets* and the *rate (percent) of return per share of common stock*:

$$\frac{\text{Net income after taxes} - \text{Preferred dividends}}{\text{Average common stockholders' equity}}$$

Of interest also is the *book value of the stock* when there is only one type of stock; this is computed as follows:

$$\frac{\text{Owners' equity}}{\text{Number of shares of stock}}$$

Any number of other ratios could be computed if it were thought they would be useful.

Although the financial statements and the ratios that can be computed from them are of great value in evaluating the overall efficiency and financial condition of a business, such reports are not foolproof. Similar information from other businesses of the same type, as well as the general economic condition of the country as a whole, must also be taken into consideration as part of any thorough analysis by management or by a potential investor. On this basis, reports which originally seemed acceptable might appear less acceptable or might indicate that the business did unusually well in comparison with competing firms.

SECTIONS 1 AND 2 PROBLEMS

Copy and complete the following financial statements, computing percents to the nearest tenth. Compute percents only at the points indicated with tinted boxes in Problems 1–4. In Problems 5–8, compute values on all rows.

1.

LEROSEN FURNITURE CO., INC.

Income Statement for Year Ending December 31, 19X2

Income from sales:			
Sales	$850,000		▓▓%
Sales discounts	50,000		▓▓
Net sales		$▓▓▓▓▓	▓▓%
Cost of goods sold:			
Inventory, January 1	$ 45,000		
Net purchases	255,000		
Goods available for sale	$▓▓▓▓		
Inventory, December 31	60,000		
Cost of goods sold		▓▓▓▓▓	▓▓%
Gross profit		$▓▓▓▓	▓▓%
Operating expenses:			
Salaries	$219,000		▓▓%
Rent	100,000		▓▓
Depreciation	52,000		▓▓
Insurance	26,000		▓▓
Utilities	20,000		▓▓
Advertising	15,000		▓▓
Office supplies	11,000		▓▓
Miscellaneous	10,000		▓▓
Total expenses		▓▓▓▓	▓▓
Net income from operations		$▓▓▓▓	▓▓%
Income taxes		32,000	▓▓
Net income after taxes		$▓▓▓▓	▓▓%

2.   **DAVIE & ASSOCIATES, INC.**

Income Statement for Year Ending December 31, 19X2

Income from sales:			
Sales	$520,000		▬%
Sales discounts	20,000		▬
Net sales		$▬	▬%
Cost of goods sold:			
Inventory, January 1	$ 82,000		
Net purchases	150,000		
Goods available for sale	$▬		
Inventory, December 31	83,000		
Cost of goods sold		▬	▬%
Gross profit		$▬	▬%
Operating expenses:			
Salaries	$200,000	▬%	
Depreciation	32,000	▬	
Utilities	18,000	▬	
Maintenance	16,500	▬	
Advertising	10,000	▬	
Insurance	8,500	▬	
Office supplies	8,000	▬	
Miscellaneous	7,000	▬	
Total expenses		▬	▬
Net income from operations		$▬	▬%
Income taxes	14,000		▬
Net income after taxes		$▬	▬%

3.

SCOTT PHARMACEUTICAL CO., INC.
Balance Sheet, June 30, 19X4

Assets				
Current assets:				
Cash		$ 5,800	▬%	
Accounts receivable		11,200	▬	
Supplies		500	▬	
Inventory		19,000	▬	
Total current assets		$▬	▬%	
Plant assets:				
Equipment	$ 44,000			
Less: Depreciation	9,000			
		$▬	▬%	
Building	$400,000			
Less: Depreciation	350,000			
		▬	▬	
Land		200,000	▬	
Total plant assets		▬		▬
Total assets		$▬		▬%
Liabilities and Equity				
Current liabilities:				
Accounts payable		$ 10,000	▬%	
Notes payable		8,000	▬	
Total current liabilities		$▬		▬%
Long-term liabilities:				
Mortgage note payable		84,000		▬
Total liabilities		$▬		▬%
Stockholders' equity:				
Common stock		$119,000	▬%	
Retained earnings		100,500	▬	
Total equity		▬		▬
Total liabilities and equity		$▬		▬%

4.

MADISON & CO.
Balance Sheet, December 31, 19X1

Assets			
Current assets:			
Cash	$ 27,000	▓%	
Accounts receivable	32,000	▓	
Notes receivable	20,000	▓	
Inventory	45,000	▓	
Total current assets	$▓		▓%
Plant assets:			
Building	$300,000		
Less: Depreciation	30,000		
	$▓	▓%	
Trucks	$38,000		
Less: Depreciation	18,000		
	▓	▓	
Land	160,000	▓	
Total plant assets	▓		▓
Total assets	$▓		▓%
Liabilities and Owner's Equity			
Current liabilities:			
Accounts payable	$ 34,000	▓%	
Notes payable	46,000	▓	
Total current liabilities	$▓		▓%
Long-term liabilities:			
Mortgage	229,000	▓	
Total liabilities	$▓		▓%
Owner's equity	265,000	▓	
Total liabilities and owner's equity	$▓		▓%

5.

KATZ ALARM SYSTEMS

Comparative Income Statement for Years Ending Dec. 31, 19X2 and 19X1

	19X2	19X1	INCREASE OR (DECREASE) AMOUNT	PERCENT	PERCENT OF NET SALES 19X2	19X1
Income:						
Net sales	$166,000	$170,000				
Cost of goods sold:						
Inventory, January 1	$ 53,000	$ 39,000				
Purchases	91,000	94,000				
Goods available for sale	$▮	$▮				
Inventory, December 31	60,000	53,000				
Cost of goods sold	▮	▮				
Gross profit	$▮	$▮				
Expenses:						
Salaries	$ 56,000	$ 60,000				
Advertising	0	1,400				
Utilities	3,000	2,800				
Depreciation	8,000	8,000				
Property taxes	5,000	5,200				
Total expenses	▮	▮				
Net profit	$▮	$▮				

6.   **NOOR FLOOR CO.**

Comparative Income Statement for Years Ending Dec. 31, 19X2 and 19X1

	19X2	19X1	INCREASE OR (DECREASE) AMOUNT	PERCENT	PERCENT OF NET SALES 19X2	19X1
Income:						
Net sales	$325,000	$300,000				
Cost of goods sold:						
Inventory, January 1	$ 90,000	$ 82,000				
Purchases	135,000	150,000				
Goods available for sale	$	$				
Inventory, December 31	81,000	90,000				
Cost of goods sold						
Gross profit	$	$				
Expenses:						
Salaries	$ 90,000	$ 60,000				
Rent	44,000	42,500				
Advertising	6,000	6,000				
Depreciation	5,000	5,500				
Utilities	3,000	2,800				
Miscellaneous	5,000	2,200				
Total expenses						
Net income	$	$				

7.

LEROSEN FURNITURE CO., INC.
Comparative Balance Sheet, December 31, 19X2 and 19X1

	19X2	19X1	INCREASE OR (DECREASE)		PERCENT OF TOTAL ASSETS	
			AMOUNT	PERCENT	19X2	19X1
Assets						
Current assets:						
Cash	$ 7,000	$ 8,000				
Marketable securities	1,500	2,000				
Accounts receivable	83,500	85,000				
Inventory	60,000	45,000				
Total current assets	$▩	$▩				
Total plant assets	600,000	610,000				
Total assets	$▩	$▩				
Liabilities and Equity						
Current liabilities	$ 37,500	$ 40,000				
Long-term liabilities	197,000	210,000				
Total liabilities	$▩	$▩				
Stockholders' equity:						
Preferred stock	$ 75,000	$ 75,000				
Common stock	330,000	325,000				
Retained earnings	112,500	100,000				
Total equity	▩	▩				
Total liabilities and equity	$▩	$▩				

8.

DAVIE & ASSOCIATES, INC.
Comparative Balance Sheet, December 31, 19X2 and 19X1

	19X2	19X1	INCREASE OR (DECREASE) AMOUNT	PERCENT	PERCENT OF TOTAL ASSETS 19X2	19X1
Assets						
Current assets:						
Cash	$ 27,000	$ 33,000				
Accounts receivable	52,000	48,000				
Inventory	83,000	82,000				
Total current assets	$	$				
Total plant assets	300,000	322,000				
Total assets	$	$				
Liabilities and Equity						
Current liabilities	$ 79,000	$ 85,000				
Long-term liabilities	150,000	172,000				
Total liabilities	$	$				
Stockholders' equity:						
Preferred stock	$ 45,000	$ 45,000				
Common stock	80,000	78,000				
Retained earnings	108,000	105,000				
Total equity						
Total liabilities and equity	$	$				

9. Compute the following ratios or percents, based on the balance sheet for Scott Pharmaceutical Co. in Problem 3. Round all answers to tenths.
   a. Working capital ratio
   b. Acid-test ratio
   c. What percent of total assets are current assets?
   d. What percent of total assets are plant assets?
   e. What percent of total assets are total liabilities?
   f. What percent of total assets is stockholders' equity (owners' equity)?

10. Determine the ratios or percents in Problem 9, based on the balance sheet for Madison & Co. in Problem 4.

11. Calculate the ratios or percents requested below, which pertain to the 19X2 business year of LeRosen Furniture Co., Inc. Use the income statement in Problem 1 and the balance sheet in Problem 7. Round all answers to tenths.
   a. Working capital ratio
   b. Acid-test ratio
   c. Rate (percent) of net income (after taxes) to average net worth
   d. Ratio of net sales to average total assets
   e. Accounts receivable turnover
   f. Average age of accounts receivable
   g. What percent of net sales was the cost of goods?

h. What percent of net sales was the gross profit?
i. What percent of net sales were the operating expenses?
j. What percent of net sales was net profit (both before and after taxes)?
k. What percent of total assets were current assets?
l. What percent of total assets were plant assets?
m. What percent of total assets were total liabilities?
n. What percent of total assets was total net worth?

12. Compute the ratios or percents listed in Problem 11 pertaining to the fiscal year ending December 31, 19X2, for Davie & Associates, Inc. The figures needed to compute these ratios are found in Problems 2 and 8. Round all answers to tenths.

SECTION 3
INVENTORY TURNOVER

One indication of a business's stability and efficiency is the number of times per year that its inventory is sold—its **inventory** (or **stock**) **turnover.** This varies greatly with different types of businesses. For instance, a florist might maintain a relatively small inventory but sell the complete stock every few days. A furniture dealer, on the other hand, might stock a rather large inventory and have a turnover only every few months.

Regardless of what might be considered a desirable turnover for a business, the methods used to calculate this turnover are similar. One business statistic that is used in this calculation is the average value of the store's inventory. Inventory may be evaluated at either its cost or its retail value, although it is probably more often valued at retail.

Average inventory for a period is found by adding the various inventories and dividing by the number of times the inventory was taken.

Example 1   Inventory at retail value at the Village Bookstore was $15,950 on January 1, $14,780 on June 30, and $17,270 on December 31. Determine the average inventory.

$$
\begin{array}{r}
\$15,950 \\
14,780 \\
17,270 \\
\hline
3)\overline{\$48,000} = \$16,000
\end{array}
$$ Average inventory (at retail) for the year

Stock turnover is found as follows:

$$\text{Inventory turnover (at retail)} = \frac{\text{Net sales}}{\text{Average inventory (at retail)}}$$

$$\text{Inventory turnover (at cost)} = \frac{\text{Cost of goods sold}}{\text{Average inventory (at cost)}}$$

Turnover is more frequently computed at retail than at cost. Many modern stores mark their goods only with retail prices, which makes it almost impossible to take inventory at cost. Another reason is that turnover at cost does not take into consideration the markdown on merchandise that was sold at a reduced price or the loss on goods that were misappropriated by shoplifters—a constant problem in many retail businesses. Turnover at retail and at cost will be the same only if all merchandise is sold at its regular price; ordinarily, the turnover at retail is slightly less than at cost.

Example 2 The bookstore in Example 1 had net sales during the year of $56,800. What was the stock turnover for the year at retail?

$$\text{Inventory turnover} = \frac{\text{Net sales}}{\text{Average inventory (at retail)}}$$

$$= \frac{\$56,800}{\$16,000}$$

$$= 3.55 \text{ times} \quad \text{(at retail)}$$

Example 3 The income statement of the store above shows that its $43,200 "cost of goods sold" is 75% of net sales. Find the bookstore's inventory turnover at cost.

The average inventory at retail (sales) value was $16,000. In general, the store's goods cost 75% of sales value. Thus, the average inventory cost is computed as follows:

$$\text{Cost} = 75\% \text{ of sales}$$

$$C = \frac{3}{4} \times \$16,000$$

$$C = \$12,000$$

The average inventory at cost was $12,000. Thus, inventory turnover at cost is

$$\text{Inventory turnover} = \frac{\text{Cost of goods}}{\text{Average inventory (at cost)}}$$

$$= \frac{\$43,200}{\$12,000}$$

$$= 3.6 \text{ times} \quad \text{(at cost)}$$

Note. When inventory is taken at retail, the firm must then calculate the cost of its inventory for the financial statements using historical percentages (similar to the illustration above).

SECTION 3 PROBLEMS

1. Inventories were taken at retail during the year at a sporting goods store as follows: $86,000, $99,000, $57,000, and $76,000. What was the average inventory at retail?

2. Determine the average inventory at retail for a convenience store. The inventory counts at four different times during the period were: $75,000, $63,000, $72,000, and $84,000.

3. A bookstore took inventory in January at cost. Later in the year, two other inventories at retail were taken. The January inventory was $20,000 and the later inventories were $35,000 and $33,000.

   a. Determine the retail value of the first inventory if the store adds 25% of cost to obtain the selling price.

   b. What is the average inventory at retail?

4. The first quarterly inventory of the year taken at cost was $74,000. The other inventories for the year were taken at retail and valued at $100,000, $104,000, and $102,000.

   a. What was the first inventory at retail if the store adds 40% of cost to obtain its selling price?

   b. What was the average inventory at retail?

5. Net sales for the store in Problem 1 were $318,000, and the cost of goods sold was $206,700. What was the inventory turnover at retail?

6. Net sales for the store in Problem 2 were $279,300, and the cost of goods sold was $176,400. What was the inventory turnover at retail?

7. An analysis of the pricing structure used by the sporting goods store in Problems 1 and 5 shows that the value of its inventory at cost is typically 66⅔% of the value of its inventory at retail. Using this information,

   a. Determine the average inventory at cost.

   b. Calculate the stock turnover at cost.

8. Refer to the company in Problems 2 and 6. The value of the store's inventory at cost is usually 60% of the value of its inventory at retail. Using this information,

   a. Determine the average inventory at cost.

   b. Calculate the inventory turnover at cost.

9. A store took the following inventories at retail: $75,000, $63,000, $66,000, and $64,000. The income statement reveals that the net sales were $281,400 and cost of goods sold was $190,950 for the same period. What was the inventory turnover at retail?

10. A business took the following inventories at retail: $116,000, $124,000, $125,000, and $126,000. Its income statement showed net sales to be $392,800 and the cost of goods sold to be $250,410. What was its inventory turnover at retail?

11. The value of the inventory at cost is usually 75% of the value of the inventory at retail for the store in Problem 9.

    a. Determine the average inventory at cost.

    b. What is the inventory turnover at cost?

12. Refer to the business in Problem 10. If its inventory value at cost is typically equal to 60% of its inventory value at retail,

   a. What is the average inventory at cost?

   b. What is the inventory turnover at cost?

SECTION 4

INVENTORY VALUATION

We have used inventory values without previously considering the process by which the inventory was evaluated. Some firms (particularly those using computers) keep a **perpetual inventory:** Each new inventory item is cost-coded and its subsequent sale is recorded, so that the firm knows at all times the exact content and value of its inventory. Many firms, however, take a **periodic inventory:** At regular intervals a physical count is made of inventory items. Four methods by which a periodic inventory may be taken are described in this section.

If each item is cost-coded, the periodic inventory may be taken by the **specific identification** method, where the inventory value is simply the total of the individual items. Evaluating items is not always an obvious process, however; inventories often contain many indistinguishable items that have been purchased at different times and for different amounts (such as valves in a plumbing supply house). If the items are small, it may be impractical to cost-code each item; thus, there would be some question about what value to attach to each. (This is a primary reason for the popularity of taking inventory at retail value—the current selling price of merchandise is always known.) One solution is to use the **weighted average method** of inventory valuation, where an average cost is computed like the weighted means we studied in Chapter 4.

The **first-in, first-out (FIFO) method** is probably most commonly used. This method assumes that those items that were bought first were also sold first and that the items still in the inventory are those that were purchased most recently. This method conforms to actual selling practice in most businesses. The number of items on hand are valued according to the invoice prices most recently paid for the corresponding number of items.

The **last-in, first-out (LIFO) method** assumes that the items most recently bought are those items that have been sold and that the items still in stock are the oldest items. Although this method is contrary to most actual selling practice, there is certain justification from an accounting standpoint: On the income statement, the LIFO method uses "current" costs (which are usually higher) for the cost of goods sold and thus produces a smaller gross profit, which many accountants argue is more realistic in an inflationary economy. Accepted accounting procedure requires that income statements specify the method(s) used to evaluate the inventories listed.

Example 1   Assume that a new valve came on the market this year and that the following purchases were made during the year: 15 at \$5, 9 at \$6, and 12 at \$8. Twenty valves were sold (leaving 16 in inventory) for a total net sales of \$300. Evaluate the ending inventory by the following methods: (a) weighted average, (b) FIFO, and (c) LIFO. (d) Compare the effect of each inventory method on the income statement.

(a) **Weighted Average Method**

$$
\begin{array}{rl}
15 \ @\ \$5 = & \$\ 75 \\
9 \ @\ \$6 = & 54 \\
+\underline{12} \ @\ \$8 = +& \underline{96} \\
36 \text{ items} & \$225 \qquad \text{Total cost} \\
\text{purchased} &
\end{array}
$$

$$
\frac{\$225}{36} = \$6.25 \qquad \text{Average cost}
$$

16 items @ \$6.25 = \$100   Inventory value at cost by weighted average method

Calculator techniques . . .   FOR EXAMPLE 1

15 $\boxed{+}$ 9 $\boxed{+}$ 12 $\boxed{=}$ $\longrightarrow$ 36 items purchased

15 $\boxed{\times}$ 5 $\boxed{M+}$ $\longrightarrow$ 75;   9 $\boxed{\times}$ 6 $\boxed{M+}$ $\longrightarrow$ 54;

12 $\boxed{\times}$ 8 $\boxed{M+}$ $\longrightarrow$ 96; $\boxed{MR}$ $\longrightarrow$ 225 total cost

(225) $\boxed{\div}$ 36 $\boxed{=}$ $\longrightarrow$ 6.25 average cost

(6.25) $\boxed{\times}$ 16 $\boxed{=}$ $\longrightarrow$ 100 inventory value

Other methods of inventory valuation use steps similar to the first three lines, where a specific number of items at some definite price(s) produce a total inventory value.

(b) **FIFO Method**

The 16 items remaining are assumed to be the last 12 plus 4 from the previous order:

$$
\begin{array}{rl}
12 \ @\ \$8 = & \$\ 96 \\
+\ \underline{4} \ @\ \$6 = +& \underline{24} \\
16 & \$120 \qquad \text{Inventory value} \\
& \qquad\qquad\ \ \text{at cost by FIFO}
\end{array}
$$

(c) **LIFO Method**

The 16 items remaining are assumed to be the first 15 plus 1 from the second order:

$$
\begin{array}{lll}
15 \ @ \ \$5 = & \$75 \\
+ \ \underline{1} \ @ \ \$6 = + & \underline{\ 6} \\
16 & \$81 & \text{Inventory value} \\
& & \text{at cost by LIFO}
\end{array}
$$

(d) **Income Statements**

	WEIGHTED AVERAGE		FIFO		LIFO	
Net sales		$300		$300		$300
Cost of goods sold:						
Beginning inventory	$ 0		$ 0		$ 0	
Purchases	+ 225		+ 225		+ 225	
Goods available	$225		$225		$225	
Ending inventory	− 100		− 120		− 81	
Cost of goods		− 125		− 105		− 144
Gross profit		$175		$195		$156
Gross profit as a % of net sales		58.3%		65%		52%

SECTION 4 PROBLEMS

Evaluate the following inventories by (1) weighted average, (2) FIFO, and (3) LIFO.

	PURCHASES	ENDING INVENTORY		PURCHASES	ENDING INVENTORY
1. a.	7 @ $ 9.00	8 units	**b.**	50 @ $6.30	52 units
	8 @ 9.75			60 @ 6.40	
	5 @ 10.00			40 @ 6.45	
2. a.	10 @ $ 5.10	15 units	**b.**	25 @ $7.40	40 units
	30 @ 5.20			45 @ 7.50	
	20 @ 5.25			30 @ 7.65	

3. A wholesaler for garden and lawn products purchased lawn chairs from a manufacturer at various times during the year at the following prices: 1,700 @ $8.30; 1,600 @ $8.40; 1,000 @ $8.45; and 1,200 @ $8.50. Evaluate the inventory cost by the following methods, assuming that 2,400 chairs are left in the inventory at the end of the accounting period.

a. Weighted average

b. FIFO

c. LIFO

4. An electronics store purchased television sets from a manufacturer for the following prices during the year: 31 @ $116; 32 @ $118; 40 @ $120; 44 @ $121; 28 @ $125; and 25 @ $126. At the end of the year, there were 10 sets left in inventory. Evaluate the inventory cost by the following methods:
   a. Weighted average
   b. FIFO
   c. LIFO

5. On April 30, the end of the quarter, Good's Bike Shop had 15 units in the inventory. The beginning inventory and purchases during the period were as follows:

   January 1 inventory   25 @ $80
   February 15 purchases   22 @ $81
   April 9 purchases   13 @ $82

   Determine the cost of the ending inventory by the following methods:
   a. Weighted average
   b. FIFO
   c. LIFO

6. On June 30, the end of Raylon Co.'s semiannual accounting period, there were 60 units in inventory. The beginning inventory and the purchases during the period were as follows:

   January 1 inventory   40 @ $35
   March 3 purchases   80 @ $34
   May 5 purchases   60 @ $32
   June 10 purchases   70 @ $30

   Determine the cost of the ending inventory by the following methods:
   a. Weighted average
   b. FIFO
   c. LIFO

7. Net sales for the lawn chairs in Problem 3 were $35,000. Complete a partial income statement that shows gross profit (dollars and percent) by each of the three inventory methods.

8. Net sales for the electronics store in Problem 4 were $40,000. Complete a partial income statement that shows gross profit (dollars and percent) by each of the three inventory methods.

CHAPTER 10 GLOSSARY

Accounts receivable (or payable). Money owed to a business (or owed by a business) for merchandise or services bought on credit.

Accounts receivable turnover. The ratio of net sales to average accounts receivable.

Acid-test (or quick) ratio. The ratio of quick assets to current liabilities; the minimum acceptable ratio is 1 : 1.

Assets. Resources owned by a business.

Average age of accounts receivable. In days: 365 divided by the accounts receivable turnover.

Balance sheet. A financial statement indicating how much a business is worth on a specific date: Assets = Liabilities + Net worth.

Contributed capital. The portion of stockholders' equity derived from the owners' investment in the business.

Current assets.   The resources that are owned by a business and that will be converted to cash or consumed within one operating cycle or one year of the balance sheet date.

Current liabilities.   The debts of a business that will be paid within one operating cycle or one year of the balance sheet date.

First-in, first-out (FIFO) method.   The inventory costing method which assumes that the first items purchased were also sold first and that present inventory items are the ones most recently purchased.

Gross profit (margin or markup).   The amount remaining when cost of merchandise is subtracted from sales income: $S - C = M$.

Horizontal analysis.   Financial statement analysis in which each item in one year is compared (amount and/or percent) with that same item in the previous year.

Income statement.   A financial statement showing a business's sales, expenses, and profit (or loss) during a specified time period: Sales − Cost of Goods − Expenses = Net profit.

Inventory turnover.   The number of times per year that the average inventory (stock) is sold:

$$\left(\begin{array}{c}\text{Turnover} \\ \text{at cost}\end{array}\right) = \frac{\text{Cost of goods sold}}{\text{Average inventory (at cost)}}$$

$$\left(\begin{array}{c}\text{Turnover} \\ \text{at retail}\end{array}\right) = \frac{\text{Net sales}}{\text{Average inventory (at retail)}}$$

Last-in, first-out (LIFO) method.   The inventory costing method which assumes that the items most recently bought are the ones that have been sold and that present inventory contains the oldest items.

Liabilities.   Money that a business owes to others.

Long-term liabilities.   Debts of a business that do not mature or come due until more than one year after the balance sheet date.

Margin or markup.   (See "Gross profit.")

Net income (or net profit).   The amount remaining after both cost of goods and operating expenses are deducted from sales income.

Net income (after taxes) to average net worth.   A ratio that indicates the owners' return on investment.

Net sales to average total assets.   A ratio indicating whether sales volume is in line with the capital assets of a business.

Net worth (net ownership or net investment).   The owners' share of a firm's assets; the amount remaining when liabilities are subtracted from assets.

Operating expenses (or overhead).   Business expenses other than the cost of materials or merchandise. (Examples: Salaries, occupancy expenses, advertising, office supplies, etc.)

Owners' equity.   (See "Net worth.")

Periodic inventory.   An inventory taken at a regular interval by a physical count of items.

Perpetual inventory.   An inventory that is continuously updated by recording the purchase and subsequent sale of each item.

Plant assets.   The long-term, tangible assets that are used repeatedly by a business in the production of income.

Quick (or liquid) assets.   Assets that can easily and quickly be converted to cash; all current assets except inventory and supplies.

Retained earnings.   That part of net profit that is kept for use in the business (rather than being distributed to the owners).

Specific identification.   The inventory costing method in which each item is cost-coded and the total inventory is the sum of these costs.

Stockholders' equity.   Net worth owned by stockholders of a corporation.

Total capital turnover.   (See "Net sales to average total assets.")

Vertical analysis.   Financial statement analysis (pertaining to 1 year only) in which each item is compared to one significant figure on the statement (that is, as a percent of net sales on the income statement or as a percent of total assets on the balance sheet).

Weighted average method.   The inventory costing method whereby each item is valued at the weighted average (or mean) of all costs paid during the time interval.

Working capital (or current) ratio.   A ratio of current assets to current liabilitites; the minimum acceptable ratio is 2:1.

11

DISTRIBUTION OF
PROFIT AND LOSS

OBJECTIVES

Upon completion of Chapter 11, you will be able to:

1. Define and use correctly the terminology associated with each topic.

2. Compute the dividend to be paid per share of common and preferred stock, given the total dividend declared and information about the number and types of stock (Section 1: Examples 1–5; Problems 1–22).

3. Compute the earnings per share on common stock, given the net income and information about the number and types of stock (Section 1: Example 5; Problems 3, 4, 21, 22).

4. Distribute profits (or losses) to the members of a partnership, given the total profit (or loss) and information about each partner's investment. Distributions will be according to (Section 2):

   a. Agreed ratios or no agreement (Examples 1–3, 5; Problems 1–12, 15–22)

   b. Average investments per month (Example 4; Problems 13–16)

   c. Salaries (Examples 5, 6; Problems 1, 2, 7, 8, 11, 12, 17–22)

   d. Interest on investments (Example 6; Problems 1, 2, 15–22)

   e. Combinations of these methods (Examples 5, 6; Problems 1, 2, 7, 8, 11, 12, 15–22).

Within our system of free enterprise, businesses are operated for the ultimate purpose of making a profit. After the amount of net profit (or loss) of a business has been determined, it is necessary to apportion that amount among the owners. Methods of distribution depend largely upon the type of business. In general, there are three categories of private business organization: *sole proprietorship, partnership,* and *corporation.*

SOLE PROPRIETORSHIP

Sole proprietorship is the simplest form of business. A single person makes the initial investment and bears the entire responsibility for the business. If a profit is earned, the entire profit belongs to the individual owner. By the same token, the sole proprietor must assume full responsibility for any loss incurred. (One principal disadvantage of this form of business is that the owner's liability extends even to personal possessions not used in business; for instance, in many states the proprietor's home might have to be sold to discharge business debts.)

Sole proprietorships are usually small businesses, since they are limited to the resources of one person. There often is no formal organization or operating procedure; an individual with sufficient skill or financial resources simply goes into business by making services available to potential clients.

Since all profit (or loss) resulting from a sole proprietorship belongs to the individual owner, there is no mathematical distribution required. Hence, the problems in this section will be limited to those concerning partnerships and corporations.

PARTNERSHIPS

A **partnership** (or general partnership) is an organization in which two or more persons engage in business as co-owners. A partnership may provide the advantages of increased investments and more diversified skills, as well as more manpower available to operate the business. (For example, individuals A, B, and C might form a partnership because A had the funds to start the business, B had the skill and experience required to manufacture their product, and C was an expert in sales and marketing. Thus, by pooling their resources, both financial and personal, a profitable business can be formed; however, none of the partners could have established a profitable business independently.) Partnerships often include a **silent partner**—one who takes no active part in the business operations, whose contribution to the partnership is entirely financial, and whose very existence as a partner may not be known outside the business.

A partnership may be entered into simply by mutual consent. To avoid misunderstanding and disagreement, however, all partners should sign a legal agreement (called *articles of partnership*) setting forth the responsibilities, restrictions, procedures, method of distribution of profits and losses, and so forth, of the partnership. (For instance, the abrupt withdrawal of a partner, if not prohibited by written agreement, might place the entire partnership in bankruptcy.)

Taking into consideration the contribution made by each partner, a partnership is permitted to distribute profits or losses in any way that is acceptable to the partners involved. In the absence of any formal written agreement, the law would require that profits be shared equally among the partners. Despite whatever agreement the partners may have among themselves, however, in most states the partners are responsible "jointly and separately" for the liabilities of the partnership. This means that if the personal resources of the partners together (jointly), when calculated according to their terms of agreement, are not sufficient to meet the partnership's liabilities, then a single partner with more personal wealth will have to bear the liabilities of the partnership individually (separately). Also, a business commitment made by any partner, even if it is not within that person's agreed area of responsibility, is legally binding upon the entire partnership.

A partnership is automatically dissolved whenever there is any change in the partners involved, such as when a partner withdraws from the business, sells to another person, or dies, or when an additional person enters the partnership. However, a new partnership may be created immediately to reflect the change, without any lapse in business operations.

Almost all states also allow a form of partnership called a *limited partnership,* which requires a legal document to establish. Certain members, known as *limited partners,* contribute only capital funds and take no active part in managing the endeavor. One or more *general partners* assume full responsibility for managing the business, as well as for its financial obligations. The distinguishing characteristic of a limited partnership is that the financial responsibility of the limited partners is restricted to the amount of capital they have invested. A common example of limited partnerships is in real estate projects, where numerous limited partners make investments and a few general partners develop, build, or sell the property. Our study, however, will concentrate on the general partnership, where all members function equally as co-owners.

CORPORATIONS

The **corporation** is the most complex form of business. It is considered to be an artificial entity, created according to state laws. Its capital is derived from the sale of **stock certificates**, which are certificates of ownership in the business. A person's ownership in a corporation is measured according to the number of shares of stock that (s)he owns. Figure 11-1 illustrates a stock certificate.

The corporation offers a distinct advantage over both the sole proprietorship and the partnership in that its liability extends only to the amount invested in the business by the stockholders. That is, the personal property of the owners (**stockholders**) may not be confiscated to meet business liabilities. (It should be noted that even small businesses may be incorporated in order to provide this protection.) The corporation also differs from the sole proprietor-

By permission of Martin Marietta Corporation.

FIGURE 11-1   Specimen Stock Certificate

ship and the partnership in that the firm must pay income taxes on its net operating income.

When a corporation is formed, it receives a *charter* from the state, authorizing the operation of business, defining the privileges and restrictions of the firm, and specifying the number of shares and type of stock that may be issued. The stockholders elect a **board of directors,** which selects the *administrative officers* of the corporation. Although it is ultimately responsible to the stockholders, the board actually makes decisions regarding policy and the direction that the business will take, and the officers are responsible for carrying out the board's decisions. The board of directors also makes the decision to distribute profits to the stockholders; the income declared on each share of stock is called a **dividend.**

When a corporation issues stock, the certificates may be assigned some value, called the **par** or **face value.** The par value may be set at any amount, although it is typically some round number. Stock is usually sold at par value when a corporation is first organized. After the firm has been in operation for some time, the *market value* (the amount for which stockholders would be willing to sell their stock) will increase or decrease, depending on the prosperity of the business, and the par value becomes insignificant. Thus, stock is often issued without any face value, in which case it is known as **non-par-value** or **no-par-value stock.**

There are two main categories of stock: *common stock* and *preferred stock*. Since there are fewer methods for distributing profits to stockholders than to partners, we shall consider first the distribution of corporate profits.

SECTION 1

DISTRIBUTION OF PROFITS IN A CORPORATION

COMMON STOCK

Common stock carries no guaranteed benefits in case a profit is made or the corporation is dissolved. It does entitle its owner to one vote in a stockholders meeting for each share (s)he owns. Many corporations issue only common stock. In this case, distributing the profits is simply a matter of dividing the profit to be distributed by the **stock outstanding** (the number of shares of stock that have been issued).

Example 1   The board of directors of the Wymore Corp. authorized $258,000 of the net profit to be distributed among its stockholders. If the corporation has issued 100,000 shares of common stock, what is the dividend per share?

$$\frac{\text{Total dividend}}{\text{Number of shares}} = \frac{\$258,000}{100,000} = \$2.58 \text{ per share}$$

A dividend of $2.58 per share of common stock will be declared.

PREFERRED STOCK

The charters of many corporations also authorize the issuance of **preferred stock.** This stock has certain advantages over common stock; most notably, preferred stock carries the provision that dividends of a specified percent of par value (or of a specified amount per share, on non-par-value stock) must be paid to preferred stockholders before any dividend is paid to common stockholders. For example, a share of 6%, $100 par-value stock would be expected to pay a $6 dividend each year (usually in quarterly payments of $1.50 each).

Also, in the event that the corporation is dissolved, the corporation's assets may first have to be applied toward refunding the par value of preferred stock, before any compensation is returned to common stockholders. Holders of preferred stock usually do not have a vote in the stockholders meeting. Also, most preferred stock is **nonparticipating,** which means it cannot earn dividends above the specified rate. (However, some preferred stock is **participating,** which permits dividends above the stated percent when sufficient profits exist.

Participating stock is frequently restricted so that additional earnings above the specified rate are permitted only after the common stockholders have received a dividend equal to the special preferred dividend.)

Even when a net profit exists, a board of directors may not think it advisable to declare a dividend. Thus, the owners of a corporation may receive no income on their investment during some years. **Cumulative preferred stock** provides an exception to this situation, however. As the name implies, unpaid dividends accumulate; that is, dividends due from previous years when no dividend was declared, as well as dividends from the current year, must first be paid to owners of cumulative preferred stock before any dividends are paid to other stockholders.

As an alternative to paying cash dividends, boards of directors may declare *stock dividends,* where each stockholder receives an additional share for each specified number of shares already owned. The stockholders thus obtain an increased investment but do not have to pay income taxes as they would on cash dividends. When a stock dividend is declared, the dividend for many stockholders would include a fractional share. Thus, the board assigns a value to each whole share (often slightly less than market value) and either pays the stockholder this value for any fractional share or else allows the stockholder to purchase the remaining portion of another whole share.

Example 2   Artcraft Sales, Inc., has 1,000 shares of 5%, $100 par-value preferred stock outstanding and 10,000 shares of common stock. The board declared a $50,000 dividend. What dividends will be paid on (a) the preferred stock and (b) the common stock?

(a)   Dividends on preferred stock are always computed first, as follows:

$$\text{Total preferred dividend} = \underbrace{\% \times \text{Par value}} \times \begin{pmatrix} \text{Number of} \\ \text{preferred shares} \end{pmatrix}$$

$$= \begin{pmatrix} \text{Dividend} \\ \text{per share} \end{pmatrix} \times \begin{pmatrix} \text{Number of} \\ \text{preferred shares} \end{pmatrix}$$

5% × $100 par value = $5   Dividend per share of preferred stock

$5 per share × 1,000 shares = $5,000   Required to pay dividends on all preferred stock

(b)
$50,000	Total dividend
− 5,000	Dividend on all preferred stock
$45,000	Dividend allotted to common stock

$$\frac{\text{Total common stock dividend}}{\text{Number of common shares}} = \frac{\$45,000}{10,000} = \$4.50 \quad \begin{array}{l}\text{Dividend} \\ \text{per share of} \\ \text{common stock}\end{array}$$

Hence, the preferred stockholders receive their full dividend of $5 per share, and a dividend of $4.50 per share will be paid to common stockholders.

Example 3   Because of low sales during the following year of operations, Artcraft Sales, Inc., declared a dividend of only $4,500. If the corporation still has 1,000 shares of 5%, $100 par-value preferred and 10,000 shares of common outstanding, what dividend per share will be paid?

Because $4,500 is not even enough to pay the required $5,000 preferred dividend (5% × $100 × 1,000 shares), the $4,500 dividend will be divided by the 1,000 shares of preferred stock.

$$\frac{\$4,500}{1,000} = \$4.50 \quad \text{Per share for preferred}$$

$$0 \quad \text{Per share for common}$$

Example 4   No dividend was paid last year by McGill, Inc. A dividend of $325,000 has now been declared for the current year. Stockholders own 100,000 shares of common stock and 25,000 shares of 6%, $50 par-value cumulative preferred stock. How much dividend will be paid on each share of stock?

Dividends for 2 years must first be paid on the cumulative preferred stock.

$$6\% \times \$50 \text{ par value} = \quad \$3 \quad \begin{array}{l}\text{Dividend per year on each share} \\ \text{of cumulative preferred stock}\end{array}$$

$$\times \underline{\quad 2} \quad \text{Years}$$

$$\$6 \quad \begin{array}{l}\text{Dividend due on each share of} \\ \text{cumulative preferred stock}\end{array}$$

$$\$6 \text{ per share} \times 25,000 \text{ shares} = \$150,000 \quad \begin{array}{l}\text{Payable to} \\ \text{preferred stockholders}\end{array}$$

$$\begin{array}{rl} \$325,000 & \text{Total dividend} \\ - \underline{\quad 150,000} & \text{Dividend on preferred stock} \\ \$175,000 & \text{Total dividend on common stock} \end{array}$$

$$\frac{\$175,000}{100,000 \text{ shares}} = \$1.75 \quad \text{Dividend per share of common stock}$$

Thus, a 2-year dividend of $6 per share will be paid to owners of the cumulative preferred stock, and $1.75 per share will go to the common stockholders.

EARNINGS PER SHARE

As a means of evaluating a firm's past and potential future performance, investors want to know the **earnings per share** of common stock. This figure represents the amount of net income earned by common shareholders during the year. The number is usually different from the cash dividend per share for common stock. If the board of directors declares a cash dividend for less than the total net income, the earnings per share and dividends per share will be different. Also, if the number of common shares outstanding changes during the year, the two numbers will be different. When the number of shares changes, a weighted-average calculation is required to determine the average number of shares. An in-depth discussion of earnings per share is beyond the scope of this text. We will limit the consideration of earnings per share to simple capital structures and situations in which the common shares outstanding do not change. The calculation for the earnings per share is:

$$\frac{\text{Net income} - \text{Preferred dividend}}{\text{Common shares outstanding}}$$

Example 5 The Powers Co. had a net income of $1,000,000 last year. The board of directors declared a cash dividend of only $600,000, retaining the remainder for future plant expansion. There were 50,000 shares of common stock outstanding and 75,000 shares of 8%, $50 par-value preferred stock.

(a) What was the dividend per share?

(b) What was the earnings per share on common stock?

(a) $8\% \times \$50 = \$4$ Dividend on each preferred share

$\$4 \times 75,000 \text{ shares} = \$300,000$ Total preferred dividend

$$
\begin{array}{rl}
\$600,000 & \text{Total dividend} \\
-\ \underline{300,000} & \text{Total preferred dividend} \\
\$300,000 & \text{Allocation for common dividend}
\end{array}
$$

$$\frac{\$300,000}{50,000} = \$6 \quad \text{Dividend on each common share}$$

(b) $\dfrac{\text{Net income} - \text{Preferred dividend}}{\text{Common shares outstanding}} = \text{Earnings per share}$

$$\frac{\$1,000,000 - \$30,000}{50,000} = \$14 \quad \text{Earnings per share}$$

MUTUAL FUNDS

Stocks, bonds, and other financial investments that are issued by corporations and government bodies and are bought and sold in securities markets are called **securities.** Instead of investing in one security, an investor may wish to diversify with ownership in many different securities by buying shares of a mutual fund. A **mutual fund** is a pooling of money from many investors into a portfolio investment that owns a wide selection of stock and often bonds or other income-producing securities. Many mutual funds specialize in a particular kind of investment, such as income-producing stocks or stocks of companies believed to have outstanding growth potential.

Today, the most popular security investment in the United States is the mutual fund. The popularity of mutual funds is due to their professional management and the reduction of the individual investor's risk, since the investment is spread among many different securities. The investor earns dividends on stocks and interest on bonds that are owned through the mutual fund. Since the investor owns only fractions of numerous stocks and bonds (instead of personally owning whole securities), the investor's return on each security in the fund is a fractional part of its total return.

SECTION 1   PROBLEMS

Determine the dividend that should be paid on each share of stock.

	NUMBER OF SHARES	TYPE OF STOCK	TOTAL DIVIDEND DECLARED
1. a.	80,000	Common	$ 94,400
b.	50,000	Common	200,000
	20,000	8%, $100 par-value preferred	
c.	25,000	Common	250,000
	30,000	7%, $50 par-value preferred, cumulative for 2 years	
2. a.	45,000	Common	$ 63,000
b.	50,000	Common	100,000
	10,000	6%, $100 par-value preferred	
c.	20,000	Common	220,000
	40,000	5%, $50 par-value preferred, cumulative for 2 years	

Determine the dividend per share on each class of stock and the earnings per share on common stock.

		NUMBER OF SHARES	TYPE OF STOCK	NET INCOME	TOTAL DIVIDEND DECLARED
3.	a.	50,000 50,000	Common 6%, $50 par-value preferred	$750,000	$400,000
	b.	40,000 70,000	Common 5%, $100 par-value preferred	800,000	600,000
4.	a.	30,000 60,000	Common 7%, $100 par-value preferred	$900,000	$450,000
	b.	10,000 20,000	Common 8%, $50 par-value preferred	200,000	150,000

5. Whitehall Publishing has 60,000 shares of common stock outstanding. At the end of the year, the board of directors voted a $123,000 dividend. What is the dividend per share?

6. The board of directors of Pape Industries, Inc., declared a $136,000 cash dividend. The firm has 80,000 shares of common stock outstanding. What will be the dividend per share?

7. The board of directors of Walton Industries voted a cash dividend of $370,000. Determine the dividend per share of stock on the 40,000 shares of common stock and 15,000 shares of 6%, $100 par-value preferred stock that are outstanding.

8. Smith Transport, Inc., declared a cash dividend of $175,000. If there are 10,000 shares of common stock and 20,000 shares of 7%, $100 par-value preferred stock outstanding, what dividend per share will be paid to both classes of stockholders?

9. The Dietrich Corporation declared a cash dividend of $300,000 this year. If there are 40,000 shares of common stock and 50,000 shares of 5%, $100 par-value preferred stock outstanding, what is the dividend per share paid to the stockholders?

10. Gofreed Products Corporation declared a cash dividend of $10,000 at the end of its first year of operations. The corporation has 500 shares of common stock and 1,500 shares of 8%, $50 par-value preferred stock outstanding. Compute the dividend on each share of stock.

11. This year a cash dividend of $14,000 has been approved for payment by the board of directors of the Anderson Group, Inc. What dividend per share will be paid if there are 7,000 shares of common and 8,000 shares of 5%, $50 par-value preferred stock outstanding?

12. The Magic Mountain Coal Co. has 100,000 shares of common stock and 65,000 shares of 6%, $100 par-value preferred stock outstanding. Compute the dividend per share on a declared dividend of $325,000.

13. Kock-Coleman Inc. did not declare or pay a cash dividend during the preceding year. This year, the board of directors voted a $55,800 cash dividend. Determine the dividend per share for each class of stock if there are 12,000 shares of common and 3,000 shares of 8%, $100 par-value cumulative preferred stock outstanding.

14. Stewart Art Distributors, Inc., did not declare or pay a cash dividend for the preceding year. This year, however, the board declared a cash dividend of $77,000. Determine the dividend

per share on 5,000 shares of common stock and 4,000 shares of 7%, $50 par-value cumulative preferred stock.

15. For the previous 2 years, the board of directors of Allison Color Lab, Inc., declared no dividends. This year, a $44,000 cash dividend was declared. What dividend per share will be paid on 8,000 shares of common stock and 2,000 shares of 6%, $100 par-value cumulative preferred stock?

16. Everett Heating and Air Conditioning, Inc., did not pay a cash dividend for the preceding two years. The board has declared a dividend of $40,000 this year. What dividend per share will be paid on 8,000 shares of common stock and 3,000 shares of 8%, $50 par-value cumulative preferred stock?

17. Determine the dividend per share this year for a corporation that has a declared dividend of $115,000. Last year no dividend was paid. There are 15,000 shares of common and 5,000 shares of 7%, $100 par-value cumulative preferred stock outstanding.

18. Determine the dividend per share this year for Midas Touch, Inc. Last year, no dividend was paid, but a $130,000 dividend was declared this year. There are 6,000 shares of common stock and 10,000 shares of 5%, $100 par-value cumulative preferred stock outstanding.

19. Good Times Video, Inc., declared a dividend of $160,000 this year, having skipped paying a dividend last year. There are 10,000 shares of common stock and 18,000 shares of 8%, $100 par-value noncumulative preferred stock outstanding. What is the dividend per share this year?

20. The Prentice Group, Inc., declared a cash dividend of $45,000 this year. The board did not pay any dividend last year. Compute the dividend per share for 6,000 shares of common and 10,000 shares of 6%, $50 par-value noncumulative preferred stock.

21. Rockville and Associates, Inc., had a net income of $80,000 last year. The board of directors declared a $50,000 cash dividend for the 6,000 shares of common stock and 8,000 shares of 4%, $100 par-value preferred stock outstanding. Calculate
    a. The dividend per share of each class of stock
    b. The earnings per share on common stock

22. Food Crown, Inc., earned $150,000 in net income last year. Its board declared a cash dividend of $90,000. The supermarket had 10,000 shares of common stock and 15,000 shares of 5%, $50 par-value preferred stock outstanding. Calculate
    a. Dividends per share paid to the stockholders
    b. Earnings per share on the common stock

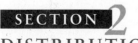

SECTION 2
DISTRIBUTION OF PROFITS IN A PARTNERSHIP

There are numerous ways in which the profits or losses of a partnership may be divided. We shall consider some of the typical methods. Recall that profits and losses are shared equally among partners unless another specific agreement is made.

Example 1 **Equal Distribution**

The partnership of Wynn, Bright & Phillips was organized without a formal agreement regarding the distribution of profits. A net profit of $39,000 was earned during the business year now ending. Their individual investments included $14,000 by Wynn, $8,000 by Bright, and $11,000 by Phillips. How much should each partner receive?

$$\frac{\$39,000}{3} = \$13,000 \text{ each}$$

In the absence of a formal agreement, each partner would receive an equal distribution of $13,000.

Example 2   **Agreed Ratio**

After considering both the capital investment and the time devoted to the business by each partner in Example 1, partners Wynn (W), Bright (B), and Phillips (P) agreed to share profits (or losses) in the ratio of 4:3:6. Divide their $39,000 profit.

The sum of $4 + 3 + 6 = 13$ indicates that the profit will be divided into 13 shares; W will receive 4 of the 13 shares, B will receive 3 shares, and P will receive 6 shares. Thus, W receives $\frac{4}{13}$ of the profit, B gets $\frac{3}{13}$, and P earns $\frac{6}{13}$ of the profit.

$$\text{W:} \qquad \tfrac{4}{13}(\$39,000) = \$12,000$$
$$\text{B:} \qquad \tfrac{3}{13}(\$39,000) = 9,000$$
$$\text{P:} \qquad \tfrac{6}{13}(\$39,000) = \underline{18,000}$$
$$\text{Total profit} \qquad\quad = \$39,000$$

Partner W will receive $12,000, B will receive $9,000, and P gets $18,000. Notice that the sum of all partners' shares equals the total profit of the partnership.

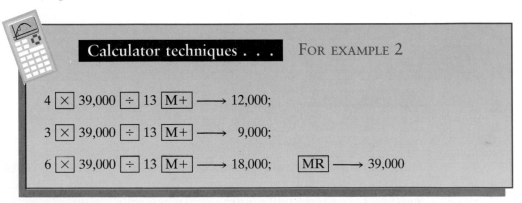

Calculator techniques . . .   FOR EXAMPLE 2

4 ⊠ 39,000 ÷ 13 M+ ⟶ 12,000;

3 ⊠ 39,000 ÷ 13 M+ ⟶ 9,000;

6 ⊠ 39,000 ÷ 13 M+ ⟶ 18,000;   MR ⟶ 39,000

Note.   Agreed distributions are also often expressed as percents, such as a 30%/20%/50% distribution. The given percents must total 100% of the profit, and finding each partner's share is simply a matter of calculating that given percent of the total profit (or loss).

Example 3   **Original Investment**

Partners A and B opened a business with a $10,000 investment by A and an $8,000 investment by B. Partner B later invested another $6,000. Their partnership agreement stipulated that profits or losses would be divided according to the ratio of their original investments. What portion of a $7,200 profit would each partner receive?

$$\$10,000 + \$8,000 = \$18,000 \qquad \text{Total original investment}$$

$$A \text{ had } \frac{\$10,000}{\$18,000} \quad \text{or} \quad \frac{5}{9} \text{ of the original investment}$$

$$B \text{ had } \frac{\$8,000}{\$18,000} \quad \text{or} \quad \frac{4}{9} \text{ of the original investment}$$

Thus, A will receive $\frac{5}{9}$ of the profit and B will receive the other $\frac{4}{9}$:

$$A: \qquad \tfrac{5}{9}(\$7,200) = \$4,000$$

$$B: \qquad \tfrac{4}{9}(\$7,200) = \underline{3,200}$$

$$\text{Total profit} = \$7,200$$

A's share is $4,000 and B has earned $3,200.

Investment at Beginning of Year

A variation of the foregoing "original-investment" distribution is to divide profits according to each partner's investment at the beginning of the year. Suppose that partners A and B (above) use this method during their second year of operations. Partner A still has $10,000 invested, while B's investment is now $8,000 + $6,000 = $14,000, giving a total investment by both partners of $24,000.

Therefore, following the second year of operations,

$$A \text{ would receive } \frac{\$10,000}{\$24,000} \quad \text{or} \quad \frac{5}{12} \text{ of the profit}$$

$$B \text{ would receive } \frac{\$14,000}{\$24,000} \quad \text{or} \quad \frac{7}{12} \text{ of the profit}$$

Example 4   **Average Investment per Month**

When partners C and D began operations on January 1, C had $4,000 invested, which did not change during the year. D had $5,800 invested on January 1; on May 1, he withdrew $200; and on August 1, he reinvested $800. If profits are to be shared according to each partner's average investment per month, how much of a $12,500 profit should each receive?

Partner C's average investment is $4,000, since his investment did not change during the year. D's average investment is computed as follows:

DATE	CHANGE	AMOUNT OF INVESTMENT		MONTHS INVESTED		
January 1	—	$5,800	×	4	=	$23,200
May 1	−$200	5,600	×	3	=	16,800
August 1	+$800	6,400	×	+ 5	=	+32,000
				12		$72,000

$$\frac{\$72,000}{12} = \$6,000 \qquad \text{D's average investment per month}$$

$$\begin{array}{ll} \$\ 4,000 & \text{C's average investment} \\ +\ \underline{6,000} & \text{D's average investment} \\ \$10,000 & \text{Total of average investments} \end{array}$$

Thus, C receives $\dfrac{\$4,000}{\$10,000}$ or $\dfrac{2}{5}$ of the profit. D's investment earns $\dfrac{\$6,000}{\$10,000}$ or $\dfrac{3}{5}$ of the profit.

$$\begin{array}{lll} \text{C:} & \tfrac{2}{5}(\$12,500) = & \$\ 5,000 \\ \text{D:} & \tfrac{3}{5}(\$12,500) = & \underline{\ 7,500} \\ & \text{Total profit} = & \$12,500 \end{array}$$

C earns $5,000 of the total profit, and D receives $7,500.

Example 5 **Salary and Agreed Ratio**

Partners X, Y, and Z agreed to pay X a salary of $10,000, to pay Y a salary of $12,000, and to divided the remaining profit or loss in the ratio of $4:3:5$. If the partnership earned a $40,000 net profit, how much will each partner receive?

The salaries should be deducted first, and the remaining profit is then divided into $4 + 3 + 5 = 12$ shares:

$$\text{Salaries:} \quad \$10,000 + \$12,000 = \$22,000$$

$$\begin{array}{ll} \$40,000 & \text{Total net profit} \\ -\ \underline{22,000} & \text{Salaries} \\ \$18,000 & \text{To be divided in the ratio } 4:3:5 \end{array}$$

X receives $\tfrac{4}{12}$ or $\tfrac{1}{3}$ of the $18,000: $\qquad \tfrac{1}{3}(\$18,000) = \quad \$\ 6,000$

Y gets $\tfrac{3}{12}$ or $\tfrac{1}{4}$ of the $18,000: $\qquad \tfrac{1}{4}(\$18,000) = \qquad 4,500$

Z has earned $\tfrac{5}{12}$ of the $18,000: $\qquad \tfrac{5}{12}(\$18,000) = +\ \underline{\ 7,500}$

$$\$18,000$$

The total net profit is distributed as follows:

	X	Y	Z	CHECK	
Salary	$10,000	$12,000	—	$16,000	X's share
Ratio	6,000	4,500	$7,500	16,500	Y's share
Total	$16,000	$16,500	$7,500	7,500	Z's share
				$40,000	Total net profit

Thus, X's share of the net profit is $16,000, Y earns $16,500, and Z receives $7,500.

Example 6   **Interest on Investment, Salary, and Agreed Ratio**

Partners R and S began operations with an investment of $15,000 by R (a silent partner) and $5,500 by S. They agreed to pay 8% interest on each partner's investment, to pay S a $10,000 salary, and to divide any remaining profit or loss equally. (a) Distribute a profit of $18,640. (b) Distribute a $10,840 net profit.

(a)   The various parts of the distribution are computed in the same order in which the agreement is stated. The interest earned by each partner is thus:

	R	S
Investment	$15,000	$5,500
Interest rate	× 8%	× 8%
Interest	$ 1,200	$ 440

The interest is deducted from the net profit, and then the salary is paid:

$1,200		$18,640	Total profit
+ 440		− 1,640	Interest
$1,640	Total interest	$17,000	
		− 10,000	Salary
		$ 7,000	To be shared equally

$$\frac{\$7,000}{2} = \$3,500 \quad \text{Additional profit due each partner}$$

Each partner thus receives the following portion of the net profit:

	R	S	CHECK	
Interest	$1,200	$ 440	$ 4,700	R's share
Salary	—	10,000	13,940	S's share
Ratio	3,500	3,500	$18,640	Total net profit
Total	$4,700	$13,940		

R receives $4,700 of the net profit, and S earns $13,940.

(b)   The same interest and salary would be due each partner as computed in part (a). The remaining calculations would then be:

$$
\begin{array}{ll}
\$10,840 & \text{Total profit} \\
-\ \ 1,640 & \text{Interest} \\
\hline
\$\ 9,200 & \\
-\ 10,000 & \text{Salary} \\
\hline
-\$\ \ \ 800 & \text{Shortage to be shared equally}
\end{array}
$$

The net profit of the partnership is $800 less than the interest and salary due. According to the terms of their agreement, this shortage will be shared equally by the partners. That is, $800 ÷ 2 or $400 of the amount due each partner will not be paid. Each partner's share is thus computed as follows:

	R	S	CHECK	
Interest	$1,200	$ 440	$ 800	R's share
Salary	—	10,000	10,040	S's share
	$1,200	$10,440	$10,840	Total profit
Ratio (shortage)	− 400	− 400		
Total	$ 800	$10,040		

R receives $800 of the net profit, and S receives $10,040.

SECTION 2   PROBLEMS

Divide the net profit among the partners according to the conditions given.

	PARTNER	INVESTMENT	METHOD OF DISTRIBUTION	NET PROFIT
1. a.	X	$28,000	Ratio of 2:1	$36,000
	Y	30,000		
b.	D	20,000	Ratio of their investment	22,500
	E	30,000		
	F	25,000		
c.	M	16,000	5% interest on investment,	44,300
	N	12,000	and ratio of 1:2:3	
	O	18,000		
d.	P	50,000	7% interest on investment,	80,300
	Q	60,000	$10,000 salary to Q, and	
	R	80,000	the remainder in ratio	
			of their investments	
2. a.	K	$12,000	Ratio of 1:3	$60,000
	L	10,000		
b.	A	5,000	Ratio of their investments	21,600
	B	6,000		
	C	7,000		
c.	S	30,000	6% interest on investment,	16,000
	T	35,000	and ratio of 2:2:1	
	U	15,000		
d.	V	30,000	5% interest on investment,	43,700
	W	28,000	$15,000 salary to V, and	
	X	24,000	remainder in ratio of	
			their investments	

3. James and Ahmed entered into a partnership without an agreement on how profits and losses would be distributed. James invested $48,000 in the business, while Ahmed invested $18,000. James spent 55 hours a week and Ahmed spent 25 hours a week in the business. How would a profit of $24,000 be divided between the two partners?

4. Catlin and Mary entered into a partnership without an agreement stipulating how profits or losses would be distributed. Catlin invested her life savings of $50,000, and Mary invested $10,000. Catlin spent 50 hours and Mary spent 40 hours a week in the business. How much of their first year's profit of $50,000 would each partner receive?

5. Partners N, O, and P agreed to share profits or losses in the ratio of 3:4:1.
   a. What is each partner's share of a $96,000 profit?
   b. How much of a $24,000 loss is each partner's share?

6. Partners Q, R, and S agreed to share profits or losses in the ratio of 2:3:1.
   a. What is each partner's share of a $42,000 profit?
   b. How much of an $18,000 loss is each partner's share?

7. Partners Jay, Kay, and Lai formed a partnership, agreeing to pay salaries of $5,000 to each partner; any additional profits or losses would be divided 20%/30%/50%, respectively. How much of a $38,000 profit would be paid to each partner?

8. A partnership was formed by Bob, Chris, and Doug. The partnership agreement stipulated salaries of $10,000 to each partner, and any additional profits or losses would be divided 25%/30%/45%, respectively. Find each person's share of a $50,000 profit.

9. Roy and Sam, two attorneys, entered into a partnership. Roy invested $20,000 and Sam invested $30,000 initially. Later in the same year, Roy increased his investment by $8,000. The partners had agreed to distribute profits and losses according to the ratio of their initial investments. What would each partner receive on a $60,000 profit?

10. Julie and Greta entered into a partnership for interior design. Julie invested $16,000 and Greta invested $24,000 initially. Later during the first year, Julie invested another $4,000. The partners agreed to distribute profits and losses according to the ratio of their initial investments. How much of a $75,000 profit would each partner receive?

11. During the following year, the partnership of Roy and Sam (in Problem 9) changed its distribution agreement as follows: Roy would draw a salary of $15,000, Sam would draw a salary of $18,000, and any remaining profits or losses would be divided according to their investment at the beginning of the current year. What would each partner receive from a $62,000 profit?

12. During the second year of operations of Julie and Greta's interior design business (Problem 10), their distribution agreement was changed as follows: Each person would draw a salary of $6,000, and the remaining profit or loss would be divided according to their investments at the beginning of the current year. What is each person's share of a $64,800 profit?

13. Partners A and K decided to distribute profits and losses from their partnership in the ratio of their average investment per month. Partner A invested $16,000 and did not change her investment during the year. K invested $16,000 on January 1 but withdrew $2,000 on April 1. She reinvested $1,500 on August 1, and did not change her investment for the remainder of the year.

    a. What was K's average investment?

    b. How much would each partner receive of a $49,800 profit?

14. Partners Steve and Tom distribute profits and losses from their construction business in the ratio of their average investment per month. Steve invested $26,000 and did not change his investment during the year. Tom invested $20,000 on January 1 but withdrew $4,000 on June 1 and reinvested $1,000 on September 1.

    a. What was Tom's average investment?

    b. How much of a $77,000 profit would each partner receive?

15. Partners M and N have agreed to distribute profits and losses as follows: 8% interest will be paid on their average investment, and the remaining profit or loss will be divided equally. M's investment of $18,000 on January 1 was unchanged. N invested $12,000 on January 1; he invested $4,000 more on May 1; and he withdrew $1,000 on November 1.

    a. What was N's average investment?

    b. How much is each partner's share of a $20,000 profit?

16. Mark and Pete became partners on January 1. Mark invested $22,000 and did not change his investment during the year. Pete invested $25,000 on January 1, withdrew $3,000 on March 1, and reinvested $6,000 on July 1. The partners have agreed to distribute profits by paying 6% interest on their average investments, with the remaining profits or losses to be divided equally.

    a. What was Pete's average investment?

    b. If the profit was $30,000, how much should each partner receive?

17. J and E formed a partnership, with J investing $17,000 and E investing $20,000. Their agreement was to pay 5% interest on investments, pay a salary of $8,000 to J, and share the remaining profit or loss equally.

   a. Distribute a net profit of $14,000.

   b. Distribute a net profit of $9,000.

18. Margaret and Katherine formed a partnership, with Margaret investing $16,000 and Katherine investing $30,000. They have agreed to pay 6% interest on investments, pay a salary of $5,000 to Margaret, and share the remaining profits or losses equally.

   a. Determine each partner's share of a $14,000 profit.

   b. Determine each partner's share of a $7,000 profit.

19. Cameron invested $24,000 and Paulson invested $18,000 in a partnership. They agreed to pay 6% interest on their investments, to pay a salary of $6,000 to Cameron and $4,000 to Paulson, and to divide the remaining profit or loss in the ratio of their investments.

   a. What is each partner's share of a $12,800 profit?

   b. What is each partner's share of an $8,320 profit?

20. Partners Tim, Chrisy, and Barbara have agreed to allocate profits and losses in the following way: 7% interest on their respective investments of $10,000, $9,000, and $8,000; salaries of $4,000 to each; and any remainder to be divided in the ratio of their investments.

   a. What is each partner's share of a $27,390 profit?

   b. What is each partner's share of a $13,350 profit?

21. Stewart, Thomas, and Udall formed a partnership and agreed to share profits and losses in the following way: 7% interest on their respective investments of $8,000, $6,000, and $11,000; salaries of $2,000 to Stewart and $3,000 to Udall; and any remainder to be divided in a 3:1:4 ratio.

   a. Distribute a profit of $16,750.

   b. Distribute a profit of $4,750.

22. Partners Richard, Sarah, and Ted have agreed to allocate profits and losses in the following way: 5% interest on their respective investments of $20,000, $40,000, and $30,000; salaries of $5,000 to be paid to Sarah and Ted; and any remainder to be divided in the ratio of 1:3:2.

   a. Distribute a profit of $16,000.

   b. Distribute a profit of $13,300.

CHAPTER *11* GLOSSARY

Board of directors.   Officials elected by stockholders to set policy and decide the direction of a corporation, including determining whether/when dividends will be paid.

Common stock.   A share of corporation ownership that includes no guarantee about the size of the dividend.

Corporation.   A business established as an artificial entity according to state law and owned by stockholders, whose liability for the business is limited to the amount invested in the stock.

Cumulative preferred stock.   Preferred stock which guarantees that dividends unpaid in any year(s) will accumulate until paid in full the next time(s) a dividend is declared.

Dividend.   The return on owner's investment (cash or additional stock) paid per share of stock in a corporation.

Earnings per share.   The amount of net income earned per share of common stock during the year: Net income minus preferred dividends divided by the number of common shares outstanding.

Limited partnership. A partnership in which the liability of the limited partners, who take no active part in the business, is restricted to the amount they invested. The business is managed by one or more general partners, who are also responsible for any financial obligations that exceed the amount invested.

Nonparticipating stock. Preferred stock that stipulates that dividends will not exceed the specified rate.

Non-par-value stock. Stock issued without a par (face) value printed on it.

Participating stock. Preferred stock that may earn dividends above the specified rate.

Partnership. A business for which two or more individuals legally are jointly and separately liable, although responsibilities and profits may be divided in any way that is mutually agreeable.

Par (or face) value. A value assigned to and printed on a share of stock when a corporation is organized (or when it issues new stock at a later time).

Preferred stock. Stock for which the dividend is preset as a certain percent of the par value, and on which dividends must be paid before dividends are paid on common stock.

Proprietorship. A business having a single (sole) owner who is also personally liable for the business's debts.

Silent partner. A member of a partnership who invests financially but takes no active part in the business.

Stock certificate. A certificate of ownership in a corporation; classified as either common or preferred stock.

Stockholder. A person who owns stock in (and thus owns a portion of) a corporation.

Stock outstanding. The number of shares of stock that a corporation has sold.

Retail Mathematics

12

COMMERCIAL DISCOUNTS

OBJECTIVES

Upon completion of Chapter 12, you will be able to:

1. Define and use correctly the terminology associated with each topic.

2. When one or more trade discount percents are given (Section 1):

   a. Find the net cost rate factor (Examples 2, 5; Problems 1, 2, 7, 8).

   b. Use the net cost rate factor to compute:

   (1) Net cost (Examples 2, 3; Problems 1–8, 19–24)

   (2) The single equivalent discount percent (Examples 4–6; Problems 1, 2, 7–10, 15–18).

3. Using the trade discount formula "% Paid × List = Net" and given two of the following amounts, find the unknown third (Section 1):

   a. Net cost (Examples 1–3; Problems 1–8, 17–24)

   b. List price (Example 5; Problems 1, 2, 11–14)

   c. Discount percent (Examples 6, 7; Problems 15–24).

4. Use sales terms given on an invoice to determine (Section 2: Examples 1–2; Problems 3–14):

   a. Cash discount   **b.** Net payment due

   c. The last day on which a cash discount may be claimed and the applicable discount rate (Example 3; Problems 1, 2).

5. Use sales terms and a variation of the formula "% Paid × List = Net" to find (Section 2: Examples 4, 5; Problems 5, 6, 13–16):

   a. The credit granted toward an account when a partial payment is made

   b. The partial payment that must be made to obtain a given credit toward an account.

A business that purchases merchandise that is then sold directly to the **consumer** (user) is known as a **retail** business or store. The retailer may have purchased this merchandise either from the manufacturer or from a **wholesaler** (a business that buys goods for distribution only to other businesses and does not sell to the general public). The purchasing, pricing, and sale of merchandise in a retail store requires a certain kind of mathematics, which is the subject of this unit.

Competition for the customer has led to widespread use of discounts, both in wholesale and retail sales. The following sections examine two of these commercial discounts—*trade discounts* and *cash discounts*.

SECTION 1
TRADE DISCOUNTS

Manufacturers and wholesalers usually issue catalogs to be used by retailers. These catalogs are often quite bulky, containing photographs, diagrams, and/ or descriptions of all items sold by the company. As would be expected, the prices of items shown in the catalogs are subject to change periodically. To reprint an entire catalog each time there is a price change would be unduly expensive; thus, several methods have been developed whereby a business can keep the customer informed of current prices without having to reprint the entire catalog.

Some catalogs are issued in looseleaf binders, and new pages are printed to replace only those pages where a price change has occurred. Other catalogs contain no prices at all; instead, a separate price list, which is revised periodically, accompanies the catalog.

We shall be concerned in this course with a third type of catalog. The prices shown in this catalog are the **list prices** or **suggested retail prices,** usually the prices that the consumer pays for the goods. The retailer buys these goods at a reduction of a certain percentage off the catalog price. This reduction is known as a **trade discount.** The actual cost to the retailer, after the trade discount has been subtracted, is the retailer's **net cost** or **net price** of the item. The percents of discount that are to be allowed (which may vary from item to item in the catalog) may be printed on a discount sheet that supplements the catalog, or they may be quoted directly to a company placing a special order.

The seller will issue to the buyer an **invoice** (see Figure 12-1), which is a bill giving a detailed list of goods shipped or services rendered, with an accounting of all costs. The invoice lists not only a description of the goods, unit prices, and the total amount of the goods, but also sales terms, trade discounts, freight charges, telephone expenses, or any other charges associated with the transaction. It should be noted that the trade discount applies only to the cost of goods or merchandise and not to the other charges. The trade discount is subtracted from the list price to give the net cost of the merchandise, then the other charges are added to determine the total invoice price.

As you can see in Figure 12-1, the quantity is multiplied by the list price to give the extension (the total price of each item). The trade discount of 20% is applied to the total list price of $264. After this discount of $52.80 is sub-

```
                    EAST STREET HARDWARE SUPPLY              Invoice
                              P.O. Box 239                   # 2414
                           Richmond, VA  23222

        TO:       Handy Hardware Co.
                  1241 Post Drive
                  Alexandria, VA 22301

        DATE:     Sept. 9, 19XX

        TERMS:    Not 30
```

Your Order # 534		Received 9 / 6 / xx	Shipped 9 / 8 / xx	Via Truck
Quantity	Cat.#	Description	List	Extension
5	S301	Lounge Chair	$18.50 ea.	$ 92.50
6	S302	Folding Chair	12.75 ea.	76.50
10	D206	Aluminum Grill	9.50 ea.	95.00
		SubTotal		$264.00
		Less 20% discount		52.80
		Net		$211.20
		Freight		37.80
		Total Due		$249.00

FIGURE 12-1   Invoice from Wholesale Supplier

tracted, the net price is $211.20. The freight charge of $37.80 is added as the last step to give the total amount due on the invoice of $249.

Example 1   A tape recorder/radio is shown at a catalog list price of $60. The discount sheet indicates that a 25% trade discount is to be allowed.

(a)   What is the net price to the retail store?

List price	$60
Less trade discount (25% × $60)	− 15
Net price	$45

Observe that when a 25% reduction is allowed, this means that the retailer still pays the other 75% of the list price. Indeed, the percent of discount subtracted from 100% (the whole list price) will always indicate what percent of the list price is still being paid for the goods. For example,

10% discount means   100% − 10% = 90% paid

35% discount means   100% − 35% = 65% paid

18% discount means   100% − 18% = 82% paid

The **complement** of any given percent (or of its decimal or fractional equivalent) is some number such that their sum equals a whole, or 100%. Each "percent paid" above is thus the complement of the corresponding discount percent. By using the complement of the given discount, the previous problem can be solved by a shorter method, applying the formula: % Paid × List = Net.

Hint. Keep in mind that percents must be converted to their decimal or fractional equivalent before being used. Many problems can be worked more conveniently when percents are changed to fractions.

(b)   A 25% discount on the tape recorder/radio above means that 100% − 25% = 75% of the list price is still being paid. (The complement of 25% is 75%.)

Using Decimals	*Using Fractions*
% Paid × List = Net	% Paid × List = Net
75% × $60 = Net	75% × $60 = Net
0.75 × 60 = Net	$\frac{3}{4}$ × 60 = Net
$45 = Net	$45 = Net

(c)   A merchant purchased 15 calculators at a list price of $7 each. The trade discount allowed on this purchase is 35%. There is a freight charge of $10 on the invoice. What is the amount due on this invoice?

The merchandise itself cost 15 × $7 = $105. Thus,

$$\% \text{ Paid} \times \text{List} = \text{Net}$$

$$0.65 \times \$105 = \text{Net}$$

$$\$68.25 = \text{Net}$$

Net price	=	$68.25
Freight	= +	10.00
Amount due	=	$78.25

Two or more successive trade discounts are often quoted on the same goods; these multiple discounts are known as **series** or **chain discounts.** If discounts of 25%, 20%, and 10% (often written 25/20/10) are offered, the net price may be found by calculating each discount in succession. (However, the order in which the discounts are taken does not affect the final result.) *The discounts in a chain are never added together.* That is, a chain discount of 25%, 20%, and 10% is *not* the same as a single discount of 55%.

Example 2   An upholstered chair is advertised at $200 less 20%, 10%, and $12\frac{1}{2}$%. Find the net cost to the furniture store.

(a)   List $\qquad\qquad$ $200
  $\qquad$ Less 20% $\qquad$ $-\underline{\quad 40}$ $\quad$ (20% × $200 = $40)
  $\qquad\qquad\qquad\quad$ $160
  $\qquad$ Less 10% $\qquad$ $-\underline{\quad 16}$ $\quad$ (10% × $160 = $16)
  $\qquad\qquad\qquad\quad$ $144
  $\qquad$ Less $12\frac{1}{2}$% $\quad$ $-\underline{\quad 18}$ $\quad$ ($12\frac{1}{2}$% × $144 = $18)
  $\qquad$ Net cost $\qquad$ $126

This net cost may be computed much more quickly using the method: % Paid × List = Net.

(b)   20%, 10%, and 12.5% discounts have complements of 80%, 90%, and 87.5% paid:

$$\% \, \text{Pd} \times L = N$$

$$(0.8)(0.9)(0.875)(\$200) = N$$

$$(0.63)200 = N$$

$$\$126 = N$$

The number that is the product of all the "percents paid" is very significant and is called the **net cost rate factor.** As demonstrated above, the net cost rate factor is a number that can be multiplied times the list price to determine the net cost. All items subject to the same discounts have the same net cost rate factor. Thus, instead of reworking the entire problem for each item, the buyer determines the net cost rate factor that applies and then uses it to find the net price of all the items.

Example 3   A coffee table subject to the same discount rates given in Example 2 lists for $80. What is the net cost?

From Example 2, the net cost rate factor = 0.8 × 0.9 × 0.875 = 0.63. Then,

$$0.63 \times \$80 = \$50.40 \qquad \text{Net cost}$$

For purposes of cost comparison, it is often desirable to know what single discount rate is equivalent (**single equivalent discount**) to a series of discounts. This is also an application of the complement, found by subtracting the percent paid (the net cost rate factor) from 100%, as follows.

Example 4   What single discount rate is equivalent to the series 20/10/12.5 offered in Example 2?

The net cost rate factor, 0.63, means that 63% of the list price was paid. Finding the complement indicates that

$$100\% - 63\% = 37\% \qquad \text{Single equivalent discount rate}$$

Observe that series discounts of 20%, 10%, and 12.5% are *not* the same as a single discount of 42.5%. If a single discount of 42.5% had been advertised on the chair in Example 2, the net cost would *not* have been $126. Rather, the percent paid would have been 57.5% and the net cost would have been:

$$\% \, \text{Pd} \times L = N$$

$$0.575 \, (\$200) = N$$

$$\$115 = N$$

Example 5   Following trade discounts of 20% and 15%, a furnace cost a dealer $374. (a) What is the list price? (b) What is the net cost rate factor? (c) What single discount percent is equivalent to discounts of 20% and 15%?

(a) $$\% \, \text{Pd} \times L = N$$

$$(0.8)(0.85)L = \$374$$

$$0.68L = 374$$

$$L = \frac{374}{0.68}$$

$$L = \$550$$

(b)   The net cost rate factor = 0.68.

(c)   The single discount equal to discounts of 20% and 15% is

$$100\% - 68\% = 32\%$$

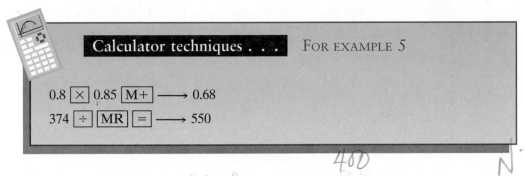

Calculator techniques . . . FOR EXAMPLE 5

$$0.8 \;\boxed{\times}\; 0.85 \;\boxed{M+} \longrightarrow 0.68$$
$$374 \;\boxed{\div}\; \boxed{MR} \;\boxed{=} \longrightarrow 550$$

Example 6 The suggested retail price of a television set is $400. If the net cost is $320, what single discount rate was offered?

The solution can be found either by the % Paid formula or the % of Change formula:

$$\% \text{ Pd} \times L = N$$

$$x \cdot 400 = 320$$

$$x = \frac{320}{400}$$

$$x = \frac{4}{5}$$

$$x = 80\%$$

But x = % Paid. Thus, a 20% discount was given.

or   The price changed by $400 − $320 = $80:

___% of Original = Change?

$$x \cdot 400 = 80$$

$$x = \frac{80}{400}$$

$$x = \frac{1}{5}$$

$$x = 20\% \text{ discount}$$

Example 7 A light fixture lists for $80. The distributor had originally allowed a 15% discount. In order to meet his competitor's price of $61.20, the distributor later offered a second discount. Find the second discount percent.

$$\% \text{ Pd} \times L = N$$

$$(0.85)(x)(\$80) = \$61.20$$

$$68x = 61.20$$

$$x = 0.9$$

But x represents 90% paid. The additional discount is thus the complement, or 100% − 90% = 10% additional discount.

SECTION 1 PROBLEMS

Complete the following.

TRADE DISCOUNT(S)	% PAID (OR NET COST) RATE FACTOR)	LIST PRICE	NET COST	SINGLE EQUIVALENT DISCOUNT PERCENT
1. a. 15%	85%	$ 54	$45.90	X
b. 30%	70%	120	84.00	X
c. 33⅓%	66⅔%	$78	$ 52	X
d. 25%	75%	$60	45	X
e. 20%, 10%	72 = 80% 90%	180	129.60	28%
f. 30%, 16⅔%	58⅓=70%, 83⅓%	132	$77	
g. 16⅔%, 25%	62½=83⅓%, 75%	$240	150	37.5%
h. 20/25/10	54 = 80/75/90	$50	27	46%
2. a. 20%	.80	$ 72	57.60	X
b. 25%	.75	$240	180.00	X
c. 40%	.60	$170	$102	X
d. 33⅓%	66⅔%	$108	72	X
e. 20%, 25%	60% = 80, 75	290	$174	40%
f. 22⅔%, 10%	70% = 77, 90	150	$105	30%
g. 37½%, 30%	43% = 62.5, 70	160	70	56¼%
h. 20/40/30	33.6% = 80, 60, 70	500	168	66⅘%

Side calculations:
52/1 ÷ 200/3
52³/1 · 3/200 = 39/50 = 78.

3. Determine the net cost for each of the following items:
   a. Quartz-tuned VCR listed at $190 less 40% $114
   b. AM/FM cassette player listed at $70 less 10%   63
   c. Answering machine listed at $40 less 25%   30
   d. Trim-style telephone listed at $15 less 16⅔%   12.50

83⅓ = 250/3 · 15/1 5

4. Compute the net cost of the following items:
   a. Stationary bike listed at $109 less 45%   59.95
   b. Treadmill listed at $500 less 35% $325.00
   c. Rollerblades listed at $40 less 37½% $ 25.00
   d. Aerobic slide listed at $18 less 11⅑% $16.00

88 8/9 = 800/9 · 18/1 = 1600

5. Complete the following invoice, determining the total list price, the trade discount amount, the net price, and the total amount due on the invoice.

The Brice Drapery Supply Co.
P.O. Box 4800
Birmingham, AL   35219

TO: The Shade Shop   INVOICE NO.:   2222
        67 Greenbriar Rd.
        Knoxville, TN 37910

DATE: Oct. 15, 19xx

TERMS: Net 30

PURCHASE ORDER NO.:		RECEIVED	SHIPPED	VIA
8976		9 / 12 / xx	9 / 15 / xx	Truck
QUANTITY	CAT NO.	DESCRIPTION	LIST	EXTENSION
10	z344	Roman shades, white	$42.50 ea.	425.00
6 doz	r458	Tassel tie backs	9.75 doz.	58.50
5 doz	r681	Brass holdbacks	26.50 doz.	132.50
12	d121	Single pull rods	7.50 ea.	90.00
8	d111	Double pull rods	10.00 ea.	80.00
			Total List	$786.00
			Less 40%	(314.4)
			Net	471.60
			Freight	$15.75
			Total Due	487.35

6. Determine the net cost of the following merchandise on a recent invoice, and the total due. All items are subject to a 25% discount. Freight charges are $27.00.

   20 boxes of formatted diskettes at $7 per box   140.00
   36 packages of double-pocket portfolios at $3 per package   108.00
    5 dozen ballpoint pens at $6 per dozen   30.00 $386 (96.5) 289.50
   10½ dozen dry chalkboard markers at $8 per dozen   84.00
   12 boxes of letter-size file folders at $2 per box   24.00   27.00
                                                                               316.5

7. Determine (1) the net cost rate factor, (2) the net cost, and (3) the single equivalent discount rate for each of the following items:
   a. A pair of pliers listed for $3, less 20% and 25%   .8 ₵ .75 = .60 · 3. = $1.80   40%
   b. A handsaw listed for $12, less 20% and 12½%   .8 ₵ .875 = .70 · 12. = $8.40   30%
   c. A toolbox listed for $30, less 20%, 30%, and 5%   .8, .7 ₵ .95 = .532 · 30 = $16.80   44%
   d. An electric drill listed for $45, less 20%, 10%, and 10%   .8, .9 ₵ .9 = .648 · $29.16   35.2%

8. Determine (1) the net cost rate factor, (2) the net cost, and (3) the single equivalent discount rate for each of the following items:
   a. A food processor listed for $45, less 20% and 20%   .8 ₵ .8 = .64 · 45 = $28.80   36%
   b. A tea kettle listed for $20, less 15% and 25%   .85 ₵ .75 = .6375 · 20. = $12.75   36¼%
   c. A toaster listed for $40, less 10%, 16⅔%, and 33⅓%   .9, .83⅓ ₵ 66⅔ = .4999 · .40 = $20.   50%
   d. A pasta maker listed for $200, less 25%, 20%, and 15%   .75, .80 ₵ .85 = .51 · 200 = $102   49%

272, 20, 200
  3   3

9. The Auto Shop offers discounts of 30% and 10%. Wheels International advertises 20%, 15%, and 5% discounts. Which company offers the better discount? *Hint:* Compare single equivalent discount rates.

10. Cotrone Co. advertises discounts of 25/10/10. Hodgkins Distributors offers 25% and 40%. Which company offers the better discount? *Hint:* Compare single equivalent discount rates.

11. After trade discounts of 30% and 10%, the net price of a tire was $50.40. What was the list price?

12. The retailer's net cost of a package of acrylic yarn was $2.04 after discounts of 20% and 15%. What was the list price?

13. A store purchased men's walking shorts for $14.40 a pair after discounts of 20% and 10%. What was the catalog list price?

14. After trade discounts of 35% and 10%, the net cost for a woman's swimsuit was $29.25. What was the catalog list price?

15. What single discount percent has been allowed if a vacuum cleaner that lists for $240 is sold for $210?

16. What single discount percent has been offered if a surge protector power outlet that lists for $10 is sold for $6?

17. The JK Market advertises a 35% discount on a $600 refrigerator.

    a. What additional discount must the store offer to meet its competitor's price of $351?

    b. What single equivalent discount would produce the same net cost?

18. The Rodgers Co. currently sells a dinnerware set for $60 less 20%. Its competitor's price is $42 for the same set.

    a. What additional discount must Rodgers offer to match its competitor's price?

    b. What single discount would produce the same net price?

19. The Home Discount Store has a coffee maker priced at $25 less 20%. Main Street Wholesalers lists the same coffee maker for $27 less 30%.

    a. What net price does each store offer now?

    b. What additional discount will Home Discount have to allow to meet Main Street's price?

20. Crabtree Pet Supplies Co. has pet beds for $30 less 16⅔%. Hannah's Pet Mart offers the same beds for $35 less 35%.

    a. What net price does each company now offer?

    b. What additional discount must Crabtree offer so its price will be the same as Hannah's?

21. A crockpot slow cooker is listed for $35 less 14²/₇% at the Arlington Cook Shop. The Dumfries Kitchen Store has the same slow cooker listed at $24 less 12½%.

    a. What is the net price at each business?

    b. What additional discount must Arlington offer to match Dumfries' price?

22. A leather chair and ottoman list for $480 less 33⅓% at Shenandoah Furniture Co., while Dickens Co. sells the same set for $400 less 25%.

    a. Find the net price at each business.

    b. What additional discount must Shenandoah offer to match Dickens' net price?

23. First Colonial Store has a stainless steel flatware set for $75 less 20% and 15%. Georgetown Homes distributes the same set for $80 less 25%.

    a. What net price does each business offer now?

b. What additional discount must be offered by the company with the higher net price in order to meet its competitor's net price?

24. Thelma's Accessories carries a wall mirror for $200 less 12½%. The same mirror is listed for $280 less 50% at Johnson Department Store.

a. What is the net price of the mirror at each business?

b. What further discount percent must be offered by the higher company in order to meet the competitor's net price?

SECTION 2

CASH DISCOUNTS

Merchants who sell goods on credit often offer the buyer a reduction on the amount due as an inducement to prompt payment. This reduction is known as a **cash discount.** The cash discount is a certain percentage of the price of the goods, but it may be deducted only if the account is paid within a stipulated time period. The invoice or the monthly statement contains **sales terms** which indicate the cash discount rate and allowable time period. There are several variations of the ordinary method, as described below.

Ordinary Dating

By the most common method, **ordinary dating,** the discount period begins with the date on the invoice. A typical example would be "Terms: 2/10, n/30." These sales terms are read "two ten, net thirty." This means that a 2% discount may be taken if payment is made within 10 days from the invoice date; from the eleventh to thirtieth days, the net amount is due; after 30 days, the account is overdue and may be subject to interest charges. (If given sales terms do not specify the day by which the net amount must be paid, which is typical in several of the following methods, it is commonly assumed that the net amount is due within 30 days after the discount period began.)

Contrary to the way it sounds, "cash discount" does not generally refer to a discount allowed only for the payment of cash on the day of purchase. On the other hand, cash discounts may be taken only on the net cost of goods, not on freight or other charges.

End of Month, or Proximo

An invoice containing terms such as "1/10, EOM" means that the 1% discount may be taken if paid within 10 days following the end of the month, that is, if paid by the tenth of the coming month. Terms of "1/10, PROX" mean essentially the same thing, as **proximo** is a Latin word indicating the next month after the present. The EOM method is often used on merchandise purchased near the end of the month, although the practice is not limited to end-of-month purchases. Some monthly statements (bills covering the whole month's purchases) often contain these sales terms. It has become common practice for merchants purchasing merchandise after the 25th of the month to add a month to the cash discount period. In other words, for merchandise purchased on February 26 with sales terms of 1/10, EOM, the cash discount period would be extended to April 10. This text will assume the more traditional approach of payment by the tenth of the coming month.

Example 1   Determine the amount due on the following statement of account.

STATEMENT

NAPIER FURNITURE DISTRIBUTORS
3608 Longworth Blvd.
Los Angeles, California 90069

Phone:
213-767-8445

Golden West Furniture Co.
Box 4018
Pasadena, California 90472

DATE: November 25, 19XX
TERMS: 3/10, E.O.M.

Date	Invoice No.	Charges	Credits	Balance
11/1				$545.50
11/8	5391	$239.20		784.70
11/12			$545.50	239.20
11/16	5437	137.35		376.55
11/20	5562	123.45		500.00
11/25	Less 20% and 10% trade discount		140.00	360.00
11/25	Total Ppd. Fgt.*	25.80		385.80

PAY LAST ↑
AMOUNT IN
THIS COLUMN

*"Ppd. Fgt." indicates "prepaid freight."

The charges shown for merchandise are $239.20 + $137.35 + $123.45, or $500 list. This is reduced by the trade discounts allowed. If the bill is paid by December 10, the 3% cash discount may be taken on the $360 net cost of the merchandise. (If no trade discount were given, the cash discount would be computed on the full $500 cost of the merchandise.) Recall that the cash discount does not apply to the freight.

Net cost of goods	$360.00		Due on goods	$349.20
Less discount (3% × $360)	− 10.80		Freight	+ 25.80
Payment due on goods	$349.20		Total amount due	$375.00

A check for $375 will be mailed with a notation or a voucher indicating that a $10.80 cash discount has been taken. The next statement will show a credit for the full $385.80 If the statement is paid December 11 or after, a check for the entire $385.80 must be sent.

Example 2   Three invoices from L.R. Hinson & Co. will be paid on May 15. Sales terms on these invoices are 4/15, 2/30, n/60. The invoices are dated April 8 for

$93.00, April 23 for $112.50, and May 9 for $297.50. What total amount must be paid to retire the entire obligation?

		Amount *Due*
No discount may be taken on the April 8 invoice, since it is over 30 days old:		$ 93.00

The April 23 invoice is less than 30 days old, so a 2% discount may be taken:

		$112.50	
(2% × $112.50)	−	2.25	
			110.25

The third invoice is less than 15 days old, so the full 4% discount may be taken:

		$297.50	
(4% × $297.50)	−	11.90	
			+ 285.60
	Total payment due		$488.85

RECEIPT OF GOODS

Sales terms such as "3/10, ROG" (**receipt of goods**) indicate that the discount period begins on the day the merchandise arrives. These terms are used especially when the goods are likely to be in transit for a long while, such as a shipment by cross-country rail freight or by boat from a foreign country. For instance, assume that an invoice dated December 15 has terms of 3/10, ROG and that the merchandise is received by the purchaser on January 9. The discount period begins January 9, the date of receipt, not December 15, the invoice date.

POSTDATING OR "AS OF"

Distributors may sometimes induce additional sales by guaranteeing that payment can be postponed without penalty. However, an invoice is dated and mailed on the same day as the order is shipped (which notifies the buyer that the order has been filled), and the distributor's ordinary sales terms (such as 2/10, 1/30, n/60) may already be printed on the invoice. Thus, another date must also be given to indicate how long the extended discount period applies, for example "Date: January 12 **AS OF** March 15." This **postdating** means that the buyer may wait until March 15 to start applying the given sales terms. (The 2% discount could be taken any time on or before March 25.)

EXTRA

The **extra** method also allows payment to be postponed and the cash discount still be taken. For example, terms of "3/10–80X" (or "3/10–80 extra," or

"3/10–80 ex.") indicate that the discount is permitted if paid within 10 days plus 80, or a total of 90 days from the date of the invoice.

Extended cash discount periods, as described in the previous two paragraphs, are particularly popular in the sale of seasonal goods, such as lawnmowers or Christmas decorations. With delayed payments possible, retailers are willing to place orders sooner. The distributor can then spread shipments to all retailers over a wider period of time, which is easier for the wholesaler. The retailer may also benefit, since in many cases the merchandise can be sold before the cash discount period expires.

Example 3   An order arrived June 4. What is the last day on which a cash discount may be claimed, given the following conditions? (a) The invoice is dated April 27, with sales terms of 3/15, ROG. (b) The invoice is dated April 27 AS OF July 7, with sales terms of 3/15, n/30. (c) The invoice is dated April 27, with sales terms of 3/15–60X.

(a) The cash discount may be taken through June 19 (or 15 days after June 4, when the goods were received).

(b) The last day for the discount will be July 22 (or 15 days after the July 7 postdate).

(c) The last day for the discount will be July 11 (or 75 days after April 27).

On occasion it may not be possible to pay an entire invoice during the discount period. Whatever payment is made, however, may be considered to be a net payment determined by calculating the discount on some part of the debt. That is, given a 3% cash discount, each 97¢ paid reduces the amount due by $1. These partial-payment problems may be solved using a variation of the % Paid formula: % Paid × Credit toward account = Net partial payment. (Some firms grant the cash discount only if the account is paid in full.)

Example 4   An invoice dated August 3 for $450 has terms of 3/20, n/60. What payment must the buyer make to reduce the obligation by $100, if paid by August 23?

The $100 portion of the account is subject to the 3% discount, as illustrated previously.

3% discount means
97% was paid:

Portion paid off	$100	*or*	% Pd × Credit = N
Less (3% × $100)	− 3		$0.97 \times \$100 = N$
Net payment	$ 97		$\$97 = N$

Thus, a payment of $97 will deduct $100 from the balance due. That is, by paying $97 of the $450 account, $450 − $100 = $350 remains to be paid.

Example 5   Sales terms of 2/15, n/30 were given on an invoice for $380 dated February 12. If a $245 payment is made on February 27, by how much has the debt been reduced? What amount is still due?

A 2% discount means that 98% was paid:

$$\% \text{ Pd} \times \text{Credit} = N$$

$$0.98C = \$245$$

$$C = \frac{245}{0.98}$$

$$C = \$250$$

Since $250 has been deducted from the debt by a $245 payment, $380 − $250 = $130 remains to be paid.

SECTION 2 ▌ PROBLEMS

Determine the applicable discount rate, if any, for the following:

	SALES TERMS	DATE OF INVOICE	DATE GOODS RECEIVED	DATE PAID	APPLICABLE DISCOUNT RATE
1. a.	4/10, 2/20, n/30	Oct. 8	Oct. 10	Oct. 28	
b.	3/10, 2/15, n/30	Nov. 11	Nov. 14	Nov. 21	
c.	3/10, PROX	Feb. 17	Feb. 21	Mar. 17	
d.	2/10, ROG	May 6	May 20	May 30	
e.	2/10, n/60	Dec. 1 AS OF Jan. 15	Dec. 2	Jan. 25	
f.	4/15–60X	Aug. 3	Aug. 25	Oct. 17	
2. a.	2/10, 1/15, n/30	Mar. 5	Mar. 18	Mar. 30	
b.	3/10, 2/20, n/30	Feb. 12	Feb. 15	Feb. 22	
c.	1/10, EOM	July 22	July 28	Aug. 10	
d.	2/15, ROG	Aug. 3	Aug. 12	Aug. 27	
e.	3/10, n/30	Sept. 4 AS OF Oct. 1	Sept. 10	Oct. 10	
f.	2/10–30X	Dec. 8	Dec. 15	Jan. 17	

Determine the amount due on each invoice if paid on the date indicated.

	DATE OF INVOICE	INVOICE AMOUNT	DATE GOODS RECEIVED	SALES TERMS	DATE PAID	AMOUNT DUE
3. a.	Jan. 2	$450	Jan. 18	1/10, n/60	Mar. 1	
b.	May 15	900	May 20	2/15, n/30	May 30	
c.	Jan. 18	390	Jan. 24	3/15, EOM	Feb. 15	
d.	July 21	826	Aug. 3	2/10, PROX	Aug. 13	
e.	Apr. 4 AS OF June 1	875	May 2	1/10, n/30	June 10	
f.	Sept. 8	498	Sept. 16	2/10, ROG	Sept. 26	
g.	Oct. 2	323	Nov. 3	4/10–60X	Dec. 11	
4. a.	Feb. 18	$824	Feb. 26	2/10, n/30	Mar. 5	
b.	Apr. 9	600	Apr. 15	3/10, 2/15, n/30	Apr. 24	
c.	May 23	325	May 28	2/10, EOM	June 10	
d.	June 15	780	June 20	1/10, PROX	July 15	
e.	Nov. 8 AS OF Dec. 10	450	Dec. 1	2/10, n/30	Dec. 20	
f.	Sept. 3	860	Oct. 15	3/10, ROG	Oct. 25	
g.	Mar. 10	550	Apr. 9	4/10–50X	May 9	

Complete the following if partial payment was made on each invoice within the discount period.

	INVOICE AMOUNT	SALES TERMS	CREDIT TOWARD ACCOUNT	NET PAYMENT AMOUNT	AMOUNT STILL DUE
5. a.	$1,200	2/10, n/30	$800		
b.	915	1/10, n/60	500		
c.	480	1/15, n/30		$237.60	
d.	650	3/10, n/60		242.50	
e.	720	4/10, n/30			$300
6. a.	$2,000	2/10, n/30	$500		
b.	860	1/10, n/30	300		
c.	670	2/10, n/60		$441	
d.	790	3/15, n/30		388	
e.	560	4/10, n/60			$200

7. Trade discounts of 33⅓% and 10% are to be taken on an invoice that totals $742.75, including a prepaid freight charge of $22.75. Sales terms on the invoice are 2/10, n/30. What is the amount due if paid within the cash discount period?

8. Determine the amount due on an invoice that totals $383.20, including a prepaid freight charge of $33.20. Trade discounts of 20% and 20% and sales terms of 3/10, n/30 are shown on the invoice. Assume that the amount due is paid within the cash discount period.

9. The merchandise below was subject to trade discounts of 15% and 10%. The invoice dated March 24 contained sales terms of 3/15, n/60 and prepaid freight of $16. How much should be remitted on April 8?

> 5 doz. bottles of aspirin at $4.50 per doz.
>
> 3 doz. boxes of adhesive tape at $4.80 per doz.
>
> ½ doz. elastic arch supports at $1.50 each
>
> 3 sets of aluminum crutches at $10.00 per set

10. The following items were purchased on a February 24 invoice. The merchandise was subject to trade discounts of 30% and 25%. The invoice contained sales terms of 2/10, n/30 and had prepaid freight charges of $12.95. How much should be paid on March 2?

> 8 doz. packages of paper towels at $8.00 per doz.
>
> 10 doz. bottles of liquid detergent at $18.00 per doz.
>
> 2½ doz. cans of furniture polish at $1.20 each
>
> 3 doz. bottles of glass cleaner at $1.50 each

11. A firm has received three invoices from Gale Enterprises, each containing sales terms of 3/10, 2/15, n/30. The invoices are $600 on November 5, $500 on November 12, and $350 on November 18. If the invoices are all paid on November 27, how much is due?

12. Three invoices from Cather Industries contained sales terms of 2/15, 1/20, n/30. The invoices were $850 on March 3, $1,000 on March 10, and $620 on March 15. If the invoices are paid on March 30, how much is due?

13. An invoice dated January 7 for $900 contains terms of 3/10, 1/20, n/30 EOM. On February 10, the purchaser wishes to make a payment of sufficient amount so that she receives credit for a third of her obligation.

   a. How much will she pay on February 10?

   b. On February 18, she finds that she can pay the remaining balance. How much is this second payment?

   c. What is the total amount that is paid to discharge the $900 obligation?

14. An invoice dated June 22 for $1,200 contains sales terms of 2/15, 1/20, n/30 PROX. On July 15, the buyer wishes to make a payment that will discharge a fourth of his obligation.

   a. How much should he remit on July 15?

   b. On July 20, the buyer is able to pay the remainder of the debt. How much is his second payment?

   c. What total amount is paid to discharge the $1,200 obligation?

15. A statement for $1,800 dated November 6 and containing terms of 3/15, 2/20, n/60 was sent by John Adams to George Moody. On November 21, Adams received an $873 check from Moody.

   a. How much credit to his account should Moody receive for this partial payment?

   b. How much does Moody still owe?

16. Creative Video Co. sent the Greenberg Connection an invoice dated August 21 AS OF September 15 for $870. Sales terms on the invoice were 3/15, n/60.

   a. If Creative Video received a check for $421.95 on September 30, how much credit should be given the Greenberg Connection?

   b. How much does the Greenberg Connection still owe?

CHAPTER *12* GLOSSARY

AS OF (or Postdating). A later date shown on an invoice, at which time the given sales terms begin.

Cash discount. A discount given for early payment of an invoice or monthly statement.

Complement (of a percent). A percent such that the sum of it and a given percent equals 100%; 100% minus the given percent. (Their fractional or decimal equivalents are also complements, equaling 1.)

Consumer. An individual who buys a product for personal use.

End of month (EOM). Sales terms that start at the beginning of the following month. (Same as "proximo" terms.)

Extra (X). Sales terms in which the discount period extends for the specified additional number of days. (Example: 2/10–50X indicates a total of 60 days.)

Invoice. A bill giving a detailed list of goods shipped or services rendered, with an accounting of all costs.

List price. A suggested retail price; price paid by the consumer.

Net cost. The total (invoice) cost after applicable trade and cash discounts.

Net cost rate factor. A factor that multiplies times the list price to produce the net cost; the product of the "percents paid" in a trade discount.

Ordinary sales terms. Sales terms in which the cash discount period begins on the date shown on the invoice.

Postdated sales terms. (See "AS OF.")

Proximo (PROX). (See "End of month.")

Receipt of goods (ROG). Sales terms in which the cash discount period begins on the day the merchandise arrives.

Retailer. A business that sells directly to consumers.

Sales terms. A code on an invoice or monthly statement that specifies the rate and the allowable time period for a cash discount, if offered.

Series (chain) discounts. Two or more trade discounts, with each succeeding discount being applied to the balance remaining after the previous discount.

Single equivalent discount. A trade discount that produces the same net cost as does a group of series discounts.

Trade discount. A discount off the list (suggested retail) price of merchandise, given to obtain the wholesale price.

Wholesaler. A business that buys merchandise from the manufacturer and sells it to retail stores.

13

MARKUP

OBJECTIVES

Upon completing Chapter 13, you will be able to:

1. Define and use correctly the terminology associated with each topic.

2. Given a percentage markup based on cost (C) or on selling price (S), use the formula "$C + M = S$" to find (C: Section 1: Examples 1–4; Problems 1–18; S: Section 2: Examples 1–3; Problems 1–18):

   a. Selling price

   b. Cost

   c. Dollar markup.

3. a. Determine the percent of markup based on cost (C) using the formula "$\underline{?}\%$ of $C = M$" (Section 1: Example 5; Problems 7, 8, 13, 14; Section 2: Example 3; Problems 1, 2, 7–14).

   b. Similarly, compute the percent of markup based on selling price (S) using the formula "$\underline{?}\%$ of $S = M$" (Section 1: Example 5; Problems 3, 4, 7–18; Section 2: Problems 9, 10).

4. Calculate the selling price factor ($\overline{Spf}$), and use "$\overline{Spf} \times C = S$" to find selling price (S) or cost (C) (C: Section 1: Example 6; Problems 3–6, 9, 10; S: Section 2: Example 2; Problems 1–4, 7, 8, 11–16).

5. a. Find the selling price required on perishable items so that the goods expected to sell will provide the desired profit on the entire purchase (Section 3: Example 1; Problems 1, 2, 5, 7–12).

   b. Similarly, determine the regular selling price when some of the items will be sold at a given reduced price (Section 3: Example 2; Problems 3, 4, 6, 13–16).

In accordance with the American economic philosophy, merchants offer their goods for sale with the intention of making a profit. If a profit is to be made, the merchandise must be marked high enough so that the cost of the goods and all selling expenses are recovered. On the other hand, merchants cannot mark their goods so as to yield an unreasonably high profit, or else they will not be able to meet the competition from other merchants in our free enterprise system. To stay within this acceptable range, therefore, it is important that business people have a thorough understanding of the factors involved and the methods used to determine the selling price of their merchandise. The first step in this direction is to become familiar with the terms involved.

The actual **cost** of a piece of merchandise includes not only its catalog price (often with trade and/or cash discounts deducted), but also charges for insurance, freight, and any other expenditures incurred during the process of getting the merchandise to the store.

The difference between this cost and the selling price is known as **gross profit**. Other names for gross profit are **markup, margin,** and **gross margin**.

Gross profit is not all clear profit, however, for out of it must be paid business expenses, or **overhead**. Examples of overhead are salaries, rent, depreciation, lights and water, heat, advertising, taxes, and insurance. Whatever remains after these expenses have been paid constitutes the clear or **net profit**. The net profit remaining is often quite small; indeed, expenses sometimes exceed the margin, in which case there is a loss. Another way of defining gross profit is to say that gross profit is overhead plus the desired net profit (assuming that the merchandise sells at the regular price).

There are many retail terms used interchangeably by business people that are confusing to business students. A term such as "cost" can mean the amount the retailer must pay to the wholesaler or manufacturer. It can also mean the amount the consumer pays to the retailer for the same goods, although, at this point in the distribution channel, it would be called the "selling price" by the retailer. Caution should be exercised to determine the position in the distribution of merchandise (manufacturer to retailer or retailer to consumer) before solving problems in Chapters 13 and 14.

Conceptually, the regular selling price of an item is determined as follows:

$$
\begin{array}{ll}
\quad \text{Catalog list price} & \\
-\ \text{Trade and/or cash discounts} & \\
+\ \underline{\text{Freight and other charges}} & \\
=\ \text{Net cost} & (C) \\
\text{Markup } (M) \begin{cases} +\ \text{Overhead} \\ +\ \underline{\text{Desired net profit}} \end{cases} & \begin{array}{l} (\overline{OH}) \\ (P) \end{array} \\
=\ \text{Regular marked selling price} & (S)
\end{array}
$$

The following diagram illustrates the breakdown of selling price. Markup (or gross profit) may be computed as a percentage of cost, or it may be based on sales. But regardless of which method is used, the same relationship between cost and markup always exists as shown in the diagram; specifically, Cost plus Markup (or margin or gross profit) always equals Selling price. From this relationship we obtain our fundamental selling price formula:

$$C + M = S$$

SELLING PRICE (*S*)

COST (*C*)	MARKUP (*M*) OVERHEAD + PROFIT

Observe that the dollar amount of markup (*M*) can be obtained in two ways:

$$S - C = M$$

or

$$\overline{OH} + P = M$$

where $\overline{OH}$ = overhead and P = profit.

Because the markup is composed of overhead and profit, we can express the basic "Cost + Markup = Selling price" formula in a slightly different manner, as follows:

$$C + (\overline{OH} + P) = S$$

MARKUP BASED ON COST

Before attempting to calculate selling price, let us consider the following examples.

Example 1
Suppose that a furniture store bought a lamp for $50 and sold it for $70. Assume that the store's expenses (overhead) run 30% of cost. (a) How much markup (or margin or gross profit) did the store obtain? (b) What were the store's expenses related to this sale? (c) What amount of profit was made?

(a)   Using the basic equation $C + M = S$ to find the markup,

$$C + M = S$$

$$\$50 + M = \$70$$

$$M = \$20$$

The markup (or margin or gross profit) on the sale was $20.

(b) Overhead averages 30% of cost; thus,

$$30\% \times C = \overline{OH}$$

$$0.3(\$50) = \overline{OH}$$

$$\$15 = \overline{OH}$$

The furniture store's expenses for this sale were $15.

(c) Since the overhead plus net profit make up the markup, then

$$\overline{OH} + P = M$$

$$\$15 + P = \$20$$

$$P = \$5$$

By using our previous diagram, we see that a net profit of $5 was made on the lamp. The $70 selling price would be broken down as shown below:

$$
\begin{array}{llr}
\text{Net cost} & (C) & = \$50 \\
+ \text{ Overhead} & (\overline{OH}) & = \ \ 15 \\
+ \text{ Profit} & (P) & = \ \ \underline{\ \ 5} \\
= \text{ Selling price} & (S) & = \$70
\end{array}
$$

Example 2   The wholesale cost of a winter coat was $80. A clothing store sold the coat for $100. According to the store's last income statement, expenses have averaged 27% of sales income. (a) What was the markup on the coat? (b) How much overhead should be charged to this sale? (c) How much profit or loss was obtained on the coat?

(a) Starting with the basic equation as before,

$$C + M = S$$

$$\$80 + M = \$100$$

$$M = \$20$$

The selling price produced a markup of $20.

(b) The expected overhead was 27% of the selling price; thus,

$$27\%S = \overline{OH}$$

$$0.27(\$100) = \overline{OH}$$

$$\$27 = \overline{OH}$$

Business expenses related to this sale would be $27.

(c) You can immediately realize that, if expenses were $27 while only $20 was brought in by the sale, there was obviously a $7 loss on the transaction. This calculation is shown in the following series of equations:

$$\overline{OH} + P = M$$

$$\$27 + P = \$20$$

$$P = \$20 - \$27$$

$$P = -\$7$$

The calculation produces a negative difference, and a negative profit ($-\$7$) indicates a $7 loss on the sale.

Example 2 illustrates a vital fact about selling—just because an item sells for more than it cost does not necessarily mean that any profit is made on the sale. Thus, it is essential that business people have the mathematical ability to take their expenses into consideration and price their merchandise to obtain the required markup.

Now let us consider some examples where the markup is based on cost.

Example 3   A merchant has found that his expenses plus the net profit he wishes to make usually run $33\frac{1}{3}\%$ of the cost of his goods. For what price should he sell a dress that cost him $36?

(a) We know that the markup equals $33\frac{1}{3}\%$ or $\frac{1}{3}$ of the cost. Since $\frac{1}{3}$ of $36 = $12, then

$$C + M = S$$

$$\$36 + \$12 = S$$

$$\$48 = S$$

(b) This same problem can be computed in another way. Since markup equals $\frac{1}{3}$ of the cost ($M = \frac{1}{3}C$), by substitution

$$C + M = S$$

$$C + \frac{1}{3}C = S$$

$$\frac{4}{3}C = S$$

But since cost equals $36,

$$\frac{4}{3}(\$36) = S$$

$$\$48 = S$$

It is customary in business to price merchandise as shown in Example 3(b), primarily because this is quicker than the method in Example 3(a), where markup was computed separately and then added to the cost (an extra step). In using method (b), you should make it a habit to *combine the Cs before substituting the dollar value of the cost.* Another advantage of method (b) is that it allows you to find the cost when the selling price is known. The following example illustrates how cost is found (as well as why the Cs should be combined before substituting for cost).

Example 4   A home improvement center marks up its merchandise 30% of cost. (a) What would be the selling price of a screen door that cost $40? (b) What was the cost of a weed trimmer selling for $65?

(a) Markup is 30% of cost, or $0.3C$.

$$C + M = S$$

$$C + 0.3C = S$$

$$1.3C = S$$

$$1.3(\$40) = S$$

$$\$52 = S$$

The door should sell for $52 in order to obtain a gross profit (markup) of 30% on cost.

(b)   Using the same 30% markup on cost when the selling price is $65,

$$C + M = S$$

$$C + 0.3C = S$$

$$1.3C = S$$

$$1.3C = \$65$$

$$C = \$50$$

The weed trimmer cost $50.

You can quickly see that markup based on cost is quite different from markup based on selling price. For instance, suppose that an article cost $1.00 and sold for $1.50. The $0.50 markup is 50% of the cost, but only $33\frac{1}{3}\%$ of the selling price.

A merchant may wish to know what margin based on selling price is equivalent to a margin based on cost, or vice versa. To find the percent margin based on cost, one would use the formula

$$\underline{\text{?}}\% \text{ of } C = M$$

where C = cost and M = markup or margin. To determine the percent of margin based on selling price, you would use the formula

$$\underline{\text{?}}\% \text{ of } S = M$$

where S = selling price and M = margin.

Example 5   A circular saw listed for $57 and required a $3 freight charge. It sold for $75. (a) What percent markup on cost did this represent? (b) What percent markup based on selling price did this represent?

The actual cost of the saw was $57 + $3 = $60. The markup is then

$$C + M = S$$

$$\$60 + M = \$75$$

$$M = \$75 - \$60$$

$$M = \$15$$

(a)   To find what percent of cost is markup,

$$\_\_\%C = M$$

$$\_\_\% \times \$60 = \$15$$

$$60x = 15$$

$$x = \frac{1}{4}$$

$$x = 25\% \text{ markup, based on cost}$$

(b)   To find what percent of selling price is markup,

$$\_\_\%S = M$$

$$\_\_\% \times \$75 = \$15$$

$$75x = 15$$

$$x = \frac{1}{5}$$

$$x = 20\% \text{ markup, based on selling price}$$

Altogether, the components of this sale can be expressed as follows:

Catalog list price		= $57
+ Freight		= $\underline{\quad 3}$
= Net cost	(C)	= $60
+ Markup	(M)	= $\underline{\quad 15}$
= Selling price	(S)	= $75

To summarize, the $75 selling price is composed of a $60 cost plus a $15 markup. This $15 markup can be expressed as a percent either of the cost or of the selling price, as shown below. Notice that, for the given dollar markup, the percent based on cost (25%C) is larger than the percent based on selling

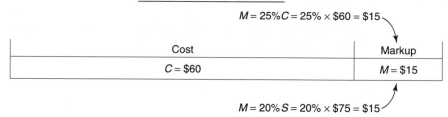

price ($20\% S$). This illustrates a general rule: The markup percent when based on cost is always larger than the markup percent when based on selling price. The next diagram will help to visualize why: Since cost is smaller than selling price, the $15 represents a larger portion of the cost than it does of the selling price.

Note. In all cases where the same markup is used, the beginning calculations would be the same. Suppose that the markup is 25% of cost ($M = \frac{1}{4}C$); then

$$C + M = S$$

$$C + \frac{1}{4}C = S$$

$$\frac{5}{4}C = S$$

When the margin is 25% of cost, the basic calculation would always yield $\frac{5}{4}C = S$. This $\frac{5}{4}$ is called the selling price factor. The **selling price factor** $\overline{(Spf)}$ is a number that can be multiplied times the cost to obtain the selling price. This definition can be expressed by the formula

$$\overline{Spf} \times C = S$$

Thus, in actual practice, a retail store would not price each item by the complete process, starting with $C + M = S$. Rather, any store that uses a markup of 25% on cost would know that its selling price factor is $\frac{5}{4}$. (The selling price factor of $\frac{5}{4}$ could be expressed in any equivalent form: $1\frac{1}{4}$, 1.25, or 125%.) Hence, the store would eliminate the first steps of the markup formula and would start with the selling price factor.

Example 6   Use $\overline{Spf} \times C = S$ to price an item that cost \$24, if the markup is 25% on cost.

To price an item that cost \$24, the store above would perform only the following calculation:

$$\overline{Spf} \times C = S$$

$$\tfrac{5}{4}C = S$$

$$\tfrac{5}{4}(\$24) = S$$

$$\$30 = S$$

The item should sell for \$30.

Note.   After trying a few examples, you will observe that, *when markup is based on cost, the selling price factor can be found simply by adding 1 to the decimal or fractional equivalent of the markup.* (This is *not* true when markup is based on selling price.)

A word to the wise: Don't forget that many percents can be handled more easily when converted to fractions instead of to decimals. If review is needed, refer to the table of Percent Equivalents of Common Fractions on page 53.

SECTION 1   PROBLEMS

Answer the following questions about each problem. (1) How much markup does each sale include? (2) How much are the overhead expenses? (3) Is a profit or loss made on the sale, and how much?

	COST	SELLING PRICE	OVERHEAD
1. a.	\$60	\$84	25% of cost
b.	30	36	30% of cost
c.	70	80	14²⁄₇% of cost
d.	18	21	33⅓% of selling price
e.	36	48	12½% of selling price
2. a.	\$80	\$95	15% of cost
b.	55	73	40% of cost
c.	75	89	20% of cost
d.	33	40	25% of selling price
e.	72	96	16⅔% of selling price

Find the missing information. (Compute the selling price before finding the markup.)

	PERCENT MARKUP ON COST	$\overline{Spf}$	COST	SELLING PRICE	MARKUP	PERCENT MARKUP ON SELLING PRICE
3. a.	25%		$60			
b.	33⅓		33			
c.	11⅑		45			
d.	20			$18		
e.	50			75		
f.	*20/50*		50	70	*20*	*20/70*
g.	*4/18*	*18*		22	$4	*4/22*
4. a.	40%		$90			
b.	20		60			
c.	12½		32			
d.	25			$15		
e.	60			64		
f.			54	66		
g.				36	$9	

5. a. A store uses a markup of 20% on cost. What would be its selling price factor?

  b. What would be the selling price of an item costing $45?

  c. How much gross profit does this yield?

6. a. What would be the selling price factor for a markup of 42% on cost?

  b. What would be the selling price of an item costing $13?

  c. What dollar markup does this yield?

7. Picture frames cost a store $60 per dozen and sold for $8 each.

  a. What percent of cost was the markup?

  b. What percent of selling price was the markup?

8. Men's socks cost $48 per dozen pairs and sold for $5 each pair.

  a. What percent of cost is the markup?

  b. What percent of selling price is the markup?

9. a. What selling price factor corresponds to a 50% markup on cost?

  b. Determine the selling price of each item listed on an invoice with the following costs: $9.00, $12.50, $7.20, and $15.40.

  c. What percent markup on selling price does this represent?

10. a. What selling price factor is equivalent to a margin of 37½% on cost?

  b. Determine the selling price of each item, having the following costs: $8.00, $4.80, $6.40, and $16.40.

  c. What percent markup on selling price does this represent?

11. a. A store using a margin of 35% on cost sold a pair of aerobic shoes for $33.75. What was the cost of these shoes?

b. A markup of 40% on cost yields a sales price of $28 on a leotard. What was the cost, and what was the equivalent percent markup on retail sales price?

12. a. After a markup of 30% on cost, an infant car seat sold for $78. What was the cost?

b. A baby stroller selling for $85 had been marked up so as to obtain a gross profit of 25% on cost. Find the cost and the equivalent percent markup on retail sales price.

13. The catalog list price of a bookcase is $150. The Miller Furniture Co. purchases this bookcase at a 60% trade discount and sells it at a 20% discount off the list price. Find

a. The cost and selling price

b. The percent markup based on cost

c. The percent markup based on selling price

14. The catalog list price of a softball glove is $70. The Sports Center buys the gloves at a 50% trade discount and sells them at a 25% discount off the list price. Find

a. The cost and selling price of a glove

b. The percent markup based on cost

c. The percent markup based on selling price

15. Eight toaster ovens listed on an invoice show a total cost of $336 plus freight of $24 on the shipment.

a. At what selling price should each toaster oven be marked to realize a gross profit of 60% on cost?

b. What percent markup on selling price does this represent?

16. A shipment of 15 smoke detectors had a catalog list price of $115. Freight on the shipment was $20.

a. At what price should each smoke detector be marked to obtain a 25% margin on cost?

b. What percent margin on selling price does this represent?

17. Fifteen espresso coffee makers were purchased by a gourmet shop for a cost of $1,050, plus shipping charges of $75 added to the invoice. The shop's markup is 20% on cost.

a. What is the selling price per espresso maker?

b. What is the markup percent on selling price?

18. A store buys 6 bottles of designer perfume for $120 plus shipping charges of $24. The store's markup is 60% on cost.

a. What is the selling price per bottle of perfume?

b. What percent markup on selling price does this represent?

SECTION 2

MARKUP BASED ON SELLING PRICE

Many business expenses are calculated as a percent of net sales. For example, sales representatives' commissions are based on their sales. Sales taxes, of course, are based on sales. When deciding how much to spend for advertising or research and development, a firm may designate a certain percent of sales. Many companies take inventory at the sales value. The income statement, which is one of two key reports used to indicate the financial condition of a company, lists the firm's sales, all its expenses, and its net profit; each of these items is then computed as a percent of net sales.

With so many other items calculated on the basis of sales, it is not surprising, then, that most businesses prefer to price their merchandise so that markup will be a certain percent of the selling price. Markup based on sales also offers a merchant the advantage of being able to refer to the daily cash register tape and immediately estimate the gross profit.

Example 1   Using a markup of 30% on sales, price an item that costs $42.

As before, we start with the equation $C + M = S$. We know that the markup equals 30% of sales, or $M = \frac{3}{10}S$. Thus, we substitute $\frac{3}{10}S$ for M in the formula $C + M = S$:

$$C + M = S$$

$$C + \frac{3}{10}S = S$$

$$C + \frac{3}{10}S - \frac{3}{10}S = S - \frac{3}{10}S$$

$$C = \frac{7}{10}S$$

It is essential that you understand the preceding calculations. When $\frac{3}{10}S$ is substituted for M, there are then Ss on both sides of the equation. When an equation contains the same variable on both sides of the equal sign, the variables cannot be combined directly. So, to combine the Ss, "$\frac{3}{10}S$" must be subtracted from both sides of the equation. (When markup is based on selling price, it will always be necessary to subtract on both sides of the equation. This should be done before the actual dollar value is substituted for the cost.)

Now we are ready to find the selling price. The $42 is substituted for the cost C:

$$C = \frac{7}{10}S$$

$$\$42 = \frac{7}{10}S$$

To find S, we must eliminate the $\frac{7}{10}$. This is accomplished by multiplying both sides of the equation by the reciprocal, $\frac{10}{7}$:

$$\$42 = \frac{7}{10}S$$

$$\frac{10}{7}(\$42) = \frac{10}{7}\left(\frac{7}{10}S\right)$$

$$\$60 = S$$

Thus, the selling price is $60. (In this example, $\frac{10}{7}$ is the selling price factor, because $\frac{10}{7}$ multiplied times the cost, $42, produces the selling price, $60.) Notice that the $60 retail price, which includes a markup of $60 − $42 = $18, meets the stated requirement that markup be 30% of sales, as illustrated below:

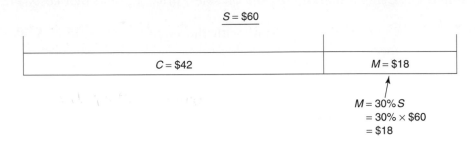

Sometimes it is more convenient to convert the markup percent to a decimal than to a fraction. The following example illustrates such a problem.

Example 2　A department store prices its merchandise so as to obtain a gross profit of 24% of the selling price. What selling price will be shown on a redwood picnic table that cost $38?

The markup is to be 24% of sales ($M = 0.24S$). Recall that $S = 1S = 1.00S$; thus,

$$C + M = S$$

$$C + 0.24S = S$$

$$C \pm 0.24S - 0.24S = S - 0.24S$$

$$C = 0.76S$$

$$\$38 = 0.76S$$

$$\frac{38}{0.76} = \frac{0.76\ S}{0.76}$$

$$\$50 = S$$

The table would sell for $50. The gross profit is $50 − $38 = $12, which is 24% of the $50 selling price, as required.

Note.　The value of the selling price factor $(\overline{Spf})$ when markup is based on retail selling price can be obtained as follows:

1. *Compute the complement of the given markup;* then
2. *Find the reciprocal of that complement.*

In Example 1, the given markup based on sales is $30\%S$ or $\frac{3}{10}S$. The complement is $\frac{7}{10}$; therefore, $\overline{Spf} = \frac{10}{7}$. Similarly in Example 2, the given markup is 24% of sales, or $0.24S$. The complement is 0.76; thus, the selling price factor would be the reciprocal of the complement, or $\overline{Spf} = \frac{1}{0.76} = 1.3158$. (That is, $\overline{Spf} \times C = 1.3158 \times \$38 = \$50$, to the nearest cent.)

Example 3 A set of patio furniture retailed for $120. The store's expenses run 19% of sales, and net profit has been 6% of sales. (a) What was the cost of the set? (b) What percent markup on cost does this represent?

(a) Recall that markup (gross profit) equals expenses plus net profit. Thus,

$$M = \overline{OH} + P$$

$$= 19\%S + 6\%S$$

$$M = 25\%S$$

Since $M = \frac{1}{4}S$, then

$$C + M = S$$

$$C + \frac{1}{4}S = S$$

$$C + \frac{1}{4}S - \frac{1}{4}S = S - \frac{1}{4}S$$

$$C = \frac{3}{4}S$$

$$= \frac{3}{4}(\$120)$$

$$C = \$90$$

The markup was $120 − $90 = $30. The selling price of $120 would then be broken down as shown below:

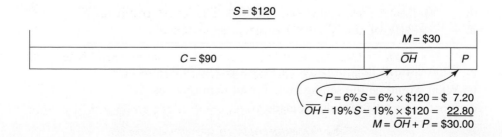

$S = \$120$

	$M = \$30$	
$C = \$90$	$\overline{OH}$	P

$P = 6\%S = 6\% \times \$120 = \$\ 7.20$
$\overline{OH} = 19\%S = 19\% \times \$120 = \underline{22.80}$
$M = \overline{OH} + P = \$30.00$

(b) Using the $90 cost and $30 markup, the markup percent (based on cost) is

$$\_\_\%C = M$$

$$\_\_\%(\$90) = \$30$$

$$90x = 30$$

$$x = \frac{30}{90}$$

$$= \frac{1}{3}$$

$$x = 33\frac{1}{3}\% \text{ markup on cost}$$

That is, $33\frac{1}{3}\%C = 33\frac{1}{3}\% \times \$90 = \$30$, which is the known markup.

SECTION 2   PROBLEMS

Complete the following.

		PERCENT MARKUP ON SELLING PRICE	$\overline{Spf}$	COST	SELLING PRICE	MARKUP	PERCENT MARKUP ON COST
1.	a.	40%		$60			
	b.	12½		42			
	c.	20			$70		
	d.	30			30		
	e.			45		$9	
2.	a.	25%		$30			
	b.	37½		35			
	c.	33⅓			$42		
	d.	50			12		
	e.			18		$2	

3. a. What is the selling price factor for a markup of 25% of sales?
   b. What would be the selling price for an item that cost $27?
   c. What gross profit would be made?

4. a. If a markup of 20% on sales is used, what would be the selling price factor?
   b. What would be the selling price for an article that cost $14?
   c. What gross profit would be made?

5. A gross profit on a clock radio was $17.50. If the markup is 35% on selling price,
   a. What was the selling price of the radio?
   b. What was the cost?

6. A store uses a markup of 37½% on selling price. If this yields a gross profit of $15,
   a. What was the selling price?
   b. What was the cost?

7. a. What is the selling price factor for a 30% markup on retail sales?
   b. Price items that cost: $2.80, $3.50, $4.90, and $5.60.
   c. What percent markup on cost does this represent?

8. a. Find the selling price factor corresponding to a 60% markup on retail sales.
   b. Price items that cost $5.00, $6.40, $7.20, and $8.50.
   c. What percent markup on cost does this represent?

9. A nursery purchased 250 clay pots at a total cost of $1,000. It sold 75 pots for $8 each, 80 for $7 each, 60 for $5 each, and 35 for $4 each.
   a. How much gross profit was made on the entire transaction?
   b. What percent markup on cost was made?
   c. What percent markup on selling price was this?

10. A store purchased 50 blankets for a total cost of $2,215. It sold 20 blankets for $60 each, 16 for $54 each, 9 for $46 each, and 5 for $36 each.
   a. How much gross profit was made on the entire transaction?
   b. What percent markup on cost was made?
   c. What percent markup on selling price was this?

11. a. Which would yield a larger gross profit, 25% on cost or 25% on selling price?
   b. Would 33⅓% on cost or 30% on selling price yield a larger gross profit?
   (Hint: Compare selling price factors.)

12. a. Would a markup of 50% on cost or 50% on selling price yield a larger gross profit?
   b. Which would yield a larger margin, 20% on cost or 28% on selling price?
   (Hint: Compare selling price factors.)

13. Using a margin of 37½% on selling price, a store sold an article for $192. Determine its
   a. Selling price factor
   b. Cost
   c. Corresponding percent markup on cost

14. A store used a markup of 33⅓% on selling price and sold an item for $33. What was the
   a. Selling price factor
   b. Cost
   c. Corresponding percent markup on cost

15. The manager of a linens department wants to buy quilts that she can retail for $200. Her company requires a 25% markup on selling price.
    a. What is the most she can pay for a quilt that she plans to resell?
    b. What is the selling price factor?

16. A home appliance manager wants to buy washing machines that can retail for $340. He must obtain a gross profit of 40% on selling price.
    a. What is the most he can pay for a washer that he plans to resell?
    b. What is the selling price factor?

17. A toy store expects its overhead to be 35% of sales and its net profit to be 15% of sales. At what price should it price a doll house that cost the store $85 less 20%?

18. A store's overhead is 25% of sales and its net profit is 5% of sales. The store paid $42 less 16⅔% for an electric chainsaw. At what price should the chainsaw be priced?

SECTION 3
MARKING PERISHABLES

A number of businesses handle products that perish in a short time if they are not sold. Examples of such businesses would be produce markets, bakeries, florists, dairies, and so forth. To a lesser extent, many other businesses would fall into this category, since styles and seasons change and the demand for a certain product diminishes. Clothing stores, appliance stores, and automobile dealers would be typical of this category.

These merchants know that not all of their goods will sell at their regular prices. The remaining items will have to be sold at a reduced price or else discarded altogether. Experienced business people know approximately how much of their merchandise will sell. They must, therefore, price the items high enough so that those which do sell will bring in the gross profit required for the entire stock.

The selling price on perishable merchandise is computed by the following basic procedure:

1. Determine the *cost* of the entire purchase.
2. Determine the *total selling price* of the entire purchase.
3. Determine the *amount of merchandise* that is expected to sell.
4. Express a *pricing equation:* Divide the total sales in item 2 by the amount of merchandise in item 3 to obtain the selling price (per item or per pound).

Example 1   A grocer bought 100 pounds of bananas at 18¢ per pound. Experience indicates that about 10% of these will spoil. At what price per pound must the bananas be priced in order to obtain a margin of 50% on cost?

The bananas cost 100 pounds × 18¢, or $18. Since the markup will be 50% on cost ($M = 0.5C$), then

$$C + M = S$$

$$C + 0.5C = S$$

$$1.5C = S$$

$$1.5(\$18) = S$$

$$\$27 = S$$

At least $27 must be made from the sale of the bananas.

The grocer expects 10% or 10 pounds of bananas to spoil; thus, his needed $27 in sales must come from the sale of the other 90 pounds. Now, if p = price per pound of bananas, then

$$90p = \$27$$

$$p = \frac{27}{90}$$

$$p = 30¢ \text{ per pound}$$

Note: When a fraction of a cent remains, retail merchants customarily mark the item up to the next penny even though the fraction may have been less than one-half cent. It should also be observed in the preceding example that, on all sales above the expected 90 pounds, the entire selling price will be additional profit.

Example 2   A delicatessen made 40 snack cakes at a cost of 25¢ each. About 15% of these will be sold as "day old" at 10¢ each. Find the regular price in order that the delicatessen can make 25% on cost.

The snack cakes cost $40 \times 25¢$ or $10 to bake, and the delicatessen uses a markup of 25% on cost $(M = 0.25C)$. Thus

$$C + M = S$$

$$C + 0.25C = S$$

$$1.25C = S$$

$$1.25(\$10) = S$$

$$\$12.50 = S$$

Now, 15% or 6 cakes will probably be sold at 10¢ each, which means that 60¢ will be earned at the reduced price. This leaves 34 cakes to be sold at the regular price. The regular sales and the "old" sales together must total $12.50; thus, if p = regular price per cake,

$$\text{Regular} + \text{Old} = \text{Total}$$

$$\text{Regular} + \$0.60 = \$12.50$$

$$\text{Regular} = \$11.90$$

$$34p = 11.90$$

$$p = \$0.35 \text{ per cake}$$

SECTION 3 PROBLEMS

Find the missing information.

	QUANTITY BOUGHT	COST PER UNIT	TOTAL COST	MARKUP ON COST	REQUIRED SALES	PERCENT TO SPOIL	AMOUNT TO SELL	SELLING PRICE
1. a.	60 lb.	$ 0.44		33⅓%		5%		
b.	25 lb.	18.00		32		8		
c.	150 doz.	3.00		50		4		
2. a.	70 lb.	$ 0.60		20%		10%		
b.	20 doz.	1.40		25		5		
c.	50	4.00		12½		4		

Complete the following, which involves part of the goods being sold at the regular price and the remaining amount at the reduced price.

	QUANTITY BOUGHT	COST PER UNIT	TOTAL COST	MARKUP ON COST	REQUIRED SALES	AMOUNT AT REGULAR PRICE	PERCENT AT REDUCED PRICE	AMOUNT AT REDUCED PRICE	REDUCED SELLING PRICE	REGULAR SELLING PRICE
3. a.	60 lb.	$0.55		28%			5%		$0.40	
b.	75 doz.	0.80		35			4		0.60	
4. a.	80 lb.	$0.40		20%			5%		$0.35	
b.	25	3.00		40			8		2.50	

5. Suppose that in Problem 1a the store actually sold 52 pounds.
   a. How much gross profit would the store earn?
   b. What percent markup on cost would that be?
   c. What percent markup on selling price would that be?

6. Suppose that in Example 2 the delicatessen actually sold 30 snack cakes at the regular price and the remainder at 10¢ each.

   a. What gross profit would be obtained?

   b. What percent markup on cost does that represent?

   c. What percent markup on selling price does that represent?

7. A produce manager at a grocery store bought 300 pounds of oranges at $0.45 per pound. Past experience indicates that about 15% of these will spoil. At what price per pound must the oranges be priced in order to obtain a margin of 30% on cost?

8. A grocery store purchased 200 pounds of grapes at $0.60 per pound. Experience shows that 10% of the grapes will spoil. At what price per pound would the store make 15% markup on cost?

9. Allen Florist bought 50 pots of red tulips at $4 each. About 10% will die and not be sold. At what price should each pot be marked to obtain a gross profit of 35% on cost?

10. A market purchased 200 pounds of tomatoes at $0.72 per pound. The manager knows from past experience that 7% of the tomatoes will not sell. At what price per pound should the manager mark the tomatoes in order to receive a gross profit of 12½% on cost?

11. A grocery store purchased 60 pounds of mushrooms at a cost of $0.60 per pound. From past experience, the produce manager knows that 10% of the mushrooms will spoil before they are sold. What must be the marked price per pound if the store wants to make a 40% markup on cost?

12. A garden shop purchased 20 dozen cut daffodils at $6 per dozen. Past experience indicates that 5% of the daffodils will not sell. In order to make a 45% markup on cost, at what price per dozen should the flowers be marked?

13. A market purchased 120 pounds of cabbage at $0.55 per pound. In the past, 5% has been sold at a reduced price of $0.49 per pound. What must be the regular price per pound in order to make a 35% markup on cost?

14. The bakery department of a supermarket bakes 50 loaves of French bread at a cost of $0.45 a loaf. The bakery's markup is 50% of cost. Approximately 10% of the loaves will sell at a reduced price of $0.40 a loaf. What must be the regular price per loaf?

15. The Beach Shop purchased 60 swimsuits at a cost of $20 each. The buyer expects that 25% of the suits will be sold at a clearance price of $23 each. What must be the regular price of the swimsuits if the store is to make 60% on cost for the entire purchase?

16. The Ski King Shop purchased 75 insulated ski jackets at $120 each. The shop has a markup of 35% on cost. About 12% of these jackets will sell at a reduced price of $118 each. What must be the regular list price of the jackets?

CHAPTER 13 GLOSSARY

Cost.   The amount paid for merchandise, including the catalog price less any trade and/or cash discounts, plus charges for insurance, freight, and any other expenses incurred during the process of getting the merchandise to the store.

Gross profit.   (See "Markup.")

Margin.   (See "Markup.")

Markup (or gross profit or margin).   The difference between cost and selling price of merchandise; the overhead and net profit combined:

$$C + M = S; \quad M = \overline{OH} + P$$

Markup is usually stated as a percent of cost:

$$\_\_\% \text{ of } C = M$$

or of selling price:

$$\text{\_\% of } S = M$$

Net profit (P). The spendable profit remaining after cost of merchandise and overhead are deducted from selling price.

Overhead ($\overline{OH}$). Business expenses other than cost of merchandise.

Selling price factor ($\overline{Spf}$). A number (usually greater than 1) that multiplies times cost to produce selling price: $\overline{Spf} \times C = S$.

14

MARKDOWN

OBJECTIVES

Upon completion of Chapter 14, you will be able to:

1. Define and use correctly the terminology associated with each topic.

2. Use "$C + M = S$," "%Pd $\times$ L = N," and "___%S = M" to find (Section 1: Examples 1–4; Problems 1–16):

   a. Regular selling price (S_1)

   b. Reduced sales price (S_2)

   c. Percent of markup (at a sale price or regular price)

   d. Net cost (when trade discounts are offered).
   (The order in which the formulas are used depends on which information is given and which must be computed.)

3. a. Using "___% of Original = Change," compute (Section 2: Examples 1, 4; Problems 1–4, 9, 10, 13–16):

      (1) Amount of markdown
      (2) Percent of markdown.

   b. When necessary, first use appropriate markup formulas to determine cost and regular selling price (Section 2: Examples 4, 5; Problems 1–4, 11–16).

4. When a markdown has occurred, find the total handling cost and use it to determine the amount and percent of (Section 2: Examples 2, 3, 5; Problems 1–18):

   a. Operating profit

   b. Breakeven

   c. Operating loss

   d. Absolute loss.

Essentially all retail concerns must sell at least some of their merchandise at reduced selling prices. Indeed, many firms must expect that some of their merchandise will not sell at all. A significant aspect of retail business thus involves pricing goods in anticipation of unsold items and computing mark-downs to minimize losses. The following sections explain the mathematics required to compute reduced selling prices and markdowns.

SECTION *1*
ACTUAL SELLING PRICE

Few, if any, retail merchants are able to sell their entire stock at its regular marked prices. Clearance sales and "storewide discounts" are common. The following problems involve markup when merchandise is sold at a sale or discount price, called the **actual selling price.**

Using the illustration from Chapter 13, Section 1 (page 336), and adding two additional lines, we see how the cost through the actual selling price flows:

$$
\begin{array}{ll}
\quad\; \text{Cost (catalog list price)} & \\
- \; \text{Trade and/or cash discount} & \\
+ \; \underline{\text{Freight and other charges}} & \\
= \; \text{Net cost} & (C) \\
+ \; \text{Overhead} & (\overline{OH}) \\
+ \; \underline{\text{Desired net profit}} & (P) \\
= \; \text{Regular marked selling price} & (S_1) \\
- \; \underline{\text{Consumer discount or markdown}} & (\overline{MD}) \\
= \; \text{Actual selling price} & (S_2)
\end{array}
$$

Both the **regular** and the **reduced prices** of merchandise are commonly known by several different terms that are used interchangeably, as follows:

REGULAR PRICE (S_1)	ACTUAL SELLING PRICE (S_2)
Regular marked price	Reduced price
Regular selling price	Sale price
Marked price	Net selling price
List price	Price after reduction

When $M = \underline{\quad}\%S$ is used for percent of markup based on sales, S may represent either the regular selling price (S_1) or the reduced selling price (S_2), depending on the information given. By using S_1 or S_2 in the formulas, confusion over which S is meant will be eliminated. There will be two Ms also, M_1 for markup at the regular price and M_2 for markup at the reduced or actual selling price. At a reduced price, markdown is represented by $\overline{MD}$. The following illustration shows the breakdown of selling price (S) both for a regular marked price (S_1) and for a reduced sale price (S_2).

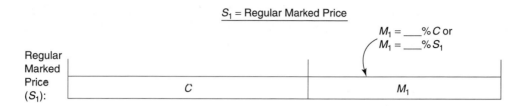

S_1 = Regular Marked Price

M_1 = ____% C or
M_1 = ____% S_1

Regular Marked Price (S_1): | C | M_1 |

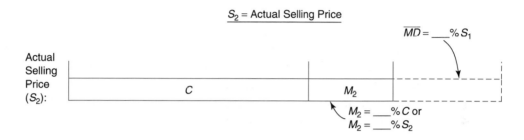

S_2 = Actual Selling Price

$\overline{MD}$ = ____% S_1

Actual Selling Price (S_2): | C | M_2 |

M_2 = ____% C or
M_2 = ____% S_2

Example 1   The regular marked price of a suitcase is $70, including a markup of 25% on *cost*. During a special sale, the case was reduced so that the store obtained a $12\frac{1}{2}\%$ markup on the clearance sale price. What was (a) the cost and (b) the sale price?

(a)   When $M_1 = 25\%$ of cost,

$70 = 25\% \, C + C$

$$C + M_1 = S_1$$

$$C + \frac{1}{4}C = S$$

$$\frac{5}{4}C = S$$

$$\frac{5}{4}C = \$70$$

$$C = \$56$$

(b)   When $M_2 = \frac{1}{8}$ of the sale price, then

$S_2 = C + 12\frac{1}{2}\% \, S_2$

$$C + M_2 = S_2$$

$$C + \frac{1}{8}S = S$$

$$C + \frac{1}{8}S - \frac{1}{8}S = S - \frac{1}{8}S$$

(handwritten annotations:)
NO %
$56 + 14 = 70$
$C + \frac{1}{4}C = S_1$
$\frac{1}{4}C \quad 70$
$\frac{5}{4}C = S_1$
$\frac{5}{4}C = 70$
$C = 56$
$C + \frac{1}{8}S = S$
$C = 7/8 \, S$
$56 = 4$

$$C = \frac{7}{8}S$$

$$\$56 = \frac{7}{8}S$$

$$\frac{8}{7}(56) = \frac{\cancel{8}}{\cancel{7}}\left(\frac{\cancel{7}}{\cancel{8}}S\right)$$

$$\$64 = S_2$$

The sale price was \$64, giving a margin of \$64 − \$56 = \$8. Notice that this \$8 markup is $12\frac{1}{2}\%S = 12\frac{1}{2}\% \times \$64$, as stipulated.

For situations in which a "sale" price is involved, you may often find it helpful to think through the entire pricing process from the time an item enters the store until it sold "on sale."

Example 2   A markup of 40% on the selling price was used to obtain the regular marked price of a suede jacket that had cost \$48. During an end-of-season clearance, the jacket was reduced by 25% of the marked price. (a) What was the regular marked price of the jacket? (b) What was the sale price? (c) What percent markup on sales was made at the clearance sale price?

(a)   First, the jacket is priced using a markup of 40% on selling price $(M_1 = \frac{2}{5}S_1)$:

$$C + M_1 = S_1$$

$$C + \frac{2}{5}S = S$$

$$C = S - \frac{2}{5}S$$

$$C = \frac{3}{5}S$$

$$\$48 = \frac{3}{5}S$$

$$\frac{5}{3}(48) = \frac{\cancel{5}}{\cancel{3}}\left(\frac{\cancel{3}}{\cancel{5}}S\right)$$

$$\$80 = S_1$$

The regular marked price (or regular selling price) of the jacket was $80.

(b) Later this regular price was reduced: A 25% discount means 75% (or $\frac{3}{4}$) of the regular marked price (or list price) was paid:

$$\% \text{ Pd} \times \text{List} = \text{Net sale price } (S_2)$$

$$\frac{3}{4}(\$80) = S_2$$

$$\$60 = S_2$$

The clearance sale price (S_2) of the jacket was $60.

(c) Before finding the percent markup based on the sale price, we must first determine the actual dollar markup at the sale price:

$$C + M_2 = S_2 \qquad\qquad \underline{\hspace{1em}}\%S_2 = M_2$$

$$M = S - C \qquad\qquad \underline{\hspace{1em}}\%(\$60) = \$12$$

$$= \$60 - \$48 \qquad\qquad 60x = 12$$

$$M_2 = \$12 \qquad\qquad\qquad x = \frac{12}{60} \text{ or } \frac{1}{5}$$

$$x = 20\%$$

Thus, 20% of the sale price (20% × $60 = $12) was gross profit.

Example 3   During a special promotion, Avery Jewelers marked a watch down 30%, selling it for $42. Even at this reduced sale price, Avery still made a gross profit (markup) of $14\frac{2}{7}\%$ on sales. What percent markup on selling price would Avery have made if the watch had sold at its regular marked price?

Before we can use the formula $\underline{\hspace{1em}}\%S = M$ to determine the percent markup on sales at the regular price, we must first know the regular markup (M_1) and the regular sales price (S_1). Thus, we must first calculate (a) the cost and (b) the regular marked price (S_1). From these we can determine (c) the regular markup (M_1). Then we can find (d) the percent markup on sales.

(a) At the clearance sale price of $42, the makup is $14\frac{2}{7}\%$ on sales (or $M_2 = \frac{1}{7}S_2$):

$$C + M_2 = S_2$$

$$C + \frac{1}{7}S = S$$

$$C = S - \frac{1}{7}S$$

$$= \frac{6}{7}S$$

$$= \frac{6}{7}(\$42)$$

$$C = \$36$$

The watch cost the jeweler $36.

(b)   The watch was reduced by 30% of the regular price to obtain the $42 sale price:

$$\% \text{ Pd} \times L = \text{Net sale price } (S_2)$$

$$0.7L = \$42$$

$$L = \frac{\$42}{0.7}$$

$$L = \$60$$

The regular marked price (S_1) was $60.

(c)   Since the watch was regularly priced at $60, the regular markup (M_1) was $60 − $36 = $24.

(d)   Thus, at the regular price,

$$\underline{\quad}\%S_1 = M_1$$

$$\underline{\quad}\%(\$60) = \$24$$

$$60x = 24$$

$$x = \frac{24}{60} \quad \text{or} \quad \frac{2}{5}$$

$$x = 40\%$$

The regular markup, based on sales, was 40% of the regular marked price (40%S = 40% × $60 = $24). The regular price and the reduced (sale) price are illustrated in the following diagram.

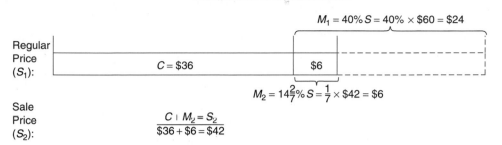

S_1 = Regular marked price = $60

$M_1 = 40\% \, S = 40\% \times \$60 = \$24$

Regular
Price
(S_1):

C = $36

$6

$M_2 = 14\frac{2}{7}\% \, S = \frac{1}{7} \times \$42 = \$6$

Sale
Price
(S_2):

$$\frac{C \mid M_2 = S_2}{\$36 + \$6 = \$42}$$

Many businesses deal in discounts every day. These merchants must price their goods very carefully, so they can allow the customer a "discount" off the marked price and still make their desired markup on the sale. The discount price thus appears to be a "sale" price (and is computed similarly), although in reality this is the price at which the goods are always expected to sell; the business never expects to obtain the "regular marked price" for its merchandise.

Example 4 Caldwell Furniture paid a wholesale distributor $750 less 20% and 20% for a dining room suite. Caldwell must obtain a markup of $33\frac{1}{3}\%$ on its actual selling price. At what price must the dining room suite be marked so that Caldwell can allow 20% and 10% off this marked price and still make its $33\frac{1}{3}\%$ markup?

This solution involves three parts: (a) to find the cost, (b) to determine what the actual selling price (the apparent "sale" price) will be, and (c) to determine what "regular" marked price, less discounts of 20% and 10%, will leave the actual (net) selling price that Caldwell needs.

(a) First, find the actual cost after discounts of 20% and 20%.

$$\% \, \text{Pd} \times L = \text{Net cost}$$

$$(0.8)(0.8)(\$750) = N$$

$$0.64(750) = N$$

$$\$480 = N$$

The dining room suite cost Caldwell $480 after trade discounts.

(b) A $33\frac{1}{3}\%$ markup on selling price (S_2) is used to determine the actual selling price (which will later be made to appear as a "sale" price):

$$C + M_2 = S_2$$

$$C + \frac{1}{3}S = S$$

$$C = S - \frac{1}{3}S$$

$$C = \frac{2}{3}S$$

$$\$480 = \frac{2}{3}S$$

$$\frac{3}{2}(480) = \frac{\cancel{3}}{\cancel{2}}\left(\frac{\cancel{2}}{\cancel{3}}\, S\right)$$

$$\$720 = S_2$$

Caldwell Furniture will actually sell the suite for $720.

(c) Finally, we compute a "regular" marked price (S_1) which, after discounts of 20% and 10%, will leave the actual $720 net selling price (the apparent "sale" price):

$$\% \, Pd \times L = \text{Net selling price}$$

$$(0.8)(0.9)L = \$720$$

$$0.72L = \$720$$

$$L = \$1,000$$

The marked price on the dining room suite will be $1,000 (although Caldwell knows the suite will never be sold at that price). These marked and actual prices are illustrated in the following diagram.

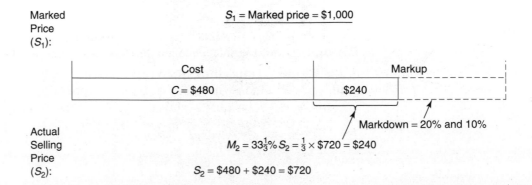

SECTION 1   PROBLEMS

Compute the missing items in the following problems.

$C + M = S$
$C + .3C = S$

	COST	PERCENT REGULAR MARKUP	REGULAR PRICE	PERCENT DISCOUNT	SALE PRICE	PERCENT SALE MARKUP
1. a.	*700*	30% of C	$910	X *875 875*		20% of S
b.	*15*	66⅔% of C	25	X *20 20*		25% of S
c.	$ 24	40% of S	*40*	20%	*32*	?% of S *25%*
d.	600	33⅓% of S	*900*	25	*675*	?% of S *11 1/9%*
e.	*315*	?% of S *37%*	*500*	30	$350	10% of S
f.	*300*	?% of S *50%*	*600*	12½	525	42 6/7% of S
g.	$620 less 30% = ? *434*	?% of S *44%*	*775*	20/20	*496*	12½% of S
h.	$800 less 25/15 = ? *510*	?% of S *49%*	*1000*	25/20	*600*	15% of S
2. a.		40% of C	$70	X		20% of S
b.		33⅓% of C	64	X		16⅔% of S
c.	$28	30% of S		20%		?% of S
d.	30	25% of S		10		?% of S
e.		?% of S		10	$ 45	22 2/9% of S
f.		?% of S		30	175	14 2/7% of S
g.	$400 less 40% = ?	?% of S		33⅓/20		25% of S
h.	$200 less 20/20 = ?	?% of S		37½/20		20% of S

3. A store sells a stationary exercise bike for a regular price of $195, which produces a 30% markup on cost. During an end-of-summer sale, the price of the bike was reduced so that the sale price produced a 20% gross profit on sales.

   a. What was the cost of the bike?

   b. What was the sale price?

4. The regular price of a gold ring was $87, which included a 45% markup on cost. During a holiday sale, the ring was marked down so that the sale price produced a 25% gross profit on sales.

   a. What was the cost of the ring? *29 06*

   b. What was the sale price?

5. Home Crafts Inc. usually sells a dining suite with a markup of 37½% of the selling price. During a sale, the suite that costs $700 received a 25% reduction off its regular price.

   a. What was the marked price?

   b. What was the sale price?

   c. What percent of the sale price was the margin?

6. Colleen's Interiors marks its furniture to obtain a markup of 33⅓% of selling price. During a sale, a curio cabinet that cost $424 received a 16⅔% reduction off its regular price.

   a. What was the regular marked price of the curio cabinet?

   b. What was the sale price?

   c. What percent of the sale price was the markup?

7. A waffle maker cost a store $21. It was priced to obtain a markup of 47½% on the regular selling price. Later, it was marked down 25% during a special sale. Find

   a. The regular marked price

   b. The sale price

   c. The percent markup on sales at the special sale price

8. A wall clock that cost a store $14 was marked to obtain a gross profit of 30% on regular selling price. During an end-of-year sale, it was marked 10% off the regular marked price.

   a. What was the regular marked price?

   b. What was the sale price?

   c. What percent of the sale price was the markup?

9. At a preholiday sale, a store marked a camera down 16⅔%, making the sale price $45. At this sales price, the store made a 20% gross profit on sales. Determine

   a. The regular price

   b. The cost

   c. The percent markup on regular selling price

10. After a markdown of 25%, a garage door opener sold for $90. At this sale price, the store made a 40% margin on sales. What had been

    a. The regular price of the system

    b. The cost

    c. The percent markup on regular selling price

11. During a sale, a retailer marked an electronic keyboard down 30%, making the sale price $210. At this price, the store still made a markup of 20% of the actual selling price. What had been

    a. The regular price

    b. The cost

    c. The percent markup on regular selling price

12. A cutlery set was marked down 12½%, making the sale price $63. At this sale price, the retailer made an 11⅑% gross profit on sales. Determine

    a. The regular price of the cutlery

    b. The cost

    c. The percent markup on regular selling price

13. A necklace cost a jeweler $100 less 30/20. After marking the necklace so that he could allow a "16⅔% discount," he still made a markup of 30% on the actual selling price. What had been

    a. The cost

    b. The actual selling price

    c. The marked price

14. A portable disc player cost a merchant $140 less 30/25. After marking the player so that he could allow a "40% discount," he still made a markup of 33⅓% on the actual selling price. What had been

    a. The cost

    b. The actual selling price

    c. The marked price

15. A roll of wallpaper cost a store $24 less 35%. The store marked the wallpaper so that it could allow a "50% discount." At this special price, the store still made a markup of 20% of the sales price. What had been

 a. The cost

 b. The sales price

 c. The marked price

16. A dress cost a store $50 less 30%. The store marked the dress so that it could allow a "22⅞%
discount." At this sale price, the store still made a gross profit of 28⁴⁄₇% of the actual selling price. Determine

 a. The cost

 b. The sales price

 c. The marked price

MARKDOWN VERSUS LOSS

In the previous section we have seen that merchandise must often be sold at a reduced price. This reduction may be expected and planned for well in advance, as in the case of seasonal clearance sales on clothing. Other reductions may be less predictable, as when the merchants' association decides to have a special citywide promotion or when the competition has unexpectedly lowered its price.

Regardless of what motivates a price reduction, there are guidelines that merchants generally follow in determining how much to mark down their merchandise. The wholesale cost plus the overhead on a sale is known as the **total handling cost.** Businesses prefer to recover this total handling cost ($C + \overline{OH}$) if possible; in this case, no net profit would be made, but there would be no loss either. This critical value between profit and loss is often called the **breakeven point.**

If the merchant fails to recover the wholesale cost plus all the operating expenses, there has been an **operating loss.** If not even the wholesale cost of the article was recovered, the transaction resulted in an **absolute** (or **gross**) **loss.** These various sales conditions are shown in the following diagram, which assumes a sale price lower than cost.

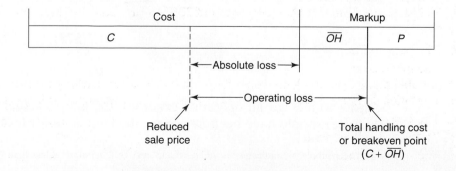

As described above, the following relationships* exist among actual selling (or sale) price ($\overline{ASP}$), total handling cost ($\overline{THC}$), and wholesale cost (C):

If $\overline{ASP} > \overline{THC}$, then a net profit exists.
If $\overline{ASP} = \overline{THC}$, then breakeven exists.
If $\overline{ASP} < \overline{THC}$, then an operating loss exists.
If $\overline{ASP} < C$, then an absolute loss exists.

We will be interested here in determining whether or not any profit was made following a markdown, as well as certain percentages related to markdown.

Example 1 A lamp that sold for $120 was reduced to $84. (a) What was the markdown on the lamp? (b) By what percent was the price reduced?

(a) The lamp was reduced by $120 − $84 = $36.

(b) ___% of Original selling price = Change?

$$\underline{\quad}\%(\$120) = \$36$$

$$120x = 36$$

$$x = \frac{36}{120} \quad \text{or} \quad \frac{3}{10}$$

$$x = 30\% \qquad \text{Reduction on selling price}$$

Example 2 The lamp in Example 1 had cost the furniture store $75. The store's operating expenses run 20% of cost. Was an operating profit (net profit) or an operating loss made on the lamp?

The breakeven point or total handling cost = $C + \overline{OH}$. Since $\overline{OH} = 20\%$ of cost and $C = \$75$, then

$$\text{Total handling cost} = C + \overline{OH}$$

$$= C + 0.2C$$

$$= 1.2C$$

$$= 1.2(\$75)$$

$$\overline{THC} = \$90$$

In order to recover all operating expenses, the lamp's actual selling price (or sale price) should have been $90. Since the lamp sold for less than this break-

*The symbol ">" indicates "is greater than," and "<" denotes "is less than."

even figure, the store's operating loss on the lamp was the difference between total handling cost and actual selling price:

Total handling cost	$90
Actual selling price	− 84
Operating loss	$ 6

Example 3 The regular marked price of a woman's suit was $144. The suit cost $120, and related selling expenses were $18. The suit was later marked down by 25% during a clearance sale. (a) What was the operating loss? (b) What was the absolute loss? (c) What was the percent of gross loss (based on wholesale cost)?

The total handling cost $(C + \overline{OH})$ of the suit was $120 + $18 = $138. The actual net selling price (or sale price) of the suit was

$$\% \, Pd \times L = Net$$

$$0.75(\$144) = Net$$

$$\$108 = Net \text{ selling price } (S_2)$$

(a) Since the suit sold for less than its total handling cost,

Total handling cost	$138
Actual selling price	− 108
Operating loss	$ 30

(b) The absolute or gross loss is the difference between wholesale cost and the actual selling price:

Wholesale cost	$120
Actual selling price	− 108
Absolute (or gross) loss	$ 12

(c) Absolute loss or gross loss is always computed as some percent of the wholesale cost, as follows:

What percent of wholesale cost was the absolute loss?

$$\_\_\%(\$120) = \$12$$

$$120x = 12$$

$$x = \frac{12}{120} \quad \text{or} \quad \frac{1}{10}$$

$$x = 10\% \text{ gross loss}$$

The operating and gross loss incurred in this sale may be visualized by referring to the following diagram.

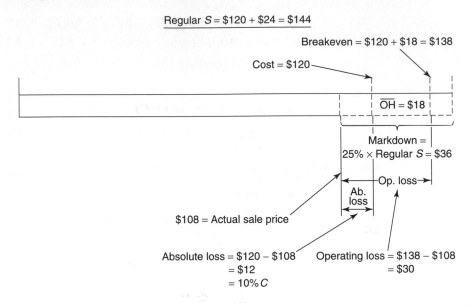

Example 4   A dress cost a department store $40. The store's operating expenses are 35% on cost, and normal profit is 15% on cost. (a) What is the regular selling price of the dress? (b) What is the maximum percent of markdown that can be offered without taking an operating loss?

(a)   The normal markup of overhead plus net profit is

$$M_1 = \overline{OH} + P$$

$$= 35\%C + 15\%C$$

$$M_1 = 50\%C$$

Therefore, using $M = 50\%C$ and a cost of $40, the regular selling price is

$$C + M_1 = S_1$$

$$C + \frac{1}{2}C = S$$

$$\frac{3}{2}C = S$$

$$\frac{3}{2}(\$40) = S$$

$$\$60 = S_1 \quad \text{(regular selling price)}$$

(b)   The operating expenses are 35% or $\frac{7}{20}$ of cost; thus,

Regular selling price	$60
Total handling cost	− 54
Maximum markdown	$ 6

Total handling cost $= C + \overline{OH}$

$$= C + 0.35C$$

$$= 1.35C$$

$$= 1.35(\$40)$$

$$\overline{THC} = \$54$$

The lowest sale price the store could offer without taking any operating loss would be the breakeven point, $54. This means that the dress could be marked down $6 from its regular selling price. The percent of reduction would be

$$\_\_\% \text{ of Original} = \text{Change}$$

$$\_\_\%(\$60) = \$6$$

$$60x = 6$$

$$x = \frac{6}{60} \quad \text{or} \quad \frac{1}{10}$$

$$x = 10\% \text{ reduction in price}$$

Example 5   A variety store bought 200 pen sets at $2 each. The regular price of $2.50 included 15% on cost for overhead. After 180 sets were sold at the regular price, the remaining 20 sets were closed out at $1 each. (a) How much net profit, operating loss, or absolute loss was made? (b) What percent of wholesale cost was it?

(a)   Cost: $200 \times \$2 = \$400$

Total sales:

$180 \times \$2.50 =$	$450
$20 \times \$1.00 = +$	20
	$470

Total handling cost:

$$\overline{THC} = C + \overline{OH}$$

$$= C + 0.15C$$

$$= 1.15C$$

$$= 1.15(\$400)$$

$$\overline{THC} = \$460$$

Total sales	$470
Total handling cost	− 460
Net profit	$ 10

(b) ___% of wholesale cost = Net profit

$$\_\_\_\%(\$400) = \$10$$

$$400x = 10$$

$$x = 2.5\% \quad \text{Net profit on cost}$$

SECTION 2   PROBLEMS

Complete the following.

REGULAR SELLING PRICE	MARKDOWN PERCENT	MARKDOWN AMOUNT	SALE PRICE	WHOLESALE COST	OVERHEAD	THC	OPERATING PROFIT OR (LOSS)
1. a. $15	20%	3	12	$ 8	$2	10	2
b. 24	25%	$ 6	18	15	4	19	-1
c. 40	40%	16	$24 =3	16 =	5 -	21	$3
d. 36	33⅓	12	24	22	6	$28	(4)
e. 50	20	10	40	32	10	42	(2)
f. 12	16⅔	2	10	7	2	9	1
2. a. $50	40%	20	30	$22	$4	26	4
b. 30	30	$ 9	21	19	3	22	-1
c. 45	22 2/9	10	$35	23	8	31	4
d. 18	11 1/9	2	16	12	7	$19	(3)
e. 60	10	6	54	50	9	59	(5)
f. 40	20	8	32	18	8	26	6

.380 x = 32 / 80 / .80

REGULAR SELLING PRICE	MARKDOWN PERCENT	MARKDOWN AMOUNT	SALE PRICE	WHOLESALE COST	OVERHEAD	THC	OPERATING LOSS	GROSS LOSS AMOUNT	GROSS LOSS PERCENT
3. a. $ 5	40	$2	3	$ 3	$1	4	-1	0	0
b. 60	40%	24	36	40		24	$10	4	10%
c. 40	25	10	$30	32	4	$36	-6	2	6¼
d. 24	33⅓	8	16	20	3	23	-7	$4	20
4. a. $36	16⅙	$6	30	$36	$3				
b. 81	11⅑	9	72	75			$ 8		
c. 40	20	8	$32		8	$43			
d. 90	30	27	63		4			$6	

5. A 33⅓% discount was offered on a tape deck that regularly sold for $96. The tape deck cost $50, and operating expenses were 30% of cost. How much operating profit or loss was made on this sale?

6. A man's suit that regularly sold for $210 was advertised for sale at a 30% discount. The suit cost the department store $130, and operating expenses were 25% of cost. How much operating profit or loss was made on the sale of the suit?

7. A $35 shirt was reduced by 20%. The shirt cost the merchant $22, and operating expenses were 40% of cost. How much operating profit or loss was made on the sale?

8. The price of a $40 pair of shoes was reduced by 25%. If the shoes cost $26 and operating expenses were 20% of cost, how much operating profit or loss was made on the sale?

9. Lambswool seat covers that usually retail for $108 were sold at a sale price of $72.
   a. What percent markdown was the store allowing?
   b. If the seat covers had cost the store $90 less 20% and 22⅔% and selling expenses had been $10, how much operating profit or loss was made?

10. A $54 brass candlestick was sold for $45.
   a. What percent markdown did this represent?
   b. If the dealer had paid $50 less 20% and 15% for the candlestick and operating expenses were $5.40, how much operating profit or loss was made?

11. A store paid $60 for a set of aluminum cookware. On regular sales, the overhead is 15% of sales and net profit is 10% of sales. At a special sale, the store sold the set at a 30% discount.
   a. What would be the regular price of the cookware?
   b. What was the operating loss?
   c. How much and what percent gross or absolute loss did the store suffer?

12. A store paid $140 for a solid wood kitchen table. The store's expenses are 20% of the regular selling price, and the expected net profit is 10% of sales. During a special sale, the table was sold at a 50% discount.
   a. What was the regular price of the table?
   b. How much operating loss did the store experience?
   c. How much and what percent gross or absolute loss did the store suffer?

13. A shoe store buys hosiery for $18 per dozen pairs, less 33⅓% and 10%. The marked price of the hose includes 40% on cost for expenses and 20% on cost for net profit.
   a. What does a pair of hose regularly sell for?
   b. What is the largest markdown that could be allowed on a pair of hose without experiencing an operating loss?
   c. What percent markdown would this represent?

14. A garden shop purchases daylily bulbs for $7.50 per dozen, less 20% and 20%. The shop marks up the bulbs to provide for expenses of 30% on cost and a net profit of 20% on cost.
   a. What is the regular selling price per bulb?
   b. What is the maximum markdown the shop could allow without taking an operating loss?
   c. What percent markdown would this be?

15. A hardware store buys pliers for $60 a dozen, less 20/10. Expenses at the store run 15% of cost, and they expect to make a net profit of 10% on cost.
   a. At what price should each pair of pliers be marked?
   b. What is the largest markdown that could be allowed on a pair of pliers without experiencing an operating loss?
   c. What percent markdown would this represent?

16. A cosmetics department purchases lipsticks for $90 per dozen, less 40/20. Expenses run 30% of cost, and net profit is expected to be 50% of cost.

    a. What should be the regular price per lipstick?

    b. What is the maximum markdown that the department could allow and still break even?

    c. What percent markdown would this represent?

17. A store purchased 200 Halloween masks for $2.80 each. The store sold 180 of the Halloween masks for $3.50 each. The last 20 of the masks were sold on Halloween Day for $2 each. The regular price included a markup of 15% on cost for overhead.

    a. What were the total wholesale cost, total handling cost, and total sales?

    b. Was a net profit, operating loss, or gross loss made on the 200 masks?

    c. How much was it, and what percent of wholesale cost did it represent (to the nearest tenth)?

18. A store paid $40 each for 50 35mm cameras. It sold 40 cameras for $55 each. The price for the last 10 cameras was $35 each. The regular markup for overhead was 20% on cost.

    a. What were the total wholesale cost, total handling cost, and total sales?

    b. Was a net profit, operating loss, or gross loss made on the 50 cameras?

    c. How much was it, and what percent of the wholesale cost did it represent?

CHAPTER 14 GLOSSARY

Absolute (or gross) loss. Loss resulting when the sale price of merchandise is less than cost: Absolute loss = Cost − Sale price.

Actual selling price. A "sale price" lower than the regular marked price; a reduced price. Actual selling price = Regular price − Markdown.

Breakeven point. A sale price that produces neither profit nor loss. (Same as "Total handling cost.")

Gross loss. (See "Absolute loss.")

Markdown. An amount or percent by which the regular selling price is reduced to obtain the actual selling price (or "sale" price).

Operating loss. Loss resulting when the sale price of merchandise is less than the total handling cost: Operating loss = Total handling cost − Sale price.

Reduced price. Selling price after a deduction or discount off the regular (higher) marked price; also called sale price, actual selling price, or net selling price.

Regular price. Selling price at which merchandise would normally be marked; also called regular marked price, regular selling price, marked price, or list price.

Total handling cost. The wholesale cost plus the related overhead on a sale ($C + \overline{OH}$) (also called the "Breakeven point").

Mathematics of Finance

SIMPLE INTEREST

OBJECTIVES

Upon completing Chapter 15, you will be able to:

1. Define and use correctly the terminology associated with each topic.
2. **a.** Use the simple interest formula $I = Prt$ to find any variable, when the other items are given (Section 1: Examples 1, 3, 4; Problems 1, 2, 5–10, 13, 14).

   b. Use the amount formula $M = P + I$ to find the amount or interest (Examples 1, 2, 4; Problems 1–4, 7–10).

   c. Use the amount formula $M = P(1 + rt)$ to find the amount or principal (Examples 2, 5; Problems 3, 4, 11, 12).
3. **a.** Compute ordinary or exact interest using exact time (Section 2: Examples 1–4; Problems 1–6; Section 3: Examples 1–3; Problems 1–8, 13–16).

   b. Predict how a given change in rate or time will affect the amount of interest (Section 3: Problems 9–12).
4. Given information about a simple interest note, use any of the above formulas to compute the unknown (Section 4: Examples 1–3; Problems 1–20):

   a. Interest

   b. Maturity value

   c. Rate

   d. Time

   e. Principal.
5. Compute the present value of a simple interest note, either

   a. On the original day (Section 5: Examples 1, 2; Problems 1–4, 7–14), or

   b. On some other given day prior to maturity (Example 3; Problems 5, 6, 15–20).

The borrowing and lending of money is a practice that dates far back into history. Never, however, has the practice of finance been more widespread than it is today. Money may be loaned at a simple interest or a simple discount rate. The loan may be repaid in a single payment or a series of payments, depending upon the type of loan. When money is invested with a financial institution, compound interest is usually earned on the deposits. The following chapters explain the basic types of loans, important methods of loan repayments, and fundamental investment procedures. Before you proceed, a review of operations with parentheses found in Chapter 1 (p. 5) would be helpful.

SECTION 1
BASIC SIMPLE INTEREST

Persons who rent buildings or equipment expect to pay for the use of someone else's property. Similarly, those who borrow money must pay rent for the use of that money. Rent paid for the privilege of borrowing another's money is called **interest.** The amount of money that was borrowed is the **principal** of a loan.

A certain percentage of the principal is charged as interest. The percent or **rate** is quoted on a yearly (per annum) basis, unless otherwise specified. The **time** is the number of days, months, or years for which the money will be loaned. In order to make rate and time correspond, time is always converted to years, since rate is always given on a yearly basis.

When **simple interest** is being charged, interest is calculated on the whole principal for the entire length of the loan. Simple interest is found using the formula

$$I = Prt$$

where I = interest, P = principal, r = rate, and t = time. The **amount** due on the ending date of the loan (also called **maturity value**) is the sum of the principal plus the interest. This is expressed by the formula

$$M = P + I$$

where M = amount (or maturity value or sum), P = principal, and I = interest.

Example 1   A loan of $900 is made at 16% for 5 months. Determine (a) the interest and (b) the maturity value of this loan.

(a) $P = \$900$ $I = Prt$

$r = 16\%$ or 0.16 or $\dfrac{16}{100}$ $= \$900 \times \dfrac{16}{100} \times \dfrac{5}{12}$

$t = 5 \text{ months}^* = \dfrac{5}{12} \text{ year}$ $I = \$60$

*Keep in mind that, for periods less than 1 year, time must be expressed as a fraction of a year.

(b) $M = P + I$

$= \$900 + \$60$

$M = \$960$

    The two formulas in Example 1 can be combined, allowing the maturity value to be found in one step. Since $I = Prt$, by substitution in the formula

$$M = P + I$$
$$\downarrow$$
$$= P + Prt$$

$$M = P(1 + rt)$$

    If the formula $M = P(1 + rt)$ is used to calculate the amount, the interest may be found by taking the difference between maturity value and principal.

Example 2   Rework Example 1, using the maturity value formula $M = P(1 + rt)$. As before, $P = \$900$, $r = \frac{16}{100}$, and $t = \frac{5}{12}$ year.

(a) $M = P(1 + rt)$

$$= \$900\left(1 + \frac{\overset{4}{\cancel{16}}}{100} \times \frac{5}{\underset{3}{\cancel{12}}}\right)$$

$$= 900\left(1 + \frac{20}{300}\right)$$

$$= 900\left(\frac{300}{300} + \frac{20}{300}\right)$$

$$= \overset{3}{\cancel{900}}\left(\frac{320}{\cancel{300}}\right)$$

$$M = \$960$$

(b) $P + I = M$

$I = M - P$

$= \$960 - \$900$

$I = \$60$

> ### Calculator techniques . . .   FOR EXAMPLE 2
>
> When a formula contains parentheses, compute that value and place it into memory; then multiply by the number preceding the parentheses.
>
> **Note.**   To store the value in parentheses, you can perform the "1 $\boxed{\text{M+}}$" step either before or after you $\boxed{\text{M+}}$ the "$r \times t$" calculation.
>
> $0.16 \boxed{\times} 5 \boxed{\div} 12 \boxed{\text{M+}} \longrightarrow 0.0666666; \qquad 1 \boxed{\text{M+}} \longrightarrow 1$
>
> $\boxed{\text{MR}} \longrightarrow 1.0666666 \boxed{\times} 900 \boxed{=} \longrightarrow 959.99994$
>
> Rounded to the nearest cent, this equals $960. This technique will apply for many problems in this chapter and the next one.
>
> **Note.**   The "$r \times t$" can be computed as: $5 \boxed{\times} 16 \boxed{\%} \longrightarrow 0.8 \boxed{\div} 12 \boxed{\text{M+}}$, if you prefer to use the $\boxed{\%}$ key rather than converting the 16% interest rate to its decimal equivalent.

Example 3   At what rate will $480 earn $28 in interest after 10 months?

$$P = \$480 \qquad\qquad\qquad I = Prt$$

$$r = ? \qquad\qquad\qquad\qquad \$28 = \$480r \times \frac{5}{6}$$

$$t = \frac{10}{12} \quad \text{or} \quad \frac{5}{6}\,\text{year} \qquad 28 = \overset{80}{\cancel{480}} \left(\frac{5}{\cancel{6}}\right) r$$

$$I = \$28 \qquad\qquad\qquad 28 = 400r$$

$$r = \frac{28}{400} \quad \text{or} \quad \frac{7}{100} \quad \text{or} \quad 0.07$$

$$r = 7\%$$

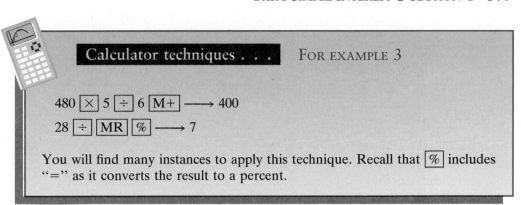

Calculator techniques . . .   FOR EXAMPLE 3

480 × 5 ÷ 6 M+ ⟶ 400

28 ÷ MR % ⟶ 7

You will find many instances to apply this technique. Recall that % includes "=" as it converts the result to a percent.

Example 4   How long will it take at an 8% simple interest rate for $950 to amount to $988?

This problem may be worked using the maturity value formula $M = P(1 + rt)$. However, the simple interest formula $I = Prt$ is easier to work with and should be used whenever possible. Notice that the formula $I = Prt$ may be used to find a missing variable whenever the interest is given or can be found quickly.

In this case, the interest may be computed easily, since

$$I = M - P \qquad\qquad\qquad I = Prt$$

$$= \$988 - \$950$$

$$\$38 = \$950\left(\frac{8}{100}\right)t$$

$$I = \$38$$

$$38 = 76t$$

$$P = \$950$$

$$\frac{38}{76} = t$$

$$r = 8\% \quad \text{or} \quad \frac{8}{100}$$

$$t = \frac{38}{76} \quad \text{or} \quad \frac{1}{2}$$

$$t = ?$$

$$t = \frac{1}{2}\text{year} \quad \text{or} \quad 6\,\text{months}$$

Example 5   What principal would have a maturity value of $636 in 8 months if the interest rate is 9%?

In this problem the interest is not given and is impossible to determine because the maturity value and principal are not both known. Thus, the amount formula $M = P(1 + rt)$ must be used to find the principal.

$$P = \$?$$

$$M = P(1 + rt)$$

$$r = 9\% \quad \text{or} \quad \frac{9}{100}$$

$$\$636 = P\left(1 + \frac{\overset{3}{\cancel{9}}}{100} \times \frac{2}{\cancel{3}}\right)$$

$$t = \frac{8}{12} \quad \text{or} \quad \frac{2}{3}\,\text{year}$$

$$= P\left(1 + \frac{6}{100}\right)$$

$$M = \$636$$

$$= P\left(\frac{100}{100} + \frac{6}{100}\right)$$

$$= P\left(\frac{106}{100}\right)$$

$$\frac{100}{\cancel{106}} \times \cancel{636}^{\;6} = P\left(\frac{\cancel{106}}{\cancel{100}}\right)\left(\frac{\cancel{100}}{\cancel{106}}\right)$$

$$\$600 = P$$

SECTION 1   PROBLEMS

Find the interest and amount.

1. a. $500 at 9% for 1 year   b. $500 at 9% for 6 months   c. $780 at 12% for 10 months
   d. $900 at 10.5% for 8 months   e. $840 at 14% for 5 months

2. a. $600 at 10% for 1 year   b. $600 at 10% for 6 months   c. $1,200 at 8% for 4 months
   d. $900 at 8.5% for 3 months   e. $2,400 at 9% for 9 months

3. Find the interest and amount for parts a, c, and e in Problem 1 using the maturity value formula $M = P(1 + rt)$.

4. Utilize the amount formula $M = P(1 + rt)$ to find the interest and amount for parts a, c, and e in Problem 2.

5. What was the interest rate on a note with a principal of $800 for 3 months if the interest was $18?

6. What was the interest rate if the interest due on a 9-month loan of $3,000 was $180?

7. How many months would it take $1,440 to amount to $1,537.20 if an interest rate of 9% was charged?

8. How many months would it take for a loan of $1,500 at 7% interest to acquire a maturity value of $1,587.50?

9. What interest rate was charged on a note with a principal of $4,000 and a time of 3 months if the maturity value was $4,160?

10. If $6,000 is worth $6,300 after 6 months, what interest rate was charged?

11. What principal will have a maturity value of $8,400 after 24 months at 10%?

12. What principal will amount to $9,680 after 18 months at 14%?

13. At what simple interest rate will an investment double itself in 10 years? (Hint: Choose any principal; what is the interest in 10 years?)

14. How long will it take for an investment to double at 16% simple interest? (Hint: Choose any principal; what will be the interest when the investment matures?)

SECTION 2

ORDINARY TIME AND EXACT TIME

The problems given in Section 1 were limited to those with time periods of even months or years. This is not always the case, as many loans are made for a certain number of days. Suppose that a loan is made on July 5 for 60 days; do we consider 60 days to be 2 months and consider the loan due on September 5, or do we take the 60 days literally and consider the amount due on September 3? Suppose this loan made on July 5 is for a time of 2 months; would we consider the time to be $\frac{2}{12} = \frac{1}{6}$ year or should we use the actual 62 days between July 5 and September 5? The answers to these questions depend upon whether ordinary time or exact time is being used.

When **ordinary time** is being used, each month is assumed to have 30 days. Thus, any 5-month period would be considered as $5 \times 30 = 150$ days. Similarly, 90 days would be considered as 3 months. In actual practice, however, ordinary time is seldom used, as we shall see later.

If **exact time** is being used, the specific number of days within the time period is calculated. Thus, 90 days would usually be slightly less than 3 months. Exact time may be calculated most conveniently using Table C-14, The Number of Each Day of the Year, in Appendix C. In this table, each day of the year is listed in consecutive order and assigned a number from 1 to 365. The following examples demonstrate use of the table.

Example 1   Using exact time, find (a) the due date of a loan made on April 15 for 180 days and (b) the exact time from March 10 to July 23.

   (a)   From the table of each day of the year, we see that April 15 is the 105th day. Thus,

$$
\begin{array}{r}
\text{April 15} = \quad 105 \text{ day} \\
+180 \text{ days} \\
\hline
285\text{th day}
\end{array}
$$
   is October 12, the due date of the note using exact time

   (b)   From March 10 to July 23:

$$
\begin{array}{r}
\text{July 23} = \quad 204 \text{ day} \\
\text{March 10} = - \ \ 69 \text{ day} \\
\hline
135 \text{ days}
\end{array}
$$
   between March 10 and July 23, using exact time

Example 2   A loan is taken out on June 1 for 5 months. Determine (a) the due date and (b) the number of days in the term of the loan, using both ordinary and exact time.

(a)   Five months after June 1 is November 1. The loan (principal plus interest) will be due on November 1 regardless of which kind of time is used.*

(b)   From June 1 to November 1:

<table>
<tr><td align="center">Ordinary Time</td><td align="center">Exact Time</td></tr>
<tr><td align="center">5 months = 5 × 30</td><td align="center">November 1 =   305 day</td></tr>
<tr><td align="center">= 150 days</td><td align="center">June 1 = −152 day</td></tr>
<tr><td></td><td align="center">153 days</td></tr>
</table>

Sometimes the time period of a loan includes a leap year. This does not affect the calculation of ordinary time, of course, since all months are considered to have 30 days. The leap year must be given consideration, however, when exact time is being used. During a leap year, February 29 becomes the 60th day of the year; thus, from March 1 on, the number of each day is one greater than is shown in Table C-14. For example, during a leap year, March 15 is the 75th day, instead of the 74th day as the table shows. This change in the number of the day must be made before the exact time is calculated.

A simple test allows you to determine whether any given year is a leap year: Leap year always falls on a year evenly divisible by 4. That is, 1996 is divisible by 4, so 1996 is a leap year. The year 1998 is not divisible by 4; thus, 1998 is not a leap year. Presidential elections also fall on leap years.

Example 3   Find the exact time from January 16, 1996 to May 7, 1996.

Since 1996 is a leap year, May 7 is the 128th day.

$$
\begin{array}{r}
\text{May 7} = \quad 128 \text{ day} \\
\text{January 16} = - \quad 16 \text{ day} \\
\hline
112 \text{ days}
\end{array}
\quad
\begin{array}{l}
\text{from January 16, 1996} \\
\text{to May 7, 1996}
\end{array}
$$

Example 4   Find the exact time between November 16 of one year and April 3 of the next (assuming no leap year is involved).

*Note.   If the given number of months would end on a day that does not exist, the due date is the last day of the month. For example, a loan on March 31 for 8 months would be due on November 30, since there is no November 31.

We must find the remaining days in the present year and then add the days in the following year:

$$\begin{array}{rl}
\text{Present year} = & 365 \text{ days} \\
\text{November 16} = & \underline{-320 \text{ day}} \\
& 45 \text{ days remaining in present year} \\
\text{April 3} = + & \underline{93 \text{ day}} \\
& 138 \text{ days from November 16 to April 3}
\end{array}$$

SECTION 2   PROBLEMS

If no year is given in the following problems, assume that it is not a leap year.

Find the exact time from the first date to the second date.

1. a. October 8 to December 5   b. April 11 to November 11
     c. May 15 to December 23   d. January 8 to September 8
     e. February 16, 1996, to June 16, 1996   f. November 23, 1997, to April 15, 1998
     g. August 16, 1998, to January 16, 1999   h. October 7, 1995, to March 7, 1996

2. a. March 6 to May 31   b. June 5 to September 15
     c. July 10 to December 10   d. January 14 to November 28
     e. February 12, 1996, to May 31, 1996   f. September 23, 1997, to April 1, 1998
     g. July 4, 1998, to March 4, 1999   h. November 15, 1999, to June 6, 2000

Determine the due date, using exact time.

3. a. 60 days after July 12   b. 90 days after September 5
     c. 180 days after May 20   d. 240 days after April 30
     e. 270 days after January 15   f. 120 days after October 31

4. a. 30 days after August 21   b. 60 days after October 14
     c. 120 days after June 18   d. 200 days after March 1
     e. 150 days after July 17   f. 180 days after November 23

Find the due date and the exact time (in days) for loans made on the given dates.

5. a. October 2 for 2 months   b. July 7 for 4 months
     c. January 12, 1996, for 8 months   d. June 18, 1997, for 4 months
     e. February 3, 1996, for 3 months

6. a. February 18 for 6 months   b. May 24 for 3 months
     c. January 26, 2000, for 9 months   d. March 30, 1998, for 4 months
     e. January 13, 1996, for 5 months

SECTION 3
ORDINARY INTEREST AND EXACT INTEREST

It has been seen that the time of a loan is frequently a certain number of days. As stated previously, a time of less than one year must be expressed as a fraction of a year. The question then arises: When converting days to fractional years, are the days placed over a denominator of 360 or 365? It is this question with which ordinary and exact interest are concerned.

The term "interest" here refers to the number of days representing one year. Just as ordinary time indicated 30 days per month, **ordinary interest** indicates interest calculated using 360 days per year. Similarly, the term **exact interest** indicates interest calculated using 365 days per year. (A 366-day year is not often used in business even for leap years.)

We have now discussed two ways of determining time (days in the loan period) and two types of interest (days per year). There are four possible combinations of these to obtain t for the interest formula:

I. Exact time over ordinary interest: $t = \dfrac{\text{Exact days}}{360}$ *(largest fraction)*

II. Exact time over exact interest: $t = \dfrac{\text{Exact days}}{365}$

III. Ordinary time over ordinary interest: $t = \dfrac{\text{Approximate days}}{360}$

IV. Ordinary time over exact interest: $t = \dfrac{\text{Approximate days}}{365}$

Type 1. The combination of exact time and ordinary interest is known as the **Banker's Rule**, a name derived from the fact that banks (as well as other financial institutions) generally compute their interest in this way:

$$\frac{\text{Exact days}}{360}$$

Example 1 A $730 loan is made on March 1 at 12% for 9 months. Compute the interest by the Bankers' Rule (that is, using exact time and ordinary interest).

Nine months after March 1 is December 1. The exact time is

$$\begin{aligned} \text{December 1} &= 335 \text{ day} \\ \text{March 1} &= \underline{60 \text{ day}} \\ & \ 275 \text{ days} \end{aligned}$$

$$P = \$730 \qquad\qquad\qquad I = Prt$$

$$r = \frac{12}{100} \qquad\qquad\qquad = \$730 \times \frac{\overset{1}{\cancel{12}}}{\cancel{100}} \times \frac{55}{\underset{6}{\cancel{72}}}$$

$$t = \frac{275}{360} \quad \text{or} \quad \frac{55}{72} \qquad\qquad I = \$66.92 \quad \text{(exact time/ordinary interest)}$$

$$I = ?$$

Calculator techniques . . .   FOR EXAMPLE 1

When using a calculator, there is no particular need to reduce the time fraction in an interest problem:

730 ⊠ 0.12 ⊠ 275 ⊡ 360 ⊟ ⟶ 66.916666

This applies throughout the simple interest and discount chapters. Also, the interest rate can be entered as 12 ⊟%, if you wish, rather than as its decimal equivalent. (Recall, however, that a percent cannot be the first value entered, since the ⊟% key also functions as an ⊟=.)

Type II.   The combination of exact time and exact interest is gaining in usage by financial institutions and has been used extensively in certain parts of the country during periods when interest rates have been quite high. This is partly to ensure that they do not exceed legal maximum (usury) rates on interest charges and partly because this method results in slightly less interest paid to depositors.

Another major application is by the Federal Reserve Bank and the United States government, which use this method when computing interest:

$$\frac{\text{Exact days}}{\text{365 or 366}}$$

This exact-time/exact-interest method applies both for interest owed to the government (such as interest for late payment of taxes) and for interest paid by the government to purchasers of certain U.S. interest-bearing certificates. (For example, earnings on U.S. Treasury Notes are computed using a half-year variation of exact interest.)

Example 2 Compute Example 1 using type II (exact time and exact interest).

$$P = \$730 \qquad\qquad\qquad I = Prt$$

$$r = \frac{12}{100} \qquad\qquad\qquad = \$\overset{10}{\cancel{730}} \times \frac{12}{\cancel{100}} \times \frac{\cancel{55}}{\cancel{73}}$$

$$t = \frac{275}{365} \quad\text{or}\quad \frac{55}{73} \qquad\qquad I = \$66.00 \quad \text{(exact time/exact interest)}$$

$$I = ?$$

Type III. The combination of ordinary time and ordinary interest is equivalent to the calculations used in the first section of this chapter. (For example, any 6-month period is equivalent to 180 days or $\frac{1}{2}$ year.) This method of interest calculation is not used commercially, but it is commonly used in private loans between individuals: $\dfrac{\text{Approx. days}}{360}$ or $\dfrac{\text{Months}}{12}$.

Example 3 Rework Example 1 using ordinary time (9 months or 270 days) and ordinary interest (360 days or 12 months).

$$P = \$730 \qquad\qquad\qquad I = Prt$$

$$r = \frac{12}{100} \qquad\qquad\qquad = \$730 \times \frac{\overset{3}{\cancel{12}}}{\cancel{100}} \times \frac{3}{\cancel{4}}$$

$$t = \frac{9}{12} \quad\text{or}\quad \frac{270}{360} \qquad\qquad I = \$65.70 \quad \text{(ordinary time/ordinary interest)}$$

$$= \frac{3}{4}$$

$$I = ?$$

Type IV. The other possible combination of ordinary time and exact interest is not customarily used. (Example 1 reworked using 270 days over a 365-day year yields interest of $64.80.)

Comparing Examples 1 through 3 reveals that type I, the combination of *exact time over ordinary interest* (a 360-day year), produces the most interest on a loan. Since this combination results in the most interest due to the lender, it is therefore the method most often used historically—which accounts for its name, the Bankers' Rule. From this point on, our problems will emphasize this type I method (Bankers' Rule), with supporting use of the type II method.

SECTION 3   PROBLEMS

Find the ordinary interest (type I) and exact interest (type II).

1. a. On $2,400 for 90 days at 12%   b. On $1,500 for 120 days at 10%
2. a. On $2,190 for 45 days at 8%   b. On $2,920 for 60 days at 9%
3. A 3-month note dated August 4 for $2,200 had an interest rate of 9%. Determine the interest due using the three combinations of time and interest (types I, II, and III).
4. A loan of $8,000 was made on June 6 for 4 months at 14%. Calculate the interest on this loan using the three combinations of time and interest (types I, II, and III).

   Use the Bankers' Rule (type I) to compute ordinary interest.

5. Find the ordinary interest on $3,000 for 60 days at each of the given interest rates. Carefully observe the relationships among your answers.
   a. 12%   b. 6%   c. 18%   d. 9%
6. Compute the ordinary interest on $3,000 for 90 days at each of the given rates. Observe carefully the relationships between rates and interest due.
   a. 5%   b. 10%   c. 15%   d. 7.5%
7. Compute the ordinary interest on a $2,800 loan at 15% for each of the given time periods. Notice how a change in time affects the interest due.
   a. 30 days   b. 90 days   c. 45 days   d. 180 days
8. Find the ordinary interest on a $4,000 loan at 12% for each of the given time periods. Carefully observe the relationships among your answers.
   a. 90 days   b. 30 days   c. 120 days   d. 270 days
9. The ordinary interest on a loan for a certain number of days at 10% was $24. What was the ordinary interest on the same principal and for the same time at the following rates? Apply what you learned in Problem 5.
   a. 5%   b. 15%   c. 18%
10. The ordinary interest on a loan for a certain number of days at 8% was $36. Using what you learned in Problem 6, compute the ordinary interest on this loan for the same time at
    a. 16%   b. 10%   c. 12%
11. The ordinary interest on a loan at a certain rate was $75 for 180 days. Applying what you learned in Problem 7, compute the interest on the same principal at the same rate if the times were as follows:
    a. 60 days   b. 45 days   c. 120 days
12. When interest was computed at a certain rate, the interest due on a 120-day loan was $40. Using your conclusions from Problem 8, determine the interest due on the same principal and at the same rate if the time periods were as follows:
    a. 60 days   b. 300 days   c. 240 days

    Use type II to compute exact interest.

13. A 3-month note was taken out on March 6 for $1,825. Calculate the exact interest due on the note
    a. At 14%   b. At 7%   c. At 10%
14. On April 5, a loan of $2,190 was taken out for 2 months. Find the exact interest due on the loan
    a. At 10%   b. At 5%   c. At 15%

15. A loan for $1,460 was made at 8%. Determine the exact interest for
   a. 120 days   b. 30 days   c. 90 days

16. A 9.5% note for $1,460 was made. Determine the exact interest for
   a. 150 days   b. 180 days   c. 270 days

SECTION 4
SIMPLE INTEREST NOTES

A person who borrows money usually signs a written promise to repay the loan; this document is called a **promissory note,** or just a **note.** The money value that is specified on a note is the **face value** of the note. If an interest rate is mentioned in the note, the face value of this simple interest note is the **principal** of the loan. Interest is computed on the whole principal for the entire length of the loan, and the **maturity value** (principal plus interest) is repaid in a single payment on the **due date** (or maturity date) stated in the note. If no interest is mentioned, the face value is the maturity value of the note. The person who borrows the money is called the **maker** of the note, and the one to whom the money will be repaid is known as the **payee.**

The use of simple interest notes varies from state to state; the banks in some states issue simple interest notes, whereas others normally use another type of note. Figure 15-1 shows a simple interest bank note. [The face value of this note is $1,200. Notice that the maturity value ($1,208.88) is also entered for convenience.] Notes between individuals are usually simple interest notes also.

Before loaning money, a bank may require the borrower to provide **collateral**—some item of value that is used to secure the note. Thus, if the loan is not repaid, the bank is entitled to obtain its money by selling the collateral. (Any excess funds above the amount owed would be returned to the borrower.) Some items commonly used as collateral are cars, real estate, insurance policies, stocks and bonds, as well as savings accounts. Figure 15-2 illustrates a simple interest note in which a certificate of deposit is the collateral.

Most promisory notes are short-term notes—for 1 year or less. The most common time period for notes is one quarter (or 90 days). When a note matures after one quarter, the interest due must be paid and the note may usually be renewed for another quarter at the rate then in effect for new loans.

In many localities, short-term loans from banks are almost exclusively either installment loans or simple discount loans. For an installment loan, the maturity value is calculated as for a simple interest loan; however, this amount is repaid in weekly or monthly payments rather than being repaid in one lump sum, as is the case for simple interest notes. (The simple discount loan and the installment loan are both discussed in detail in later topics.)

The bank is required by law to provide the borrower with information about the annual percentage rate and the finance charge. Some banks provide this information on the promissory note itself, as shown in Figure 15-1. Other

(FIXED OR VARIABLE RATE/SINGLE PAYMENT OR INSTALLMENTS) (365) (NATIONAL OR STATE BANK)   NOTE NO. X4-10436

Anything appearing below this block shall control if it conflicts with anything above or in this block.

| MATURITY: Sept. 29, 19X4 | RATE: 9% | TERM: 30 days | COLLATERAL: none |

| **PROMISSORY NOTE** | NORTH AUSTIN STATE BANK | DATE OF NOTE Aug. 30, 19X4 |
| AMOUNT OF NOTE $ 1,200.00 | TOTAL OF PAYMENTS $ 1,208.88 | AUSTIN, TEXAS | OFFICER GKW |

MAKER: Stanley L. Warren
ADDRESS: 11208 Apache Trail
Round Rock, Texas 78626

ACCT. NO. OR CUST. NO. 029-75-3643

TELEPHONE NO. 528-4117

For value received, each of the undersigned (called Maker whether one or more) jointly and severally promises to pay to the order of the bank named above (called Bank) at its office in the city named above in immediately available current funds of the United States of America the principal sum of One thousand two hundred and no/100 -- Dollars, together with interest on such principal sum, or so much thereof as may be advanced and outstanding, at the rate checked below:

☑ A fixed rate of _____9%_____ percent per annum.

☐ A variable rate (called the Variable Rate) based on the prime or base percent per annum rate of interest (called the Base Rate) charged by _____ for short-term loans to substantial and responsible commercial borrowers, each change in the Variable Rate to be effective, without notice to Maker, on the effective date of each change in the Base Rate. The Variable Rate shall be:

　☐ The Base Rate plus _____ percentage points.
　☐ _____

Interest charges will be calculated for actual days elapsed on the basis of a 365-day year. In no event shall the rate charged hereunder exceed the maximum rate of interest permitted by applicable law, and if application of any variable rate or any other circumstance would cause the rate of interest hereunder to exceed such maximum rate, the rate of interest hereunder automatically shall be reduced to such maximum rate.

Payment shall be due as checked below:

☑ Principal and interest shall be due on demand or, if no demand is made, on Sept. 29, 19X4.

☐ Principal shall be due _____. Interest shall be due _____.

☐ Payment shall be due in _____ installments, with the first installment of $_____ due _____, and installments of the same amount due on the _____ day of each _____ thereafter until a final installment equal to the total unpaid balance is due on _____.

☐ _____

Interest:
　☐ Is included in above installments.
　☐ Is in addition to above installments and shall be due _____, beginning _____.
　☐ _____

Purpose: _____

Bank may require payment by Maker and any surety, indorser or guarantor without first resorting to any security. Maker and all other parties liable hereon consent to the release or discharge of any party liable hereon (including any of the undersigned) and to the release, impairment or substitution of any collateral for this note by Bank.

If default occurs in the payment of any principal or interest when due hereunder, or upon the occurrence of any default or failure to perform any covenant, agreement or obligation to be performed under any document or instrument executed in connection with or as security for this note, or if Bank in good faith believes that the prospect of payment of this note is impaired, Bank may declare the entirety of this note, principal and interest, immediately due and payable, but failure to do so at any time shall not constitute a waiver of Bank's right to do so at any other time.

All past due principal and interest shall bear interest from maturity of such principal or interest at the maximum rate of interest permitted by applicable law. If default occurs in the payment of this note at maturity, whether maturity may occur by acceleration or otherwise, or this note is collected through probate, bankruptcy or other proceedings, Maker promises to pay all costs and expenses of collection and enforcement. If this note is placed in the hands of an attorney for collection, Maker promises to pay, in addition to all other costs and expenses of collection and enforcement, an additional amount equal to 15% of the principal and interest then due, as attorney's fees.

Maker and every surety, indorser and guarantor of this note waive presentment for payment, protest, notice of dishonor, grace and notice of acceleration and all other notice, filing of suit and diligence in collecting this note and the enforcing of any of the security rights of Bank, and consent and agree that time of payment hereof may be extended without notice at any time and from time to time, and for periods of time whether or not for a term or terms in excess of the original term hereof, without notice or consideration to, or consent from, any of them.

Stanley L. Warren

Courtesy of North Austin State Bank, Austin, Texas

FIGURE 15-1   Interest-Bearing Note

Consumer Note - Crestar Bank

CRESTAR

William J. McDaniel
Borrower

May 1, 19x4
Date

Old Town
Originating Office

4651-9121
Account No.

A2-22341
Note No.

Co-Borrower
3001 Dawes Avenue; Alexandra, Va 22311
Borrower's Address

Five thousand and no/100 dollars
Loan Amount
Dollars ($ 5,000.00   )

In this Note, the words "you" and "your" mean the Borrower and any Co-Borrower. "We," "our," "us" and "the Bank" mean Crestar Bank. "Scheduled Payment" means any payment specified in this Note, whether it is a payment of principal plus interest or a payment of interest only.

This Note covers your loan with the Bank. When you sign it, each one of you is fully responsible for fulfilling all of the promises you make in this Note. By signing this Note, you acknowledge that you received a loan from us in the Loan Amount shown above and agree to pay to us at any of our offices, or at such place as the Bank may in writing designate, the Loan Amount shown above, plus or including interest, and any other amounts due, upon the terms described below.

Terms of Note

☐ **Simple Interest Instalment Loan**

You will make a total of _____ monthly payments of principal and interest beginning on _____, 19 _____ and continuing on the same day of each succeeding month until this Note is paid in full. You will make _____ monthly payments of principal and interest equal to $ _____ and then a final payment equal to the unpaid principal balance plus interest and any other amounts due.
☐ Payment will be by the Alternative Payment Schedule specified below.

☐ **Principal Plus Instalment Loan**

You will make a total of _____ monthly payments of principal and interest beginning on _____, 19 _____ and continuing on the same day of each succeeding month until this Note is paid in full. Each month for _____ months you will make a payment equal to $_____ in principal plus all interest accrued on the unpaid balance, and then a final payment equal to $_____ in principal plus all accrued interest and any other amounts due.
☐ Payment will be by the Alternative Payment Schedule specified below.

☒ **Single Payment Loan** due on May 31,   , 19 x4 (30 days from the date of this Note).

You will make one payment in full of principal plus accrued interest from the date of this Note until this Note is paid in full plus any other amounts due.
☒ New obligation ☐ Renewal - New disclosure required ☐ Renewal - Same terms and no new disclosure required.

Interest

Interest on an Instalment Loan will accrue on a 30/360 day basis. On all other loan types, interest will accrue daily on the basis of a 360-day year. Interest will accrue at the stated interest rate on the unpaid balance from the date of this Note until paid in full. Interest will continue to accrue after maturity, whether by acceleration or otherwise, at the stated interest rate until this Note has been paid in full.
Subject to the above, the interest rate applicable to this Note (the "Rate") is:

☒ 9.0 _____ % per annum, fixed for the term of this Note.
☐ The initial Rate is _____ % per annum, subject to change with changes in the Index identified below. This Rate is based upon the Index plus _____ % per annum.

☐ The Bank's prime rate, which is the rate established from time to time by the Bank and recorded in its Credit Administration Division as a reference for fixing the lending rate on commercial loans. It is not necessarily the lowest rate charged by the Bank for commercial borrowings.
☐ The Bank's personal rate, which is the rate established from time to time by the Bank and recorded in its Credit Administration Division as a reference for fixing the lending rate on personal loans. It is not necessarily the lowest rate charged by the Bank for personal borrowings.

Collateral

We will have a security interest in the property described below (the "Collateral"). The Collateral is described more fully in a
☐ Security Agreement dated _____ ☐ Assignment of Deposit dated _____
☐ Deed of Trust dated _____ ☐ Credit Line Deed of Trust dated _____
☒ Other Certificate of Deposit for $5,000.00 at Crestar   dated March 1, 19x4
Covering: _____

You also give us the right of set-off, which means we can apply any funds you have on deposit with us to any amount due and unpaid on this Note. We may use the Collateral (except your dwelling(s)) to secure other loans you have with us and the collateral we are holding for your other loans (except your dwelling(s) and household goods) may also secure this Note. If credit insurance is obtained with this loan, we will have a security interest in all unearned credit insurance premiums.

There Are Important Additional Terms And Conditions On The Reverse Side Of This Note. Be Sure To Read Them Before You Sign.

By signing below, you agree to all the terms of this Note and acknowledge receipt of a completed copy.

William J. McDaniel   (Seal)
Borrower's Signature

_____ (Seal)
Co-Borrower's Signature

By signing below, you agree to be bound by all the terms of this Note and acknowledge receipt of a copy of the "Notice to Co-Signer" form.

_____ (Seal)
Endorser's Signature

Endorser's Address

_____ (Seal)
Endorser's Signature

Endorser's Address

CRE-0301 VA (1/91) For use by Crestar Bank only

Distribution: Original - Note File; Canary - Quality Control; Pink - Customer Copy

Courtesy of Crestar Bank, Alexandria, Virginia

FIGURE 15-2   Note with Collateral

Truth-in-Lending Disclosure Statement   **CRESTAR**

Creditor: ☒ Crestar Bank   Borrower: William J. McDaniel
☐ Crestar Bank N.A.
☐ Crestar Bank MD
Co-Borrower:

Loan Type: ☐ Variable Rate Instalment Loan
☐ Fixed Rate Instalment Loan
☐ Single Payment Loan   Originating Office: Old Town
☐ Demand Loan

Use of Terms: In this Disclosure, the words "you" and "your" mean any Borrower or Co-Borrower. "We," "our," and "us" mean the Creditor. "Scheduled Payment"
means any payment disclosed in this disclosure or specified in the Note to which this disclosure applies, whether such payment is a payment of
principal and interest or a payment of interest only.

Estimates: The Finance Charge and Total of Payments disclosed below are estimates. These amounts have been computed on the assumption that all
payments will be received on their scheduled due date. Any other estimate is noted with an asterisk (*) and is based on the best information
reasonably available.

ANNUAL PERCENTAGE RATE The cost of your credit as a yearly rate	FINANCE CHARGE The dollar amount the credit will cost you	AMOUNT FINANCED The amount of credit provided to you or on your behalf	TOTAL OF PAYMENTS Amount you will have paid after you have made all payments as scheduled
9.0%	$ 37.50	$ 5,000.00	$ 5,037.50

Payment Schedule:

Number of Payments	Amount of Payments	When Payments Are Due	
		Monthly Beginning	19

☒ In one payment of $ 5,037.50   * due on May 31   19 x4
☒ This obligation is payable on demand.
   ☐ All disclosures are based on an "assumed maturity" of one year. Scheduled interest payments for the first year: _____

   ☐ Alternate payment schedule for a Demand Loan: _____

Variable Rate: ☐ The Annual Percentage Rate will increase during the term of the loan if the Index identified below increases:
   ☐ The Creditor's Prime Rate, which is the rate established from time to time by the Creditor and recorded in its Credit Administration
      Division as a reference for fixing the lending rate on commercial loans.
   ☐ The Creditor's personal rate, which is the rate established from time to time by the Creditor and recorded in its Credit
      Administration Division as a reference for fixing the lending rate on personal loans.
   ☐ The lowest prime rate published in the "Money Rates" section of *The Wall Street Journal*.

   ☐ _____
   The maximum interest rate will not exceed the maximum permitted by law or 24%, whichever is less. If the specified Index ceases to be
   published, we will substitute a similar index and send you notice of it. On an Instalment Loan, the effect of an increase in the Annual
   Percentage Rate may be more payments of the same amount, an increased final payment, or an increase in the amount of your payments.
   For example, if your loan were for $5,000 at 12% for three years, your monthly payment would be $166.07. If after six months your rate
   increased to 13% for the remainder of the loan term, your final payment would increase to $237.62; or you would have to make one additional
   partial payment of $72.33; or your regular monthly payments would increase to $168.10. On a Single Payment or Demand Loan, the effect
   of an increase in the Annual Percentage Rate would be an increase in the amount of your payments. For example, if your loan were for
   $10,000 at 14% payable on demand with an alternate payment schedule of $1,000 per month plus interest and the rate increased to 16%
   at the end of the first month, the payment due at the end of the second month would increase from $1,103.56 to $1,118.36.

Late Charge: If any portion of a Scheduled Payment is late, you will have to pay us a late charge of 5% of the amount which is late.

Security: ☐ You are giving a security interest in ☐ the goods or property being purchased. ☐ other property described as follows:

   Collateral securing other loans with us (except your consumer dwelling(s),) may also secure this loan.

   Assumption (applicable only to loans secured by a principal dwelling): Someone buying your house cannot assume the remainder of
   the loan on the original terms.

Filing Fees: $_____
Property Insurance: You may obtain any required property insurance from an insurance company of your choice.
Prepayment: If you pay off early, you will not have to pay a penalty and will not be entitled to a rebate of any prepaid Finance Charge.
   Read all your loan documents for additional information about nonpayment, default, late charges, any required payment in full before the scheduled date, security
   interests and other matters pertaining to this credit transaction.

Credit Insurance - CREDIT INSURANCE IS NOT REQUIRED FOR THIS
LOAN. If you would like Credit Life insurance on your Single Payment or
Instalment Loan, or Credit Disability for your Instalment Loan, we can provide it
for you. You must specify any insurance protection you want by checking the
appropriate blocks below. No insurance protection will be provided unless you
sign your request on this disclosure. **Credit insurance is not available on
demand loans.**

☐ Credit Life for: ☐ Borrower ☐ Co-Borrower

   The premium is $_____ for the term of the loan.
☐ Credit Disability for: ☐ Borrower ☐ Co-Borrower.

   The premium is $_____ for the term of the loan.
   Credit Life coverage must be provided for you to be eligible for Credit
   Disability protection

Term Insurance - TERM INSURANCE IS NOT REQUIRED FOR THIS LOAN.
If term insurance is desired, you must specify any insurance protection you
want by checking the appropriate blocks below and complete an application.
The coverage will not be effective unless and until your application is accepted
by the insurance company. **This coverage is not offered in the District of
Columbia or Maryland.**

☐ Single-Premium Term Life for: ☐ Borrower ☐ Co-Borrower.

   The premium is $_____ for _____ months.
☐ Single-Premium Disability for: ☐ Borrower ☐ Co-Borrower.

   The premium is $_____ for _____ months.

SIGNATURE OF PARTY REQUESTING INSURANCE _____

☐ If checked, a RESPA Estimated Closing Costs form was provided in place
of the Itemization of the Amount Financed.

ITEMIZATION OF AMOUNT FINANCED

Paid directly to you or deposited in
your account #_____ $ 5,000.00
Used to pay off other loans with us:
_____ $_____
_____ $_____
_____ $_____
_____ $_____

Paid to others on your behalf:
Credit and/or term insurance premium(s)
paid to insurance company $_____
_____ $_____
_____ $_____
_____ $_____
_____ $_____

Amount Financed $ 5,000.00
Prepaid Finance Charge $_____

Acknowledgement of Receipt: The undersigned acknowledge receipt of a completed copy of this disclosure before entering into any agreement with Creditor
concerning this credit transaction.

William J. McDaniel 5/1/x4
Borrower   Date   Co-Borrower   Date   Date
CRE-0279 SYS (1/91)   Distribution Instructions: White To Borrower; Canary To Disclosure File; Pink To Quality Control

Courtesy of Crestar Bank, Alexandria, Virginia

FIGURE 15-3   Truth-in-Lending Disclosure Statement

banks provide a separate statement, the "Truth-in-Lending Disclosure Statement" (Figure 15-3).

Example 1   A note for $450 was dated June 12 and repaid in 3 months with interest at 12%. Find (a) the due date and (b) the amount repaid.

(a)   The due date was September 12.
(b)   Using the Bankers' Rule, interest and maturity value were as follows:

$$\text{Sept. } 12 = 255 \text{ day}$$
$$\underline{\text{June } 12 = 163 \text{ day}}$$

$$t = \quad 92 \text{ days} \quad \text{or} \quad \tfrac{92}{360} \text{ year}$$

$$P = \$450$$
$$r = 12\%$$

Interest

$$I = Prt$$

$$= \overset{15}{\cancel{450}} \times \frac{\cancel{12}}{\cancel{100}} \times \frac{\overset{.92}{\cancel{.92}}}{\underset{20}{\cancel{360}}}$$

$$I = \$13.80$$

Maturity Value

$$M = P + I$$

$$= \$450 + \$13.80$$

$$M = \$463.80$$

Example 2   The maturity value of a 60-day simple interest note was $976, including interest at 10%. What was the face value of the note?

$$M = \$976$$

$$r = \frac{10}{100}$$

$$t = \frac{60}{360} \quad \text{or} \quad \frac{1}{6}$$

$$P = ?$$

$$M = P(1 + rt)$$

$$\$976 = P\left(1 + \frac{\overset{5}{\cancel{10}}}{100} \cdot \frac{1}{\underset{3}{\cancel{6}}}\right)$$

$$976 = P\left(1 + \frac{5}{300}\right)$$

$$976 = P\left(\frac{305}{300}\right)$$

$$\left(\frac{300}{305}\right)976 = P\left(\frac{\cancel{305}}{\cancel{300}}\right)\left(\frac{\cancel{300}}{\cancel{305}}\right)$$

$$\$960 = P$$

Calculator techniques . . .   FOR EXAMPLE 2

Here, since you can't multiply the value in parentheses times P, you must divide by the parenthetical value (using $\boxed{MR}$) to find P:

$$0.1 \boxed{\div} 6 \boxed{M+} \longrightarrow 0.0166666; \quad 1 \boxed{M+} \longrightarrow 1$$

$$976 \boxed{\div} \boxed{MR} \boxed{=} \longrightarrow 960.00006$$

Recall that the "$1 \boxed{M+}$" can be done first, if you prefer.

Example 3   A \$750 note drawn at 15% exact interest had a maturity value of \$795. What was the exact time of the note?

$$
\begin{aligned}
M &= \ \ \$795 \\
P &= - \ \ \underline{750} \\
I &= \ \ \ \$ \ 45
\end{aligned}
\qquad\qquad I = Prt
$$

$$
r = \frac{15}{100}
\qquad\qquad \$45 = \$750 \times \frac{15}{100} \times t
$$

$$
\qquad\qquad\qquad 45 = 112.5t
$$

$$
t = ?
\qquad\qquad \frac{45}{112.5} = t
$$

$$
0.4 \text{ year} = t
$$

$$
t = 0.4 \text{ year} \times 365 \text{ days per year}
$$

$$
t = 146 \text{ days}
$$

SECTION 4 PROBLEMS

Note.   You should use exact days in all problems unless exact time is impossible to determine from the information given. For example, if a problem involves a 3-month loan but no date is included, it is impossible to compute exact time; thus, ordinary time would have to be used. However, if a problem states that a 3-month loan began on some specific date, then the exact days during that particular 3-month interval would be used.

Further, ordinary interest should normally be computed, using the Bankers' Rule (exact days over 360). However, a 365-day year should be observed when the loan problem specifies "exact interest."

1. Identify each part of the note shown in Figure 15-1.

   a. Face value   b. Maker   c. Payee   d. Date

   e. Due date   f. Principal   g. Rate   h. Time

   i. Interest   j. Maturity value

2. Answer Problem 1 for the note shown in Figure 15-2 and the Truth-in-Lending Disclosure Statement shown in Figure 15-3.

   Complete the following simple interest problems, using the Bankers' Rule whenever possible.

		PRINCIPAL	RATE	DATE	DUE DATE	TIME	INTEREST	MATURITY VALUE
3.	a.	$ 400	15%	4/12	5/27			
	b.	650	12	9/1		120 days		
	c.	720	10	7/13		4 months		
	d.		8		6/15	150 days	$36	
4.	a.	$4,800	9%	2/12	11/9			
	b.	900	11	4/22		60 days		
	c.	4,500	12	8/21		3 months		
	d.		10		7/13	100 days	$20	

   In completing the following problems, use exact days if possible. Find ordinary interest (360-day year) unless exact interest (365-day year) is indicated.

5. A 3-month note dated January 16, 1997, for $1,460 had interest at 15%. Find the exact interest and the maturity value of the note.

6. A note dated January 24, 1997, was made for 4 months at 9%. If the principal of the note was $6,000, find the exact interest and maturity value of the note.

7. A note dated June 6, 1996, reads: "One month from date, I promise to pay $2,400 with interest at 8.5%." Find the maturity value.

8. A note dated October 5, 1996, read: "Two months from date, I promise to pay $400 with interest at 10.5%." What was the maturity value of the note?

9. An 8-month note is drawn for $4,800 with interest at 10%. What maturity value will be paid?

10. Find the maturity value of a 9%, 10-month note with a face value of $2,000.

11. A note dated August 21, 1998, will be due in 3 months. If the face value of the note is $1,095 and $1,128.12 will be paid to discharge the debt, what exact interest rate will be charged?

12. A 6-month note dated February 14, 1998, has a face value of $2,190 and a maturity value of $2,298.60. Determine the exact interest rate charged.

13. How long will it take $1,800 to earn $72 interest at a 10% exact interest rate?

14. How long will it take $730 to earn $18 interest at 10% exact interest?

15. How long will it take $1,600 to earn $64 interest at an 8% interest rate?

16. What time is required for $400 to earn $36 at 12% interest?

17. The interest on a 150-day note is $48.75. If interest was computed at 9%, find the principal of the note.

18. Find the principal of a 180-day note made at 14% if the interest was $105.

19. If $5,230 is paid on November 10 to discharge a note dated May 10, find the face value of the 9% note.

20. Find the face value of a 10% note dated September 7 that matures on December 6. The maturity value on the note is $3,075.

SECTION 5
PRESENT VALUE

In Problems 19 and 20 of Section 4, we were told the maturity value of a note and asked to find what principal had been loaned. When the principal is being found, this is often referred to as finding the **present value.** Present value can also be explained by answering the question: What amount would have to be invested today (at the present) in order to obtain a given maturity value? This "X" amount that would have to be invested is the present value.

Present value at simple interest can be found using the same maturity value formula that has been used previously: $M = P(1 + rt)$. However, when present value is to be found repeatedly, the formula is usually altered slightly by dividing by the quantity in parentheses.

$$M = P(1 + rt)$$

$$\frac{M}{1 + rt} = \frac{P(1 + rt)}{(1 + rt)}$$

$$\frac{M}{1 + rt} = P$$

or

$$P = \frac{M}{1 + rt}$$

The formula can be used more conveniently in this form, since it is set up to solve for the principal or present value. However, Example 1 illustrates a present-value problem solved with both versions of the formula, and either form may be used for this type of problem.

Example 1   If the maturity value of a 12%, 3-month note dated March 11 was $618.40, find the present value (or principal, or face value) of the note.

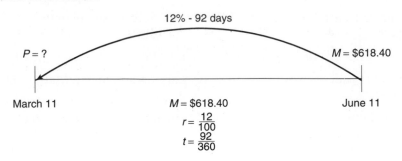

$$M = P(1 + rt) \qquad\qquad P = \frac{M}{1 + rt}$$

$$\$618.40 = P\left(1 + \frac{\cancel{12}}{\cancel{100}} \times \frac{.92}{\cancel{360}\,_{30}}\right) \qquad = \frac{\$618.40}{1 + \frac{\cancel{12}}{\cancel{100}} \times \frac{.92}{\cancel{360}\,_{30}}}$$

$$618.40 = P\left(1 + \frac{.92}{30}\right) \qquad\qquad = \frac{618.40}{\frac{30}{30} + \frac{.92}{30}}$$

$$618.40 = P\left(\frac{30}{30} + \frac{.92}{30}\right) \qquad\qquad = \frac{618.40}{\frac{30.92}{30}}$$

$$618.40 = P\left(\frac{30.92}{30}\right)$$

$$\frac{30}{\cancel{30.92}}(\cancel{618.40})^{20} = P\left(\frac{\cancel{30.92}}{\cancel{30}}\right)\left(\frac{\cancel{30}}{\cancel{30.92}}\right) \qquad = \cancel{618.40}^{\,20}\left(\frac{30}{\cancel{30.92}}\right)$$

$$\$600 = P \qquad\qquad\qquad\qquad P = \$600$$

![calculator] **Calculator techniques . . .**   FOR EXAMPLE 1

This technique illustrates the equation on the right, where the formula is set up to solve for the present value, P:

$0.12\ \boxed{\times}\ 92\ \boxed{\div}\ 360\ \boxed{M+} \longrightarrow 0.0306666; \qquad 1\ \boxed{M+} \longrightarrow 1$

$618.40\ \boxed{\div}\ \boxed{MR}\ \boxed{=} \longrightarrow 600.00003$

You will use this technique for many subsequent problems.

Example 1 involves only one interest rate. Many investments, however, involve two interest rates. To illustrate present value at two interest rates, consider the following case.

Suppose that I find a place where I can earn 20% on my investment. The most I have been offered any other place is 12%, so I immediately invest $1,000 in this "gold mine." On the way home, I meet a friend who offers to buy this investment for the same $1,000 I just deposited. Would I sell the investment for $1,000, knowing I could not earn 20% interest on any other investment I might make? No! My $1,000 is really worth more than $1,000 to me, because I would have to deposit more than $1,000 elsewhere in order to earn the same amount of interest.

This case demonstrates that an investment is not always worth its exact face value; it may be worth more or less than its face value when compared to the average rate of interest being paid by most financial institutions. This typical, or average, interest rate is referred to as the **rate money is worth.** Thus, if most financial institutions are paying 8%, then money is worth 8%.

The rate money is worth varies considerably, depending on whether one is borrowing or depositing—and on the kind of deposit made. An ordinary **savings account,** where the money can be withdrawn at any time, earns less interest than a **certificate of deposit** (or CD). A CD usually requires a minimum deposit (often $1,000), and the money must remain on deposit for a specified period of time (usually a 3-month minimum) in order to earn interest at the higher rate. CD rates increase when more money is deposited and/or longer time periods are established.

Another popular account, the **money-market account,** shares some characteristics with each of the others. The money-market account may also require a minimum deposit (often $500), but it allows a limited number of withdrawals or checks (usually three per month) without penalty so long as the minimum balance remains met. Its rate of interest usually falls between that of the other two accounts.

As would be expected, institutions must charge higher rates for notes and other loans than they pay on deposits, or they could not cover expenses and return a profit to their owners.

If an investment is made at a rate that is not the prevailing rate money is worth, then the present value of the investment is different—either greater or less—than the principal. We therefore find what investment, made at the rate money is worth, would have the same maturity value that the actual investment has. The size of this "rate-money-is-worth investment" is the present or true value of the actual investment.

Example 2 A 90-day note for $1,000 is drawn on November 12 at 8%. If money is worth 12%, find the present value of the note on the day it is drawn.

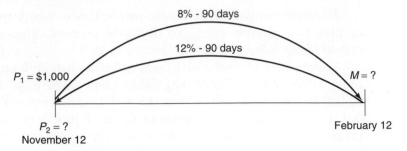

The solution to this kind of problem involves two steps: (a) Find the *actual maturity value* of the note, and (b) find the *present value at the rate money is worth* of that calculated amount.

(a) $P_1 = \$1,000$

$r = 8\%$

$t = \dfrac{90}{360}$ or $\dfrac{1}{4}$

$M = ?$

$M = P(1 + rt)$

$= \$1,000\left(1 + \dfrac{\cancel{8}^{2}}{100} \times \dfrac{1}{\cancel{4}}\right)$

$= 1,000\left(1 + \dfrac{2}{100}\right)$

$= \cancel{1,000}^{10}\left(\dfrac{102}{\cancel{100}}\right)$

$M = \$1,020$

(b) $M = \$1,020$

$r = 12\%$

$t = \dfrac{90}{360}$ or $\dfrac{1}{4}$

$P_2 = ?$

$P_2 = \dfrac{M}{1 + rt}$

$= \dfrac{\$1,020}{1 + \dfrac{\cancel{12}^{3}}{100} \times \dfrac{1}{\cancel{4}}}$

$= \dfrac{1,020}{1 + \dfrac{3}{100}}$

$= \dfrac{1,020}{\dfrac{103}{100}}$

$= 1,020\left(\dfrac{100}{103}\right)$

$P_2 = \$990.29$

The present value of the note on the day it was drawn was $990.29. This means that the lender (payee) could have invested only $990.29 at the rate money is worth (12%) and would achieve the same maturity value ($1,020) that will be obtained on the $1,000 note at only 8% interest. That is, $990.29 invested at 12% would earn $29.71 interest ($990.29 + $29.71 = $1,020), producing the same $1,020 maurity value as $1,000 plus $20 interest. Thus, the $1,000 is really worth only $990.29 to the lender.

When computing present value of an investment where two interest rates are involved, you should know beforehand whether the present value is more or less than the principal. In general, the present value is less than the principal when the investment rate is less than the rate money is worth. That is, when the actual rate is less, then the present value is less. The reverse is also true: If the actual rate is more than the rate money is worth, then the present value is more than the principal.

Previous examples have required finding the present value on the day a note was drawn. It is often desirable to know the worth of an investment on some day nearer the due date. The worth of an investment on any day prior to the due date is also called present value. The procedure for finding this present value is the same as that already discussed, except that in the second step the time as well as the rate will be different. It should be emphasized that present value is

1.   Computed using the maturity value.
2.   Computed for an exact number of days *prior* to the due date.
3.   Computed using the rate money is worth.

Example 3   Find the (present) value on October 11 of a $720, 4-month note taken out on July 10 at 15%, if money is worth 12%.

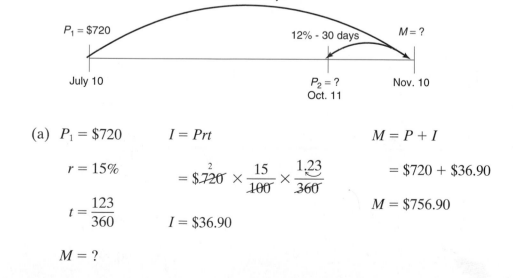

(a) $P_1 = \$720$ $I = Prt$ $M = P + I$

$r = 15\%$ $= \$\overset{2}{720} \times \dfrac{15}{100} \times \dfrac{1.23}{360}$ $= \$720 + \$36.90$

$t = \dfrac{123}{360}$ $I = \$36.90$ $M = \$756.90$

$M = ?$

(b) $M = \$756.90$

$r = 12\%$

$t = \dfrac{30}{360}$ or $\dfrac{1}{12}$

$P_2 = ?$

$P_2 = \dfrac{M}{1 + rt}$

$= \dfrac{\$756.90}{1 + \dfrac{\cancel{12}}{100} \times \dfrac{1}{\cancel{12}}}$

$= \dfrac{756.90}{1 + \dfrac{1}{100}}$

$= \dfrac{756.90}{\dfrac{101}{100}}$

$= 756.90\left(\dfrac{100}{101}\right)$

$P_2 = \$749.41$

This means that if the note were sold on October 11, it would be sold for $749.41; this money could then be invested at 12% interest, and on November 10 the maturity value of the new investment ($749.41 plus $7.49 interest) would also be $756.90.

SECTION 5   PROBLEMS

Any convenient method may be used to find the amount in the following problems: $I = Prt$ and $M = P + I$, or $M = P(1 + rt)$.

Determine the present value, using the Bankers' Rule.

	MATURITY VALUE	RATE	TIME (DAYS)	PRESENT VALUE
1. a.	$ 936	12%	120	
b.	812	9	60	
2. a.	$ 735	12%	150	
b.	2,484	14	90	

Complete the following, finding the present value on the day of investment.

	PRINCIPAL	RATE	TIME (DAYS)	MATURITY VALUE	RATE MONEY IS WORTH	PRESENT VALUE
3. a.	$1,200	10%	180		8%	
b.	1,500	12	30		9	
4. a.	$6,000	10%	90		9%	
b.	8,000	12	120		15	

Complete the following, finding the present value on the day indicated.

	PRINCIPAL	RATE	TIME (DAYS)	MATURITY VALUE	RATE MONEY IS WORTH	DAYS BEFORE MATURITY	PRESENT VALUE
5. a.	$2,000	9%	270		10%	90	
b.	2,500	15	180		12	45	
6. a.	$9,000	12%	300		10%	180	
b.	7,500	15	120		18	60	

7. On June 3, $4,084 was paid in settlement of a note dated March 23. If interest was charged at 10.5%, what was the principal of the note?

8. The maturity value was $4,062 on a 9% note dated July 15. If the note was due on September 15, what had been the face value?

9. On September 6, $2,080 was paid at the maturity of a note dated May 9. If money is worth 12%, what was the present value of the note on the date it was made?

10. The maturity value of $6,165 was paid on April 15 to discharge a simple interest note dated January 15. If money is worth 11%, what was the present value of the note on the day it was made?

11. A 3-month note for $3,800 was drawn on January 13 at 8%. If money is worth 10%, find the present value of the note on the day it was made.

12. A 4-month note for $3,500 was drawn on February 8 at 15%. If money is worth 18%, find the present value of the note on the day it was issued.

13. Determine the present value of a 6%, 5-month note with a face value of $15,000 made on February 13, if money is worth 7.5%.

14. A 4-month note dated February 12 was made at 11% for $6,000. What is the present value of the note on the day it was made, if money is worth 12%?

15. A 9-month note for $1,440 was made at 9% on March 23. If money is worth 12%, find the present value of the note on November 23.

16. A 3-month note dated May 16 was made at 15% for $12,000. Find the present value of the note on July 17 if money is worth 12%.

17. A 6-month, 12% note for $6,000 was dated March 5. What is the present value on August 16 if money is worth 10%?

18. A 4-month, 9% note dated January 17 had a face value of $2,500. Find the present value on April 17 if money is worth 8%.

19. The following offers were made on a house: $123,000 cash, $128,000 in 3 months, or $133,000 in 9 months. Which offer should be accepted if money is worth 10%? Hint: Compare the present value of the three amounts.

20. A contractor is considering three payment options for a contract: $30,000 on signing the contract today, $30,700 in 3 months, or $31,500 on the completion of the contract in 6 months. Money is worth 12%. Which is the best offer? Hint: Compare the present value of each payment.

CHAPTER 15 GLOSSARY

Amount. (See "Maturity value at simple interest.")

Bankers' Rule. The most commonly used method of interest calculation, where time is expressed in exact days over 360 days per year.

Certificate of deposit (CD). A savings investment that earns a higher interest rate than savings accounts because the money is committed for a set period of time (usually a 3-month minimum); may also require a minimum investment (often $1,000).

Collateral. An item of value that is pledged by a borrower in order to insure that a note will be repaid or equivalent value received.

Date (of a note). The date on which a note is drawn (or signed).

Due date (of a note). The date on which a note is to be repaid; the maturity date.

Exact interest. Interest calculated using 365 days per year.

Exact time. The time period of a loan expressed in the specific number of calendar days.

Face value. The money value specified on a promissory note; the principal of a simple interest note.

Interest. Rent paid for the privilege of borrowing money: $I = Prt$, where $I =$ interest, $P =$ principal, $r =$ rate, and $t =$ time.

Maker. A person who signs a note promising to repay a loan.

Maturity value (at simple interest). The total amount (principal plus interest) due on a simple interest loan; also called "amount" or "sum": $M = P + I$, or $M = P(1 + rt)$, where $M =$ maturity value, $P =$ principal, $I =$ interest, $r =$ rate, and $t =$ time.

Maturity value (of a note). The amount to be repaid on the due date of a note (which may or may not include interest).

Money-market account. A combined savings/CD account that requires a minimum balance (often $500) but allows a limited number of checks or withdrawals. Pays higher interest than regular savings, lower than CDs.

Note. A written promise to repay a loan (either with or without interest, and with or without collateral).

Ordinary interest. Interest calculated using 360 days per year.

Ordinary time. The time period of a loan where each month is assumed to have 30 days.

Payee. The person (or bank or firm) to whom the maturity value of a note will be paid.

Present value. (1) Principal; also (2) the value of another investment which, if made (on any given date) at the rate money is worth, would have the same maturity value as has an actual, given investment:

$$P = \frac{M}{1 + rt}$$

where $P =$ principal or present value, $M =$ maturity value, $r =$ rate, and $t =$ time.

Principal.   The original amount of money that is loaned or borrowed; the investment value on which interest is computed.

Promissory note.   (See "Note.")

Rate.   An annual percent at which simple interest is computed.

Rate money is worth.   An average or typical rate currently being used at most financial institutions.

Savings account.   A low-interest account where any amount may be deposited or withdrawn at any time; also called an open account.

Simple interest.   Interest that is computed on the original principal of a loan for the entire time of the loan, and is then added to the principal to obtain the maturity value. (See also "Interest.")

Sum.   (See "Maturity value at simple interest.")

Time.   The duration of a simple-interest investment. (Time must be expressed as a fractional year, in order to correspond with the annual percentage rate at which interest is calculated.)

16

BANK DISCOUNT

OBJECTIVES

Upon completion of Chapter 16, you will be able to:

1. Define and use correctly the terminology associated with each topic.

2.   **a.** Determine the proceeds of a simple discount note, using the simple discount formulas (Section 1: Examples 1, 2; Problems 1–24):
   (1) $D = Mdt$
   (2) $p = M - D$.

       b. Also, apply $p = M(1 - dt)$ to compute
   (1) Proceeds (Section 1: Examples 2, 3; Problems 9, 10, 17–20)
   (2) Maturity value (Section 2: Example 1; Problems 1, 2, 5–10).

3.   **a.** Determine the equivalent simple interest rate charged on a simple discount note (Section 1: Example 4; Problems 7, 8).

       b. Find the unknown discount rate or time of a simple discount note (Section 1: Example 5; Problems 13–16, 21, 22).

4.   **a.** Rediscount a simple discount note (or a simple interest note) on a given day prior to the maturity date (Section 2: Examples 2, 3; Problems 3–6, 11–18).

       b. Also, determine the amount of interest made
   (1) By the original lender
   (2) By the second lender.

5. Compute the savings gained by signing a discount note on the last day allowed by invoice sales terms, in order to take advantage of the cash discount (assuming that the bank note will be repaid on the same due date specified by the sales terms) (Section 3: Example 1; Problems 1–10).

6. Explain the similarities and differences of simple interest and simple discount notes (Section 4: Problems 1–22).

In the preceding chapter we discussed the simple interest note. Some banks, however, use the simple discount note (illustrated in the following section) for short-term loans of 1 year or less. There are some similarities and several major differences between the two types which it is imperative that you understand. (The two notes will also be summarized in Section 4.)

SECTION 1

SIMPLE DISCOUNT NOTES

Recall that the face value of the simple interest note is the principal of the loan. Interest is computed on this principal and added to it to obtain the maturity value.

The **face value** of a discounted note, however, is the **maturity value**—the amount that will be repaid to the payee. As in the case of the simple interest note, the maturity value is repaid in a single payment at the end of the loan.

The interest on a discounted note is called **bank discount.** Bank discount is computed on the maturity value (face value) of the note and is then *subtracted* from the maturity value (as opposed to simple interest, which is computed on the *principal* and then *added* to the principal to *obtain* the maturity value). The amount remaining after the bank discount has been subtracted from the maturity value of the note is called the **proceeds.** The proceeds is the actual amount that the borrower will receive from the bank and that can then be spent. Because the bank discount is subtracted from the face value before the borrower receives any money, this is often referred to as paying "interest in advance."

To reemphasize: The distinguishing aspect of the simple discount note is that interest is charged on the amount that is to be paid back, rather than on the amount that was actually borrowed.

Figures 16-1 and 16-2 illustrate typical simple discount notes. Figure 16-1 is an unsecured or signature note; that is, the loan is made simply on the basis of the borrower's good credit standing. Banks often establish some maximum amount that they will lend on an unsecured note. Figure 16-2 illustrates a collateral note, which requires the maker (borrower) to secure the note with some item of value.

Bank discount is computed using the formula

$$D = Mdt$$

where D = discount, M = maturity value (face value) of the discount note, d = discount rate, and t = time. We will utilize the Bankers' Rule for most discount notes, although a 365-day year is sometimes used, as described later.

Proceeds p are then found using the formula

$$p = M - D$$

NOTE NO_____   OFFICER APPROVAL_____

UNSECURED PROMISSORY NOTE

$ 2,400.00 _____   January 30   19XX

For value received, the undersigned Borrower(s) jointly and severally promise to pay to North Star Savings Association (hereinafter called "NORTH STAR"), or order, at the main banking office of NORTH STAR in Richmond _____ ,VA, the sum of Two thousand four hundred and no/100------------Dollars _____ in lawful money of the United States of America with principal and interest payable as stated in the schedule of payments.

SCHEDULE OF PAYMENTS   Select one (terms not selected no not apply)

☒ DISCOUNT NOTE

The above stated amount shall be due and payable on March 31 _____ , 19XX , 60 days from the date of this Note, and includes interest in the amount of $ 40.00 _____ , discounted at the rate of 10 %.

☐ SINGLE MATURITY NOTE WITH INTEREST FROM DATE

The above stated amount shall be due and payable on _____ , 19____ , _____ days from the date of this Note, plus interest in the amount of $_____ computed at the rate of _____ %.

☐ INSTALLMENT TERM NOTE

The above stated amount shall be payable in _____ consecutive _____ installments of $_____ each, beginning on _____ , 19____ , and one final installment or balloon payment of $_____ , payable on _____ , 19____ . The installment payment ☐ includes or ☐ is in addition to interest payable at the rate of _____ %.

If final payment is a balloon payment, NORTH STAR makes no commitment to refinance same if not paid when due. If default be made on any payment of this Note, or any other obligation to NORTH STAR, the maturity of this Note may be accelerated and the entire balance shall become due and payable at the option of NORTH STAR. In the event of any default or failure to pay by maturity, interest thereafter shall be payable at the maximum contract rate then allowed by law. In the event that any payment shall be past due for 10 days or more, there will be a delinquency charge of the lesser of 12% of the unpaid payment or $25, which is in addition to any interest owed.

THIS NOTE IS SUBJECT TO ANY ADDITIONAL PROVISIONS, WARRANTIES, AND RIGHTS SET FORTH ON THE REVERSE SIDE HEREOF.

In Witness Whereof, the Borrower has caused this Note to be executed by its duly authorized officers (if a corporation), or has hereunto set hand and seal (if an individual) to be affixed hereto on the day and year first written above.

CORPORATE BORROWER: INDIVIDUAL BORROWER(S):

_____   _A. Sample Borrower_ (Seal)
Name of Corporation

By_____   _____ (Seal)
President

By_____   _____ (Seal)
Secretary

Corporate Seal: (Seal)   _____ (Seal)

Address:_____

_____

Form 7126   NOTICE: SEE OTHER SIDE FOR IMPORTANT INFORMATION

FIGURE 16-1   Simple Discount Note

$1,500 _____   July 15 _____, 19 X1

   90 days _____after date, for value received and with interest discounted to maturity _____

at 12 percent _____ per annum,

        the undersigned promise(s) to pay to **NORTH CAROLINA NATIONAL BANK** or order
   One thousand five hundred and no/100--Dollars

payable at any office of the North Carolina National Bank in North Carolina. Interest shall be computed on the basis of a 360 day year for the actual number of days in the interest period and interest shall accrue after maturity or demand, until paid, at the rate stated above.

   To secure the payment of this note and liabilities as herein defined, the parties hereto hereby pledge and grant to said Bank (the word "Bank" wherever used herein shall include any holder or assignee of this note) a security interest in the collateral described as follows:

   25 shares - IBM stock _____

and any collateral added thereto or substituted therefor, including shares issued as stock dividends and stock splits and dividends representing distribution of capital assets. The Bank is hereby authorized at any time to charge against any deposit accounts of any party hereto any and all liabilities whether due or not. The Bank may declare all liabilities due at once in the event any party hereto becomes subject to any proceedings for the relief of creditors including but not limited to proceedings under the Bankruptcy Act or otherwise, or if in the judgment of Bank the collateral decreases in value so as to render Bank insecure and Bank demands additional collateral which is not furnished, or if Bank at any time otherwise deems itself insecure. In the event the indebtedness evidenced hereby or liabilities as defined herein be collected by or through an attorney at law, the holder shall be entitled to collect reasonable attorneys' fees.

   Upon failure to pay any liability when due, Bank may sell the collateral at public or private sale, for cash or on credit, as a whole or in parcels, without notice, and Bank may at any such sale purchase the collateral or any part thereof for its own account, and the proceeds of any such sale shall be applied first to the costs of such sale and the expenses of collection, including reasonable attorneys' fees, and then to the outstanding balance due on said liabilities, the application to be made in the manner and proportions as Bank elects. The Bank may forbear from realizing on the collateral or any part thereof, by sale or otherwise, all as the Bank may decide, and the liabilities of the parties hereto shall not be released, discharged or in any way affected by any such forbearance, nor shall any of the parties hereto have any rights or recourse against the Bank by reason of any action the Bank may take or omit to take under this note, by reason of any deterioration, waste, or loss of any of the collateral unless such deterioration, waste, or loss be caused by the willful act or willful failure to act of the Bank. Upon payment of this note the Bank may release the collateral but shall have the right to retain the same to secure any unpaid liabilities. Upon any transfer of this note and the collateral, the Bank shall be fully relieved of responsibility with reference thereto. "Liabilities" or "Liability" as herein used, shall include this note and all obligations of every kind of any party hereto in whatever capacity to Bank, now or hereafter existing, whether arising directly or acquired from others as collateral or otherwise, whether absolute or contingent, joint or several, joint and several, secured or unsecured, due or not due, direct or indirect, including, but not limited to, liabilities arising by operation of law, contractual or tortious, liquidated or unliquidated or otherwise.

   All persons bound on this obligation, whether primarily or secondarily liable as principals, sureties, guarantors, endorsers or otherwise, hereby waive presentment, protest, notice of dishonor and of acceleration of maturity and any right to require the Bank to retain any collateral pledged as security for this note or any other liabilities and agree that any extension of time for payment with or without notice shall not affect their joint and several liabilities.

   Witness our/my hand(s) and seal(s).

Address   2008 E. Corning Dr. _____   _Mason R. Weatherford_ (Seal)

                                                          Joan D. Weatherford (Seal)
NCNB 2158 Rev. 2/75 Due   Oct. 13   No. 73885 _____

Courtesy of North Carolina National Bank, Statesville

FIGURE 16-2   Collateral Discount Note

Let us consider the difference between the amount of money one would be able to spend if (s)he borrows at a discount rate versus the amount available to spend when borrowing at an interest rate. (The following discount computation illustrates Figure 16-2.)

Example 1   The face values of two notes are $1,500 each. The discount rate and the interest rate are both 12%, and the time of each is 90 days.

Interest Note	*Discount Note*
$P = \$1,500$	$M = \$1,500$
$r = 12\%$	$d = 12\%$

	Interest Note			*Discount Note*	

$$t = \frac{90}{360} \quad \text{or} \quad \frac{1}{4} \qquad\qquad\qquad t = \frac{90}{360} \quad \text{or} \quad \frac{1}{4}$$

$$I = Prt \qquad\qquad\qquad\qquad\qquad D = Mdt$$

$$= \$1{,}500 \times \frac{12}{100} \times \frac{1}{4} \qquad\qquad = \$1{,}500 \times \frac{12}{100} \times \frac{1}{4}$$

$$I = \$45 \qquad\qquad\qquad\qquad\qquad D = \$45$$

$$M = P + I \qquad\qquad\qquad\qquad p = M - D$$

$$= \$1{,}500 + \$45 \qquad\qquad\qquad = \$1{,}500 - \$45$$

$$M = \$1{,}545 \qquad\qquad\qquad\qquad p = \$1{,}455$$

Notice on the simple interest note that the borrower pays $45 for the use of $1,500. On the simple discount note, however, the borrower pays $45 for the use of only $1,455. Thus, based on the actual amount of money the borrower has available to spend, more interest is paid at a discount rate than at an interest rate. Or, stated another way: The borrower is really paying slightly more than 12% of the money (s)he actually uses, because 12% of $1,455 for 90 days would be only $43.65 interest due, not the $45 actually charged.

From this example we can conclude that, when a borrower obtains money at a discount rate, (s)he will always be paying somewhat more than the stated rate on the money actually received. In addition, the **Truth-in-Lending Law** enables the borrower to be more informed about the true rate being paid. This true rate is the **equivalent simple interest rate.** The law requires the lender to reveal both the amount of interest and the annual percentage rate charged on the money received, correct to the nearest $\frac{1}{4}$%. Banks have available tables or computer programs that enable them to determine this true rate very easily.

It will be of interest here to know the true rate required under the Truth-in-Lending Law for the examples above. Figure 16-3 shows the Truth-in-Lending Disclosure Statement that corresponds to the note illustrated in Figure 16-2 and Example 1. We see that the true annual percentage rate paid for the $1,455 that the borrower received is actually 12.25% (correct to the nearest $\frac{1}{4}$%), rather than 12% (the discount rate). For the note in Figure 16-1, the true rate is 10.25%, correct to the nearest $\frac{1}{4}$%, as compared to the stated 10% discount rate. This procedure is described later in the chapter.

TRUTH IN LENDING DISCLOSURE STATEMENT
NORTH CAROLINA NATIONAL BANK
COMMERCIAL BANKING DEPARTMENT

DATE OF NOTE (Date on which finance charge begins to accrue)
July 15, 19X1

BORROWER'S NAME
Mason R. & Joan D. Weatherford

AMOUNT OF NOTE
$1,500.00

FINANCE CHARGES			AMOUNT FINANCED		
Interest	2	$ 45.00	Paid to Borrower	11	$ 1,455.00
Fees	3	$	Credit Life Insurance Prem. (if any)	12	$ –
Other (specify):	4	$	Financing Statement Filing Fee (if any)	13	$ –
	5	$	Other Amounts Paid on Behalf of Borrower (specify):	14	$
	6	$		15	$
Total FINANCE CHARGE	7	$ 45.00		16	$
ANNUAL PERCENTAGE RATE	8	12.25 %	AMOUNT FINANCED	17	$ 1,455.00

No. of Payments	Due Dates or Periods	Amount of Regular Payment	Amount of Final Payment	Balloon Payment (if any) *	Total of Payments
18	19	20	21	22	23
1	October 13, 19X1	–	–	–	1,500.00

24 To secure the obligation borrower has granted NCNB a Security Interest in the following collateral (if applicable)

25 shares - IBM stock

25
NOTE: The Security Agreement will secure future or other indebtedness and will cover after-acquired property.

☐ If checked, the collateral includes the principal residence of the borrower, and two (2) copies of the Notice of Right to Rescind are attached hereto.

26 In the event of prepayment of the obligation before maturity, borrower shall receive a rebate of unearned Finance Charge as follows:

☐ Not applicable

27
ADDITIONAL INFORMATION

Property insurance, if required, may be purchased from any reputable insurer selected by the borrower.
If the obligation is collected by or through an attorney at law after maturity, borrower shall be required to pay all collection costs and reasonable attorney's fees.

28 *Conditions (if any) under which Balloon Payment may be refinanced if not paid when due

Receipt of the foregoing statement fully completed and any attachments referred to therein is hereby acknowledged.

29 Mason R. Weatherford
Borrower's Signature

Credit Life Insurance Election (if applicable):
I understand that Credit Life Insurance is not required as a condition of this loan but may be purchased through NCNB at a cost of $ _____ for the term of the loan. I hereby affirm my desire to purchase such insurance.

30
Borrower's Signature for Credit Life Insurance

NCNB 2260 Rev 12/77

Courtesy of North Carolina National Bank, Statesville

FIGURE 16-3   Truth-in-Lending Disclosure Statement

The bank discount formulas given above can be combined into a single formula for finding proceeds. By substituting Mdt for D, we have

$$p = M - D$$

$$= M - Mdt$$

$$p = M(1 - dt)$$

Example 2   A \$3,000 note dated January 31 is to be discounted at 10% for 5 months. Find the bank discount and the proceeds using (a) $D = Mdt$ and (b) $p = M(1 - dt)$.

The Bankers' Rule is applied below when computing bank discount, as follows:

$$
\begin{array}{rl}
\text{June 30} = & 181 \text{ day} \\
\text{January 31} = - & 31 \text{ day} \\
\hline
\text{time} = & 150 \text{ days}
\end{array}
$$

$M = \$3,000$

$d = 10\%$

$t = \dfrac{150}{360}$ or $\dfrac{5}{12}$

(a) $D = Mdt$

$= \$3,000 \times \dfrac{10}{100} \times \dfrac{5}{12}$

$D = \$125$

$p = M - D$

$p = \$3,000 - \$125$

$p = \$2,875$

(b) $p = M(1 - dt)$

$= \$3,000 \left(1 - \dfrac{\overset{5}{\cancel{10}}}{100} \times \dfrac{5}{\underset{6}{\cancel{12}}}\right)$

$= 3,000 \left(1 - \dfrac{25}{600}\right)$

$= 3,000 \left(\dfrac{600}{600} - \dfrac{25}{600}\right)$

$= 3,000 \left(\dfrac{575}{600}\right)$

$p = \$2,875$

$D = M - p$

$= \$3,000 - \$2,875$

$D = \$125$

Although bank discount and proceeds may be found in either of the above ways, you would probably use method (a), since the computation would be easier.

It was shown in Example 1 that when interest and bank discount are equal, the borrower has less spendable money from a discounted note. In that example, however, the maturity values of the two notes were not the same. We will now consider a simple interest note and a simple discount note that both have the same maturity value, to compare the amounts that the borrower would have available to spend.

Example 3   A simple interest note and a simple discount note both have a maturity value of $1,000. If the notes both had interest computed at 15% for 120 days, find (a) the principal of the simple interest note and (b) the proceeds of the discount note.

(a)   *Simple Interest Note*

$$M = \$1,000$$

$$r = 15\%$$

$$t = \frac{120}{360} \quad \text{or} \quad \frac{1}{3}$$

$$P = ?$$

$$P = \frac{M}{1 + rt}$$

$$= \frac{\$1,000}{1 + \frac{\overset{}{\cancel{15}}^{\,5}}{100} \times \frac{1}{\cancel{3}}}$$

$$= \frac{1,000}{\frac{100}{100} + \frac{5}{100}}$$

$$= \frac{1,000}{\frac{105}{100}}$$

$$= 1,000\left(\frac{100}{105}\right)$$

$$P = \$952.38$$

(b)   *Simple Discount Note*

$$M = \$1,000$$

$$d = 15\%$$

$$t = \frac{120}{360} \quad \text{or} \quad \frac{1}{3}$$

$$p = ?$$

$$p = M(1 - dt)$$

$$= \$1,000\left(1 - \frac{\overset{5}{\cancel{15}}}{100} \times \frac{1}{\cancel{3}}\right)$$

$$= 1,000\left(1 - \frac{5}{100}\right)$$

$$= 1,000\left(\frac{100}{100} - \frac{5}{100}\right)$$

$$= \overset{10}{\cancel{1,000}}\left(\frac{95}{\cancel{100}}\right)$$

$$p = \$950$$

In this case, the maker of the note would have $2.38 more to spend if the note were a simple interest note. If $952.38 were invested at 15% for 120 days, the maturity value would be $1,000.

Calculator techniques . . .   FOR EXAMPLE 3(B)

For simple discount, notice that you must subtract the "$d \times t$" calculation using M−; the "1 M+" can be entered either before or after the M− operation.

0.15 ÷ 3 M− ⟶ $0.05;$ 1 M+ ⟶ 1

$1,000$ × MR = ⟶ 950

As you press MR, notice that the value 0.95 appears, which verifies that the calculations on line one did in fact subtract 0.05 from 1.

As mentioned earlier, borrowing money by signing a simple discount note is more expensive than by signing a simple interest note. To determine the equivalent simple interest rate on the simple discount note in Example 3, use the simple interest formula, $I = Prt$.* The I is the discount or the difference between the maturity value and the proceeds. The proceeds will be substituted for the P in the formula, because the amount of money available to spend is the proceeds of a discount note. The t will be the same, but r is not 15% because 15% is the discount rate on a maturity value of $1,000, not 15% of the proceeds. The r will be the unknown.

Example 4   Determine the equivalent simple interest rate being charged on the simple discount note with a maturity value of $1,000, discount rate of 15%, and time of 120 days (Example 3).

$$M = \$1,000 \qquad\qquad I = Prt$$

$$p = \$950$$
$$\$50 = \$950 \times r \times \frac{1}{3}$$

$$I = \$1,000 - \$950$$

$$I = \$50 \qquad\qquad 50 = \frac{950}{3} r$$

$$t = \frac{120}{360} \quad \text{or} \quad \frac{1}{3} \qquad \left(\frac{3}{950}\right) 50 = r$$

$$r = 15.79\% \quad \text{or} \quad 15.75\%$$

*Another formula for finding the equivalent simple interest rate is as follows: $r = \dfrac{D}{P} \times$ reciprocal of time. In Example 4, this is $\dfrac{\$50}{\$950} \times \dfrac{3}{1} = 0.1578947$ or 15.75%, correct to the nearest ¼%.

The rate of interest disclosed to the borrower would be 15.75%, correct to the nearest $\frac{1}{4}$%.

As was true for simple interest notes, the use of a 365-day year has become more widespread with bank discounts. The trend began during a time of higher rates of interest and discount, which made the difference between a 360-day and a 365-day year more significant. In some parts of the country, bank discount is thus routinely calculated using an "exact" discount rate (exact days over 365 or 366). This exact discount method is also applied when a commercial bank borrows funds from a Federal Reserve bank.

Example 5 A bank received proceeds of $14,150 on a 90-day discounted note from its Federal Reserve bank. The face value of the note was $14,600 (using as collateral certain notes owed to the bank). What exact discount rate was charged?

$$M = \$14,600 \qquad D = Mdt$$
$$p = -\;\underline{14,150}$$
$$D = \quad\$ \quad 450 \qquad \$450 = \$14,600 \times d \times \frac{90}{365}$$

$$t = \frac{90}{365} \qquad\qquad 450 = 3,600d$$

$$d = ? \qquad\qquad \frac{450}{3,600} = d$$

$$d = 0.125 \quad \text{or} \quad 12.5\%$$

Another type of investment that utilizes discount calculations is U.S. Treasury Bills, which may be purchased in either 13-week or 26-week categories. The price that an investor pays for a Treasury Bill (or "T-bill," as it is often called) is the proceeds remaining after the discount calculation. The face of the Treasury Bill is the amount that the investor will receive from the government when the bill matures. (It should be noted, however, that *discounts on Treasury Bills are calculated by the Bankers' Rule*—that is, $\dfrac{13 \times 7}{360}$ or $\dfrac{26 \times 7}{360}$.)

SECTION 1 PROBLEMS

1. Identify the parts of the discounted note shown in Figure 16-1.

a. Face value b. Maker c. Payee d. Date
e. Due date f. Rate g. Time h. Bank discount
i. Proceeds j. Maturity value

2. Answer Problem 1 for the discounted note shown in Figures 16-2 and 16-3.

Supply the missing entries, using the Bankers' Rule.

		MATURITY VALUE	DISCOUNT RATE	DATE	DUE DATE	TIME (DAYS)	BANK DISCOUNT	PROCEEDS
3.	a.	$1,800	12%	6/4		60		
	b.	2,400	11	4/28		120		
	c.	3,200	9	1/16	4/16			
	d.	1,200	10		9/18	75		
4.	a.	$9,000	8%	10/5		30		
	b.	7,600	15	2/25		270		
	c.	5,000	10	5/14	8/12			
	d.	3,000	9		9/3	60		

Use an exact bank discount (365-day year) to find the proceeds.

5. a. $1,825 at 13% for 60 days   b. $2,190 at 8% for 180 days   c. $1,460 at 10% for 120 days

6. a. $3,650 at 9% for 30 days   b. $7,300 at 11% for 120 days   c. $1,095 at 15% for 90 days

Use the Bankers' Rule and supply the missing parts. Round the equivalent simple interest rate correct to the nearest $\frac{1}{4}$%.

		MATURITY VALUE	DISCOUNT RATE	TIME (DAYS)	BANK DISCOUNT	PROCEEDS	EQUIVALENT SIMPLE INTEREST RATE
7.	a.	$5,400	10%	60			
	b.	3,200	11	90			
	c.	4,500	8	210			
	d.	1,600	9	180			
8.	a.	$4,000	12%	90			
	b.	8,000	15	150			
	c.	7,200	13	120			
	d.	5,400	10	60			

Complete the following, using the Bankers' Rule.

		MATURITY VALUE	DISCOUNT RATE	DATE	DUE DATE	TIME	PROCEEDS
9.	a.	$2,160	9.5%	6/19		4 months	
	b.	1,080	10	8/13		3 months	
	c.	1,440	12	1/5		120 days	
	d.	2,520	11	2/9		30 days	
10.	a.	$4,320	14%	1/9		5 months	
	b.	6,000	9	10/3		2 months	
	c.	6,500	12	3/17		90 days	
	d.	9,000	10	5/20		180 days	

11. A florist signed a note with a face value of $5,000 discounted at 9% for 180 days. How much did the business receive from the bank?

12. A 90-day note for $10,000 was discounted by a bank at 11%. What amount will the borrower receive from the bank?

13. The bank discount was $60 on a note with a maturity value of $1,500. If the note was discounted at 12%, what was the time of the note?

14. The face value of an 8% note was $7,800. If the bank discount was $156, what was the time of the note?

15. The face value of a note was $1,160 and the proceeds were $1,131. What was the discount rate on this 90-day note?

16. A borrower received $8,554 when his $9,400 note was discounted at the bank. If the note had a time of 270 days, what discount rate was charged?

17. A 5-month note dated March 10 had a maturity value of $2,880. If the note was discounted at 10%, what were the proceeds?

18. A 3-month note having a maturity value of $6,000 was dated January 4. The bank charged a discount rate of 13%. What were the proceeds?

19. Two 2-month notes both have a maturity value of $4,000. Compare the present value at 12% interest with the proceeds at 12% discount.

20. The maturity values of two 6-month notes are both $3,000. Compare the present value at 9% interest with the proceeds at 9% discount.

21. A bank discounts a 180-day note at its Federal Reserve bank, receiving $31,230. Collateral for the note was $32,850 in loans owed to the bank. At what exact discount rate was the note computed?

22. A bank used $54,750 of loans as collateral and received $53,940 when it discounted a note at its Federal Reserve bank. What exact discount rate was charged on the 60-day note?

23. An investor purchased a $10,000 U.S. Treasury Bill at a 7.5% discount rate for 13 weeks.
   a. How much was invested in the T-bill?
   b. What is the maturity value?
   c. How much interest is earned on the investment?

24. A \$10,000 U.S. Treasury Bill was purchased at a 10% discount rate for 26 weeks.
   a. How much did the investor pay for the T-bill?
   b. What is the maturity value?
   c. How much interest is earned on the T-bill?

SECTION 2
MATURITY VALUE (AND REDISCOUNTING NOTES)

It often happens that the proceeds of a simple discount note are known and one wishes to find the maturity value of the note. When problems of this type are to be computed, it is customary to use an altered version of the basic discount formula $p = M(1 - dt)$. The formula is altered by dividing both sides of the equation by the expression in parentheses:

$$p = M(1 - dt)$$

$$\frac{p}{1 - dt} = \frac{M(\cancel{1 - dt})}{\cancel{(1 - dt)}}$$

$$\frac{p}{1 - dt} = M \quad \text{or} \quad M = \frac{p}{1 - dt}$$

Thus, the formula is set up to solve for the maturity value, M. Example 1, however, shows a problem worked by both versions of the discount formula, and either form may be used to solve the assignment problems.

Example 1   The proceeds were \$290.80 on a 3-month note dated August 29 and discounted at 12%. What is the maturity value?

$p = \$290.80$

$d = 12\%$

$t = \dfrac{92}{360}$

$M = ?$

$p = M(1 - dt)$

$$\$290.80 = M\left(1 - \frac{\cancel{12}}{\cancel{100}} \times \frac{.92}{\cancel{360}_{30}}\right)$$

$$= M\left(1 - \frac{.92}{30}\right)$$

$$= M\left(\frac{30}{30} - \frac{.92}{30}\right)$$

$$= M\left(\frac{29.08}{30}\right)$$

$M = \dfrac{p}{1 - dt}$

$$= \frac{\$290.80}{1 - \dfrac{\cancel{12}}{\cancel{100}} \times \dfrac{.92}{\cancel{360}_{30}}}$$

$$= \frac{290.80}{1 - \dfrac{.92}{30}}$$

$$= \frac{290.80}{\dfrac{30}{30} - \dfrac{.92}{30}}$$

$$\left(\frac{30}{\cancel{29.08}}\right)\,\cancel{290.80}^{10} = M\left(\frac{\cancel{29.08}}{\cancel{30}}\right)\left(\frac{\cancel{30}}{\cancel{29.08}}\right)$$

$$= \frac{\dfrac{290.80}{29.08}}{30}$$

$$\$300 = M$$

$$= \cancel{290.80}^{10}\left(\frac{30}{\cancel{29.08}}\right)$$

$$M = \$300$$

The maturity value is \$300 on November 29.

Calculator techniques . . .   FOR EXAMPLE 1

This technique illustrates the equation on the right, where the formula is set up to solve for the maturity value, M.

1 [M+];   0.12 [×] 92 [÷] 360 [M−] → 0.0306666

290.80 [÷] [MR] [=] → 299.99997

This is \$300, to the nearest cent. As you press [MR], notice that 0.9693334 appears briefly as the value of $(1 - dt)$.

A promissory note, like a check or currency, is a negotiable instrument. That is, it can be sold to another party (a person, a business, or a bank), or it can be used to purchase something or to pay a debt.

If the payee of a discounted note (usually some financial institution) sells the note to another bank, the note has been **rediscounted**. In that case, the second bank also uses the maturity value of the note as the basis for determining the proceeds that it will pay for the note. It should be observed that the discount rates that financial institutions charge each other are somewhat lower than those that banks ordinarily charge businesses and individuals.

When a note is rediscounted, the actual amount of interest earned by the original lender is the difference between the rediscounted note's proceeds (p_2) and the original note's proceeds (p_1).

$$\begin{array}{r} \text{Proceeds}_2\,(p_2) \\ -\text{Proceeds}_1\,(p_1) \\ \hline \text{Net interest earned} \end{array} \quad \text{(by original lender)}$$

Example 2   City National Bank was the holder (payee) of a $1,200, 120-day note discounted at 12%. Thirty days before the note was due, City Bank rediscounted (or sold) the note to Capital Mortgage and Trust at 9%. (a) How much did City Bank receive for the note? (b) How much did City Bank actually make on the transaction?

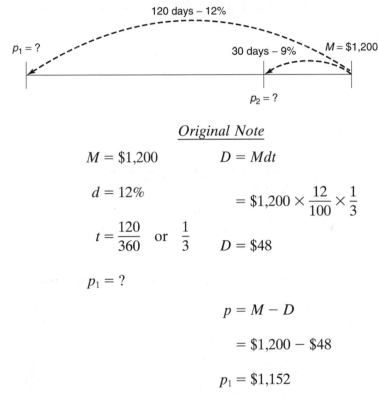

Original Note

$M = \$1,200$ 　　　　　$D = Mdt$

$d = 12\%$ 　　　　　　　$= \$1,200 \times \dfrac{12}{100} \times \dfrac{1}{3}$

$t = \dfrac{120}{360}$　or　$\dfrac{1}{3}$ 　　$D = \$48$

$p_1 = \ ?$

　　　　　　　　　　$p = M - D$

　　　　　　　　　　　$= \$1,200 - \$48$

　　　　　　　　　　$p_1 = \$1,152$

The proceeds of the original note (p_1), the amount that City Bank loaned to the maker, were $1,152.

Rediscounted Note

$M = \$1,200$ 　　　　　$p = M(1 - dt)$

$d = 9\%$ 　　　　　　　$= \$1,200 \left(1 - \dfrac{\overset{3}{\cancel{9}}}{100} \times \dfrac{1}{\underset{4}{\cancel{12}}}\right)$

$t = \dfrac{30}{360}$　or　$\dfrac{1}{12}$ 　　$= 1,200 \left(1 - \dfrac{3}{400}\right)$

$p_2 = \ ?$ 　　　　　　　$= \overset{3}{\cancel{1,200}} \left(\dfrac{397}{\cancel{400}}\right)$

　　　　　　　　　　$p_2 = \$1,191$

(a) The proceeds of the rediscounted note (p_2) were \$1,191. That is, City Bank received \$1,191 when it sold the note to Capital Mortgage and Trust.

(b) City Bank made \$1,191 − \$1,152 = \$39 on the transaction. This is only \$9 less than the entire \$48 interest that City Bank would have earned if it had kept the note, although only ¾ of the time had elapsed.

DISCOUNTING SIMPLE INTEREST NOTES

It is also possible that the payee of a simple interest note may wish to sell the note at a bank before the due date. The amount that the bank would pay is determined as follows:

1. Find the maturity value of the simple interest note.

2. Discount the maturity value of the note using the bank's rate and the time *until the due date.* The proceeds remaining is the amount that the bank would pay for the note.

3. The interest that the original payee has made is the difference between what the note was sold for (the discounted proceeds) and the original amount loaned (the principal of the simple interest note):

$$
\begin{array}{r}
\text{Proceeds} \\
-\text{Principal} \\
\hline
\text{Net interest earned}
\end{array}\quad\text{(by original lender)}
$$

Example 3   A-1 Appliance Sales was the payee of a \$900, 80-day note with interest at 10%. Forty days before the due date, A-1's bank discounted the note at 9%. (a) How much did A-1 receive for the note? (b) How much did A-1 make on the transaction?

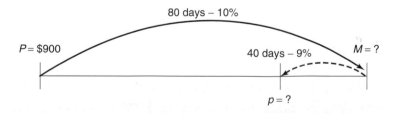

Simple Interest Note
(Finding Maturity Value)

$P = \$900$ $I = Prt$

$r = 10\%$ $= \cancel{\$900}^{9} \times \dfrac{10}{\cancel{100}} \times \dfrac{2}{\cancel{9}}$

$t = \dfrac{80}{360}$ or $\dfrac{2}{9}$ $I = \$20$

$M = \ ?$

$$M = P + I$$

$$= \$900 + \$20$$

$$M = \$920$$

Discount Note
(Finding Proceeds)

$M = \$920$ $D = Mdt$

$d = 9\%$ $= \cancel{\$920}^{9.20} \times \dfrac{\cancel{9}}{\cancel{100}} \times \dfrac{1}{\cancel{9}}$

$t = \dfrac{40}{360}$ or $\dfrac{1}{9}$ $D = \$9.20$

$p = \ ?$

$$p = M - D$$

$$= \$920 - \$9.20$$

$$p = \$910.80$$

(a)   A-1 received $910.80 when the note was sold.

(b)   Since A-1 had loaned $900 originally, it made $10.80 in interest, as shown below. This is approximately ½ of the $20 interest that the maker will pay on the note, although ⅝ of the time has passed.

Proceeds	$910.80
Principal	− 900.00
Net interest	$ 10.80

SECTION 2   PROBLEMS

Complete the following.

	PROCEEDS	DISCOUNT RATE	DATE	DUE DATE	TIME	MATURITY VALUE
1. a.	$1,173	9%	9/1		90 days	
b.	2,232	14	4/1		180 days	
c.	1,785	10	4/4		1 month	
d.	2,688	12	1/15		4 months	
2. a.	$6,860	12%	10/18		60 days	
b.	1,925	9	6/1		150 days	
c.	8,721	18	7/9		2 months	
d.	4,275	15	2/25		4 months	

Find the additional information.

	ORIGINAL NOTE				REDISCOUNTED NOTE			
	MATURITY VALUE	RATE	TIME (DAYS)	PROCEEDS	RATE	TIME (DAYS)	PROCEEDS	NET INTEREST EARNED
3. a.	$4,000	9%	120		10%	45		
b.	3,600	13	150		12	60		
c.	2,500	12	90		10	45		
4. a.	$7,600	13%	270		14%	90		
b.	8,500	9	120		10	30		
c.	6,400	14	90		12	60		

Find the net interest made on the following.

	SIMPLE INTEREST NOTE				DISCOUNTED NOTE			
	PRINCIPAL	RATE	TIME (DAYS)	MATURITY VALUE	DISCOUNT RATE	TIME (DAYS)	PROCEEDS	NET INTEREST EARNED
5. a.	$4,800	10%	120		12%	90		
b.	2,400	14	270		9	180		
6. a.	$1,600	15%	90		12%	30		
b.	8,000	12	150		10	45		

7. On February 15, the proceeds of a 5-month note were $1,140. If the bank discount had been computed at 12%, what was the face value of the note?

8. The proceeds of a 4-month note dated January 10 were $3,800. If the discount rate was 15%, what was the face value of the note?

9. The proceeds of a 3-month note dated January 27 were $778. If the discount rate was 11%, what was the maturity value of the note?

10. The maker of a 5-month note dated February 20 received proceeds of $8,470. If the discount rate had been 9%, how much did the maker pay at maturity?

11. First National Bank was the payee of a $2,700, 60-day note dated September 12 and discounted at 10%. On October 22, the note was rediscounted at Madison Bank at 12%.

    a. What was the amount originally loaned by First National Bank?

    b. What was the amount received when the note was sold?

    c. What was the amount that First National Bank made on the transaction?

    d. How much less interest was this than First National would have earned had it kept the note to maturity?

12. Cullman National Bank was the payee of a $20,000 note dated June 3. The note was made for 90 days and was discounted at 16%. On August 2, the note was rediscounted at Decatur Bank at 15%.

    a. What amount had Cullman National Bank loaned the maker of the note?

    b. How much had Cullman received when it rediscounted the note?

    c. How much did Cullman make on the transaction?

    d. How much more would Cullman have received if it had kept the note to maturity?

13. Calvert Office Furniture Stores was the payee of a $3,240, 90-day note dated June 12, with interest at 10%. On August 11, Calvert discounted this note at its bank at 12%.

    a. How much did Calvert receive from discounting the note?

    b. How much interest did Calvert earn on the note?

    c. How much interest did Calvert lose by not holding the note until maturity?

14. Professional Office Movers, Inc., was the payee of a $9,000, 120-day note dated March 16, with simple interest at 14%. On May 15, the business discounted the note at its bank at 12%.

    a. How much did Professional Office Movers receive from the note?

    b. How much interest did Professional Office Movers make on the note?

    c. How much more interest would the business have made by keeping the note until maturity?

15. The Annandale City Bank was the payee of a $6,000, 90-day note dated May 11 and discounted at 15%. On July 10, the note was rediscounted with the Burke National Bank at 12%.

    a. What amount did the Annandale City Bank lend to the maker of the note?

    b. What amount did the Annandale City Bank receive when it sold the note?

    c. How much interest did Annandale earn on the note?

    d. How much more interest would Annandale have earned had it kept the note to maturity?

16. Radford National Bank was the payee of a 120-day note dated June 25. The face value was $6,300, and the discount rate was 10%. On July 25, Radford National rediscounted the note at 9% at First Georgia State Bank.

    a. What amount did Radford National lend to the maker of the note?

    b. How much did Radford National receive when it rediscounted the note?

    c. How much interest did Radford National earn from these transactions?

    d. What interest did Radford National forgo by not holding the note to maturity?

17. The A & M Bank was the payee of a $20,000, 180-day note dated March 15, earning 10% simple interest. On July 28, the bank used the maturity value as collateral to arrange a note with its Federal Reserve bank at an exact discount rate of 9%.
    a. What amount did A & M Bank receive from the Federal Reserve bank?
    b. How much interest did A & M Bank make on the funds it loaned?
    c. How much interest did A & M Bank forfeit by taking out the Federal Reserve note?

18. A 12%, 90-day simple interest note dated May 13 for $4,200 was payable at Sheraton National Bank. On June 22, Sheraton National Bank discounted the face value of the note at the Federal Reserve bank at an exact discount rate of 11%.
    a. How much did Sheraton National receive from the Federal Reserve bank?
    b. How much interest did Sheraton National earn from these transactions?
    c. How much interest did Sheraton National lose by not holding the original note to maturity?

SECTION 3
DISCOUNTING NOTES TO TAKE ADVANTAGE OF CASH DISCOUNT

In Chapter 12, we learned that many merchants offer cash discounts as inducements to buyers to pay their accounts promptly. These discounts typically are a full 1% to 2% in only 10 days. When businesses do not have sufficient cash available on the last day of a cash discount period, they often sign a bank note in order to take advantage of the cash discount. Although the bank charges interest on the note, the business still "comes out ahead" if it borrows for only a short term. Remember that a bank rate of 13.5% means 13.5% for an entire year; for a 20-day note, the actual interest charge would be only $\frac{3}{4}$%. Thus, the business would still save $1\frac{1}{4}$% by borrowing to take advantage of a 2% cash discount.

The business would save money by borrowing provided that it borrowed for a period of time short enough so that the bank discount it must pay is less than the cash discount. However, we will consider only the situation when the maximum amount will be saved: On the *last* day of the cash discount period, a note is discounted so that the proceeds exactly equal the amount required to take advantage of the cash discount. The note will be for only the time remaining until the net amount of the invoice would have been due.

Example 1   An invoice dated July 2 covers merchandise of $665 and freight of $18.80. Sales terms on the invoice are 2/10, n/30. A note is discounted at 12% in order to take advantage of the cash discount. (a) What will be the face value of the note? (b) How much will be saved by borrowing the necessary cash?

The last day of the discount period is July 12; the full net amount will be due 20 days later on August 1. Therefore, to obtain maximum savings, a discounted note will be signed on July 12 for 20 days.

(a) $665.00 Goods $665.00 Goods
     ×   0.02   −   13.30 Cash discount
    $ 13.30 Cash discount $651.70
                    +   18.80 Freight
                    $670.50 Amount needed to take advantage of cash discount (proceeds needed)

$$p = \$670.50$$

$$d = 12\%$$

$$t = 20 \text{ days}$$

$$= \frac{20}{360} \quad \text{or} \quad \frac{1}{18}$$

$$M = ?$$

$$M = \frac{p}{1 - dt}$$

$$= \frac{\$670.50}{1 - \frac{\cancel{12}^{\,2}}{100} \times \frac{1}{\cancel{18}_{\,3}}}$$

$$= \frac{670.50}{1 - \frac{2}{300}}$$

$$= \frac{670.50}{\dfrac{298}{300}}$$

$$= 670.50 \left(\frac{300}{298}\right)$$

$$M = \$675$$

The required note will have a face (maturity) value of $675.

(b) $665.00 Goods
   +   18.80 Freight
    $683.80 Amount due August 1 if a note is not signed
  − 675.00 Maturity value of note due August 1
   $ 8.80 Amount saved by discounting a note

SECTION 3 PROBLEMS

Complete the following.

	AMOUNT OF INVOICE	SALES TERMS	AMOUNT NEEDED (PROCEEDS)	TIME OF NOTE (DAYS)	DISCOUNT RATE	FACE VALUE OF NOTE	AMOUNT SAVED
1. a.	$909.18	2/10, n/50			9%		
b.	758.90	2/15, n/30			12		
c.	873.00	3/15, n/30			9		
2. a.	$6,404.64	3/10, n/30			9%		
b.	7,614.80	2/15, n/30			12		
c.	4,242.86	2/15, n/60			8		

3. An invoice for $2,615.72 has sales terms of 3/15, n/30. A 12% note was discounted in order to take advantage of the cash discount.

   a. What was the face value of the note?

   b. How much was saved by borrowing to take advantage of the cash discount?

4. A bank note was discounted at 18% in order to take advantage of the cash discount on an invoice for $5,613.40. The terms of the invoice were 3/10, n/30.

   a. What was the maturity value of the note?

   b. How much was saved by borrowing to take advantage of the cash discount?

5. An invoice for $1,626.53 has terms of 2/15, n/30. If a bank note is discounted at 9% in order to take advantage of the cash discount,

   a. What will be the face value of the note?

   b. How much will be saved by borrowing the necessary cash?

6. An invoice for $3,073.45 has terms of 3/15, n/30. A note is discounted at 15% in order to take advantage of the cash discount.

   a. What will be the face value of the note?

   b. How much will be saved by borrowing the necessary cash?

7. An invoice for $1,498 has terms of 2/10, n/30 and freight charges of $23.63. If a bank note is discounted at 10% in order to take advantage of the cash discount,

   a. What will be the face value of the note?

   b. How much will be saved by borrowing to take advantage of the cash discount?

8. A note is discounted at 9% in order to take advantage of a cash discount on an invoice listing merchandise of $9,259.11 plus freight charges of $21.25. The cash discount is 4/10, n/50.

   a. What will be the face value of the note?

   b. How much will be saved by signing the note?

9. A business takes out a bank note discounted at 8% in order to take advantage of a cash discount on an invoice listing merchandise of $923.71 with terms of 3/10, n/30. The invoice also shows shipping charges of $18.00.

   a. What is the maturity value of the note?

   b. By borrowing the necessary cash, how much will be saved?

10. An invoice has terms of 3/15, n/30. Merchandise is listed for $5,680, and there is a freight charge of $48.40. If a bank note is discounted at 18%,

    a. What is the maturity value of the note?

    b. By borrowing the necessary cash, how much is saved?

SECTION 4

SUMMARY OF SIMPLE INTEREST AND SIMPLE DISCOUNT

It is essential that the student understand the identifying characteristics of both the simple interest and simple discount notes as well as the differences between the two. Both kinds of *promissory notes* are

1. Generally for 1 year or less.
2. Repaid in a single payment at the end of the time.

As for their differences, the following are characteristic of the *simple interest note* (a sample of which is shown in Figure 16-4):

1. The face value is the principal of the note. The face value is the actual amount that was loaned.
2. Interest is computed on the face value of the note at the rate and for the time stated on the note.
3. The maturity value of the simple interest note is the sum of the face value plus interest.
4. A simple interest problem will contain one or more of these identifying characteristics: "*interest* rate of $x\%$," "with *interest* at $x\%$," a maturity value which exceeds the face value, "*amount*."

Face value	(P)	$400
Interest rate	(r)	12%
Time	(t)	60 days
Interest	(I)	$8
Maturity value	(M)	$408

FIGURE 16-4   Simple Interest Note

5. The solution of simple interest problems requires one or more of the following formulas:

(a) $I = Prt$ (b) $M = P + I$ (c) $M = P(1 + rt)$

(d) $P = \dfrac{M}{1 + rt}$

The following are characteristic of the *simple discount note* (a sample of which is shown in Figure 16-5):

1. The face value of the simple discount note is the maturity value—the amount that will be repaid.

2. Discount notes may contain a rate quoted as a certain percent "discounted to maturity." The interest (called "bank discount") is computed on the face value of the note. (That is, the borrower pays interest on the amount that will be repaid, rather than on the amount actually borrowed.)

3. The actual amount that the borrower (maker) receives from the lender (payee) is called the "proceeds." The proceeds are found by subtracting the bank discount from the face value. This is often called paying interest in advance.

4. Simple discount problems may be distinguished from simple interest problems by one or more of the following: "*discount* rate of $x\%$," "*discounted* at $x\%$," "*proceeds*," maturity value equals the face value, "interest in advance."

5. The solution of simple discount problems requires one or more of the following formulas:

(a) $D = Mdt$ (b) $p = M - D$ (c) $p = M(1 - dt)$

(d) $M = \dfrac{p}{1 - dt}$

Face value	(M)	$400
Discount rate	(d)	12%
Time	(t)	60 days
Discount	(D)	$8
Proceeds	(p)	$392
Maturity value	(M)	$400

FIGURE 16-5   Simple Discount Note

SECTION 4 PROBLEMS

Solve the following simple interest and simple discount problems.

1. How much interest is due on a 60-day, 9.5% note with a face value of $1,080? What is the maturity value of the note?

2. A $5,000 note is taken out for 45 days at 11%. Find the interest and the maturity value of the note.

3. The face value of a 120-day note is $1,440. If the bank discount rate is 12%, what is the discount and what are the proceeds?

4. The face value of a $3,960 note is discounted for 40 days at 10%. What is the discount and what are the proceeds?

5. A simple interest note and a simple discount note both have face values of $3,600. Both notes are for 90 days and both have interest charged at 10%.
   a. Find the interest that will be paid on each.
   b. Find the amount the borrower will actually receive on each note.
   c. Find the maturity value of each note.

6. Compare a simple interest note and a simple discount note. Both notes have face values of $1,600, interest charged at 15%, and a time of 120 days.
   a. How much interest will be charged on each note?
   b. How much will the maker receive on each note?
   c. What is the maturity value of each note?

7. The proceeds of a 9% note for 150 days were $693. What was
   a. The maturity value?
   b. The face value?

8. a. Determine the maturity value of a 60-day note that has proceeds of $2,352. The discount rate is 12%.
   b. What is the face value of the note?

9. The maturity value is $1,875 on a 150-day note with interest at 10%.
   a. What is the face value of this note?
   b. How much did the maker receive on the loan?

10. A 180-day note with interest at 9.5% has a maturity value of $4,190.
    a. What is the principal of the note?
    b. What amount did the maker receive on the loan?

11. The Liberty Trust Co. discounted a 12%, 120-day note for $8,500 from Tri-Cities Cable Co. What equivalent simple interest rate (to the nearest ¼%) was charged by Liberty Trust?

12. The Regis Services Co. borrowed $7,000 from the bank by discounting a 90-day, 10% note. What equivalent simple interest rate (to the nearest ¼%) was charged by the bank?

13. Compare the present value at 14% interest with the proceeds at 14% discount on two notes with maturity values of $1,400 and times of 90 days.

14. Compare the present value at 15% interest with the proceeds at 15% discount on two notes with maturity values of $8,400 and terms of 120 days.

15. On June 10, a $9,000, 3-month note was drawn with interest at 12%. If money is worth 9%, what was the present value of the note on the day it was drawn?

16. On August 22, a 4-month note with interest at 10% was made. The face value of the note was $1,080. If most investments are earning 9%, what was the present value of the note on the day it was made?

17. On January 6, a 4-month note was signed for $3,240 at 10% interest. What is the present value of the note 45 days before it is due, if money is worth 8%?

18. On February 5, a 5-month note with a face value of $5,200 was drawn at 12% interest. What is the present value of the note 90 days before maturity, if most investments are earning 10% interest?

19. Massey Manufacturing Co. was the payee of a 4-month note dated February 1 for $3,000 at 12% interest. On May 2, Massey discounted the note at 9% at the bank.

    a. What did the business receive for the note?

    b. How much did Massey make on the note?

    c. How much interest did the bank earn?

20. On January 8, Cook Construction Co. received a 6-month note for $8,000 at 18% interest for a remodeling job. On April 9, Cook discounted the $8,000 note at 15% at the bank.

    a. What did Cook Construction Co. receive from the bank?

    b. How much did Cook make on the note?

    c. How much interest did the bank make?

21. Jefferson Bank was the holder of a 3-month note dated January 10 with a discount rate of 12%. The face value was $15,000. On March 11, Jefferson Bank rediscounted the note at Washington Bank at 10%.

    a. How much did Jefferson Bank lend to the maker of the note?

    b. How much did Jefferson Bank receive when it rediscounted the note?

    c. How much interest did Jefferson Bank earn?

    d. How much interest did Jefferson Bank lose by not holding the note to maturity?

22. American Bank was the payee of a $6,000, 4-month note dated August 15, discounted at 9%. American Bank rediscounted the note at 8.5% at another bank on October 16.

    a. What amount had American Bank loaned the maker?

    b. How much did American Bank receive when it sold the note?

    c. How much interest did American Bank earn?

    d. How much interest did the second bank make?

CHAPTER 16 GLOSSARY

Bank (or simple) discount.   Interest that is computed on the maturity value of a loan and is then subtracted to obtain the proceeds; often called "interest in advance":

$$D = Mdt$$

where D = bank discount, M = maturity value, d = discount rate, and t = time.

Equivalent simple interest rate.   The true rate of interst charged on a simple discount note; the interest rate which, if computed on the money the borrower

actually received (the proceeds) and for the same time, would produce the same maturity value.

Face value.   The money value specified on any promissory note; the maturity value of a simple discount note.

Maturity value.   The total amount (Proceeds plus Bank discount) due after a simple discount loan; the value on which simple discount is computed.

Proceeds.   The amount remaining after the bank discount is subtracted from the maturity value; the amount actually received by a borrower:

$$p = M - D \quad \text{or} \quad p = M(1 - dt)$$

where p = proceeds, M = maturity value, D = bank discount, d = discount rate, and t = time.

Rediscounted note. A simple discount note that is sold for a value equal to the proceeds of the note at the buyer's discount rate.

Simple discount. (See "Bank discount.")

Truth-in-Lending Law. A federal law requiring lenders to disclose both their total finance charge and corresponding annual percentage rate (based on the amount actually received by a borrower).

MULTIPLE PAYMENT PLANS

OBJECTIVES

Upon completion of Chapter 17, you will be able to:

1. Define and use correctly the terminology associated with each topic.
2. Apply the United States rule to
   a. Find the credit due for partial payments made on a note, and
   b. Find the balance due on the maturity date (Section 1: Example 1; Problems 1–6).
3. On an open-end charge account,
   a. Compute the interest due on adjusted monthly balances, and
   b. Determine the total amount required to pay off the account (Section 2: Example 1; Problems 1, 2, 5, 6).
   c. Compute the interest due on average daily balances, and
   d. Find the final balance (Example 2; Problems 3, 4, 7, 8).
4. Determine for an installment plan purchase (Section 3: Examples 1, 2; Problems 1, 2, 7–20, 23–26):
   a. The down payment and the finance charge
   b. The regular monthly or weekly payment
   c. The total cost on the time payment plan.
5. For loans or purchases repaid on an installment plan (Section 3: Examples 2–4; Problems 3, 4, 11–24):
   a. Compute "finance charge per $100" (or "rate × time").
   b. Consult a table to determine the annual percentage rate charged on an installment purchase.
6. Construct a table to verify that the annual percentage rate (Objective 5b) would actually result in the amount of interest charged (Section 3: Examples 3, 4; Problems 5, 6, 23–26, 28).
7. When prepayment occurs in an installment plan with set payments, determine the interest saved and the balance due
   a. Using the actuarial method (Section 4: Examples 1, 3; Problems 1, 2, 5–8).
   b. Using the rule of 78s (Example 2; Problems 3, 4, 9–12).

The simple interest and simple discount notes studied earlier are normally repaid in a single payment at the end of the time period. However, a borrower may prefer to repay part of a debt before the entire loan is due. The lender, in order to give the borrower proper credit for payments made prior to the maturity date, should apply the United States rule. Also, many firms, when extending credit, feel that the entire debt will more likely be repaid if the borrower is required to make regular installment payments that begin immediately, rather than a single payment that is not due for some time. Partial payments may thus be made under the *U.S. rule* or on basic *installment plans,* both of which are discussed in the following sections.

SECTION *1*

UNITED STATES RULE

A person who has borrowed money on a note may sometimes wish to pay off part of the debt before the due date, in order to reduce the amount of interest paid. Such partial payments are not required (as they would be for an installment loan), but may be made whenever the borrower's financial condition permits. The United States rule, which we shall study, is used to calculate the credit that should be given for partial payments and to determine how much more is owed on the debt.

The **United States rule,** or **U.S. rule,** derives its name from the fact that this method has been upheld by the U.S. Supreme Court as well as by a number of state courts. It is the method used by the federal government in its financial transactions.* The Bankers' Rule should be used to determine "time" for the interest calculations in this section.

Basically, the U.S. rule is applied in the following manner:

1. Interest is computed on the principal from the first day until the date of the first partial payment.

2. The partial payment is used first to pay the interest due, and the remaining part of the payment is deducted from the principal.

3. The next time an amount is paid, interest is calculated on the adjusted principal from the date of the previous payment up to the present. The partial payment is then used to pay the interest due and to reduce the principal, as before. (This step may be repeated as often as additional payments are made.)

4. The balance due on the maturity date is found by computing interest due since the last partial payment and adding this interest to the unpaid principal.

---

*It might also be noted that interest calculations by the U.S. government use exact interest: $t =$ exact days over 365 or 366. However, exact interest is not an essential party of the U.S. rule.

Example 1   On April 1, Mary Carrington Interiors took out a 90-day note for $1,500 bearing interest at 10%. On May 1, Carrington Interiors paid $700 toward the obligation, and on May 16, another $300. How much remains to be paid on the due date?

(a) Interest is computed on the original $1,500 from April 1 until the day of the first partial payment (May 1):

$$\begin{array}{ll}
\text{May 1} = & 121 \text{ day} \\
\text{April 1} = & -\ 91 \text{ day} \\
\hline
& 30 \text{ days}
\end{array} \qquad I = Prt$$

$$= \$1{,}500 \times \frac{10}{100} \times \frac{1}{12}$$

$$t = \frac{30}{360} \quad \text{or} \quad \frac{1}{12} \qquad I = \$12.50$$

(b) The $700 is first applied to pay the interest, and the remaining amount ($700 − $12.50 = $687.50) is then deducted from the principal:

Principal	$1,500.00
Less: Payment to principal	− 687.50
Adjusted principal	$ 812.50

(c) Interest is computed on the $812.50 adjusted principal from May 1 until May 16. The $300 payment first pays this interest, and the remainder of the payment is then subtracted from the current principal:

$$\begin{array}{ll}
\text{May 16} = & 136 \text{ day} \\
\text{May 1} = & -121 \text{ day} \\
\hline
& 15 \text{ days}
\end{array} \qquad I = Prt$$

$$= \$812.50 \times \frac{10}{100} \times \frac{1}{24}$$

$$t = \frac{15}{360} = \frac{1}{24} \qquad I = \$3.39$$

Partial payment	$300.00
Less: Interest	− 3.39
Payment to principal	$296.61
Current principal	$812.50
Less: Payment to principal	− 296.61
Adjusted principal	$515.89

(d)   No other payment is made until the maturity date; therefore, interest is computed on $515.89 from May 16 until the 90-day period ends on June 30. The maturity value $(P + I)$ is the final payment.

$$\begin{aligned}
\text{June 30} &= 181 \text{ day} \\
\text{May 16} &= \underline{-136} \text{ day} \\
&\ \ 45 \text{ days}
\end{aligned}$$

$$I = Prt$$
$$= \$515.89 \times \frac{10}{100} \times \frac{1}{8}$$

$$t = \frac{45}{360} = \frac{1}{8}$$

$$I = \$6.45$$

$$\begin{aligned}
M &= P + I \\
&= \$515.89 + \$6.45 \\
M &= \$522.34
\end{aligned}$$

Thus, $522.34 remains to be paid on June 30.

The total interest paid was $12.50 + $3.39 + $6.45 = $22.34. With no partial payments, interest on the note would have been $1,500 × $10\% \times \dfrac{90}{360} = \$37.50$. Thus, $37.50 − $22.34 = $15.16 was saved by making early payments.

The entire process required in the foregoing application of the U.S. rule is summarized as follows:

Original principal, 4/1		$1,500.00
First partial payment, 5/1	$700.00	
Less: Interest ($1,500, 10%, 30 days)	− 12.50	
Payment to principal		− 687.50
Adjusted principal		$ 812.50
Second partial payment, 5/16	$300.00	
Less: Interest ($812.50, 10%, 15 days)	− 3.39	
Payment to principal		− 296.61
Adjusted principal		$ 515.89
Interest due ($515.89, 10%, 45 days)		+ 6.45
Balance due, 6/30		$ 522.34

Complications arise from the U.S. rule if the partial payment is not large enough to pay the interest due. The unpaid interest cannot be added to the principal, for then interest would be earned on interest; this would constitute compound interest, and the charging of compound interest on loans is illegal. This problem is solved simply by holding the payment or payments (without giving credit for them) until the sum of these partial payments is large enough to pay the interest due to that date. The procedure then continues as usual.

(Note that the borrower does not save any interest unless the partial payment is sufficiently large to pay the interest due and reduce the principal somewhat.)

SECTION 1   PROBLEMS

Using the U.S. rule, determine the balance due on the maturity date of each of the following notes. (Use the Bankers' Rule.)

		NOTE			PARTIAL PAYMENTS	
		PRINCIPAL	RATE	TIME (DAYS)	AMOUNT	DAY
1.	a.	$2,000	10%	180	$ 550	90th
					825	150th
	b.	1,800	8	120	612	30th
					304	45th
	c.	3,000	9	240	690	120th
					918	150th
2.	a.	$6,000	12	150	$2,040	20th
					1,500	80th
	b.	3,600	15	90	2,030	20th
					1,030	65th
	c.	8,000	10	270	2,200	90th
					2,875	135th

3. Griffin Parts Co. borrowed $5,000 by signing a 9% note for 240 days on March 6. A payment of $625 was made toward the note on March 26, and $2,066 was paid on May 25. Find the balance due on the day the note matures.

4. Carlton Interiors borrowed $10,000 by signing a 12%, 180-day note on April 5. The business made a $2,200 partial payment on June 4 and paid $3,240 on September 2. Determine the amount still owed at the end of the loan period.

5. Mitchell Candy Co. borrowed $4,000 by signing a 9% note dated June 15 for 150 days. The firm made a $520 partial payment on July 5 and another payment of $1,070 on September 23. What payment will be required when the note is due?

6. On June 10, Perez Beauty Supplies borrowed $12,000 by signing an 18%, 90-day note. A partial payment of $2,120 was made on June 30, and a second partial payment of $5,225 was made on August 14. How much will be owed on the due date?

SECTION 2
INSTALLMENT PLANS: OPEN END

Undoubtedly, you already have some knowledge of that uniquely American institution, the **installment plan.** Almost all department stores, appliance stores, and furniture stores, as well as finance companies and banks, offer some form of installment plan whereby a customer may take possession of a purchase immediately and, for an additional charge, pay for it later by a series of regular payments. Americans pay exhorbitant rates, as we shall see, for the privilege of buying on credit, but, because the dollar amounts involved are not extremely large, many people are willing to pay them.

The installment plan actually provides a great boost to the American economy, because many families buy things on the spur of the moment on the installment plan that they would never buy were they compelled to save the money required for a cash purchase. The lure of the sales pitch—"nothing down; low monthly payments," "budget terms," "buy now; no interest or payment for 60 days," "small down payment; 3 years to pay," "consolidate all your bills into one low, monthly payment"—has caused many families to feel that they will never miss that small extra amount each month. The result is that many families soon find themselves saddled with so many of these "small monthly payments" that they cannot make ends meet and are in danger of bankruptcy. The Family Service Agency or similar organizations in many cities provide inexpensive counseling to help such families learn to budget their incomes and "get back on their feet" financially.

The **Truth-in-Lending Law** (officially entitled Regulation Z), which took effect in 1969, enables consumers to determine how much the privilege of credit buying actually costs, thereby allowing them to make intelligent decisions when faced with the question of when to buy on the installment plan. Indeed, the question has become "when to buy" rather than "whether to buy," because almost all Americans take advantage of time-payment plans at one time or another. The single item most often purchased in this way (other than homes) is the automobile, which is followed by other durable items such as furniture and appliances. Even realizing the expense of these plans, most people feel that the added convenience at times justifies the additional cost.

The high rate of interest charged by companies offering installment plans is not without justification. The merchant incurs numerous additional operating expenses as a result of the installment plan: the costs of investigating the customer's credit standing, discounting bank notes, buying extra insurance, paying cashiers and bookkeepers for the additional work, collection costs necessitated by buyers who do not keep up with payments, as well as "bad debt" loss from customers who never finish paying. And, of course, the merchant is entitled to interest, since the business has capital invested in the merchandise being bought on the time-payment plan. Thus, the installment plan is expensive for the seller as well as the buyer.

Basically, there are two types of installment plans: open-end accounts (this section) and accounts with a set number of payments (next section).

The first type of installment plan is called **open-end credit** because the time period of the credit account is not definite and the customer may receive additional credit before the first credit is entirely repaid. Under this plan, interest is computed each month on the unpaid balance (or sometimes on an average balance). Each payment is applied first toward the interest and then toward the balance due, in accordance with the U.S. rule of partial payments.

This plan is used by most department stores and other retail businesses offering "charge accounts" and "credit cards." The monthly interest rate is usually $1\frac{1}{4}$% (= 15% per year) to $1\frac{1}{2}$% (= 18% per year). Customer accounts at these stores usually vary quite a bit from month to month, because the account results from a number of small purchases rather than one large purchase. Many accounts of this type are never actually paid off, because new purchases continue to be made. These accounts are therefore often called **revolving charge accounts.** There is usually a limit to the amount a customer is entitled to charge, and a prescribed minimum monthly payment may be required according to the amount owed. (For instance, suppose the maximum balance that a charge account may reach is $500. The customer may be required to pay at least $15 per month if his balance is under $200 and pay at least $25 per month if he owes $200 or more.)

The following example illustrates the payment of a single purchase on an installment plan. This procedure would be used also to pay off an account balance when no other purchases are charged to the account.

Example 1　**Installment Plan Having Monthly Rate**

Mrs. Cohen's charge account at Butler-Stohr has a balance of $100. Interest of $1\frac{1}{2}$% per month is charged on unpaid customer accounts. Mrs. Cohen has decided to pay off her account by making monthly payments of $25. (a) How much of each payment is interest, and how much will apply toward her account balance? (b) What is the total amount Mrs. Cohen will pay, and how much of this is interest?

(a)　The following schedule computes the interest and the payment that applies toward the balance due each month. "Interest" is found by taking $1\frac{1}{2}$% of the "balance due." This interest must then be deducted from the "monthly payment" in order to determine the "payment (that applies) toward the balance due." The "payment toward the balance due" is then subtracted from the "balance due" to find the "adjusted balance due," which is carried forward to the next line as the "balance due."

PAYMENT NUMBER	BALANCE DUE	$1\frac{1}{2}$% INTEREST PAYMENT	MONTHLY PAYMENT	PAYMENT TOWARD BALANCE DUE	ADJUSTED BALANCE DUE
1	$100.00	$1.50	$ 25.00	$ 23.50	$76.50
2	76.50	1.15	25.00	23.85	52.65
3	52.65	0.79	25.00	24.21	28.44
4	28.44	0.43	25.00	24.57	3.87
5	3.87	0.06	3.93	3.87	0.00
		$3.93	$103.93	$100.00	

(b) During the last month, Mrs. Cohen paid only the remaining balance due plus the interest due. She thus paid a total of $103.93 in order to discharge her $100 obligation, which included total interest of $3.93.

Calculator techniques . . .   For example 1

Set up a table using the format shown in the example; fill in the values as you perform the following calculations. Start by entering the $100 original balance; round each month's 1.5% interest to the nearest cent. Recall that numbers shown here in parentheses mean to use the value currently displayed, without reentering it.

100 [M+]

(100) [×] 1.5 [%] ——→ 1.5

   25 [−] 1.5 [M−] ——→ 23.5;   [MR] ——→ 76.5

(76.5) [×] 1.5 [%] ——→ 1.1475

   25 [−] 1.15 [M−] ——→ 23.85;   [MR] ——→ 52.65

(52.65) [×] 1.5 [%] ——→ 0.78975

   25 [−] 0.79 [M−] ——→ 24.21;   [MR] ——→ 28.44   and so forth.

Finally, add the middle columns to verify that your figures produce the correct amounts of interest, total monthly payments, and payments of the original balance due.

There are a number of variations in the way that interest may be computed on a revolving charge account. Example 1 illustrates interest computed on the *adjusted balance* remaining after payment was received during the month. By contrast, when the *previous balance* method is used, this month's interest is based on the balance due at the end of the previous month.

The most common interest method, the *average daily balance* method, computes a weighted average that is a combination of the two preceding methods. In this case, the previous month's balance is multiplied by the number of days until a payment is received. After that payment has been deducted, the adjusted balance is multiplied by the number of days remaining in the month (or other payment cycle). Their sum is then divided by 30 (or by 31 or other total days in the cycle) to obtain the average daily balance for the billing period, and interest is calculated on this average.

Ordinarily, interest is not charged on this month's "current" purchases until after the customer's statement is mailed. This is illustrated in the following example, which utilizes the average daily balance method. Example 2 also illustrates a more realistic revolving charge account, where each month's transactions include additional credit purchases as well as the required amount of payment.

Example 2   **Revolving Charge Account**

Michael Boyer's charge account at Wymer's had a $50 balance at the beginning of the 6-month period illustrated below. Payments and additional charges occurred on the indicated dates of each 30-day month. Interest is computed by the average daily balance method at 2% per month. (No interest is charged on current purchases.) Each month's ending balance is indicated by an "*". Calculation of three of the average balances is shown in footnotes. Notice also that a check confirms that the totals are correct. (Refer to the table on the following page for the monthly calculations of the revolving charge account.)

Calculator techniques . . .   FOR EXAMPLE 2

There are many balances to keep up with in this problem, so ordinary addition/subtraction is suggested for most lines. Calculations using memory are suggested only to determine the average balance and its corresponding 2% interest charge. Here is the average-balance calculation of $59.23 illustrated in footnote b of the associated table, with its resulting $1.18 interest charge:

10 ☒ 65.90 ☐M+ ⟶ 659;   20 ☒ 55.90 ☐M+ ⟶ 1,118

☐MR ⟶ 1,777 ☐÷ 30 ☐= ⟶ 59.233333 ☒ 2 ☐% ⟶ 1.1846666

| | | REVOLVING ACCOUNT | | | | | CURRENT ACCOUNT | |
MO.	DATE	PREV. BAL.	PAYMT.	ADJ. BAL.	AVG. BAL.	2% INT. CHARGE	PURCHASES	CUR. BAL.
1	15	$ 50.00	$ 10	$ 40.00				$ 40.00
	20						$ 25	65.00
	30				$ 45.00[a]	$ 0.90		65.90*
2	5						30	95.90
	10	65.90	10	55.90				85.90
	14						20	105.90
	30				59.23[b]	1.18		107.08*
3	20	107.08	20	87.08				87.08
	22						15	102.08
	28						10	112.08
	30				100.41	2.01		114.09*
4	10	114.09	20	94.09				94.09
	18						15	109.09
	30				100.76	2.02		111.11*
5	8						25	136.11
	15	111.11	20	91.11				116.11
	20						5	121.11
	30				101.11	2.02		123.13*
6	2						24	147.13
	12	123.13	20	103.13				127.13
	22						15	142.13
	30				111.13[c]	2.22		144.35*
	Totals		$100			$10.35	$184	

*Denotes average daily balance

$$^a\frac{(15 \times 50) + (15 \times 40)}{30} = \$45$$

$$^b\frac{(10 \times 65.90) + (20 \times 55.90)}{30} = \$59.23$$

$$^c\frac{(12 \times 123.13) + (18 \times 103.13)}{30} = \$111.13$$

Check:

$ 50.00	Beginning balance
184.00	Purchases
+ 10.35	Interest
$244.35	
− 100.00	Payments
$144.35	Ending balance

SECTION 2 PROBLEMS

For each account, complete a table (as in Example 1) that shows both the monthly interest due on the adjusted balance and the amount of each payment that applies to the balance due, until the account is paid in full.

	AMOUNT OF ACCOUNT	MONTHLY PAYMENT	INTEREST PER MONTH
1. a.	$350	$50	1½%
b.	220	60	1
2. a.	$175	$25	1¼%
b.	250	50	1

Use the average daily balance method (as in Example 2) to compute 2% monthly interest for the following charge accounts. No interest is charged on current purchases. (Assume a 30-day billing cycle.) Compute total interest and final balance.

	MONTH	PREVIOUS BALANCE	PAYMENTS	PURCHASES	TOTAL INTEREST	FINAL BALANCE
3.	1	$ 80	$10 on 12th	$30 on 16th		
	2		$15 on 15th	$20 on 18th; $10 on 25th		
	3		$15 on 5th; $30 on 20th	$10 on 12th	▬▬	▬▬
4.	1	$300	$50 on 18th	$20 on 20th		
	2		$75 on 15th	$25 on 19th; $10 on 25th		
	3		$45 on 12th	$30 on 28th	▬▬	▬▬

5. Anita Alfred's account balance at Blanchard Department Store is $500. The store charges 1½% per month interest on adjusted charge account balances. If she agrees to pay $60 monthly toward her account, how much interest will she pay?

6. Nicole Nelson charged a video camera for $300 to her account at Herman's Department Store. The store charges 1% per month interest on adjusted charge account balances. If she agrees to pay $50 monthly toward her account, how much interest will she pay?

7. Van Lamar's charge account reflects the following purchases. Interest at 1% is computed on his average daily balance (but not on current purchases). A $15 payment is required when the preceding month's ending balance is under $100, or $20 when the balance exceeds $100. Assuming 30-day months, find

a. The total payments   b. The total interest   c. The ending balance

MONTH	PREVIOUS BALANCE	DATE OF PAYMENT	PURCHASES
1	$90	10th	$20 on 15th
2		5th	$30 on 10th
3		20th	$25 on 13th; $15 on 24th; $10 on 25th

8. Sarah Castleberry's purchases and payments at Wilder Department Store are shown below. Interest at 1% is computed on her average daily balance but not on current purchases. A $10 payment is required when the preceding month's ending balance is under $100 or $20 when the balance exceeds $100. Assuming 30-day months, find

a. The total payments   b. The total interest   c. The ending balance

MONTH	PREVIOUS BALANCE	DATE OF PAYMENT	PURCHASES
1	$85	15th	$30 on 20th
2		18th	$45 on 25th
3		12th	$25 on 21st

SECTION 3
INSTALLMENT PLANS: SET PAYMENTS

A second type of installment plan—a set number of payments—is used to finance a single purchase (automobile, suite of furniture, washing machine, and so forth). In this case, the account will be paid in full in a specified number of payments; hence, finance charges for the entire time are computed and added to the cost at the time of purchase.

The typical procedure for an installment purchase is as follows:

1. The customer makes a **down payment** (sometimes including a trade-in) which is subtracted from the cash price to obtain the **outstanding balance.**

2. A **carrying charge** (or **time-payment differential**) is then added to the outstanding balance.

3. The resulting sum is divided by the number of payments (usually weekly or monthly) to obtain the amount of each installment payment.

This type of time-payment loan is also used by finance companies and by banks in their installment loan, personal loan, and/or automobile loan departments. The only variations in the plan as described above are that there is no down payment made (the amount of the loan becomes the outstanding balance) and that the interest charged corresponds to the carrying charge. Bank interest rates are usually lower, because banks will lend money only to people with acceptable credit ratings. Finance companies cater to persons whose credit ratings would not qualify them for loans at more selective institutions. Many states place essentially no restrictions on these small-loan agencies, allowing rates of 100% or more on small loans of 12 months or less. Their profits are not so high as their rates might indicate, however, for these companies have a high percentage of bad debts from uncollected loans.

As noted previously, high-powered advertising has often misled buyers into making financial commitments that were substantially greater than they realized beforehand. The Truth-in-Lending Law protects against false impressions by requiring that if a business mentions one feature of credit in its advertising (such as the amount of down payment), it must also mention all other important features (such as the number, amount, and frequency of payments that follow). If an advertisement states "only $20 down," for example, it must also state that the buyer will have to pay $10 a week for the next 2 years.

The principal accomplishment of the law is that it enables the buyer to know both the **finance charge** (the total amount of extra money paid for buying on credit) and the **annual percentage rate** to which this is equivalent. Some merchants may require a carrying charge, a service charge, interest, insurance, or other special charges levied only upon credit sales; the finance charge is the sum of all these items. The annual percentage rate, correct to the nearest quarter of a percent, must then be determined by one of the two methods approved in the law. Tables are available from the government which enable the merchant to determine the correct percentage rate without extensive mathematical calculations.

It is important to understand the distinction between the interest rates required under the law and simple interest rates such as we studied previously. Formerly, merchants often advertised a simple interest rate, and interest was computed on the original balance for the entire payment period. Consider the following, however: On a simple interest note, the borrower keeps all the money until the due date; thus, the borrower rightfully owes interest on the original amount for the entire period. On an installment purchase at simple interest, on the other hand, the buyer must immediately begin making payments. As the credit period expires, therefore, the buyer has repaid most of the obligation and then owes very little, yet the latter payments would still include just as much interest as the early payments did. This means that the buyer would be paying at a much higher interest rate than on the early payments. A simple interest rate would thus be deceiving; however, the Truth-in-Lending Law permits the buyer to know the true rate that is paid. (Examples in this section will illustrate the differences in these rates.) Since the rates required under the law are quite complicated to compute, the Federal Reserve System publishers tables that merchants can use to determine annual percentage rates. These tables contain rates corresponding to both monthly and weekly payments; however, this text will include tables only for monthly payments.

It should be pointed out that the Truth-in-Lending Law does not establish any maximum interest rates or maximum finance charges or otherwise restrict what a seller may charge for credit. Neither are there restrictions on what methods the seller may use to determine the finance charge. The law only requires that the merchant must fully inform buyers what they are paying for the privilege of credit buying. In addition to the finance charge and the annual percentage rate, other important information must also be itemized for the buyers, including penalties they must pay for not conforming to the provisions of the credit contract.

Note. The term "installment plan," as used in this section, describes high-interest payment plans that run for a limited period. (The maximum time is usually 4 years.) Chapter 20 includes the periodic payments made to repay home loans or other real estate loans, to repay long-term personal bank loans, or to repay other loans when a comparatively lower interest rate is computed only on the balance due at each payment (such as credit union loans or loans on insurance policies).

The following examples all pertain to installment plans of a set period of time, where the finance charge (carrying charge, interest, and similar items) is calculated at the time the contract is made.

Example 1   The Appliance Warehouse will sell a garden weeder for $64 cash or on the following "easy-payment" terms: a $\frac{1}{4}$ down payment, a full 10% carrying charge, and monthly payments for 6 months. (a) Find the monthly payment. (b) What is the total cost of the weeder on the time-payment plan? (c) How much more than the cash price will be paid on the installment plan?

(a)

Cash price	$64.00	
Down payment	− 16.00	($\frac{1}{4}$ of $64)
Outstanding balance	$48.00	
Carrying charge	+ 4.80	(10% of $48)
Total of payments	$52.80	

$$\frac{\$52.80}{6} = \$8.80 \qquad \text{Monthly payment}$$

A payment of $8.80 per month is required. Notice that the finance (carrying) charge was a full 10% of the outstanding balance (without multiplying by $\frac{1}{2}$ for the 6 months' time period). The merchant may use this or any convenient method for determining the finance charge; however, this 10% rate is much lower than the rate the seller is required to disclose to the credit buyer. Since these payments cover only half a year, this rate is equivalent to simple interest at 20% per year. However, the rate that the merchant must reveal to the buyer is the rate that applies when interest is computed only on the balance due at the time each payment is made. Thus, the merchant must tell the buyer that an annual percentage rate of 33.5% is being charged. (Later examples will further illustrate these rates.)

Be aware of the correct procedure: If the finance charge is given as a rate (either a fraction or a percent), the finance charge is found by multiplying this rate times the outstanding balance remaining *after* the down payment has been made.

(b)   The total cost of the weeder, when bought on the installment plan, is $68.80, as follows:

Down payment	$16.00
Total of monthly payments	+ 52.80
Total installment plan cost	$68.80

(c)   The $4.80 carrying charge represents the additional cost to the credit customer:

Total installment plan price	$68.80
Cash price	− 64.00
Time-payment differential	$ 4.80

This $4.80 finance charge, along with the 33.5% annual percentage rate, represent the two most important items the merchant must disclose to the credit buyer under the Truth-in-Lending Law, although several other things are also required.

Example 2   Gerald Wood bought a new color television that sold for $675 cash. He received a $35 trade-in allowance on his old set and also paid $40 cash. The store arranged a 2-year payment plan, including a finance charge computed at regular 8% simple interest. (a) What is Wood's monthly payment? (b) What will be the total cost of the TV? (c) How much extra will he pay for the convenience of installment buying? (d) What annual percentage rate must be disclosed under the Truth-in-Lending Law?

(a)   Wood's regular payment of $29 per month is computed as follows:

Cash price	$675	
Down payment	− 75	($35 trade-in + $40 cash)
Outstanding balance	$600	
Finance charge (I)	+ 96	$\left(\begin{array}{l} I = Prt \\ = \$600 \times \frac{8}{100} \times 2 \end{array}\right)$
Total of payments	$696	

$$\frac{\$696}{24} = \$29 \text{ per month}$$

(b)   The total cost of the TV will be $771:

Down payment:	Trade-in	$ 35
	Cash	40
Total of monthly payments		696
Total cost		$771

(c)   The $771 total cost includes an extra $96 interest (or finance or carrying) charge above the cash price.

(d)   The annual percentage rate that the dealer must disclose is found using Table C-15, Appendix C. First, the finance charge should be divided by the amount financed (the outstanding balance), and the result multiplied by 100. This gives the "finance charge per $100 of amount financed":

$$\frac{\text{Finance charge}}{\text{Amount financed}} \times 100 = \frac{\$96}{\$600} \times 100 = 0.16 \times 100 = \$16$$

Thus, the credit buyer pays $16 interest for each $100 being financed. Note that this result is also the total simple interest percent charged; that is, the "rate × time": 8% × 2 years = 16%. This relationship will always exist, so either method could be used to determine the table value.

To use Table C-15, look in the left-hand column headed "Number of Payments" and find line 24, since this problem includes 24 monthly payments. Look to the right until you find the column containing the value nearest to the $16 we just computed. The percent at the top of that column will then be the correct annual percentage rate. In this example, the value nearest to our 16 is the entry 16.08. (In case a computed value is midway between two table values, the larger table value should be used.) Thus, the percent at the top of the column indicates that the annual percentage rate is 14.75%.

For practical purposes, this means that if interest were computed at 14.75% only on the remaining balance each month and the U.S. rule were used so that the $29 applied first to the interest due and then to reduce the principal, then 24 payments of $29 would exactly repay the $600 and the $96 interest.* Thus, it is the 14.75% rate that the merchant must state is being charged.

When simple interest is calculated on the *original balance* for the entire length of a loan, this rate is often called a **nominal interest rate.** The interest computed in this way is frequently called **add-on interest.** By contrast, an annual rate that is applied only to the *balance due at the time of each payment* is called an **effective interest rate.** The annual rates determined for the Truth-in-Lending Law by Table C-15 are known as **actuarial rates,** which are almost identical to effective rates. It is this actuarial (or effective) rate that must be disclosed to the buyer by the merchant.

Example 3   (a) What actuarial (or effective) interest rate is equivalent to a nominal (or simple) interest rate of 9%, if there are six monthly installments? (b) How

---

*Strictly speaking, there is a mathematical distinction between the annual percentage rates in Table C-15 and the annual rates that one should use to apply the U.S. rule (as was done here), although either rates are acceptable under the Truth-in-Lending Law. The U.S. rule computes interest on the amount financed from the original date until the day payments are made. The rates in Table C-15, however, are applied in the opposite direction: The present value on the original day of all the payments must equal the amount financed. Although this is a distinct difference in mathematical procedure, the difference in the actual percentage rates is insignificant for our purposes. Thus, since a table of rates applicable to the U.S. rule was not available, we shall use the rates from Table C-15 in applying the U.S. rule.

much simple interest would be required on a $480 loan at a nominal 9% for 6 months? What is the monthly payment? (c) Verify that the actuarial rate would produce the same amount of interest, if six monthly payments were made.

(a) The value we look for in Table C-15 is found by multiplying "rate × time":

$$9\% \times \frac{1}{2}\,\text{year} = 4.5$$

This indicates that the interest due is 4.5% of the amount borrowed. It also means that the borrower pays $4.50 interest for each $100 borrowed.

Because this is a 6-month loan, we look on line 6 for the column containing the value nearest 4.50. The nearest value is 4.49. We thus find that a simple (nominal) interest rate of 9% for 6 months corresponds to an actuarial (effective) rate of 15.25% correct to the nearest $\frac{1}{4}$%. It is this actuarial rate* that is required under the Truth-in-Lending Law.

(b) A loan for $480 at 9% simple interest for 6 months would require interest as follows:

$$I = Prt$$

$$= \$480 \times \underbrace{9\% \times \tfrac{1}{2}}$$

$$= \$480 \times \quad 4.5\% \quad [\text{as in part (a)}]$$

$$I = \$21.60$$

or at $4.50 per $100 financed:

$$
\begin{array}{rl}
4.8 & \text{hundreds} \\
\times\$\ \underline{\quad 4.5} & \text{per \$100} \\
\$21.60 & \text{interest}
\end{array}
$$

If the $480 principal plus $21.60 interest were repaid in six monthly payments, then

$$\frac{\$501.60}{6} = \$83.60\,\text{per month}$$

---

*A very similar rate can be computed without tables using the formula $e = \dfrac{2nr}{n+1}$, where e = effective rate, r = annual simple interest rate, and n = number of payments (either monthly or weekly). Thus, for Example 3, $e = \dfrac{2(6)(9)}{6+1} = 15.43\%$.

(c) The U.S. rule is used to verify the 15.25% actuarial rate. Each month, the interest due is subtracted from the $83.60 payment. The remainder of the payment is then deducted from the outstanding principal to obtain the adjusted principal upon which interest will be calculated at the time of the next payment.

MONTH	$Prt = I$	PAYMENT TO PRINCIPAL
1	$480.00 \times \dfrac{15.25}{100} \times \dfrac{1}{12} = \$\ 6.10$	$ 77.50
2	$402.50 \times \dfrac{15.25}{100} \times \dfrac{1}{12} =\ \ 5.12$	78.48
3	$324.02 \times \dfrac{15.25}{100} \times \dfrac{1}{12} =\ \ 4.12$	79.48
4	$244.54 \times \dfrac{15.25}{100} \times \dfrac{1}{12} =\ \ 3.11$	80.49
5	$164.05 \times \dfrac{15.25}{100} \times \dfrac{1}{12} =\ \ 2.08$	81.52
6	$82.53 \times \dfrac{15.25}{100} \times \dfrac{1}{12} =\ \ \underline{1.05}$	$\underline{82.55}$
	$\$21.58\ \ \ +$	$\$480.02 = \$501.60$

Thus we see that a rate of 15.25%, applied only to the unpaid balance each month, would result in interest of $21.58 over 6 months' time. This is $0.02 less than the $21.60 simple interest that would be due at a 9% nominal rate. (The difference results from rounding and from the use of Table C-15 in applying the U.S. rule. Recall that Table C-15 is correct only to the nearest quarter of a percent.)

Calculator techniques . . .   FOR EXAMPLE 3(c)

You could use the techniques shown for Example 1 of Section 2. However, since this "rate × time" calculation produces such a long decimal, you may prefer to keep that in memory, as shown here (rather than keeping the running balance in memory). Recall that $83.60 is the monthly payment. (Parentheses appear around values you use without reentering.)

.1525 $\boxed{\div}$ 12 $\boxed{M+}$ ⟶ 0.0127083

480 $\boxed{\times}$ $\boxed{MR}$ $\boxed{=}$ ⟶ 6.099984

83.60 $\boxed{-}$ 6.10 $\boxed{=}$ ⟶ 77.5;   480 $\boxed{-}$ 77.5 $\boxed{=}$ ⟶ 402.5

(402.5) $\boxed{\times}$ $\boxed{MR}$ $\boxed{=}$ ⟶ 5.11509075

83.60 $\boxed{-}$ 5.12 $\boxed{=}$ ⟶ 78.48;   402.5 $\boxed{-}$ 78.48 $\boxed{=}$ ⟶ 324.02

(324.02) $\boxed{\times}$ $\boxed{MR}$ $\boxed{=}$ ⟶ 4.117743366

83.60 $\boxed{-}$ 4.12 $\boxed{=}$ ⟶ 79.48;   324.02 $\boxed{-}$ 79.48 $\boxed{=}$ ⟶ 244.54

and so on. You will use this technique again for several subsequent examples.

Example 4   To repay a loan of $160 from IOU Finance Co. requires six payments of $28 each. Determine (a) the amount of interest that will be paid and (b) the annual (actuarial) percentage rate charged. (c) Verify that this annual percentage rate would actually result in the amount of interest charged.

(a)   The six payments of $28 will total $168 altogether. The interest charged will thus be $8:

$ 28	Total of payments	$168
× 6	Amount of loan	− 160
$168	Interest (finance charge)	$ 8

(b)   Before finding the actuarial rate, we first find the finance charge per $100 of amount financed:

$$\frac{\text{Finance charge}}{\text{Amount financed}} \times 100 = \frac{\$8}{\$160} \times 100 = 0.05 \times 100 = \$5.00$$

Then on line 6 of the table, we find that the value nearest to 5.00 is the entry 5.02. Thus, IOU Finance Co. must inform the borrower that (s)he is paying interest at an annual percentage rate of 17%.

(c) As before, the monthly payment applies first to the interest due, and the remainder of the payment is then deducted from the outstanding principal.

Month	$Prt = I$	Payment to Principal
1	$\$160.00 \times \dfrac{17}{100} \times \dfrac{1}{12} = \$2.27$	$ 25.73
2	$134.27 \times \dfrac{17}{100} \times \dfrac{1}{12} = 1.90$	26.10
3	$108.17 \times \dfrac{17}{100} \times \dfrac{1}{12} = 1.53$	26.47
4	$81.70 \times \dfrac{17}{100} \times \dfrac{1}{12} = 1.16$	26.84
5	$54.86 \times \dfrac{17}{100} \times \dfrac{1}{12} = 0.78$	27.22
6	$27.64 \times \dfrac{17}{100} \times \dfrac{1}{12} = \underline{\;0.39\;}$	$\underline{\;27.61\;}$
	$\$8.03 \quad + $	$\$159.97 = \$168$

Thus, we find that an effective rate of 17%, applied only to the unpaid balance each month, would require $8.03 in interest on a $160.00 loan. (This calculation contains a $0.03 rounding error.) The $8.00 interest on this loan would be equivalent to simple interest at the rate of only 10%, which illustrates well how misleading advertised installment rates could be before the Truth-in-Lending Law was passed.

Note. In verifying rates as in part (c), the "rate × time" will be the same each month (in this case, $\frac{17}{100} \times \frac{1}{12}$). In doing calculations of this kind, therefore, it is easier to find the product ($\frac{17}{100} \times \frac{1}{12} = 0.01417$) and use this factor to multiply times the principal each month. (Be sure you carry the factor out to enough places to obtain correct cents. Review Section 1, "Accuracy of Computation," in Chapter 1.)

As a general rule, the shorter the period of time, the higher installment rates tend to be. Customers seem to be willing to pay these very high rates, because the actual dollar amounts for shorter periods are not very large.

Section 3   Problems

Find the regular (monthly or weekly) payment required to pay for the following installment purchases.

	CASH PRICE	DOWN PAYMENT REQUIRED	FINANCE CHARGE	TIME PERIOD	REGULAR PAYMENT REQUIRED
1. a.	$600	$50	$30	5 months	
b.	825	20%	$40	16 weeks	
c.	300	¼	8% simple interest	18 months	
2. a.	$500	$50	$20	10 months	
b.	960	25%	$30	12 weeks	
c.	645	⅓	10% simple interest	24 months	

For the following installment problems, determine the annual (actuarial) percentage rate charged.

	AMOUNT FINANCED	FINANCE CHARGE	MONTHS	ANNUAL PERCENTAGE RATE
3. a.	$ 80	$5	8	
b.	240	$20	10	
c.	—	10% simple interest	48	
d.	—	9% simple interest	4	
4. a.	$120	$20	9	
b.	300	$30	12	
c.	—	14% simple interest	36	
d.	—	12% simple interest	5	

Compute monthly interest and payment to principal to verify that the following annual percentage rates would produce the given interest.

	PRINCIPAL	INTEREST	MONTHLY PAYMENTS	MONTHS	ANNUAL RATE	MONTHLY INTEREST	PAYMENT TO PRINCIPAL
5.	$500	$30	$106	5	23.75%		
6.	800	40	140	6	17.00		

7. A camcorder is priced at $900 cash, or it can be purchased with a ¼ down payment, a finance charge of $45, and the balance in 36 monthly installments. Compute the monthly payment.

8. A man's suit sells for $336 cash or for ⅓ down, a finance charge of $40, and the balance in 12 monthly installments. What is the monthly payment?

9. A fax machine is marked $470. The time-payment plan requires a 20% down payment, 24 monthly payments, and a finance charge of $50. What is the amount of each payment?

10. A multimedia computer system can be purchased for $2,500 cash. On the installment plan, a 30% down payment, 18 monthly payments, and a finance charge of $50 are required. Compute the monthly payment.

11. A couple purchased a washer and dryer costing $800. The appliances may be purchased for a 25% down payment and 24 monthly payments. For convenience, the dealer computes the finance charge at 9% nominal (simple) interest.

    a. What is the finance charge on the washer and dryer?

    b. How much is the monthly payment on the purchase?

    c. Determine the total cost of the appliances on the installment plan.

    d. What annual percentage rate must the dealer disclose under the Truth-in-Lending Law?

12. Four radial tires can be purchased for $520 cash. The tires can be purchased on the time-payment plan with a 25% down payment and 10 monthly payments that include a finance charge computed at 12% nominal (simple) interest.

    a. What is the finance charge?

    b. How much is the monthly payment?

    c. What is the total cost of the tires on the installment plan?

    d. Compute the annual percentage rate that must be disclosed under the Truth-in-Lending Law.

13. Eighteen monthly payments of $57.50 are required to repay a $900 loan. Determine

    a. The finance charge the borrower pays

    b. The annual percentage rate charged on the loan

14. Monthly payments of $125 are required for 24 months to repay a loan of $2,600. Determine

    a. The finance charge the borrower pays

    b. The annual percentage rate the seller must reveal

15. A $600 loan is repaid in 24 monthly payments of $31 each.

    a. What is the finance charge on the loan?

    b. At what annual percentage rate is interest paid?

16. A $580 recliner chair was purchased by making payments of $35 for 18 months. Determine

    a. The finance charge on the chair

    b. The annual percentage rate charged on the installment plan

17. A sofa costs $600 cash. If purchased on the installment plan, it can be purchased for $100 down and 12 monthly payments of $46.25 each. Find

    a. The total cost of the sofa when purchased on credit

    b. The finance charge on the purchase

    c. The annual percentage rate the lender must disclose

18. A laser printer costs $1,400 cash. It also can be purchased for $200 down and 24 monthly payments of $56.

    a. How much is the total cost of the printer when purchased on credit?

    b. What is the finance charge on the purchase?

    c. At what annual percentage rate is interest paid?

19. George Sawyer can purchase a cross-training system listing for $800 by making a down payment of $100. He will make 15 monthly payments of $50.40 each.

    a. What is the total cost of the system?

    b. What is the finance charge for this time-payment plan?

    c. What annual percentage rate must be disclosed?

20. The Smiths can purchase a lawn tractor for $1,300 by making a down payment of $250 and 36 monthly payments of $35. Determine

   a. The total cost of the tractor on the installment plan.

   b. The finance charge.

   c. The annual percentage rate that must be disclosed to the Smiths.

21. For simplicity, the Supply Market computes interest at 12% simple interest. What annual percentage rate must the store reveal under the Truth-in-Lending Law on the following monthly installment loans?

   a. A 1-year loan

   b. A 24-month loan

   c. A 10-month loan

22. Jerry's PC Services computes interest at 14% simple interest on installment purchases. Determine the annual percentage rate the business must disclose under the Truth-in-Lending Law on the following monthly installment purchases.

   a. A 12-month installment purchase

   b. An 18-month installment purchase

   c. A 9-month installment purchase

23. a. What amount of interest would be due on a 4-month loan of $1,000 if a nominal rate of 9% was charged?

   b. How much will the monthly payments be?

   c. What annual actuarial rate is charged on the loan?

   d. Verify that this actuarial rate is correct.

24. Decor Furniture Co. computes 9% nominal interest on an installment purchase of $2,400 to be repaid in 6 monthly payments.

   a. What amount of interest will be due?

   b. What monthly payment is required?

   c. What annual actuarial rate is charged?

   d. Verify that this actuarial rate is correct.

25. Suppose that $500 was borrowed at the rate and time in part d of Problem 3.

   a. How much interest would be charged?

   b. How much is the monthly payment?

   c. Verify that the annual rate you found in part d of Problem 3 would actually result in the correct amount of interest.

26. Assume that $800 was borrowed at the rate and time of part d of Problem 4.

   a. How much interest would be charged?

   b. What is the monthly payment?

   c. Verify that the annual rate you found in part d of Problem 4 would actually result in the correct amount of interest.

27. Look for newspaper advertisements showing installment purchases. What terms are given? Do these merchants adhere to the requirements of the Truth-in-Lending Law?

28. Find an installment plan in a newspaper or other periodical. Verify the actuarial rate given (as in Problems 25 and 26).

SECTION 4
INSTALLMENT PLAN PREPAYMENTS

Like the borrower on a promissory note, the borrower on an installment plan may sometimes wish to pay off the remaining balance early, in order to avoid further payments and to save interest. In this case, the lender deducts from the remaining payments the interest that would otherwise have been due on the remaining principal.

Since lenders often do not keep schedules that show the remaining principal, however, the unearned interest must usually be determined by some computation. As noted earlier, the Truth-in-Lending Law places no restrictions on how the lender computes interest, so long as the borrower is clearly informed of the rate actually paid. The examples in this section illustrate the **actuarial method,** which utilizes the Truth-in-Lending tables, as well as the **rule of 78s,** a traditional method that computes a similar amount of unearned interest without tables.

Unearned interest is computed by the actuarial method using the formula

$$\frac{n \times \text{Pmt} \times \text{Value}}{100 + \text{Value}}$$

where n is the number of monthly installment payments *remaining* in the established time period, Pmt is the regular monthly payment, and Value is the value from the Truth-in-Lending table that corresponds to the annual (actuarial) percentage rate for n payments.

Example 1   Mason Jansen borrowed $600 at 13% simple interest, to be repaid in 12 monthly payments of $56.50. When making his eighth payment, Jansen also wishes to pay his remaining balance. (a) What annual (actuarial) percentage rate was charged on the loan? (b) How much interest will Jansen save, computed by the actuarial method? (c) What is the remaining balance on that day?

(a)   As presented in the preceding section, the value we look for in the table is the product of "rate × time":

13% × 1 year = 13   Corresponding to a 23.25% annual percentage rate

(b)   Since 8 payments have been made, there are 4 payments remaining. Referencing the annual percentage rate table for 4 periods at 23.25%, we find a value of 4.89. Thus, the unearned interest by the formula is as follows:

$$\frac{n \times \text{Pmt} \times \text{Value}}{100 + \text{Value}} = \frac{4 \times \$56.50 \times 4.89}{100 + 4.89}$$

$$= \frac{\$1,105.14}{104.89}$$

$$= \$10.54$$

Jansen will save $10.54 interest by the actuarial method.

(c) Since the remaining payments total 4 × $56.50 = $226, Jansen's remaining balance is as follows:

$226.00   Total of remaining payments
−   10.54   Interest saved
$215.46   Balance due

A payment of $215.46 will terminate Jansen's installment loan. (This is in addition to the regular $56.50 payment paid on that day.)

The rule of 78s for interest is similar to the sum-of-the-years'-digits method of calculating depreciation. Originally, when the concept of installment loans first began, most of the loans were for a period of just one year—12 monthly installments. Over the period of a 12-month installment loan, the borrower has the use of $\frac{12}{12}$ of the money during the first month, $\frac{11}{12}$ of the money during the second month, and so on. The lender, the bank, has more money at risk during the early part of the loan period than it does toward the end of the loan period. Thus, the bank claims it has the right to earn a greater amount of interest in the earlier months of the loan than in the later months.

A formula is derived from allocating interest over the term of a one-year loan. Using the numbers 12, 11, 10, 9, 8, 7, 6, 5, 4, 3, 2, and 1 to represent the months involved in a one-year loan, the digits are added together to obtain a total of 78. During the first month of the loan, when the borrower has the use of $\frac{12}{12}$ of the principal, the bank is entitled to received $\frac{12}{78}$ of the total interest; during the second month, the bank is entitled to receive $\frac{11}{78}$ of the principal, and so forth.

The steps to follow for the rule of 78s are illustrated in the following example.

Example 2   Refer to Mason Jansen's loan in Example 1, in which $600 was repaid in 12 monthly payments of $56.50. The loan is paid in full after the 8th installment payment, with his final payment computed here using the rule of 78s.

Step 1   Determine the total amount of *interest* on the loan.

The installment payments for all 12 months total $12 \times \$56.50 = \$678$, which includes \$78 interest:

$$
\begin{array}{rl}
\$678 & \text{Total payments} \\
-\ \underline{600} & \text{Principal} \\
\$\ 78 & \text{Total interest}
\end{array}
$$

Step 2   Determine the *numerator* for the remaining period by applying the formula from sum-of-the-years'-digits method, $\dfrac{n(n+1)}{2}$.

There are four payments remaining, so *n* is 4:

$$\frac{4(4+1)}{2} = \frac{20}{2} = 10$$

Step 3   Determine the *denominator,* which is the sum of the years' digits for the entire loan period:

$$\frac{12(12+1)}{2} = \frac{156}{2} = 78$$

Step 4   Multiply the fraction $\dfrac{n}{d}$ times the total interest (step 1). The result is the amount of *interest saved.*

$$\frac{10}{78} \times \$78 = \$10 \text{ interest saved}$$

Step 5   Determine the total of the *remaining regular payments.*

Jansen has four remaining payments:

$$4 \times \$56.50 = \$226$$

Step 6   Subtract the amount of interest saved (step 4) from the total of the remaining installment payments (step 5), to determine the total *payment necessary to pay off the loan.*

$$
\begin{array}{rl}
\$226.00 & \text{Total of remaining payments} \\
-\ \underline{10.00} & \text{Interest saved} \\
\$216.00 & \text{Balance due}
\end{array}
$$

Jansen will owe \$216 under the rule of 78s (as compared with \$215.46 using the actuarial method).

Example 3   Refer again to Jansen's $600 loan in Examples 1 and 2, with monthly payments of $56.50. Verify that a 23.25% annual percentage rate produces the correct balance remaining after the 8th payment.

MONTH	$Prt = I$	PAYMENT TO PRINCIPAL
1	$\$600.00 \times \dfrac{23.25}{100} \times \dfrac{1}{12} = \$11.63$	$ 44.87
2	$555.13 \times \dfrac{23.25}{100} \times \dfrac{1}{12} = 10.76$	45.74
3	$509.39 \times \dfrac{23.25}{100} \times \dfrac{1}{12} = 9.87$	46.63
4	$462.76 \times \dfrac{23.25}{100} \times \dfrac{1}{12} = 8.97$	47.53
5	$415.23 \times \dfrac{23.25}{100} \times \dfrac{1}{12} = 8.05$	48.45
6	$366.78 \times \dfrac{23.25}{100} \times \dfrac{1}{12} = 7.11$	49.39
7	$317.39 \times \dfrac{23.25}{100} \times \dfrac{1}{12} = 6.15$	50.35
8	$267.04 \times \dfrac{23.25}{100} \times \dfrac{1}{12} = 5.18$	51.32
	215.72	0.00 215.72
		$67.72 $600.00

By this calculation, we obtain $215.72 of the principal remaining after the 8th payment. This compares with $215.46 by the actuarial method and $216 by the rule of 78s. (Differences in the first two figures are due to rounding.)

The rule of 78s, although named on the basis of 12-month installments, also applies for installment periods of other lengths. For convenience, the sums of numbers representing other time periods are given in Table 17-1.

TABLE 17-1	SUM OF MONTHS' DIGITS
NUMBER OF MONTHS	SUM: 1 THROUGH LARGEST MONTH
6	21
9	45
10	55
12	78
15	120
18	171
24	300
36	666

Thus, if a 24-month installment plan is paid off 4 months early, $\frac{10}{300}$ of the total interest is saved. Similarly, if an 18-month plan is terminated 6 months early, $\frac{21}{171}$ of the interest is avoided.

As a reminder, for convenience in finding the sum of a series of consecutive integers for months 1 through n, the following formula may be applied (as illustrated for the sum $8 + 7 + 6 + 5 + 4 + 3 + 2 + 1 = 36$):

$$\frac{n(n + 1)}{2} = \frac{8(9)}{2} = \frac{72}{2} = 36$$

Section 4   Problems

Based on the information given, determine the interest saved and the balance due according to the actuarial method.

		Months in Plan	Months Early	Monthly Payment	Annual Rate	Total Interest	Interest Saved	Balance Due
1.	a.	10	4	$167.50	24.75%	$210.00		
	b.	9	3	61.39	24.5	52.50		
2.	a.	24	6	$110.00	18.25%	$440.00		
	b.	15	5	237.50	26.75	562.50		

3. Repeat Problem 1 using the rule of 78s.

4. Repeat Problem 2 using the rule of 78s.

5. Janice Jeffers bought furniture for $2,000 under a finance plan of 12 monthly payments at 7% simple interest.
   a. What is the total finance charge for the plan?
   b. What will be her monthly payment?
   c. What annual percentage rate must the lender reveal?
   d. As Jeffers makes her 8th payment, she also wants to pay her remaining balance. How much interest will her prepayment save by the actuarial method?
   e. What is her remaining balance?

6. George Lunt purchased a diamond ring that had a cash price of $2,800. He purchased it on an installment plan specifying 24 monthly payments at 10% simple interest.
   a. What amount of interest was computed for the 24-month plan?
   b. What was his monthly payment?
   c. What was the annual percentage rate that must be disclosed by the seller?
   d. How much interest would be saved by the actuarial method if he pays the balance off after the 10th payment?
   e. How much is the remaining balance after the 10th payment?

7. Edwin King borrowed $1,500 at 8% simple interest to be repaid in 10 monthly payments. When making his 6th payment, King decided to pay his remaining balance.
   a. How much interest will be included in the 10 payments?
   b. What is his monthly payment?

c. What is the actuarial percentage rate charged?

d. How much interest will King save by the actuarial method when he pays his remaining balance early?

e. What is his remaining balance?

8. Ruth Blake borrowed $7,500 at 11% simple interest to be repaid in 15 monthly payments. After making her 9th payment, she decided to pay her remaining balance. Determine

   a. The interest that would be included in the 15-month plan

   b. The monthly payment

   c. The actuarial percentage rate charged

   d. The interest saved under the actuarial method by paying the remaining balance early

   e. The remaining balance after the 9th payment

9. Tony Mendoza is buying a $700 CD/stereo system by making 24 monthly payments at 10% simple interest. After the 18th payment, he decides to pay off his balance.

   a. What is the total interest included in the 24 payments?

   b. How much does Mendoza pay each month?

   c. How much interest is saved under the rule of 78s?

   d. What amount is required to pay the remaining balance due?

10. Janet Hunter borrowed $3,600 by signing a 14% simple interest note requiring 9 monthly payments.

    a. What is the total interest due on the 9-month note?

    b. How much does Hunter pay each month?

    c. After making her 4th payment, Hunter pays off her balance. How much interest does she save under the rule of 78s?

    d. What amount is required to pay the remaining balance due?

11. Chris Watson purchased a $1,800 health club membership under a plan requiring 24 monthly payments at 9% simple interest.

    a. How much interest was included in the 24 payments?

    b. How much did Watson pay each month?

    c. Watson paid off the balance after making his 14th payment. How much interest did he save under the rule of 78s?

    d. What was the remaining balance at that time?

12. An oil painting costing $1,200 was purchased on an installment plan permitting 15 monthly payments at 12% simple interest.

    a. How much interest was included in the 15 payments?

    b. What was the monthly payment?

    c. After making the 6th payment, the purchaser paid off the balance. How much interest was saved under the rule of 78s?

    d. What was the remaining balance required to pay for the painting?

CHAPTER *17* GLOSSARY

Actuarial (or effective) interest rate. The annual percentage rate required by the Truth-in-Lending Law; a rate computed on the balance due at the time of each installment payment.

Actuarial method. A method of determining the interest savings when a set installment plan is paid in full early, by applying an actuarial interest rate.

Add-on interest.   Simple interest.

Annual percentage rate.   The annual rate required to be disclosed by the Truth-in-Lending Law; an actuarial or effective rate.

Carrying charge.   An extra charge paid for the privilege of buying on the installment plan; a time-payment differential or interest.

Down payment.   An initial payment made at the time of purchase and deducted from the cash price before an installment payment is computed.

Effective interest rate.   (See "Actuarial interest rate.")

Finance charge.   The total amount of extra money paid for credit buying—carrying charge plus service charge, insurance, or any other required charges; the amount that must be dislcosed under the Truth-in-Lending Law.

Installment plan.   A payment plan that requires periodic payments for a limited time, usually at a relatively high interest rate; either (1) an open-end account or (2) an account with a set number of payments.

Nominal (or simple) interest rate.   An interest rate computed on the original balance for the entire time, without regard to the decreasing balance due after each payment to an installment account.

Open-end credit.   An account without a definite repayment period and to which additional credit may be added before the first credit is entirely repaid; a revolving charge account.

Outstanding balance.   On an installment purchase, the portion of the cash price remaining after the down payment and before the carrying charge is added.

Revolving charge account.   An open-end charge account offered by many retail businesses, where new charges may be added up to a set maximum, and minimum monthly payments (depending on the balance due) are made, with interest usually at $1\frac{1}{4}\%$ to $1\frac{1}{2}\%$ monthly computed on the unpaid balance.

Rule of 78s.   A method of determining the interest savings when a set installment plan is paid in full early. The savings is that fraction of the total interest equal to the sum of the digits of the early months over the sum of the digits of the total months.

Simple interest rate.   (See "Nominal interest rate.")

Time-payment differential.   (See "Carrying charge.")

Truth-in-Lending Law.   A federal regulation requiring sellers to inform buyers of (1) the finance charge and (2) the effective annual percentage rate paid on any credit account (installment purchase, installment loan, promissory note, etc.).

U.S. rule.   A standard method used to give credit for partial payments; each payment is first used to pay the interest due, and the remainder is then deducted from the principal, to obtain an adjusted principal upon which future interest will be computed.

18

COMPOUND INTEREST

OBJECTIVES

Upon completion of Chapter 18, you will be able to:

1. Define and use correctly the terminology associated with each topic.

2. Compute compound amount without use of a table for short periods (Section 1: Examples 1–4; Problems 1–10).

3. Compute compound interest and amount using the formula $M = P(1 + i)^n$ and the compound amount table (Section 2: Examples 1–3; Problems 1–20).

4.   **a.** Find compound interest and amount at institutions paying interest compounded daily from date of deposit to date of withdrawal (Section 3: Examples 1–4; Problems 1–10).

   b. Also, compare this with interest and amount that would be earned if interest were not compounded daily (Problems 3, 4).

   c. Use the formulas $I = P \times$ Dep. tab. and $I = W \times$ W/D tab. for computing interest compounded daily (Examples 1–4; Problems 7–10).

5. Compute present value at compound interest on investments made at either simple or compound interest. The formula $P = M(1 + i)^{-n}$ and the present value table will be used (Section 4: Examples 1–3; Problems 1–14).

It was previously noted that money invested with a financial institution earns compound interest. Compound interest is more profitable than simple interest to the investor, because, at compound interest, "interest is earned on interest." That is, interest is earned, not only on the principal, but also on all interest accumulated since the original deposit. Most compound interest is calculated using prepared tables or computer programs. To be certain that you clearly understand compound interest, however, you should compute a few problems yourself.

Note. Pages 521–522 contain a summary of Chapters 18 through 20, which all contain topics involving interest at a compound rate. As you study the forthcoming chapters, you may also refer to the summary to identify the characteristics of each topic.

SECTION 1

COMPOUND INTEREST (BY COMPUTATION)

Recall that for a simple interest investment, interest is paid on the *original principal* only; at the end of the time, the maturity value is the total of the principal plus the simple interest. Now consider the following example.

Example 1   Suppose that Mr. A makes a $1,000 investment for 3 years at 7% simple interest. Then

$$P = \$1,000 \qquad I = Prt \qquad\qquad M = P + I$$

$$r = 7\% \qquad\qquad = \$1,000 \times \frac{7}{100} \times \frac{3}{1} \qquad = \$1,000 + \$210$$

$$t = 3 \text{ years} \qquad I = \$210 \qquad\qquad M = \$1,210$$

Mr. A would earn $210 interest on this investment, making the total maturity value $1,210.

Now suppose that Ms. B invests $1,000 for only 6 months, also at 7% interest. Then

$$P = \$1,000 \qquad\qquad I = Prt \qquad\qquad M = P + I$$

$$r = 7\% \qquad\qquad = \$1,000 \times \frac{7}{100} \times \frac{1}{2} \qquad = \$1,000 + \$35$$

$$t = 6 \text{ months or} \qquad I = \$35 \qquad\qquad M = \$1,035$$

$$\tfrac{1}{2} \text{ year}$$

Ms. B would have $1,035 at the end of this 6-month investment.

Assume that Ms. B then reinvests this $1,035 for another 6 months at 7%; she would earn $36.23 on her second investment, making the total amount $1,071.23. If this total were then deposited for another 6 months, the interest earned would be $37.49 and Ms. B would have $1,108.72. If this procedure were repeated each 6 months until 3 years had passed, Ms. B would have made six investments, and the computations would be as follows:

First 6 Months

$$I = Prt$$

$$= \$1,000 \times \frac{7}{100} \times \frac{1}{2}$$

$$I = \$35$$

$$M = \$1,035$$

Second 6 Months

$$I = Prt$$

$$= \$1,035 \times \frac{7}{100} \times \frac{1}{2}$$

$$I = \$36.23$$

$$M = \$1,071.23$$

Third 6 Months

$$I = Prt$$

$$= \$1,071.23 \times \frac{7}{100} \times \frac{1}{2}$$

$$I = \$37.49$$

$$M = \$1,108.72$$

Fourth 6 Months

$$I = Prt$$

$$= \$1,108.72 \times \frac{7}{100} \times \frac{1}{2}$$

$$I = \$38.81$$

$$M = \$1,147.53$$

Fifth 6 Months

$$I = Prt$$

$$= \$1,147.53 \times \frac{7}{100} \times \frac{1}{2}$$

$$I = \$40.16$$

$$M = \$1,187.69$$

Sixth 6 Months

$$I = Prt$$

$$= \$1,187.69 \times \frac{7}{100} \times \frac{1}{2}$$

$$I = \$41.57$$

$$M = \$1,229.26$$

Thus, after 3 years, Ms. B's original principal would have amounted to $1,229.26. Because Mr. A had only $1,210 after his single, 3-year investment, Ms. B made $1,229.26 − $1,210.00, or $19.26 more interest by making successive, short-term investments.

The above example illustrates the idea of **compound interest:** Each time that interest is computed, the interest is added to the previous principal; that total then becomes the principal for the next interest period. Thus, money accumulates faster at compound interest because *interest is earned on interest* as well as on the principal.

Interest is said to be "compounded" whenever interest is computed and added to the previous principal. This is done at regular intervals known as **conversion periods** (or just **periods**). Interest is commonly compounded annually (once a year), semiannually (twice a year), quarterly (four times a year), or monthly. The total value at the end of the investment (original principal plus all interest) is the **compound amount.** The *compound interest* earned is the difference between the compound amount and the original principal. The length of the investment is known as the **term.** The quoted interest rate is always the nominal (or yearly) rate.

Before compound interest can be computed, (1) the term must be expressed as its total number of periods and (2) the interest rate must be converted to its corresponding rate per period.

Example 2   Determine the number of periods for each of the following investments: (a) 7 years compounded annually, (b) 5 years compounded semiannually, (c) 6 years compounded quarterly, and (d) 3 years compounded monthly.

In general, the number of periods is found in this way:

Years × Number of periods per year = Total number of periods

Thus,

(a)   7 years compounded annually = 7 years × 1 period per year
= 7 periods

(b)   5 years compounded semiannually = 5 years × 2 periods per year
= 10 periods

(c)   6 years compounded quarterly = 6 years × 4 periods per year
= 24 periods

(d)   3 years compounded monthly = 3 years × 12 periods per year
= 36 periods

Example 3   Determine the rate per period for each investment: (a) 6% compounded annually, (b) $8\frac{1}{2}$% compounded semiannually, (c) 5% compounded quarterly, and (d) 4% compounded monthly.

Keep in mind that the stated interest rate is always the yearly rate. Thus, if the rate is 8% compounded quarterly, the rate per period is 2% (since 2% paid four times during the year is equivalent to 8% annually).

You can find rate per period as follows:

$$\frac{\text{Yearly rate}}{\text{Number of periods per year}} = \text{Rate per period}$$

Hence,

(a)   6% compounded annually $= \dfrac{6\%}{1 \text{ period per year}}$

$= 6\%$ per period (or 6% each year)

(b) $8\frac{1}{2}\%$ compounded semiannually $= \dfrac{8\frac{1}{2}\%}{2 \text{ periods per year}}$

$= 4\frac{1}{4}\%$ per period (or $4\frac{1}{4}\%$ each 6 months)

(c)   5% compounded quarterly $= \dfrac{5\%}{4 \text{ periods per year}}$

$= 1\frac{1}{4}\%$ per period (or $1\frac{1}{4}\%$ each quarter)

(d)   4% compounded monthly $= \dfrac{4\%}{12 \text{ periods per year}}$

$= \frac{1}{3}\%$ per period (or $\frac{1}{3}\%$ each month)

Now let us compute a problem at compound interest:

Example 4   Find the compound amount and the compound interest for the following investments:

(a) $1,000 for 3 years at 8% compounded annually:

3 years $\times$ 1 period per year = 3 periods

$$\dfrac{8\%}{1 \text{ period per year}} = 8\% \text{ per period}$$

First period:	Principal	$1,000.00	
	Interest	+ 80.00	(8% of $1,000)
Second period:	Principal	$1,080.00	
	Interest	+ 86.40	(8% of $1,080)
Third period:	Principal	$1,166.40	
	Interest	+ 93.31	(8% of $1,166.40)
	Compound amount	$1,259.71	

Compound amount	$1,259.71
Less: Original principal	− 1,000.00
Compound interest	$ 259.71

(b) $1,000 for 1 year at 7% compounded quarterly:

1 year $\times$ 4 periods per year = 4 periods

$$\dfrac{7\%}{4 \text{ periods per year}} = 1\frac{3}{4}\% \text{ per period}$$

First period:	Principal	$1,000.00	
	Interest	+ 17.50	($1\frac{3}{4}$% of $1,000)
Second period:	Principal	$1,017.50	
	Interest	+ 17.81	($1\frac{3}{4}$% of $1,017.50)
Third period:	Principal	$1,035.31	
	Interest	+ 18.12	($1\frac{3}{4}$% of $1,035.31)
Fourth period:	Principal	$1,053.43	
	Interest	+ 18.44	($1\frac{3}{4}$% of $1,053.43)
	Compound amount	$1,071.87	

Compound amount	$1,071.87
Less: Original principal	− 1,000.00
Compound interest	$ 71.87

Note. Compare this $1,071.87 compound amount with Example 1, in which the compound amount after 1 year at 7% compounded *semiannually* was $1,071.23.

SECTION 1 PROBLEMS

Determine the number of periods and the rate per period for each of the following.

	RATE	COMPOUNDED	YEARS	NO. OF PERIODS	RATE PER PERIOD
1. a.	7%	Annually	5	5	7%
b.	6	Monthly	3	36	1/2 %
c.	8.5	Semiannually	6	12	4.25%
d.	8	Quarterly	9	36	2%
e.	5.5	Semiannually	8	16	2.75%
2. a.	4.5%	Annually	6	6	4.5%
b.	9	Monthly	2	24	.75
c.	7.5	Semiannually	5	10	3.75
d.	6	Quarterly	8	32	1.5
e.	5	Semiannually	9	18	2.5

Find the compound amount and the compound interest for each of the following.

3. $5,000 invested for 2 years at 8% compounded
   a. Semiannually 4/2%   b. Quarterly 8/1%

4. $4,000 invested for 4 years at 7% compounded
   a. Annually 4/1.75   b. Semiannually 8/.875

5. $2,000 invested for 3 years at 9% compounded
   a. Annually 3/32%   b. Semiannually 6/1.5%

6. $3,000 invested for 1 year at 5% compounded
   a. Semiannually   b. Quarterly
7. $8,000 invested for 6 months at 6% compounded
   a. Quarterly   b. Monthly
8. $7,000 invested for 9 months at 6% compounded
   a. Quarterly   b. Monthly
9. Study carefully your answers to parts a and b of the preceding problems. What conclusion seems to be indicated?
10. Suppose you are given the principal, rate, and years of a compound interest problem. What else must you know in order to work the problem?

SECTION 2
COMPOUND AMOUNT (USING TABLES)

Example 1 in Section 1 illustrated the advantage of a compound interest investment over a simple interest investment. All financial institutions in the United States pay compound interest on savings. There are several factors that an investor should consider before opening an account, however, as various types of accounts are available, often at considerably different rates and under different conditions.

In an **open account,** the owner may deposit or withdraw funds at any time. The most common type of open account is the **statement account.** The owner of a statement account receives a statement periodically from the bank to show deposits, withdrawals, and interest credited to the account for the period. (Deposits to the bank savings account are made using deposit slips similar to those for checking accounts.) Today, financial institutions pay interest on savings accounts from the date of deposit to the date of withdrawal.

Funds deposited for a specific period of time may be used to purchase a **certificate of deposit** (commonly called a **CD**). CDs offer higher rates than open accounts, and their rates increase as funds are committed for longer periods of time. CDs are commonly purchased for 30 days, 90 days (one quarter), 6 months, and 1 year. Periods of 2 or 3 years are available but are not popular during times of low interest rates, as investors hope rates will rise before then. Minimum deposits of $1,000 or more are usually required for CDs. If funds are withdrawn early from a CD, the owner forfeits the higher interest rate and is usually paid only at the institution's open-account rate on the withdrawn amount. Some institutions offer a no-penalty CD at a slightly lower interest rate than the traditional CD, which allows the owner to withdraw funds early without being assessed a penalty. These CDs are attractive to investors since a low interest rate is not locked in on a long-term investment. Figure 18-1 illustrates a certificate of deposit.

Money-market accounts offer interest rates comparable to certificates of deposits. Unlike CDs, money can be withdrawn from the money-market ac-

Certificate of Deposit   *First National Bank*

May 02, 19X1   May 02, 19X2
Issue Date   Maturity Date

Virginia A. Jenkins   111111111
Name   Taxpayer Number

7000006570746   12 MONTHS $2,800.00
Account Number   Initial Term   Amount

TWO THOUSAND EIGHT HUNDRED DOLLARS AND 00 CENTS
Amount Written Out

4.000%   4.06%   TO THIS ACCOUNT QUARTERLY
Interest Rate   Annual Percentage Yield Payable

SEMINARY PLAZA   6471   ALEXANDRIA VA
Branch Name   RU#   City/State

The Annual Percentage Yield assumes interest remains on deposit until maturity. A withdrawal will reduce earnings.

By signing this:
- You acknowledge the receipt of a copy of the Rules and Regulations For Savings Certificates and accept the terms described therein.
- You understand this time deposit is subject to such Rules and Regulations and as amended from time to time.
- You acknowledge that the Bank's statement of early withdrawal penalities for time deposits was called to your attention. If this Savings Certificate is designated as a No Penalty certificate, one penalty-free withdrawal of all or part of your deposit may be made after funds have been on deposit seven (7) calendar days.
- You understand that this Savings Certificate will renew automatically for like successive periods unless you redeem this certificate on the maturity date or within ten (10) calendar days beginning with the maturity date. Certificates which earn a fixed rate of interest will renew at the interest rate in effect on the maturity date. For variable rate certificates which have a floor rate, the rate in effect on the maturity date will be the floor rate for the renewal term.
- If joint, this Savings Certificate shall be a (choose one):

  (   ) Joint account with survivorship (See Rules and Regulations)   (   ) Joint account with no survivorship

Virginia A. Jenkins
Signature   Signature

Signature   Signature

Estate or Trust Account Certification

I hereby certify that I am the executor/executrix or administrator/administratrix or trustee of the Estate/Trust of

_____ ("Estate"/"Trust"). I also hereby certify

that all beneficiaries of the Estate/Trust, irrespective of any possible remainder interests or powers of appointments,

are natural persons. In Witness thereof, this _____ day of _____, 19 _____.

Signature   Signature

Prepared By   Authorized By

Non-Transferable - Initial Deposit Receipt
Member FDIC

SCT-0200 PC (7/93)   COPY 1: CUSTOMER

Courtesy of First Virginia Bank, Arlington

FIGURE 18-1   Certificate of Deposit

count at any time without penalty, although normally a limit of only two or three checks per month can be written on the account without charge. The interest rate on the money-market account may change daily, weekly, or monthly, whereas the rate on the CD remains the same throughout the term.

Historically, the rates that financial institutions paid to depositors were controlled by federal regulations. The phasing out of those rates caused increased competition among financial institutions for investors' deposits, and the rates now fluctuate with the general economic climate. The mid-1980s and early 1990s experienced an extreme drop in interest rates overall, however, despite competition, and rates still remain low compared to earlier years. The standard by which interest rates are compared is known as the **prime rate**—the lowest rate that large financial institutions charge their "best" customers for loans. Investors earn somewhat less than the prime rate on deposits, and small borrowers pay considerably more than the prime rate when they take out loans or mortgages.

Problems in the preceding section of this chapter demonstrated the advantage of more frequent compounding. Thus, if two accounts offer $5\frac{1}{4}\%$ interest, but one compounds daily and the other compounds quarterly, the depositor would earn slightly more interest in the account compounded daily. (Most CDs have their interest compounded quarterly, even though the CD is for a longer period of time.)

While computing the problems in Section 1, it no doubt became obvious to you that this procedure can become quite long and tedious. Compound amount can also be found using the formula

$$M = P(1 + i)^n$$

where M = compound amount, P = original principal, i = interest rate per period, and n = number of periods.

Example 1

(a) Using the formula for compound amount, compute the compound amount and compound interest on $1,000 invested for 3 years at 8% compounded semiannually.

$$P = \$1,000 \qquad\qquad M = P(1 + i)^n$$

$$i = 4\% \text{ per period} \qquad\qquad = \$1,000(1 + 4\%)^6$$

$$n = 6 \text{ periods} \qquad\qquad = 1,000(1 + 0.04)^6$$

$$M = ? \qquad\qquad M = 1,000(1.04)^6$$

Recall that the exponent (here, "6") tells how many times the factor 1.04 should be written down before being multiplied. Thus, 1.04 should be used as a factor 6 times:

$$M = \$1,000(1.04)^6$$

$$= 1,000\underbrace{(1.04)(1.04)(1.04)(1.04)(1.04)(1.04)}$$

$$= 1,000 \quad \times \quad (1.2653190\ldots)$$

$$= \cancel{1,000} \, (1.2653190\ldots)$$

$$M = \$1,265.32$$

Therefore, the maturity value (compound amount) would be $1,265.32. The compound interest is $1,265.32 − $1,000, or $265.32.

Even using this formula, however, the calculation of compound amount would still be quite tedious if the number of periods (the exponent) were very large. The computation is greatly simplified through the use of a compound amount table—a list of the values obtained when the parenthetical expression $(1 + i)$ is used as a factor for the indicated numbers of periods.

Example 1
(cont.)

(b)   Using the compound amount formula and the table, rework part (a).

$$P = \$1,000 \qquad M = P(1 + i)^n$$

$$i = 4\% \qquad\qquad = \$1,000(1 + 4\%)^6$$

$$n = 6$$

To find compound amount, turn to Table C-18, Amount of 1 (at Compound Interest); Appendix C. Various interest rates per period are given at the top left-hand margin of each page; find the page headed by 4%. The lines of the columns correspond to the number of periods, and these are numbered on both the right-hand and left-hand sides of the page. Go down to line 6 (for 6 periods) of the Amount column on the 4% page and there read "1.2653190185"; this is the value of $(1.04)^6$. Now,

$$M = P(1 + i)^n \qquad\qquad I = M - P$$

$$= \$1,000(1 + 4\%)^6 \qquad = \quad \$1,265.32$$
$$\qquad\qquad\qquad\qquad\qquad - \quad 1,000.00$$
$$= 1,000(1.2653190185) \qquad I = \quad \$\ \ 265.32$$

$$M = \$1,265.32$$

Note.   The value in the table includes the "1" from the parenthetical expression $(1 + i)$. It is *not* correct to add "1" to the tabular value before multiplying by the principal. (Students should now review Section 1, "Accuracy of Computation," in Chapter 1, which demonstrates how many digits from the table must be used in order to ensure an answer correct to the nearest penny.)

Example 2 Find the compound amount and compound interest on $2,000 invested at 7% compounded quarterly for 6 years.

$$P = \$2,000 \qquad M = P(1 + i)^n \qquad\qquad I = M - P$$

$$i = 1\frac{3}{4}\% \qquad = \$2,000\left(1 + 1\frac{3}{4}\%\right)^{24} \qquad\begin{array}{r} = \quad \$3,032.89 \\ - \quad 2,000.00 \\ \hline I = \quad \$1,032.89 \end{array}$$

$$n = 24 \qquad = \ 2,000(1.516443)$$

$$= \ 3,032.886$$

$$M = \$3,032.89$$

After 6 years, the $2,000 investment would thus be worth $3,032.89, of which $1,032.89 is interest.

Example 3 Ray Copeland opened a savings account on April 1, 19X1, with a deposit of $800. The account paid 4% compounded quarterly. On October 1, 19X1, Ray closed that account and added enough additional money to purchase a $1,000 6-month CD earning interest at 6% compounded monthly. (a) How much more did Ray deposit on October 1? (b) What was the maturity value of his CD on April 1, 19X2? (c) How much total interest was earned?

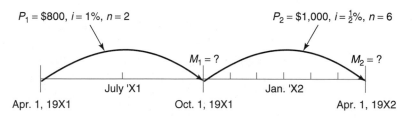

(a) Banking quarters usually begin on the first days of January, April, July, and October. This means that Ray's $800 deposit earned interest for two quarters. Hence,

$$P_1 = \$800 \qquad M_1 = P(1 + i)^n \qquad\qquad \begin{array}{r} \$1,000.00 \\ - \quad 816.08 \\ \hline \$ \ \ 183.92 \end{array} \text{ Additional}$$

$$i = 1\% \qquad = \$800(1 + 1\%)^2 \qquad\qquad\qquad\qquad\qquad \text{deposit}$$

$$n = 2 \qquad = \ 800(1.020100)$$

$$= \ 816.080$$

$$M_1 = \$816.08$$

(b) On October 1, Ray's savings account had a closing balance of $816.08. An additional $183.92 was required to purchase a $1,000 CD. There were then 6 months (two quarters) until the CD matured on April 1, 19X2, during which time the $1,000 principal earned interest at 6% compounded monthly.

$$P_2 = \$1,000 \qquad M_2 = P(1 + i)^n$$

$$i = \frac{1}{2}\% \qquad = \$1,000\left(1 + \frac{1}{2}\%\right)^6$$

$$n = 6 \qquad = 1,000(1.030378)$$

$$= 1,030.378$$

$$M_2 = \$1,030.38$$

The maturity value of Ray's CD was $1,030.38 on April 1, 19X2.

(c) The total interest may be found in either of two ways:

1. By adding the interest paid on each of the principals.
2. By finding the difference between the final balance and the total of all deposits.

(1)

$I = M - P$	$I = M - P$	$16.08 Interest₁
= $816.08	= $1,030.38	+ 30.38 Interest₂
− 800.00	− 1,000.00	$46.46 Total interest
I_1 = $ 16.08	I_2 = $ 30.38	

$$(1) \quad I = M - P \qquad I = M - P$$
$$= \$816.08 \qquad = \$1,030.38$$
$$-\ 800.00 \qquad -\ 1,000.00$$
$$I_1 = \$\ 16.08 \qquad I_2 = \$\ 30.38$$

$16.08 Interest₁
+ 30.38 Interest₂
$46.46 Total interest

(2)

$800.00 Deposit₁
+ 183.92 Deposit₂
$983.92 Total deposits

$1,030.38 Final balance
− 983.92 Total deposits
$ 46.46 Total interest

How Long Does It Take to Double an Investment?

For a sum of money invested at various interest rates compounded quarterly, the following table shows how long it takes to double (to the nearest quarter after the money doubles):

QUARTERLY RATE	YEARS TO DOUBLE	QUARTERLY RATE	YEARS TO DOUBLE	QUARTERLY RATE	YEARS TO DOUBLE
5%	14	$7\frac{1}{2}\%$	$9\frac{1}{2}$	10%	$7\frac{1}{4}$
$5\frac{1}{2}$	$12\frac{3}{4}$	8	9	$10\frac{1}{2}$	$6\frac{3}{4}$
6	$11\frac{3}{4}$	$8\frac{1}{2}$	$8\frac{1}{4}$	11	$6\frac{1}{2}$
$6\frac{1}{2}$	11	9	8	$11\frac{1}{2}$	$6\frac{1}{4}$
7	10	$9\frac{1}{2}$	$7\frac{1}{2}$	12	6

SECTION 2 PROBLEMS

1–6. Rework Problems 3–8 of Section 1 using the compound amount formula and Table C-18. (Some answers may vary by a few cents because the tables are not rounded off after each period.)

Compute the following compound amounts and compound interest, using Table C-18.

		PRINCIPAL	RATE	COMPOUNDED	YEARS
7.	a.	$ 300	7%	Monthly	2
	b.	6,400	8	Quarterly	4
	c.	1,500	5	Semiannually	10
	d.	4,800	9	Monthly	8
	e.	900	10	Quarterly	6
8.	a.	$ 500	5%	Monthly	3
	b.	1,700	6	Quarterly	5
	c.	3,200	7	Semiannually	6
	d.	4,500	8	Monthly	4
	e.	8,100	9	Quarterly	7

9. Compute compound amount and interest on the following investments at 6% compounded monthly for 2 years.

a. $800   b. $1,600   c. $3,200   d. What conclusion can be reached about the principal?

10. Compute compound amount and interest on the following deposits at 5% compounded monthly for 3 years.

a. $500   b. $1,000   c. $2,000   d. What can be concluded regarding the principal?

11. Determine compound amount and interest on $1,000 earning interest compounded quarterly for 4 years at the following rates:

a. 3%   b. 6%   c. 12%   d. Does doubling the rate exactly double the interest?

12. Calculate compound amount and interest on $1,000 drawing interest compounded semiannually for 2 years at the following rates:

a. $3\frac{1}{2}$%   b. 7%   c. 14%   d. Does doubling the rate exactly double the interest?

13. Compute compound amount and interest on $1,000 at 9% compounded semiannually for

a. 5 years   b. 10 years   c. 20 years   d. Does the interest exactly double when the time is doubled?

14. Determine compound amount and interest on $1,000 at 6% compounded semiannually for

a. 2 years   b. 4 years   c. 8 years   d. Does the interest exactly double when the time is doubled?

15. Mike Bracy opened a savings account on January 1, 19X1, with a deposit of $800. The account earned interest at 5% compounded quarterly.

a. What was the value of his account on October 1, 19X1?

b. On October 1, 19X1, Mike deposited enough money to make his account total $1,000. What amount did he deposit?

c. How much was Mike's account worth on July 1, 19X2?

d. How much total interest did Mike earn?

16. On January 1, 19X6, Patricia Humble opened a savings account with a deposit of $1,000. Her bank paid 4% interest compounded quarterly.

a. What was the value of Humble's account on July 1, 19X6?

b. On that day, Humble made a deposit sufficient to bring the value of the account to $1,200. What amount did she deposit?

c. How much was her account worth on July 1, 19X7?

d. How much total interest did her account earn?

17. On January 2, Tom Ford purchased a $1,000, 90-day CD earning interest at 6% compounded monthly.

a. What was the maturity value of the CD?

b. On April 2, Ford added enough money to buy a $1,500, 6-month CD that paid interest at 7% compounded quarterly. How much was his additional deposit?

c. What was the maturity value of the second CD?

d. What total amount of interest did the two CDs earn?

18. On March 1, Elena Ticer purchased a $1,000, 90-day CD earning interest at 5% compounded monthly.

a. What was the maturity value of the CD?

b. On June 1, Ticer added enough money to purchase a $1,400, 1-year CD that paid interest at 6% compounded quarterly. How much was her additional deposit?

c. What was the maturity value of this 1-year CD?

d. How much total interest did Ticer earn?

19. On July 3, 19X1, Sarah Hodges purchased a $1,000, 6-month CD that earned 7% compounded monthly.

a. What was the maturity value of this CD?

b. On January 3, 19X2, Sarah added enough money to purchase a $1,300, 2-year CD paying 8% compounded quarterly. How much additional money was deposited?

c. What was the maturity value of this 2-year CD?

d. How much interest was earned on the two CDs?

20. On May 1, Ed Grant purchased a $1,000, 6-month CD that earned interest at 5% compounded monthly.

a. What was the maturity value of the CD?

b. On November 1, Grant added enough funds to purchase a $1,400 2-year CD paying 7% compounded quarterly. How much additional deposit was made?

c. What was the maturity value of this 2-year CD?

d. How much interest was earned on the two CDs?

SECTION 3
INTEREST COMPOUNDED DAILY

Financial institutions also pay "daily interest" (interest compounded daily) on all open accounts where deposits or withdrawals may be made at any time.

That is, financial institutions pay interest for the exact number of days that money has been on deposit.

Interest on deposits is compounded daily, but to eliminate excessive bookkeeping, most institutions enter interest in the depositor's account only once each quarter. For this reason, a daily interest table contains factors for one quarter. In accordance with the practice followed by most savings institutions, it is a 90-day quarter; deposits made on the 31st of any month earn interest as if they were made on the 30th. Recall that quarters begin in January, April, July, and October.

INTEREST ON DEPOSITS

As pointed out in the preceding section, interest rates on deposits in savings accounts and regular CDs have remained low in recent years, despite the termination of federal regulations controlling them.

Our study will be limited to $4\frac{1}{2}\%$ interest compounded daily, which approximates the historic passbook interest rate. Interest is found by multiplying the principal (deposit) times the appropriate value from Table C-16, Interest from Day of Deposit, Appendix C. That is,

$$\text{Interest} = \text{Principal} \times \text{Deposit table}$$

which might be abbreviated

$$I = P \times \text{Dep. tab.}$$

In the daily interest deposit table, there is a column for each of the three months of the quarter; each column (month) contains entries for 30 days. To use the table, you look for the date on which a deposit was made; the factor beside that date is the number to be multiplied by the amount of deposit in order to obtain the interest (provided the money remained on deposit until the quarter ended).

Example 1   Find the interest that would be earned at $4\frac{1}{2}\%$ compounded daily, if \$1,000 were deposited in a savings and loan association on July 17.

July is the first month of the quarter; therefore, we refer to the "1st Month" column of the deposit table and to the 17th line under that heading:

$$I = P \times \text{Dep. tab.}$$

$$= \$1,000(0.0092923)$$

$$= 9.2923$$

$$I = \$9.29$$

When the quarter ends, interest of $9.29 will be added to the depositor's account, bringing the total to $1,009.29.

Example 2   Richard Chiles has an account in a financial institution where deposits earn interest at $4\frac{1}{2}\%$ compounded daily. When the quarter began on January 1, Chiles's account contained $1,000. During the quarter, he made deposits of $200 on February 7 and $300 on March 13. (a) How much interest will the account earn during the quarter? (b) What will be the balance in Chiles's account at the end of the quarter?

For any deposits made after a quarter begins, interest must be computed separately for each deposit. The total interest for the quarter is the sum of all interest for the various deposits.

(a)   Chiles's initial balance of $1,000 will earn interest for the entire quarter:

$$I = P \times \text{Dep. tab.}$$

$$= \$1,000(0.0113128)$$

$$= 11.3128$$

$$I = \$11.31$$

The $200 deposit on February 7 would earn interest from the 7th day of the second month:

$$I = P \times \text{Dep. tab.}$$

$$= \$200(0.0067724)$$

$$= 1.35448$$

$$I = \$1.35$$

Interest on the $300 deposit made March 13 (the third month) is

$$I = P \times \text{Dep. tab.}$$

$$= \$300(0.0022524)$$

$$= 0.67572$$

$$I = \$0.68$$

The transactions for the entire quarter would thus be summarized as follows:

Principal		Interest
$1,000		$11.31
200		1.35
+ 300		+ 0.68
$1,500	+	$13.34 = $1,513.34 Balance, end of quarter

The account would earn total interest of $13.34 during the quarter.

(b)   Chiles's account would have a balance of $1,513.34 at the end of the quarter.

INTEREST ON WITHDRAWALS

When withdrawals are involved, interest may be calculated in the following manner:

1.   Subtract the withdrawals from the opening principal. Compute interest on this remaining balance for the entire quarter.

2.   Determine the interest that would be earned on the withdrawn funds until the date they were withdrawn. This interest is found as follows, using Table C-17, Interest to Day of Withdrawal, Appendix C:

$$Interest = Withdrawal \times Withdrawal\ table$$

which may be abbreviated

$$I = W \times W/D\ tab.$$

(The daily interest withdrawal table is similar to the daily interest deposit table in that it is divided into the 3 months of the quarter, and you use the factor beside the appropriate date.)

3.   Total interest for the quarter is the sum of the various amounts of interest found in steps 1 and 2.

Example 3   On July 1, a savings account in a credit union contained $1,250. A withdrawal of $250 was made on September 15. (a) How much interest did the account earn for the quarter, if interest was paid at $4\frac{1}{2}\%$ compounded daily? (b) What was the ending balance in the account?

(a)   Since $250 was withdrawn from the $1,250 account, only $1,000 earned interest for the entire quarter. In Example 2 we found that the interest on $1,000 for one quarter is $11.31.

Next, we compute interest on the $250 between the beginning of the quarter and September 15, when the funds were withdrawn. Since September is the third month of the quarter, we use the 15th line in the "3rd Month" column in the withdrawal table:

$$I = W \times \text{W/D tab.}$$

$$= \$250(0.0094185)$$

$$= 2.3546$$

$$I = \$2.35$$

The $250 will earn interest of $2.35 before it is withdrawn on September 15. Total interest for the quarter is thus

Interest

$11.31
+ 2.35
$13.66   Total interest for the quarter

(b)   The balance in this account at the end of the quarter was

Opening balance $1,250.00
Withdrawal   −   250.00
                      $1,000.00
Interest   +   13.66
                      $1,013.66   Balance, end of quarter

Example 4   Compute the (a) interest and (b) balance at the end of the quarter after the following transactions, when daily interest is compounded at $4\frac{1}{2}\%$.

Balance   April 1 $1,800
Withdrawal   May 18   200
Deposit   June 14   400
Withdrawal   June 21   200

(a)   The account would earn interest on $1,800 for the entire quarter, if there had been no withdrawals. Since there were two withdrawals of $200, however, only $1,800 − $400, or $1,400 will earn interest for the whole quarter:

Withdrawals

May 18	$200	Opening balance	$1,800
June 21	200	Withdrawals	− 400
	$400	Principal earning interest for entire quarter	$1,400

Interest for the quarter is thus computed as follows:

$1,400 principal for entire quarter:

$I = P \times$ Dep. tab.

$= \$1,400(0.0113128)$

$= 15.8379$

$I = \$15.84$

$400 deposit on June 14 (third month):

$I = P \times$ Dep. tab.

$= \$400(0.0021271)$

$= 0.8508$

$I = \$0.85$

The following interest was earned prior to the two withdrawals:

$200 withdrawal on May 18 (second month):

$I = W \times$ W/D tab.

$= \$200(0.0060177)$

$= 1.2035$

$I = \$1.20$

$200 withdrawal on June 21 (third month):

$I = W \times$ W/D tab.

$= \$200(0.0101758)$

$= 2.0352$

$I = \$2.04$

Thus, total interest for the quarter is

	Interest
Interest on funds on deposit for entire quarter	$15.84
Interest on $400 deposited June 14	0.85
Interest on $200 withdrawn May 18	1.20
Interest on $200 withdrawn June 21	2.04
Total interest for quarter	$19.93

(b) The balance after the quarter ended would be

Deposits *Withdrawals*

$400	$200	Opening balance	$1,800.00
	200	Deposits	+ 400.00
	$400		$2,200.00
		Withdrawals	− 400.00
			$1,800.00
		Interest	+ 19.93
		Balance, end of quarter	$1,819.93

The savings account would contain $1,819.93 after the quarter ended.

SECTION 3 PROBLEMS

The following problems are all for interest paid at $4\frac{1}{2}$% compounded daily. Find (1) the amount of interest and (2) the balance at the end of the quarter for the following deposits.

1. a. $1,000 on April 15   b. $500 on September 5   c. $1,500 on February 21
   d. $2,000 on October 10
2. a. $1,000 on May 3   b. $800 on July 6   c. $4,000 on September 16   d. $600 on October 8

Compute the interest that would be earned for one quarter (1) at $4\frac{1}{2}$% compounded daily and (2) at $4\frac{1}{2}$% compounded quarterly on the given principals. The factor for $4\frac{1}{2}$% compounded quarterly for one quarter is 0.01125.

3. a. $10,000   b. $4,500
4. a. $7,500   b. $12,000

Find (1) the total amount of interest and (2) the balance after the quarter ended for the following accounts.

OPENING BALANCE	DEPOSITS
5. a. $1,000 on July 1	$ 300 on July 5 $ 200 on August 21 $ 400 on September 9
b. $3,700 on April 1	$ 500 on April 6 $1,000 on May 15 $ 600 on June 24
6. a. $700 on October 1	$ 500 on October 22 $ 300 on November 15 $ 400 on December 5
b. $600 on January 1	$ 200 on January 30 $ 400 on February 13 $ 300 on March 9

Compute (1) the interest for the quarter and (2) the balance when the quarter ended for the following problems.

	OPENING BALANCE	WITHDRAWALS
7. a.	$7,400 on January 1	$300 on March 8
b.	$6,800 on October 1	$400 on October 15
		$200 on November 20
8. a.	$10,000 on April 1	$1,000 on May 4
b.	$5,500 on July 1	$100 on August 10
		$400 on September 1

Transactions for an entire quarter are given below. Calculate (1) the total interest each account would earn and (2) the balance in each account after the quarter ends.

	OPENING BALANCE	DEPOSITS	WITHDRAWALS
9. a.	$3,500 on January 1	$ 400 on January 18	$ 200 on March 3
b.	$5,500 on October 1	$ 200 on November 6	$ 400 on October 8
		$ 100 on December 8	$ 300 on December 23
c.	$8,800 on April 1	$ 300 on April 8	$ 200 on April 29
		$ 100 on May 26	$1,000 on June 1
10. a.	$5,600 on October 1	$ 500 on October 30	$1,000 on November 11
b.	$4,900 on January 1	$ 800 on February 5	$ 600 on January 15
		$1,000 on March 12	$ 400 on March 20
c.	$5,100 on April 1	$ 400 on April 8	$ 200 on April 27
		$ 100 on May 9	$ 900 on June 20

PRESENT VALUE (AT COMPOUND INTEREST)

It often happens that someone wishes to know how much would have to be deposited now (at the present) in order to obtain a certain maturity value. The principal that would have to be deposited is the **present value.**

The formula for obtaining present value at compound interest is a variation of the compound amount formula. The formula is rearranged to solve for P rather than M, by dividing both sides of the equation by the parenthetical expression $(1 + i)^n$:

$$M = P(1 + i)^n$$

$$\frac{M}{(1 + i)^n} = \frac{P\cancel{(1 + i)^n}}{\cancel{(1 + i)^n}}$$

$$\frac{M}{(1 + i)^n} = P$$

or, reversing the order,

$$P = \frac{M}{(1 + i)^n}$$

Therefore, present value could be found by dividing the known maturity value M by the appropriate value from the compound amount table (Table C-18) that we have already been using. However, division using such long decimal numbers would be extremely tedious (without a calculator); hence, present value tables have been developed which allow present value to be computed by multiplication.

Just as $\frac{12}{3} = 12 \cdot \frac{1}{3}$, so

$$\frac{M}{(1 + i)^n} = M \cdot \frac{1}{(1 + i)^n}$$

The entries in Table C-21, Present Value, Appendix C, are the quotients obtained when the numbers from the compound amount table are divided into 1. These quotients (the present value entries) can then be multiplied by the appropriate maturity value to obtain the present value.

Since present value is usually computed using multiplication, the formula is commonly written in a form that indicates multiplication and uses a negative exponent. Recall that a negative exponent indicates that the factor actually belongs in the opposite part of the fraction. $\left(\text{That is, } 3x^{-2} \text{ means } \dfrac{3}{x^2}.\right)$ Thus,

$$P = \frac{M}{(1 + i)^n}$$

is usually written

$$P = M(1 + i)^{-n}$$

The following examples will illustrate that present value may be found using either the compound amount table (Table C-18) or the present value table (Table C-21). As a general rule, however, you should use the present value table when computing present value at compound interest.

Example 1   An investment made for 6 years at 9% compounded monthly is to have a maturity value of $1,000. Determine the present value using (a) the compound amount table (Table C-18) and (b) the present value table (Table C-21). Also, (c) find how much interest will be included.

(a) $M = \$1,000$ $P = \dfrac{M}{(1 + i)^n}$

$i = \dfrac{3}{4}\%$ $= \dfrac{\$1,000}{\left(1 + \dfrac{3}{4}\%\right)^{72}}$

$n = 72$

$P = ?$ $= \dfrac{\$1,000}{1.712553}$

$= 583.924$

$P = \$583.92$

(b) $P = M(1 + i)^{-n}$ (c) $I = M - P$

$= \$1,000\left(1 + \dfrac{3}{4}\%\right)^{-72}$

$\begin{aligned} &= \quad \$1,000.00 \\ &- \quad\underline{583.92} \\ I &= \quad \$ \;\; 416.08 \end{aligned}$

$= 1,000(0.583924)$

$P = \$583.92$

Thus, $583.92 invested now at 9% compounded monthly will earn $416.08 interest and mature to $1,000 after 6 years.

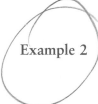

Example 2   Charles and Sue Baker would like to have $6,000 in 4 years for a down payment on a condominium. (a) What single deposit would have this maturity value if CDs earn 7% compounded quarterly? (b) How much of the final amount will be interest?

(a) $M = \$6,000$ $P = M(1 + i)^{-n}$ (b) $I = M - P$

$n = 16$ $= \$6,000\left(1 + 1\dfrac{3}{4}\%\right)^{-16}$

$\begin{aligned} &= \quad \$6,000.00 \\ &- \quad\underline{4,545.70} \\ I &= \quad \$1,454.30 \end{aligned}$

$i = 1\dfrac{3}{4}\%$ $= 6,000(0.7576163)$

$P = ?$ $= 4,545.698$

$P = \$4,545.70$

If the Bakers purchase a $4,545.70 CD now, it will earn interest of $1,454.30 during 4 years at 7% compounded quarterly and have a $6,000 maturity value.

As was true of simple interest problems, present value (or present worth) may on occasion differ from principal. The actual amount of money that is invested is always the principal; it may or may not be invested at the interest rate currently being paid by most financial institutions. The present value of the investment is the amount that would have to be invested at the rate money is worth (the rate being paid by most financial institutions) in order to obtain the same maturity value that the actual investment will have. If an investment is sold before its maturity date, it should theoretically be sold for its present value at that time.

Example 3   Arthur Levy made a $12,000 real estate investment that he expects will have a maturity value equivalent to interest at 8% compounded monthly for 5 years. If most savings institutions are currently paying 6% compounded quarterly on 5-year CDs, what is the least amount for which Levy should sell his property?

This is a two-part problem. We must first determine the maturity value of the $12,000 investment. Then, using the rate money is actually worth, we must compute the present worth of the maturity value obtained by step 1.

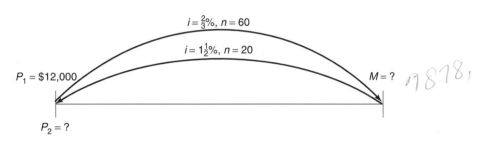

(a) $P = \$12,000$ $M = P(1 + i)^n$

$i = \dfrac{2}{3}\%$ $= \$12,000\left(1 + \dfrac{2}{3}\%\right)^{60}$

$n = 60$ $= 12,000(1.4898457)$

$M = ?$ $= 17,878.148$

$M = \$17,878.15$

(b) $M = \$17,878.15$ $P = M(1 + i)^{-n}$

$i = 1\dfrac{1}{2}\%$ $= \$17,878.15\left(1 + 1\dfrac{1}{2}\%\right)^{-20}$

$n = 20$ $= 17,878.15(0.74247042)$

$P_2 = ?$ $= 13,273.997$

$P_2 = \$13,274.00$

Levy should not sell the $12,000 property for less than $13,274, because he would have to purchase a CD of this value, paying interest at 6% compounded quarterly, in order to have $17,878.15 in 5 years.

Calculator Tip.   For most calculators, the value of i that you use from the interest tables must be rounded to a maximum of 7 decimal places (or a maximum of 8 digits altogether). For instance, in part (b) of this example you would use (0.7424704). In most problems, your result will be within a cent or two of the correct answer. Frequently you will obtain exactly the same amount, to the nearest cent, which happens here.

SECTION 4   PROBLEMS

Compute present value at compound interest, using the data given in Problems 1 and 2. Also determine the amount of interest that will be earned before maturity.

	MATURITY VALUE	RATE	COMPOUNDED	YEARS
1. a.	$1,000	4%	Quarterly	9
b.	4,100	5	Monthly	6
c.	8,000	6	Semiannually	4
d.	1,500	7	Quarterly	15
e.	3,000	6	Monthly	8
2. a.	$3,200	4%	Semiannually	2
b.	2,500	5	Quarterly	5
c.	6,800	7	Monthly	4
d.	1,800	5	Semiannually	1
e.	4,400	6	Quarterly	3

3. Bradford and Maria Quinn want to have $50,000 in savings when their son graduates from high school in 12 years. If their bank will pay 7% compounded quarterly on long-term CDs,

   a. What single deposit now will obtain their goal?

   b. How much interest will their investment earn during this time?

4. Bill and Beth Reno want to have $5,000 in savings in 1 year for a family vacation. If they can make an investment that pays 6% compounded monthly,

   a. How much should the Renos invest now in order to have this maturity value? *PV*

   b. How much interest will their investment earn?

5. The Chavez Co. plans to replace its office furniture in 3 years. The business wants to have $30,000 available for this project.

   a. What single deposit must be made now in a 3-year CD that earns 6% interest compounded monthly?

   b. How much interest will the CD earn?

6. The Capitol Plumbing Co. plans to increase the size of its warehouse in 2 years. The business will need $75,000 for the project.

    a. What single deposit must be made now in an investment earning 8% interest compounded quarterly?

    b. How much interest will the investment earn?

7. Kent Construction Co. expects to expand its operations into other cities in 3 years at a cost of $80,000. Its bank pays 8% interest compounded quarterly.

    a. What single deposit now would produce this maturity value?

    b. How much interest will be included?

8. Dominion Rentals, Inc., expects to remodel its offices in 6 years at a cost of $100,000. If the company can make an investment paying 9% compounded semiannually,

    a. What single deposit made now will produce this maturity value?

    b. How much interest will be included?

9. A $6,000 investment is made at 5% compounded semiannually for a 5-year period.

    a. What is the maturity value of the investment? (Round the maturity value to the nearest $10 for use in part b.)

    b. If money is generally worth 4% compounded quarterly, what is the present value of the investment?

10. A $7,000 investment is made at 6% compounded quarterly for a 4-year period.

    a. What is the maturity value of the investment? (Round the maturity value to the nearest $10 for use in part b.)

    b. If money is generally worth 5% compounded monthly, what is the present value of the investment?

11. Simon Chong is the payee of a 5-year, 10% simple interest note for $5,400. Money invested in a 5-year CD is worth 6% compounded quarterly.

    a. How much will be due when the note matures?

    b. If Simon had sold the note on the day it was drawn, how much would he have received?

12. Maggie Clifford is the payee of a 2-year, 11% simple interest note for $6,000. Most financial institutions are currently paying 7% compounded monthly.

    a. How much will be the maturity value of the note?

    b. How much would Maggie have received if she had sold the note on the day it was originated?

13. A 9% simple interest rate was earned on a 6-month CD for $10,000. Savings accounts are earning 5% compounded monthly.

    a. What will be the maturity value of the CD?

    b. Using the savings account rate, what is the present worth of the CD?

14. A $5,000, 1-year CD was purchased at a 9% simple interest rate.

    a. What maturity value will the CD have?

    b. If inflation is increasing at a rate equivalent to 6% compounded monthly (the rate money is worth), what was the value of the CD on the day it was purchased?

CHAPTER 18 GLOSSARY

Certificate of deposit (CD). A savings contract that offers a slightly higher rate when a (usually $1,000 minimum) deposit is made for a specified time period. (An early withdrawal would forfeit the higher rate.)

Compound amount.   The total value at the end of a compound interest investment (original principal plus all interest).

Compound interest.   Interest computed periodically and then added to the previous principal, so that this total then becomes the principal for the next interest period. "Interest earned on (previous) interest." The difference between final maturity value and original principal.

Conversion period.   A regular time interval at which interest is compounded (or computed and added to the previous principal). Interest is commonly compounded daily, monthly, quarterly, semiannually, or annually.

Money-market account.   A savings contract similar to a CD that pays higher interest rates than open accounts, permits the withdrawal of funds at any time, and allows up to three checks per month to be written.

Open account.   A savings account upon which deposits or withdrawals may be made at any time.

Period.   (See "Conversion period.")

Present value.   The amount of money that would have to be deposited (on some given date prior to maturity) in order to obtain a specified maturity value.

Prime rate.   The interest rate banks charge on loans to their "best" customers.

Statement account.   An open account for which the bank sends periodic statements showing transactions that have occurred since the last statement.

Term.   The length of time for a compound interest investment.

19

ANNUITIES

OBJECTIVES

Upon completion of Chapter 19, you will be able to:

1. Define and use correctly the terminology associated with each topic.

2. Determine amount and compound interest earned on an annuity, using the procedure $M = \text{Pmt.} \times \text{Amt. ann. tab.}_{\overline{n}|i}$ and the amount of annuity table (Section 1: Example 1; Problems 1–10).

3.   **a.** Compute present value of an annuity, using the procedure P.V. $= \text{Pmt.} \times \text{P.V. ann. tab.}_{\overline{n}|i}$ and the present value of annuity table. (That is, determine the original value required in order for one to withdraw the given annuity payments). (Section 2: Examples 1, 2; Problems 1–12).

   b. Also, determine the total amount received and the interest included.

4. Use the same procedure (Objective 3) to determine (Section 2: Example 3, Problems 13–16):

   a. The total amount paid for a real estate purchase

   b. The equivalent cash price.

Notice that the compound interest problems in the previous chapter basically involved making a *single deposit* which remained invested for the entire time. There are few people, however, who have large sums available to invest in this manner. Most people must attain their savings goals by making a series of regular deposits. This leads to the idea of annuities.

An **annuity** is a series of payments (normally equal in amount) that are made at regular intervals of time. Most people think of an annuity as the regular payment received from an insurance policy when it is cashed in after retirement. This is one good example, but there are many other everyday examples that are seldom thought of as annuities. Besides savings deposits, other common examples are rent, salaries, Social Security payments, installment plan payments, loan payments, insurance payments—in fact, any payment made at regular intervals of time.

There are several time variables that may affect an annuity. For instance, some annuities have definite beginning and ending dates; such an annuity is called an **annuity certain.** Examples of an annuity certain are installment plan payments or the payments from a life insurance policy converted to an annuity of a specified number of years.

If the beginning and/or ending dates are uncertain, the annuity is called a **contingent annuity.** Monthly Social Security retirement benefits and the payments on an ordinary life insurance policy are examples of contingent annuities for which the ending dates are unknown, because both will terminate when the person dies. If a person provides in a will that following death a beneficiary is to receive an annuity for a fixed number of years, this is a contingent annuity for which the beginning date is uncertain. A man with a large estate might provide that his surviving wife receive a specified yearly income for the remainder of her life and that the balance then be donated to some charity; this contingent annuity would then be uncertain on both the beginning and ending dates.

Another factor affecting annuities is whether the payment is made at the beginning of each time interval (such as rent and insurance premiums, which are normally paid in advance) or at the end of the period (such as salaries and Social Security retirement benefits). An annuity for which payments are made at the beginning of each period is known as an **annuity due.** When payments come at the end of each period, the annuity is called an **ordinary annuity.**

We shall be studying *investment annuities*—annuities that earn compound interest (rather than rent or installment payments, for example, which earn no interest). An annuity is said to be a **simple annuity** when the date of payment coincides with the conversion date of the compound interest.

The study of contingent annuities requires some knowledge of probability, which is not within the scope of this text. Thus, since our purpose is just to give you a basic introduction to annuities, our study of annuities will be limited to simple, ordinary annuities certain—annuities for which both the beginning and ending dates are fixed, and for which the payments are made on the conversion date at the end of each period.

SECTION 1

AMOUNT OF AN ANNUITY

The **amount of an annuity** is the maturity value that an account will have after a series of equal payments into it. As in the case of other compound interest problems, amount of an annuity is usually found by using the appropriate table for the appropriate rate per period and the number of periods. Thus, we shall compute amount of an annuity by using Table C-19, Amount of Annuity, Appendix C, and the following procedure:

$$\text{Amount} = \text{Payment} \times \text{Amount of annuity table}_{\overline{n}|i}$$

where n = number of periods and i = interest rate per period. For simplicity, the procedure* might be abbreviated as

$$M = \text{Pmt.} \times \text{Amt. ann. tab.}_{\overline{n}|i}$$

The total amount of deposits is found by multiplying the periodic payment times the number of periods. Total interest earned is then the difference between the maturity value and the total deposits made into the account.

Example 1   (a) Determine the amount (maturity value) of an annuity if Leonard Feldman deposits $100 each quarter for 6 years into an account earning 8% compounded quarterly. (b) How much of this total will Feldman deposit himself? (c) How much of the final amount is interest?

(a)   (Notice that deposits are made each quarter to coincide with the interest conversion date at the bank; we would not be able to compute the annuity if they were otherwise.)

Pmt. = $100	$M = \text{Pmt.} \times \text{Amt. ann. tab.}_{\overline{n}|i}$
$n = 24$	$= \$100 \times \text{Amt. ann. tab.}_{\overline{24}|2\%}$
$i = 2\%$	$= 100(30.42186)$
	$= 3{,}042.186$
	$M = \$3{,}042.19$

Feldman's account will contain $3,042.19 after 6 years.

---

* Many texts give the procedure for amount of an annuity in the form $M = Pm_{\overline{n}|i}$, where M = maturity value, P = payment, and $m_{\overline{n}|i}$ indicates use of the amount of annuity table. Strictly speaking, $M = Pm_{\overline{n}|i}$ is also a procedure, not a formula. The actual formula is $M = P \times \dfrac{(1+i)^n - 1}{i}$.

(b)   There will be 24 deposits of $100 each, totaling 24 × $100 or $2,400.

(c)   The interest earned during this annuity period is

Amount of annuity	$3,042.19
Total deposits	− 2,400.00
Interest	$ 642.19

SECTION 1   PROBLEMS

Using the amount-of-annuity procedure, find the maturity value that would be obtained if one makes the payments given below. Also determine the total deposits and the total interest earned. Assume that all problems are ordinary annuities, where payments are made at the end of the period, as in Table C-19.

		PERIODIC PAYMENT	RATE	COMPOUNDED	YEARS
1.	a.	$ 200	7%	Semiannually	4
	b.	400	6	Monthly	5
	c.	400	8	Quarterly	8
	d.	1,000	9	Monthly	6
	e.	1,200	5	Semiannually	10
2.	a.	$ 500	6%	Quarterly	7
	b.	800	7	Quarterly	3
	c.	300	8	Semiannually	8
	d.	1,500	5	Quarterly	6
	e.	1,100	6	Monthly	3

3.   Edwin Brice deposited $200 each quarter for 3 years into his savings account which paid 6% compounded quarterly.

a. What was the value of his account after 3 years?

b. How much did Mr. Brice invest?

c. How much interest was earned?

4.   Erin Anderson deposited $500 each quarter for 2 years into a savings account that paid 7% compounded quarterly.

a. What was the value of her account after 2 years?

b. How much had actually been deposited?

c. How much interest was earned?

5.   Maria James deposits $100 every month into an account that earns 5% compounded monthly.

a. What will be the value of her account after 6 years?

b. How much of this total will Ms. James have invested herself?

c. How much interest will her account have earned?

6. Paul Smith Mattress Co. invested $500 each month into an account that earned 6% compounded monthly.

   a. How much was the company's account worth in 5 years?

   b. How much of the total had the company deposited?

   c. How much was interest?

7. The grandparents of Lizzy Grieves deposited $500 into her savings account each year for 15 years. The account earned 5% compounded annually.

   a. How much did they deposit altogether?

   b. What was the amount in the account after the 15th deposit?

   c. How much interest was earned during this time?

8. Strickland Printing Co. invested $6,000 each year for 5 years into an account earning 8% compounded annually.

   a. How much did the company invest in 5 years?

   b. What was the maturity value of the annuity?

   c. How much interest did the annuity earn?

9. The Waters Corp. invested $5,000 each quarter for 4 years into an account paying 8% compounded quarterly.

   a. How much did Waters invest during this time?

   b. What was the amount in the account after the 4th year?

   c. How much interest did they earn?

10. Jones Telecommunications, Inc., invested $8,000 each quarter for 5 years into an account paying 6% compounded quarterly.

   a. How much did the company invest?

   b. What was the amount in the account after 5 years?

   c. How much interest did the maturity value include?

SECTION 2
PRESENT VALUE OF AN ANNUITY

Observe that our study of amount of an annuity involved starting with an empty account and making payments into it so that the account contained its largest amount at the end of the term. The study of present value of an annuity is exactly the reverse: The account contains its largest balance at the beginning of the term and someone receives payments from the account until it is empty. This balance which an account must contain at the beginning of the term is the **present value** or **present worth of an annuity.**

   Rather than studying the actual formula for present value of an annuity, we will use an informal procedure as we did for amount of an annuity. By consulting Table C-22, Present Value of Annuity, Appendix C, present value can be computed as follows:

$$\text{Present value} = \text{Payment} \times \text{Present value of annuity table}_{\overline{n}|i}$$

where, as before, n = number of payments and i = interest rate per period. The procedure* can be abbreviated

$$\text{P.V.} = \text{Pmt.} \times \text{P.V. ann. tab.}_{\overline{n}|i}$$

The total amount to be received from an annuity is found by multiplying the payment times the number of periods. As long as there are still funds in the account, it will continue to earn interest; thus, even though the balance of the account is declining during the term of the annuity, the account will still continue to earn some interest until the final payment is received. The total interest that the account will earn is found by subtracting the beginning balance (the present value) from the total payments to be received.

Example 1

Elaine Shaw wishes to receive a $100 annuity each quarter for 6 years while attending college and graduate school. Her account earns 8% compounded quarterly. (a) What must be the (present) value of Elaine's account when she starts college? (b) How much total will Elaine actually receive? (c) How much interest will these annuity payments include?

(a)   Pmt. = $100 $\text{P.V.} = \text{Pmt.} \times \text{P.V. ann. tab.}_{\overline{n}|i}$

$n = 24$ $= \$100 \times \text{P.V. ann. tab.}_{\overline{24}|2\%}$

$i = 2\%$ $= 100(18.91393)$ *Chart #18 PV ANN*

$= 1,891.393$

$\text{P.V.} = \$1,891.39$

Elaine's account must contain $1,891.39 when she enters college.

(b) She will receive 24 annuity payments of $100 each, for a total of $2,400.

(c) The account will earn interest of

Total payments	$2,400.00
Present value	− 1,891.39
Interest	$ 508.61

Thus, from an account containing $1,891.39, Elaine may withdraw $100 each quarter until a total of $2,400 has been withdrawn. During this time, the declining fund will have earned $508.61 interest. After the final payment is received, the balance in her account will be exactly $0.

---

* The procedure for present value of an annuity is often indicated as $A = Pa_{\overline{n}|i}$, where A = present value, P = payment, and $a_{\overline{n}|i}$ indicates use of the present value of annuity table.

Example 2 Grover Campbell would like to receive an annuity of $5,000 semiannually for 10 years after he retires; he will retire in 18 years. Money is worth 6% compounded semiannually. (a) How much must Campbell have when he retires in order to finance this annuity? (b) What single deposit made now would provide the funds for the annuity? (c) How much will Campbell actually receive in payments from the annuity? (d) How much interest will the single deposit earn before the annuity ends?

This is a two-part problem: (1) to find the beginning balance (present value) required for the 10-year annuity, and (2) to find what single deposit made now (present value at compound interest) would produce a maturity value equal to the answer obtained in step 1.

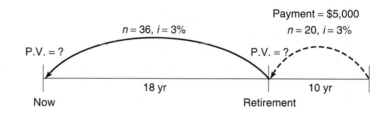

(a) $\text{Pmt.} = \$5,000$ $\text{P.V.} = \text{Pmt.} \times \text{P.V. ann. tab.}_{\overline{n}|i}$

   $n = 20$ $= \$5,000 \times \text{P.V. ann. tab.}_{\overline{20}|3\%}$

   $i = 3\%$ $= 5,000(14.877475)$ *Chart # 18*

      PV-ann

   $= 74,387.375$

   $\text{P.V.} = \$74,387.38$

The account must contain $74,387.38 at Campbell's retirement, in order for him to receive $5,000 each 6 months for 10 years.

(b) We next find what single deposit, made now, would reach a maturity value of $74,387.38 when Campbell retires in 18 years.

   $M = \$74,387.38$ $P = M(1 + i)^{-n}$

   $n = 36$ $= 74,387.38(1 + 3\%)^{-36}$

   $i = 3\%$ $= 74,387.38(0.3450324)$ *Chart #17*

      PV-SS

   $= 25,666.056$

   $P = \$25,666.06$

A single deposit of $25,666.06 now will be worth $74,387.38 after 18 years at 6% compounded semiannually. Notice that $i = 0.3450324251$ was rounded to 7 decimal places in accordance with the Calculator Tip on p. 487.

(c) Campbell will receive 20 payments of $5,000 each, for a total of $100,000 during the term of his annuity.

(d) The total interest is the difference between what he will actually receive from the annuity ($100,000) and the original amount invested ($25,666.06):

Total received	$100,000.00
Principal invested	− 25,666.06
Interest	$ 74,333.94

The original investment of $25,666.06 would produce interest of $74,333.94 before the annuity expires. (Campbell will receive nearly 4 times the original amount invested.)

Equivalent Cash Price.   Keep in mind that "present value" often refers to the value of an investment on the first day of the term. (It always refers to the value on some day prior to the maturity date.) This fact will help clarify the following example.

Example 3   A homebuyer made a down payment of $20,000 and will make payments of $7,500 semiannually for 25 years. Money is worth 7% compounded semiannually. (a) What would have been the equivalent cash price of the house? (b) How much will the buyer actually pay for the house?

(a) The "cash price" would have been the cost on the original day. This value on the beginning day of the term indicates that present value is required.

Cash price = Down payment + Present value of periodic payments

$$\text{Pmt.} = \$7,500 \qquad \text{P.V.} = \text{Pmt.} \times \text{P.V. ann. tab.}_{\overline{n}|i}$$

$$n = 50 \qquad\qquad = \$7,500 \times \text{P.V. ann. tab.}_{\overline{50}|3\frac{1}{2}\%}$$

$$i = 3\frac{1}{2}\% \qquad\qquad = 7,500(23.455618)$$

$$= 175,917.13$$

$$\text{P.V.} = \$175,917.13$$

Thus,

Down payment	$ 20,000.00
Present value of periodic payments	+ 175,917.13
Cash price	$195,917.13

(b) The buyer wll actually make a down payment of $20,000 plus 50 payments of $7,500 each, for a total cost of $395,000:

Down payment	$ 20,000
Total of periodic payments	+ 375,000
Total cost	$395,000

The equivalent cash price was $195,917.13 because that sum, invested in an account earning 7% semiannual interest, would have the same maturity value ($1,094,182.80) as if the $20,000 down payment and each semiannual payment were deposited at the same rate. Notice that the buyer's total cost ($395,000) will be about 2 times the equivalent cash price. This is not an uncommon ratio for the payment of home loans. This subject is discussed in more detail in Section 2 on "Amortization" in the next chapter.

SECTION 2 PROBLEMS

Hint. Remember that the basic problems of amount and present value at compound interest involve a *single* deposit for some period of time, whereas the annuity problems involve a *series* of payments.

Using the present-value-of-annuity procedure, determine (1) the present value required in order to receive each of the annuities given below, (2) the total amount that each annuity would pay, and (3) how much interest would be included.

		PAYMENT	RATE	COMPOUNDED	YEARS
1.	a.	$1,000	10%	Semiannually	10
	b.	1,800	5	Monthly	4
	c.	2,500	6	Quarterly	5
	d.	1,200	4	Quarterly	7
	e.	2,000	7	Semiannually	9
2.	a.	$1,000	15%	Semiannually	8
	b.	1,500	9	Monthly	3
	c.	3,000	8	Quarterly	10
	d.	1,700	12	Quarterly	6
	e.	2,600	6	Semiannually	5

(1) *Calculate the* amount *of an annuity (as in Section 1) and the interest included if one* makes deposits *as given below. Also,* (2) *determine the* present value *(as in Section 2) required in order to* receive *the given* annuity payments, *and the total interest that would be included in the annuity.*

	PAYMENT	RATE	COMPOUNDED	YEARS
3. a.	$3,000	6%	Quarterly	6
b.	4,000	5	Semiannually	8
c.	5,000	7	Monthly	5
4. a.	$6,000	8	Quarterly	8
b.	300	7	Semiannually	12
c.	4,900	6	Monthly	4

5. Janice Strauss wants to receive a quarterly annuity of $900 while she takes a 2-year nursing course. She has an investment paying 9% compounded quarterly.

   a. How much must she have in her investment account when the annuity begins?

   b. How much will the annuity pay?

   c. How much interest will she earn before the account is exhausted?

6. Bill Jefferson wishes to purchase an annuity of $450 quarterly for 3 years. If the current interest rate is 6% compounded quarterly,

   a. How much would be required to finance the annuity?

   b. How much will Jefferson actually receive from the annuity?

   c. How much total interest will the annuity include?

7. Susan Moody wants to receive an annuity of $300 a month during the first two years of her child's life. If money is compounded at 5% per month,

   a. How much must she have in a savings account to finance the annuity?

   b. How much will she receive over the life of the annuity?

   c. How much interest will the annuity payments include?

8. Sam Wallace wants to receive an annuity of $600 per quarter for a 2-year period. If his investment account pays 5% compounded quarterly,

   a. How much must Wallace have in his account to finance the annuity?

   b. How much will he actually receive from the annuity?

   c. How much interest will his investment earn?

   In problems 9–12, round your solution to each part to the nearest $100 for use in succeeding parts.

9. Jake and Greta Hogan want to receive a $6,000 annuity quarterly for 5 years after they retire. They can invest money at 6% compounded quarterly.

   a. How much money must the Hogans have when they retire in order to receive the annuity?

   b. The Hogans retire in 8 years. What single deposit could they make today in order to finance the annuity?

   c. How much interest will the account earn before the annuity ends?

10. Eric Walton wants to receive a $700 annuity semiannually for 4 years after he retires. He can invest his money at 8% compounded semiannually.

   a. What amount must be on deposit when he starts receiving the annuity?

   b. Walton will retire in 10 years. What single deposit must be made today to provide the funds for the annuity?

   c. How much interest will his deposit earn until the annuity ends?

11. Linn Yann wants to take a 2-year interior design course at the local college, starting in 3 years. During the time she is in school, she wants to receive an annuity of $200 per month. Money can be invested at 8% compounded monthly.

   a. What amount must be on deposit when she starts the courses in order to receive this annuity?

   b. How much would Linn Yann have to deposit today in order to provide for the annuity?

   c. How much interest will she earn before the annuity expires?

12. Jenny Miller plans to send her daughter to college in 6 years. Miller wants to provide for an annuity of $350 monthly during her daughter's 4 years in college. Her money can be invested at 6% compounded monthly.

   a. What amount must be on deposit when the daughter enters college in order to receive this annuity?

   b. How much would Miller have to deposit today in order to provide for the annuity?

   c. How much total interest will this deposit earn during the 10-year period?

13. The MacArthur Group, Inc., bought land for $50,000 down and monthly payments of $1,500 for 7 years.

   a. How much will MacArthur pay altogether for the land?

   b. What cash price invested at 9% compounded monthly would give the seller of the land the same total after 7 years that (s)he would have by depositing MacArthur's payments?

14. Cavett Enterprises purchased property for $25,000 down and monthly payments of $3,000 for 6 years.

   a. What was the total cost to the business?

   b. What cash price deposited at 10% compounded monthly would yield the seller of the property the same maturity value (s)he would have after 6 years of depositing Cavett's payments?

15. A family purchased a home by making a $10,000 down payment and payments of $6,000 quarterly for 20 years.

   a. What was the total cost of the home under this plan?

   b. What would have been the equivalent cash price, if money were worth 8% compounded quarterly?

16. A local area network (LAN) system was purchased by making a $5,000 down payment and quarterly payments of $2,000 for 3 years.

   a. What was the total cost of the LAN system?

   b. What would have been the equivalent cash price, if money were worth 9% compounded quarterly?

CHAPTER 19 GLOSSARY

Amount (of an annuity). The maturity value of an account after a series of equal (annuity) payments into it.

Annuity. A series of equal payments made at regular intervals of time (either with or without interest).

Annuity certain.   An annuity with definite beginning and ending dates.

Annuity due.   An annuity paid (or received) at the beginning of each time interval. (Example: Rent or insurance premiums.)

Contingent annuity.   An annuity for which the beginning and/or ending dates are uncertain.

Ordinary annuity.   An annuity paid (or received) at the end of each time period. (Example: Salaries or stock dividends.)

Present value (of an annuity).   The beginning balance an account must contain in order to receive a given annuity from it.

Simple annuity.   An annuity in which the date of payment coincides with the conversion date of the compound interest.

SINKING FUNDS AND AMORTIZATION

OBJECTIVES

Upon completion of Chapter 20, you will be able to:

1. Define and use correctly the terminology associated with each topic.

2. Using the procedure Pmt. = $M \times$ S.F. tab.$_{\overline{n}|i}$ and the sinking fund table, determine (Section 1: Example 1; Problems 1, 2, 5–10):

   a. The regular payment required to finance a sinking fund
   b. The total amount deposited
   c. The interest earned.

3. Prepare a sinking fund schedule to verify that the payments (Objective 2) will result in the required maturity value (Section 1: Example 2; Problems 13, 14).

4. Given the quoted price of a bond and its interest rate, determine its (Section 1: Examples 3, 4; Problems 3, 4, 11, 12):

   a. Purchase price   b. Premium or discount
   c. Annual interest   d. Current yield.

5. Using the procedure Pmt. = P.V. $\times$ Amtz. tab.$_{\overline{n}|i}$ and the amortization table, compute (Section 2: Examples 1, 2; Problems 1–14):

   a. The periodic payment required to amortize a loan
   b. The total amount paid
   c. The interest included.

6. Prepare an amortization schedule to verify that the payments (Objective 5) pay off the loan correctly (Section 2: Example 3; Problems 15, 16).

7. Explain the characteristics of the six basic types of compound interest and annuity problems (Section 3: Problems 1–24).

Far-sighted investors may wish to establish a savings plan whereby regular deposits will achieve a specified savings goal by a certain time. An account of this type is known as a *sinking fund*.

In the loan plans studied in previous chapters, either simple interest or simple discount was charged. For large loans that take many years to repay, however, ordinary simple interest would not be profitable to lenders, as later examples will show. Thus, long-term loans must be repaid in a series of payments that include all interest due since the previous payment. This loan repayment procedure is known as **amortization.**

Sinking funds and amortization are the subjects of the following sections; both topics involve finding a required periodic payment (or annuity).

SECTION *1*

SINKING FUNDS AND BONDS

Section 1 of Chapter 19, on the amount of an annuity, dealt with making given periodic payments into an account (at compound interest) and finding what the maturity value of that account would be. It frequently happens that businesses know what amount will be needed on some future date and are concerned with determining what periodic payment (annuity) would have to be invested in order to obtain this amount.

When a special account is established so that the maturity value (equal periodic deposits plus compound interest) will exactly equal a specific amount, such a fund is called a **sinking fund.** Sinking funds are often used to finance the replacement of machinery, equipment, facilities, and so forth, or to finance the redemption of bonds.

Bonds are somewhat similar to promissory notes in that they are written promises to repay a specified debt on a certain date. Both notes and bonds earn simple interest. The primary difference is that interest on bonds is typically paid periodically, whereas the interest on a note is all repaid on the maturity date. Large corporations, municipalities, and state governments usually finance long-term improvements by selling bonds. (A term of 10 years or more is typical.) We shall be concerned first with setting up a sinking fund to provide the *face value* (or **par value** or redemption value) of the bonds at maturity, as well as sinking funds to provide significant maturity values for other purposes.

SINKING FUND PROCEDURES

The object of a sinking fund problem, then, is to determine what periodic payment, invested at compound interest, will produce a given maturity value. Since amount-of-annuity problems also involve building a fund that would contain its largest amount at maturity, we can see that sinking fund problems are a variation of amount-of-annuity problems. The difference is that in this case we are concerned with finding the periodic payment of the annuity rather than the maturity value. (The periodic payment to a sinking fund is often called the **rent.**)

The period payment to a sinking fund can be found using the amount-of-annuity procedure $M = \text{Pmt.} \times \text{Amt. ann. tab.}_{\overline{n}|i}$ and solving for Pmt. Since use of this procedure would involve long and tedious division, however, you will normally find the periodic payment by using Table C-20, Sinking Fund, Appendix C, and the following procedure:

$$\text{Payment} = \text{Maturity value} \times \text{Sinking fund table}_{\overline{n}|i}$$

where n = number of payments and i = interest rate per period. This procedure* may be abbreviated

$$\text{Pmt.} = M \times \text{S.F. tab.}_{\overline{n}|i}$$

Example 1   Brandon County issued bonds totaling $1,000,000 in order to build an addition to the courthouse. The county commissioners set up a sinking fund at 9% compounded quarterly in order to redeem the 5-year bonds. (a) What quarterly rent must be deposited to the sinking fund? (b) How much of the maturity value will be deposits? (c) How much interest will the sinking fund earn?

(a) $M = \$1,000,000$ 　　　　　　　　　 $\text{Pmt.} = M \times \text{S.F. tab.}_{\overline{n}|i}$

　　　$n = 20$ 　$(5\text{yrs} \times Q)$ 　　　　　　 $= \$1,000,000 \times \text{S.F. tab.}_{\overline{20}|2\frac{1}{4}\%}$

　　　$i = 2\frac{1}{4}\%$ 　$(9 \div Q)$ 　　　　　　　 $= 1,000,000(0.04014207)$

　　　　　　　　　　　　　　　　　　　　 $\text{Pmt.} = \$40,142.07$

A payment of $40,142.07 must be deposited into the sinking fund quarterly in order to have $1,000,000 at maturity.

(b)   There will be 20 deposits of $40,142.07 each, making the total deposits $20 \times \$40,142.07$, or $802,841.40.

(c)   The interest earned is thus

Maturity value	$1,000,000.00
Total deposits	− 802,841.40
Interest	$ 197,158.60

*The procedure for finding the payment to a sinking fund is often given as $P = M \times \dfrac{1}{m_{\overline{n}|i}}$, where P = payment, M = maturity value, and $\dfrac{1}{m_{\overline{n}|i}}$ indicates use of the sinking fund table. The procedure $P = M \times \dfrac{1}{m_{\overline{n}|i}}$ is a variation of the amount-of-annuity procedure $M = Pm_{\overline{n}|i}$.

When a sinking fund is in progress, businesses often keep a **sinking fund schedule.** The schedule shows how much interest the fund has earned during each period and what the current balance is. The schedule also verifies that the periodic payments will result in the desired maturity value.

Example 2   Prepare a sinking fund schedule to show that semiannual payments of $949 for 2 years will amount to $4,000 when invested at 7% compounded semiannually.

The periodic "interest" earned is found by multiplying the previous "balance at end of period" by the periodic interest rate i. (Example: $949 $\times$ $3\frac{1}{2}$% = $33.22)

Payment	Periodic Interest ($i = 3\frac{1}{2}$%)	Periodic Payment	Total Increase	Balance at End of Period
1	$ 0	$ 949.00	$ 949.00	$ 949.00
2	33.22	949.00	982.22	1,931.22
3	67.59	949.00	1,016.59	2,947.81
4	103.17	949.00	1,052.17	3,999.98
Totals	$203.98	$3,796.00		

$3,999.98   Final balance

Notice that the final balance in the sinking fund will be $0.02 less than the $4,000 needed, because of rounding.

BOND PROCEDURES

As described at the beginning of this topic, bonds are similar to promissory notes. The issuers of bonds (usually governmental bodies or large corporations) must pay periodic interest on the face (par) value at a rate specified on the bonds. For *registered bonds,* the bond owners are recorded, and checks are mailed directly. For *coupon bonds,* the owner collects the interest by clipping an attached coupon and submitting it (to the company or specified trustee) on the indicated date. A registered bond is shown in Figure 20-1.

If the Brandon County bonds in Example 1 pay 7% semiannual interest, then each 6 months the county must pay the bond owners interest that collectively totals (by $I = Prt$)

$$\$1,000,000 \times \frac{7}{100} \times \frac{1}{2} = \$35,000 \quad \text{Semiannual interest}$$

This $35,000 interest (the total for all bonds) is in addition to the semiannual payment Brandon County makes to the sinking fund in order to redeem the bonds at maturity.

Most bonds are issued with a par (redemption) value of $1,000 each. Buyers are willing to pay more or less than this face value, depending on the issuer's

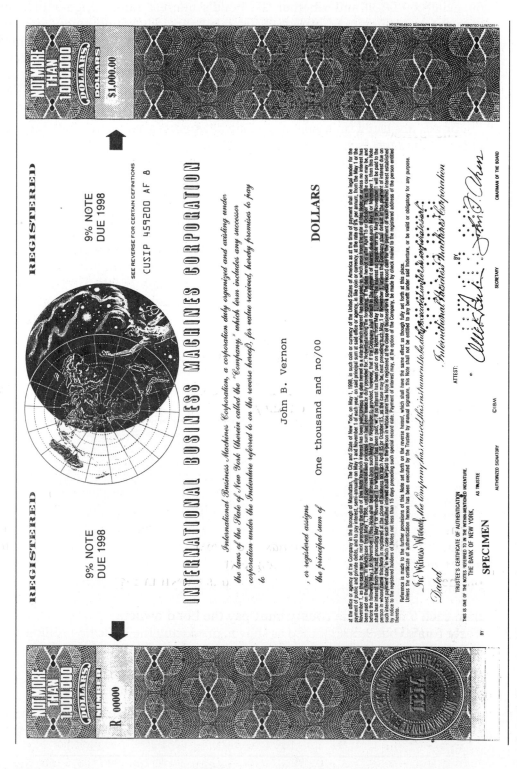

Courtesy of International Business Machines Corporation

FIGURE 20-1   Registered Bond

507

financial condition and whether the bond's interest rate is higher or lower than the prevailing market rate. Thus, if a financially sound company issues bonds at a higher-than-normal interest rate, the bonds may sell at a **premium** (more than face value). Low-interest bonds would probably sell at a **discount** (less than redemption value). If sold for face value, the bonds are sold at **par.** Generally, as interest rates increase, bond prices decrease, and as interest rates decrease, bond prices increase.

Bonds are listed on a sales exchange by company name, interest rate, and redemption date, as shown in the box below. For instance, "IBM $6\frac{3}{8}00$" indicates that IBM bonds pay $6\frac{3}{8}\%$ interest and mature in the year 2000. The optional "s" in "SouBell 6s04" is merely a plural form of reference. The initials "cv" in "PacSci $7\frac{3}{4}03$cv" indicate that these bonds are convertible to common stock under certain terms. A "current yield" of 6.5 is read as 6.5%. The "volume" indicates the number of bonds sold per $1,000 of face value. A volume of 328 thus stands for $328,000 in face value. Bonds are quoted as a percentage of face value, which also indicates whether they are selling at a premium or a discount. Thus, the "close" figure, such as $98\frac{1}{2}$ (or $985), is the bond price at the close of the previous business day. The "net change," such as $\frac{3}{8}$ (or $3.75), is the change in the price of the bond from the previous trading day. The "vj" preceding a bond name indicates that trading has been suspended by the Securities and Exchange Commission and that the company is in bankruptcy, receivership, or being reorganized under the Bankruptcy Act.

CORPORATION BONDS

BONDS	CURRENT YIELD	VOLUME	CLOSE	NET CHANGE
vjColuG $10\frac{1}{4}11$	. . .	50	$145\frac{1}{8}$	. . .
IBM $6\frac{3}{8}00$	6.5	328	$98\frac{1}{2}$	$+\frac{3}{8}$
Kroger 9s99	8.8	90	$101\frac{7}{8}$	. . .
MobilCp $8\frac{3}{8}01$	7.8	35	108	$+1$
PacSci $7\frac{3}{4}03$cv	7.2	5	108	-1
Revlon $10\frac{7}{8}10$	11.0	474	$98\frac{1}{2}$	$-1\frac{1}{2}$
SouBell 6s04	6.5	6	$92\frac{3}{4}$	$+1\frac{3}{4}$

Example 3   Three $1,000 bonds are quoted as shown, producing the indicated selling price and premium (or discount):

QUOTE	FACE VALUE	SELLING PRICE	PREMIUM/DISCOUNT
104	$1,000	104% × $1,000 = $1,040	$40 premium
97	1,000	97% × 1,000 = 970	30 discount
$108\frac{1}{2}$	1,000	$108\frac{1}{2}$% × 1,000 = 1,085	85 premium

The buyer of a bond normally also pays the previous owner for interest accrued (earned) since the last interest payment. This amount will be reimbursed to the new owner, who will receive full interest at the end of the period. A small (insignificant for our purposes) brokerage commission is also added to a bond's quoted price when the bond is purchased.

Bond purchases are made to provide income for the investor. Certain government bonds pay tax-free income, which makes these bonds desirable even at reduced interest rates. When bonds sell at a discount or a premium, the stated interest rate will differ from the **current yield** (the rate of interest based on the amount actually paid).

Example 4   Two $1,000 bonds each pay 6% simple interest annually. The quoted price of bond A was 98 and of bond B was 114. Find the current annual yield on each.

Annual interest on each bond, by $I = Prt$, is

$$\$1,000 \times \frac{6}{100} \times 1 = \$60$$

Bond A cost $980 and bond B was $1,140. Thus,

$$\text{Current yield} = \frac{\text{Annual interest}}{\text{Market or purchase price}}$$

$$\text{Bond A} = \frac{\$60}{\$980} = 6.12\% \text{ current yield}$$

$$\text{Bond B} = \frac{\$60}{\$1,140} = 5.26\% \text{ current yield}$$

Section 1   Problems

Using the sinking fund table, determine the periodic payment necessary to finance each sinking fund. Also calculate the total deposits and the total amount of interest contained in the maturity value.

	MATURITY VALUE	RATE	COMPOUNDED	YEARS
1. a.	$ 20,000	6%	Quarterly	8
b.	10,000	5	Semiannually	10
c.	40,000	8	Monthly	5
d.	130,000	9	Quarterly	15
e.	400,000	7	Monthly	7
2. a.	$ 30,000	7%	Quarterly	5
b.	50,000	8	Semiannually	6
c.	75,000	6	Monthly	7
d.	100,000	5	Quarterly	4
e.	200,000	9	Monthly	3

For each bond below, assume that the face value is $1,000. Compute the additional information related to each. Express current yield as a percent correct to 2 decimal places.

		PRICE QUOTE	PURCHASE PRICE	PREMIUM (P) OR DISCOUNT (D)	INTEREST RATE	ANNUAL INTEREST	CURRENT YIELD
3.	a.	105	*1000*		$5\frac{1}{4}\%$		
	b.	92			$5\frac{3}{4}$		
	c.	100			$7\frac{1}{4}$		
	d.	$96\frac{1}{4}$			7.7		
	e.	$101\frac{1}{4}$			$10\frac{1}{8}$		
4.	a.	101			$6\frac{1}{4}\%$		
	b.	88			$5\frac{1}{8}$		
	c.	100			$9\frac{3}{4}$		
	d.	$95\frac{1}{2}$			$8\frac{1}{2}$		
	e.	$106\frac{1}{4}$			10		

(handwritten next to row e of problem 3: $81/8 \times \frac{1000}{1}$)

5. Yucci Enterprises plans to spend $50,000 in 3 years for new factory equipment. To finance the purchase, the firm will establish a sinking fund at 8% compounded quarterly.

   a. What periodic payment must Yucci make to the sinking fund each quarter?

   b. What will be the total of these payments?

   c. How much interest will the fund contain at maturity?

6. The Osteopathic Medical Center plans to remodel its offices in 4 years. The directors expect to need $75,000.

   a. What quarterly payment is required to be invested in a sinking fund earning 9% compounded quarterly?

   b. What part of the maturity value will be deposits?

   c. How much interest will the fund draw?

7. Miller Shopping Center plans in 4 years to add a parking deck that will cost $1,000,000. Miller can set up a sinking fund at 8% compounded semiannually.

   a. What periodic payment will be made? *Chart 16 × MV*

   b. What will be the total of these payments? *#a × N*

   c. What will be the total amount of interest earned? *MV − #b (m) = I*

8. The First Edition Bookstore wants $25,000 to upgrade its computer system in 2 years. The store can set up a sinking fund at 7% compounded semiannually.

   a. What semiannual payment must be made to finance the fund?

   b. What total amount will the store invest?

   c. How much interest will the account accumulate?

9. The city of Pleasant Grove voted to issue bonds totaling $700,000 to build an addition to the community center. The bonds will mature in 5 years. If a sinking fund is established earning 6% compounded monthly,

   a. How much must be deposited each month to finance the fund?

   b. How much will the city invest?

   c. How much of the maturity value will be interest?

10. The city of Boise sold bonds totaling $1,000,000 to finance an addition to a recreation center. The bonds mature in 8 years. To provide for redemption of the bonds, a sinking fund was established at 9% compounded monthly.

   a. What will be the monthly payment to the fund?

   b. How much will the city invest?

   c. How much interest will the mature fund include?

11. Boyd Corp. sold 300 bonds of $1,000 par value at 94. The bonds pay 9.4% interest annually to the bondholder.

   a. How much did each bond cost? *(94% × 1000) = .94 × 1000 = $940*

   b. What was the premium or discount? *P + /₂ (1000 − 940) = $60 disc*

   c. What total amount did Boyd receive from the sale? *300 × 940 = 282,000*

   d. How much interest is paid on each bond each year? *9.4% × 1000 = (.094 × 1000) = $94*

   e. What is the current yield to the bondholder? *$94 ÷ $940 = .1 = 10%*

   f. How much total interest must Boyd pay to all bondholders each year? *√10% × 282,000 = 28,200*

12. Austin Enterprises, Inc., sold 500 bonds of $1,000 par value at 104. The bonds pay $8\frac{1}{4}$% interest annually to the bondholders.

   a. What was the cost of each bond? *1.04 × 1000 = 1040.*

   b. What was the premium or discount? *1000 − 1040 = +40 prem*

   c. What total amount did Austin receive from the sale? *1040 × qty 500 = 520,000*

   d. How much interest is paid on each bond each year? *1000 × .0825 = 82.50*

   e. What is the current yield for each bond? *104 ÷ 1040 = .1 = 10%*

   f. How much total interest must the corporation pay to all bondholders each year? *.10 × 520,000 = 52,000*

Find the periodic payment necessary to finance each of the following sinking funds. Then prepare a sinking fund schedule to verify that these periodic payments will result in the desired maturity value.

(52% ÷ 5)

		MATURITY VALUE	RATE	COMPOUNDED	YEARS
13.	a.	$ 5,000	5%	Semiannually	4
	b.	20,000	6	Annually	5
14.	a.	$ 8,000	8%	Quarterly	2
	b.	50,000	9	Semiannually	3

SECTION 2

AMORTIZATION

BANK LOANS

In our study of simple interest, it was pointed out that simple interest notes are usually for short periods of time—a year or less. Example 1 in Section 1 of Chapter 18 (p. 464) was given to illustrate the advantage to the depositor of compound interest over simple interest. The same example can be used to illustrate that simple interest notes for long periods are not profitable to bankers. Suppose that the investors in that example are replaced by bankers who are each lending $1,000. Banker A lends $1,000 at 7% simple interest for 3 years, for which he would receive $210 interest.

Banker B, on the other hand, lends $1,000 at 7% for only 6 months. When the borrower repays the loan after 6 months, banker B then lends both the principal and the interest for another 6 months. This process is repeated for 3 years; so, in effect, banker B earns compound interest on the $1,000 she originally lent.

That is, banker B would earn $229.26 total interest, or $19.26 more than banker A. You can easily understand that banks would prefer to invest their money in successive short-term loans rather than tying it up in single long-term loans. If carried to its logical conclusion, the result would be that no one could borrow money for long-term projects such as building a home or starting a business.

Therefore, a borrower who receives a large bank loan makes periodic, partial payments. That is, at regular intervals payments are made that include both a payment to reduce the principal and also the interest due on the principal still owed. This procedure is the same as paying on the installment plan at an annual percentage rate required by the Truth-in-Lending Law (that is, an effective interest rate—interest paid only on the balance due). At the same time, it enables the bank, in effect, to earn compound interest (by reloaning the interest just repaid).

Note. It should be emphasized that the interest rates associated with amortization are effective rates; that is, the interest rate each period is applied to the outstanding balance only. To put it another way: The borrower receives credit for the principal that has been repaid, and interest is paid only on the amount still owed.

Since the bulk of the debt exists at the beginning of the time period, and the problem is to determine what periodic payment is necessary to discharge the debt, amortization is thus a variation of the present-value-of-annuity problem already studied. The debt itself is the present (beginning) value of the annuity. The payment may be found by using the present-value-of-annuity procedure, P.V. $=$ Pmt. $\times$ P.V. ann. tab.$_{\overline{n}|i}$, and solving for Pmt.

This procedure, however, necessitates dividing by the number from the present-value-of-annuity table. To simplify the calculations, financial institutions usually use Table C-23, Amortization, Appendix C, which is an amortization table for regular payments scheduled each period. This table can be used in the following procedure:

$$\text{Payment} = \text{Present value} \times \text{Amortization table}_{\overline{n}|i}$$

where, as before, n = number of periods and i = interest rate per period. The procedure* may be abbreviated

$$\text{Pmt.} = \text{P.V.} \times \text{Amtz. tab.}_{\overline{n}|i}$$

The total amount paid to amortize a loan is found by multiplying the periodic payment by the number of payments. The total amount of interest is then the difference between the total paid and the original loan (present value).

Example 1   Marge and Gary Thomas bought a \$30,000 residential lot by making a \$3,000 down payment and making equal semiannual payments for 4 years. (a) How much was each payment if they had a 9% loan? (b) What was the total of their payments? (c) How much interest did the payments include? (d) How much did they pay altogether for the lot?

     (a)   Following the \$3,000 down payment, the principal (or present value) of the loan was \$27,000.

$$\text{P.V.} = \$27,000 \qquad \text{Pmt.} = \text{P.V.} \times \text{Amtz. tab.}_{\overline{n}|i}$$

$$n = 8 \qquad\qquad\qquad = \$27,000 \times \text{Amtz. tab.}_{\overline{8}|4\frac{1}{2}\%}$$

$$i = 4\tfrac{1}{2}\% \qquad\qquad\quad = \$27,000(0.1516097)$$

$$\text{Pmt.} = \$4,093.46$$

A payment of \$4,093.46 each 6 months would repay both the principal and the interest due on a \$27,000 loan.

     (b)   There were 8 payments of \$4,093.46 each, making a total (principal plus interest) of \$32,747.68.

     (c)   Interest would be computed each 6 months on the balance still owed (balance times $4\tfrac{1}{2}\%$). The total interest would be

Total payments	\$32,747.68
Principal	− 27,000.00
Interest	\$ 5,747.68

*This procedure may also be indicated by $P = A \times \dfrac{1}{a_{\overline{n}|i}}$, where P = payment, A = present value,

and $\dfrac{1}{a_{\overline{n}|i}}$ denotes use of the amortization table. (This table is often entitled "Annuity Whose Present

Value is 1.") The form $P = A \times \dfrac{1}{a_{\overline{n}|i}}$ is a variation of the present-value-of-annuity procedure $A = Pa_{\overline{n}|i}$.

(d)   The total cost of the lot, including the down payment, was thus

Total cost of loan	$32,747.68
Down payment	+ 3,000.00
Total cost of lot	$35,747.68

MORTGAGES

Monthly payments on home mortgages are probably the most familiar amortization payment. Since our amortization table (Table C-23) contains only 100 periods, it can only be used to determine monthly payments for loans with terms no longer than $8\frac{1}{3}$ years. However, tables similar to the one below are used by real estate agents and mortgage bankers to determine monthly mortgage payments required at various prevailing rates on home loans of 20, 25, or 30 years.

MONTHLY PAYMENT PER $1,000
OF MORTGAGE[a]

RATE	20 YEARS	25 YEARS	30 YEARS
8%	$ 8.37	$7.72	$7.34
$8\frac{1}{4}$	8.53	7.89	7.52
$8\frac{1}{2}$	8.68	8.06	7.69
$8\frac{3}{4}$	8.84	8.23	7.87
9	9.00	8.40	8.05
$9\frac{1}{4}$	9.16	8.57	8.23
$9\frac{1}{2}$	9.33	8.74	8.41
$9\frac{3}{4}$	9.49	8.92	8.60
10	9.66	9.09	8.78
11	10.33	9.81	9.53

[a] Monthly payments including principal and interest.

The following example illustrates the cost of purchasing a home. As in Example 1, these mortgage payments include interest computed only on the balance due.

Example 2   The Townsends purchased a $98,000 home by making an $8,000 down payment and signing a 10% mortgage with monthly payments for 25 years. (a) Determine the Townsends' monthly payment. (b) What was the total of their payments, and how much of this was interest? (c) What was the total cost of their home?

(a)   After their $8,000 down payment, the mortgage principal was $90,000. Their payment was thus

$ 9.09	per $1,000 at 10% for 25 years
× 90	
$818.10	monthly payment on $90,000 mortgage

(b)   Twelve payments per year for 25 years equal 300 payments required to repay the mortgage. Their total payments and interest included would be

Monthly payment	$ 818.10
	× 300
Total payments	$245,430
Principal	− 90,000
Interest	$155,430

(c)   Including their down payment, the Townsends' $98,000 home cost them

Total cost of loan	$245,430
Down payment	+ 8,000
Total cost of home	$253,430

The **fixed-rate mortgage** has been the traditional home mortgage in the United States. This mortgage has a fixed or set rate of interest, such as 8%, for a fixed time period, such as 20, 25, or 30 years. During the late 1970s and early 1980s, a period of rapid increases in interest rates, the fixed-rate mortgage lost popularity to the **adjustable-rate mortgage (ARM)** or **renegotiable-rate mortgage.** Typically, the interest rate on these mortgages was below market rates initially but then rose after a few years to a maximum rate that remained in force thereafter.

By the mid-1980s, interest rates decreased substantially throughout the economy, and fixed-rate mortgages once more became standard on new loans. As interest rates continued to fall through the early 1990s, many people who had taken out the adjustable-rate mortgages refinanced their homes at the lower interest rate, thereby reducing their monthly payments.

About the same time, the price of new homes increased, making it more difficult for buyers to qualify for financing. Some building contractors arranged financing plans which for the first two or three years were 2% to 3% below the conventional rates. When the rates and monthly payments of these mortgages subsequently increased, and when the economy moved into the recession of 1990–1991, many families were unable to meet their obligations, and the number of foreclosures rose dramatically. Attempting to halt this trend, the Federal Housing Administration (FHA), which underwrites the loans for a large portion of U.S. veterans, greatly tightened their requirements for applicants' proof of financial ability—including proof that the applicants had saved their own down payment and had not borrowed it or even accepted it as a gift.

At the time this edition went to press, mortgage interest rates were approximately 8% for an average 30-year fixed-rate mortgage.

AMORTIZATION SCHEDULES

Persons making payments to amortize a loan are often given an **amortization schedule,** which is a period-by-period breakdown showing how much of each payment goes toward the principal, how much is interest, the total amount of principal and interest that has been paid, the principal still owed, and so forth. Figure 20-2 illustrates an actual amortization schedule of a $50,000 loan at 10% interest for 10 years, which will be repaid in monthly payments of $633.38. Observe that a total of $76,005.46 will be repaid altogether—the $50,000 principal plus $26,005.46 interest.

Example 3   Prepare a simplified amortization schedule for Example 1, which found that a semiannual payment of $4,093.46 would amortize a $27,000 loan at 9% for 4 years.

The "interest" due each period is found by multiplying the "principal owed" by the periodic interest rate i. (Example: $27,000 $\times$ $4\frac{1}{2}$% = $1,215.) The next period's principal is found by subtracting the "payment to principal" from the "principal owed." (Example: $27,000 $-$ $2,878.46 = $24,121.54)

PAYMENT	PRINCIPAL OWED	INTEREST $(i = 4\frac{1}{2}\%)$	PAYMENT TO PRINCIPAL ($4,093.46 $-$ INTEREST)
1	$27,000.00	$1,215.00	$ 2,878.46
2	24,121.54	1,085.47	3,007.99
3	21,113.55	950.11	3,143.35
4	17,970.20	808.66	3,284.80
5	14,685.40	660.84	3,432.62
6	11,252.78	506.38	3,587.08
7	7,665.70	344.96	3,748.50
8	3,917.20	176.27	3,917.19
Totals		$5,747.69	$26,999.99

Total cost of loan $32,747.68

Since interest is rounded to the nearest cent each period, the totals on an amortization schedule like this are often off by a few cents. In actual practice, the final loan payment is usually slightly different from the regular payment in order to have the total amounts come out exact.

The discussion at the beginning of this section pointed out that long-term simple interest notes (with a single payment at maturity) would not be profitable to bankers. We now see that such notes would also be unreasonably expensive to borrowers. Thus, effective interest with periodic payments works to the advantage of both the borrower and the lender. If financial institutions could charge simple interest and also receive periodic payments—as is frequently done on installment purchases and loans (although the Truth-in-Lending Law does require that the equivalent effective annual percentage rate be dis-

```
                         MORTGAGE AMORTIZATION SCHEDULE

   PRIN. 50000              PERIOD RATE      .0075      AMOUNT OF
   APR.  .0900              NO. OF PYTS      120         PAYMENT      633.38
   TERM  10
   PPA   12              TOTAL AMOUNT PAID                           76005.46
   MONTH 10              TOTAL INTEREST PAID                         26005.46
   YEAR  1996
```

PAYMENT	INTEREST	PRINCIPAL	PRINCIPAL BALANCE
#	PYT	PYT	
			50000.00
1	375.00	258.38	49741.62
2	373.06	260.32	49481.30
3	371.11	262.27	49219.04
4	369.14	264.24	48954.80
5	367.16	266.22	48688.58
6	365.16	268.21	48420.37
7	363.15	270.23	48150.14
8	361.13	272.25	47877.89
9	359.08	274.29	47603.59
10	357.03	276.35	47327.24
11	354.95	278.42	47048.82
12	352.87	280.51	46768.30
22	331.10	302.28	43844.82
23	328.84	304.54	43540.28
24	326.55	306.83	43233.45
25	324.25	309.13	42924.32
37	295.25	338.13	39028.88
38	292.72	340.66	38688.22
39	290.16	343.22	38345.00
40	287.59	345.79	37999.21
41	284.99	348.38	37650.83
42	282.38	351.00	37299.83
60	231.85	401.53	30512.00
61	228.84	404.54	30107.46
62	225.81	407.57	29699.88
63	222.75	410.63	29289.26
64	219.67	413.71	28875.55
65	216.57	416.81	28458.73
66	213.44	419.94	28038.80
67	210.29	423.09	27615.71
68	207.12	426.26	27189.45
69	203.92	429.46	26759.99
70	200.70	432.68	26327.31
101	87.92	545.46	11177.13
102	83.83	549.55	10627.58
103	79.71	553.67	10073.91
104	75.55	557.82	9516.08
105	71.37	562.01	8954.07
106	67.16	566.22	8387.85
107	62.91	570.47	7817.38
108	58.63	574.75	7242.63
109	54.32	579.06	6663.57
113	36.75	596.63	4303.58
114	32.28	601.10	3702.48
115	27.77	605.61	3096.87
116	23.23	610.15	2486.72
117	18.65	614.73	1871.99
118	14.04	619.34	1252.65
119	9.39	623.98	628.66
120	4.71	628.66	0.00

FIGURE 20-2   Schedule of Direct-Reduction Loan

closed)—this would be the most profitable arrangement for the lender; however, this is not done on long-term loans from banking institutions.

Calculator techniques . . .   FOR EXAMPLE 3

This is a technique you have seen before, for paying off a revolving charge account. Loan amortization is an identical process, illustrated here for the first two payments of $4,903.46 each, made on a loan with a beginning principal of $27,000. (As previously, parentheses appear around displayed values that you will use without reentering.)

27,000 $\boxed{M+}$

(27,000) $\boxed{\times}$ 4.5 $\boxed{\%}$ $\longrightarrow$ 1215

   4,093.46 $\boxed{-}$ 1215 $\boxed{M-}$ $\longrightarrow$ 2,878.46; $\boxed{MR}$ $\longrightarrow$ 24,121.54

(24,121.54) $\boxed{\times}$ 4.5 $\boxed{\%}$ $\longrightarrow$ 1,085.47

   4,093.46 $\boxed{-}$ 1,085.47 $\boxed{M-}$ $\longrightarrow$ 3,007.99; $\boxed{MR}$ $\longrightarrow$ 21,113.55

and so forth.

The six remaining payments are calculated in a similar manner. Then, total the columns of interest and the payments to principal, to verify your calculations.

EFFECTIVE VS. SIMPLE INTEREST

The advantage to the borrower of effective interest over simple interest is obvious from the following example, which compares 12% simple interest ($I = Prt$) with effective interest at 12% compounded semiannually ($i = 6\%$) on a $25,000 loan for 10, 20, and 30 years.

Effective versus Simple Interest on a $25,000 Loan at 12%

TERM (YEARS)	TOTAL EFFECTIVE INTEREST ($i = 6\%$)	TOTAL SIMPLE INTEREST (12%)	EXTRA INTEREST AT SIMPLE RATE
10	$18,592.20	$30,000	$11,407.80
20	41,461.60	60,000	18,538.40
30	67,813.40	90,000	22,186.60

Observe on the 10-year loan that there would be over 60% more interest at the simple rate. The percentage difference decreases with longer terms, but on the 30-year loan there would still be 45% more interest at the simple rate. The simple interest alone on the 30-year loan would be 3.6 times as much as the principal that was borrowed. It is easy to see that no one could afford to buy a home or start a business if it were necessary to borrow money at simple interest.

SECTION 2   PROBLEMS

Using the amortization table, C-23, find the payment necessary to amortize each loan. Also compute the total amount that will be repaid, and determine how much interest is included.

		PRINCIPAL (P.V.)	RATE	COMPOUNDED (PAID)	TERM (YEARS)
1.	a.	$400,000	9%	Semiannually	15
	b.	180,000	8	Quarterly	20
	c.	70,000	10	Monthly	6
	d.	200,000	7	Semiannually	12
	e.	60,000	6	Monthly	5
2.	a.	$300,000	8%	Semiannually	15
	b.	150,000	9	Quarterly	10
	c.	80,000	7	Monthly	5
	d.	500,000	10	Semiannually	7
	e.	20,000	9	Monthly	8

Using the table of Monthly Payments per $1,000 of Mortgage, determine the monthly payment required for each mortgage below. Also find the total amount of those payments and the amount of interest included.

		MORTGAGE PRINCIPAL	RATE	TERM (YEARS)				MORTGAGE PRINCIPAL	RATE	TERM (YEARS)
3.	a.	$150,000	$8\frac{1}{4}$%	20		4.	a.	$200,000	8%	30
	b.	80,000	$9\frac{1}{4}$	25			b.	300,000	$8\frac{3}{4}$	20
	c.	250,000	9	30			c.	60,000	$9\frac{1}{4}$	25
	d.	400,000	10	30			d.	90,000	9	20

5. Jack Daniels borrowed $75,000 in order to go into partnership with his brother. His loan was made at 7% with monthly payments for 5 years.

   a. What is his monthly payment?

   b. How much total interest will Daniels pay?

6. Paul Martin borrowed $30,000 to build an addition to his home. He will repay the 9% loan with monthly payments for 3 years.

   a. What is Martin's monthly payment?

   b. How much total interest will be paid?

7. One Day Cleaners, Inc., borrowed $50,000 to purchase new drycleaning equipment. Quarterly payments will be made for 8 years in order to amortize the 8% loan.

   a. What is the amount of each payment?

   b. How much total interest will be charged?

8. Wooten Veterinarian Supply signed a $60,000 loan to purchase a new inventory system. The company will make quarterly payments at 8% interest for 7 years.

   a. What quarterly payment is necessary to discharge the loan?

   b. How much interest will be included in the payments?

9. Mr. and Mrs. Talkington purchased a $98,000 townhouse by paying $8,000 down and signing a 9% mortgage with quarterly payments for 20 years.

   a. How much is each quarterly payment? *Chart 19 × (cost−down)*

   b. How much interest will the Talkingtons pay? *TTL cost − price = I*

   c. What will be the total cost of the townhouse? *(N × PP + down = TTL cost*

10. The Dumas Co. purchased real property for a building site for $145,000. It paid $25,000 down and financed the balance at 8% quarterly for 5 years.

    a. What quarterly payment was necessary to repay the loan?

    b. How much interest was included in the payments?

    c. What was the actual total cost of the property?

11. Additions to the offices of Hodges Construction Co. cost $300,000. The firm paid $50,000 down and financed the balance with monthly payments for 6 years at 8%.

    a. What monthly payment was necessary to repay the loan?

    b. How much interest was included in the payments?

    c. What was the actual total cost to Hodges Construction?

12. A new automated distribution system cost Davis Delivery Co. $175,000. Davis paid $25,000 down and financed the remainder at 7% monthly for 5 years.

    a. What monthly payment is required to repay the loan?

    b. How much interest will Davis pay?

    c. What will be the total cost of the system?

13. Mr. and Mrs. Blood purchased a $320,000 home by making a $20,000 down payment. The interest rate was $9\frac{1}{4}\%$.

    a. Compare monthly payments for 25 years to those for 30 years.

    b. What would be the total payments in each case?

    c. Including down payment, how much would the house cost altogether in each case?

    d. Determine how much extra the Bloods would pay under the 30-year mortgage.

14. Jane and Tom Brown purchased a $245,000 home by making a $15,000 down payment. Their mortgage rate is $9\frac{1}{2}\%$.

    a. Compare the monthly payment required for a 20-year mortgage with that for a 25-year mortgage.

    b. How much would each mortgage cost altogether?

    c. What would be the total cost of the house in each case?

    d. How much more would the house cost with the 25-year mortgage?

Find the periodic payment required for each of the following loans. Then verify your answer by preparing an amortization schedule similar to the one in Example 3, showing how much of each payment is interest and how much applies toward principal.

		PRINICPAL (P.V.)	RATE	COMPOUNDED (PAID)	TERM (YEARS)
15.	a.	$8,000	6%	Annually	4
	b.	4,000	9	Semiannually	2
16.	a.	$5,000	7%	Annually	5
	b.	6,000	8	Semiannually	3

SECTION 3
REVIEW

The following criteria may help enable you to distinguish between the basic types of problems studied in Chapters 18 through 20.

1. Determine whether the problem is a compound interest problem or an annuity problem: If a series of *regular payments* is involved, it is an *annuity;* otherwise, it is a compound interest problem.

2. Then, decide whether amount or present value is required: In general, if the question in the problem refers to the *end* of the time period, it is an *amount* problem; if the question relates to the *beginning* of the time period, it is a *present value* problem.

Note. This review includes only the basic problems of each type so that students can confidently identify each without its being listed under a section heading. This is not intended to be a complete review of the entire unit; your personal review should also include the variations of the basic problems studied in each section. The accompanying chart identifies each basic type of problem and the corresponding procedure for each.

① compound interest OR annuity
(single sum (regular payments)

② amount (mat.)? OR present value?
(end of time) (beginning of time)

OR
③ periodic payment
3A) sinking 3B) amortization
(Build up) (discharge a debt)
(PAYOFF

Retire payments = 496

COMPOUND INTEREST	ANNUITY
The basic problem involves a *single* deposit invested for the entire time.	The problem involves a *series* of *regular* payments (or deposits).

M = compound amount
P = original principal
i = Rate per period
N = # of periods

P = present value
M = compound amt
i = rate per period
—N = # of periods

1. *Amount* at compound interest: A single deposit is made, and you wish to know how much it will be worth at the *end* of the time. Chart # 14 pg 471

$$M = P(1 + i)^n$$

1. *Amount* of an annuity: Regular deposits are *made*, and you wish to know the value of the account at the *end* of the time. Chart # 15 pg 493

$$M = \text{Pmt.} \times \text{Amt. ann. tab.}_{\overline{n}|i}$$

2. *Present value* at compound interest: You wish to find what single deposit must be invested at the *beginning* of the time period in order to obtain a given maturity value. Chart # 17 pg 485

$$P = M(1 + i)^{-n}$$

2. *Present value* of an annuity: Regular payments are to be *received*, and you want to find how much must be on deposit at the *beginning* from which to withdraw the payments. Chart # 18 pg 496

$$\text{P.V.} = \text{Pmt.} \times \text{P.V. ann. tab.}_{\overline{n}|i}$$

If the problem involves *finding a periodic payment,* one of the following applies:

3. *Sinking fund:* You are asked to find what regular payment must be made to *build up* an account to a given amount. Chart # 16 pg 505

$$\text{Pmt.} = M \times \text{S.F. tab.}_{\overline{n}|i}$$

4. *Amortization:* You are asked to determine what regular payment must be made to *discharge* a debt. Chart # 19 pg 513

$$\text{Pmt.} = \text{P.V.} \times \text{Amtz. tab.}_{\overline{n}|i}$$

SECTION 3 PROBLEMS

Use the following information to complete Problems 1, 3, 5, and 7.

	VALUE	RATE	COMPOUNDED	TERM (YEARS)
a.	$10,000	8%	Semiannually	8
b.	7,000	7	Monthly	5

Use the following information to complete Problems 2, 4, 6, and 8.

a.	$ 5,000	9	Quarterly	7
b.	20,000	10	Monthly	4

1–2. Assume that each value given above is the principal of an investment. Find the (1) compound amount and (2) compound interest.

3–4. Assume that each value given above is the maturity value of an investment. Compute the (1) present value and (2) compound interest.

5–6. Assume that each value given above is the desired maturity value of a sinking fund. (1) Find the periodic payment required to finance the fund. (2) How much of the final value will be actual deposits? (3) How much interest will be included?

7–8. Assume that each value given above is the principal (present value) of a loan. (1) Compute the periodic payment required to amortize each debt. (2) What total amount will be paid to discharge each obligation? (3) How much interest will be included?

Use the following information to complete Problems 9 and 11.

	PAYMENTS	PERIODIC RATE	COMPOUNDED	TERM (YEARS)
a.	$1,000	6%	Quarterly	20
b.	100	8	Monthly	6

Use the following information to complete Problems 10 and 12.

a.	$100,000	5%	Quarterly	5
b.	10,000	7	Monthly	3

9–10. (1) Determine the maturity value if one makes (invests) each annuity payment above. (2) How much would be deposited during the time period? (3) How much of the maturity value is interest?

11–12. (1) Find the present value required in order to receive each annuity payment above. (2) How much would be received during the term of the annuity? (3) How much interest would be included?

13. Mariana Anderson graduates from college in 2 years. She would like to have $5,000 at that time for a vacation. Money is worth 6% compounded monthly.

  a. How much must she invest now in order to achieve her savings goal?

  b. How much interest will she earn?

14. Sam Levett wants to have $15,000 in 4 years to make a down payment on a house. His savings can earn 6% compounded monthly.

  a. How much must he deposit now to achieve his goal?

  b. How much of the $15,000 will be interest?

15. Rich Dairy, Inc., is investing $6,000 each quarter to enlarge its processing plant in 5 years. The dairy's investment earns 8% compounded quarterly.

  a. What amount will the account contain in 4 years?

  b. How much total interest will the account earn?

16. BBB Trucking Co. is investing $7,000 each quarter to purchase a new refrigerated truck in 4 years. The investment earns 7% compounded quarterly.

  a. What amount will the company have after 4 years?

  b. How much of this amount will be interest?

17. William Bingham borrowed $70,000 to buy his sisters' shares of stock in the family business. He signed a 9%, 10-year loan, making semiannual payments.

  a. How much does he pay each 6 months?

  b. How much total interest will he pay on the loan?

18. Lock Manufacturing Co. borrowed $50,000 to purchase furniture for its new headquarters. The company signed a 10%, 3-year loan, making semiannual payments.

    a. What was the semiannual payment?

    b. How much total interest did the company pay on the loan?

19. Ken Atwater would like to receive $500 monthly for 8 years after he retires. He can invest in savings certificates earning 7% compounded monthly.

    a. How much must Atwater's account contain at retirement in order to provide these payments?

    b. How much will he receive before the fund is exhausted?

    c. How much interest will be earned on the account?

20. Instructor George Hamid plans to return to graduate school for 2 years, and he would like to receive $400 each month during this time. He can invest his money at 6% compounded monthly.

    a. What amount must Hamid's account contain when he returns to school?

    b. How much will he receive while he is in school?

    c. What part of these payments will be interest?

21. The Tyler family wants to travel around the United States beginning in 3 years. To make this trip, they anticipate a need for $15,000. Money is currently worth 8% compounded monthly.

    a. How much must the Tylers save each month to reach their goal?

    b. How much of the $15,000 will be their deposits?

    c. How much interest will the account contain?

22. Beltway Equipment Co. sold bonds worth $800,000 at maturity in order to expand its operations. Beltway will make deposits each 6 months for 10 years to provide the funds to redeem the bonds. It can invest money at 9% compounded semiannually.

    a. What semiannual payment will be required to finance the fund?

    b. How much of the $800,000 will be actual deposits?

    c. How much interest will the deposits earn?

23. Beth Coleman invested $5,000 in a savings account which earns 6% compounded quarterly.

    a. How much will the account be worth in 2 years?

    b. How much interest will she earn?

24. Maria Rodman purchased a $1,000 CD that pays 5% compounded quarterly when held for 3 years.

    a. How much will the CD be worth at maturity?

    b. How much interest will the certificate earn?

CHAPTER 20 GLOSSARY

Adjustable-rate mortgage (ARM). A mortgage in which the interest rate may change annually but the amount and time are fixed for the entire loan period.

Amortization. The process of repaying a loan (principal plus interest) by equal periodic payments.

Amortization schedule. A listing of the principal and interest included in each periodic loan payment, as well as a statement of the balance still owed.

Bond. A written promise to repay a specified debt on a certain date, with periodic interest to be paid during the term of the bond. Bonds are typically sold by large corporations and by city and state governments when they borrow money.

Current yield.   A rate of interest based on the amount actually paid for a bond (rather than its stated rate based on the par value).

Discount.   The amount by which the purchase price of a bond is less than its par value.

Fixed-rate mortgage.   A mortgage in which the amount, time, and interest rate are fixed at the beginning of the loan and do not change.

Par value.   Face value or redemption value of a bond. The value upon which interest is computed.

Premium.   The excess amount above par value that is paid for a bond.

Renegotiable-rate mortgage.   (See "Adjustable-rate mortgage.")

Rent.   The periodic payment to a sinking fund.

Sinking fund.   An annuity account established so that the maturity value (equal periodic deposits plus interest) will exactly equal a specific amount.

Sinking fund schedule.   A periodic listing of the growth (in principal and interest) of a sinking fund.

APPENDIX A

ARITHMETIC

OBJECTIVES

Upon completing Appendix A, you will be able to:

1. Define and use correctly the terminology associated with each topic.

2.
   a. Identify the place value associated with each digit in a whole number or a decimal number (Section 1: Problems 1, 2; Section 4: Problem 1).

   b. Pronounce and write out the complete number (Section 1: Problems 3, 4; Section 4: Problem 2).

3. Perform accurately the arithmetic operations (addition, subtraction, multiplication, and division) using whole numbers, common fractions, and decimal fractions (Section 2: Examples 1, 2; Problems 1–16; Section 3: Examples 1–4; Problems 6–10; Section 4: Examples 3–9; Problems 5–9).

4. Manipulate common fractions accurately (Section 3):

   a. Reduce (Problems 1, 5–10)

   b. Change to higher terms (least common denominator) (Problems 2, 3, 6, 7)

   c. Convert mixed numbers to improper fractions and vice versa (Problems 4, 5)

   d. Convert common fractions to their equivalent decimal value (Section 4: Example 1; Problem 3).

5. Manipulate decimal fractions accurately (Section 4)

   a. Convert to common fractions (including decimals with fractional remainders) (Example 2; Problem 4).

   b. Multiply and divide by powers of 10 by moving the decimal point (Examples 5, 6, 8, 9; Problem 5).

Although calculators or computers are used in businesses to perform most arithmetic computations, it is still advantageous for everyone to have a certain facility with computation. The following topics are presented to give you an opportunity to develop or practice the necessary skills.

SECTION 1
READING NUMBERS

The Hindu-Arabic number system that we use is known as the "base 10" system. This means that it is organized by 10 and powers of 10. Any number can be written using combinations of 10 basic characters called **digits:** 0, 1, 2, 3, 4, 5, 6, 7, 8, and 9. When combined in a number, each digit represents a particular value according to its rank in the order of digits and to the position it holds in the number (the latter is known as its **place value**). The following chart shows what place value (and power of 10) is represented by each place in a number (the number 1,376,049,528 is illustrated). Larger place values are on the left and smaller values are toward the right.

Note.   Each power of 10 is indicated (or abbreviated) by an **exponent,** which is a small number written after the 10 in a raised position. The exponent indicates the number of tens that would have to be multiplied together in order to obtain the unabbreviated place value. Thus, in the expression 10^4, the exponent "4" indicates that the complete number is found by multiplying together four 10s: $10^4 = 10 \times 10 \times 10 \times 10 = 10,000$. It should also be observed that the exponent indicates the number of zeros that the complete number contains. That is, 10^5 is equal to a "1" followed by 5 zeros:

$$10^5 = 10 \times 10 \times 10 \times 10 \times 10 = 100,000$$

BILLIONS			MILLIONS			THOUSANDS			HUNDREDS		
Hundred billions: 10^{11}	Ten billions: 10^{10}	Billions: 10^9	Hundred millions: 10^8	Ten millions: 10^7	Millions: 10^6	Hundred thousands: 10^5	Ten thousands: 10^4	Thousands: 10^3	Hundreds: 10^2	Tens: 10^1	Ones: 10^0
		1	3	7	6	0	4	9	5	2	8

The "2" in the number in the chart, because of the place it occupies, represents "two 10s" or 20. Similarly, the "4" represents "four 10,000s" or 40,000. The last three digits in the chart, "528" are read "five hundred twenty-eight."

For convenience in reading, large numbers are usually separated with commas into groups of three digits, starting at the right. Each three-digit group is then read as the appropriate number of hundreds, tens, and units followed by the family name of the groupings. Thus, the digits "376" above are read "three hundred seventy-six million." The entire number, written with commas, would appear 1,376,049,528 and would be read "one billion, three hundred seventy-six million, forty-nine thousand, five hundred twenty-eight." Commas in a number have no mathematical significance and are used merely to separate the family groupings and indicate the point at which you should pronounce the family name.

When preparing to read a large number, you should never have to name every place value (smallest to largest) in order to determine the largest value; rather, you should point off the groups of three, reading the family names as you proceed, until the largest family grouping is reached. Thus, the largest value of the number 207,XXX,XXX,XXX is quickly identified, reading right to left:

$$\text{B} \quad \text{M} \quad \text{T} \quad \text{H}$$
$$207,\text{XXX},\text{XXX},\text{XXX}$$

The 207, in the billions grouping, thus represents "two hundred seven billion," and succeeding values would be read in decreasing order.

In reading the value of a number, it is incorrect to use the word "and" except to designate the location of a decimal point. The number 16,000,003 is pronounced "sixteen million, three."

SECTION 1   PROBLEMS

In the number 3,851,469,720, identify the place value of each digit.

1. a. 1   b. 2   c. 3   d. 4   e. 0
2. a. 5   b. 6   c. 7   d. 8   e. 9

Write in words the values of the following numbers:

3. a. 633,520,481   b. 25,543,128   c. 150,286,413   d. 6,046,125   e. 812,344,601,022
4. a. 842,416,375   b. 2,655,120,688   c. 48,136,472   d. 8,108,027   e. 762,300

SECTION *2*
WHOLE NUMBERS

The simplest set of numbers in our number system is those used to identify a single object or a group of objects. These are the **counting numbers** or **natural**

numbers or positive **whole numbers.** The counting numbers, along with zero and the negative whole numbers (whole numbers less than zero), comprise the set known as the **integers.** Earliest "mathematics" consisted simply of using the integers to count one's possessions—sheep, cattle, tents, wives, and so on. As civilization advanced and became more complex, the need for more efficient mathematics also increased. This led to the perfection of the four arithmetic **operations:** addition, subtraction, multiplication, and division. (This review includes operations only with nonnegative numbers.)

ADDITION

This operation provides a shortcut that eliminates having to count each item consecutively until the total is reached. The names of the numbers in an operation of addition are as follows:

$$
\begin{array}{ll}
\text{Addend} & 23 \\
\text{Addend} & +16 \\
\hline
\text{Sum} & 39
\end{array}
$$

Addition functions under two basic mathematical laws, expressed as (1) the commutative property of addition and (2) the associative property of addition. The **commutative property of addition** means that the *order* in which *two* addends are taken does not affect the sum. More simply, if two numbers are to be added, it does not matter which is written down first. The commutative property is usually written in general terms (terms which apply to all numbers) as

$$\text{CPA:} \quad a + b = b + a$$

The commutative property of addition can be illustrated using the numbers 2 and 3:

$$2 + 3 \stackrel{?}{=} 3 + 2$$

$$5 = 5$$

The **associative property of addition** applies to the *grouping* of *three* numbers; it means that, when three numbers are to be added, the sum will be the same regardless of which two are grouped together to be added first. The associative property is expressed in the following general terms:

$$\text{APA:} \quad (a + b) + c = a + (b + c)$$

The associative property can be verified as follows. Suppose that the numbers 2, 3, and 4 are to be added. By the associative property,

$$(2 + 3) + 4 \overset{?}{=} 2 + (3 + 4)$$

$$5 \quad + 4 \overset{?}{=} 2 + \quad 7$$

$$9 = 9$$

By applying both the commutative and associative properties of addition, one can verify other groupings or the addition of more addends.

An excellent way in which you can speed addition is by adding in groups of numbers that total 10. You should become thoroughly familiar with the combinations of two numbers that total 10 (1 + 9; 2 + 8; 3 + 7; 4 + 6; 5 + 5) and should be alert for these combinations in problems. For example,

$$\begin{array}{r} 3 \\ 7 \\ 6 \\ +\ 4 \end{array}$$ This should not be added as "3 plus 7 is 10, plus 6 is 16, plus 4 is 20." Rather, you should immediately recognize the two groups of 10 and add "10 plus 10 is 20," as

$$\begin{array}{r} 3 \\ 7 \end{array} \Big]\ 10$$
$$\begin{array}{r} 6 \\ +\ 4 \end{array} \Big]\ 10$$
$$\overline{\quad 20}$$

Most problems do not consist of such obvious combinations, but the device may frequently be employed if you stay alert for it. For instance,

$$\begin{array}{r} 8 \\ 4 \end{array} \Big\} \ 10$$
$$\begin{array}{r} 2 \\ +\ 6 \end{array} \Big\} \ 10$$
$$\overline{\quad 20}$$

$$\begin{array}{r} 4 \\ 5 \\ 9 \\ +\ 5 \end{array}$$ (added "4 plus 10 is 14, plus 9 is 23" or "4 plus 9 is 13, plus 10 is 23")

$$\overline{\quad 23}$$

$$\begin{array}{r} 9 \\ 6 \\ 1 \\ 4 \\ +\ 6 \end{array}$$ (added "10 plus 6 is 16, plus 10 is 26")

$$\overline{\quad 26}$$

$$\begin{array}{r} 8 \\ 6 \\ 4 \\ 3 \\ 5 \\ +\ 7 \end{array}$$ (added "8 plus 10 is 18, plus 10 is 28, plus 5 is 33")

$$\overline{\quad 33}$$

This technique, like many others, can be carried to extremes. It is intended to be a timesaver; it will be, provided that you stay alert for combinations adding up to 10 when they are conveniently located near each other. However, a problem may take longer to add if you waste time searching for widely separated combinations. Also, it is easy to miss a number altogether if you are adding numbers from all parts of the problem rather than adding in nearly consecutive order. Therefore, you should use the device when convenient but should not expect it to apply to every problem.

If a first glance at a problem reveals that the same digit appears several times in a column, the following device is useful: Count the number of times the digit appears and multiply this number by the digit itself; the remaining digits are then added to this total. For example,

$$
\begin{array}{r}
7 \\
4 \\
7 \\
7 \\
8 \\
+\ 7 \\
\hline
40
\end{array}
\quad \text{(added ``four 7s are 28, plus 4 is 32, plus 8 is 40'')}
$$

As you probably recall, addition problems may be checked either by adding upward in reverse order or simply by re-adding the sum.

Horizontal addition problems seem harder because the numbers being added are so far apart. More mistakes are made in horizontal problems because a digit from the wrong place is often added. However, many business forms require horizontal addition. The following hint may make such addition easier:

1. A right-handed person should use the left index finger to cover all the higher-place digits in the first number, over to the digit that is currently being added. (Unfortunately, left-handed persons will have to forego this aid.)
2. As digits from succeeding numbers are added, place the pencil point beneath the digit currently being added. Mentally repeat the current subtotal frequently as you proceed.

For example, while the tens'-place digits are being added in the following problem, the left index finger should cover the digits "162" of the first number. The pencil point should be placed under the 5, 1, 2, and 7 as each is added.

$$16,2)83 + 22,054 + 40,611 + 7,927 + 19,670 = \underline{\ .\ .\ .\ 45}$$

SUBTRACTION

This operation is the opposite of addition. (Since subtraction produces the same result as when a negative number is added to a positive, subtraction is sometimes defined as an extension of addition.) The parts of a subtraction operation have the following names:

Minuend	325
Subtrahend	-198
Difference or Remainder	127

A subtraction problem can be checked mentally by adding together the subtrahend and difference; if the resultant sum equals the minuend, the operation is verified.

You can easily determine that neither the commutative nor the associative property applies to subtraction. For example, it makes a great deal of difference whether we find $17 − $8 or $8 − $17. And any example will show that the associative property does not hold:

$$(7 - 5) - 2 \overset{?}{=} 7 - (5 - 2)$$

$$2 \quad - 2 \overset{?}{=} 7 - \quad 3$$

$$0 \neq 4$$

As you know, subtraction often necessitates "borrowing," or regrouping of the place values. For example, the number 325 in our first example of subtraction—which represents three 100s, two 10s, and five 1s—had to be regrouped into two 100s, eleven 10s, and fifteen 1s. Both groupings equal 325, as the following example shows:

Original	*Regrouped*		*Subtraction*	
300	200	(two 100s)		(two 100s)
20	110	(eleven 10s)		(eleven 10s)
5	15	(fifteen 1s)		(fifteen 1s)
325	325			

$$\begin{array}{r} {}^{2}\cancel{3}\ {}^{11}\cancel{2}\ {}^{1}5 \\ -1\ \ 9\ \ 8 \\ \hline 1\ \ 2\ \ 7 \end{array}$$

You are also aware that addition problems often contain many numbers, whereas subtraction problems are done only two numbers at a time. Subtraction problems could just as well contain more numbers were it not for the fact that regrouping becomes so complicated. Calculators, which can regroup repeatedly when necessary, can perform a series of subtractions before indicating a total.

MULTIPLICATION

This operation was developed as a shortcut for addition. Suppose that a student has three classes on each of five days; this week's total classes could be found by addition:

$$M \quad T \quad W \quad T \quad F \quad TOTAL$$
$$3 + 3 + 3 + 3 + 3 = \quad 15$$

This approach to the problem would become quite tedious, however, if we wanted to know the total receipts at a grocery store from the sale of 247 cartons of soft drinks at $2.49 each.

Thus, multiplication originated when persons learned from experience what sums to expect following repeated additions of small numbers, and the actual addition process then became unnecessary. Perfection of the operation led to the knowledge that any two numbers, regardless of size, could be multiplied if one had memorized the products of all combinations of the digits 0–9. The frequent occurrence in business situations of the number 12 makes knowledge of its multiples quite valuable also.

The numbers of an operation of multiplication have the following names:

Multiplicand (or Factor)	18
Multiplier (or Factor)	× 4
Product	72

MULTIPLICATION TABLE

The following table lists the products of the digits 2–9, as well as the numbers 10–12. You should be able to recite instantly all products up through the multiples of 9. A knowledge of products through the multiples of 12 is also very useful.

	2	3	4	5	6	7	8	9	10	11	12
2	4	6	8	10	12	14	16	18	20	22	24
3	6	9	12	15	18	21	24	27	30	33	36
4	8	12	16	20	24	28	32	36	40	44	48
5	10	15	20	25	30	35	40	45	50	55	60
6	12	18	24	30	36	42	48	54	60	66	72
7	14	21	28	35	42	49	56	63	70	77	84
8	16	24	32	40	48	56	64	72	80	88	96
9	18	27	36	45	54	63	72	81	90	99	108
10	20	30	40	50	60	70	80	90	100	110	120
11	22	33	44	55	66	77	88	99	110	121	132
12	24	36	48	60	72	84	96	108	120	132	144

There are three properties affecting the operation of multiplication: (1) the commutative property of multiplication, (2) the associative property of multiplication, and (3) the distributive property.

The **commutative property of multiplication,** like the commutative property of addition, applies to the *order* in which *two* numbers are taken. The commutative property asserts that either factor can be written first, and the product will be the same. In general terms, the commutative property of multiplication is written

$$\text{CPM:} \qquad a{\cdot}b = b{\cdot}a$$

It can be illustrated using the factors 4 and 7:

$$4{\cdot}7 \overset{?}{=} 7{\cdot}4$$

$$28 = 28$$

The **associative property of multiplication** is similar to the associative property of addition in that it applies to the *grouping* of any *three* numbers. That is, if three numbers are to be multiplied, the product will be the same regardless of which two are multiplied first. The associative property of multiplication is expressed in general terms as

$$\text{APM:} \quad (a{\cdot}b)c = a(b{\cdot}c)$$

Using the numbers 2, 5, and 3, the associative property is verified as follows:

$$(2{\cdot}5)3 \overset{?}{=} 2(5{\cdot}3)$$

$$(10)3 \overset{?}{=} 2(15)$$

$$30 = 30$$

The third property applies to the combination of multiplication and addition. It is the **distributive property,** sometimes called the **distributive property of multiplication over addition.** This property means that the product of a factor times the sum of two numbers equals the sum of the individual products. Stated differently, this means that if two numbers are added and their sum is multiplied by some other number, the final total is the same as if the two numbers were first separately multiplied by the third number and the products then added. The distributive property is expressed in general terms as

$$\text{DP:} \qquad a(b + c) = ab + ac$$

Using 2 for the multiplier and the numbers 3 and 5 for the addends, the distributive property can be verified as follows:

$$2(3 + 5) \overset{?}{=} (2 \cdot 3) + (2 \cdot 5)$$

$$2(8) \overset{?}{=} 6 + 10$$

$$16 = 16$$

Note. The distributive property has frequent application in business. One of the best examples is sales tax. When several items are purchased, the sales tax is found by adding all the prices and then multiplying only once by the applicable sales tax percent. This is much simpler than multiplying the tax rate by the price of each separate item and then adding to find the total sales tax, although the total would be the same by either method.

While other methods are sometimes used, the easiest, quickest, and most reliable check of a multiplication problem is simply to repeat the multiplication. (If convenient, the factors may be reversed for the checking operation.)

Students often waste time and do much unnecessary work when *multiplying by numbers that contain zeros.* When multiplying by 10, 100, 1,000, etc., the product should be written immediately simply by affixing to the original factor the same number of zeros as contained in the multiplier. Thus,

$$54 \times 10 = 540; \quad 2{,}173 \times 100 = 217{,}300; \quad \text{and} \quad 145 \times 1{,}000 = 145{,}000$$

When multiplying numbers ending in zeros, the factors should be written with the zeros to the right of the problem as it will actually be performed. After multiplication with the other digits has been completed, affix all these zeros to the basic product, as in Example 1.

Example 1   (a) 148×60 (b) $6{,}700 \times 52$

```
      148 |                      5 2 |
    ×   6 | 0              ×   6 7 | 00
    ——————                  ——————————
    8,88 | 0                  36 4 |
                              312   |
                            ——————————
                            348,4 | 00
```

(c) $130 \times 1{,}500$

```
      13 | 0          Note here that all three zeros are affixed to
    × 15 | 00          the product
    ——————
      65 |
      13 |
    ——————
    195, | 000
```

When a multiplier contains zeros within the number, many people write whole rows of useless zeros in order to be certain that the significant digits remain properly aligned. These unnecessary zeros should be omitted; the other digits will still be in correct order if you practice this basic rule of multiplication: On each line of multiplication, the first digit to be written down is placed directly beneath the digit in the multiplier (bottom factor) being multiplied at the time.

Example 2   (a)   671 × 305

Right		*Inefficient*
671		671
× 305		× 305
3 355		3 355
201 3 (because 300 × 1 = 300)		0 00
204,655		201 3
		204,655

(b)   4,382 × 3,004

$$
\begin{array}{r}
4{,}382 \\
\times\ 3{,}004 \\
\hline
17\ 528 \\
13\ 146 \qquad \text{(because 3,000 × 2 = 6,000)} \\
\hline
13{,}163{,}528
\end{array}
$$

DIVISION

This operation is the reverse of multiplication. The numbers in an operation of division are identified as follows:

$$
\begin{array}{r}
4 \quad \text{Quotient} \\
\text{Divisor} \quad 18\overline{)77} \quad \text{Dividend} \\
\underline{72} \\
5 \quad \text{Remainder}
\end{array}
$$

That is,

$$\frac{\text{Quotient}}{\text{Divisor}\,)\,\text{Dividend}} \quad \text{or} \quad \text{Dividend} \div \text{Divisor} = \text{Quotient}$$

Division is neither commutative, associative, nor distributive, as can be determined by brief experiments. Obviously, $4\overline{)16}$ yields quite a different quotient from $16\overline{)4}$. An example to test the associative property reveals different quotients:

$$(36 \div 6) \div 3 \overset{?}{=} 36 \div (6 \div 3)$$

$$6 \quad \div 3 \overset{?}{=} 36 \div \quad 2$$

$$2 \neq 18$$

Similarly, a test of the distributive property reveals that it does not apply either:

$$60 \div (4 + 6) \overset{?}{=} (60 \div 4) + (60 \div 6)$$

$$60 \div \quad 10 \quad \overset{?}{=} \quad 15 \quad + \quad 10$$

$$6 \neq 25$$

Division by one-digit divisors should be carried out mentally and the quotient written directly; this method is commonly called *short division.* You may find it helpful to write each remainder in front of the next digit as the operation progresses. For example,

$$7\overline{)9422} \qquad \text{written} \qquad \begin{array}{c} 1\,3\,4\,6 \\ 7\overline{)9^2 4^3 2^4 2} \end{array}$$

Division involving divisors of two or more digits is usually performed by writing each step completely. Division performed in this manner is known as *long division.* The following general rules apply to both long and short division and cover aspects of division where mistakes are often made:

1. The first digit of the quotient should be written directly above the last digit of the partial dividend which the divisor goes into. For example, when dividing $25\overline{)17628}$, the first digit of the quotient should be written above the 6:

$$\begin{array}{r} 7 \\ 25\overline{)17628} \\ \underline{175} \end{array}$$

2. The amount remaining after each succeeding step in division must always be smaller than the divisor. Otherwise, the divisor would have divided into the dividend at least one more time. If a remainder is larger than the divisor, the previous step should be repeated to correct

the corresponding digit in the quotient. For example, if the first step had been

$$
\begin{array}{r}
6 \\
25\overline{)17,628} \\
15\ 0 \\
\hline
2\ 6
\end{array}
$$

⟵ Remainder is more than 25

the quotient should be corrected:

$$
\begin{array}{r}
7 \\
25\overline{)17,628} \\
17\ 5 \\
\hline
1
\end{array}
$$

⟵ Remainder is less than 25

3. Succeeding digits of the dividend should be brought down one at a time. For example,

$$
\begin{array}{r}
7 \\
25\overline{)17,628} \\
17\ 5\downarrow \\
\hline
1\overset{\downarrow}{2}
\end{array}
$$

4. Each time a digit is brought down, a new digit must be affixed to the quotient directly over the digit that was just brought down. Thus, a zero is affixed to the quotient when a brought-down digit does not create a number large enough for the divisor to divide into it at least one whole time. For example,

$$
\begin{array}{r}
70 \\
25\overline{)17,628} \\
17\ 5 \\
\hline
12
\end{array}
$$

The next digit is then brought down immediately and the division process continued as follows:

$$
\begin{array}{r}
705 \\
25\overline{)17,628} \\
17\ 5 \\
\hline
128 \\
125 \\
\hline
3
\end{array}
$$

5. The final remainder may be written directly after the quotient as follows:

```
        705   R3
25)17,628
    17 5
       128
       125
         3
```

Division problems can be checked by multiplying the divisor times the quotient and then adding the remainder (if any). This result should equal the original dividend. For example, for the above division problem,

```
      705
×      25
  17,625
      +3
  17,628
```

SECTION 2   PROBLEMS

The following problems were especially designed to provide practice in adding by combinations that total 10.

1. a. 7 b. 4 c. 9 d. 6 e. 4
        2   8   7   5   5
        3   6   1   2   1
       +8   +1   5   4   8
                    +5   +8   +9

   f. 6 g. 2 h. 32 i. 64
       1   5   54   23
       9   3   48   33
       2   6   66   75
      +2   +3   +25   +11

2. a. 3 b. 6 c. 8 d. 9 e. 5
        2   4   2   4   5
        7   5   6   3   8
       +8   +5   3   1   4
                    +7   +6   +2

   f. 7 g. 3 h. 13 i. 48
       9   2   25   16
       1   4   77   52
       3   6   65   84
      +8   3   +32   +25
             3
            +4

Add, using time-saving techniques when possible. Check your answers.

3. a.
```
   6
   6
   5
   4
  +1
```
b.
```
   8
   7
   2
   3
  +4
```
c.
```
   6
   7
   5
   5
   3
  +2
```
d.
```
   4
   4
   8
   3
   4
   6
  +8
```
e.
```
  23
  14
  53
  63
  42
  33
 +56
```

f.
```
  42
  48
  44
  46
 +41
```
g.
```
  12
  16
  46
  56
  17
 +66
```
h.
```
  88
  14
  76
  54
  32
 +51
```
i.
```
  67
  13
  44
  27
  63
 +54
```

4. a.
```
   123
   986
   290
  +451
```
b.
```
   802
   463
   623
   440
  +716
```
c.
```
   528
    73
   265
   939
  + 80
```
d.
```
  7,132
  3,040
  6,355
 +8,098
```
e.
```
  3,613
  4,872
  7,520
  5,638
 +2,516
```

f.
```
    186
  4,351
    265
     68
 +3,081
```
g.
```
  61,046
  42,844
  36,545
 +80,622
```
h.
```
  44,315
  34,045
  25,438
  66,055
 +47,681
```
i.
```
  $5,808.25
      49.18
     219.15
   2,363.75
       7.19
 +   340.93
```

5. a. 41 + 39 + 22 + 16 + 88
   b. 78 + 24 + 45 + 52 + 63
   c. 154 + 238 + 448 + 261 + 311

6. a.   63 +   41 +   22 +   18 +   73
   b.   240 +   313 +   467 +   243 +   324
   c. 1,206 + 1,344 + 2,281 + 3,468 + 2,615

Subtract. Check by adding.

7. a.
```
   36
  -28
```
b.
```
   863
  -199
```
c.
```
   3,468
  -1,547
```
d.
```
   6,002
  -5,679
```

8. a.
```
   43
  -17
```
b.
```
   655
  -458
```
c.
```
   5,044
  -3,950
```
d.
```
   1,824
  -  469
```

Multiply.

9. a. 466 × 10   b. 64 × 100   c. 553 × 100
   d. 56 × 200   e. 1,800 × 46   f. 25,000 × 83

10. a. 312 × 100   b. 88 × 1,000   c. 65 × 30
   d. 525 × 300   e. 3,300 × 74   f. 2,700 × 44

11. a. 76   b. 714   c. 1,203   d. 6,890
       ×24   × 53   × 317   × 254

12. a. 47   b. 669   c. 3,365   d. 5,724
       ×35   × 31   × 842   × 864

Divide. Check using multiplication.

13. a. $6\sqrt{3,588}$ b. $8\sqrt{5,968}$ c. $36\sqrt{648}$

    d. $48\sqrt{1,248}$ e. $63\sqrt{18,333}$

14. a. $5\sqrt{13,545}$ b. $3\sqrt{38,232}$ c. $537\sqrt{13,425}$

    d. $627\sqrt{183,711}$ e. $52\sqrt{13,416}$

15. The following table shows the sales made by various departments of a discount store during one week. Find the total sales of the store for each day and the total sales of each department for the entire week. (This final total serves as a check of your addition: The sum of the weekly totals for each register must equal the sum of the daily totals for the entire store. That is, the sum of the totals in the right-hand column must equal the sum of the totals across the bottom.)

THE DISCOUNT MART
Sales for week of August 4

DEPT.	MONDAY	TUESDAY	WEDNESDAY	THURSDAY	FRIDAY	SATURDAY	DEPT. TOTALS
#1	$345.68	$406.29	$323.76	$386.91	$459.85	$481.86	$
#2	562.09	438.88	475.45	424.18	582.37	519.12	
#3	445.21	459.16	398.06	437.75	493.46	418.25	
#4	293.34	302.28	285.37	356.11	374.04	380.84	
#5	322.66	289.55	314.42	349.23	352.58	334.17	
#6	488.49	419.37	392.76	407.33	469.87	493.66	
Daily Totals	$	$	$	$	$	$	$

16. Complete the following sales invoice. Multiply the cost of each item by the number of items purchased. Add to find the total cost of the purchase.

OFFICE SUPPLY HOUSE

Invoice

No. PURCHASED	DESCRIPTION	COST PER ITEM	TOTAL
14	X-22 pencil sharpener	$ 8.12	$
12	X-25 pencil sharpener	10.55	
25	B-13 ream bond stationery	4.95	
15	K-46 ream fax paper	2.95	
26	R-12 ream computer paper	2.85	
6	TC-32 multimedia drawer	8.95	
72	C-28 pencil	.10	
46	L-4 ballpoint pen	.55	
8	RF-10 color inkjet printer	215.00	
5	P-7 computer speaker	24.95	
3	A-IV calculator	8.38	
2	HD-5 CD-ROM drive	142.95	
7	RP-34 notebook computer	2,142.00	
		Total	$

SECTION 3

COMMON FRACTIONS

The system of fractions was developed to meet the need for measuring quantities that are not whole amounts. The parts of a fraction are identified by the following terms:

Numerator   3
Denominator   5

A **common fraction** is an indicated division—a short method of writing a division problem. That is, if $3 is to be divided among five persons, each person's share is indicated as

$$\$3 \div 5 \quad \text{or} \quad \$\frac{3}{5}$$

Thus, the fraction line indicates division (the numerator is to be divided by the denominator), and the entire fraction represents each person's share. We now see why common fractions are often known as "rational numbers." A **rational number** is any number that can be expressed as the quotient of two integers. The fraction $\frac{3}{5}$ is a rational number because it is expressed as the quotient of the integers (whole numbers) 3 and 5. The entire fraction represents the quotient—the value of 3 divided by 5.

A common fraction is called a **proper fraction** if its numerator is smaller than its denominator (for example, $\frac{4}{7}$) and if the indicated division would result in a value less than 1. The numerator of an **improper fraction** is equal to or greater than the denominator (for example, $\frac{7}{7}$ or $\frac{15}{7}$) and thus has a value equal to or greater than 1.

REDUCING FRACTIONS

Two fractions are equal if they represent the same quantity. When writing fractions, however, it is customary to use the smallest possible numbers. Thus, it is often necessary to *reduce* a fraction—to change it to a fraction of equal value but written with smaller numbers.

The fraction $\frac{3}{3}$ represents $3 \div 3$ or 1. Obviously, any number written over itself as a fraction equals 1 $\left(\dfrac{x}{x} = 1 \right)$. We also know from multiplication that 1 times any number equals that same number ($1 \cdot x = x$). These facts are the mathematical basis for reducing fractions. A fraction can be reduced if the same number is a factor of both the numerator and the denominator. For example,

$$\frac{6}{15} = \frac{2 \times 3}{5 \times 3} \qquad \text{(the factors of 6 are 2 and 3)}$$
$$\text{(the factors of 15 are 5 and 3)}$$

But since $\frac{3}{3} = 1$, then

$$\frac{6}{15} = \frac{2 \times 3}{5 \times 3} = \frac{2}{5} \times \frac{3}{3} = \frac{2}{5} \times 1 = \frac{2}{5}$$

Hence, $\frac{2}{5}$ represents the same quantity as $\frac{6}{15}$.

In actual practice, you reduce a fraction by mentally testing the numerator and denominator to determine the *greatest* factor that they have in common (that is, to determine the largest number that will divide into both evenly). The reduced fraction is then indicated by writing these quotients in fraction form:

$\dfrac{20}{15}$ (Think: 5 divides into 20 *four* times and into 15 *three* times)   Write: $\dfrac{20}{15} = \dfrac{4}{3}$

Sometimes, however, the greatest common factor may not be immediately obvious. When the greatest common factor is not apparent, a fraction may be reduced to its lowest terms by performing a series of reductions:

$$\frac{60}{84} = \frac{2}{2} \times \frac{30}{42} = \frac{30}{42} = \frac{2}{2} \times \frac{15}{21} = \frac{15}{21} = \frac{3}{3} \times \frac{5}{7} = 1 \times \frac{5}{7} = \frac{5}{7}$$

The end result will be the same whether the reduction is performed in one step or in several steps. It should be noted that if a series of reductions is used, the product of these several common factors equals the number that would have been the greatest single common factor. For example, the fraction $\frac{60}{84}$ above was reduced by 2, 2, and 3. The product of these factors is 12, which is the greatest common factor:

$$\frac{60}{84} = \frac{2}{2} \times \frac{2}{2} \times \frac{3}{3} \times \frac{5}{7} = 1 \times 1 \times 1 \times \frac{5}{7} = \frac{5}{7} \quad \text{or} \quad \frac{60}{84} = \frac{12}{12} \times \frac{5}{7} = 1 \times \frac{5}{7} = \frac{5}{7}$$

Therefore, you should divide by the largest number that you recognize as a factor of both the numerator and denominator. However, you should always examine the result to determine whether it still contains another common factor.

CHANGING TO HIGHER TERMS

Suppose that we wish to know whether $\frac{5}{6}$ or $\frac{7}{9}$ is larger. To compare the size of fractions, as a practical matter we must rewrite them using the same denominator. The process of rewriting fractions so that they have a common denominator (which is usually larger than either original denominator) is called *changing to higher terms.*

The product of the various denominators will provide a common denominator. However, it is customary to use the **least common denominator,** which is the smallest number that all the denominators will divide into evenly. Thus, the least (or lowest) common denominator is often a number that is smaller than the product of all the denominators. In fact, one of the given denominators may be the least common denominator.

When finding the least common denominator, you should first consider the multiples of the largest given denominator. If this method does not reveal the least common denominator quickly, then you should find the product of the denominators. Before using this product as the common denominator, though, first consider $\frac{1}{2}$, $\frac{1}{3}$, or even $\frac{1}{4}$ of this product to determine if that might be the lowest common denominator. If not, then the product itself probably represents the lowest common denominator.

Consider the fractions, $\frac{2}{3}$, $\frac{1}{6}$, and $\frac{3}{8}$. To find the least common denominator, examine the multiples of 8 (the largest denominator). The multiples of 8 are 8, 16, 24, 32, 40, etc. Neither 8 nor 16 is divisible by 3 or 6. However, 24 is divisible by both 3 and 6 (and, of course, by 8). Hence, 24 is the least common denominator of $\frac{2}{3}$, $\frac{1}{6}$, and $\frac{3}{8}$.

Next, consider the fractions $\frac{5}{6}$, $\frac{3}{7}$, and $\frac{5}{8}$. The product of the denominators is $6 \times 7 \times 8$, or 336. However, $\frac{1}{2}$ of 336, or 168, represents the smallest number that 6, 7, and 8 will all divide into evenly. Thus, 168 is the least common denominator. Also consider the fractions $\frac{1}{6}$, $\frac{2}{7}$, and $\frac{2}{9}$. The product of these denominators is $6 \times 7 \times 9$, or 378. One-half of 378 is 189, which is not divisible by 6. However, $\frac{1}{3}$ of 378 is 126, which is divisible by 6, 7, and 9 all. Hence, 126 is the least common denominator.

Changing to higher terms is based on the same principles as reducing fractions, but applied in reverse order. That is, the procedure is based on the facts that any number over itself equals 1 $\left(\dfrac{x}{x} = 1\right)$ and that any number multiplied by 1 equals that same number ($x \cdot 1 = x$). Thus, once the least common denominator has been determined, you must then find how many times the given denominator divides into the common denominator. This result is used as both the numerator and denominator to create a fraction equal to 1, which is then used to convert the fraction to higher terms. For example, suppose that we have already determined that 12 is the least common denominator for $\frac{3}{4}$ and $\frac{1}{6}$. To convert $\frac{3}{4}$ to higher terms, think "4 divides into 12 *three* times." Thus, $\frac{3}{3}$ is used to change $\frac{3}{4}$ to higher terms:

$$\frac{3}{4} = \frac{3}{4} \times 1 = \frac{3}{4} \times \frac{3}{3} = \frac{3 \times 3}{4 \times 3} = \frac{9}{12}$$

We have multiplied $\frac{3}{4}$ by a value, $\frac{3}{3}$, equal to 1. Therefore, we know that the value of the fraction remains unchanged and that $\frac{3}{4} = \frac{9}{12}$. Similarly, 6 divides into 12 *two* times. Thus, $\frac{2}{2}$ is used to change $\frac{1}{6}$ to higher terms:

$$\frac{1}{6} = \frac{1}{6} \times 1 = \frac{1}{6} \times \frac{2}{2} = \frac{1 \times 2}{6 \times 2} = \frac{2}{12}$$

Hence, $\frac{1}{6} = \frac{2}{12}$.

In actual practice, it is unnecessary to write out this entire process. Rather, you mentally divide the denominator into the lowest common denominator and multiply that quotient times the numerator to obtain the numerator of the converted fraction. Therefore, to change $\frac{3}{4}$ to twelfths, think: "4 divides into 12 *three* times; that *three* times the numerator 3 equals 9; thus $\frac{3}{4} = \frac{9}{12}$." Or, when changing $\frac{1}{6}$ to twelfths, think: "6 divides into 12 *two* times; *two* times 1 equals 2; therefore, $\frac{1}{6} = \frac{2}{12}$."

Now let us return to $\frac{5}{6}$ and $\frac{7}{9}$ to determine which is larger.

$$\frac{5}{6} = \frac{15}{18} \qquad \text{and} \qquad \frac{7}{9} = \frac{14}{18}$$

Thus, comparing the fractions on the same terms, $\frac{15}{18}$ is larger than $\frac{14}{18}$. Hence, $\frac{5}{6}$ is larger than $\frac{7}{9}$.

ADDITION AND SUBTRACTION

The parts of each operation have the same terminology that they did with integers, as follows:

Addend	$\dfrac{3}{7}$	Minuend	$\dfrac{3}{7}$
Addend	$+\dfrac{2}{7}$	Subtrahend	$-\dfrac{2}{7}$
Sum	$\dfrac{5}{7}$	Difference or remainder	$\dfrac{1}{7}$

Before common fractions can be either added or subtracted, the fractions must first be written with a common denominator. The numerators are then added or subtracted (whichever is indicated). The result is then written over the *same* common denominator. (Be certain that this fact is firmly impressed in your mind, for mistakes are frequently made at this point.) The answer should then be reduced, if possible. Consider the following examples.

Example 1

(a)
$$\frac{3}{5} = \frac{9}{15}$$
$$+\frac{2}{3} = \frac{10}{15}$$
$$\frac{19}{15}$$

(b)
$$\frac{3}{4} = \frac{9}{12}$$
$$\frac{2}{3} = \frac{8}{12}$$
$$+\frac{5}{6} = \frac{10}{12}$$
$$\frac{27}{12} = \frac{9}{4}$$

(c)
$$\frac{1}{4} + \frac{3}{8} + \frac{1}{6} = \frac{6}{24} + \frac{9}{24} + \frac{4}{24}$$
$$= \frac{6+9+4}{24}$$
$$\frac{1}{4} + \frac{3}{8} + \frac{1}{6} = \frac{19}{24}$$

(d)
$$\frac{9}{16} = \frac{9}{16}$$
$$-\frac{3}{8} = -\frac{6}{16}$$
$$\frac{3}{16}$$

(e)
$$\frac{5}{7} = \frac{15}{21}$$
$$-\frac{1}{3} = -\frac{7}{21}$$
$$\frac{8}{21}$$

(f)
$$\frac{1}{2} - \frac{1}{6} = \frac{3}{6} - \frac{1}{6}$$
$$= \frac{3-1}{6}$$
$$= \frac{2}{6}$$
$$\frac{1}{2} - \frac{1}{6} = \frac{1}{3}$$

MULTIPLICATION AND DIVISION

As would be expected, the terms previously used to identify the parts of multiplication and division with whole numbers also apply to those operations with fractions.

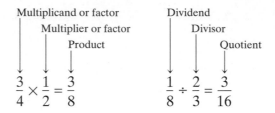

When multiplying fractions, we multiply the various numerators together to obtain the numerator of the product and then multiply the denominators to obtain the denominator of the product. Where possible, the product should be reduced.

$$\frac{2}{5} \times \frac{2}{3} = \frac{2}{5} \times \frac{2}{3} = \frac{2 \times 2}{5 \times 3} = \frac{4}{15}$$

When one fraction (dividend) is to be divided by another (divisor), the divisor (first fraction) must first be inverted (that is, the numerator and denominator must be interchanged). Then, proceed as in multiplication. When a fraction is inverted, the new fraction is called the **reciprocal** of the original fraction. Hence, the reciprocal of $\frac{3}{5}$ is $\frac{5}{3}$. Thus, it is often said that to divide by a fraction we "multiply by the reciprocal."

$$\frac{3}{4} \div \frac{2}{3} = \frac{3}{4} \times \frac{3}{2} = \frac{3 \times 3}{4 \times 2} = \frac{9}{8}$$

$$\left(\text{To divide by } \tfrac{2}{3}, \textit{multiply} \text{ by the reciprocal } \tfrac{3}{2}. \right)$$

Note.   With common fractions, the number properties apply exactly as they did for whole numbers. That is, the *commutative* and *associative* properties apply for both addition and multiplication. Then, the *distributive* property of multiplication over addition illustrates how the two operations interact. You might pick some fractions and try confirming that these properties do work, using the examples from the previous section as a guide. (As before, none of these properties holds for subtraction or division.)

You may be accustomed to "canceling" when multiplying or dividing fractions. When you use cancellation, this is actually just reducing the problem rather than waiting and reducing the answer. We know that the value of a fraction remains unchanged when it is reduced. Thus, cancellation is permissible because it is a reduction process and does not change the value of the answer.

Consider the problem $\frac{3}{4} \times \frac{1}{3}$. If multiplied in the standard manner, the product would be $\frac{3}{12}$, which reduces to $\frac{1}{4}$. The following operations illustrate how the problem itself can be reduced before the answer is obtained:

$$\frac{3}{4} \times \frac{1}{3} = \frac{3 \times 1}{4 \times 3}$$

Since multiplication is commutative, we can reverse the order of the numbers:

$$\frac{3 \times 1}{4 \times 3} = \frac{1 \times 3}{4 \times 3}$$

$$= \frac{1}{4} \times \frac{3}{3}$$

$$= \frac{1}{4} \times 1$$

$$\frac{3 \times 1}{4 \times 3} = \frac{1}{4}$$

Thus, $\frac{3}{4} \times \frac{1}{3} = \frac{1}{4}$. However, you would not usually reduce by rewriting the problems to make the 3s form the fraction $\frac{3}{3}$ and thereby reduce to 1. Rather, seeing that there is one 3 in the numerator and another in the denominator, you would cancel the 3s (or reduce $\frac{3}{3}$ to 1) as follows:

$$\frac{3}{4} \times \frac{1}{3} = \frac{\overset{1}{\cancel{3}}}{4} \times \frac{1}{\underset{1}{\cancel{3}}} = \frac{1}{4}$$

Two numbers can be canceled even though they are widely separated in the problem, provided that one is in the numerator and the other in the denominator.

$$\frac{4}{5} \times \frac{3}{5} \times \frac{7}{8} = \frac{\overset{1}{\cancel{4}}}{5} \times \frac{3}{5} \times \frac{7}{\underset{2}{\cancel{8}}} = \frac{1}{5} \times \frac{3}{5} \times \frac{7}{2} = \frac{21}{50}$$

Now consider the division problem $\frac{4}{5} \div \frac{8}{15}$. If this problem were divided before it is reduced, the operation would be as follows:

$$\frac{4}{5} \div \frac{8}{15} = \frac{4}{5} \times \frac{15}{8}$$

$$= \frac{60}{40}$$

$$\frac{4}{5} \div \frac{8}{15} = \frac{3}{2}$$

The problem could also be reduced prior to the actual division:

$$\frac{4}{5} \div \frac{8}{15} = \frac{4}{5} \times \frac{15}{8}$$

$$= \frac{4}{5} \times \frac{5 \times 3}{4 \times 2}$$

$$= \frac{4 \times 5 \times 3}{5 \times 4 \times 2}$$

Because multiplication is commutative and associative, we may rearrange the numbers:

$$= \frac{5 \times 4 \times 3}{5 \times 4 \times 2}$$

$$= \frac{5}{5} \times \frac{4}{4} \times \frac{3}{2}$$

$$= 1 \times 1 \times \frac{3}{2}$$

$$\frac{4}{5} \div \frac{8}{15} = \frac{3}{2}$$

Hence, $\frac{4}{5} \div \frac{8}{15} = \frac{3}{2}$ by either method. In actual practice, however, neither method is normally used. Rather, after the divisor has been inverted, cancellation would be used to reduce the fractions. That is, the implied fraction $\frac{4}{8}$ would be reduced to $\frac{1}{2}$, and $\frac{15}{5}$ would be reduced to $\frac{3}{1}$ as follows:

$$\frac{4}{5} \div \frac{8}{15} = \frac{4}{5} \times \frac{15}{8}$$

$$= \frac{\overset{1}{\cancel{4}}}{\underset{1}{\cancel{5}}} \times \frac{\overset{3}{\cancel{15}}}{\underset{2}{\cancel{8}}}$$

$$= \frac{1}{1} \times \frac{3}{2}$$

$$\frac{4}{5} \div \frac{8}{15} = \frac{3}{2}$$

Thus, we see that cancellation is a method of reducing fractions and therefore produces the same answers as ordinary reduction. However, cancellation may be used only under the following conditions: Cancellation may be performed only in a multiplication problem, or in a division problem *after* the divisor has been inverted; and a number in the *numerator* and a number in the *denominator* must contain a common factor (that is, must comprise a fraction that can be reduced). The numbers being reduced are slashed with a diagonal mark; the quotient obtained when dividing by the common factor is written above the number in the numerator and below the number in the denominator. These reduced values are then used in performing the multiplication.

MIXED NUMBERS

A number composed of both a whole number and a fraction is known as a **mixed number** (for example, $3\frac{1}{2}$). Multiplication and division involving mixed numbers is easier if the mixed number is first converted to a common fraction.

Mixed numbers can be converted to fractions by applying the same basic principles that we have used to reduce fractions or to change to higher terms— namely, that any number over itself equals 1 $\left(\dfrac{x}{x} = 1\right)$ and that multiplying any number by 1 equals that same number ($x \cdot 1 = x$). You will recall that a whole number can be written in fraction form using a denominator of 1. (That is, $5 = \frac{5}{1}$. This is correct because $\frac{5}{1}$ means $5 \div 1$, which equals the original 5.) Now consider the mixed number $3\frac{1}{2}$:

$$3\frac{1}{2} = \quad 3 \quad + \frac{1}{2}$$

$$= (3 \times 1) + \frac{1}{2}$$

$$= \left(3 \times \frac{2}{2}\right) + \frac{1}{2}$$

$$= \left(\frac{3}{1} \times \frac{2}{2}\right) + \frac{1}{2}$$

$$= \frac{6}{2} + \frac{1}{2}$$

$$3\frac{1}{2} = \frac{7}{2}$$

Thus, we see that $3\frac{1}{2}$ can be converted to the common fraction $\frac{7}{2}$ by multiplying the whole-number part by a fraction equal to 1 and then adding the fraction from the mixed number. (The "fraction equal to 1" is always composed of the denominator from the fraction in the mixed number over itself—in this case, $\frac{2}{2}$. Thus, the mixed number $4\frac{2}{5}$ would be converted by substituting $\frac{5}{5}$ for 1.) Any mixed number always equals an improper fraction.

The familiar procedure used to convert a mixed number to a common fraction is to "multiply the denominator times the whole number and add the numerator." This total over the original denominator (from the fraction part) is the value of the common fraction that is equivalent to the mixed number. Thus,

$$3\frac{1}{2} = 3 \overset{+\;1}{\underset{\times\;2}{\frown}} 2 = \begin{cases} 3 \\ \times\, \underline{2} \text{ halves} \\ 6 \text{ halves} + 1 \text{ half} = \dfrac{7}{2} \end{cases}$$

If the whole number is so large that the product is not immediately obvious, the multiplication can be performed digit by digit and the numerator added in the same way as are numbers that have been "carried." For example,

$$16\frac{3}{8} = \frac{?}{8} \qquad \text{Think: "8} \times \text{6 is 48, plus 3 is 51."}$$

$$\text{Write 1 and carry the 5.} \quad \left(16\frac{3}{8} = \frac{\ldots\,1}{8}\right)$$

Then, "8 × 1 is 8, plus 5 is 13." So,

$$16\frac{3}{8} = \frac{131}{8}$$

Example 2 demonstrates how mixed numbers are handled in multiplication and division problems.

Example 2

(a) $\quad 4\dfrac{1}{2} \times 3\dfrac{1}{3} = \dfrac{9}{2} \times \dfrac{10}{3}$

$$= \dfrac{\overset{3}{\cancel{9}}}{\underset{1}{\cancel{2}}} \times \dfrac{\overset{5}{\cancel{10}}}{\underset{1}{\cancel{3}}}$$

$$= \dfrac{3}{1} \times \dfrac{5}{1}$$

$$= \dfrac{15}{1}$$

$$4\dfrac{1}{2} \times 3\dfrac{1}{3} = 15$$

(b) $\quad 5\dfrac{2}{5} \div 3\dfrac{3}{5} = \dfrac{27}{5} \div \dfrac{18}{5}$

$$= \dfrac{27}{5} \times \dfrac{5}{18}$$

$$= \dfrac{\overset{3}{\cancel{27}}}{\underset{1}{\cancel{5}}} \times \dfrac{\overset{1}{\cancel{5}}}{\underset{2}{\cancel{18}}}$$

$$= \dfrac{3}{1} \times \dfrac{1}{2}$$

$$5\dfrac{2}{5} \div 3\dfrac{3}{5} = \dfrac{3}{2}$$

(c) $\quad 4\dfrac{7}{8} \div 1\dfrac{11}{16} = \dfrac{39}{8} \div \dfrac{27}{16}$

$$= \dfrac{39}{8} \times \dfrac{16}{27}$$

$$= \dfrac{\overset{13}{\cancel{39}}}{\underset{1}{\cancel{8}}} \times \dfrac{\overset{2}{\cancel{16}}}{\underset{9}{\cancel{27}}}$$

$$= \dfrac{13}{1} \times \dfrac{2}{9}$$

$$4\dfrac{7}{8} \div 1\dfrac{11}{16} = \dfrac{26}{9}$$

Notice in division that a mixed-number divisor should be converted to an improper fraction *before* the divisor is inverted.

In certain instances, it is necessary to change an improper fraction to a mixed number. This operation simply reverses the former procedure. Suppose that $\frac{19}{8}$ is to be converted to a mixed number. The numerator can first be broken down into a multiple of the denominator plus a remainder. The familiar properties of 1 $\left(\text{that is,}\ \dfrac{x}{x} = 1\ \text{and}\ x \cdot 1 = x\right)$ are then applied as before:

$$\dfrac{19}{8} = \dfrac{2(8) + 3}{8}$$

$$= \dfrac{2(8)}{8} + \dfrac{3}{8}$$

$$= \left(\dfrac{2}{1} \times \dfrac{8}{8}\right) + \dfrac{3}{8}$$

$$= \left(\dfrac{2}{1} \times 1\right) + \dfrac{3}{8}$$

$$= 2 + \dfrac{3}{8}$$

$$\dfrac{19}{8} = 2\dfrac{3}{8}$$

Thus, $\frac{19}{8} = 2\frac{3}{8}$. In actual practice, however, the procedure used to convert improper fractions to mixed numbers applies the fact that a fraction is an indicated division. That is, $\frac{19}{8}$ means $19 \div 8$, which equals 2 with a remainder of 3. Hence, $\frac{19}{8} = 2\frac{3}{8}$.

Addition may be performed using mixed numbers (without changing them to improper fractions). Although it is generally considered acceptable to leave answers as improper fractions, it is not correct to leave a mixed number that includes an improper fraction. The improper fraction should be changed to a mixed number and this amount added to the whole number to obtain a mixed number that includes a proper fraction (reduced to lowest terms). Thus, if the sum of an addition problem is $7\frac{14}{6}$, this result must be converted as follows:

$$7\frac{14}{6} = 7 + \frac{14}{6}$$

$$= 7 + 2\frac{2}{6}$$

$$= 9\frac{2}{6}$$

$$7\frac{14}{6} = 9\frac{1}{3}$$

Example 3 shows cases of addition with mixed numbers.

Example 3

(a)
$$7\frac{5}{9}$$
$$+ \ 3\frac{1}{9}$$
$$\overline{10\frac{6}{9} = 10\frac{2}{3}}$$

(b)
$$6\frac{5}{8} = \ 6\frac{15}{24}$$
$$+ \ 9\frac{2}{3} = \ 9\frac{16}{24}$$
$$\overline{15\frac{31}{24} = 16\frac{7}{24}}$$

(c)
$$16\frac{2}{3} = 16\frac{4}{6}$$
$$+ 17\frac{5}{6} = 17\frac{5}{6}$$
$$\overline{33\frac{9}{6} = 34\frac{3}{6} = 34\frac{1}{2}}$$

Subtraction may also be performed using mixed numbers. You will recall that subtraction usually requires regrouping (or borrowing). This procedure, when applied to fractions, means that improper fractions must often be created where they did not exist.

Consider the problem $7\frac{3}{8} - 4\frac{7}{8}$. Although the mixed number $4\frac{7}{8}$ is smaller than $7\frac{3}{8}$, the fraction $\frac{7}{8}$ cannot be subtracted from $\frac{3}{8}$. In order to perform the operation, we must regroup the $7\frac{3}{8}$. Mixed numbers are regrouped by borrowing one unit from the whole number, converting this unit to a fraction equal to 1 (the common denominator over itself), and adding this fractional "1" to the fraction from the mixed number. Using this procedure, we regroup $7\frac{3}{8}$ as follows:

$$7\frac{3}{8} = 7 + \frac{3}{8}$$

$$= 6 + 1 + \frac{3}{8}$$

$$= 6 + \frac{8}{8} + \frac{3}{8}$$

$$= 6 + \frac{11}{8}$$

$$7\frac{3}{8} = 6\frac{11}{8}$$

Thus, $7\frac{3}{8}$ equals $6\frac{11}{8}$. In actual practice, the procedure of borrowing 1 from the whole number and adding it to the existing fraction may be done mentally, with only the result being written. The preceding subtraction problem and other problems are illustrated in Example 4.

Example 4

(a)
$$5\frac{5}{6}$$
$$-1\frac{1}{6}$$
$$\overline{4\frac{4}{6} = 4\frac{2}{3}}$$

(b)
$$7\frac{3}{8} = 6\frac{11}{8}$$
$$-4\frac{7}{8} = -4\frac{7}{8}$$
$$\overline{2\frac{4}{8} = 2\frac{1}{2}}$$

(c)
$$15\frac{1}{3} = 15\frac{4}{12} = 14\frac{16}{12}$$
$$-8\frac{3}{4} = -8\frac{9}{12} = -8\frac{9}{12}$$
$$\overline{6\frac{7}{12}}$$

Notice that the fractions in mixed numbers are first changed to fractions with a common denominator before regrouping for subtraction, as illustrated by (c).

SECTION 3 PROBLEMS

1. Reduce the following fractions to lowest terms (that is, reduce as much as possible).

   a. $\dfrac{16}{24}$ b. $\dfrac{14}{6}$ c. $\dfrac{28}{36}$ d. $\dfrac{48}{84}$ e. $\dfrac{34}{51}$

   f. $\dfrac{75}{45}$ g. $\dfrac{57}{76}$ h. $\dfrac{108}{90}$ i. $\dfrac{189}{321}$ j. $\dfrac{168}{288}$

2. Convert the following fractions to fractions with least common denominators.

   a. $\dfrac{1}{3}$ and $\dfrac{3}{4}$ b. $\dfrac{1}{2}$ and $\dfrac{4}{7}$ c. $\dfrac{4}{9}$ and $\dfrac{2}{3}$ d. $\dfrac{3}{8}$ and $\dfrac{1}{16}$

   e. $\dfrac{5}{8}$ and $\dfrac{1}{6}$ f. $\dfrac{1}{5}, \dfrac{1}{4},$ and $\dfrac{2}{3}$ g. $\dfrac{2}{7}, \dfrac{1}{6},$ and $\dfrac{1}{2}$ h. $\dfrac{7}{10}, \dfrac{2}{5},$ and $\dfrac{1}{3}$

   i. $\dfrac{2}{9}, \dfrac{3}{4},$ and $\dfrac{5}{6}$ j. $\dfrac{3}{5}, \dfrac{1}{4},$ and $\dfrac{5}{8}$

3. Compare the following fractions by converting to a common denominator. Then arrange the original fractions in order of size, starting with the smallest.

   a. $\dfrac{3}{7}, \dfrac{3}{4}, \dfrac{5}{14}, \dfrac{1}{2}, \dfrac{5}{8}, \dfrac{5}{7}$ b. $\dfrac{7}{12}, \dfrac{5}{6}, \dfrac{3}{4}, \dfrac{7}{9}, \dfrac{5}{8}, \dfrac{13}{24}$

4. Change the following mixed numbers to improper fractions.

   a. $3\dfrac{1}{4}$ b. $6\dfrac{5}{8}$ c. $8\dfrac{2}{7}$ d. $5\dfrac{2}{9}$ e. $12\dfrac{2}{5}$ f. $20\dfrac{1}{3}$

   g. $26\dfrac{3}{7}$ h. $32\dfrac{1}{2}$ i. $18\dfrac{5}{6}$ j. $9\dfrac{2}{11}$ k. $15\dfrac{2}{7}$ l. $7\dfrac{7}{12}$

5. Convert the following improper fractions (and improper mixed numbers) to mixed numbers. Reduce if possible.

   a. $\dfrac{17}{6}$ b. $\dfrac{46}{5}$ c. $\dfrac{34}{4}$ d. $\dfrac{24}{7}$ e. $\dfrac{49}{3}$ f. $\dfrac{37}{12}$

   g. $\dfrac{92}{6}$ h. $\dfrac{54}{4}$ i. $4\dfrac{5}{3}$ j. $8\dfrac{16}{5}$ k. $9\dfrac{7}{3}$ l. $5\dfrac{9}{2}$

6. Find the sums, reducing when possible.

   a. $\dfrac{1}{4}$ b. $\dfrac{1}{6}$ c. $\dfrac{3}{5}$ d. $\dfrac{1}{4}$ e. $\dfrac{7}{12}$ f. $\dfrac{5}{6}$
   $+\dfrac{2}{5}$ $+\dfrac{3}{8}$ $\dfrac{2}{3}$ $\dfrac{1}{8}$ $\dfrac{4}{9}$ $\dfrac{2}{3}$
   $+\dfrac{1}{10}$ $+\dfrac{5}{6}$ $\dfrac{1}{4}$ $\dfrac{4}{9}$
   $+\dfrac{1}{6}$ $+\dfrac{3}{18}$

   g. $6\dfrac{2}{5}$ h. $3\dfrac{3}{4}$ i. $7\dfrac{3}{8}$ j. $22\dfrac{5}{6}$
   $+5\dfrac{1}{2}$ $+6\dfrac{1}{9}$ $+7\dfrac{2}{5}$ $+14\dfrac{4}{5}$

7. Subtract, reducing when possible.

a. $\dfrac{5}{6}$ b. $\dfrac{3}{4}$ c. $\dfrac{7}{8}$ d. $\dfrac{7}{12}$ e. $\dfrac{2}{3}$

$\underline{-\dfrac{1}{3}}$ $\underline{-\dfrac{1}{5}}$ $\underline{-\dfrac{3}{7}}$ $\underline{-\dfrac{1}{2}}$ $\underline{-\dfrac{3}{5}}$

f. $15\dfrac{1}{4}$ g. $24\dfrac{1}{6}$ h. 33 i. 56 j. 28

$\underline{-\ 3\dfrac{1}{3}}$ $\underline{-\ 5\dfrac{3}{4}}$ $\underline{-12\dfrac{1}{6}}$ $\underline{-40\dfrac{5}{7}}$ $\underline{-\ 7\dfrac{1}{2}}$

8. Multiply, reducing whenever possible.

a. $\dfrac{2}{7} \times \dfrac{5}{6}$ b. $\dfrac{3}{8} \times \dfrac{7}{12}$ c. $\dfrac{5}{9} \times \dfrac{9}{10}$ d. $\dfrac{14}{20} \times \dfrac{5}{28}$ e. $\dfrac{4}{15} \times \dfrac{3}{24}$

f. $\dfrac{6}{7} \times 2\dfrac{1}{2}$ g. $6\dfrac{2}{3} \times \dfrac{1}{5}$ h. $4\dfrac{1}{5} \times \dfrac{4}{7}$ i. $12\dfrac{1}{4} \times 5\dfrac{1}{7}$ j. $3\dfrac{3}{8} \times 3\dfrac{5}{9}$

9. Multiply, reducing whenever possible.

a. $\dfrac{4}{14} \times \dfrac{7}{15} \times \dfrac{3}{16}$ b. $\dfrac{3}{8} \times \dfrac{5}{6} \times \dfrac{7}{10}$ c. $4\dfrac{1}{2} \times \dfrac{2}{3} \times 8\dfrac{1}{6}$ d. $2\dfrac{5}{8} \times 2\dfrac{2}{7} \times 4\dfrac{2}{3}$

10. Divide, reducing to lowest terms.

a. $\dfrac{2}{9} \div \dfrac{2}{3}$ b. $\dfrac{4}{9} \div \dfrac{4}{5}$ c. $\dfrac{3}{8} \div \dfrac{6}{15}$ d. $\dfrac{5}{6} \div \dfrac{3}{18}$ e. $\dfrac{6}{7} \div \dfrac{18}{21}$

f. $\dfrac{3}{5} \div \dfrac{6}{11}$ g. $\dfrac{9}{10} \div \dfrac{1}{2}$ h. $3\dfrac{1}{5} \div \dfrac{2}{5}$ i. $\dfrac{3}{8} \div 2\dfrac{1}{2}$ j. $1\dfrac{1}{3} \div 7\dfrac{1}{5}$

SECTION 4
DECIMAL FRACTIONS

As the developing number system became more sophisticated, a method was devised for expressing parts of a whole without writing a complete common fraction; this is the system of **decimal notation.** Decimal values are actually abbreviated fractions. That is, the written number is only the numerator of an indicated fraction. The denominator of the indicated fraction is some power of 10. This denominator is not written but is indicated by the number of digits to the right of the decimal point. For example, 0.7 indicates $\frac{7}{10}$, or 0.43 indicates $\frac{43}{100}$, without the 10 or 100 being written. Thus, since decimal values represent fractional parts of a whole, they are known as **decimal fractions.**

Decimal fractions are an extension of our whole number (and integer) system, both because they are based on powers of 10 and also because the value of any digit is determined by the position it holds in the number. The following chart shows the value and power of 10 that is associated with each place in a decimal number. (Some whole-number places are included to show that decimal values are an extension of whole-number place values.) Notice that the names for decimal place values repeat, in reverse order, the names of integer place values—except that the letters "-ths" are added to indicate that these are decimal-place values.

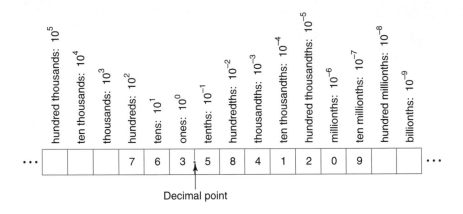

Decimal point

A decimal point (or period) is used in a number to show where the whole amount ends and the fractional portion begins. A whole number is usually written without a decimal point, but every integer is understood to have a decimal point after the units place. Thus, 15 could be written "15." (or "15.0," or "15.00," etc.) and would be just as correct.

You will recall from our study of integers that 10^2 represents 100, and the digit 7 in the chart above thus has the value 7×100, or 700. Similarly, 10^{-2} represents $\frac{1}{100}$, and the digit 8 has the value $8 \times \frac{1}{100}$, or $\frac{8}{100}$. In the same manner, the digit 5 represents "5 tenths," or $\frac{5}{10}$; the digit 4 represents "4 thousandths," or $\frac{4}{1000}$. Hence, any digit in the first place following a decimal represents that number divided by 10 $\left(\dfrac{x}{10}\right)$; any number in the second place represents that number over 100 $\left(\dfrac{y}{100}\right)$; any number in the third decimal place represents that number over 1,000 $\left(\dfrac{z}{1000}\right)$; and so on.

As pointed out previously, a decimal is actually an abbreviated fraction. Thus, the decimal fraction 0.475 (the zero is placed before the decimal point purely for clarity) represents a common fraction with a numerator of 475 and (since the decimal contains three digits to the right of the decimal point) a denominator of 1,000. Therefore, the decimal fraction 0.475 could be written $\frac{475}{1000}$. This illustrates why decimal fractions, like common fractions, are known as **rational numbers,** since they can be expressed as the quotient of two integers.

Decimal fractions are spoken (or written) by saying the digits as if they represented a whole number and then saying the place value associated with the farthest place to the right of the decimal point. Thus,

NUMBER	SPOKEN (OR WRITTEN) FORM
0.9	"Nine *tenths*"
0.14	"Fourteen *hundredths*"
0.007	"Seven *thousandths*"
0.3024	"Three thousand twenty-four *ten-thousandths*"
0.00508	"Five hundred eight *hundred-thousandths*"
25.6	"Twenty-five and six *tenths*"
2.03	"Two and three *hundredths*"
4,070.035	"Four thousand seventy and thirty-five *thousandths*"

Notice that zeros between the decimal point and the first significant digit do not affect the spoken form. (For example, 0.007 is said "*seven* thousandths.") Also, the word "and" is spoken (or written) solely to indicate the location of the decimal point. (Thus, 2.03 is pronounced "two *and* three hundredths.") Zeros to the right of a decimal fraction (with no other digits following) do not affect its value. Thus, 0.4, 0.40, 0.400, and 0.4000 are equal in value, because

$$\frac{4}{10} = \frac{40}{100} = \frac{400}{1000} = \frac{4,000}{10,000}$$

CONVERTING COMMON FRACTIONS TO DECIMAL FRACTIONS

Manual computation can often be accomplished efficiently using common fractions, although at times common fractions become cumbersome. Most computation in modern offices, however, is done on electronic machines, which are equipped to handle decimal values rather than common fractions. Thus, business efficiency requires a knowledge not only of common fractions and decimal fractions separately, but also of how to convert each form to the other.

You will recall that a common fraction is an indicated division. Hence, $\frac{2}{5}$ means $2 \div 5$. This principle is the basis for converting common fractions to decimal fractions. That is, a common fraction is converted to a decimal fraction by dividing the denominator into the numerator:

$$\frac{n}{d} = n \div d = d\overline{)n}$$

This procedure applies to improper as well as to proper fractions. If a mixed number is to be converted to a decimal number, the fractional part of the mixed number is converted in this same manner, and the decimal equivalent is then affixed to the whole number of the mixed number. Example 1 illustrates these procedures.

Example 1   (a) $\frac{2}{5} = 5\overline{)2.0}$ $.4 = 0.4$

$\underline{2\,0}$

(b) $\frac{3}{40} = 40\overline{)3.000}$ $.075 = 0.075$

$\underline{2\,80}$

200

$\underline{200}$

(c) $\dfrac{9}{8} = 8\overline{)9.000}^{\,1.125} = 1.125$

(d) $2\dfrac{8}{25}$: $\dfrac{8}{25} = 25\overline{)8.00}^{\,.32} = 0.32$

so $2\dfrac{8}{25} = 2.32$

The student is no doubt aware that many fractions do not have an exact decimal equivalent. This will always be the case (unless the fraction represents a whole number) if the denominator is 3 or is divisible by 3, or if the denominator is a prime number larger than five. (A **prime number** is any number that is divisible by no numbers except itself and 1—for example, 7, 11, 17, 23, 31, to name but a few.)

You may not always recognize whether a fraction has an exact decimal equivalent. The *recommended procedure for converting fractions* is first to divide to the hundredths place (two decimal places). If further inspection reveals that the quotient will come out to an exact digit after one more division, then include the third decimal place. If it is obvious that the quotient would not be exact even with three decimal places, then stop dividing after two decimal places and express the remainder as a fraction. (Students using calculators may round quotients to the nearest thousandth, or third decimal place. Rounding is reviewed in Chapter 1.) Consider the following examples.

Example 1 (cont.)

(e) $\dfrac{3}{8} = 8\overline{)3.000}^{\,.375} = 0.375$

(f) $\dfrac{13}{200} = 200\overline{)13.000}^{\,.065} = 0.065$

(g) $\dfrac{3}{7} = 7\overline{)3.00}^{\,.42\frac{6}{7}} = 0.42\frac{6}{7}$ (or 0.429)

(h) $\dfrac{16}{29} = 29\overline{)16.00}^{\,.55\frac{5}{29}} = 0.55\frac{5}{29}$ (or 0.552)

CONVERTING DECIMAL FRACTIONS TO COMMON FRACTIONS

The standard procedure for converting a decimal fraction to a common fraction is to pronounce the value of the decimal, write this value in fraction form, and then reduce this fraction. This is shown in Example 2.

Example 2

(a)   0.2   =   2 tenths $= \dfrac{2}{10}$ $= \dfrac{1}{5}$

(b)   0.12   =   12 hundredths $= \dfrac{12}{100}$ $= \dfrac{3}{25}$

(c)   0.336   = 336 thousandths $= \dfrac{336}{1,000}$ $= \dfrac{42}{125}$

(d)   0.0045 =   45 ten-thousandths $= \dfrac{45}{10,000}$ $= \dfrac{9}{2,000}$

If a decimal fraction also contains a common fraction, the conversion process begins in a similar manner but requires additional steps. First, the given decimal is rewritten as a fraction, and the numerator is then converted to an improper fraction. Next, we multiply by an appropriate fraction equal to 1 $\left(\dfrac{x}{x} = 1\right)$ and then reduce. The following examples show how $0.83\frac{1}{3}$ and $0.31\frac{1}{4}$ are converted to fractions.

Example 2 (cont.)

(e) $0.83\dfrac{1}{3} = \dfrac{83\frac{1}{3}}{100}$

$= \dfrac{\frac{250}{3}}{100}$

$= \dfrac{\frac{250}{3}}{100} \times \dfrac{\cancel{3}}{3}$

$= \dfrac{250}{300}$

$0.83\dfrac{1}{3} = \dfrac{5}{6}$

(f) $0.31\dfrac{1}{4} = \dfrac{31\frac{1}{4}}{100}$

$= \dfrac{\frac{125}{4}}{100}$

$= \dfrac{\frac{125}{4}}{100} \times \dfrac{\cancel{4}}{4}$

$= \dfrac{125}{400}$

$0.31\dfrac{1}{4} = \dfrac{5}{16}$

As noted previously, many computations can be done more easily by using fractional equivalents for decimal numbers, especially if the decimals are as complicated as those above. To save time, however, you should be able to recognize when a decimal is the equivalent of one of the common fractions most often used—without having to actually convert it. These equivalents are listed on page 53, and you should memorize all that you do not already know.

ADDITION AND SUBTRACTION

Addition and subtraction with decimals are very similar to addition and subtraction with integers. The same names apply to the parts of each problem:

Addend	0.36	Minuend	0.73
Addend	+0.58	Subtrahend	−0.46
Sum	0.94	Difference or remainder	0.27

As in addition and subtraction with whole numbers, decimal numbers must also be arranged so that all the digits of the same place are in the same column. That is, all the tenths places must be in a vertical line, all the hundredths places in one column, and so forth. In so arranging decimal numbers, the decimal points will always fall in a straight line. Therefore, in actual practice, we arrange the numbers by aligning the decimal points.

The actual addition or subtraction of decimal fractions is performed in exactly the same way as with integers; the decimal points have no effect on the operations. The decimal point in the answer is located directly under the decimal points of the problem.

It was pointed out earlier that appending zeros to the right of a decimal number does not alter its value (that is, $3.7 = 3.70 = 3.700$, and so on). Thus, if the numbers of a problem contain different numbers of decimal places, you may wish to add zeros so that all will contain the same number of decimal places. This is particularly helpful in subtraction problems where the subtrahend contains more decimal places than the minuend.

Example 3   (a) Add: $0.306 + 9.8 + 14.65 + 6.775 + 24 + 0.09$. (b) Subtract: $18.6 - 5.897$.

$$
\begin{array}{rl}
\text{(a)} & 0.306 \\
 & 9.8 \\
 & 14.65 \\
 & 6.775 \\
 & 24. \\
+ & 0.09 \\
\hline
 & 55.621
\end{array}
\qquad
\begin{array}{rl}
\text{(b)} & 18.600 \\
- & 5.897 \\
\hline
 & 12.703
\end{array}
$$

Notice that zeros were inserted in the minuend to facilitate the subtraction.

MULTIPLICATION

The standard terminology is used to identify decimal numbers in a multiplication problem:

$$
\begin{array}{ll}
\text{Multiplicand or factor} & 0.3 \\
\text{Multiplier or factor} & \times\ 0.4 \\
\hline
\text{Product} & 0.12
\end{array}
$$

Decimal numbers are written down for multiplication in the same arrangement that would be used if there were no decimal points. The actual multiplication of digits is also performed in the same way as for integers. The only variation comes in the placing of the decimal point in the product. To determine how this is done, consider the problem 12.4×3.13:

$$
\begin{array}{r}
12.4 \\
\times\ 3.13 \\
\hline
372 \\
124 \\
372 \\
\hline
38812
\end{array}
$$
Observe that this multiplication problem is approximately 12×3; therefore, the product must be approximately 36. This indicates that the decimal point must be located between the 8s. Thus,
$$
\begin{array}{r}
12.4 \\
\times\ 3.13 \\
\hline
372 \\
124 \\
372 \\
\hline
38.812
\end{array}
$$

The correct position of the decimal point in every multiplication product could thus be determined by approximating the answer and locating the decimal point accordingly. However, you would notice that the product invariably contains the same number of decimal places as the total number of decimal places in the factors. Therefore, as a time-saving device, it is customary to count the total number of decimal places in the factors and place the decimal point accordingly, rather than to estimate the product. (In the preceding problem, 12.4 contained one decimal place and 3.13 contained two decimal places; thus, the product contains three decimal places—38.812.) This example can also be explained as $\frac{124}{10} \times \frac{313}{100} = \frac{38,812}{1,000}$. Observe that the denominator of 1,000 indicates three decimal places.

Sometimes basic multiplication produces a number containing fewer digits than the total number of decimal places required for the product. In this case, you must insert a zero or zeros between the decimal point and the first significant digit so that the product will contain the correct number of decimal places. Consider the example 0.04×0.2:

$$
\begin{array}{r}
0.04 \\
\times\ 0.2 \\
\hline
8
\end{array}
\qquad
\begin{array}{l}
\text{The factors require a three-digit product. Thus, two} \\
\text{zeros must be inserted after the decimal point:}
\end{array}
\qquad
\begin{array}{r}
0.04 \\
\times\ 0.2 \\
\hline
0.008
\end{array}
$$

You can readily understand that 0.008 is the correct answer, because 0.04×0.2 is equal to $\frac{4}{100} \times \frac{2}{10} = \frac{8}{1,000}$.

Some additional examples are given below.

Example 4

$$
\text{(a)}\quad
\begin{array}{r}
41.8173 \\
\times\ \ \ \ 2.36 \\
\hline
2509038 \\
1254519\ \ \\
836346\ \ \ \ \\
\hline
98.688828
\end{array}
\qquad
\text{(b)}\quad
\begin{array}{r}
0.0547 \\
\times\ \ \ 3.6 \\
\hline
3282 \\
1641\ \ \\
\hline
0.19692
\end{array}
\qquad
\text{(c)}\quad
\begin{array}{r}
7.38 \\
\times\ 0.004 \\
\hline
0.02952
\end{array}
\qquad
\text{(d)}\quad
\begin{array}{r}
0.24 \\
\times\ 0.03 \\
\hline
0.0072
\end{array}
$$

Multiplication by powers of 10 presents a special case. Study the operations in Example 5.

Example 5

$$
\text{(a)}\quad
\begin{array}{r}
25.13 \\
\times\ \ 10 \\
\hline
251.30
\end{array}
\qquad
\text{(b)}\quad
\begin{array}{r}
65.8 \\
\times\ \ 100 \\
\hline
6,580.0
\end{array}
\qquad
\text{(c)}\quad
\begin{array}{r}
78 \\
\times\ 100 \\
\hline
7,800
\end{array}
\qquad
\text{(d)}\quad
\begin{array}{r}
3.9471 \\
\times\ \ \ 1,000 \\
\hline
3,947.1000
\end{array}
$$

Observe that each product contains the same digits as the multiplicand (along with some zeros appended) and that each decimal point has been moved to the right the same number of places as there were zeros in the power of 10. By applying this information, you should never actually have to multiply by a power of 10. Instead, you can obtain the product simply by moving the decimal point as many places to the right as there are zeros in the power of 10. This is shown in Example 6.

Example 6

(a) $25.13 \times 10 = 25.1.3$ (b) $65.8 \times 100 = 65.80.$

            $25.13 \times 10 = 251.3$ $65.8 \times 100 = 6{,}580$

(c) $78 \times 100 = 78.00.$ (d) $3.9471 \times 1{,}000 = 3.947.1$

         $78 \times 100 = 7{,}800$ $3.9471 \times 1{,}000 = 3{,}947.1$

DIVISION

Decimal numbers in a division problem are identified by the terms previously learned:

$$\text{Divisor} \quad 1.5)\overline{7.95} \quad \begin{array}{l} 5.3 \quad \text{Quotient} \\ \text{Dividend} \end{array}$$

The basic operation of division is performed in the same manner for decimal numbers as for whole numbers. The decimal points do not affect the division process itself, but only the location of the decimal point in the quotient. To understand how the decimal points are handled, consider the following example:

$$\begin{array}{r} 51 \\ 3.02)\overline{15.402} \\ \underline{15\ 10} \\ 302 \\ \underline{302} \end{array}$$

Notice that this division is approximately $3)\overline{15}$; therefore, the quotient must be approximately 5. This indicates that the decimal point in the quotient must be located after the 5.

$$\begin{array}{r} 5.1 \\ 3.02)\overline{15.402} \\ \underline{15\ 10} \\ 302 \\ \underline{302} \end{array}$$

It would be possible to position the decimal point in every quotient by first estimating the result and then placing the decimal point accordingly. However, this requires additional time and effort; hence, some quick, easy method for determining the correct location of the quotient decimal is necessary. You are no doubt familiar with the method commonly used to position the decimal quickly; that is, to "move the decimal point to the end of the divisor and move the decimal in the dividend the same number of places." Let us see why this procedure is permissible.

It was observed that the decimal point of the quotient is always located directly above the decimal point of the dividend if the divisor is a whole number. For example,

$$\begin{array}{r} 2\ 13 \\ 12)\overline{25.56} \\ \underline{24} \\ 1\ 5 \\ \underline{1\ 2} \\ 36 \\ \underline{36} \end{array}$$

Since this problem is approximately $12)\overline{25}$, the result must be approximately 2. Thus, the decimal point of the quotient should be located directly above the decimal point in the operation:

$$\begin{array}{r} 2.13 \\ 12)\overline{25.56} \\ \underline{24} \\ 1\ 5 \\ \underline{1\ 2} \\ 36 \\ \underline{36} \end{array}$$

Thus, you know exactly where the decimal point belongs if the divisor is a whole number. With this in mind, let us apply three familiar principles—namely, that a fraction is an indicated division, that any number over itself equals 1 $\left(\text{that is,} \frac{x}{x} = 1\right)$, and that multiplying any number by 1 equals that same number $(x \cdot 1 = x)$. Now suppose that we have the division problem $3.2\overline{)9.28}$. This is the division indicated by the fraction $\frac{9.28}{3.2}$. Then,

$$\frac{9.28}{3.2} = \frac{9.28}{3.2} \times 1$$

$$= \frac{9.28}{3.2} \times \frac{10}{10}$$

$$\frac{9.28}{3.2} = \frac{92.8}{32}$$

Thus the divisions indicated by each fraction are equal, or

$$3.2\overline{)9.28} = 32\overline{)92.8}$$

But we know the correct location for the decimal point when dividing by a whole number:

$$= \begin{array}{r} 2.9 \\ 32\overline{)92.8} \\ \underline{64} \\ 28\ 8 \\ \underline{28\ 8} \end{array}$$

Therefore, the other quotient must also be 2.9.

$$\begin{array}{r} 2.9 \\ 3.2\overline{)9.28} \\ \underline{6\ 4} \\ 2\ 88 \\ \underline{2\ 88} \end{array} = \begin{array}{r} 2.9 \\ 32\overline{)92.8} \end{array}$$

Thus, we can always position the decimal easily by "making the divisor a whole number," so to speak. But when we "move the decimal point to the end of the divisor and move the decimal point in the dividend a corresponding number of places," we are, in effect, multiplying by a fraction equal to 1, which does not change the value. Stated differently, moving both decimals one place to the right is the same as multiplying by $\frac{10}{10}$; moving the decimals two places is equivalent to multiplying by $\frac{100}{100}$; moving the decimals three digits is equivalent to multiplying by $\frac{1,000}{1,000}$; and so forth.

Having determined why it is permissible to "move the decimal points," let us now consider some special cases. The decimal point in a dividend must often be moved more places to the right than there are digits. In this case, we must append zeros to the dividend in order to fill in the required number of places. (Suppose that the decimal point in the dividend 14.3 must be moved three places to the right. This is the same as multiplying 14.3 by 1,000; so, 14.3 _ _. = 14,300.) If the dividend is a whole number, we obviously must first insert a decimal point before moving it to the right. (For example, if a divisor must be multiplied by 10 to become a whole number and the dividend is 17, then 17 must also be multiplied by 10. That is, the decimal point must be moved one place to the right: $17 \times 10 = 17. \times 10 = 17$ _. = 170.)

It has been pointed out that appending zeros to the right of a decimal fraction does not change its value. (That is, $0.52 = 0.520 = 0.5200 = 0.52000$, and so forth). When a division problem does not terminate (or end) within the number of digits contained in the dividend, the operation may be continued by appending zeros to the dividend. The process may be continued until the quotient terminates or until it contains as many decimal places as required.

Once the decimal point has been located in the quotient, each succeeding place must contain a digit. If normal division skips some of the first decimal places, these places must be held by zeros. (This procedure is similar to that in multiplication, when zeros must be inserted after the decimal point in order to complete the required number of decimal places.)

Example 7 shows how these procedures apply.

Example 7

(a) $3.02 \overline{)15.402}$

$$
\begin{array}{r}
5.1 \\
3.02 \overline{)15.40.2} \\
15\ 10 \\
\hline
30\ 2 \\
30\ 2 \\
\hline
\end{array}
$$

(b) $4.725 \overline{)37.8}$

$$
\begin{array}{r}
8. \\
4.725 \overline{)37.800.} \\
37\ 800 \\
\hline
\end{array}
$$

(c) $0.72 \overline{)36.}$

$$
\begin{array}{r}
50. \\
0.72 \overline{)36.00.} \\
36\ 0 \\
\hline
0 \\
0 \\
\hline
\end{array}
$$

(d) $6.4 \overline{)0.3968}$

$$
\begin{array}{r}
.062 \\
6.4 \overline{)0.3.968} \\
3\ 84 \\
\hline
128 \\
128 \\
\hline
\end{array}
$$

A special case develops when the divisor of a problem is a power of 10, as shown in Example 8.

Example 8

(a) 10)13.6

(b) 100)245.3

```
        1.36
   10)13.60
      10
       3 6
       3 0
         60
         60
```

```
          2.453
   100)245.300
       200
        45 3
        40 0
          5 30
          5 00
            300
            300
```

(c) 100)7.31

(d) 1,000)629

```
         .0731
   100)7.3100
      7 00
        310
        300
        100
        100
```

```
           .629
   1,000)629.000
         600 0
          29 00
          20 00
           9 000
           9 000
```

In each operation in Example 8, the quotient contains the same digits as the dividend, but the decimal point is moved to the left the same number of places as there were zeros in the divisor. Knowing this, students should never actually have to perform a division by a power of 10. Rather, you can obtain the answer simply by moving the decimal point to the left. This is shown in Example 9.

Example 9

(a) $13.6 \div 10 = 1.3.6 = 1.36$

(b) $245.3 \div 100 = 2.45.3 = 2.453$

(c) $7.31 \div 100 = .\_7.31 = 0.0731$

(d) $629 \div 1,000 = .629 = 0.629$

Hint. Students often confuse multiplication and division by powers of 10 and forget which way to move the decimal point. If this happens, you can quickly determine the correct direction by making up an example using small numbers. For example. $4 \times 10 = 40$ (that is, $4.0 \times 10 = 40$). The decimal point of the 4 moved to the right; therefore, the decimal point moves toward the right for *all* multiplication by powers of 10. Hence, the decimal point obviously moves to the left for division.

Note. Addition with decimal fractions is both commutative and associative; multiplication is similarly commutative and associative. Further, the distributive property of multiplication over addition also applies. You can verify these properties for yourself by selecting some decimal values and following the examples for whole numbers in Section 2 of this Appendix.

SECTION 4   PROBLEMS

1. In the number 6,432.85179, identify the place value associated with each of the following digits.
   a. 1   b. 2   c. 3   d. 4   e. 5   f. 6   g. 7   h. 8   i. 9

2. Write in words the values of the following numbers.
   a. 0.5   b. 0.06   c. 0.018   d. 0.32   e. 8.6   f. 17.44   g. 500.103   h. 9.0015

3. Convert the following common fractions (and mixed numbers) to their decimal fraction equivalents.

   a. $\dfrac{2}{5}$ b. $\dfrac{1}{4}$ c. $\dfrac{6}{20}$ d. $\dfrac{21}{30}$ e. $\dfrac{16}{50}$ f. $\dfrac{15}{200}$ g. $\dfrac{1}{7}$ h. $\dfrac{4}{9}$ i. $\dfrac{5}{12}$

   j. $\dfrac{6}{11}$ k. $\dfrac{4}{13}$ l. $\dfrac{7}{26}$ m. $\dfrac{18}{3}$ n. $\dfrac{163}{4}$ o. $\dfrac{108}{20}$ p. $5\dfrac{1}{9}$ q. $3\dfrac{5}{6}$ r. $9\dfrac{107}{250}$

4. Find the common fraction equivalent to each of the following decimals.
   a. 0.06   b. 0.4   c. 0.25   d. 0.325   e. 0.155

   f. 0.008   g. 0.036   h. 0.45½   i. 0.04¼   j. 0.036⅘

5. Multiply or divide, as indicated, by moving the decimal points.
   a. 356.2 ÷ 10   b. 2.645 × 10   c. 556.63 ÷ 100   d. 104.88 ÷ 1,000

   e. 0.88452 × 1,000   f. 16 ÷ 10   g. 26.5 × 10   h. 8.812 × 100

   i. 2.8 × 100   j. 3.651 ÷ 1,000   k. 525 × 100   l. 0.644 ÷ 100

   m. 54.6 × 10,000   n. 99.45 ÷ 10,000

6. Add. [Arrange parts j–m in columns before adding.]

a.	4.96	b.	0.164	c.	0.1887	d.	12.713	e.	175.024
	6.01		0.917		0.5295		81.591		716.054
	4.53		0.244		0.4897		20.307		505.149
	+8.70		+0.415		+0.0972		+81.633		+347.664

f.	18.131	g.	1.02	h.	6.3	i.	52.7
	4.01		62.6		341		149.66
	215.8		7.482		8.98		62.435
	0.145		138.5		12.455		9.15
	42.5		16.933		0.138		317.8
	+ 0.005		+541.87		+530.68		+274.511

   j. 123.7 + 0.515 + 22.7 + 156.68 + 7.37 + 120.64

   k. 54.88 + 6.83 + 328.6 + 330 + 5.81 + 146.9

   l. 3.691 + 304.84 + 26.641 + 31.106 + 0.482 + 56.2

   m. 94.171 + 135 + 8.06 + 26.71 + 125.7 + 206.728

7. Subtract. [Arrange parts n–q in columns before subtracting.]

a. 135.86
    − 88.91

b. 5.224
    −1.499

c. 38.267
    − 9.408

d. 385.002
    −144.686

e. 56.1146
    −48.3036

f. 15.001
    −11.156

g. 634.703
    − 75.881

h. 27.549
    −14.62

i. 34.1006
    − 5.091

j. 265.4
    −123.056

k. 108.6
    56.321

l. 543.4
    − 88.355

m. 64
    −26.303

n. 54.2 − 6.18   o. 314 − 28.88   p. 105.5 − 36.606   q. 5.4 − 0.8212

8. Multiply.

a. 6.24
    × 6.4

b. 18.3
    × 5.4

c. 6.06
    ×0.48

d. 352.8
    ×0.015

e. 0.124
    × 5.25

f. 0.311
    × 0.47

g. 0.514
    × 0.82

h. 0.005
    ×0.029

i. 0.3551
    × 2.64

j. 0.418
    ×0.033

9. Divide. (When quotients are not exact whole numbers, follow the procedure recommended in this section for changing fractions to decimals.)

a. $6.5\overline{)507}$ b. $5.17\overline{)635.91}$ c. $0.125\overline{)5.75}$ d. $0.122\overline{)8.296}$

e. $5.62\overline{)764.32}$ f. $1.742\overline{)5,696.34}$ g. $0.64\overline{)0.8128}$ h. $3.32\overline{)0.7968}$

i. $13.5\overline{)4.644}$ j. $0.724\overline{)23.58792}$ k. $5.6\overline{)4.368}$ l. $28.44\overline{)1.83438}$

m. $25.8\overline{)11,352}$ n. $5.62\overline{)230.42}$ o. $0.285\overline{)91.485}$ p. $4.132\overline{)1.15696}$

APPENDIX A   GLOSSARY

Addend.   Any of the numbers being combined in addition. (Example: In 2 + 3 = 5, the 2 and 3 are addends.)

Associative property.   Pertains to the grouping of three numbers for addition or multiplication; states that the result will be the same regardless of which two numbers are added (or multiplied) first. That is, $(a + b) + c = a + (b + c)$ for addition; $(a \cdot b)c = a(b \cdot c)$ for multiplication.

Common fraction.   An indicated division, using a horizontal line. (Examples: $\frac{3}{5}$ or $\frac{7}{4}$)

Commutative property.   Pertains to the order of two numbers for addition or multiplication; states that the result will be the same regardless of which number is taken first. That is, $a + b = b + a$ for addition; $a \cdot b = b \cdot a$ for multiplication.

Counting numbers.   (See "Natural numbers.")

Decimal fraction.   An abbreviated fraction whose denominator is a certian power of 10 indicated by the location of a dot called the decimal point. (Example: 0.37 means $\frac{37}{100}$.)

Decimal notation.   The system used to denote decimal fractions.

Denominator.   The number below the division line of a common fraction. (Example: The denominator of $\frac{3}{5}$ is 5.)

Difference.   The result in a subtraction operation.

Digit.   Any of the 10 characters 0–9 with which decimal (or base 10) numbers are formed.

Distributive property.   States that a factor times the sum of two numbers equals the sum of the individual products; that is, $a(b + c) = ab + ac$.

Dividend.   In a division operation, the number that is being divided. (Dividend ÷ Divisor = Quotient)

Divisor.   In a division operation, the number that divides into another.

(Example: In $4\overline{)12}^{\,3}$, the divisor is 4.)

Exponent.   A superscript that indicates how many times another number, called the base, should be written for repeated multiplication times itself. (Example: The exponent 4 denotes that $10^4 = 10 \cdot 10 \cdot 10 \cdot 10 = 10,000$.)

Factor.   Any number (multiplicand or multiplier) used in multiplication.

Fraction.   (See "Common fraction.")

Higher terms.   An equivalent fraction with a larger denominator. (Example: $\frac{3}{5}$ may be expressed in higher terms as $\frac{9}{15}$.)

Improper fraction.   A common fraction whose numerator is equal to or larger than the denominator. (Examples: $\frac{7}{7}$ or $\frac{15}{8}$)

Integer.   Any number, positive or negative, that can be expressed without a decimal or fraction. Any number from the set {. . . , −3, −2, −1, 0, 1, 2, 3, . . .}.)

Least (or lowest) common denominator.   The smallest number (denominator) into which all given denominators will divide.

Minuend.   In a subtraction operation, the number from which another number is subtracted. (Minuend − Subtrahend = Difference)

Mixed number.   A number composed of a whole number and a fraction. (Example: $12\frac{3}{7}$)

Multiplicand.   The first (or top) factor in a multiplication operation. (Multiplicand × Multiplier = Product)

Multiplier.   The second (or bottom) factor in a multiplication operation. (Example: In 8×5, the multiplier is 5.)

Natural (or counting) number.   Any number that may be used in counting objects. (Any number from the set {1, 2, 3, 4, . . .}.)

Numerator.   The number above the division line of a common fraction. (Example: The numerator of $\frac{4}{9}$ is 4.)

Operation.   Any of the arithmetic processes of addition, subtraction, multiplication, and division.

Place value.   The value associated with a digit because of its position in a whole number or decimal number.

Prime number.   A number (larger than 1) that can be evenly divided only by itself and 1. (Examples: 2, 3, 5, 7, 11, . . .)

Product.   The result in a multiplication operation.

Proper fraction.   A fraction whose numerator is smaller than the denominator. (Example: $\frac{4}{5}$)

Quotient.   The result in a division operation.

Rational number.   Any number that can be expressed as a common fraction (the quotient of two integers).

Reciprocal (of a fraction).   A fraction formed by interchanging the original numerator and denominator. (Example: The reciprocal of $\frac{5}{8}$ is $\frac{8}{5}$.)

Subtrahend.   In a subtraction operation, the number being subtracted from another. (Example: In $10 − 7 = 3$, the subtrahend is 7.)

Sum.   The result in an addition operation.

Whole number.   Any number zero or larger that can be expressed without a decimal or fraction. (Any number from the set {0, 1, 2, 3, 4, . . .}.)

B

METRIC AND CURRENCY CONVERSIONS

OBJECTIVES

Upon completion of Appendix B, you will be able to:

1. Define and use correctly the terminology associated with each topic.

2. Given any multiple of a metric unit (Section 1: Examples 1, 2; Problems 1–4):

   a. Name its single equivalent metric unit

   b. Convert the given measurement to its equivalent measurement in another given metric unit.

3. Given a metric measurement, find its equivalent in a given U.S. unit (Section 1: Example 3; Problems 5, 6, 11, 12, 17, 18).

4. Given a U.S. measurement, determine its equivalent in a given metric unit (Section 1: Example 4; Problems 7, 8, 11–16).

5. Given temperature readings in either Celsius or Fahrenheit, convert to the other temperature scale (Section 1: Example 5; Problems 9, 10, 19, 20).

6. Given any multiple of a U.S. dollar, find its equivalent in the currency of another country (Section 2: Example 1; Problems 1, 2, 5–8, 13, 14).

7. Given any multiple of a foreign currency, determine its equivalent in U.S. dollars (Section 2: Example 2; Problems 3, 4, 9–12).

There is practically no individual, business, or industry that is not affected by international trade and commerce. Many of the employees who are hired by corporations and the customers who purchase goods come from culturally diverse backgrounds. Similarly, many products sold in the United States today come from El Salvador, Saudi Arabia, or Turkey, to name but a few countries of origination. U. S.-made products are sold worldwide; our economy is indeed global. To recognize the importance of the global marketplace, two international topics—metric conversions and currency conversions—are covered in this Appendix.

SECTION 1

METRIC CONVERSION

In 1971, a National Bureau of Standards report to Congress, entitled "A Metric America: A Decision Whose Time Has Come," described the United States as an isolated island in a metric world. The report recommended the passage of the Metric Conversion Act. Although the act was passed, it lacked clearly stated objectives and a timetable for implementation. In 1988, a metric usage provision was included in the Omnibus Trade and Competitiveness Act. The legislation designated the metric system as "the preferred system of weights and measurements for United States trade and commerce." It required federal agencies to use the metric system in procurements, grants, and other business-related activities. An executive order was signed by President Bush in 1991 that gave specific management authority to the Secretary of Commerce to coordinate implementation of the metric usage provisions of the Omnibus Trade Act.

The goal of these legislative and executive acts is to help U.S. industry and business compete more successfully in the global marketplace. The metric system (SI—International Systems of Units) is the international standard of measurement. The United States is the only developed country in the world that has not completely converted. Many U.S. industries, however, have converted to metrics. Today, we purchase photographic film in millimeters, medicines in milligrams, and soft drinks in liters.

Metric measurement is a decimal system, based on the number 10. The metric system was established with the **meter** (m) as the basic unit for length, the **gram** (g) as the basic unit for weight, and the **liter** (l) as the basic unit for liquid volume or capacity. Prefixes before each basic unit are used to name the other units of the system and to identify its multiple of the basic unit. The principal prefixes and indicated multiples of the basic unit are as follows:

METRIC PREFIX	MULTIPLE OF BASIC UNIT	METRIC PREFIX	MULTIPLE OF BASIC UNIT
deca (dk)	10	deci (d)	0.1 or 1/10
hecto (h)	100	centi (c)	0.01 or 1/100
kilo (k)	1,000	milli (m)	0.001 or 1/1000
mega	1,000,000	micro	0.000001 or 1/1000000

Using the information in the table, we see that

$$1\ kilo\text{meter} = 1000\ \text{meters}$$

$$1\ deca\text{liter} = 10\ \text{liters}$$

$$1\ centi\text{gram} = 0.01\ \text{gram}$$

The following chart, which is similar to a place-value table for numbers, will help identify the power of 10 associated with each prefix.

* Meter, gram, or liter

Example 1   Identify the unit name for the given metric measurements.

(a)   10 meters = 1 ____   (b) 1000 liters = 1 ____

(c)   0.1 gram = 1 ____   (d) 100 grams = 1 ____

(e)   0.001 liter = 1 ____   (f) 0.01 meter = 1 ____

Using the basic units and prefix definitions above,

(a)   10 meters = 1 decameter   (b) 1000 liters = 1 kiloliter

(c)   0.1 gram = 1 decigram   (d) 100 grams = 1 hectogram

(e)   0.001 liter = 1 milliliter   (f) 0.01 meter = 1 centimeter

Conversion to larger or smaller metric units is accomplished by first moving to the right or left in the chart of prefixes, then moving the decimal in the value the same number of places. To convert meters to decameters: Since decameters is 1 place to the left of meters, move the decimal 1 place to the left in the value in meters. Consider 40 meters as an example:

			4.0	40.			
km	hm	dkm	m	dm	cm	mm	

So, 40 meters = 4 decameters. To convert 40 meters to millimeters, move the decimal three places to the right in the value in meters, since millimeters is three places to the right of meters:

			40.			40 000.
km	hm	dkm	m	dm	cm	mm

So, 40 meters = 40 000 millimeters.

Example 2

(a) Convert 10 grams to centigrams. To convert grams to centigrams, move the decimal in the grams value to the right two places.

			10.		1 000.	
kg	hg	dkg	g	dg	cg	mg

$$10.0 \text{ g} = 10.00$$

$$= 1000 \text{ cg}$$

(b) Convert 65 deciliters to hectoliters. To convert deciliters to hectoliters, move the decimal in the deciliters value to the left three places.

	.065			65.		
kl	hl	dkl	l	dl	cl	ml

$$65 \text{ dl} = 0\ 065.$$

$$= 0.065 \text{ hl}$$

The major challenge when anyone first uses the metric system is to develop a mental picture of the quantities that metric measures represent. By rough approximation, 1 meter is slightly longer than 1 yard; 1 liter is nearly equivalent to 1 quart. The gram is so small that it is used to weigh only small amounts, such as food items. Larger amounts are weighed using the kilogram, which is somewhat more than 2 pounds. Distances are expressed in kilometers, each of which is about ⅝ mile. As the following illustration shows, 1 inch is equivalent to 2.54 centimeters.

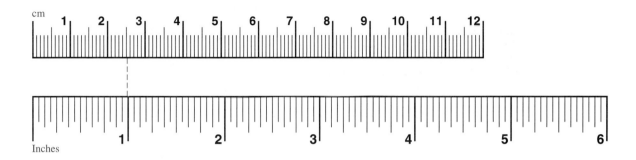

Example 3   Convert the following metric measurements to their U.S. equivalents, using the equivalents given in Table B-1.

(a)   10 meters = ? feet   (b) 5 liters = ? quarts

(c)   200 grams = ? ounces   (d) 500 kilograms = ? pounds

(a)   1 meter =   3.28 feet
      × 10   ×   10
      ─────────   ─────────
      10 meters =   32.8 feet

(b)   1 liter =   1.06 quarts
      × 5   ×   5
      ─────────   ─────────
      5 liters =   5.3 quarts

(c)   1 gram   =   0.035 ounce
      × 200   ×   200
      ──────────   ──────────
      200 grams =   7.0 ounces

(d)   1 kilogram =   2.20 pounds
      × 500   ×   500
      ────────────   ───────────
      500 kilograms =   1,100 pounds

TABLE B-1	METRIC MEASURES WITH U.S. EQUIVALENTS		
LENGTH	**WEIGHT**		**LIQUID CAPACITY**

LENGTH	WEIGHT	LIQUID CAPACITY
1 millimeter = 0.039 inch 1 centimeter = 0.394 inch 1 decimeter = 3.94 inches 1 meter = 39.37 inches = 3.28 feet = 1.09 yards 1 kilometer = 0.621 mile	1 gram[a] = 0.035 ounce 1 kilogram = 2.20 pounds **TEMPERATURE** $C = \frac{5}{9}(F - 32)$ or $C = (F - 32) \div 1.8$	1 liter[b] = 33.8 ounces = 2.11 pints = 1.06 quarts = 0.264 gallon

[a]1 gram is defined as the weight of 1 cubic centimeter of pure water at 4° Celsius.
[b]1 liter is defined as the capacity of a cube that is 1 decimeter long on each edge.

Example 4 Use the equivalents given in Table B-2 to find the metric equivalents of the following U.S. measurements.

    (a) 50 yards = ? meters   (b) 4 gallons = ? liters

    (c) 6 inches = ? centimeters   (d) 150 pounds = ? kilograms

(a)   1 yard = 0.914 meter
    × 50   × 50
     50 yards = 45.7 meters

(b)   1 gallon = 3.79 liters
    × 4   × 4
     4 gallons = 15.16 liters

(c)   1 inch = 2.54 centimeters
    × 6   × 6
     6 inches = 15.24 centimeters

(d)   1 pound = 0.454 kilogram
    × 150   × 150
     150 pounds = 68.1 kilograms

Metric quantities will develop meaning to you as you use them, including the **Celsius** (or **centigrade**) temperature scale. You will learn, for example,

TABLE B-2	U.S. MEASURES WITH METRIC EQUIVALENTS		

LENGTH	WEIGHT	LIQUID CAPACITY
1 inch = 2.54 centimeters 1 foot = 0.305 meter 1 yard = 0.914 meter 1 mile = 1.61 kilometers	1 ounce = 28.4 grams 1 pound = 454 grams = 0.454 kilograms **TEMPERATURE** $F = \frac{9}{5}C + 32$ or $F = 1.8C + 32$	1 ounce = 0.030 liter 1 pint = 0.473 liter 1 quart = 0.946 liter 1 gallon = 3.79 liters

United States Department of Commerce
Technology Administration
National Institute of Standards and Technology
Metric Program, Gaithersburg, MD 20899

All You Will Need to Know About Metric
(For Your Everyday Life)

10

Metric is based on the Decimal system

The metric system is simple to learn. For use in your everyday life you will need to know only ten units. You will also need to get used to a few new temperatures. Of course, there are other units which most persons will not need to learn. There are even some metric units with which you are already familiar; those for time and electricity are the same as you use now.

BASIC UNITS

METER: a little longer than a yard (about 1.1 yards)
LITER: a little larger than a quart (about 1.06 quarts)
GRAM: a little more than the weight of a paper clip

(comparative sizes are shown)

1 METER

1 YARD

25 DEGREES FARENHEIT

COMMON PREFIXES
(to be used with basic units)

milli: one-thousandth (0.001)
centi: one-hundredth (0.01)
kilo: one-thousand times (1000)

For example
1000 millimeters = 1 meter
100 centimeters = 1 meter
1000 meters = 1 kilometer

1 LITER

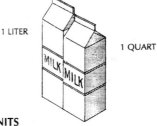

1 QUART

OTHER COMMONLY USED UNITS

millimeter:	0.001 meter	diameter of a paper clip wire
centimeter:	0.01 meter	a little more than the width of a paper clip (about 0.4 inch)
kilometer:	1000 meters	somewhat further than 1/2 mile (about 0.6 mile)
kilogram:	1000 grams	a little more than 2 pounds (about 2.2 pounds)
millimeter:	0.001 liter	five of them make a teaspoon

25 DEGREES CELSIUS

OTHER USEFUL UNITS
hectare: about 2 1/2 acres
metric ton: about one ton

WEATHER UNITS: **FOR TEMPERATURE** degrees celsius **FOR PRESSURE** kilopascals are used
100 kilopascals = 29.5 inches of Hg (14.5 psi)

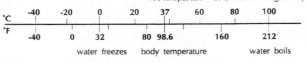

1 POUND

1 KILOGRAM

°C −40 −20 0 20 37 60 80 100
°F −40 0 32 80 98.6 160 212
water freezes body temperature water boils

Courtesy of U.S. Department of Commerce Metric Program

that 10°C is a cool day, that 20°–25°C is the comfortable indoor range, and that 35°C is a hot summer day.

Temperature conversions may be done using the formulas in Tables B-1 and B-2. Either the decimal or fractional forms of each formula may be used, as convenient.

Example 5 Find the following temperature equivalents.

(a) 32°F = ?°C   (b) 35°C = ?°F

(c) 212°F = ?°C   (d) 10°C = ?°F

(a) $C = \frac{5}{9}(F - 32)$

$= \frac{5}{9}(32 - 32)$

$= \frac{5}{9}(0)$

$C = 0°$ (32°F = 0°C)

(b) $F = \frac{9}{5}C + 32$

$= \frac{9}{5}(35) + 32$

$= 63 + 32$

$F = 95°$ (35°C = 95°F)

(c) $C = (F - 32) \div 1.8$

$= (212 - 32) \div 1.8$

$= 180 \div 1.8$

$C = 100°$ (212°F = 100°C)

(d) $F = 1.8C + 32$

$= 1.8(10) + 32$

$= 18 + 32$

$F = 50°$ (10°C = 50°F)

If you once become accustomed to the metric system, the present U.S. system of weights and measures will seem as inefficient as any foreign monetary system that is not based on dollars and 100 cents.

Section 1   Problems

Name each metric unit or indicate the multiple of each given unit.

1. a. 1 decameter = ? meters

c. 0.001 gram = 1 __?__

e. 1 meter = ? centimeters

g. 1000 meters = 1 __?__

i. 1 hectogram = ? decigrams

k. 1 kilometer = ? decameters

b. 100 liters = 1 __?__

d. 1 deciliter = ? liter

f. 0.1 decagram = 1 __?__

h. 1 milliliter = ? liter

j. 0.1 centiliter = 1 __?__

l. 100 decigrams = 1 __?__

2. a. 1 kilogram = ? grams

c. 0.01 liter = 1 __?__

e. 1 meter = ? kilometer

g. 10 decigrams = 1 __?__

i. 1 hectometer = ? decameters

k. 0.01 deciliter = 1 __?__

b. 10 meters = 1 __?__

d. 1 millimeter = ? meter

f. 0.01 hectoliter = 1 __?__

h. 1 liter = ? centiliters

j. 1000 centigrams = 1 __?__

l. 1 hectoliter = ? kiloliter

Convert the following metric measurements as indicated.

3. a. 3.6 hectoliters = ? liters   b. 45 decimeters = ? meters
c. 2150 milligrams = ? grams   d. 0.8 kilometers = ? meters
e. 4.25 grams = ? centigrams   f. 188 liters = ? hectoliters
g. 30 meters = ? decimeters   h. 1200 grams = ? kilograms
i. 2.9 centigrams = ? milligrams   j. 76 meters = ? hectometers

4. a. 1.5 hectometers = ? meters   b. 125 centimeters = ? meters
c. 36 decagrams = ? grams   d. 2137 milliliters = ? liters
e. 1475 grams = ? kilograms   f. 17.4 grams = ? centigrams
g. 24 liters = ? decaliters   h. 50 meters = ? millimeters
i. 75 hectometers = ? kilometers   j. 362 grams = ? kilograms

Determine the U.S. equivalent of the following metric measurements.

5. a. 6 meters = ? yards   b. 50 kilograms = ? pounds
c. 10 liters = ? quarts   d. 80 kilometers = ? miles
e. 20 centimeters = ? inches   f. 0.5 liter = ? ounces
g. 5 meters = ? feet   h. 400 grams = ? ounces

6. a. 10 centimeters = ? inches   b. 3 meters = ? inches
c. 4 kilograms = ? pounds   d. 6 liters = ? ounces
e. 20 meters = ? yards   f. 5 decimeters = ? inches
g. 8 liters = ? quarts   h. 50 kilometers = ? miles

Find the metric equivalent of the following U.S. measurements.

7. a. 8 feet = ? meters   b. 100 pounds = ? kilograms
c. 4 quarts = ? liters   d. 20 pints = ? liters
e. 80 miles = ? kilometers   f. 6 inches = ? centimeters
g. 9 ounces = ? grams   h. 5 gallons = ? liters

8. a. 8 inches = ? centimeters   b. 10 miles = ? kilometers
c. 8 ounces = ? grams   d. 6 quarts = ? liters
e. 3 pounds = ? grams   f. 20 ounces = ? liters
g. 6 feet = ? meters   h. 50 gallons = ? liters

Convert each temperature measurement as indicated.

9. a. 77°F = ?°C   b. 95°F = ?°C   c. 59°F = ?°C   d. 14°F = ?°C
e. 20°C = ?°F   f. 80°C = ?°F   g. 55°C = ?°F   h. −5°C = ?°F

10. a. 104°F = ?°C   b. 41°F = ?°C   c. 68°F = ?°C   d. 5°F = ?°C
e. 35°C = ?°F   f. 45°C = ?°F   g. 10°C = ?°F   h. 0°C = ?°F

Find each measurement below as indicated.

11. An American-made car has a 120-inch wheelbase, and a European car has a 250-centimeter wheelbase.
a. How many centimeters is the wheelbase of the American car?
b. How many inches is the European car's wheelbase?
c. Which wheelbase is longer, and by how many centimeters?

12. Track sprinters run the 100-yard dash, while swimmers compete in 100-meter races.
    a. How many meters is 100 yards?
    b. How many yards is 100 meters?
    c. Which race is longer, and by how many meters?

13. Elaine Stevens is 5 feet, 6 inches tall and weighs 120 pounds. Express her (a) height in centimeters and (b) weight in kilograms. (Hint: First convert her height to inches.)

14. Bill Gray is 5 feet, 10 inches tall and weighs 170 pounds. Express his (a) height in centimeters and (b) weight in kilograms.

15. The United States is approximately 3,000 miles wide. The earth is approximately 25,000 miles in circumference. Express each distance in kilometers.

16. Tasty Food Processors markets a pancake mix with a net weight of 40 ounces and a pancake syrup that contains 30 fluid (liquid) ounces. What metric equivalents should the labels contain for (a) the mix in grams and (b) the syrup in liters?

17. The Import House offers a gold pendant that weighs 20 grams and a silver tea service that weighs 1.75 kilograms.
    a. How many ounces of gold are in the pendant?
    b. How many pounds of silver are in the tea service?

18. The gas tank of a European car holds 50 liters, and 4 liters of oil are required in the motor. What is the U.S. equivalent of (a) the gas tank capacity in gallons and (b) the oil in quarts?

19. The high temperature in New York City was 95°F. The high in Madrid that same day was 30°C. Determine (a) the New York temperature in Celsius and (b) the Madrid temperature in Fahrenheit. (c) Which city was warmer, and by how many degrees Celsius?

20. Miami had a low temperature of 50°F on a day when the low in Madrid was 5°C. What were (a) the Miami temperature in Celsius and (b) the Madrid temperature in Fahrenheit? (c) Where was the air colder, and by how many degrees Celsius?

SECTION 2
FOREIGN CURRENCY CONVERSION

Although the economy is global, there is not a universal currency used worldwide. Before international transactions can be conducted, money must be converted or exchanged into the local currency. An **exchange rate** is the relative value of the currency of one country to that of another. For example, the exchange rate of U.S. dollars to Japanese yen is equal to the number of yen a dollar will purchase. Under the flexible (floating) exchange rate system, which has been in use since March 1973, the rate is determined by supply and demand in the marketplace for a particular country's currency. Just as prices will vary for any good or service, a currency will fluctuate as the supply and demand for it changes over time.

Foreign exchange rates are printed daily in newspapers. An illustration is shown in Table B-3, which includes conversions for the British pound (£), the Canadian dollar (C$), the French franc (F), the German mark or Deutsche mark (DM), the Japanese yen (¥), and the Mexican (new) peso (N$).

According to the illustration in Table B-3, one U.S. dollar would be exchanged for 86.55 (or 87) yen. Likewise, a U.S. dollar could be exchanged for 5.92 Mexican pesos. When the dollar is strong, it will buy more foreign goods and investments (imports). When the dollar is weak, U.S. dollars will purchase less in foreign goods; conversely, U.S. products (exports) will be cheaper, relatively speaking, in other countries with stronger currencies. Thus, a weak dollar generally means fewer imports and more exports.

The question frequently arises as to rounding. It is common practice to convert American dollars to whole Japanese yen, whereas it is common to convert American dollars to other countries' currencies by rounding to two decimal places. When converting currencies to American dollars, round to two decimal places (or correct to cents).

Example 1 Sam Perdue, on an exploratory business trip to Germany, exchanges $1,500 to German marks. How many marks does he receive?

Using the sample chart in Table B-3, U.S. $1 = DM1.4380. Thus, for $1,500,

$$\$1,500 \times DM1.4380 = DM2,157$$

Or, from the chart, since DM1 = $0.6954,

$$\$1,500 \div 0.6954 = DM2,157.03$$

Note. Most people use the first approach (multiplication) when converting U.S. dollars to a foreign currency.

Example 2 While in Japan, Terry Daniels paid 17,850 yen for dinner and 44,625 yen for a hotel room one night. How much did Ms. Daniels pay for the dinner and room in U.S. dollars?

From the sample chart in Table B-3, ¥1 = U.S. $0.011554. Thus,

$$\begin{aligned}
\yen17,850 \times 0.011554 &= \$206.24 \quad \text{for dinner} \\
\yen44,625 \times 0.011554 &= \underline{\$515.60} \quad \text{for the room} \\
&\ \$721.84 \quad \text{total}
\end{aligned}$$

TABLE B-3 SAMPLE FOREIGN CURRENCY EXCHANGE RATES

COUNTRY AND CURRENCY	$1 U.S. IN FOREIGN CURRENCY	FOREIGN CURRENCY IN U.S. DOLLARS
Britain (£)	0.6371	1.5695
Canada (C$)	1.3570	0.7369
France (F)	5.0800	0.1969
Germany (DM)	1.4380	0.6954
Japan (¥)	86.5500	0.011554
Mexico (N$)	5.9200	0.168919

SECTION 2 PROBLEMS

For Problems 1–14, determine the currency equivalencies, using the sample chart in Table B-3. Round all answers to 2 decimal places except for answers in yen, which should be rounded to whole numbers.

	U.S. ($)	GERMAN (DM)	JAPANESE (¥)	MEXICAN (N$)
1. a.	$ 10			
b.	600			
c.	1,800			
2. a.	$ 20			
b.	500			
c.	2,000			
3. a.		2,812	X	X
b.		X	53,550	X
c.		X	X	3,390
4. a.		3,515	X	X
b.		X	12,495	X
c.		X	X	1,356

5. How much will Marcia Murphy receive when she exchanges $8,000 into Canadian dollars?

6. Phil Kelly exchanges $6,000 for Canadian dollars. How much will he receive?

7. During a trip to the United States, a French family purchased 4 tickets to a theme park for $80. How much did they pay for the tickets in French francs?

8. Before traveling to France, Sarah Delano exchanged $2,000 into French francs. How much did she receive in francs?

9. On a recent trip to Great Britain, Charles Ness purchased a sweater for £50.144. How many U.S. dollars did it cost?

10. While in Great Britain, Lucy and John Goodwin attended a stage production. The tickets cost £18.804 each. How much did they pay in U.S. dollars for two tickets?

11. While visiting Germany, Elizabeth Smith purchased a crystal necklace for DM158.18. How much had the necklace cost in U.S. dollars, rounded to the nearest dollar?

12. On a recent trip to Germany, Bob Glazer purchased gasoline for his car at DM3.60 a liter. What was the equivalent price per liter in U.S. dollars to the nearest cent?

13. If a Big Mac meal cost $3.99 in the United States, what would it cost in
   a. Germany
   b. Japan
   c. Canada

14. A Kentucky Fried Chicken dinner cost $4.29 in the United States. How much would the meal cost in
   a. Great Britain
   b. Mexico
   c. France

APPENDIX *B*   GLOSSARY

Celsius (centigrade).   The metric temperature scale, using 100 degrees for the range from the freezing point of water (0°C) to the boiling point (100°C).

Centi-.   Prefix indicating 0.01 of a basic metric unit. (Example: 1 centimeter = 0.01 meter)

Centigrade.   (See "Celsius.")

Deca-.   Prefix indicating 10 times a basic metric unit. (Example: 1 decaliter = 10 liters)

Deci-.   Prefix indicating 0.1 of a basic metric unit. (Example: 1 decigram = 0.1 gram)

Exchange rate.   The relative value of one country's currency to that of another.

Gram.   The basic metric unit of weight; defined as the weight of 1 cubic centimeter of pure water at 4°C.

Hecto-.   Prefix indicating 100 times a basic metric unit. (Example: 1 hectometer = 100 meters)

Kilo-.   Prefix indicating 1000 times a basic metric unit. (Example: 1 kilogram = 1000 grams)

Liter.   The basic metric unit of liquid volume (or capacity); defined as the capacity of a cube that is 1 decimeter long on each edge.

Mega-.   Prefix indicating 1 million times a basic metric unit.

Meter.   The basic metric unit of length; originally defined (erroneously) as 1 ten-millionth of a quadrant ($\frac{1}{4}$) of the earth's circumference; now officially defined as a multiple of a certain wavelength of krypton 86, a rare gas.

Metric system.   The decimal system of weights and measures, where each unit is 10 times the size of the next smaller unit. Its basic units are the meter (length), the gram (weight), and the liter (capacity).

Micro-.   Prefix indicating 1 millionth of a basic metric unit.

Milli-.   Prefix indicating 0.001 of a basic metric unit. (Example: 1 milliliter = 0.001 liter)

METRIC TALK

When the United States goes to the metric system, perhaps some of our favorite sayings will have to be changed thus:

- Traffic was 2.54 centimetering along the freeway.
- It hit me like 907 kilograms of bricks.
- Cried 3.79 liters of tears.
- A miss is as good as 1.61 kilometers.
- A decigram of salt.
- Beat him within 2.54 centimeters of his life.
- All wool and 91.4 centimeters wide.
- Give her 2.54 centimeters and she'll take 1.61 kilometers.
- Give him 454 grams of flesh.
- Missed it by 1.61 country kilometers.

TABLES

LIST OF TABLES

TABLE C-1 CONSUMER PRICE INDEX OF SELECTED EXPENSES IN SELECTED URBAN AREAS

(1982–84 = 100)

EXPENSE	U.S. URBAN AVERAGE	ATLANTA	CHICAGO	DALLAS	LOS ANGELES	NEW YORK	SEATTLE
Food	147.2	141.1	147.0	142.2	148.5	151.9	146.9
Housing	145.4	140.1	144.8	129.0	151.0	159.9	147.9
Utilities	122.0	132.2	110.7	126.3	143.1	112.4	112.7
Transportation	137.1	123.8	130.3	134.5	140.5	141.8	135.0
Medical care	215.3	227.0	213.2	205.6	215.2	217.6	199.8

Source: U.S. Department of Labor.

TABLE C-2 COMPARATIVE INDEX OF SELECTED EXPENSES IN SELECTED URBAN AREAS AT INTERMEDIATE STANDARD OF LIVING

U.S. Urban Average = 100

EXPENSE	ATLANTA	CHICAGO	DALLAS	LOS ANGELES	NEW YORK	SEATTLE
Food	98.6	99.3	98.1	110.3	112.8	97.3
Housing	86.8	107.5	93.8	147.9	105.6	132.8
Utilities	110.4	113.2	124.0	91.0	131.0	57.1
Transportation	98.3	108.0	103.2	106.8	95.5	94.7
Medical care	110.4	105.8	113.8	145.5	106.2	124.8
Combined index	97.0	103.5	102.0	121.1	106.0	106.2

Source: American Chamber of Commerce Researchers Association.

TABLE C-3 ANNUAL FIRE INSURANCE PREMIUMS FOR DWELLINGS

Per $100 of Face Value

TERRITORY	STRUCTURE CLASS A	B	C	D	CONTENTS CLASS A	B	C	D
1	$0.45	$0.56	$0.66	$0.75	$0.51	$0.79	$0.79	$0.98
2	0.58	0.64	0.78	0.80	0.68	0.86	0.91	1.03
3	0.63	0.72	0.85	0.93	0.71	0.89	0.95	1.12

TABLE C-4 SHORT-TERM RATES INCLUDING CANCELLATION BY THE INSURED

MONTHS OF COVERAGE	PERCENT OF ANNUAL PREMIUM CHARGED	MONTHS OF COVERAGE	PERCENT OF ANNUAL PREMIUM CHARGED
1	20%	7	75%
2	30	8	80
3	40	9	85
4	50	10	90
5	60	11	95
6	70	12	100

TABLE C-5 COMPREHENSIVE AND COLLISION INSURANCE

Base Annual Premiums

MODEL CLASS	AGE GROUP	TERRITORY 1 COMPRE-HENSIVE	$250 DEDUCTIBLE COLLISION	$500 DEDUCTIBLE COLLISION	TERRITORY 2 COMPRE-HENSIVE	$250 DEDUCTIBLE COLLISION	$500 DEDUCTIBLE COLLISION	TERRITORY 3 COMPRE-HENSIVE	$250 DEDUCTIBLE COLLISION	$500 DEDUCTIBLE COLLISION
(1) A-G	1	$55	$ 82	$ 76	$59	$ 92	$ 80	$73	$100	$ 91
	2,3	52	77	73	56	86	76	58	94	85
	4	49	71	67	51	79	70	54	85	78
(3) J-K	1	63	111	101	69	128	108	75	141	127
	2,3	59	103	95	64	118	101	68	131	118
	4	54	93	86	57	106	91	61	116	105
(4) L-M	1	68	123	112	76	143	120	83	169	142
	2,3	64	125	104	70	133	112	75	138	132
	4	57	102	94	62	117	100	66	129	117
(5) N-O	1	77	140	126	86	164	136	95	183	162
	2,3	70	130	117	77	151	126	85	168	150
	4	62	115	105	68	133	112	73	147	132

TABLE C-6 DRIVER CLASSIFICATIONS

Multiples of Base Annual Automobile Insurance Premiums

			PLEASURE; LESS THAN 3 MILES TO WORK EACH WAY	DRIVES TO WORK, 3 TO 9 MILES EACH WAY	DRIVES TO WORK, 10 MILES OR MORE EACH WAY	USED IN BUSINESS
No young operators	Only operator is female, age 30–64		0.90	1.00	1.30	1.40
	One or more operators age 65 or over		1.00	1.10	1.40	1.50
	All others		1.00	1.10	1.40	1.50
Young females	Age 16	DT[a]	1.40	1.50	1.80	1.90
		No DT	1.55	1.65	1.95	2.05
	Age 20	DT	1.05	1.15	1.45	1.55
		No DT	1.10	1.20	1.50	1.60
Young males (married)	Age 16	DT	1.60	1.70	2.00	2.10
		No DT	1.80	1.90	2.20	2.30
	Age 20	DT	1.45	1.55	1.85	1.95
		No DT	1.50	1.60	1.90	2.00
	Age 21		1.40	1.50	1.80	1.90
	Age 24		1.10	1.20	1.50	1.60
Young unmarried males (not principal operator)	Age 16	DT	2.05	2.15	2.45	2.55
		No DT	2.30	2.40	2.70	2.80
	Age 20	DT	1.60	1.70	2.00	2.10
		No DT	1.70	1.80	2.10	2.20
	Age 21		1.55	1.65	1.95	2.05
	Age 24		1.10	1.20	1.50	1.60
Young unmarried males (owner or principal operator)	Age 16	DT	2.70	2.80	3.10	3.20
		No DT	3.30	3.40	3.70	3.80
	Age 20	DT	2.55	2.65	2.95	3.05
		No DT	2.70	2.80	3.10	3.20
	Age 21		2.50	2.60	2.90	3.00
	Age 24		1.90	2.00	2.30	2.40
	Age 26		1.50	1.60	1.90	2.00
	Age 29		1.10	1.20	1.50	1.60

[a] "DT" indicates completion of a certified driver training course.

TABLE C-7 AUTOMOBILE LIABILITY AND MEDICAL PAYMENT INSURANCE

Base Annual Premiums

	BODILY INJURY				PROPERTY DAMAGE		
COVERAGE	TERRITORY 1	TERRITORY 2	TERRITORY 3	COVERAGE	TERRITORY 1	TERRITORY 2	TERRITORY 3
15/30	$ 81	$ 91	$112	$ 5,000	$83	$ 95	$100
25/25	83	94	115	10,000	85	97	103
25/50	86	97	120	25,000	86	99	104
50/50	88	101	125	50,000	87	101	107
50/100	90	103	129	100,000	90	103	108
100/100	91	104	131	MEDICAL PAYMENT			
100/200	94	108	136				
100/300	95	110	139	$ 1,000	$62	$63	$64
200/300	98	112	141	2,500	65	66	67
300/300	100	115	144	5,000	67	68	69
				10,000	70	72	74

TABLE C-8 ANNUAL LIFE INSURANCE PREMIUMS

Per $1,000 of Face Value for Male Applicants[a]

AGE ISSUED (YEARS)	TERM 10-YEAR	WHOLE LIFE	VARIABLE UNIVERSAL	LIMITED PAYMENT 20-YEAR
18	$ 6.81	$14.77	$18.36	$24.69
20	6.88	15.46	19.81	25.59
22	6.95	16.12	20.28	26.53
24	7.05	16.82	21.33	27.53
25	7.10	17.22	22.17	28.06
26	7.18	17.67	23.43	28.63
28	7.35	18.64	24.79	29.84
30	7.59	19.73	26.05	31.12
35	8.68	23.99	30.87	35.80
40	10.64	28.26	35.55	40.34
45	14.52	33.79	41.24	46.01
50	22.18	40.77	48.11	53.24
55	32.93	51.38	57.62	64.77
60	—	59.32	—	70.86

[a] Because of women's longer life expectancy, premiums for women approximately equal those of men who are 5 years younger.

TABLE C-9 NONFORFEITURE OPTIONS[a] ON TYPICAL LIFE
INSURANCE POLICIES

Issued at Age 25

| YEARS IN FORCE | WHOLE LIFE | | | | 20-PAYMENT LIFE | | | | VARIABLE UNIVERSAL |
| | CASH VALUE | PAID-UP INSURANCE | EXT. TERM | | CASH VALUE | PAID-UP INSURANCE | EXT. TERM | | CASH VALUE |
			YEARS	DAYS			YEARS	DAYS	
3	$ 9	$ 19	1	190	$ 32	$ 94	10	84	$ 44
5	31	87	9	200	74	218	19	184	95
10	93	248	18	91	187	507	28	186	321
15	162	387	20	300	319	768	32	164	595
20	251	535	22	137	470	1,000	Life		1,030
40	576	827	—	—	701	—	—	—	2,979

[a] The "cash value" and "paid-up insurance" nonforfeiture values are per $1,000 of life insurance coverage. The time period for "extended term insurance" applies as shown to all policies, regardless of face value. The cash value for variable universal life represents a reasonable estimate of the increase in value, rather than a guaranteed amount.

TABLE C-10 SETTLEMENT OPTIONS

Monthly Installments per $1,000 of Face Value

| OPTIONS 1 AND 2: FIXED AMOUNT OR FIXED NUMBER OF YEARS | | OPTIONS 3 AND 4: INCOME FOR LIFE | | | | |
| | | AGE WHEN ANNUITY BEGINS | | LIFE ANNUITY | LIFE WITH 10 YEARS CERTAIN | LIFE WITH 20 YEARS CERTAIN |
YEARS	AMOUNT	MALE	FEMALE			
10	$9.60	40	45	$4.60	$4.56	$4.44
12	8.52	45	50	5.16	5.07	4.80
14	7.71	50	55	5.30	5.28	5.00
15	6.89	55	60	6.13	6.00	5.46
16	6.43	60	65	6.56	6.31	5.65
18	6.08	65	70	7.22	6.70	5.78
20	5.66					

TABLE C-11 SOCIAL SECURITY EMPLOYEE TAX TABLE

7.65% Employee Tax Deductions

Wages at least	But less than	Tax to be withheld	Wages at least	But less than	Tax to be withheld	Wages at least	But less than	Tax to be withheld	Wages at least	But less than	Tax to be withheld
199.94	200.07	15.30	212.75	212.88	16.28	225.56	225.69	17.26	238.37	238.50	18.24
200.07	200.20	15.31	212.88	213.01	16.29	225.69	225.82	17.27	238.50	238.63	18.25
200.20	200.33	15.32	213.01	213.14	16.30	225.82	225.95	17.28	238.63	238.76	18.26
200.33	200.46	15.33	213.14	213.27	16.31	225.95	226.08	17.29	238.76	238.89	18.27
200.46	200.59	15.34	213.27	213.40	16.32	226.08	226.21	17.30	238.89	239.02	18.28
200.59	200.72	15.35	213.40	213.53	16.33	226.21	226.34	17.31	239.02	239.15	18.29
200.72	200.85	15.36	213.53	213.66	16.34	226.34	226.48	17.32	239.15	239.29	18.30
200.85	200.98	15.37	213.66	213.80	16.35	226.48	226.61	17.33	239.29	239.42	18.31
200.98	201.12	15.38	213.80	213.93	16.36	226.61	226.74	17.34	239.42	239.55	18.32
201.12	201.25	15.39	213.93	214.06	16.37	226.74	226.87	17.35	239.55	239.68	18.33
201.25	201.38	15.40	214.06	214.19	16.38	226.87	227.00	17.36	239.68	239.81	18.34
201.38	201.51	15.41	214.19	214.32	16.39	227.00	227.13	17.37	239.81	239.94	18.35
201.51	201.64	15.42	214.32	214.45	16.40	227.13	227.26	17.38	239.94	240.07	18.36
201.64	201.77	15.43	214.45	214.58	16.41	227.26	227.39	17.39	240.07	240.20	18.37
201.77	201.90	15.44	214.58	214.71	16.42	227.39	227.52	17.40	240.20	240.33	18.38
201.90	202.03	15.45	214.71	214.84	16.43	227.52	227.65	17.41	240.33	240.46	18.39
202.03	202.16	15.46	214.84	214.97	16.44	227.65	227.78	17.42	240.46	240.59	18.40
202.16	202.29	15.47	214.97	215.10	16.45	227.78	227.91	17.43	240.59	240.72	18.41
202.29	202.42	15.48	215.10	215.23	16.46	227.91	228.04	17.44	240.72	240.85	18.42
202.42	202.55	15.49	215.23	215.36	16.47	228.04	228.17	17.45	240.85	240.98	18.43
202.55	202.68	15.50	215.36	215.49	16.48	228.17	228.31	17.46	240.98	241.12	18.44
202.68	202.82	15.51	215.49	215.63	16.49	228.31	228.44	17.47	241.12	241.25	18.45
202.82	202.95	15.52	215.63	215.76	16.50	228.44	228.57	17.48	241.25	241.38	18.46
202.95	203.08	15.53	215.76	215.89	16.51	228.57	228.70	17.49	241.38	241.51	18.47
203.08	203.21	15.54	215.89	216.02	16.52	228.70	228.83	17.50	241.51	241.64	18.48
203.21	203.34	15.55	216.02	216.15	16.53	228.83	228.96	17.51	241.64	241.77	18.49
203.34	203.47	15.56	216.15	216.28	16.54	228.96	229.09	17.52	241.77	241.90	18.50
203.47	203.60	15.57	216.28	216.41	16.55	229.09	229.22	17.53	241.90	242.03	18.51
203.60	203.73	15.58	216.41	216.54	16.56	229.22	229.35	17.54	242.03	242.16	18.52
203.73	203.86	15.59	216.54	216.67	16.57	229.35	229.48	17.55	242.16	242.29	18.53
203.86	203.99	15.60	216.67	216.80	16.58	229.48	229.61	17.56	242.29	242.42	18.54
203.99	204.12	15.61	216.80	216.93	16.59	229.61	229.74	17.57	242.42	242.55	18.55
204.12	204.25	15.62	216.93	217.06	16.60	229.74	229.87	17.58	242.55	242.68	18.56
204.25	204.38	15.63	217.06	217.19	16.61	229.87	230.00	17.59	242.68	242.82	18.57
204.38	204.51	15.64	217.19	217.32	16.62	230.00	230.14	17.60	242.82	242.95	18.58
204.51	204.65	15.65	217.32	217.46	16.63	230.14	230.27	17.61	242.95	243.08	18.59
204.65	204.78	15.66	217.46	217.59	16.64	230.27	230.40	17.62	243.08	243.21	18.60
204.78	204.91	15.67	217.59	217.72	16.65	230.40	230.53	17.63	243.21	243.34	18.61
204.91	205.04	15.68	217.72	217.85	16.66	230.53	230.66	17.64	243.34	243.47	18.62
205.04	205.17	15.69	217.85	217.98	16.67	230.66	230.79	17.65	243.47	243.60	18.63
205.17	205.30	15.70	217.98	218.11	16.68	230.79	230.92	17.66	243.60	243.73	18.64
205.30	205.43	15.71	218.11	218.24	16.69	230.92	231.05	17.67	243.73	243.86	18.65
205.43	205.56	15.72	218.24	218.37	16.70	231.05	231.18	17.68	243.86	243.99	18.66
205.56	205.69	15.73	218.37	218.50	16.71	231.18	231.31	17.69	243.99	244.12	18.67
205.69	205.82	15.74	218.50	218.63	16.72	231.31	231.44	17.70	244.12	244.25	18.68
205.82	205.95	15.75	218.63	218.76	16.73	231.44	231.57	17.71	244.25	244.38	18.69
205.95	206.08	15.76	218.76	218.89	16.74	231.57	231.70	17.72	244.38	244.51	18.70
206.08	206.21	15.77	218.89	219.02	16.75	231.70	231.83	17.73	244.51	244.65	18.71
206.21	206.34	15.78	219.02	219.15	16.76	231.83	231.97	17.74	244.65	244.78	18.72
206.34	206.48	15.79	219.15	219.29	16.77	231.97	232.10	17.75	244.78	244.91	18.73
206.48	206.61	15.80	219.29	219.42	16.78	232.10	232.23	17.76	244.91	245.04	18.74
206.61	206.74	15.81	219.42	219.55	16.79	232.23	232.36	17.77	245.04	245.17	18.75
206.74	206.87	15.82	219.55	219.68	16.80	232.36	232.49	17.78	245.17	245.30	18.76
206.87	207.00	15.83	219.68	219.81	16.81	232.49	232.62	17.79	245.30	245.43	18.77
207.00	207.13	15.84	219.81	219.94	16.82	232.62	232.75	17.80	245.43	245.56	18.78
207.13	207.26	15.85	219.94	220.07	16.83	232.75	232.88	17.81	245.56	245.69	18.79
207.26	207.39	15.86	220.07	220.20	16.84	232.88	233.01	17.82	245.69	245.82	18.80
207.39	207.52	15.87	220.20	220.33	16.85	233.01	233.14	17.83	245.82	245.95	18.81
207.52	207.65	15.88	220.33	220.46	16.86	233.14	233.27	17.84	245.95	246.08	18.82
207.65	207.78	15.89	220.46	220.59	16.87	233.27	233.40	17.85	246.08	246.21	18.83
207.78	207.91	15.90	220.59	220.72	16.88	233.40	233.53	17.86	246.21	246.34	18.84
207.91	208.04	15.91	220.72	220.85	16.89	233.53	233.66	17.87	246.34	246.48	18.85
208.04	208.17	15.92	220.85	220.98	16.90	233.66	233.80	17.88	246.48	246.61	18.86
208.17	208.31	15.93	220.98	221.12	16.91	233.80	233.93	17.89	246.61	246.74	18.87
208.31	208.44	15.94	221.12	221.25	16.92	233.93	234.06	17.90	246.74	246.87	18.88
208.44	208.57	15.95	221.25	221.38	16.93	234.06	234.19	17.91	246.87	247.00	18.89
208.57	208.70	15.96	221.38	221.51	16.94	234.19	234.32	17.92	247.00	247.13	18.90
208.70	208.83	15.97	221.51	221.64	16.95	234.32	234.45	17.93	247.13	247.26	18.91
208.83	208.96	15.98	221.64	221.77	16.96	234.45	234.58	17.94	247.26	247.39	18.92
208.96	209.09	15.99	221.77	221.90	16.97	234.58	234.71	17.95	247.39	247.52	18.93
209.09	209.22	16.00	221.90	222.03	16.98	234.71	234.84	17.96	247.52	247.65	18.94
209.22	209.35	16.01	222.03	222.16	16.99	234.84	234.97	17.97	247.65	247.78	18.95
209.35	209.48	16.02	222.16	222.29	17.00	234.97	235.10	17.98	247.78	247.91	18.96
209.48	209.61	16.03	222.29	222.42	17.01	235.10	235.23	17.99	247.91	248.04	18.97
209.61	209.74	16.04	222.42	222.55	17.02	235.23	235.36	18.00	248.04	248.17	18.98
209.74	209.87	16.05	222.55	222.68	17.03	235.36	235.49	18.01	248.17	248.31	18.99
209.87	210.00	16.06	222.68	222.82	17.04	235.49	235.63	18.02	248.31	248.44	19.00
210.00	210.14	16.07	222.82	222.95	17.05	235.63	235.76	18.03	248.44	248.57	19.01
210.14	210.27	16.08	222.95	223.08	17.06	235.76	235.89	18.04	248.57	248.70	19.02
210.27	210.40	16.09	223.08	223.21	17.07	235.89	236.02	18.05	248.70	248.83	19.03
210.40	210.53	16.10	223.21	223.34	17.08	236.02	236.15	18.06	248.83	248.96	19.04
210.53	210.66	16.11	223.34	223.47	17.09	236.15	236.28	18.07	248.96	249.09	19.05
210.66	210.79	16.12	223.47	223.60	17.10	236.28	236.41	18.08	249.09	249.22	19.06
210.79	210.92	16.13	223.60	223.73	17.11	236.41	236.54	18.09	249.22	249.35	19.07
210.92	211.05	16.14	223.73	223.86	17.12	236.54	236.67	18.10	249.35	249.48	19.08
211.05	211.18	16.15	223.86	223.99	17.13	236.67	236.80	18.11	249.48	249.61	19.09
211.18	211.31	16.16	223.99	224.12	17.14	236.80	236.93	18.12	249.61	249.74	19.10
211.31	211.44	16.17	224.12	224.25	17.15	236.93	237.06	18.13	249.74	249.87	19.11
211.44	211.57	16.18	224.25	224.38	17.16	237.06	237.19	18.14	249.87	250.00	19.12
211.57	211.70	16.19	224.38	224.51	17.17	237.19	237.32	18.15	250.00	250.14	19.13
211.70	211.83	16.20	224.51	224.65	17.18	237.32	237.46	18.16	250.14	250.27	19.14
211.83	211.97	16.21	224.65	224.78	17.19	237.46	237.59	18.17	250.27	250.40	19.15
211.97	212.10	16.22	224.78	224.91	17.20	237.59	237.72	18.18	250.40	250.53	19.16
212.10	212.23	16.23	224.91	225.04	17.21	237.72	237.85	18.19	250.53	250.66	19.17
212.23	212.36	16.24	225.04	225.17	17.22	237.85	237.98	18.20	250.66	250.79	19.18
212.36	212.49	16.25	225.17	225.30	17.23	237.98	238.11	18.21	250.79	250.92	19.19
212.49	212.62	16.26	225.30	225.43	17.24	238.11	238.24	18.22	250.92	251.05	19.20
212.62	212.75	16.27	225.43	225.56	17.25	238.24	238.37	18.23	251.05	251.18	19.21

TABLE C-11 (cont.) SOCIAL SECURITY EMPLOYEE TAX TABLE

7.65% Employee Tax Deductions

Wages at least	But less than	Tax to be withheld	Wages at least	But less than	Tax to be withheld	Wages at least	But less than	Tax to be withheld	Wages at least	But less than	Tax to be withheld
251.18	251.31	19.22	263.99	264.12	20.20	276.80	276.93	21.18	289.61	289.74	22.16
251.31	251.44	19.23	264.12	264.25	20.21	276.93	277.06	21.19	289.74	289.87	22.17
251.44	251.57	19.24	264.25	264.38	20.22	277.06	277.19	21.20	289.87	290.00	22.18
251.57	251.70	19.25	264.38	264.51	20.23	277.19	277.32	21.21	290.00	290.14	22.19
251.70	251.83	19.26	264.51	264.65	20.24	277.32	277.46	21.22	290.14	290.27	22.20
251.83	251.97	19.27	264.65	264.78	20.25	277.46	277.59	21.23	290.27	290.40	22.21
251.97	252.10	19.28	264.78	264.91	20.26	277.59	277.72	21.24	290.40	290.53	22.22
252.10	252.23	19.29	264.91	265.04	20.27	277.72	277.85	21.25	290.53	290.66	22.23
252.23	252.36	19.30	265.04	265.17	20.28	277.85	277.98	21.26	290.66	290.79	22.24
252.36	252.49	19.31	265.17	265.30	20.29	277.98	278.11	21.27	290.79	290.92	22.25
252.49	252.62	19.32	265.30	265.43	20.30	278.11	278.24	21.28	290.92	291.05	22.26
252.62	252.75	19.33	265.43	265.56	20.31	278.24	278.37	21.29	291.05	291.18	22.27
252.75	252.88	19.34	265.56	265.69	20.32	278.37	278.50	21.30	291.18	291.31	22.28
252.88	253.01	19.35	265.69	265.82	20.33	278.50	278.63	21.31	291.31	291.44	22.29
253.01	253.14	19.36	265.82	265.95	20.34	278.63	278.76	21.32	291.44	291.57	22.30
253.14	253.27	19.37	265.95	266.08	20.35	278.76	278.89	21.33	291.57	291.70	22.31
253.27	253.40	19.38	266.08	266.21	20.36	278.89	279.02	21.34	291.70	291.83	22.32
253.40	253.53	19.39	266.21	266.34	20.37	279.02	279.15	21.35	291.83	291.97	22.33
253.53	253.66	19.40	266.34	266.48	20.38	279.15	279.29	21.36	291.97	292.10	22.34
253.66	253.80	19.41	266.48	266.61	20.39	279.29	279.42	21.37	292.10	292.23	22.35
253.80	253.93	19.42	266.61	266.74	20.40	279.42	279.55	21.38	292.23	292.36	22.36
253.93	254.06	19.43	266.74	266.87	20.41	279.55	279.68	21.39	292.36	292.49	22.37
254.06	254.19	19.44	266.87	267.00	20.42	279.68	279.81	21.40	292.49	292.62	22.38
254.19	254.32	19.45	267.00	267.13	20.43	279.81	279.94	21.41	292.62	292.75	22.39
254.32	254.45	19.46	267.13	267.26	20.44	279.94	280.07	21.42	292.75	292.88	22.40
254.45	254.58	19.47	267.26	267.39	20.45	280.07	280.20	21.43	292.88	293.01	22.41
254.58	254.71	19.48	267.39	267.52	20.46	280.20	280.33	21.44	293.01	293.14	22.42
254.71	254.84	19.49	267.52	267.65	20.47	280.33	280.46	21.45	293.14	293.27	22.43
254.84	254.97	19.50	267.65	267.78	20.48	280.46	280.59	21.46	293.27	293.40	22.44
254.97	255.10	19.51	267.78	267.91	20.49	280.59	280.72	21.47	293.40	293.53	22.45
255.10	255.23	19.52	267.91	268.04	20.50	280.72	280.85	21.48	293.53	293.66	22.46
255.23	255.36	19.53	268.04	268.17	20.51	280.85	280.98	21.49	293.66	293.80	22.47
255.36	255.49	19.54	268.17	268.31	20.52	280.98	281.12	21.50	293.80	293.93	22.48
255.49	255.63	19.55	268.31	268.44	20.53	281.12	281.25	21.51	293.93	294.06	22.49
255.63	255.76	19.56	268.44	268.57	20.54	281.25	281.38	21.52	294.06	294.19	22.50
255.76	255.89	19.57	268.57	268.70	20.55	281.38	281.51	21.53	294.19	294.32	22.51
255.89	256.02	19.58	268.70	268.83	20.56	281.51	281.64	21.54	294.32	294.45	22.52
256.02	256.15	19.59	268.83	268.96	20.57	281.64	281.77	21.55	294.45	294.58	22.53
256.15	256.28	19.60	268.96	269.09	20.58	281.77	281.90	21.56	294.58	294.71	22.54
256.28	256.41	19.61	269.09	269.22	20.59	281.90	282.03	21.57	294.71	294.84	22.55
256.41	256.54	19.62	269.22	269.35	20.60	282.03	282.16	21.58	294.84	294.97	22.56
256.54	256.67	19.63	269.35	269.48	20.61	282.16	282.29	21.59	294.97	295.10	22.57
256.67	256.80	19.64	269.48	269.61	20.62	282.29	282.42	21.60	295.10	295.23	22.58
256.80	256.93	19.65	269.61	269.74	20.63	282.42	282.55	21.61	295.23	295.36	22.59
256.93	257.06	19.66	269.74	269.87	20.64	282.55	282.68	21.62	295.36	295.49	22.60
257.06	257.19	19.67	269.87	270.00	20.65	282.68	282.82	21.63	295.49	295.63	22.61
257.19	257.32	19.68	270.00	270.14	20.66	282.82	282.95	21.64	295.63	295.76	22.62
257.32	257.46	19.69	270.14	270.27	20.67	282.95	283.08	21.65	295.76	295.89	22.63
257.46	257.59	19.70	270.27	270.40	20.68	283.08	283.21	21.66	295.89	296.02	22.64
257.59	257.72	19.71	270.40	270.53	20.69	283.21	283.34	21.67	296.02	296.15	22.65
257.72	257.85	19.72	270.53	270.66	20.70	283.34	283.47	21.68	296.15	296.28	22.66
257.85	257.98	19.73	270.66	270.79	20.71	283.47	283.60	21.69	296.28	296.41	22.67
257.98	258.11	19.74	270.79	270.92	20.72	283.60	283.73	21.70	296.41	296.54	22.68
258.11	258.24	19.75	270.92	271.05	20.73	283.73	283.86	21.71	296.54	296.67	22.69
258.24	258.37	19.76	271.05	271.18	20.74	283.86	283.99	21.72	296.67	296.80	22.70
258.37	258.50	19.77	271.18	271.31	20.75	283.99	284.12	21.73	296.80	296.93	22.71
258.50	258.63	19.78	271.31	271.44	20.76	284.12	284.25	21.74	296.93	297.06	22.72
258.63	258.76	19.79	271.44	271.57	20.77	284.25	284.38	21.75	297.06	297.19	22.73
258.76	258.89	19.80	271.57	271.70	20.78	284.38	284.51	21.76	297.19	297.32	22.74
258.89	259.02	19.81	271.70	271.83	20.79	284.51	284.65	21.77	297.32	297.46	22.75
259.02	259.15	19.82	271.83	271.97	20.80	284.65	284.78	21.78	297.46	297.59	22.76
259.15	259.29	19.83	271.97	272.10	20.81	284.78	284.91	21.79	297.59	297.72	22.77
259.29	259.42	19.84	272.10	272.23	20.82	284.91	285.04	21.80	297.72	297.85	22.78
259.42	259.55	19.85	272.23	272.36	20.83	285.04	285.17	21.81	297.85	297.98	22.79
259.55	259.68	19.86	272.36	272.49	20.84	285.17	285.30	21.82	297.98	298.11	22.80
259.68	259.81	19.87	272.49	272.62	20.85	285.30	285.43	21.83	298.11	298.24	22.81
259.81	259.94	19.88	272.62	272.75	20.86	285.43	285.56	21.84	298.24	298.37	22.82
259.94	260.07	19.89	272.75	272.88	20.87	285.56	285.69	21.85	298.37	298.50	22.83
260.07	260.20	19.90	272.88	273.01	20.88	285.69	285.82	21.86	298.50	298.63	22.84
260.20	260.33	19.91	273.01	273.14	20.89	285.82	285.95	21.87	298.63	298.76	22.85
260.33	260.46	19.92	273.14	273.27	20.90	285.95	286.08	21.88	298.76	298.89	22.86
260.46	260.59	19.93	273.27	273.40	20.91	286.08	286.21	21.89	298.89	299.02	22.87
260.59	260.72	19.94	273.40	273.53	20.92	286.21	286.34	21.90	299.02	299.15	22.88
260.72	260.85	19.95	273.53	273.66	20.93	286.34	286.48	21.91	299.15	299.29	22.89
260.85	260.98	19.96	273.66	273.80	20.94	286.48	286.61	21.92	299.29	299.42	22.90
260.98	261.12	19.97	273.80	273.93	20.95	286.61	286.74	21.93	299.42	299.55	22.91
261.12	261.25	19.98	273.93	274.06	20.96	286.74	286.87	21.94	299.55	299.68	22.92
261.25	261.38	19.99	274.06	274.19	20.97	286.87	287.00	21.95	299.68	299.81	22.93
261.38	261.51	20.00	274.19	274.32	20.98	287.00	287.13	21.96	299.81	299.94	22.94
261.51	261.64	20.01	274.32	274.45	20.99	287.13	287.26	21.97			
261.64	261.77	20.02	274.45	274.58	21.00	287.26	287.39	21.98			
261.77	261.90	20.03	274.58	274.71	21.01	287.39	287.52	21.99			
261.90	262.03	20.04	274.71	274.84	21.02	287.52	287.65	22.00			
262.03	262.16	20.05	274.84	274.97	21.03	287.65	287.78	22.01			
262.16	262.29	20.06	274.97	275.10	21.04	287.78	287.91	22.02			
262.29	262.42	20.07	275.10	275.23	21.05	287.91	288.04	22.03			
262.42	262.55	20.08	275.23	275.36	21.06	288.04	288.17	22.04			
262.55	262.68	20.09	275.36	275.49	21.07	288.17	288.31	22.05			
262.68	262.82	20.10	275.49	275.63	21.08	288.31	288.44	22.06			
262.82	262.95	20.11	275.63	275.76	21.09	288.44	288.57	22.07			
262.95	263.08	20.12	275.76	275.89	21.10	288.57	288.70	22.08			
263.08	263.21	20.13	275.89	276.02	21.11	288.70	288.83	22.09			
263.21	263.34	20.14	276.02	276.15	21.12	288.83	288.96	22.10			
263.34	263.47	20.15	276.15	276.28	21.13	288.96	289.09	22.11			
263.47	263.60	20.16	276.28	276.41	21.14	289.09	289.22	22.12			
263.60	263.73	20.17	276.41	276.54	21.15	289.22	289.35	22.13			
263.73	263.86	20.18	276.54	276.67	21.16	289.35	289.48	22.14			
263.86	263.99	20.19	276.67	276.80	21.17	289.48	289.61	22.15			

TABLE C-11 (cont.) SOCIAL SECURITY EMPLOYEE TAX TABLE

7.65% Employee Tax Deductions

Wages at least	But less than	Tax to be withheld	Wages at least	But less than	Tax to be withheld	Wages at least	But less than	Tax to be withheld	Wages at least	But less than	Tax to be withheld
299.94	300.07	22.95	312.75	312.88	23.93	325.56	325.69	24.91	338.37	338.50	25.89
300.07	300.20	22.96	312.88	313.01	23.94	325.69	325.82	24.92	338.50	338.63	25.90
300.20	300.33	22.97	313.01	313.14	23.95	325.82	325.95	24.93	338.63	338.76	25.91
300.33	300.46	22.98	313.14	313.27	23.96	325.95	326.08	24.94	338.76	338.89	25.92
300.46	300.59	22.99	313.27	313.40	23.97	326.08	326.21	24.95	338.89	339.02	25.93
300.59	300.72	23.00	313.40	313.53	23.98	326.21	326.34	24.96	339.02	339.15	25.94
300.72	300.85	23.01	313.53	313.66	23.99	326.34	326.48	24.97	339.15	339.20	25.05
300.85	300.98	23.02	313.66	313.80	24.00	326.48	326.61	24.98	339.29	339.42	25.96
300.98	301.12	23.03	313.80	313.93	24.01	326.61	326.74	24.99	339.42	339.55	25.97
301.12	301.25	23.04	313.93	314.06	24.02	326.74	326.87	25.00	339.55	339.68	25.98
301.25	301.38	23.05	314.06	314.19	24.03	326.87	327.00	25.01	339.68	339.81	25.99
301.38	301.51	23.06	314.19	314.32	24.04	327.00	327.13	25.02	339.81	339.94	26.00
301.51	301.64	23.07	314.32	314.45	24.05	327.13	327.26	25.03	339.94	340.07	26.01
301.64	301.77	23.08	314.45	314.58	24.06	327.26	327.39	25.04	340.07	340.20	26.02
301.77	301.90	23.09	314.58	314.71	24.07	327.39	327.52	25.05	340.20	340.33	26.03
301.90	302.03	23.10	314.71	314.84	24.08	327.52	327.65	25.06	340.33	340.46	26.04
302.03	302.16	23.11	314.84	314.97	24.09	327.65	327.78	25.07	340.46	340.59	26.05
302.16	302.29	23.12	314.97	315.10	24.10	327.78	327.91	25.08	340.59	340.72	26.06
302.29	302.42	23.13	315.10	315.23	24.11	327.91	328.04	25.09	340.72	340.85	26.07
302.42	302.55	23.14	315.23	315.36	24.12	328.04	328.17	25.10	340.85	340.98	26.08
302.55	302.68	23.15	315.36	315.49	24.13	328.17	328.31	25.11	340.98	341.12	26.09
302.68	302.82	23.16	315.49	315.63	24.14	328.31	328.44	25.12	341.12	341.25	26.10
302.82	302.95	23.17	315.63	315.76	24.15	328.44	328.57	25.13	341.25	341.38	26.11
302.95	303.08	23.18	315.76	315.89	24.16	328.57	328.70	25.14	341.38	341.51	26.12
303.08	303.21	23.19	315.89	316.02	24.17	328.70	328.83	25.15	341.51	341.64	26.13
303.21	303.34	23.20	316.02	316.15	24.18	328.83	328.96	25.16	341.64	341.77	26.14
303.34	303.47	23.21	316.15	316.28	24.19	328.96	329.09	25.17	341.77	341.90	26.15
303.47	303.60	23.22	316.28	316.41	24.20	329.09	329.22	25.18	341.90	342.03	26.16
303.60	303.73	23.23	316.41	316.54	24.21	329.22	329.35	25.19	342.03	342.16	26.17
303.73	303.86	23.24	316.54	316.67	24.22	329.35	329.48	25.20	342.16	342.29	26.18
303.86	303.99	23.25	316.67	316.80	24.23	329.48	329.61	25.21	342.29	342.42	26.19
303.99	304.12	23.26	316.80	316.93	24.24	329.61	329.74	25.22	342.42	342.55	26.20
304.12	304.25	23.27	316.93	317.06	24.25	329.74	329.87	25.23	342.55	342.68	26.21
304.25	304.38	23.28	317.06	317.19	24.26	329.87	330.00	25.24	342.68	342.82	26.22
304.38	304.51	23.29	317.19	317.32	24.27	330.00	330.14	25.25	342.82	342.95	26.23
304.51	304.65	23.30	317.32	317.46	24.28	330.14	330.27	25.26	342.95	343.08	26.24
304.65	304.78	23.31	317.46	317.59	24.29	330.27	330.40	25.27	343.08	343.21	26.25
304.78	304.91	23.32	317.59	317.72	24.30	330.40	330.53	25.28	343.21	343.34	26.26
304.91	305.04	23.33	317.72	317.85	24.31	330.53	330.66	25.29	343.34	343.47	26.27
305.04	305.17	23.34	317.85	317.98	24.32	330.66	330.79	25.30	343.47	343.60	26.28
305.17	305.30	23.35	317.98	318.11	24.33	330.79	330.92	25.31	343.60	343.73	26.29
305.30	305.43	23.36	318.11	318.24	24.34	330.92	331.05	25.32	343.73	343.86	26.30
305.43	305.56	23.37	318.24	318.37	24.35	331.05	331.18	25.33	343.86	343.99	26.31
305.56	305.69	23.38	318.37	318.50	24.36	331.18	331.31	25.34	343.99	344.12	26.32
305.69	305.82	23.39	318.50	318.63	24.37	331.31	331.44	25.35	344.12	344.25	26.33
305.82	305.95	23.40	318.63	318.76	24.38	331.44	331.57	25.36	344.25	344.38	26.34
305.95	306.08	23.41	318.76	318.89	24.39	331.57	331.70	25.37	344.38	344.51	26.35
306.08	306.21	23.42	318.89	319.02	24.40	331.70	331.83	25.38	344.51	344.65	26.36
306.21	306.34	23.43	319.02	319.15	24.41	331.83	331.97	25.39	344.65	344.78	26.37
306.34	306.48	23.44	319.15	319.29	24.42	331.97	332.10	25.40	344.78	344.91	26.38
306.48	306.61	23.45	319.29	319.42	24.43	332.10	332.23	25.41	344.91	345.04	26.39
306.61	306.74	23.46	319.42	319.55	24.44	332.23	332.36	25.42	345.04	345.17	26.40
306.74	306.87	23.47	319.55	319.68	24.45	332.36	332.49	25.43	345.17	345.30	26.41
306.87	307.00	23.48	319.68	319.81	24.46	332.49	332.62	25.44	345.30	345.43	26.42
307.00	307.13	23.49	319.81	319.94	24.47	332.62	332.75	25.45	345.43	345.56	26.43
307.13	307.26	23.50	319.94	320.07	24.48	332.75	332.88	25.46	345.56	345.69	26.44
307.26	307.39	23.51	320.07	320.20	24.49	332.88	333.01	25.47	345.69	345.82	26.45
307.39	307.52	23.52	320.20	320.33	24.50	333.01	333.14	25.48	345.82	345.95	26.46
307.52	307.65	23.53	320.33	320.46	24.51	333.14	333.27	25.49	345.95	346.08	26.47
307.65	307.78	23.54	320.46	320.59	24.52	333.27	333.40	25.50	346.08	346.21	26.48
307.78	307.91	23.55	320.59	320.72	24.53	333.40	333.53	25.51	346.21	346.34	26.49
307.91	308.04	23.56	320.72	320.85	24.54	333.53	333.66	25.52	346.34	346.48	26.50
308.04	308.17	23.57	320.85	320.98	24.55	333.66	333.80	25.53	346.48	346.61	26.51
308.17	308.31	23.58	320.98	321.12	24.56	333.80	333.93	25.54	346.61	346.74	26.52
308.31	308.44	23.59	321.12	321.25	24.57	333.93	334.06	25.55	346.74	346.87	26.53
308.44	308.57	23.60	321.25	321.38	24.58	334.06	334.19	25.56	346.87	347.00	26.54
308.57	308.70	23.61	321.38	321.51	24.59	334.19	334.32	25.57	347.00	347.13	26.55
308.70	308.83	23.62	321.51	321.64	24.60	334.32	334.45	25.58	347.13	347.26	26.56
308.83	308.96	23.63	321.64	321.77	24.61	334.45	334.58	25.59	347.26	347.39	26.57
308.96	309.09	23.64	321.77	321.90	24.62	334.58	334.71	25.60	347.39	347.52	26.58
309.09	309.22	23.65	321.90	322.03	24.63	334.71	334.84	25.61	347.52	347.65	26.59
309.22	309.35	23.66	322.03	322.16	24.64	334.84	334.97	25.62	347.65	347.78	26.60
309.35	309.48	23.67	322.16	322.29	24.65	334.97	335.10	25.63	347.78	347.91	26.61
309.48	309.61	23.68	322.29	322.42	24.66	335.10	335.23	25.64	347.91	348.04	26.62
309.61	309.74	23.69	322.42	322.55	24.67	335.23	335.36	25.65	348.04	348.17	26.63
309.74	309.87	23.70	322.55	322.68	24.68	335.36	335.49	25.66	348.17	348.31	26.64
309.87	310.00	23.71	322.68	322.82	24.69	335.49	335.63	25.67	348.31	348.44	26.65
310.00	310.14	23.72	322.82	322.95	24.70	335.63	335.76	25.68	348.44	348.57	26.66
310.14	310.27	23.73	322.95	323.08	24.71	335.76	335.89	25.69	348.57	348.70	26.67
310.27	310.40	23.74	323.08	323.21	24.72	335.89	336.02	25.70	348.70	348.83	26.68
310.40	310.53	23.75	323.21	323.34	24.73	336.02	336.15	25.71	348.83	348.96	26.69
310.53	310.66	23.76	323.34	323.47	24.74	336.15	336.28	25.72	348.96	349.09	26.70
310.66	310.79	23.77	323.47	323.60	24.75	336.28	336.41	25.73	349.09	349.22	26.71
310.79	310.92	23.78	323.60	323.73	24.76	336.41	336.54	25.74	349.22	349.35	26.72
310.92	311.05	23.79	323.73	323.86	24.77	336.54	336.67	25.75	349.35	349.48	26.73
311.05	311.18	23.80	323.86	323.99	24.78	336.67	336.80	25.76	349.48	349.61	26.74
311.18	311.31	23.81	323.99	324.12	24.79	336.80	336.93	25.77	349.61	349.74	26.75
311.31	311.44	23.82	324.12	324.25	24.80	336.93	337.06	25.78	349.74	349.87	26.76
311.44	311.57	23.83	324.25	324.38	24.81	337.06	337.19	25.79	349.87	350.00	26.77
311.57	311.70	23.84	324.38	324.51	24.82	337.19	337.32	25.80	350.00	350.14	26.78
311.70	311.83	23.85	324.51	324.65	24.83	337.32	337.46	25.81	350.14	350.27	26.79
311.83	311.97	23.86	324.65	324.78	24.84	337.46	337.59	25.82	350.27	350.40	26.80
311.97	312.10	23.87	324.78	324.91	24.85	337.59	337.72	25.83	350.40	350.53	26.81
312.10	312.23	23.88	324.91	325.04	24.86	337.72	337.85	25.84	350.53	350.66	26.82
312.23	312.36	23.89	325.04	325.17	24.87	337.85	337.98	25.85	350.66	350.79	26.83
312.36	312.49	23.90	325.17	325.30	24.88	337.98	338.11	25.86	350.79	350.92	26.84
312.49	312.62	23.91	325.30	325.43	24.89	338.11	338.24	25.87	350.92	351.05	26.85
312.62	312.75	23.92	325.43	325.56	24.90	338.24	338.37	25.88	351.05	351.18	26.86

TABLE C-11 (cont.) SOCIAL SECURITY EMPLOYEE TAX TABLE

7.65% Employee Tax Deductions

Wages at least	But less than	Tax to be withheld	Wages at least	But less than	Tax to be withheld	Wages at least	But less than	Tax to be withheld	Wages at least	But less than	Tax to be withheld
351.18	351.31	26.87	363.99	364.12	27.85	376.80	376.93	28.83	389.61	389.74	29.81
351.31	351.44	26.88	364.12	364.25	27.86	376.93	377.06	28.84	389.74	389.87	29.82
351.44	351.57	26.89	364.25	364.38	27.87	377.06	377.19	28.85	389.87	390.00	29.83
351.57	351.70	26.90	364.38	364.51	27.88	377.19	377.32	28.86	390.00	390.14	29.84
351.70	351.83	26.91	364.51	364.65	27.89	377.32	377.46	28.87	390.14	390.27	29.85
351.83	351.97	26.92	364.65	364.78	27.90	377.46	377.59	28.88	390.27	390.40	29.86
351.97	352.10	26.93	364.78	364.91	27.91	377.59	377.72	28.89	390.40	390.53	29.87
352.10	352.23	26.94	364.91	365.04	27.92	377.72	377.85	28.90	390.53	390.66	29.88
352.23	352.36	26.95	365.04	365.17	27.93	377.85	377.98	28.91	390.66	390.79	29.89
352.36	352.49	26.96	365.17	365.30	27.94	377.98	378.11	28.92	390.79	390.92	29.90
352.49	352.62	26.97	365.30	365.43	27.95	378.11	378.24	28.93	390.92	391.05	29.91
352.62	352.75	26.98	365.43	365.56	27.96	378.24	378.37	28.94	391.05	391.18	29.92
352.75	352.88	26.99	365.56	365.69	27.97	378.37	378.50	28.95	391.18	391.31	29.93
352.88	353.01	27.00	365.69	365.82	27.98	378.50	378.63	28.96	391.31	391.44	29.94
353.01	353.14	27.01	365.82	365.95	27.99	378.63	378.76	28.97	391.44	391.57	29.95
353.14	353.27	27.02	365.95	366.08	28.00	378.76	378.89	28.98	391.57	391.70	29.96
353.27	353.40	27.03	366.08	366.21	28.01	378.89	379.02	28.99	391.70	391.83	29.97
353.40	353.53	27.04	366.21	366.34	28.02	379.02	379.15	29.00	391.83	391.97	29.98
353.53	353.66	27.05	366.34	366.48	28.03	379.15	379.29	29.01	391.97	392.10	29.99
353.66	353.80	27.06	366.48	366.61	28.04	379.29	379.42	29.02	392.10	392.23	30.00
353.80	353.93	27.07	366.61	366.74	28.05	379.42	379.55	29.03	392.23	392.36	30.01
353.93	354.06	27.08	366.74	366.87	28.06	379.55	379.68	29.04	392.36	392.49	30.02
354.06	354.19	27.09	366.87	367.00	28.07	379.68	379.81	29.05	392.49	392.62	30.03
354.19	354.32	27.10	367.00	367.13	28.08	379.81	379.94	29.06	392.62	392.75	30.04
354.32	354.45	27.11	367.13	367.26	28.09	379.94	380.07	29.07	392.75	392.88	30.05
354.45	354.58	27.12	367.26	367.39	28.10	380.07	380.20	29.08	392.88	393.01	30.06
354.58	354.71	27.13	367.39	367.52	28.11	380.20	380.33	29.09	393.01	393.14	30.07
354.71	354.84	27.14	367.52	367.65	28.12	380.33	380.46	29.10	393.14	393.27	30.08
354.84	354.97	27.15	367.65	367.78	28.13	380.46	380.59	29.11	393.27	393.40	30.09
354.97	355.10	27.16	367.78	367.91	28.14	380.59	380.72	29.12	393.40	393.53	30.10
355.10	355.23	27.17	367.91	368.04	28.15	380.72	380.85	29.13	393.53	393.66	30.11
355.23	355.36	27.18	368.04	368.17	28.16	380.85	380.98	29.14	393.66	393.80	30.12
355.36	355.49	27.19	368.17	368.31	28.17	380.98	381.12	29.15	393.80	393.93	30.13
355.49	355.63	27.20	368.31	368.44	28.18	381.12	381.25	29.16	393.93	394.06	30.14
355.63	355.76	27.21	368.44	368.57	28.19	381.25	381.38	29.17	394.06	394.19	30.15
355.76	355.89	27.22	368.57	368.70	28.20	381.38	381.51	29.18	394.19	394.32	30.16
355.89	356.02	27.23	368.70	368.83	28.21	381.51	381.64	29.19	394.32	394.45	30.17
356.02	356.15	27.24	368.83	368.96	28.22	381.64	381.77	29.20	394.45	394.58	30.18
356.15	356.28	27.25	368.96	369.09	28.23	381.77	381.90	29.21	394.58	394.71	30.19
356.28	356.41	27.26	369.09	369.22	28.24	381.90	382.03	29.22	394.71	394.84	30.20
356.41	356.54	27.27	369.22	369.35	28.25	382.03	382.16	29.23	394.84	394.97	30.21
356.54	356.67	27.28	369.35	369.48	28.26	382.16	382.29	29.24	394.97	395.10	30.22
356.67	356.80	27.29	369.48	369.61	28.27	382.29	382.42	29.25	395.10	395.23	30.23
356.80	356.93	27.30	369.61	369.74	28.28	382.42	382.55	29.26	395.23	395.36	30.24
356.93	357.06	27.31	369.74	369.87	28.29	382.55	382.68	29.27	395.36	395.49	30.25
357.06	357.19	27.32	369.87	370.00	28.30	382.68	382.82	29.28	395.49	395.63	30.26
357.19	357.32	27.33	370.00	370.14	28.31	382.82	382.95	29.29	395.63	395.76	30.27
357.32	357.46	27.34	370.14	370.27	28.32	382.95	383.08	29.30	395.76	395.89	30.28
357.46	357.59	27.35	370.27	370.40	28.33	383.08	383.21	29.31	395.89	396.02	30.29
357.59	357.72	27.36	370.40	370.53	28.34	383.21	383.34	29.32	396.02	396.15	30.30
357.72	357.85	27.37	370.53	370.66	28.35	383.34	383.47	29.33	396.15	396.28	30.31
357.85	357.98	27.38	370.66	370.79	28.36	383.47	383.60	29.34	396.28	396.41	30.32
357.98	358.11	27.39	370.79	370.92	28.37	383.60	383.73	29.35	396.41	396.54	30.33
358.11	358.24	27.40	370.92	371.05	28.38	383.73	383.86	29.36	396.54	396.67	30.34
358.24	358.37	27.41	371.05	371.18	28.39	383.86	383.99	29.37	396.67	396.80	30.35
358.37	358.50	27.42	371.18	371.31	28.40	383.99	384.12	29.38	396.80	396.93	30.36
358.50	358.63	27.43	371.31	371.44	28.41	384.12	384.25	29.39	396.93	397.06	30.37
358.63	358.76	27.44	371.44	371.57	28.42	384.25	384.38	29.40	397.06	397.19	30.38
358.76	358.89	27.45	371.57	371.70	28.43	384.38	384.51	29.41	397.19	397.32	30.39
358.89	359.02	27.46	371.70	371.83	28.44	384.51	384.65	29.42	397.32	397.46	30.40
359.02	359.15	27.47	371.83	371.97	28.45	384.65	384.78	29.43	397.46	397.59	30.41
359.15	359.29	27.48	371.97	372.10	28.46	384.78	384.91	29.44	397.59	397.72	30.42
359.29	359.42	27.49	372.10	372.23	28.47	384.91	385.04	29.45	397.72	397.85	30.43
359.42	359.55	27.50	372.23	372.36	28.48	385.04	385.17	29.46	397.85	397.98	30.44
359.55	359.68	27.51	372.36	372.49	28.49	385.17	385.30	29.47	397.98	398.11	30.45
359.68	359.81	27.52	372.49	372.62	28.50	385.30	385.43	29.48	398.11	398.24	30.46
359.81	359.94	27.53	372.62	372.75	28.51	385.43	385.56	29.49	398.24	398.37	30.47
359.94	360.07	27.54	372.75	372.88	28.52	385.56	385.69	29.50	398.37	398.50	30.48
360.07	360.20	27.55	372.88	373.01	28.53	385.69	385.82	29.51	398.50	398.63	30.49
360.20	360.33	27.56	373.01	373.14	28.54	385.82	385.95	29.52	398.63	398.76	30.50
360.33	360.46	27.57	373.14	373.27	28.55	385.95	386.08	29.53	398.76	398.89	30.51
360.46	360.59	27.58	373.27	373.40	28.56	386.08	386.21	29.54	398.89	399.02	30.52
360.59	360.72	27.59	373.40	373.53	28.57	386.21	386.34	29.55	399.02	399.15	30.53
360.72	360.85	27.60	373.53	373.66	28.58	386.34	386.48	29.56	399.15	399.29	30.54
360.85	360.98	27.61	373.66	373.80	28.59	386.48	386.61	29.57	399.29	399.42	30.55
360.98	361.12	27.62	373.80	373.93	28.60	386.61	386.74	29.58	399.42	399.55	30.56
361.12	361.25	27.63	373.93	374.06	28.61	386.74	386.87	29.59	399.55	399.68	30.57
361.25	361.38	27.64	374.06	374.19	28.62	386.87	387.00	29.60	399.68	399.81	30.58
361.38	361.51	27.65	374.19	374.32	28.63	387.00	387.13	29.61	399.81	399.94	30.59
361.51	361.64	27.66	374.32	374.45	28.64	387.13	387.26	29.62	399.94	400.07	30.60
361.64	361.77	27.67	374.45	374.58	28.65	387.26	387.39	29.63	400.07	400.20	30.61
361.77	361.90	27.68	374.58	374.71	28.66	387.39	387.52	29.64	400.20	400.33	30.62
361.90	362.03	27.69	374.71	374.84	28.67	387.52	387.65	29.65	400.33	400.46	30.63
362.03	362.16	27.70	374.84	374.97	28.68	387.65	387.78	29.66	400.46	400.59	30.64
362.16	362.29	27.71	374.97	375.10	28.69	387.78	387.91	29.67	400.59	400.72	30.65
362.29	362.42	27.72	375.10	375.23	28.70	387.91	388.04	29.68	400.72	400.85	30.66
362.42	362.55	27.73	375.23	375.36	28.71	388.04	388.17	29.69	400.85	400.98	30.67
362.55	362.68	27.74	375.36	375.49	28.72	388.17	388.31	29.70	400.98	401.12	30.68
362.68	362.82	27.75	375.49	375.63	28.73	388.31	388.44	29.71	401.12	401.25	30.69
362.82	362.95	27.76	375.63	375.76	28.74	388.44	388.57	29.72	401.25	401.38	30.70
362.95	363.08	27.77	375.76	375.89	28.75	388.57	388.70	29.73	401.38	401.51	30.71
363.08	363.21	27.78	375.89	376.02	28.76	388.70	388.83	29.74	401.51	401.64	30.72
363.21	363.34	27.79	376.02	376.15	28.77	388.83	388.96	29.75	401.64	401.77	30.73
363.34	363.47	27.80	376.15	376.28	28.78	388.96	389.09	29.76	401.77	401.90	30.74
363.47	363.60	27.81	376.28	376.41	28.79	389.09	389.22	29.77	401.90	402.03	30.75
363.60	363.73	27.82	376.41	376.54	28.80	389.22	389.35	29.78	402.03	402.16	30.76
363.73	363.86	27.83	376.54	376.67	28.81	389.35	389.48	29.79	402.16	402.29	30.77
363.86	363.99	27.84	376.67	376.80	28.82	389.48	389.61	29.80	402.29	402.42	30.78

TABLE C-12 FEDERAL INCOME TAX WITHHOLDING TABLE—
SINGLE PERSONS

Weekly Payroll Period

If the wages are—		And the number of withholding allowances claimed is—										
At least	But less than	0	1	2	3	4	5	6	7	8	9	10
		The amount of income tax to be withheld is—										
150	155	15	8	1	0	0	0	0	0	0	0	0
155	160	16	9	2	0	0	0	0	0	0	0	0
160	165	17	10	2	0	0	0	0	0	0	0	0
165	170	18	10	3	0	0	0	0	0	0	0	0
170	175	18	11	4	0	0	0	0	0	0	0	0
175	180	19	12	5	0	0	0	0	0	0	0	0
180	185	20	13	5	0	0	0	0	0	0	0	0
185	190	21	13	6	0	0	0	0	0	0	0	0
190	195	21	14	7	0	0	0	0	0	0	0	0
195	200	22	15	8	0	0	0	0	0	0	0	0
200	210	23	16	9	2	0	0	0	0	0	0	0
210	220	25	18	10	3	0	0	0	0	0	0	0
220	230	26	19	12	5	0	0	0	0	0	0	0
230	240	28	21	13	6	0	0	0	0	0	0	0
240	250	29	22	15	8	0	0	0	0	0	0	0
250	260	31	24	16	9	2	0	0	0	0	0	0
260	270	32	25	18	11	3	0	0	0	0	0	0
270	280	34	27	19	12	5	0	0	0	0	0	0
280	290	35	28	21	14	6	0	0	0	0	0	0
290	300	37	30	22	15	8	1	0	0	0	0	0
300	310	38	31	24	17	9	2	0	0	0	0	0
310	320	40	33	25	18	11	4	0	0	0	0	0
320	330	41	34	27	20	12	5	0	0	0	0	0
330	340	43	36	28	21	14	7	0	0	0	0	0
340	350	44	37	30	23	15	8	1	0	0	0	0
350	360	46	39	31	24	17	10	2	0	0	0	0
360	370	47	40	33	26	18	11	4	0	0	0	0
370	380	49	42	34	27	20	13	5	0	0	0	0
380	390	50	43	36	29	21	14	7	0	0	0	0
390	400	52	45	37	30	23	16	8	1	0	0	0
400	410	53	46	39	32	24	17	10	3	0	0	0
410	420	55	48	40	33	26	19	11	4	0	0	0
420	430	56	49	42	35	27	20	13	6	0	0	0
430	440	58	51	43	36	29	22	14	7	0	0	0
440	450	59	52	45	38	30	23	16	9	2	0	0
450	460	61	54	46	39	32	25	17	10	3	0	0
460	470	62	55	48	41	33	26	19	12	5	0	0
470	480	64	57	49	42	35	28	20	13	6	0	0
480	490	66	58	51	44	36	29	22	15	8	0	0
490	500	69	60	52	45	38	31	23	16	9	2	0
500	510	72	61	54	47	39	32	25	18	11	3	0
510	520	75	63	55	48	41	34	26	19	12	5	0
520	530	78	64	57	50	42	35	28	21	14	6	0
530	540	80	67	58	51	44	37	29	22	15	8	1
540	550	83	70	60	53	45	38	31	24	17	9	2
550	560	86	73	61	54	47	40	32	25	18	11	4
560	570	89	75	63	56	48	41	34	27	20	12	5
570	580	92	78	65	57	50	43	35	28	21	14	7
580	590	94	81	68	59	51	44	37	30	23	15	8
590	600	97	84	70	60	53	46	38	31	24	17	10
600	610	100	87	73	62	54	47	40	33	26	18	11
610	620	103	89	76	63	56	49	41	34	27	20	13
620	630	106	92	79	65	57	50	43	36	29	21	14
630	640	108	95	82	68	59	52	44	37	30	23	16
640	650	111	98	84	71	60	53	46	39	32	24	17
650	660	114	101	87	74	62	55	47	40	33	26	19
660	670	117	103	90	76	63	56	49	42	35	27	20
670	680	120	106	93	79	66	58	50	43	36	29	22
680	690	122	109	96	82	69	59	52	45	38	30	23
690	700	125	112	98	85	71	61	53	46	39	32	25
700	710	128	115	101	88	74	62	55	48	41	33	26
710	720	131	117	104	90	77	64	56	49	42	35	28
720	730	134	120	107	93	80	66	58	51	44	36	29
730	740	136	123	110	96	83	69	59	52	45	38	31
740	750	139	126	112	99	85	72	61	54	47	39	32
750	760	142	129	115	102	88	75	62	55	48	41	34
760	770	145	131	118	104	91	78	64	57	50	42	35
770	780	148	134	121	107	94	80	67	58	51	44	37
780	790	150	137	124	110	97	83	70	60	53	45	38
790	800	153	140	126	113	99	86	72	61	54	47	40

TABLE C-12 (cont.) FEDERAL INCOME TAX WITHHOLDING TABLE—
MARRIED PERSONS

Weekly Payroll Period

If the wages are—		And the number of withholding allowances claimed is—										
At least	But less than	0	1	2	3	4	5	6	7	8	9	10
		The amount of income tax to be withheld is—										
145	150	4	0	0	0	0	0	0	0	0	0	0
150	155	4	0	0	0	0	0	0	0	0	0	0
155	160	5	0	0	0	0	0	0	0	0	0	0
160	165	6	0	0	0	0	0	0	0	0	0	0
165	170	7	0	0	0	0	0	0	0	0	0	0
170	175	7	0	0	0	0	0	0	0	0	0	0
175	180	8	1	0	0	0	0	0	0	0	0	0
180	185	9	2	0	0	0	0	0	0	0	0	0
185	190	10	2	0	0	0	0	0	0	0	0	0
190	195	10	3	0	0	0	0	0	0	0	0	0
195	200	11	4	0	0	0	0	0	0	0	0	0
200	210	12	5	0	0	0	0	0	0	0	0	0
210	220	14	7	0	0	0	0	0	0	0	0	0
220	230	15	8	1	0	0	0	0	0	0	0	0
230	240	17	10	2	0	0	0	0	0	0	0	0
240	250	18	11	4	0	0	0	0	0	0	0	0
250	260	20	13	5	0	0	0	0	0	0	0	0
260	270	21	14	7	0	0	0	0	0	0	0	0
270	280	23	16	8	1	0	0	0	0	0	0	0
280	290	24	17	10	3	0	0	0	0	0	0	0
290	300	26	19	11	4	0	0	0	0	0	0	0
300	310	27	20	13	6	0	0	0	0	0	0	0
310	320	29	22	14	7	0	0	0	0	0	0	0
320	330	30	23	16	9	1	0	0	0	0	0	0
330	340	32	25	17	10	3	0	0	0	0	0	0
340	350	33	26	19	12	4	0	0	0	0	0	0
350	360	35	28	20	13	6	0	0	0	0	0	0
360	370	36	29	22	15	7	0	0	0	0	0	0
370	380	38	31	23	16	9	2	0	0	0	0	0
380	390	39	32	25	18	10	3	0	0	0	0	0
390	400	41	34	26	19	12	5	0	0	0	0	0
400	410	42	35	28	21	13	6	0	0	0	0	0
410	420	44	37	29	22	15	8	1	0	0	0	0
420	430	45	38	31	24	16	9	2	0	0	0	0
430	440	47	40	32	25	18	11	4	0	0	0	0
440	450	48	41	34	27	19	12	5	0	0	0	0
450	460	50	43	35	28	21	14	7	0	0	0	0
460	470	51	44	37	30	22	15	8	1	0	0	0
470	480	53	46	38	31	24	17	10	2	0	0	0
480	490	54	47	40	33	25	18	11	4	0	0	0
490	500	56	49	41	34	27	20	13	5	0	0	0
500	510	57	50	43	36	28	21	14	7	0	0	0
510	520	59	52	44	37	30	23	16	8	1	0	0
520	530	60	53	46	39	31	24	17	10	3	0	0
530	540	62	55	47	40	33	26	19	11	4	0	0
540	550	63	56	49	42	34	27	20	13	6	0	0
550	560	65	58	50	43	36	29	22	14	7	0	0
560	570	66	59	52	45	37	30	23	16	9	1	0
570	580	68	61	53	46	39	32	25	17	10	3	0
580	590	69	62	55	48	40	33	26	19	12	4	0
590	600	71	64	56	49	42	35	28	20	13	6	0
600	610	72	65	58	51	43	36	29	22	15	7	0
610	620	74	67	59	52	45	38	31	23	16	9	2
620	630	75	68	61	54	46	39	32	25	18	10	3
630	640	77	70	62	55	48	41	34	26	19	12	5
640	650	78	71	64	57	49	42	35	28	21	13	6
650	660	80	73	65	58	51	44	37	29	22	15	8
660	670	81	74	67	60	52	45	38	31	24	16	9
670	680	83	76	68	61	54	47	40	32	25	18	11
680	690	84	77	70	63	55	48	41	34	27	19	12
690	700	86	79	71	64	57	50	43	35	28	21	14
700	710	87	80	73	66	58	51	44	37	30	22	15
710	720	89	82	74	67	60	53	46	38	31	24	17
720	730	90	83	76	69	61	54	47	40	33	25	18
730	740	92	85	77	70	63	56	49	41	34	27	20
740	750	93	86	79	72	64	57	50	43	36	28	21
750	760	95	88	80	73	66	59	52	44	37	30	23
760	770	96	89	82	75	67	60	53	46	39	31	24
770	780	98	91	83	76	69	62	55	47	40	33	26
780	790	99	92	85	78	70	63	56	49	42	34	27
790	800	101	94	86	79	72	65	58	50	43	36	29
800	810	102	95	88	81	73	66	59	52	45	37	30

TABLE C-13 MACRS COST RECOVERY FACTORS

For Property Placed into Service in 1987 and Thereafter

RECOVERY YEAR	3-YR. CLASS	5-YR. CLASS	7-YR. CLASS	10-YR. CLASS
1	0.333333	0.200000	0.142857	0.100000
2	0.444444	0.320000	0.244898	0.180000
3	0.148148	0.192000	0.174927	0.144000
4	0.074074	0.115200	0.124948	0.115200
5	—	0.115200	0.089249	0.092160
6	—	0.057600	0.089249	0.073728
7	—	—	0.089249	0.065536
8	—	—	0.044624	0.065536
9	—	—	—	0.065536
10	—	—	—	0.065536
	—	—	—	0.032768

RECOVERY YEAR(S)	15-YR. CLASS	20-YR. CLASS	27.5-YR. REAL ESTATE	31.5-YR. REAL ESTATE
1	0.050000	0.037500	0.034848	0.030423
2	0.095000	0.072188	0.036364	0.031746
3	0.085500	0.066773	0.036364	0.031746
4	0.076950	0.061765	0.036364	0.031746
5	0.069255	0.057133	0.036364	0.031746
6	0.062330	0.052848	0.036364	0.031746
7	0.059049	0.048884	0.036364	0.031746
8	0.059049	0.045218	0.036364	0.031746
9–15	0.059049	0.044615	0.036364	0.031746
16	0.029525	0.044615	0.036364	0.031746
17–20	—	0.044615	0.036364	0.031746
21	—	0.022308	0.036364	0.031746
22–27	—	—	0.036364	0.031746
28	—	—	0.019697	0.031746
29–31	—	—	—	0.031746
32	—	—	—	0.017196

EMPLOYEE

FICA - check YTD : under $68,400, then gross × 7.65% (.0765)
over $68,400, then gross × 1.45% (.0145)

FED W/T - check table : single w' or w/o dependents pg. 595
married w' or w/o dependents pg. 596

STATE W/T - gross times 3% (.03)

NET - gross minus FICA, FED W/T and STATE W/T = NET

EMPLOYER

FICA - match employee (TTL) paid amount

FED UNEM. - YTD (each employee) under $7000 × .8% (.008)

ST. UNEM. - YTD (each employee) under $9000 × 3.5% (.035)

BALANCE - ADD columns down & total
- EMPLOYEE GROSS - FICA, FED W/T and STATE W/T = NET

TABLE C-14 THE NUMBER OF EACH DAY OF THE YEAR*

DAY OF MONTH	JAN.	FEB.	MAR.	APR.	MAY	JUNE	JULY	AUG.	SEPT.	OCT.	NOV.	DEC.	DAY OF MONTH
1	1	32	60	91	121	152	182	213	244	274	305	335	1
2	2	33	61	92	122	153	183	214	245	275	306	336	2
3	3	34	62	93	123	154	184	215	246	276	307	337	3
4	4	35	63	94	124	155	185	216	247	277	308	338	4
5	5	36	64	95	125	156	186	217	248	278	309	339	5
6	6	37	65	96	126	157	187	218	249	279	310	340	6
7	7	38	66	97	127	158	188	219	250	280	311	341	7
8	8	39	67	98	128	159	189	220	251	281	312	342	8
9	9	40	68	99	129	160	190	221	252	282	313	343	9
10	10	41	69	100	130	161	191	222	253	283	314	344	10
11	11	42	70	101	131	162	192	223	254	284	315	345	11
12	12	43	71	102	132	163	193	224	255	285	316	346	12
13	13	44	72	103	133	164	194	225	256	286	317	347	13
14	14	45	73	104	134	165	195	226	257	287	318	348	14
15	15	46	74	105	135	166	196	227	258	288	319	349	15
16	16	47	75	106	136	167	197	228	259	289	320	350	16
17	17	48	76	107	137	168	198	229	260	290	321	351	17
18	18	49	77	108	138	169	199	230	261	291	322	352	18
19	19	50	78	109	139	170	200	231	262	292	323	353	19
20	20	51	79	110	140	171	201	232	263	293	324	354	20
21	21	52	80	111	141	172	202	233	264	294	325	355	21
22	22	53	81	112	142	173	203	234	265	295	326	356	22
23	23	54	82	113	143	174	204	235	266	296	327	357	23
24	24	55	83	114	144	175	205	236	267	297	328	358	24
25	25	56	84	115	145	176	206	237	268	298	329	359	25
26	26	57	85	116	146	177	207	238	269	299	330	360	26
27	27	58	86	117	147	178	208	239	270	300	331	361	27
28	28	59	87	118	148	179	209	240	271	301	332	362	28
29	29		88	119	149	180	210	241	272	302	333	363	29
30	30		89	120	150	181	211	242	273	303	334	364	30
31	31		90		151		212	243		304		365	31

* In leap years, after February 28, add 1 to the tabular number.

TABLE C-15 ANNUAL PERCENTAGE RATE TABLE FOR MONTHLY PAYMENT PLANS

Number of Payments	ANNUAL PERCENTAGE RATE															
	10.00%	10.25%	10.50%	10.75%	11.00%	11.25%	11.50%	11.75%	12.00%	12.25%	12.50%	12.75%	13.00%	13.25%	13.50%	13.75%
	Finance charge per $100 of amount financed															
1	0.83	0.85	0.87	0.90	0.92	0.94	0.96	0.98	1.00	1.02	1.04	1.06	1.08	1.10	1.12	1.15
2	1.25	1.28	1.31	1.35	1.38	1.41	1.44	1.47	1.50	1.53	1.57	1.60	1.63	1.66	1.69	1.73
3	1.67	1.71	1.76	1.80	1.84	1.88	1.92	1.96	2.01	2.05	2.09	2.13	2.17	2.22	2.26	2.30
4	2.09	2.14	2.20	2.25	2.30	2.35	2.41	2.46	2.51	2.57	2.62	2.67	2.72	2.78	2.83	2.88
5	2.51	2.58	2.64	2.70	2.77	2.83	2.89	2.96	3.02	3.08	3.15	3.21	3.27	3.34	3.40	3.46
6	2.94	3.01	3.08	3.16	3.23	3.31	3.38	3.45	3.53	3.60	3.68	3.75	3.83	3.90	3.97	4.05
7	3.36	3.45	3.53	3.62	3.70	3.78	3.87	3.95	4.04	4.12	4.21	4.29	4.38	4.47	4.55	4.64
8	3.79	3.88	3.98	4.07	4.17	4.26	4.36	4.46	4.55	4.65	4.74	4.84	4.94	5.03	5.13	5.22
9	4.21	4.32	4.43	4.53	4.64	4.75	4.85	4.96	5.07	5.17	5.28	5.39	5.49	5.60	5.71	5.82
10	4.64	4.76	4.88	4.99	5.11	5.23	5.35	5.46	5.58	5.70	5.82	5.94	6.05	6.17	6.29	6.41
11	5.07	5.20	5.33	5.45	5.58	5.71	5.84	5.97	6.10	6.23	6.36	6.49	6.62	6.75	6.88	7.01
12	5.50	5.64	5.78	5.92	6.06	6.20	6.34	6.48	6.62	6.76	6.90	7.04	7.18	7.32	7.46	7.60
13	5.93	6.08	6.23	6.38	6.53	6.68	6.84	6.99	7.14	7.29	7.44	7.59	7.75	7.90	8.05	8.20
14	6.36	6.52	6.69	6.85	7.01	7.17	7.34	7.50	7.66	7.82	7.99	8.15	8.31	8.48	8.64	8.81
15	6.80	6.97	7.14	7.32	7.49	7.66	7.84	8.01	8.19	8.36	8.53	8.71	8.88	9.06	9.23	9.41
16	7.23	7.41	7.60	7.78	7.97	8.15	8.34	8.53	8.71	8.90	9.08	9.27	9.46	9.64	9.83	10.02
17	7.67	7.86	8.06	8.25	8.45	8.65	8.84	9.04	9.24	9.44	9.63	9.83	10.03	10.23	10.43	10.63
18	8.10	8.31	8.52	8.75	8.93	9.14	9.35	9.56	9.77	9.98	10.19	10.40	10.61	10.82	11.03	11.24
19	8.54	8.76	8.98	9.20	9.42	9.64	9.86	10.08	10.30	10.52	10.74	10.96	11.18	11.41	11.63	11.85
20	8.98	9.21	9.44	9.67	9.90	10.13	10.37	10.60	10.83	11.06	11.30	11.53	11.76	12.00	12.23	12.46
21	9.42	9.66	9.90	10.15	10.39	10.63	10.88	11.12	11.36	11.61	11.85	12.10	12.34	12.59	12.84	13.08
22	9.86	10.12	10.37	10.62	10.88	11.13	11.39	11.64	11.90	12.16	12.41	12.67	12.93	13.19	13.44	13.70
23	10.30	10.57	10.84	11.10	11.37	11.63	11.90	12.17	12.44	12.71	12.97	13.24	13.51	13.78	14.05	14.32
24	10.75	11.02	11.30	11.58	11.86	12.14	12.42	12.70	12.98	13.26	13.54	13.82	14.10	14.38	14.66	14.95
25	11.19	11.48	11.77	12.06	12.35	12.64	12.93	13.22	13.52	13.81	14.10	14.40	14.69	14.98	15.28	15.57
26	11.64	11.94	12.24	12.54	12.85	13.15	13.45	13.75	14.06	14.36	14.67	14.97	15.28	15.59	15.89	16.20
27	12.09	12.40	12.71	13.03	13.34	13.66	13.97	14.29	14.60	14.92	15.24	15.56	15.87	16.19	16.51	16.83
28	12.53	12.86	13.18	13.51	13.84	14.16	14.49	14.82	15.15	15.48	15.81	16.14	16.47	16.80	17.13	17.46
29	12.98	13.32	13.66	14.00	14.33	14.67	15.01	15.35	15.70	16.04	16.38	16.72	17.07	17.41	17.75	18.10
30	13.43	13.78	14.13	14.48	14.83	15.19	15.54	15.89	16.24	16.60	16.95	17.31	17.66	18.02	18.38	18.74
31	13.89	14.25	14.61	14.97	15.33	15.70	16.06	16.43	16.79	17.16	17.53	17.90	18.27	18.63	19.00	19.38
32	14.34	14.71	15.09	15.46	15.84	16.21	16.59	16.97	17.35	17.73	18.11	18.49	18.87	19.25	19.63	20.02
33	14.79	15.18	15.57	15.95	16.34	16.73	17.12	17.51	17.90	18.29	18.65	19.08	19.47	19.87	20.26	20.66
34	15.25	15.65	16.05	16.44	16.85	17.25	17.65	18.05	18.46	18.86	19.27	19.67	20.08	20.49	20.90	21.31
35	15.70	16.11	16.53	16.94	17.35	17.77	18.18	18.60	19.01	19.43	19.85	20.27	20.69	21.11	21.53	21.95
36	16.16	16.58	17.01	17.43	17.86	18.29	18.71	19.14	19.57	20.00	20.43	20.87	21.30	21.73	22.17	22.60
37	16.62	17.06	17.49	17.93	18.37	18.81	19.25	19.69	20.13	20.58	21.02	21.46	21.91	22.36	22.81	23.25
38	17.08	17.53	17.98	18.43	18.88	19.33	19.78	20.24	20.69	21.15	21.61	22.07	22.52	22.99	23.45	23.91
39	17.54	18.00	18.46	18.93	19.39	19.86	20.32	20.79	21.26	21.73	22.20	22.67	23.14	23.61	24.09	24.56
40	18.00	18.48	18.95	19.43	19.90	20.38	20.86	21.34	21.82	22.30	22.79	23.27	23.76	24.25	24.73	25.22
41	18.47	18.95	19.44	19.93	20.42	20.91	21.40	21.89	22.39	22.88	23.38	23.88	24.38	24.88	25.38	25.88
42	18.93	19.43	19.93	20.43	20.93	21.44	21.94	22.45	22.96	23.47	23.98	24.49	25.00	25.51	26.03	26.55
43	19.40	19.91	20.42	20.94	21.45	21.97	22.49	23.01	23.53	24.05	24.57	25.10	25.62	26.15	26.68	27.21
44	19.86	20.39	20.91	21.44	21.97	22.50	23.03	23.57	24.10	24.64	25.17	25.71	26.25	26.79	27.33	27.88
45	20.33	20.87	21.41	21.95	22.49	23.03	23.58	24.12	24.67	25.22	25.77	26.32	26.88	27.43	27.99	28.55
46	20.80	21.35	21.90	22.46	23.01	23.57	24.13	24.69	25.25	25.81	26.37	26.94	27.51	28.08	28.65	29.22
47	21.27	21.83	22.40	22.97	23.53	24.10	24.68	25.25	25.82	26.40	26.98	27.56	28.14	28.72	29.31	29.89
48	21.74	22.32	22.90	23.48	24.06	24.64	25.23	25.81	26.40	26.99	27.58	28.18	28.77	29.37	29.97	30.57
49	22.21	22.80	23.39	23.99	24.58	25.18	25.78	26.38	26.98	27.59	28.19	28.80	29.41	30.02	30.63	31.34
50	22.69	23.29	23.89	24.50	25.11	25.72	26.33	26.95	27.56	28.18	28.80	29.42	30.04	30.67	31.29	31.92
51	23.16	23.78	24.40	25.02	25.64	26.26	26.89	27.52	28.15	28.78	29.41	30.05	30.68	31.32	31.96	32.60
52	23.64	24.27	24.90	25.53	26.17	26.81	27.45	28.09	28.73	29.38	30.02	30.67	31.32	31.98	32.63	33.29
53	24.11	24.76	25.40	26.05	26.70	27.35	28.00	28.66	29.32	29.98	30.64	31.30	31.97	32.63	33.30	33.97
54	24.59	25.25	25.91	26.57	27.23	27.90	28.56	29.23	29.91	30.58	31.25	31.93	32.61	33.29	33.98	34.66
55	25.07	25.74	26.41	27.09	27.77	28.44	29.13	29.81	30.50	31.18	31.87	32.56	33.26	33.95	34.65	35.35
56	25.55	26.23	26.92	27.61	28.30	28.99	29.69	30.39	31.09	31.79	32.49	33.20	33.91	34.62	35.33	36.04
57	26.03	26.73	27.43	28.13	28.84	29.54	30.25	30.97	31.68	32.39	33.11	33.83	34.56	35.28	36.01	36.74
58	26.51	27.23	27.94	28.66	29.37	30.10	30.82	31.55	32.27	33.00	33.74	34.47	35.21	35.95	36.69	37.43
59	27.00	27.72	28.45	29.18	29.91	30.65	31.39	32.13	32.87	33.61	34.36	35.11	35.86	36.62	37.37	38.13
60	27.48	28.22	28.96	29.71	30.45	31.20	31.96	32.71	33.47	34.23	34.99	35.75	36.52	37.29	38.06	38.83

TABLE C-15 (cont.) ANNUAL PERCENTAGE RATE TABLE FOR MONTHLY PAYMENT PLANS

Number of Payments	ANNUAL PERCENTAGE RATE															
	14.00%	14.25%	14.50%	14.75%	15.00%	15.25%	15.50%	15.75%	16.00%	16.25%	16.50%	16.75%	17.00%	17.25%	17.50%	17.75%
	Finance charge per $100 of amount financed															
1	1.17	1.19	1.21	1.23	1.25	1.27	1.29	1.31	1.33	1.35	1.37	1.40	1.42	1.44	1.46	1.48
2	1.75	1.78	1.82	1.85	1.88	1.91	1.94	1.97	2.00	2.04	2.07	2.10	2.13	2.16	2.19	2.22
3	2.34	2.38	2.43	2.47	2.51	2.55	2.59	2.64	2.68	2.72	2.76	2.80	2.85	2.89	2.93	2.97
4	2.93	2.99	3.04	3.09	3.14	3.20	3.25	3.30	3.36	3.41	3.46	3.51	3.57	3.62	3.67	3.73
5	3.53	3.59	3.65	3.72	3.78	3.84	3.91	3.97	4.04	4.10	4.16	4.23	4.29	4.35	4.42	4.48
6	4.12	4.20	4.27	4.35	4.42	4.49	4.57	4.64	4.72	4.79	4.87	4.94	5.02	5.09	5.17	5.24
7	4.72	4.81	4.89	4.98	5.06	5.15	5.23	5.32	5.40	5.49	5.58	5.66	5.75	5.83	5.92	6.00
8	5.32	5.42	5.51	5.61	5.71	5.80	5.90	6.00	6.09	6.19	6.29	6.38	6.48	6.58	6.67	6.77
9	5.92	6.03	6.14	6.25	6.35	6.46	6.57	6.68	6.78	6.89	7.00	7.11	7.22	7.32	7.43	7.54
10	6.53	6.65	6.77	6.88	7.00	7.12	7.24	7.36	7.48	7.60	7.72	7.84	7.96	8.08	8.19	8.31
11	7.14	7.27	7.40	7.53	7.66	7.79	7.92	8.05	8.18	8.31	8.44	8.57	8.70	8.83	8.96	9.09
12	7.74	7.89	8.03	8.17	8.31	8.45	8.59	8.74	8.88	9.02	9.16	9.30	9.45	9.59	9.73	9.87
13	8.36	8.51	8.66	8.81	8.97	9.12	9.27	9.43	9.58	9.73	9.89	10.04	10.20	10.35	10.50	10.66
14	8.97	9.13	9.30	9.46	9.63	9.79	9.96	10.12	10.29	10.45	10.62	10.78	10.95	11.11	11.28	11.45
15	9.59	9.76	9.94	10.11	10.29	10.47	10.64	10.82	11.00	11.17	11.35	11.51	11.71	11.88	12.06	12.24
16	10.20	10.39	10.58	10.77	10.95	11.14	11.33	11.52	11.71	11.90	12.09	12.28	12.46	12.65	12.84	13.03
17	10.82	11.02	11.22	11.42	11.62	11.82	12.02	12.22	12.42	12.62	12.83	13.03	13.23	13.43	13.63	13.83
18	11.45	11.66	11.87	12.08	12.29	12.50	12.72	12.93	13.14	13.35	13.57	13.78	13.99	14.21	14.42	14.64
19	12.07	12.30	12.52	12.74	12.97	13.19	13.41	13.64	13.86	14.09	14.31	14.54	14.76	14.99	15.22	15.44
20	12.70	12.93	13.17	13.41	13.64	13.88	14.11	14.35	14.59	14.82	15.06	15.30	15.54	15.77	16.01	16.25
21	13.33	13.58	13.82	14.07	14.32	14.57	14.82	15.06	15.31	15.56	15.81	16.05	16.31	16.56	16.81	17.07
22	13.96	14.22	14.48	14.74	15.00	15.26	15.52	15.78	16.04	16.30	16.57	16.83	17.09	17.36	17.62	17.88
23	14.59	14.87	15.14	15.41	15.68	15.96	16.23	16.50	16.78	17.05	17.32	17.60	17.88	18.15	18.43	18.70
24	15.23	15.51	15.80	16.08	16.37	16.65	16.94	17.22	17.51	17.80	18.09	18.37	18.65	18.95	19.24	19.53
25	15.87	16.17	16.46	16.76	17.06	17.35	17.65	17.95	18.25	18.55	18.85	19.15	19.45	19.75	20.05	20.36
26	16.51	16.82	17.13	17.44	17.75	18.06	18.37	18.68	18.99	19.30	19.62	19.93	20.24	20.56	20.87	21.19
27	17.15	17.47	17.80	18.12	18.44	18.76	19.09	19.41	19.74	20.06	20.39	20.71	21.04	21.37	21.69	22.02
28	17.80	18.13	18.47	18.80	19.14	19.47	19.81	20.15	20.48	20.82	21.16	21.50	21.84	22.18	22.52	22.86
29	18.45	18.79	19.14	19.49	19.83	20.18	20.53	20.88	21.23	21.58	21.94	22.29	22.64	22.99	23.35	23.70
30	19.10	19.45	19.81	20.17	20.54	20.90	21.26	21.62	21.99	22.35	22.72	23.08	23.45	23.81	24.18	24.55
31	19.75	20.12	20.49	20.87	21.24	21.61	21.99	22.37	22.74	23.12	23.50	23.88	24.26	24.64	25.02	25.40
32	20.40	20.79	21.17	21.56	21.95	22.33	22.72	23.11	23.50	23.89	24.28	24.68	25.07	25.46	25.86	26.25
33	21.06	21.46	21.85	22.25	22.65	23.06	23.46	23.86	24.26	24.67	25.07	25.48	25.88	26.29	26.70	27.11
34	21.72	22.13	22.54	22.95	23.37	23.78	24.19	24.61	25.03	25.44	25.86	26.28	26.70	27.12	27.54	27.97
35	22.38	22.80	23.23	23.65	24.08	24.51	24.94	25.36	25.79	26.23	26.66	27.09	27.52	27.96	28.39	28.83
36	23.04	23.48	23.92	24.35	24.80	25.24	25.68	26.12	26.57	27.01	27.46	27.90	28.35	28.80	29.25	29.70
37	23.70	24.16	24.61	25.06	25.51	25.97	26.42	26.88	27.34	27.80	28.26	28.72	29.18	29.64	30.10	30.57
38	24.37	24.84	25.30	25.77	26.24	26.70	27.17	27.64	28.11	28.59	29.06	29.53	30.01	30.49	30.96	31.44
39	25.04	25.52	26.00	26.48	26.96	27.44	27.92	28.41	28.89	29.38	29.87	30.36	30.85	31.34	31.83	32.32
40	25.71	26.20	26.70	27.19	27.69	28.18	28.68	29.18	29.68	30.18	30.68	31.18	31.68	32.19	32.69	33.20
41	26.39	26.89	27.40	27.91	28.41	28.92	29.44	29.95	30.46	30.97	31.49	32.01	32.52	33.04	33.56	34.08
42	27.06	27.58	28.10	28.62	29.15	29.67	30.19	30.72	31.25	31.78	32.31	32.84	33.37	33.90	34.44	34.97
43	27.74	28.27	28.81	29.34	29.88	30.42	30.96	31.50	32.04	32.58	33.13	33.67	34.22	34.76	35.31	35.86
44	28.42	28.97	29.52	30.07	30.62	31.17	31.72	32.28	32.83	33.39	33.95	34.51	35.07	35.63	36.19	36.76
45	29.11	29.67	30.23	30.79	31.36	31.92	32.49	33.06	33.63	34.20	34.77	35.35	35.92	36.50	37.08	37.66
46	29.79	30.36	30.94	31.52	32.10	32.68	33.26	33.84	34.43	35.01	35.60	36.19	36.78	37.37	37.96	38.56
47	30.48	31.07	31.66	32.25	32.84	33.44	34.03	34.63	35.23	35.83	36.43	37.04	37.64	38.25	38.86	39.46
48	31.17	31.77	32.37	32.98	33.59	34.20	34.81	35.42	36.03	36.65	37.27	27.88	38.50	39.13	39.75	40.37
49	31.86	32.48	33.09	33.71	34.34	34.96	35.59	36.21	36.84	37.47	38.10	38.74	39.37	40.01	40.65	41.29
50	32.55	33.18	33.82	34.45	35.09	35.73	36.37	37.01	37.65	38.30	38.94	39.59	40.24	40.89	41.55	42.20
51	33.25	33.89	34.54	35.19	35.84	36.49	37.15	37.81	38.46	39.12	39.79	40.45	41.11	41.78	42.45	43.12
52	33.95	34.61	35.27	35.93	36.60	37.27	37.94	38.61	39.28	39.96	40.63	41.31	41.99	42.67	43.36	44.04
53	34.65	35.32	36.00	36.68	37.36	38.04	38.72	39.41	40.10	40.79	41.48	42.17	42.87	43.57	44.27	44.97
54	35.35	36.04	36.73	37.42	38.12	38.82	39.52	40.22	40.92	41.63	42.33	43.04	43.75	44.47	45.18	45.90
55	36.05	36.76	37.46	38.17	38.88	39.60	40.31	41.03	41.74	42.47	43.19	43.91	44.64	45.37	46.10	46.83
56	36.76	37.48	38.20	38.92	39.65	40.38	41.11	41.84	42.57	43.31	44.05	44.79	45.53	46.27	47.02	47.77
57	37.47	38.20	38.94	39.68	40.42	41.16	41.91	42.65	53.40	44.15	44.91	45.66	46.42	47.18	47.94	48.71
58	38.18	38.93	39.68	40.43	41.19	41.95	42.71	43.47	44.23	45.00	45.77	46.54	47.32	48.09	48.87	49.65
59	38.89	39.66	40.42	41.19	41.96	42.74	43.51	44.29	45.07	45.85	46.64	47.42	48.41	49.01	49.80	50.60
60	39.61	40.39	41.17	41.95	42.74	43.53	44.32	45.11	45.91	46.71	47.51	48.31	49.12	49.92	50.73	51.55

TABLE C-15 (cont.) ANNUAL PERCENTAGE RATE TABLE FOR MONTHLY PAYMENT PLANS

Number of Payments	ANNUAL PERCENTAGE RATE															
	18.00%	18.25%	18.50%	18.75%	19.00%	19.25%	19.50%	19.75%	20.00%	20.25%	20.50%	20.75%	21.00%	21.25%	21.50%	21.75%
	Finance charge per $100 of amount financed															
1	1.50	1.52	1.54	1.56	1.58	1.60	1.62	1.65	1.67	1.69	1.71	1.73	1.75	1.77	1.79	1.81
2	2.26	2.29	2.32	2.35	2.38	2.41	2.44	2.48	2.51	2.54	2.57	2.60	2.63	2.66	2.70	2.73
3	3.01	3.06	3.10	3.14	3.18	3.23	3.27	3.31	3.35	3.39	3.44	3.48	3.52	3.56	3.60	3.65
4	3.78	3.83	3.88	3.94	3.99	4.04	4.10	4.15	4.20	4.25	4.31	4.36	4.41	4.47	4.52	4.57
5	4.54	4.61	4.67	4.74	4.80	4.86	4.93	4.99	5.06	5.12	5.18	5.25	5.31	5.37	5.44	5.50
6	5.32	5.39	5.46	5.54	5.61	5.69	5.76	5.84	5.91	5.99	6.06	6.14	6.21	6.29	6.36	6.44
7	6.09	6.18	6.26	6.35	6.43	6.52	6.60	6.69	6.78	6.86	6.95	7.04	7.12	7.21	7.29	7.38
8	6.87	6.96	7.06	7.16	7.26	7.35	7.45	7.55	7.64	7.74	7.84	7.94	8.03	8.13	8.23	8.33
9	7.65	7.76	7.87	7.97	8.08	8.19	8.30	8.41	8.52	8.63	8.73	8.84	8.95	9.06	9.17	9.28
10	8.43	8.55	8.67	8.79	8.91	9.03	9.15	9.27	9.39	9.51	9.63	9.75	9.88	10.00	10.12	10.24
11	9.22	9.35	9.49	9.62	9.75	9.88	10.01	10.14	10.28	10.41	10.54	10.67	10.80	10.94	11.07	11.20
12	10.02	10.16	10.30	10.44	10.59	10.73	10.87	11.02	11.16	11.31	11.45	11.59	11.74	11.88	12.02	12.17
13	10.81	10.97	11.12	11.28	11.43	11.59	11.74	11.90	12.05	12.21	12.36	12.52	12.67	12.83	12.99	13.14
14	11.61	11.78	11.95	12.11	12.28	12.45	12.61	12.78	12.95	13.11	13.28	13.45	13.62	13.79	13.95	14.12
15	12.42	12.59	12.77	12.95	13.13	13.31	13.49	13.67	13.85	14.03	14.21	14.39	14.57	14.75	14.93	15.11
16	13.22	13.41	13.60	13.80	13.99	14.18	14.37	14.56	14.75	14.94	15.13	15.33	15.52	15.71	15.90	16.10
17	14.04	14.24	14.44	14.64	14.85	15.05	15.25	15.46	15.66	15.86	16.07	16.27	16.48	16.68	16.89	17.09
18	14.85	15.07	15.28	15.49	15.71	15.93	16.14	16.36	16.57	16.79	17.01	17.22	17.44	17.66	17.88	18.09
19	15.67	15.90	16.12	16.35	16.58	16.81	17.03	17.26	17.49	17.72	17.95	18.18	18.41	18.64	18.87	19.10
20	16.49	16.73	16.97	17.21	17.45	17.69	17.93	18.17	18.41	18.66	18.90	19.14	19.38	19.63	19.87	20.11
21	17.32	17.57	17.82	18.07	18.33	18.58	18.83	19.09	19.34	19.60	19.85	20.11	20.36	20.62	20.87	21.13
22	18.15	18.41	18.68	18.94	19.21	19.47	19.74	20.01	20.27	20.54	20.81	21.08	21.34	21.61	21.88	22.15
23	18.98	19.26	19.54	19.81	20.09	20.37	20.65	20.93	21.21	21.49	21.77	22.05	22.33	22.61	22.90	23.18
24	19.82	20.11	20.40	20.69	20.98	21.27	21.56	21.86	22.15	22.44	22.74	23.03	23.33	23.62	23.92	24.21
25	20.66	20.96	21.27	21.57	21.87	22.18	22.48	22.79	23.10	23.40	23.71	24.02	24.32	24.63	24.94	25.25
26	21.50	21.82	22.14	22.45	22.77	23.09	23.41	23.73	24.04	24.36	24.68	25.01	25.33	25.65	25.97	26.29
27	22.35	22.68	23.01	23.34	23.67	24.00	24.33	24.67	25.00	25.33	25.67	26.00	26.34	26.67	27.01	27.34
28	23.20	23.55	23.89	24.23	24.58	24.92	25.27	25.61	25.96	26.30	26.65	27.00	27.35	27.70	28.05	28.40
29	24.06	24.41	24.77	25.13	25.49	25.84	26.20	26.56	26.92	27.28	27.64	28.00	28.37	28.73	29.09	29.46
30	24.92	25.29	25.66	26.03	26.40	26.77	27.14	27.52	27.89	28.26	28.64	29.01	29.39	29.77	30.14	30.52
31	25.78	26.16	26.55	26.93	27.32	27.70	28.09	28.47	28.86	29.25	29.64	30.03	30.42	30.81	31.20	31.59
32	26.65	27.04	27.44	27.84	28.24	28.64	29.04	29.44	29.84	30.24	30.64	31.05	31.45	31.85	32.26	32.67
33	27.52	27.93	28.34	28.75	29.16	29.57	29.99	30.40	30.82	31.23	31.65	32.07	32.49	32.91	33.33	33.75
34	28.39	28.81	29.24	29.66	30.09	30.52	30.95	31.37	31.80	32.23	32.67	33.10	33.53	33.96	34.40	34.83
35	29.27	29.71	30.14	30.58	31.02	31.47	31.91	32.35	32.79	33.24	33.68	34.13	34.58	35.03	35.47	35.92
36	30.15	30.60	31.05	31.51	31.96	32.42	32.87	33.33	33.79	34.25	34.71	35.17	35.63	36.09	36.56	37.02
37	31.03	31.50	31.97	32.43	32.90	33.37	33.84	34.32	34.79	35.26	35.74	36.21	36.69	37.16	37.64	38.12
38	31.92	32.40	32.88	33.37	33.85	34.33	34.82	35.30	35.79	36.28	36.77	37.26	37.75	38.24	38.73	39.23
39	32.81	33.31	33.80	34.30	34.80	35.30	35.80	36.30	36.80	37.30	37.81	38.31	38.82	39.32	39.83	40.34
40	33.71	34.22	34.73	35.24	35.75	36.26	36.78	37.29	37.81	38.33	38.85	39.37	39.89	40.41	40.93	41.46
41	34.61	35.13	35.66	36.18	36.71	37.24	37.77	38.30	38.83	39.36	39.89	40.43	40.96	41.50	42.04	42.58
42	35.51	36.05	36.59	37.13	37.67	38.21	38.76	39.30	39.85	40.40	40.95	41.50	42.05	42.60	43.15	43.71
43	36.42	36.97	37.52	38.08	38.63	39.19	39.75	40.31	40.87	41.44	42.00	42.57	43.13	43.70	44.27	44.84
44	37.33	37.89	38.46	39.03	39.60	40.18	40.75	41.33	41.90	42.48	43.06	43.64	44.22	44.81	45.39	45.98
45	38.24	38.82	39.41	39.99	40.58	41.17	41.75	42.35	42.94	43.53	44.13	44.72	45.32	45.92	46.52	47.12
46	39.16	39.75	40.35	40.95	41.55	42.16	42.76	43.37	43.98	44.58	45.20	45.81	46.42	47.03	47.65	48.27
47	40.08	40.69	41.30	41.92	42.54	43.15	43.77	44.40	45.02	45.64	46.27	46.90	47.53	48.16	48.79	49.42
48	41.00	41.63	42.26	42.89	43.52	44.15	44.79	45.43	46.07	46.71	47.35	47.99	48.64	49.28	49.93	50.58
49	41.93	42.57	43.22	43.86	44.51	45.16	45.81	46.46	47.12	47.77	48.43	49.09	49.75	50.41	51.08	51.74
50	42.86	43.52	44.18	44.84	45.50	46.17	46.83	47.50	48.17	48.84	49.52	50.19	50.87	51.55	52.23	52.91
51	43.79	44.47	45.14	45.82	46.50	47.18	47.86	48.55	49.23	49.92	50.61	51.30	51.99	52.69	53.38	54.08
52	44.73	45.42	46.11	46.80	47.50	48.20	48.89	49.59	50.30	51.00	51.71	52.41	53.12	53.83	54.55	54.26
53	45.67	46.38	47.08	47.79	48.50	49.22	49.93	50.65	51.37	52.09	52.81	53.53	54.26	54.98	55.71	56.44
54	46.62	47.34	48.06	48.79	49.51	50.24	50.97	51.70	52.44	53.17	53.91	54.65	55.39	56.14	56.88	57.63
55	47.57	48.30	49.04	49.78	50.52	51.27	52.02	52.76	53.52	54.27	55.02	55.78	56.54	57.30	58.06	58.82
56	48.52	49.27	50.03	50.78	51.54	52.30	53.06	53.83	54.60	55.37	56.14	56.91	57.68	58.46	59.24	60.02
57	49.47	50.24	51.01	51.79	52.56	53.34	54.12	54.90	55.68	56.47	57.25	58.04	58.84	59.63	60.43	61.22
58	50.43	51.22	52.00	52.79	53.58	54.38	55.17	55.97	56.77	57.57	58.38	59.18	59.99	60.80	61.62	62.43
59	51.39	52.20	53.00	53.80	54.61	55.42	56.23	57.05	57.87	58.68	59.51	60.33	61.15	61.98	62.81	63.64
60	52.36	53.18	54.00	54.82	55.64	56.47	57.30	58.13	58.96	59.80	60.64	61.48	62.32	63.17	64.01	64.86

TABLE C-15 (cont.) ANNUAL PERCENTAGE RATE TABLE FOR MONTHLY PAYMENT PLANS

Number of Payments	ANNUAL PERCENTAGE RATE															
	22.00%	22.25%	22.50%	22.75%	23.00%	23.25%	23.50%	23.75%	24.00%	24.25%	24.50%	24.75%	25.00%	25.25%	25.50%	25.75%
	Finance charge per $100 of amount financed															
1	1.83	1.85	1.87	1.90	1.92	1.94	1.96	1.98	2.00	2.02	2.04	2.06	2.08	2.10	2.12	2.15
2	2.76	2.79	2.82	2.85	2.88	2.92	2.95	2.98	3.01	3.04	3.07	3.10	3.14	3.17	3.20	3.23
3	3.69	3.73	3.77	3.82	3.86	3.90	3.94	3.98	4.03	4.07	4.11	4.15	4.20	4.24	4.28	4.32
4	4.62	4.68	4.73	4.78	4.84	4.89	4.94	5.00	5.05	5.10	5.16	5.21	5.26	5.32	5.37	5.42
5	5.57	5.63	5.69	5.76	5.82	5.89	5.95	6.02	6.08	6.14	6.21	6.27	6.34	6.40	6.46	6.53
6	6.61	6.59	6.66	6.74	6.81	6.89	6.96	7.04	7.12	7.19	7.27	7.34	7.42	7.49	7.57	7.64
7	7.47	7.55	7.64	7.73	7.81	7.90	7.99	8.07	8.16	8.24	8.33	8.42	8.51	8.59	8.68	8.77
8	8.42	8.52	8.62	8.72	8.82	8.91	9.01	9.11	9.21	9.31	9.40	9.50	9.60	9.70	9.80	9.90
9	9.39	9.50	9.61	9.72	9.83	9.94	10.04	10.15	10.26	10.37	10.48	10.59	10.70	10.81	10.92	11.03
10	10.36	10.48	10.60	10.72	10.84	10.96	11.08	11.21	11.33	11.45	11.57	11.69	11.81	11.93	12.06	12.18
11	11.33	11.47	11.60	11.73	11.86	12.00	12.13	12.26	12.40	12.53	12.66	12.80	12.93	13.06	13.20	13.33
12	12.31	12.46	12.60	12.75	12.89	13.04	13.18	13.33	13.47	13.62	13.76	13.91	14.05	14.20	14.34	14.49
13	13.30	13.46	13.61	13.77	13.93	14.08	14.24	14.40	14.55	14.71	14.87	15.03	15.18	15.34	15.50	15.66
14	14.29	14.46	14.63	14.80	14.97	15.13	15.30	15.47	15.64	15.81	15.98	16.15	16.32	16.49	16.66	16.83
15	15.29	15.47	15.65	15.83	16.01	16.19	16.37	16.56	16.74	16.92	17.10	17.28	17.47	17.65	17.83	18.02
16	16.29	16.48	16.68	16.87	17.06	17.26	17.45	17.65	17.84	18.03	18.23	18.42	18.62	18.81	19.01	19.21
17	17.30	17.50	17.71	17.92	18.12	18.33	18.53	18.74	18.95	19.16	19.36	19.57	19.78	19.99	20.20	20.40
18	18.31	18.53	18.75	18.97	19.19	19.41	19.62	19.84	20.06	20.28	20.50	20.72	20.95	21.17	21.39	21.61
19	19.33	19.56	19.79	20.02	20.26	20.49	20.72	20.95	21.19	21.42	21.65	21.89	22.12	22.35	22.59	22.82
20	20.35	20.60	20.84	21.09	21.33	21.58	21.82	22.07	22.31	22.56	22.81	23.05	23.30	23.55	23.79	24.04
21	21.38	21.64	21.90	22.16	22.41	22.67	22.93	23.19	23.45	23.71	23.97	24.23	24.49	24.75	25.01	25.27
22	22.42	22.69	22.96	23.23	23.50	23.77	24.04	24.32	24.59	24.86	25.13	25.41	25.68	25.96	26.23	26.50
23	23.46	23.74	24.03	24.31	24.60	24.88	25.17	15.45	25.74	26.02	26.31	26.60	26.88	27.17	27.46	27.75
24	24.51	24.80	25.10	25.40	25.70	25.99	26.29	26.59	26.89	27.19	27.49	27.79	28.09	28.39	28.69	29.00
25	25.56	25.87	26.18	26.49	26.80	27.11	27.43	27.74	28.05	28.36	28.68	28.99	29.31	29.62	29.94	30.25
26	26.62	26.94	27.26	27.59	27.91	28.24	28.56	28.89	29.22	29.55	29.87	30.20	30.53	30.86	31.19	31.52
27	27.68	28.02	28.35	28.69	29.03	29.37	29.71	30.05	30.39	30.73	31.07	31.42	31.76	32.10	32.45	32.79
28	28.75	29.10	29.45	29.80	30.15	30.51	30.86	31.22	31.57	31.93	32.28	32.64	33.00	33.35	33.71	34.07
29	29.82	30.19	30.55	30.92	31.28	31.65	32.02	32.39	32.76	33.13	33.50	33.87	34.24	34.61	34.98	35.36
30	30.90	31.28	31.66	32.04	32.42	32.80	33.18	33.57	33.95	34.33	34.72	35.10	35.49	35.88	36.26	36.65
31	31.98	32.38	32.77	33.17	33.56	33.96	34.35	34.75	35.15	35.55	35.95	36.35	36.75	37.15	37.55	37.95
32	33.07	33.48	33.89	34.30	34.71	35.12	35.53	35.94	36.35	36.77	37.18	37.60	38.01	38.43	38.84	39.26
33	34.17	34.59	35.01	35.44	35.86	36.29	36.71	37.14	37.57	37.99	38.42	38.85	39.28	39.71	40.14	40.58
34	35.27	35.71	36.14	36.58	37.02	37.46	37.90	38.34	38.78	39.23	39.67	40.11	40.56	41.01	41.45	41.90
35	36.37	36.83	37.28	37.73	38.18	38.64	39.09	39.55	40.01	40.47	40.92	41.38	41.84	42.31	42.77	43.23
36	37.49	37.95	38.42	38.89	39.35	39.82	40.29	40.77	41.24	41.71	42.19	42.66	43.14	43.61	44.09	44.57
37	38.60	39.08	39.56	40.05	40.53	41.02	41.50	41.99	42.48	42.96	43.45	43.94	44.43	44.92	45.42	45.91
38	39.72	40.22	40.72	41.21	41.71	42.21	42.71	43.22	43.72	44.22	44.73	45.23	45.74	46.25	46.75	47.26
39	40.85	41.36	41.87	42.39	42.90	43.42	43.93	44.45	44.97	45.49	46.01	46.53	47.05	47.57	48.10	48.62
40	41.98	42.51	43.04	43.56	44.09	44.62	45.16	45.69	46.22	46.76	47.29	47.83	48.37	48.91	49.45	49.99
41	43.12	43.66	44.20	44.75	45.29	45.84	46.39	46.94	47.48	48.04	48.59	49.14	49.69	50.25	50.80	51.36
42	44.26	44.82	45.38	45.94	46.50	47.06	47.62	48.19	48.75	49.32	49.89	50.46	51.03	51.60	52.17	52.74
43	45.41	45.98	46.56	47.13	47.71	48.29	48.87	49.45	50.03	50.61	51.19	51.78	52.36	52.95	53.54	54.13
44	46.56	47.15	47.74	48.33	48.93	49.52	50.11	50.71	51.31	51.91	52.51	53.11	53.71	54.31	54.92	55.52
45	47.72	48.33	48.93	49.54	50.15	50.76	51.37	51.98	52.59	53.21	53.82	54.44	55.06	55.68	56.30	56.92
46	48.89	49.51	50.13	50.75	51.37	52.00	52.63	53.26	53.89	54.52	55.15	55.78	56.42	57.05	57.69	58.33
47	50.06	50.69	51.33	51.97	52.61	53.25	53.89	54.54	55.18	55.83	56.48	57.13	57.78	58.44	59.09	59.75
48	51.23	51.88	52.54	53.19	53.85	54.51	55.16	55.83	56.49	57.15	57.82	58.49	59.15	59.82	60.50	61.17
49	52.41	53.08	53.75	54.42	55.09	55.77	56.44	57.12	57.80	58.48	59.16	59.85	60.53	61.22	61.91	62.60
50	53.59	54.28	54.96	55.65	56.34	57.03	57.73	58.42	59.12	59.81	60.51	61.21	61.92	62.62	63.33	64.03
51	54.78	55.48	56.19	56.89	57.60	58.30	59.01	59.73	60.44	61.15	61.87	62.59	63.31	64.03	64.75	65.47
52	55.98	56.69	57.41	58.13	58.86	59.58	60.31	61.04	61.77	62.50	63.23	63.97	64.70	65.44	66.18	66.92
53	57.18	57.91	58.65	59.38	60.12	60.87	61.61	62.35	63.10	63.85	64.60	65.35	66.11	66.86	67.62	68.38
54	58.38	59.13	59.88	60.64	61.40	62.16	62.92	63.68	64.44	65.21	65.98	66.75	67.52	68.29	69.07	69.84
55	59.59	60.36	61.13	61.90	62.67	63.45	64.23	65.01	65.79	66.57	67.36	68.14	68.93	69.72	70.52	71.31
56	60.80	61.59	62.38	63.17	63.96	64.75	65.54	66.34	67.14	67.94	68.74	69.55	70.36	71.16	71.97	72.79
57	62.02	62.83	63.63	64.44	65.25	66.06	66.87	67.68	68.50	69.32	70.14	70.96	71.78	72.61	73.44	74.27
58	63.25	64.07	64.89	65.71	66.54	67.37	68.20	69.03	69.86	70.70	71.54	72.38	73.22	74.06	74.91	75.76
59	64.48	65.32	66.15	67.00	67.84	68.68	69.53	70.38	71.23	72.09	72.94	73.80	74.66	75.52	76.39	77.25
60	65.71	66.57	67.42	68.28	69.14	70.01	70.87	71.74	72.61	73.48	74.35	75.23	76.11	76.99	77.87	78.76

TABLE C-15 (cont.) ANNUAL PERCENTAGE RATE TABLE FOR MONTHLY PAYMENT PLANS

Number of Payments	ANNUAL PERCENTAGE RATE															
	26.00%	26.25%	26.50%	26.75%	27.00%	27.25%	27.50%	27.75%	28.00%	28.25%	28.50%	28.75%	29.00%	29.25%	29.50%	29.75%
	Finance charge per $100 of amount financed															
1	2.17	2.19	2.21	2.23	2.25	2.27	2.29	2.31	2.33	2.35	2.37	2.40	2.42	2.44	2.46	2.48
2	3.26	3.29	3.32	3.36	3.39	3.42	3.45	3.48	3.51	3.54	3.58	3.61	3.64	3.67	3.70	3.73
3	4.36	4.41	4.45	4.49	4.53	4.58	4.62	4.66	4.70	4.74	4.79	4.83	4.87	4.91	4.96	5.00
4	5.47	5.53	5.58	5.63	5.69	5.74	5.79	5.85	5.90	5.95	6.01	6.06	6.11	6.17	6.22	6.27
5	6.59	6.66	6.72	6.79	6.85	6.91	6.98	7.04	7.11	7.17	7.24	7.30	7.37	7.43	7.49	7.56
6	7.72	7.79	7.87	7.95	8.02	8.10	8.17	8.25	8.32	8.40	8.48	8.55	8.63	8.70	8.78	8.85
7	8.85	8.94	9.03	9.11	9.20	9.29	9.37	9.46	9.55	9.64	9.72	9.81	9.90	9.98	10.07	10.16
8	9.99	10.09	10.19	10.29	10.39	10.49	10.58	10.68	10.78	10.88	10.98	11.08	11.18	11.28	11.38	11.47
9	11.14	11.25	11.36	11.47	11.58	11.69	11.80	11.91	12.03	12.14	12.25	12.36	12.47	12.58	12.69	12.80
10	12.30	12.42	12.54	12.67	12.79	12.91	13.03	13.15	13.28	13.40	13.52	13.64	13.77	13.89	14.01	14.14
11	13.46	13.60	13.73	13.87	14.00	14.13	14.27	14.40	14.54	14.67	14.81	14.94	15.08	15.21	15.35	15.48
12	14.64	14.78	14.93	15.07	15.22	15.37	15.51	15.66	15.81	15.95	16.10	16.25	16.40	16.54	16.69	16.84
13	15.82	15.97	16.13	16.29	16.45	16.61	16.77	16.93	17.09	17.24	17.40	17.56	17.72	17.88	18.04	18.20
14	17.00	17.17	17.35	17.52	17.69	17.86	18.03	18.20	18.37	18.54	18.72	18.89	19.06	19.23	19.41	19.58
15	18.20	18.38	18.57	18.75	18.93	19.12	19.30	19.48	19.67	19.85	20.04	20.22	20.41	20.59	20.78	20.96
16	19.40	19.60	19.79	19.99	20.19	20.38	20.58	20.78	20.97	21.17	21.37	21.57	21.76	21.96	22.16	22.36
17	20.61	20.82	21.03	21.24	21.45	21.66	21.87	22.08	22.29	22.50	22.71	22.92	23.13	23.34	23.55	23.77
18	21.83	22.05	22.27	22.50	22.72	22.94	23.16	23.39	23.61	23.83	24.06	24.28	24.51	24.73	24.96	25.18
19	23.06	23.29	23.53	23.76	24.00	24.23	24.47	24.71	24.94	25.18	25.42	25.65	25.89	26.13	26.37	26.61
20	24.29	24.54	24.79	25.04	25.28	25.53	25.78	26.03	26.28	26.53	26.78	27.04	27.29	27.54	27.79	28.04
21	25.53	25.79	26.05	26.32	26.58	26.84	27.11	27.37	27.63	27.90	28.16	28.43	28.69	28.96	29.22	29.49
22	26.78	27.05	27.33	27.61	27.88	28.16	28.44	28.71	28.99	29.27	29.55	29.82	30.10	30.38	30.66	30.94
23	28.04	28.32	28.61	28.90	29.19	29.48	29.77	30.07	30.36	30.65	30.94	31.23	31.53	31.82	32.11	32.41
24	29.30	29.60	29.90	30.21	30.51	30.82	31.12	31.43	31.73	32.04	32.34	32.65	32.96	33.27	33.57	33.88
25	30.57	30.89	31.20	31.52	31.84	32.16	32.48	32.80	33.12	33.44	33.76	34.08	34.40	34.72	35.04	35.37
26	31.85	32.18	32.51	32.84	33.18	33.51	33.84	34.18	34.51	34.84	35.18	35.51	35.85	36.19	36.52	36.86
27	33.14	33.48	33.83	34.17	34.52	34.87	35.21	35.56	35.91	36.26	36.61	36.96	37.31	37.66	38.01	38.36
28	34.43	34.79	35.15	35.51	35.87	36.23	36.59	36.96	37.32	37.68	38.05	38.41	38.78	39.15	39.51	39.88
29	35.73	36.10	36.48	36.85	37.23	37.61	37.98	38.36	38.74	39.12	39.50	39.88	40.26	40.64	41.02	41.40
30	37.04	37.43	37.82	38.21	38.60	38.99	39.38	39.77	40.17	40.56	40.95	41.35	41.75	42.14	42.54	42.94
31	38.35	38.76	39.16	39.57	39.97	40.38	40.79	41.19	41.60	42.01	42.42	42.83	43.24	43.65	44.06	44.48
32	39.68	40.10	40.52	40.94	41.36	41.78	42.20	42.62	43.05	43.47	43.90	44.32	44.75	45.17	45.60	46.03
33	41.01	41.44	41.88	42.31	42.75	43.19	43.62	44.06	44.50	44.94	45.38	45.82	46.26	46.70	47.15	47.59
34	42.35	42.80	43.25	43.70	44.15	44.60	45.05	45.51	45.96	46.42	46.87	47.33	47.79	48.24	48.70	49.16
35	43.69	44.16	44.62	45.09	45.56	46.02	46.49	46.96	47.43	47.90	48.37	48.85	49.32	49.79	50.27	50.74
36	45.05	45.53	46.01	46.49	46.97	47.45	47.94	48.42	48.91	49.40	49.88	50.37	50.86	51.35	51.84	52.33
37	46.41	46.90	47.40	47.90	48.39	48.89	49.39	49.89	50.40	50.90	51.40	51.91	52.41	52.92	53.42	53.93
38	47.77	48.29	48.80	49.31	49.82	50.34	50.86	51.37	51.89	52.41	52.93	53.45	53.97	54.49	55.02	55.54
39	49.15	49.68	50.20	50.73	51.26	51.79	52.33	52.86	53.39	53.93	54.46	55.00	55.54	56.08	56.62	57.16
40	50.53	51.07	51.62	52.16	52.71	53.26	53.81	54.35	54.90	55.46	56.01	56.56	57.12	57.67	58.23	58.79
41	51.92	52.48	53.04	53.60	54.16	54.73	55.29	55.86	56.42	56.99	57.56	58.13	58.70	59.28	59.85	60.42
42	53.32	53.89	54.47	55.05	55.63	56.21	56.79	57.37	57.95	58.54	59.12	59.71	60.30	60.89	61.48	62.07
43	54.72	55.31	55.90	56.50	57.09	57.69	58.29	58.89	59.49	60.09	60.69	61.30	61.90	62.51	63.11	63.72
44	56.13	56.74	57.35	57.96	58.57	59.19	59.80	60.42	61.03	61.65	62.27	62.89	63.51	64.14	64.76	65.39
45	57.55	58.17	58.80	59.43	60.06	60.69	61.32	61.95	62.59	63.22	63.86	64.50	65.13	65.77	66.42	67.06
46	58.97	59.61	60.26	60.90	61.55	62.20	62.84	63.49	64.15	64.80	65.45	66.11	66.76	67.42	68.08	68.74
47	60.40	61.06	61.72	62.38	63.05	63.71	64.38	65.05	65.71	66.38	67.06	67.73	68.40	69.08	69.75	70.43
48	61.84	62.52	63.20	63.87	64.56	65.24	65.92	66.60	67.29	67.98	68.67	69.36	70.05	70.74	71.44	72.13
49	63.29	63.98	64.68	65.37	66.07	66.77	67.47	68.17	68.87	69.58	70.29	70.99	71.70	72.41	73.13	73.84
50	64.74	65.45	66.16	66.88	67.59	68.31	69.03	69.75	70.47	71.19	71.91	72.64	73.37	74.10	74.83	75.56
51	66.20	66.93	67.66	68.39	69.12	69.86	70.59	71.33	72.07	72.81	73.55	74.29	75.04	75.78	76.53	77.28
52	67.67	68.41	69.16	69.91	70.66	71.41	72.16	72.92	73.67	74.43	75.19	75.95	76.72	77.48	78.25	79.02
53	69.14	69.90	70.67	71.43	72.20	72.97	73.74	74.52	75.29	76.07	76.85	77.62	78.41	79.19	79.97	80.76
54	70.62	71.40	72.18	72.97	73.75	74.54	75.33	76.12	76.91	77.71	78.50	79.30	80.10	80.90	81.71	82.51
55	72.11	72.91	73.71	74.51	75.31	76.12	76.92	77.73	78.55	79.36	80.17	80.99	81.81	82.63	83.45	84.27
56	73.60	74.42	75.24	76.06	76.88	77.70	78.53	79.35	80.18	81.02	81.85	82.68	83.52	84.36	85.20	86.04
57	75.10	75.94	76.77	77.61	78.45	79.29	80.14	80.98	81.83	82.68	83.53	84.39	85.24	86.10	86.96	87.82
58	76.61	77.46	78.32	79.17	80.03	80.89	81.75	82.62	83.48	84.35	85.22	86.10	86.97	87.85	88.72	89.60
59	78.12	78.99	79.87	80.74	81.62	82.50	83.38	84.26	85.15	86.03	86.92	87.81	88.71	89.60	90.50	91.40
60	79.64	80.53	81.42	82.32	83.21	84.11	85.01	85.91	86.81	87.72	88.63	89.54	90.45	91.37	92.28	93.20

TABLE C-15 °(cont.) ANNUAL PERCENTAGE RATE TABLE FOR MONTHLY PAYMENT PLANS

Number of Payments	ANNUAL PERCENTAGE RATE															
	30.00%	30.25%	30.50%	30.75%	31.00%	31.25%	31.50%	31.75%	32.00%	32.25%	32.50%	32.75%	33.00%	33.25%	33.50%	33.75%
	Finance charge per $100 of amount financed															
1	2.50	2.52	2.54	2.56	2.58	2.60	2.62	2.65	2.67	2.69	2.71	2.73	2.75	2.77	2.79	2.81
2	3.77	3.80	3.83	3.86	3.89	3.92	3.95	3.99	4.02	4.05	4.08	4.11	4.14	4.18	4.21	4.24
3	5.04	5.08	5.13	5.17	5.21	5.25	5.30	5.34	5.38	5.42	5.46	5.51	5.55	5.59	5.63	5.68
4	6.33	6.38	6.43	6.49	6.54	6.59	6.65	6.70	6.75	6.81	6.86	6.91	6.97	7.02	7.08	7.13
5	7.62	7.69	7.75	7.82	7.88	7.95	8.01	8.08	8.14	8.20	8.27	8.33	8.40	8.46	8.53	8.59
6	8.93	9.01	9.08	9.16	9.23	9.31	9.39	9.46	9.54	9.61	9.69	9.77	9.84	9.92	9.99	10.07
7	10.25	10.33	10.42	10.51	10.60	10.68	10.77	10.86	10.95	11.03	11.12	11.21	11.30	11.39	11.47	11.56
8	11.57	11.67	11.77	11.87	11.97	12.07	12.17	12.27	12.37	12.47	12.57	12.67	12.77	12.87	12.97	13.07
9	12.91	13.02	13.13	13.24	13.36	13.47	13.58	13.69	13.80	13.91	14.02	14.14	14.25	14.36	14.47	14.58
10	14.26	14.38	14.50	14.63	14.75	14.87	15.00	15.12	15.24	15.37	15.49	15.62	15.74	15.86	15.99	16.11
11	15.62	15.75	15.89	16.02	16.16	16.29	16.43	16.56	16.70	16.84	16.97	17.11	17.24	17.38	17.52	17.65
12	16.98	17.13	17.28	17.43	17.58	17.72	17.87	18.02	18.17	18.32	18.47	18.61	18.76	18.91	19.06	19.21
13	18.36	18.52	18.68	18.84	19.00	19.16	19.33	19.49	19.65	19.81	19.97	20.13	20.29	20.45	20.62	20.78
14	19.75	19.92	20.10	20.27	20.44	20.62	20.79	20.96	21.14	21.13	21.49	21.66	21.83	22.01	22.18	22.36
15	21.15	21.34	21.52	21.71	21.89	22.08	22.27	22.45	22.64	22.83	23.01	23.20	23.39	23.58	23.76	23.95
16	22.56	22.76	22.96	23.16	23.35	23.55	23.75	23.95	24.15	24.35	24.55	24.75	24.96	25.16	25.36	25.56
17	23.98	24.19	24.40	24.61	24.83	25.04	25.25	25.47	25.68	25.89	26.11	26.32	26.53	26.75	26.96	27.18
18	25.41	25.63	25.86	26.08	26.31	26.54	26.76	26.99	27.22	27.44	27.67	27.90	28.13	28.35	28.58	28.81
19	26.85	27.08	27.32	27.56	27.80	28.04	28.28	28.52	28.76	29.00	29.25	29.49	29.73	29.97	30.21	30.45
20	28.29	28.55	28.80	29.05	29.31	29.56	29.81	30.07	30.32	30.58	30.83	31.09	31.34	31.60	31.86	32.11
21	29.75	30.02	30.29	30.55	30.82	31.09	31.36	31.62	31.89	32.16	32.43	32.70	32.97	33.24	33.51	33.78
22	31.22	31.50	31.78	32.06	32.35	32.63	32.91	33.19	33.48	33.76	34.04	34.33	34.61	34.89	35.18	35.46
23	32.70	33.00	33.29	33.59	33.88	34.18	34.48	34.77	35.07	35.37	35.66	35.96	36.26	36.56	36.86	37.16
24	34.19	34.50	34.81	35.12	35.43	35.74	36.05	36.26	36.67	36.99	37.30	37.61	37.92	38.24	38.55	38.87
25	35.69	36.01	36.34	36.66	36.99	37.31	37.64	37.96	38.29	38.62	38.94	39.27	39.60	39.93	40.26	40.59
26	37.20	37.54	37.88	38.21	38.55	38.89	39.23	39.58	39.92	40.26	40.60	40.94	41.29	41.63	41.97	42.32
27	38.72	39.07	39.42	39.78	40.13	40.49	40.84	41.20	41.56	41.91	42.27	42.63	42.99	43.34	43.70	44.06
28	40.25	40.61	40.98	41.35	41.72	42.09	42.46	42.83	43.20	43.58	43.95	44.32	44.70	45.07	45.45	45.82
29	41.78	42.17	42.55	42.94	43.32	43.71	44.09	44.48	44.87	45.25	45.64	46.03	46.42	46.81	47.20	47.59
30	43.33	43.73	44.13	44.53	44.93	45.33	45.73	46.13	46.54	46.94	47.34	47.75	48.15	48.56	48.96	49.37
31	44.89	45.30	45.72	46.13	46.55	46.97	47.38	47.80	48.22	48.64	49.06	49.48	49.90	50.32	50.74	51.17
32	46.46	46.89	47.32	47.75	48.18	48.61	49.05	49.48	49.91	50.35	50.78	51.22	51.66	52.09	52.53	52.97
33	48.04	48.48	48.93	49.37	49.82	50.27	50.72	51.17	51.62	52.07	52.52	52.97	53.43	53.88	54.33	54.79
34	49.62	50.08	50.55	51.01	51.47	51.94	52.40	52.87	53.33	53.80	54.27	54.74	55.21	55.68	56.15	56.62
35	51.22	51.70	52.17	52.65	53.13	53.61	54.09	54.58	55.06	55.54	56.03	56.51	57.00	57.48	57.97	58.46
36	52.83	53.32	53.81	54.31	54.80	55.30	55.80	56.30	56.80	57.30	57.80	58.30	58.80	59.30	59.81	60.31
37	54.44	54.95	55.46	55.97	56.49	57.00	57.51	58.03	58.54	59.06	59.58	60.10	60.62	61.14	61.66	62.18
38	56.07	56.59	57.12	57.65	58.18	58.71	59.24	59.77	60.30	60.84	61.37	61.90	62.44	62.98	63.52	64.06
39	57.70	58.24	58.79	59.33	59.88	60.42	60.97	61.52	62.07	62.62	63.17	63.72	64.28	64.83	65.39	65.94
40	59.34	59.90	60.47	61.03	61.59	62.15	62.72	63.28	63.85	64.42	64.99	65.56	66.13	66.70	67.27	67.84
41	61.00	61.57	62.15	62.73	63.31	63.89	64.47	65.06	65.64	66.22	66.81	67.40	67.99	68.57	69.16	69.76
42	62.66	63.25	63.85	64.44	65.04	65.64	66.24	66.84	67.44	68.04	68.65	69.25	69.86	70.46	71.07	71.68
43	64.33	64.94	65.56	66.17	66.78	67.40	68.01	68.63	69.25	69.87	70.49	71.11	71.74	72.36	72.99	73.61
44	66.01	66.64	67.27	67.90	68.53	69.17	69.80	70.43	71.07	71.71	72.35	72.99	73.63	74.27	74.91	75.56
45	67.70	68.35	69.00	69.64	70.29	70.94	71.60	72.25	72.90	73.56	74.21	74.87	75.53	76.19	76.85	77.52
46	69.40	70.07	70.73	71.40	72.06	72.73	73.40	74.07	74.74	75.42	76.09	76.77	77.44	78.12	78.80	79.48
47	71.11	71.79	72.47	73.16	73.84	74.53	75.22	75.90	76.60	77.29	77.98	78.67	79.37	80.07	80.76	81.46
48	72.83	73.53	74.23	74.93	75.63	76.34	77.04	77.75	78.46	79.17	79.88	80.59	81.30	82.02	82.74	83.45
49	74.55	75.27	75.99	76.71	77.43	78.15	78.88	79.60	80.33	81.06	81.79	82.52	83.25	83.98	84.72	84.45
50	76.29	77.02	77.76	78.50	79.24	79.98	80.72	81.46	82.21	82.96	83.70	84.45	85.20	85.96	86.71	87.47
51	78.03	78.79	79.54	80.30	81.06	81.81	82.58	83.34	84.10	84.87	85.63	86.40	87.17	87.94	88.71	89.49
52	79.79	80.56	81.33	82.11	82.88	83.66	84.44	85.22	86.00	86.79	87.57	88.36	89.15	89.94	90.73	91.52
53	81.55	82.34	83.13	83.92	84.72	85.51	86.31	87.11	87.91	88.72	89.52	90.33	91.13	91.94	92.75	93.57
54	83.32	84.13	84.94	85.75	86.56	87.38	88.19	89.01	89.83	90.66	91.48	92.30	92.13	93.96	94.79	95.62
55	85.10	85.93	86.75	87.58	88.42	89.25	90.09	90.92	91.76	92.60	93.45	94.29	95.14	95.99	96.83	97.69
56	86.89	87.73	88.58	89.43	90.28	91.13	91.99	92.84	93.70	94.56	95.43	96.29	97.15	98.02	98.89	99.76
57	88.68	89.55	90.41	91.28	92.15	93.02	93.90	94.77	95.65	96.53	97.41	98.30	99.18	100.07	100.96	101.85
58	90.49	91.37	92.26	93.14	94.03	94.92	95.82	96.71	97.61	98.51	99.41	100.31	101.22	102.12	103.03	103.94
59	92.30	93.20	94.11	95.01	95.92	96.83	97.75	98.66	99.58	100.50	101.42	102.34	103.26	104.19	105.12	106.05
60	94.12	95.04	95.97	96.89	97.82	98.75	99.68	100.62	101.56	102.49	103.43	104.38	105.32	106.27	107.21	108.16

TABLE C-15 (cont.) ANNUAL PERCENTAGE RATE TABLE FOR
MONTHLY PAYMENT PLANS

Number of Payments	ANNUAL PERCENTAGE RATE															
	34.00%	34.25%	34.50%	34.75%	35.00%	35.25%	35.50%	35.75%	36.00%	36.25%	36.50%	36.75%	37.00%	37.25%	37.50%	37.75%
	Finance charge per $100 of amount financed															
1	2.83	2.85	2.87	2.90	2.92	2.94	2.96	2.98	3.00	3.02	3.04	3.06	3.08	3.10	3.12	3.15
2	4.27	4.30	4.33	4.36	4.40	4.43	4.46	4.49	4.52	4.55	4.59	4.62	4.65	4.68	4.71	4.74
3	5.72	5.76	5.80	5.85	5.89	5.93	5.97	6.02	6.06	6.10	6.14	6.19	6.23	6.27	6.31	6.36
4	7.18	7.24	7.29	7.34	7.40	7.45	7.50	7.56	7.61	7.66	7.72	7.77	7.83	7.88	7.93	7.99
5	8.66	8.72	8.79	8.85	8.92	8.98	9.05	9.11	9.18	9.24	9.31	9.37	9.44	9.50	9.57	9.63
6	10.15	10.22	10.30	10.38	10.45	10.53	10.61	10.68	10.76	10.83	10.91	10.99	11.06	11.14	11.22	11.29
7	11.65	11.74	11.83	11.91	12.00	12.09	12.18	12.27	12.35	12.44	12.53	12.62	12.71	12.80	12.88	12.97
8	13.17	13.27	13.36	13.46	13.56	13.66	13.76	13.86	13.97	14.07	14.17	14.27	14.37	14.47	14.57	14.67
9	14.69	14.81	14.92	15.03	15.14	15.25	15.37	15.48	15.59	15.70	15.82	15.93	16.04	16.15	16.27	16.38
10	16.24	16.36	16.48	16.61	16.73	16.86	16.98	17.11	17.23	17.36	17.48	17.60	17.73	17.85	17.98	18.10
11	17.79	17.93	18.06	18.20	18.34	18.47	18.61	18.75	18.89	19.02	19.16	19.30	19.43	19.57	19.71	19.85
12	19.36	19.51	19.66	19.81	19.96	20.11	20.25	20.40	20.55	20.70	20.85	21.00	21.15	21.31	21.46	21.61
13	20.94	21.10	21.26	21.43	21.59	21.75	21.91	22.08	22.24	22.40	22.56	22.73	22.89	23.05	23.22	23.38
14	22.53	22.71	22.88	23.06	23.23	23.41	23.59	23.76	23.94	24.11	24.29	24.47	24.64	24.82	25.00	25.17
15	24.14	24.33	24.52	24.71	24.89	25.08	25.27	25.46	25.65	25.84	26.03	26.22	26.41	26.60	26.79	26.98
16	25.76	25.96	26.16	26.37	26.57	26.77	26.97	27.17	27.38	27.58	27.78	27.99	28.19	28.39	28.60	28.80
17	27.39	27.61	27.82	28.04	28.25	28.47	28.69	28.90	29.12	29.34	29.55	29.77	29.99	30.20	30.42	30.64
18	29.04	29.27	29.50	29.73	29.96	30.19	30.42	30.65	30.88	31.11	31.34	31.57	31.80	32.03	32.26	32.49
19	30.70	30.94	31.18	31.43	31.67	31.91	32.16	32.40	32.65	32.89	33.14	33.38	33.63	33.87	34.12	34.36
20	32.37	32.63	32.88	33.14	33.40	33.66	33.91	34.17	34.43	34.69	34.95	35.21	35.47	35.73	35.99	36.25
21	34.05	34.32	34.60	34.87	35.14	35.41	35.68	35.96	36.23	36.50	36.78	37.05	37.33	37.60	37.88	38.15
22	35.75	36.04	36.32	36.61	36.89	37.18	37.47	37.76	38.04	38.33	38.62	38.91	39.20	39.49	39.78	40.07
23	37.46	37.76	38.06	38.36	38.66	38.96	39.27	39.57	39.87	40.18	40.48	40.78	41.09	41.39	41.70	42.00
24	39.18	39.50	39.81	40.13	40.44	40.76	41.08	41.40	41.71	42.03	42.35	42.67	42.99	43.31	43.63	43.95
25	40.92	41.25	41.58	41.91	42.24	42.57	42.90	43.24	43.57	43.90	44.24	44.57	44.91	45.24	45.58	45.91
26	42.66	43.01	43.36	43.70	44.05	44.40	44.74	45.09	45.44	45.79	46.14	46.49	46.84	47.19	47.54	47.89
27	44.42	44.78	45.15	45.51	45.87	46.23	46.60	46.96	47.32	47.69	48.05	48.42	48.78	49.15	49.52	49.88
28	46.20	46.57	46.95	47.33	47.70	48.08	48.46	48.84	49.22	49.60	49.98	50.36	50.75	51.13	51.51	51.89
29	47.98	48.37	48.77	49.16	49.55	49.95	50.34	50.74	51.13	51.53	51.93	52.32	52.72	53.12	53.52	53.92
30	49.78	50.19	50.60	51.00	51.41	51.82	52.23	52.65	53.06	53.47	53.88	54.30	54.71	55.13	55.54	55.96
31	51.59	52.01	52.44	52.86	53.29	53.71	54.14	54.57	55.00	55.43	55.85	56.28	56.72	57.15	57.58	58.01
32	53.41	53.85	54.29	54.73	55.17	55.62	56.06	56.50	56.95	57.39	57.84	58.29	58.73	59.18	59.63	60.08
33	55.24	55.70	56.16	56.62	57.07	57.53	57.99	58.45	58.92	59.38	59.84	60.30	60.77	61.23	61.70	62.16
34	57.09	57.56	58.04	58.51	58.99	59.46	59.94	60.42	60.89	61.37	61.85	62.33	62.81	63.30	63.78	64.26
35	58.95	59.44	59.93	60.42	60.91	61.40	61.90	62.39	62.89	63.38	63.88	64.38	64.88	65.37	65.87	66.37
36	60.82	61.33	61.83	62.34	62.85	63.36	63.87	64.38	64.89	65.41	65.92	66.43	66.95	67.47	67.98	68.50
37	62.70	63.22	63.75	64.27	64.80	65.33	65.85	66.38	66.91	67.44	67.97	68.51	69.04	69.57	70.11	70.64
38	64.59	65.14	65.68	66.22	66.76	67.31	67.85	68.40	68.95	69.49	70.04	70.59	71.14	71.69	72.25	72.80
39	66.50	67.06	67.62	68.18	68.74	69.30	69.86	70.43	70.99	71.56	72.12	72.69	73.26	73.83	74.40	74.97
40	68.42	68.99	69.57	70.15	70.13	71.31	71.89	72.47	73.05	73.63	74.22	74.80	75.39	75.98	76.56	77.15
41	70.35	70.94	71.53	72.13	72.73	73.32	73.92	74.52	75.12	75.72	76.32	76.93	77.53	78.14	78.74	79.35
42	72.29	72.90	73.51	74.12	74.74	75.35	75.97	76.59	77.20	77.82	78.44	79.07	79.69	80.31	80.94	81.56
43	74.24	74.87	75.50	76.13	76.76	77.40	78.03	78.67	79.30	79.94	80.58	81.22	81.86	82.50	83.14	83.79
44	76.20	76.85	77.50	78.15	78.80	79.45	80.10	80.76	81.41	82.07	82.72	83.38	84.04	84.70	85.36	86.03
45	78.18	78.84	79.51	80.18	80.85	81.52	82.19	82.86	83.53	84.21	84.88	85.56	86.24	86.92	87.60	88.28
46	80.17	80.85	81.53	82.22	82.91	83.60	84.28	84.98	85.67	86.36	87.06	87.75	88.45	89.15	89.85	90.55
47	82.16	82.87	83.57	84.27	84.98	85.69	86.39	87.10	87.81	88.53	89.24	89.95	90.67	91.39	92.11	92.83
48	84.17	84.89	85.61	86.34	87.06	87.79	88.52	89.24	89.97	90.70	91.44	92.17	92.91	93.64	94.38	95.12
49	86.19	86.93	87.67	88.41	89.16	89.90	90.65	91.40	92.14	92.89	93.65	94.40	95.15	95.91	96.67	97.42
50	88.22	88.98	89.74	90.50	91.26	92.03	92.79	93.56	94.33	95.10	95.87	96.64	97.41	98.19	98.96	99.74
51	90.26	91.04	91.82	92.60	93.38	94.16	94.95	95.74	96.52	97.31	98.10	98.89	99.69	100.48	101.28	102.07
52	92.32	93.11	93.91	94.71	95.51	96.31	97.12	97.92	98.73	99.54	100.35	101.16	101.97	102.79	103.60	104.42
53	94.38	95.20	96.01	96.83	97.65	98.47	99.30	100.12	100.95	101.78	102.61	103.44	104.27	105.10	105.94	106.78
54	96.45	97.29	98.13	98.96	99.80	100.64	101.49	102.33	103.18	104.03	104.87	105.73	106.58	107.43	108.29	109.14
55	98.54	99.39	100.25	101.11	101.97	102.83	103.69	104.55	105.42	106.29	107.16	108.03	108.90	109.77	110.65	111.53
56	100.63	101.51	102.38	103.26	104.14	105.02	105.90	106.79	107.67	108.56	109.45	110.34	111.23	112.13	113.02	113.92
57	102.74	103.63	104.53	105.43	106.32	107.22	108.13	109.03	109.94	110.85	111.75	112.67	113.58	114.49	115.41	116.33
58	104.85	105.77	106.68	107.60	108.52	109.44	110.36	111.29	112.21	113.14	114.07	115.00	115.93	116.87	117.81	118.74
59	106.98	107.91	108.85	109.79	110.73	111.67	112.61	113.55	114.50	115.45	116.40	117.35	118.30	119.26	120.22	121.17
60	109.12	110.07	111.02	111.98	112.94	113.90	114.87	115.83	116.80	117.77	118.74	119.71	120.68	121.66	122.64	123.62

Table C-15 (cont.) Annual percentage rate table for monthly payment plans

Number of Payments	ANNUAL PERCENTAGE RATE															
	38.00%	38.25%	38.50%	38.75%	39.00%	39.25%	39.50%	39.75%	40.00%	40.25%	40.50%	40.75%	41.00%	41.25%	41.50%	41.75%
	Finance charge per $100 of amount financed															
1	3.17	3.19	3.21	3.23	3.25	3.27	3.29	3.31	3.33	3.35	3.37	3.40	3.42	3.44	3.46	3.48
2	4.77	4.81	4.84	4.87	4.90	4.93	4.96	5.00	5.03	5.06	5.09	5.12	5.15	5.19	5.22	5.25
3	6.40	6.44	6.48	6.53	6.57	6.61	6.65	6.70	6.74	6.78	6.82	6.87	6.91	6.95	7.00	7.04
4	8.04	8.09	8.15	8.20	8.25	8.31	8.36	8.42	8.47	8.52	8.58	8.63	8.69	8.74	8.79	8.85
5	9.70	9.76	9.83	9.89	9.96	10.02	10.09	10.15	10.22	10.28	10.35	10.41	10.48	10.54	10.61	10.68
6	11.37	11.45	11.52	11.60	11.68	11.75	11.83	11.91	11.99	12.06	12.14	12.22	12.29	12.37	12.45	12.52
7	13.06	13.15	13.24	13.33	13.42	13.50	13.59	13.68	13.77	13.86	13.95	14.04	14.13	14.21	14.30	14.39
8	14.77	14.87	14.97	15.07	15.17	15.27	15.37	15.47	15.57	15.67	15.77	15.88	15.98	16.08	16.18	16.28
9	16.49	16.60	16.72	16.83	16.94	17.05	17.17	17.28	17.39	17.51	17.62	17.73	17.85	17.96	18.07	18.19
10	18.23	18.35	18.48	18.61	18.73	18.86	18.98	19.11	19.23	19.36	19.48	19.61	19.74	19.86	19.99	20.12
11	19.99	20.12	20.26	20.40	20.54	20.68	20.81	20.95	21.09	21.23	21.37	21.51	21.65	21.78	21.92	22.06
12	21.76	21.91	22.06	22.21	22.36	22.51	22.66	22.81	22.97	23.12	23.27	23.42	23.57	23.72	23.88	24.03
13	23.54	23.71	23.87	24.04	24.20	24.37	24.53	24.69	24.86	25.02	25.19	25.35	25.52	25.68	25.85	26.01
14	25.35	25.53	25.70	25.88	26.06	26.24	26.41	26.59	26.77	26.95	27.13	27.30	27.48	27.66	27.84	28.02
15	27.17	27.36	27.55	27.74	27.93	28.12	28.32	28.51	28.70	28.89	29.08	29.27	29.47	29.66	29.85	30.04
16	29.01	29.21	29.41	29.62	29.82	30.03	30.23	30.44	30.65	30.85	31.06	31.26	31.47	31.68	31.88	32.09
17	30.86	31.08	31.29	31.51	31.73	31.95	32.17	32.39	32.61	32.83	33.05	33.27	33.49	33.71	33.93	34.15
18	32.73	32.96	33.19	33.42	33.66	33.89	34.12	34.36	34.59	34.83	35.06	35.29	35.53	35.76	36.00	36.23
19	34.61	34.86	35.10	35.35	35.60	35.85	36.09	36.34	36.59	36.84	37.09	37.34	37.59	37.84	38.09	38.34
20	36.51	36.77	37.03	37.30	37.56	37.82	38.08	38.35	38.61	38.87	39.14	39.40	39.66	39.93	40.19	40.46
21	38.43	38.70	38.98	39.26	39.53	39.81	40.09	40.36	40.64	40.92	41.20	41.48	41.76	42.04	42.32	42.60
22	40.36	40.65	40.94	41.23	41.52	41.82	42.11	42.40	42.69	42.99	43.28	43.58	43.87	44.16	44.46	44.75
23	42.31	42.61	42.92	43.23	43.53	43.84	44.15	44.46	44.76	45.07	45.38	45.69	46.00	46.31	46.62	46.93
24	44.27	44.59	44.91	45.23	45.56	45.88	46.20	46.53	46.85	47.17	47.50	47.82	48.15	48.48	48.80	49.13
25	46.25	46.59	46.92	47.26	47.60	47.94	48.28	48.61	48.95	49.29	49.63	49.98	50.32	50.66	51.00	51.34
26	48.24	48.60	48.95	49.30	49.66	50.01	50.36	50.72	51.07	51.43	51.79	52.14	52.50	52.86	53.22	53.58
27	50.25	50.62	50.99	51.36	51.73	52.10	52.47	52.84	53.21	53.58	53.96	54.33	54.70	55.08	55.45	55.83
28	52.28	52.66	53.05	53.43	53.82	54.20	54.59	54.98	55.37	55.76	56.14	56.53	56.92	57.31	57.70	58.10
29	54.32	54.72	55.12	55.52	55.92	56.33	56.73	57.13	57.54	57.94	58.35	58.75	59.16	59.57	59.97	60.38
30	56.37	56.79	57.21	57.63	58.05	58.46	58.88	59.30	59.73	60.15	60.57	60.99	61.42	61.84	62.26	62.69
31	58.44	58.88	59.31	59.75	60.18	60.62	61.05	61.49	61.93	62.37	62.81	63.25	63.69	64.13	64.57	65.01
32	60.53	60.98	61.43	61.88	62.34	62.79	63.24	63.70	64.15	64.61	65.06	65.52	65.98	66.43	66.89	67.35
33	62.63	63.10	63.57	64.03	64.50	64.97	65.44	65.92	66.39	66.86	67.33	67.81	68.28	68.76	69.23	69.71
34	64.75	65.23	65.72	66.20	66.69	67.18	67.66	68.15	68.64	69.13	69.62	70.11	70.61	71.10	71.59	72.09
35	66.88	67.38	67.88	68.38	68.89	69.39	69.90	70.40	70.91	71.42	71.93	72.44	72.95	73.46	73.97	74.48
36	69.02	69.54	70.06	70.58	71.10	71.62	72.15	72.67	73.20	73.72	74.25	74.78	75.30	75.83	76.36	76.89
37	71.18	71.72	72.25	72.79	73.33	73.87	74.41	74.96	75.50	76.04	76.59	77.13	77.68	78.22	78.77	79.32
38	73.35	73.91	74.46	75.02	75.58	76.14	76.69	77.25	77.81	78.38	78.94	79.50	80.07	80.63	81.20	81.76
39	75.54	76.11	76.69	77.26	77.84	78.41	78.99	79.57	80.15	80.73	81.31	81.89	82.47	83.06	83.64	84.23
40	77.74	78.33	78.93	79.52	80.11	80.71	81.30	81.90	82.50	83.09	83.69	84.29	84.90	85.50	86.10	86.70
41	79.96	80.57	81.18	81.79	82.40	83.01	83.63	84.24	84.86	85.48	86.09	86.71	87.33	87.95	88.58	89.20
42	82.19	82.82	83.45	84.07	84.71	85.34	85.97	86.60	87.24	87.87	88.51	89.15	89.79	90.43	91.07	91.71
43	84.43	85.08	85.73	86.37	87.02	87.67	88.33	88.98	89.63	90.29	90.94	91.60	92.26	92.92	93.58	94.24
44	86.69	87.36	88.02	88.69	89.36	90.03	90.70	91.37	92.04	92.72	93.39	94.07	94.74	95.42	96.10	96.78
45	88.96	89.65	90.33	91.02	91.70	92.39	93.08	93.77	94.47	95.16	95.85	96.55	97.24	97.94	98.64	99.34
46	91.25	91.95	92.65	93.36	94.07	94.77	95.48	96.19	96.90	97.62	98.33	99.04	99.76	100.48	101.20	101.91
47	93.55	94.27	94.99	95.72	96.44	97.17	97.90	98.63	99.36	100.09	100.82	101.56	102.29	103.03	103.77	104.50
48	95.86	96.60	97.34	98.09	98.83	99.58	100.33	101.07	101.82	102.58	103.33	104.08	104.84	105.59	106.35	107.11
49	98.18	98.94	99.71	100.47	101.23	102.00	102.77	103.54	104.31	105.08	105.85	106.62	107.40	108.18	108.95	109.73
50	100.52	101.30	102.08	102.87	103.65	104.44	105.22	106.01	106.80	107.59	108.39	109.18	109.98	110.77	111.57	112.37
51	102.87	103.67	104.47	105.28	106.08	106.89	107.69	108.50	109.31	110.12	110.94	111.75	112.57	113.38	114.20	115.02
52	105.24	106.06	106.88	107.70	108.53	109.35	110.18	111.01	111.84	112.67	113.50	114.33	115.17	116.01	116.85	117.69
53	107.61	108.45	109.29	110.14	110.98	111.83	112.68	113.52	114.37	115.23	116.08	116.93	117.79	118.65	119.51	120.37
54	110.00	110.86	111.72	112.59	113.45	114.32	115.19	116.05	116.93	117.80	118.67	119.55	120.42	121.30	122.18	123.06
55	112.40	113.28	114.17	115.05	115.94	116.82	117.71	118.60	119.49	120.38	121.28	122.17	123.07	123.97	124.87	125.77
56	114.82	115.72	116.62	117.53	118.43	119.34	120.25	121.16	122.07	122.98	123.90	124.81	125.73	126.65	127.57	128.49
57	117.25	118.17	119.09	120.01	120.94	121.87	122.80	123.73	124.66	125.59	126.53	127.47	128.40	129.34	130.29	131.23
58	119.68	120.63	121.57	122.51	123.46	124.41	125.36	126.31	127.26	128.22	129.17	130.13	131.09	132.05	133.02	133.98
59	122.13	123.10	124.06	125.03	125.99	126.96	127.93	128.91	129.88	130.86	131.83	132.81	133.79	134.78	135.76	136.75
60	124.60	125.58	126.56	127.55	128.54	129.53	130.52	131.51	132.51	133.51	134.51	135.51	136.51	137.51	138.52	139.52

TABLE C-15 (cont.) ANNUAL PERCENTAGE RATE TABLE FOR
MONTHLY PAYMENT PLANS

Number of Payments	ANNUAL PERCENTAGE RATE															
	42.00%	42.25%	42.50%	42.75%	43.00%	43.25%	43.50%	43.75%	44.00%	44.25%	44.50%	44.75%	45.00%	45.25%	45.50%	45.75%
	Finance charge per $100 of amount financed															
1	3.50	3.52	3.54	3.56	3.58	3.60	3.62	3.65	3.67	3.69	3.71	3.73	3.75	3.77	3.79	3.81
2	5.28	5.31	5.34	5.37	5.41	5.44	5.47	5.50	5.53	5.56	5.60	5.63	5.66	5.69	5.72	5.75
3	7.08	7.12	7.17	7.21	7.25	7.29	7.34	7.38	7.42	7.46	7.51	7.55	7.59	7.63	7.68	7.72
4	8.90	8.95	9.01	9.06	9.12	9.17	9.22	9.28	9.33	9.39	9.44	9.49	9.55	9.60	9.66	9.71
5	10.74	10.81	10.87	10.94	11.00	11.07	11.13	11.20	11.26	11.33	11.39	11.46	11.53	11.59	11.66	11.72
6	12.60	12.68	12.76	12.83	12.91	12.99	13.06	13.14	13.22	13.30	13.37	13.45	13.53	13.60	13.68	13.76
7	14.48	14.57	14.66	14.75	14.84	14.93	15.02	15.10	15.19	15.28	15.37	15.46	15.55	15.64	15.73	15.82
8	16.38	16.48	16.58	16.69	16.79	16.89	16.99	17.09	17.19	17.29	17.40	17.50	17.60	17.70	17.80	17.90
9	18.30	18.42	18.53	18.64	18.76	18.87	18.98	19.10	19.21	19.33	19.44	19.55	19.67	19.78	19.90	20.01
10	20.24	20.37	20.49	20.62	20.75	20.87	21.00	21.13	21.25	21.38	21.51	21.63	21.76	21.89	22.02	22.14
11	22.20	22.34	22.48	22.62	22.76	22.90	23.04	23.18	23.32	23.46	23.60	23.74	23.88	24.02	24.16	24.30
12	24.18	24.33	24.49	24.64	24.79	24.94	25.10	25.25	25.40	25.55	25.71	25.86	26.01	26.17	26.32	26.48
13	26.18	26.35	26.51	26.68	26.84	27.01	27.18	27.34	27.51	27.67	27.84	28.01	28.18	28.34	28.51	28.68
14	28.20	28.38	28.56	28.74	28.92	29.10	29.28	29.46	29.64	29.82	30.00	30.18	30.36	30.54	30.72	30.90
15	30.24	30.43	30.62	30.82	31.01	31.20	31.40	31.59	31.79	31.98	32.17	32.37	32.56	32.76	32.95	33.15
16	32.30	32.50	32.71	32.92	33.12	33.33	33.54	33.75	33.96	34.17	34.37	34.58	34.79	35.00	35.21	35.42
17	34.37	34.59	34.82	35.04	35.26	35.48	35.70	35.93	36.15	36.37	36.60	36.82	37.04	37.27	37.49	37.71
18	36.47	36.71	36.94	37.18	37.41	37.65	37.89	38.13	38.36	38.60	38.84	39.08	39.31	39.55	39.79	40.03
19	38.59	38.84	39.09	39.34	39.59	39.84	40.09	40.34	40.60	40.85	41.10	41.35	41.61	41.86	42.11	42.37
20	40.72	40.99	41.25	41.52	41.79	42.05	42.32	42.59	42.85	43.12	43.39	43.66	43.92	44.19	44.46	44.73
21	42.88	43.16	43.44	43.72	44.00	44.28	44.56	44.85	45.13	45.41	45.69	45.98	46.26	46.55	46.83	47.11
22	45.05	45.35	45.64	45.94	46.24	46.53	46.83	47.13	47.43	47.72	48.02	48.32	48.62	48.92	49.22	49.52
23	47.24	47.55	47.87	48.18	48.49	48.80	49.12	49.43	49.74	50.06	50.37	50.69	51.00	51.32	51.63	51.95
24	49.45	49.78	50.11	50.44	50.77	51.09	51.42	51.75	52.08	52.41	52.74	53.07	53.41	53.74	54.07	54.40
25	51.69	52.03	52.37	52.72	53.06	53.40	53.75	54.10	54.44	54.79	55.13	55.48	55.83	56.18	56.53	56.87
26	53.93	54.29	54.65	55.01	55.37	55.73	56.10	56.46	56.82	57.18	57.55	57.91	58.27	58.64	59.00	59.37
27	56.20	56.58	56.95	57.33	57.71	58.08	58.46	58.84	59.22	59.60	59.98	60.36	60.74	61.12	61.50	61.89
28	58.49	58.88	59.27	59.67	60.06	60.45	60.85	61.24	61.64	62.04	62.43	62.83	63.23	63.63	64.02	64.42
29	60.79	61.20	61.61	62.02	62.43	62.84	63.25	63.67	64.08	64.49	64.91	65.32	65.73	66.15	66.57	66.98
30	63.11	63.54	63.97	64.39	64.82	65.25	65.68	66.11	66.54	66.97	67.40	67.83	68.26	68.70	69.13	69.56
31	65.45	65.90	66.34	66.79	67.23	67.68	68.12	68.57	69.02	69.46	69.91	70.36	70.81	71.26	71.71	72.16
32	67.81	68.27	68.73	69.20	69.66	70.12	70.59	71.05	71.51	71.98	72.45	72.91	73.38	73.85	74.32	74.79
33	70.19	70.67	71.15	71.63	72.11	72.59	73.07	73.55	74.03	74.52	75.00	75.48	75.97	76.45	76.94	77.43
34	72.58	73.08	73.58	74.07	74.57	75.07	75.57	76.07	76.57	77.07	77.57	78.07	78.58	79.08	79.59	80.09
35	74.99	75.51	76.02	76.54	77.05	77.57	78.09	78.61	79.12	79.64	80.16	80.68	81.21	81.73	82.25	82.78
36	77.42	77.95	78.49	79.02	79.55	80.09	80.62	81.16	81.70	82.24	82.77	83.31	83.85	84.39	84.94	85.48
37	79.87	80.42	80.97	81.52	82.07	82.63	83.18	83.74	84.29	84.85	85.40	85.96	86.52	87.08	87.64	88.20
38	82.33	82.90	83.47	84.04	84.61	85.18	85.76	86.33	86.90	87.48	88.05	88.63	89.21	89.79	90.37	90.95
39	84.81	85.40	85.99	86.58	87.17	87.76	88.35	88.94	89.53	90.13	90.72	91.32	91.91	92.51	93.11	93.71
40	87.31	87.91	88.52	89.13	89.74	90.35	90.96	91.57	92.18	92.79	93.41	94.02	94.64	95.25	95.87	96.49
41	89.82	90.45	91.07	91.70	92.33	92.96	93.58	94.22	94.85	95.48	96.11	96.75	97.38	98.02	98.65	99.29
42	92.35	93.00	93.64	94.29	94.93	95.58	96.23	96.88	97.53	98.18	98.83	99.49	100.14	100.80	101.45	102.11
43	94.90	95.56	96.23	96.89	97.56	98.22	98.89	99.56	100.23	100.90	101.57	102.25	102.92	103.60	104.27	104.95
44	97.46	98.14	98.83	99.51	100.20	100.88	101.57	102.26	102.95	103.64	104.33	105.03	105.72	106.41	107.11	107.81
45	100.04	100.74	101.45	102.15	102.85	103.56	104.27	104.98	105.69	106.40	107.11	107.82	108.53	109.25	109.96	110.68
46	102.63	103.36	104.08	104.80	105.53	106.25	106.98	107.71	108.44	109.17	109.90	110.63	111.37	112.10	112.84	113.58
47	105.25	105.99	106.73	107.47	108.22	108.96	109.71	110.46	111.21	111.96	112.71	113.46	114.22	114.97	115.73	116.49
48	107.87	108.63	109.39	110.16	110.92	111.69	112.46	113.23	113.99	114.77	115.54	116.31	117.09	117.86	118.64	119.42
49	110.51	111.29	112.08	112.86	113.64	114.43	115.22	116.01	116.80	117.59	118.38	119.17	119.97	120.77	121.56	122.36
50	113.17	113.97	114.77	115.58	116.38	117.19	118.00	118.81	119.62	120.43	121.24	122.06	122.87	123.69	124.51	125.32
51	115.84	116.66	117.48	118.31	119.14	119.96	120.79	121.62	122.45	123.28	124.12	124.95	125.79	126.63	127.46	128.30
52	118.53	119.37	120.21	121.06	121.90	122.75	123.60	124.45	125.30	126.16	127.01	127.87	128.72	129.58	130.44	131.30
53	121.23	122.09	122.95	123.82	124.69	125.56	126.43	127.30	128.17	129.04	129.92	130.80	131.67	132.55	133.43	134.32
54	123.94	124.83	125.71	126.60	127.49	128.38	129.27	130.16	131.05	131.95	132.84	133.74	134.64	135.54	136.44	137.35
55	126.67	127.58	128.48	129.39	130.30	131.21	132.12	133.03	133.95	134.87	135.78	136.70	137.62	138.54	139.47	140.39
56	129.42	130.34	131.27	132.20	133.13	134.06	134.99	135.93	136.86	137.80	138.74	139.68	140.62	141.56	142.51	143.45
57	132.17	133.12	134.07	135.02	135.97	136.92	137.88	138.83	139.79	140.75	141.71	142.67	143.63	144.60	145.56	146.53
58	134.95	135.91	136.88	137.85	138.83	139.80	140.78	141.75	142.73	143.71	144.69	145.68	146.66	147.65	148.63	149.62
59	137.73	138.72	139.71	140.70	141.70	142.69	143.69	144.69	145.69	146.69	147.69	148.70	149.70	150.71	151.72	152.73
60	140.53	141.54	142.55	143.57	144.58	145.60	146.62	147.64	148.66	149.68	150.71	151.73	152.76	153.79	154.82	155.85

"How To"

477

TABLE C-16 INTEREST FROM DAY OF DEPOSIT

At 4.5% Compounded Daily (for One Quarter)

QUARTER	1ST MONTH	2ND MONTH	3RD MONTH
1st	January	February	March
2nd	April	May	June
3rd	July	August	September
4th	October	November	December

1ST MONTH		2ND MONTH		3RD MONTH	
DEP. DATE	FACTOR	DEP. DATE	FACTOR	DEP. DATE	FACTOR
1	.0113128	1	.0075277	1	.0037568
2	.0111864	2	.0074018	2	.0036314
3	.0110600	3	.0072759	3	.0035059
4	.0109337	4	.0071500	4	.0033805
5	.0108073	5	.0070241	5	.0032551
6	.0106810	6	.0068983	6	.0031297
7	.0105547	7	.0067724	7	.0030043
8	.0104284	8	.0066466	8	.0028790
9	.0103021	9	.0065208	9	.0027536
10	.0101758	10	.0063950	10	.0026283
11	.0100495	11	.0062692	11	.0025030
12	.0099233	12	.0061434	12	.0023777
13	.0097971	13	.0060177	13	.0022524
14	.0096709	14	.0058919	14	.0021271
15	.0095447	15	.0057662	15	.0020019
16	.0094185	16	.0056405	16	.0018766
17	.0092923	17	.0055148	17	.0017514
18	.0091662	18	.0053891	18	.0016262
19	.0090401	19	.0052635	19	.0015010
20	.0089139	20	.0051378	20	.0013759
21	.0087878	21	.0050122	21	.0012507
22	.0086618	22	.0048866	22	.0011256
23	.0085357	23	.0047610	23	.0010004
24	.0084096	24	.0046354	24	.0008753
25	.0082836	25	.0045099	25	.0007502
26	.0081576	26	.0043843	26	.0006252
27	.0080316	27	.0042588	27	.0005001
28	.0079056	28	.0041333	28	.0003751
29	.0077796	29	.0040078	29	.0002500
30	.0076537	30	.0038823	30	.0001250

TABLE C-17	INTEREST TO DAY OF WITHDRAWAL

At 4.5% Compounded Daily (for One Quarter)

1ST MONTH		2ND MONTH		3RD MONTH	
W/D DATE	FACTOR	W/D DATE	FACTOR	W/D DATE	FACTOR
1	.0001250	1	.0038823	1	.0076537
2	.0002500	2	.0040078	2	.0077796
3	.0003751	3	.0041333	3	.0079056
4	.0005001	4	.0042588	4	.0080316
5	.0006252	5	.0043843	5	.0081576
6	.0007502	6	.0045099	6	.0082836
7	.0008753	7	.0046354	7	.0084096
8	.0010004	8	.0047610	8	.0085357
9	.0011256	9	.0048866	9	.0086618
10	.0012507	10	.0050122	10	.0087878
11	.0013759	11	.0051378	11	.0089139
12	.0015010	12	.0052635	12	.0090401
13	.0016262	13	.0053891	13	.0091662
14	.0017514	14	.0055148	14	.0092923
15	.0018766	15	.0056405	15	.0094185
16	.0020019	16	.0057662	16	.0095447
17	.0021271	17	.0058919	17	.0096709
18	.0022524	18	.0060177	18	.0097971
19	.0023777	19	.0061434	19	.0099233
20	.0025030	20	.0062692	20	.0100495
21	.0026283	21	.0063950	21	.0101758
22	.0027536	22	.0065208	22	.0103021
23	.0028790	23	.0066466	23	.0104284
24	.0030043	24	.0067724	24	.0105547
25	.0031297	25	.0068983	25	.0106810
26	.0032551	26	.0070241	26	.0108073
27	.0033805	27	.0071500	27	.0109337
28	.0035059	28	.0072759	28	.0110600
29	.0036314	29	.0074018	29	.0111864
30	.0037568	30	.0075277	30	.0113128

Tables C-18 through C-23 are printed courtesy of the Financial Publishing Company, Boston, Massachusetts.

		XIV COMPOUND AMOUNT	XV AMOUNT OF ANNUITY	XVI SINKING FUND	XVII PRESENT VALUE	XVIII PRESENT VALUE OF ANNUITY	XIX AMORTIZATION	
RATE 5/12%	**PERIODS**	AMOUNT OF 1 How $1 left at compound interest will grow.	AMOUNT OF 1 PER PERIOD How $1 deposited periodically will grow.	SINKING FUND Periodic deposit that will grow to $1 at future date.	PRESENT WORTH OF 1 What $1 due in the future is worth today.	PRESENT WORTH OF 1 PER PERIOD What $1 payable periodically is worth today.	PARTIAL PAYMENT Annuity worth $1 today. Periodic payment necessary to pay off a loan of $1.	**PERIODS**
	1	1.004 166 6667	1.000 000 0000	1.000 000 0000	.995 850 6224	.995 850 6224	1.004 166 6667	1
	2	1.008 350 6944	2.004 166 6667	.498 960 4990	.991 718 4621	1.987 569 0846	.503 127 1656	2
	3	1.012 552 1557	3.012 517 3611	.331 948 2944	.987 603 4478	2.975 172 5323	.336 114 9611	3
	4	1.016 771 1230	4.025 069 5168	.248 442 9140	.983 505 5082	3.958 678 0405	.252 609 5807	4
	5	1.021 007 6693	5.041 840 6398	.198 340 2633	.979 424 5724	4.938 102 6129	.202 506 9300	5
	6	1.025 261 8680	6.062 848 3091	.164 938 9774	.975 360 5701	5.913 463 1830	.169 105 6440	6
	7	1.029 533 7924	7.088 110 1771	.141 081 3285	.971 313 4308	6.884 776 6138	.145 247 9951	7
	8	1.033 823 5165	8.117 643 9695	.123 188 4527	.967 283 0846	7.852 059 6984	.127 355 1193	8
	9	1.038 131 1145	9.151 467 4860	.109 272 0923	.963 269 4618	8.815 329 1602	.113 438 7590	9
	10	1.042 456 6608	10.189 598 6005	.098 139 2927	.959 272 4931	9.774 601 6533	.102 305 9593	10
	11	1.046 800 2303	11.232 055 2614	.089 030 9010	.955 292 1093	10.729 893 7626	.093 197 5677	11
	12	1.051 161 8979	12.278 855 4916	.081 440 8151	.951 328 2416	11.681 222 0043	.085 607 4818	12
	13	1.055 541 7391	13.330 017 3895	.075 018 6568	.947 380 8216	12.628 602 8259	.079 185 3235	13
	14	1.059 939 8297	14.385 559 1286	.069 514 1559	.943 449 7808	13.572 052 6067	.073 680 8226	14
	15	1.064 356 2457	15.445 498 9583	.064 743 7809	.939 535 0514	14.511 587 6581	.068 910 4475	15
	16	1.068 791 0633	16.509 855 2040	.060 569 8831	.935 636 5657	15.447 224 2238	.064 736 5498	16
	17	1.073 244 3594	17.578 646 2673	.056 887 2019	.931 754 2563	16.378 978 4802	.061 053 8686	17
	18	1.077 716 2109	18.651 890 6268	.053 613 8679	.927 888 0561	17.306 866 5363	.057 780 5346	18
	19	1.082 206 6952	19.729 606 8377	.050 685 2472	.924 037 8982	18.230 904 4344	.054 851 9139	19
	20	1.086 715 8897	20.811 813 5329	.048 049 6329	.920 203 7160	19.151 108 1505	.052 216 2996	20
	21	1.091 243 8726	21.898 529 4226	.045 665 1669	.916 385 4434	20.067 493 5938	.049 831 8335	21
	22	1.095 790 7221	22.989 773 2952	.043 497 6016	.912 583 0141	20.980 076 6080	.047 664 2683	22
	23	1.100 356 5167	24.085 564 0173	.041 518 6457	.908 796 3626	21.888 872 9706	.045 685 3124	23
	24	1.104 941 3356	25.185 920 5340	.039 704 7231	.905 025 4234	22.793 898 3940	.043 871 3897	24
	25	1.109 545 2578	26.290 861 8696	.038 036 0296	.901 270 1311	23.695 168 5251	.042 202 6963	25
	26	1.114 168 3630	27.400 407 1273	.036 495 8081	.897 530 4211	24.592 698 9462	.040 662 4748	26
	27	1.118 810 7312	28.514 575 4904	.035 069 7839	.893 806 2284	25.486 505 1746	.039 236 4506	27
	28	1.123 472 4426	29.633 386 2216	.033 745 7215	.890 097 4889	26.376 602 6635	.037 912 3882	28
	29	1.128 153 5778	30.756 858 6642	.032 513 0733	.886 404 1383	27.263 006 8018	.036 679 7400	29
	30	1.132 854 2177	31.885 012 2419	.031 362 6977	.882 726 1129	28.145 732 9147	.035 529 3644	30
	31	1.137 574 4436	33.017 866 4596	.030 286 6329	.879 063 3489	29.024 796 2636	.034 453 2995	31
	32	1.142 314 3371	34.155 440 9032	.029 277 9122	.875 415 7831	29.900 212 0467	.033 444 5789	32
	33	1.147 073 9802	35.297 755 2403	.028 330 4135	.871 783 3525	30.771 995 3992	.032 497 0801	33
	34	1.151 853 4551	36.444 829 2205	.027 438 7347	.868 165 9942	31.640 161 3934	.031 605 4014	34
	35	1.156 652 8445	37.596 682 6756	.026 598 0913	.864 563 6457	32.504 725 0391	.030 764 7580	35
	36	1.161 472 2313	38.753 335 5200	.025 804 2304	.860 976 2447	33.365 701 2837	.029 970 8971	36
	37	1.166 311 6990	39.914 807 7514	.025 053 3588	.857 403 7291	34.223 105 0129	.029 220 0255	37
	38	1.171 171 3310	41.081 119 4503	.024 342 0825	.853 846 0373	35.076 951 0501	.028 508 7492	38
	39	1.176 051 2116	42.252 290 7814	.023 667 3558	.850 303 1077	35.927 254 1578	.027 834 0225	39
	40	1.180 951 4250	43.428 341 9930	.023 026 4374	.846 774 8790	36.774 029 0368	.027 193 1041	40
	41	1.185 872 0559	44.609 293 4179	.022 416 8536	.843 261 2903	37.617 290 3271	.026 583 5203	41
	42	1.190 813 1895	45.795 165 4738	.021 836 3661	.839 762 2808	38.457 052 6079	.026 003 0328	42
	43	1.195 774 9111	46.985 978 6633	.021 282 9450	.836 277 7900	39.293 330 3979	.025 449 6117	43
	44	1.200 757 3066	48.181 753 5744	.020 754 7448	.832 807 7577	40.126 138 1556	.024 921 4115	44
	45	1.205 760 4620	49.382 510 8810	.020 250 0841	.829 352 1238	40.955 490 2795	.024 416 7508	45
	46	1.210 784 4639	50.588 271 3430	.019 767 4278	.825 910 8287	41.781 401 1082	.023 934 0944	46
	47	1.215 829 3992	51.799 055 8069	.019 305 3712	.822 483 8128	42.603 884 9210	.023 472 0379	47
	48	1.220 895 3550	53.014 885 2061	.018 862 6269	.819 071 0169	43.422 955 9379	.023 029 2936	48
	49	1.225 982 4190	54.235 780 5611	.018 438 0125	.815 672 3820	44.238 628 3199	.022 604 6792	49
	50	1.231 090 6791	55.461 762 9801	.018 030 4402	.812 287 8493	45.050 916 1692	.022 197 1069	50
	51	1.236 220 2236	56.692 853 6592	.017 638 9075	.808 917 3603	45.859 833 5295	.021 805 5741	51
	52	1.241 371 1412	57.929 073 8828	.017 262 4890	.805 560 8567	46.665 394 3862	.021 429 1557	52
	53	1.246 543 5209	59.170 445 0240	.016 900 3292	.802 218 2806	47.467 612 6668	.021 066 9959	53
	54	1.251 737 4523	60.416 988 5449	.016 551 6360	.798 889 5740	48.266 502 2408	.020 718 3026	54
	55	1.256 953 0250	61.668 725 9972	.016 215 6747	.795 574 6795	49.062 076 9203	.020 382 3414	55
	56	1.262 190 3293	62.925 679 0222	.015 891 7634	.792 273 5397	49.854 350 4600	.020 058 4300	56
	57	1.267 449 4556	64.187 869 3514	.015 579 2677	.788 986 0977	50.643 336 5577	.019 745 9344	57
	58	1.272 730 4950	65.455 318 8071	.015 277 5973	.785 712 2964	51.429 048 8542	.019 444 2639	58
	59	1.278 033 5388	66.728 049 3021	.014 986 2016	.782 452 0794	52.211 500 9336	.019 152 8683	59
	60	1.283 358 6785	68.006 082 8408	.014 704 5670	.779 205 3903	52.990 706 3239	.018 871 2336	60
	61	1.288 706 0063	69.289 441 5193	.014 432 2133	.775 972 1729	53.766 678 4969	.018 598 8800	61
	62	1.294 075 6147	70.578 147 5257	.014 168 6915	.772 752 3714	54.539 430 8682	.018 335 3582	62
	63	1.299 467 5964	71.872 223 1404	.013 913 5810	.769 545 9300	55.308 976 7982	.018 080 2477	63
	64	1.304 882 0447	73.171 690 7368	.013 666 4875	.766 352 7934	56.075 329 5916	.017 833 1542	64
	65	1.310 319 0533	74.476 572 7815	.013 427 0410	.763 172 9063	56.838 502 4979	.017 593 7077	65
	66	1.315 778 7160	75.786 891 8348	.013 194 8939	.760 006 2137	57.598 508 7116	.017 361 5606	66
	67	1.321 261 1273	77.102 670 5508	.012 969 7194	.756 852 6609	58.355 361 3725	.017 136 3860	67
	68	1.326 766 3820	78.423 931 6781	.012 751 2097	.753 712 1935	59.109 073 5660	.016 917 8764	68
	69	1.332 294 5753	79.750 698 0600	.012 539 0752	.750 584 7570	59.859 658 3230	.016 705 7419	69
	70	1.337 845 8026	81.082 992 6353	.012 333 0426	.747 470 2974	60.607 128 6204	.016 499 7092	70
	71	1.343 420 1602	82.420 838 4379	.012 132 8540	.744 368 7609	61.351 497 3813	.016 299 5207	71
	72	1.349 017 7442	83.764 258 5981	.011 938 2660	.741 280 0939	62.092 777 4752	.016 104 9327	72
	73	1.354 638 6514	85.113 276 3423	.011 749 0484	.738 204 2428	62.830 981 7180	.015 915 7150	73
	74	1.360 282 9791	86.467 914 9937	.011 564 9834	.735 141 1547	63.566 122 8727	.015 731 6500	74
	75	1.365 950 8249	87.828 197 9728	.011 385 8649	.732 090 7765	64.298 213 6492	.015 552 5316	75
	76	1.371 642 2867	89.194 148 7977	.011 211 4978	.729 053 0554	65.027 266 7046	.015 378 1644	76
	77	1.377 357 4629	90.565 791 0844	.011 041 6967	.726 027 9390	65.753 294 6435	.015 208 3634	77
	78	1.383 096 4523	91.943 148 5472	.010 876 2862	.723 015 3749	66.476 310 0185	.015 042 9529	78
	79	1.388 859 3542	93.326 244 9995	.010 715 0995	.720 015 3111	67.196 325 3296	.014 881 7662	79
	80	1.394 646 2681	94.715 104 3537	.010 557 9781	.717 027 6957	67.913 353 0253	.014 724 6448	80
	81	1.400 457 2943	96.109 750 6218	.010 404 7716	.714 052 4771	68.627 405 5024	.014 571 4382	81
	82	1.406 292 5330	97.510 207 9161	.010 255 3366	.711 089 6037	69.338 495 1061	.014 422 0032	82
	83	1.412 152 0852	98.916 500 4490	.010 109 5368	.708 139 0245	70.046 634 1306	.014 276 2035	83
	84	1.418 036 0522	100.328 652 5342	.009 967 2424	.705 200 6883	70.751 834 8188	.014 133 9091	84
	85	1.423 944 5358	101.746 688 5865	.009 828 3297	.702 274 5443	71.454 109 3632	.013 994 9964	85
	86	1.429 877 6380	103.170 633 1223	.009 692 6807	.699 360 5421	72.153 469 9052	.013 859 3473	86
	87	1.435 835 4615	104.600 510 7603	.009 560 1828	.696 458 6311	72.849 928 5363	.013 726 8494	87
	88	1.441 818 1093	106.036 346 2218	.009 430 7286	.693 568 7613	73.543 497 2976	.013 597 3952	88
	89	1.447 825 6847	107.478 164 3310	.009 304 2155	.690 690 8826	74.234 188 1802	.013 470 8821	89
	90	1.453 858 2917	108.925 990 0157	.009 180 5454	.687 824 9453	74.922 013 1255	.013 347 2121	90
	91	1.459 916 0346	110.379 848 3075	.009 059 6247	.684 970 8999	75.606 984 0254	.013 226 2914	91
	92	1.465 999 0181	111.839 764 3421	.008 941 3636	.682 128 6970	76.289 112 7224	.013 108 0303	92
	93	1.472 107 3473	113.305 763 3602	.008 825 6764	.679 298 2875	76.968 411 0098	.012 992 3431	93
	94	1.478 241 1279	114.777 870 7075	.008 712 4808	.676 479 6224	77.644 890 6322	.012 879 1475	94
	95	1.484 400 4660	116.256 111 8355	.008 601 6983	.673 672 6530	78.318 563 2852	.012 768 3650	95
	96	1.490 585 4679	117.740 512 3014	.008 493 2533	.670 877 3308	78.989 440 6159	.012 659 9200	96
	97	1.496 796 2407	119.231 097 7694	.008 387 0737	.668 093 6074	79.657 534 2233	.012 553 7403	97
	98	1.503 032 8917	120.727 894 0101	.008 283 0899	.665 321 4348	80.322 855 6581	.012 449 7566	98
	99	1.509 295 5288	122.230 926 9018	.008 181 2355	.662 560 7649	80.985 416 4230	.012 347 9022	99
	100	1.515 584 2601	123.740 222 4305	.008 081 4466	.659 811 5501	81.645 227 9731	.012 248 1133	100

	XIV COMPOUND AMOUNT	XV AMOUNT OF ANNUITY	XVI SINKING FUND	XVII PRESENT VALUE	XVIII PRESENT VALUE OF ANNUITY	XIX AMORTIZATION	
RATE 1/2% PERIODS	AMOUNT OF 1 How $1 left at compound interest will grow.	AMOUNT OF 1 PER PERIOD How $1 deposited periodically will grow.	SINKING FUND Periodic deposit that will grow to $1 at future date.	PRESENT WORTH OF 1 What $1 due in the future is worth today.	PRESENT WORTH OF 1 PER PERIOD What $1 payable periodically is worth today.	PARTIAL PAYMENT Annuity worth $1 today. Periodic payment necessary to pay off a loan of $1.	PERIODS
1	1.005 000 0000	1.000 000 0000	1.000 000 0000	.995 024 8756	.995 024 8756	1.005 000 0000	1
2	1.010 025 0000	2.005 000 0000	.498 753 1172	.990 074 5031	1.985 099 3787	.503 753 1172	2
3	1.015 075 1250	3.015 025 0000	.331 672 2084	.985 148 7593	2.970 248 1380	.336 672 2084	3
4	1.020 150 5006	4.030 100 1250	.248 132 7930	.980 247 5217	3.950 495 6597	.253 132 7930	4
5	1.025 251 2531	5.050 250 6256	.198 009 9750	.975 370 6684	4.925 866 3281	.203 009 9750	5
6	1.030 377 5094	6.075 501 8788	.164 595 4556	.970 518 0780	5.896 384 4061	.169 595 4556	6
7	1.035 529 3969	7.105 879 3881	.140 728 5355	.965 689 6298	6.862 074 0359	.145 728 5355	7
8	1.040 707 0439	8.141 408 7851	.122 828 8649	.960 885 2038	7.822 959 2397	.127 828 8649	8
9	1.045 910 5791	9.182 115 8290	.108 907 3606	.956 104 6804	8.779 063 9201	.113 907 3606	9
10	1.051 140 1320	10.228 026 4082	.097 770 5727	.951 347 9407	9.730 411 8608	.102 770 5727	10
11	1.056 395 8327	11.279 166 5402	.088 659 0331	.946 614 8664	10.677 026 7272	.093 659 0331	11
12	1.061 677 8119	12.335 562 3729	.081 066 4297	.941 905 3397	11.618 932 0668	.086 066 4297	12
13	1.066 986 2009	13.397 240 1848	.074 642 2387	.937 219 2434	12.556 151 3103	.079 642 2387	13
14	1.072 321 1319	14.464 226 3857	.069 136 0060	.932 556 4611	13.488 707 7714	.074 136 0060	14
15	1.077 682 7376	15.536 547 5176	.064 364 3640	.927 916 8768	14.416 624 6482	.069 364 3640	15
16	1.083 071 1513	16.614 230 2552	.060 189 3669	.923 300 3749	15.339 925 0231	.065 189 3669	16
17	1.088 486 5070	17.697 301 4065	.056 505 7902	.918 706 8407	16.258 631 8637	.061 505 7902	17
18	1.093 928 9396	18.785 787 9135	.053 231 7305	.914 136 1599	17.172 768 0236	.058 231 7305	18
19	1.099 398 5843	19.879 716 8531	.050 302 5273	.909 588 2188	18.082 356 2424	.055 302 5273	19
20	1.104 895 5772	20.979 115 4373	.047 666 4520	.905 062 9043	18.987 419 1467	.052 666 4520	20
21	1.110 420 0551	22.084 011 0145	.045 281 6293	.900 560 1037	19.887 979 2504	.050 281 6293	21
22	1.115 972 1553	23.194 431 0696	.043 113 7973	.896 079 7052	20.784 058 9556	.048 113 7973	22
23	1.121 552 0161	24.310 403 2250	.041 134 6530	.891 621 5972	21.675 680 5529	.046 134 6530	23
24	1.127 159 7762	25.431 955 2411	.039 320 6103	.887 185 6689	22.562 866 2218	.044 320 6103	24
25	1.132 795 5751	26.559 115 0173	.037 651 8570	.882 771 8098	23.445 638 0316	.042 651 8570	25
26	1.138 459 5530	27.691 910 5924	.036 111 6289	.878 379 9103	24.324 017 9419	.041 111 6289	26
27	1.144 151 8507	28.830 370 1453	.034 685 6456	.874 009 8610	25.198 027 8029	.039 685 6456	27
28	1.149 872 6100	29.974 521 9961	.033 361 6663	.869 661 5532	26.067 689 3561	.038 361 6663	28
29	1.155 621 9730	31.124 394 6060	.032 129 1390	.865 334 8788	26.933 024 2349	.037 129 1390	29
30	1.161 400 0829	32.280 016 5791	.030 978 9184	.861 029 7302	27.794 053 9651	.035 978 9184	30
31	1.167 207 0833	33.441 416 6620	.029 903 0394	.856 746 0002	28.650 799 9653	.034 903 0394	31
32	1.173 043 1187	34.608 623 7453	.028 894 5324	.852 483 5823	29.503 283 5475	.033 894 5324	32
33	1.178 908 3343	35.781 666 8640	.027 947 2727	.848 242 3704	30.351 525 9179	.032 947 2727	33
34	1.184 802 8760	36.960 575 1983	.027 055 8560	.844 022 2591	31.195 548 1771	.032 055 8560	34
35	1.190 726 8904	38.145 378 0743	.026 215 4958	.839 823 1434	32.035 371 3205	.031 215 4958	35
36	1.196 680 5248	39.336 104 9647	.025 421 9375	.835 644 9188	32.871 016 2393	.030 421 9375	36
37	1.202 663 9274	40.532 785 4895	.024 671 3861	.831 487 4814	33.702 503 7207	.029 671 3861	37
38	1.208 677 2471	41.735 449 4170	.023 960 4464	.827 350 7278	34.529 854 4484	.028 960 4464	38
39	1.214 720 6333	42.944 126 6640	.023 286 0714	.823 234 5550	35.353 089 0034	.028 286 0714	39
40	1.220 794 2365	44.158 847 2974	.022 645 5186	.819 138 8607	36.172 227 8641	.027 645 5186	40
41	1.226 898 2077	45.379 641 5338	.022 036 3133	.815 063 5430	36.987 291 4070	.027 036 3133	41
42	1.233 032 6987	46.606 539 7415	.021 456 2163	.811 008 5005	37.798 299 9075	.026 456 2163	42
43	1.239 197 8622	47.839 572 4402	.020 903 1969	.806 973 6323	38.605 273 5398	.025 903 1969	43
44	1.245 393 8515	49.078 770 3024	.020 375 4086	.802 958 8381	39.408 232 3779	.025 375 4086	44
45	1.251 620 8208	50.324 164 1539	.019 871 1696	.798 964 0180	40.207 196 3959	.024 871 1696	45
46	1.257 878 9249	51.575 784 9747	.019 388 9439	.794 989 0727	41.002 185 4686	.024 388 9439	46
47	1.264 168 3195	52.833 663 8996	.018 927 3264	.791 033 9031	41.793 219 3717	.023 927 3264	47
48	1.270 489 1611	54.097 832 2191	.018 485 0290	.787 098 4111	42.580 317 7828	.023 485 0290	48
49	1.276 841 6069	55.368 321 3802	.018 060 8690	.783 182 4986	43.363 500 2814	.023 060 8690	49
50	1.283 225 8149	56.645 162 9871	.017 653 7580	.779 286 0683	44.142 786 3497	.022 653 7580	50
51	1.289 641 9440	57.928 388 8020	.017 262 6931	.775 409 0231	44.918 195 3728	.022 262 6931	51
52	1.296 090 1537	59.218 030 7460	.016 886 7486	.771 551 2668	45.689 746 6396	.021 886 7486	52
53	1.302 570 6045	60.514 120 8997	.016 525 0686	.767 712 7033	46.457 459 3429	.021 525 0686	53
54	1.309 083 4575	61.816 691 5042	.016 176 8606	.763 893 2371	47.221 352 5800	.021 176 8606	54
55	1.315 628 8748	63.125 774 9618	.015 841 3897	.760 092 7732	47.984 445 3532	.020 841 3897	55
56	1.322 207 0192	64.441 403 8366	.015 517 9735	.756 311 2171	48.737 756 5704	.020 517 9735	56
57	1.328 818 0543	65.763 610 8558	.015 205 9777	.752 548 4748	49.490 305 0452	.020 205 9777	57
58	1.335 462 1446	67.092 428 9100	.014 904 8114	.748 804 4525	50.239 109 4977	.019 904 8114	58
59	1.342 139 4553	68.427 891 0546	.014 613 9240	.745 079 0572	50.984 188 5549	.019 613 9240	59
60	1.348 850 1525	69.770 030 5099	.014 332 8015	.741 372 1962	51.725 560 7511	.019 332 8015	60
61	1.355 594 4033	71.118 880 6624	.014 060 9637	.737 683 7774	52.463 244 5285	.019 060 9637	61
62	1.362 372 3753	72.474 475 0657	.013 797 9613	.734 013 7088	53.197 258 2373	.018 797 9613	62
63	1.369 184 2372	73.836 847 4411	.013 543 3735	.730 361 8993	53.927 620 1366	.018 543 3735	63
64	1.376 030 1584	75.206 031 6783	.013 296 8058	.726 728 2580	54.654 348 3946	.018 296 8058	64
65	1.382 910 3092	76.582 061 8366	.013 057 8882	.723 112 6946	55.377 461 0892	.018 057 8882	65
66	1.389 824 8607	77.964 972 1458	.012 826 2728	.719 515 1190	56.096 976 2082	.017 826 2728	66
67	1.396 773 9850	79.354 797 0066	.012 601 6326	.715 935 4418	56.812 911 6499	.017 601 6326	67
68	1.403 757 8550	80.751 570 9916	.012 383 6600	.712 373 5739	57.525 285 2238	.017 383 6600	68
69	1.410 776 6442	82.155 328 8466	.012 172 0650	.708 829 4267	58.234 114 6505	.017 172 0650	69
70	1.417 830 5275	83.566 105 4908	.011 966 5742	.705 302 9122	58.939 417 5627	.016 966 5742	70
71	1.424 919 6801	84.983 936 0182	.011 766 9297	.701 793 9425	59.641 211 5052	.016 766 9297	71
72	1.432 044 2785	86.408 855 6983	.011 572 8879	.698 302 4303	60.339 513 9355	.016 572 8879	72
73	1.439 204 4999	87.840 899 9768	.011 384 2185	.694 828 2889	61.034 342 2244	.016 384 2185	73
74	1.446 400 5224	89.280 104 4767	.011 200 7037	.691 371 4317	61.725 713 6561	.016 200 7037	74
75	1.453 632 5250	90.726 504 9991	.011 022 1374	.687 931 7729	62.413 645 4290	.016 022 1374	75
76	1.460 900 6876	92.180 137 5241	.010 848 3240	.684 509 2267	63.098 154 6557	.015 848 3240	76
77	1.468 205 1911	93.641 038 2117	.010 679 0785	.681 103 7082	63.779 258 3639	.015 679 0785	77
78	1.475 546 2170	95.109 243 4028	.010 514 2252	.677 715 1325	64.456 973 4964	.015 514 2252	78
79	1.482 923 9481	96.584 789 6198	.010 353 5971	.674 343 4154	65.131 316 9118	.015 353 5971	79
80	1.490 338 5678	98.067 713 5679	.010 197 0359	.670 988 4731	65.802 305 3849	.015 197 0359	80
81	1.497 790 2607	99.558 052 1357	.010 044 3910	.667 650 2220	66.469 955 6069	.015 044 3910	81
82	1.505 279 2120	101.055 842 3964	.009 895 5189	.664 328 5791	67.134 284 1859	.014 895 5189	82
83	1.512 805 6080	102.561 121 6084	.009 750 2834	.661 023 4618	67.795 307 6477	.014 750 2834	83
84	1.520 369 6361	104.073 927 2164	.009 608 6545	.657 734 7878	68.453 042 4355	.014 608 6545	84
85	1.527 971 4843	105.594 296 8525	.009 470 2084	.654 462 4754	69.107 504 9110	.014 470 2084	85
86	1.535 611 3417	107.122 268 3368	.009 335 1272	.651 206 4432	69.758 711 3542	.014 335 1272	86
87	1.543 289 3984	108.657 879 6784	.009 203 1982	.647 966 6102	70.406 677 9644	.014 203 1982	87
88	1.551 005 8454	110.201 169 0768	.009 074 3139	.644 742 8957	71.051 420 8601	.014 074 3139	88
89	1.558 760 8746	111.752 174 9222	.008 948 3717	.641 535 2196	71.692 956 0797	.013 948 3717	89
90	1.566 554 6790	113.310 935 7968	.008 825 2735	.638 343 5021	72.331 299 5818	.013 825 2735	90
91	1.574 387 4524	114.877 490 4758	.008 704 9255	.635 167 6638	72.966 467 2455	.013 704 9255	91
92	1.582 259 3896	116.451 877 9282	.008 587 2381	.632 007 6256	73.598 474 8712	.013 587 2381	92
93	1.590 170 6866	118.034 137 3178	.008 472 1253	.628 863 3091	74.227 338 1803	.013 472 1253	93
94	1.598 121 5400	119.624 308 0044	.008 359 5050	.625 734 6359	74.853 072 8162	.013 359 5050	94
95	1.606 112 1477	121.222 429 5445	.008 249 2984	.622 621 5283	75.475 694 3445	.013 249 2984	95
96	1.614 142 7085	122.828 541 6922	.008 141 4302	.619 523 9087	76.095 218 2532	.013 141 4302	96
97	1.622 213 4220	124.442 684 4006	.008 035 8279	.616 441 7002	76.711 659 9535	.013 035 8279	97
98	1.630 324 4891	126.064 897 8226	.007 932 4222	.613 374 8261	77.325 034 7796	.012 932 4222	98
99	1.638 476 1116	127.695 222 3118	.007 831 1466	.610 323 2101	77.935 357 9896	.012 831 1466	99
100	1.646 668 4921	129.333 698 4233	.007 731 9369	.607 286 7762	78.542 644 7658	.012 731 9369	100

	PERIODS	XIV COMPOUND AMOUNT — AMOUNT OF 1 — How $1 left at compound interest will grow.	XV AMOUNT OF ANNUITY — AMOUNT OF 1 PER PERIOD — How $1 deposited periodically will grow.	XVI SINKING FUND — SINKING FUND — Periodic deposit that will grow to $1 at future date.	XVII PRESENT VALUE — PRESENT WORTH OF 1 — What $1 due in the future is worth today.	XVIII PRESENT VALUE OF ANNUITY — PRESENT WORTH OF 1 PER PERIOD — What $1 payable periodically is worth today.	XIX AMORTIZATION — PARTIAL PAYMENT — Annuity worth $1 today. Periodic payment necessary to pay off a loan of $1.	PERIODS
RATE 7/12%	1	1.005 833 3333	1.000 000 0000	1.000 000 0000	.994 200 4971	.994 200 4971	1.005 833 3333	1
	2	1.011 700 6944	2.005 833 3333	.498 545 9078	.988 434 6284	1.982 635 1255	.504 379 2411	2
	3	1.017 602 2818	3.017 534 0278	.331 396 4286	.982 702 1989	2.965 337 3245	.337 229 7619	3
	4	1.023 538 2951	4.035 136 3096	.247 823 1027	.977 003 0147	3.942 340 3392	.253 656 4360	4
	5	1.029 508 9352	5.058 674 6047	.197 680 2380	.971 336 8829	4.913 677 2220	.203 513 5714	5
	6	1.035 514 4040	6.088 183 5399	.164 252 6040	.965 703 6118	5.879 380 8338	.170 085 9373	6
	7	1.041 554 9047	7.123 697 9439	.140 376 5303	.960 103 0109	6.839 483 8447	.146 209 8636	7
	8	1.047 630 6416	8.165 252 8486	.122 470 1817	.954 534 8907	7.794 018 7355	.128 303 5150	8
	9	1.053 741 8204	9.212 883 4902	.108 543 6499	.948 999 0628	8.743 017 7983	.114 376 9832	9
	10	1.059 888 6476	10.266 625 3106	.097 402 9898	.943 495 3400	9.686 513 1383	.103 236 3231	10
	11	1.066 071 3314	11.326 513 9582	.088 288 4181	.938 023 5361	10.624 536 6744	.094 121 7514	11
	12	1.072 290 0809	12.392 585 2896	.080 693 4128	.932 583 4658	11.557 120 1402	.086 526 7461	12
	13	1.078 545 1063	13.464 875 3705	.074 267 3046	.927 174 9453	12.484 295 0856	.080 100 6379	13
	14	1.084 836 6194	14.543 420 4768	.068 759 6155	.921 797 7916	13.406 092 8771	.074 592 9488	14
	15	1.091 164 8331	15.628 257 0963	.063 986 6617	.916 451 8226	14.322 544 6997	.069 819 9950	15
	16	1.097 529 9613	16.719 421 9293	.059 810 6803	.911 136 8576	15.233 681 5573	.065 644 0136	16
	17	1.103 932 2194	17.816 951 8906	.056 126 3232	.905 852 7167	16.139 534 2740	.061 959 6565	17
	18	1.110 371 8240	18.920 884 1100	.052 851 6529	.900 599 2213	17.040 133 4953	.058 684 9863	18
	19	1.116 848 9929	20.031 255 9339	.049 921 9821	.895 376 1935	17.935 509 6888	.055 755 3154	19
	20	1.123 363 9454	21.148 104 9269	.047 285 5607	.890 183 4567	18.825 693 1454	.053 118 8941	20
	21	1.129 916 9018	22.271 468 8723	.044 900 4960	.885 020 8351	19.710 713 9805	.050 733 8294	21
	22	1.136 508 0837	23.401 385 7740	.042 732 5121	.879 888 1542	20.590 602 1348	.048 565 8454	22
	23	1.143 137 7142	24.537 893 8577	.040 753 2939	.874 785 2403	21.465 387 3751	.046 586 6272	23
	24	1.149 806 0175	25.681 031 5719	.038 939 2458	.869 711 9208	22.335 099 2958	.044 772 5791	24
	25	1.156 513 2193	26.830 837 5894	.037 270 5472	.864 668 0240	23.199 767 3198	.043 103 8806	25
	26	1.163 259 5464	27.987 350 8087	.035 730 4272	.859 653 3793	24.059 420 6991	.041 563 7605	26
	27	1.170 045 2271	29.150 610 3550	.034 304 5990	.854 667 8170	24.914 088 5161	.040 137 9324	27
	28	1.176 870 4909	30.320 655 5821	.032 980 8172	.849 711 1685	25.763 799 6846	.038 814 1506	28
	29	1.183 735 5688	31.497 526 0730	.031 748 5252	.844 783 2661	26.608 582 9507	.037 581 8585	29
	30	1.190 640 6929	32.681 261 6418	.030 598 5739	.839 883 9431	27.448 466 8938	.036 431 9072	30
	31	1.197 586 0970	33.871 902 3347	.029 522 9949	.835 013 0338	28.283 479 9276	.035 356 3282	31
	32	1.204 572 0159	35.069 488 4316	.028 514 8157	.830 170 3732	29.113 650 3008	.034 348 1491	32
	33	1.211 598 6859	36.274 060 4475	.027 567 9091	.825 355 7978	29.939 006 0986	.033 401 2424	33
	34	1.218 666 3449	37.485 659 1334	.026 676 8685	.820 569 1444	30.759 575 2430	.032 510 2019	34
	35	1.225 775 2320	38.704 325 4784	.025 836 9055	.815 810 2513	31.575 385 4943	.031 670 2388	35
	36	1.232 925 5875	39.930 100 7103	.025 043 7635	.811 078 9574	32.386 464 4516	.030 877 0969	36
	37	1.240 117 6534	41.163 026 2978	.024 293 6463	.806 375 1026	33.192 839 5542	.030 126 9796	37
	38	1.247 351 6730	42.403 143 9512	.023 583 1570	.801 698 5279	33.994 538 0821	.029 416 4903	38
	39	1.254 627 8911	43.650 495 6243	.022 909 2413	.797 049 0749	34.791 587 1570	.028 742 5807	39
	40	1.261 946 5538	44.905 123 5154	.022 269 1738	.792 426 5865	35.584 013 7435	.028 102 5071	40
	41	1.269 307 9087	46.167 070 0692	.021 660 4606	.787 830 9062	36.371 844 6497	.027 493 7939	41
	42	1.276 712 2049	47.436 377 9780	.021 080 8675	.783 261 8786	37.155 106 5283	.026 914 2009	42
	43	1.284 159 6927	48.713 090 1829	.020 528 3630	.778 719 3490	37.933 825 8773	.026 361 6964	43
	44	1.291 650 6243	49.997 249 8756	.020 001 1001	.774 203 1639	38.708 029 0413	.025 834 4343	44
	45	1.299 185 2529	51.288 900 4999	.019 497 3959	.769 713 1704	39.477 742 2117	.025 330 7293	45
	46	1.306 763 8336	52.588 085 7528	.019 015 7140	.765 249 2167	40.242 991 4284	.024 849 0474	46
	47	1.314 386 6226	53.894 849 5863	.018 554 6487	.760 811 1516	41.003 802 5800	.024 387 9820	47
	48	1.322 053 8779	55.209 236 2089	.018 112 9113	.756 398 8251	41.760 201 4051	.023 946 2447	48
	49	1.329 765 8588	56.531 290 0868	.017 689 3186	.752 012 0880	42.512 213 4931	.023 522 6519	49
	50	1.337 522 8263	57.861 055 9456	.017 282 7817	.747 650 7917	43.259 864 2848	.023 116 1151	50
	51	1.345 325 0428	59.198 578 7720	.016 892 2974	.743 314 7887	44.003 179 0735	.022 725 6308	51
	52	1.353 172 7723	60.543 903 8148	.016 516 9396	.739 003 9325	44.742 183 0060	.022 350 2729	52
	53	1.361 066 2801	61.897 076 5871	.016 155 8519	.734 718 0770	45.476 901 0830	.021 989 1852	53
	54	1.369 005 8334	63.258 142 8672	.015 808 2415	.730 457 0774	46.207 358 1604	.021 641 5748	54
	55	1.376 991 7008	64.627 148 7006	.015 473 3733	.726 220 7895	46.933 578 9498	.021 306 7067	55
	56	1.385 024 1523	66.004 140 4013	.015 150 5647	.722 009 0699	47.655 588 0197	.020 983 8980	56
	57	1.393 103 4599	67.389 164 5537	.014 839 1808	.717 821 7762	48.373 409 7959	.020 672 5142	57
	58	1.401 229 8667	68.782 268 0136	.014 538 6308	.713 658 7667	49.087 068 5626	.020 371 9641	58
	59	1.409 403 7378	70.183 497 9103	.014 248 3636	.709 519 9006	49.796 588 4633	.020 081 6970	59
	60	1.417 625 2596	71.592 901 6481	.013 967 8652	.705 405 0379	50.501 993 5012	.019 801 1985	60
	61	1.425 894 7403	73.010 526 9077	.013 696 6550	.701 314 0393	51.203 307 5405	.019 529 9884	61
	62	1.434 212 4596	74.436 421 6480	.013 434 2836	.697 246 7665	51.900 554 3071	.019 267 6170	62
	63	1.442 578 6990	75.870 634 1076	.013 180 3301	.693 203 0819	52.593 757 3890	.019 013 6634	63
	64	1.450 993 7414	77.313 212 8066	.012 934 3997	.689 182 8486	53.282 940 2376	.018 767 7331	64
	65	1.459 457 8715	78.764 206 5480	.012 696 1223	.685 185 9307	53.968 126 1683	.018 529 4556	65
	66	1.467 971 3758	80.223 664 4195	.012 465 1499	.681 212 1929	54.649 338 3611	.018 298 4832	66
	67	1.476 534 5421	81.691 635 7953	.012 241 1553	.677 261 5008	55.326 599 8619	.018 074 4886	67
	68	1.485 147 6603	83.168 170 3374	.012 023 8307	.673 333 7208	55.999 933 5827	.017 857 1640	68
	69	1.493 811 0217	84.653 317 9977	.011 812 8861	.669 428 7199	56.669 362 3026	.017 646 2194	69
	70	1.502 524 9193	86.147 129 0194	.011 608 0479	.665 546 3661	57.334 908 6687	.017 441 3812	70
	71	1.511 289 6480	87.649 653 9387	.011 409 0582	.661 686 5280	57.996 595 1967	.017 242 3915	71
	72	1.520 105 5043	89.160 943 5866	.011 215 6731	.657 849 0751	58.654 444 2718	.017 049 0065	72
	73	1.528 972 7864	90.681 049 0909	.011 027 6625	.654 033 8775	59.308 478 1493	.016 860 9958	73
	74	1.537 891 7943	92.210 021 8772	.010 844 8082	.650 240 8061	59.958 718 9554	.016 678 1415	74
	75	1.546 862 8298	93.747 913 6715	.010 666 9040	.646 469 7327	60.605 188 6880	.016 500 2374	75
	76	1.555 886 1963	95.294 776 5013	.010 493 7546	.642 720 5296	61.247 909 2176	.016 327 0879	76
	77	1.564 962 1991	96.850 662 6975	.010 325 1746	.638 993 0700	61.886 902 2876	.016 158 5079	77
	78	1.574 091 1452	98.415 624 8966	.010 160 9882	.635 287 2278	62.522 189 5154	.015 994 3215	78
	79	1.583 273 3436	99.989 716 0418	.010 001 0285	.631 602 8777	63.153 792 3931	.015 834 3618	79
	80	1.592 509 1047	101.572 989 3854	.009 845 1370	.627 939 8950	63.781 732 2881	.015 678 4704	80
	81	1.601 798 7412	103.165 498 4902	.009 693 1631	.624 298 1557	64.406 030 4438	.015 526 4964	81
	82	1.611 142 5672	104.767 297 2314	.009 544 9632	.620 677 5368	65.026 707 9806	.015 378 2966	82
	83	1.620 540 8988	106.378 439 7985	.009 400 4011	.617 077 9156	65.643 785 8962	.015 233 7344	83
	84	1.629 994 0541	107.998 980 6974	.009 259 3466	.613 499 1704	66.257 285 0667	.015 092 6800	84
	85	1.639 502 3527	109.628 974 7514	.009 121 6761	.609 941 1802	66.867 226 2469	.014 955 0094	85
	86	1.649 066 1164	111.268 477 1041	.008 987 2714	.606 403 8246	67.473 630 0715	.014 820 6047	86
	87	1.658 685 6688	112.917 543 2206	.008 856 0198	.602 886 9838	68.076 517 0553	.014 689 3531	87
	88	1.668 361 3352	114.576 228 8894	.008 727 8139	.599 390 5390	68.675 907 5944	.014 561 1472	88
	89	1.678 093 4430	116.244 590 2246	.008 602 5509	.595 914 3719	69.271 821 9662	.014 435 8842	89
	90	1.687 882 3214	117.922 683 6675	.008 480 1327	.592 458 3647	69.864 280 3310	.014 313 4660	90
	91	1.697 728 3016	119.610 565 9889	.008 360 4654	.589 022 4007	70.453 302 7317	.014 193 7987	91
	92	1.707 631 7167	121.308 294 2905	.008 243 4594	.585 606 3636	71.038 909 0953	.014 076 7927	92
	93	1.717 592 9017	123.015 926 0072	.008 129 0288	.582 210 1378	71.621 119 2331	.013 962 3621	93
	94	1.727 612 1936	124.733 518 9089	.008 017 0912	.578 833 6084	72.199 952 8415	.013 850 4246	94
	95	1.737 689 9314	126.461 131 1026	.007 907 5681	.575 476 6631	72.775 429 5028	.013 740 9014	95
	96	1.747 826 4560	128.198 821 0340	.007 800 3837	.572 139 1827	73.347 568 6854	.013 633 7171	96
	97	1.758 022 1104	129.946 647 4900	.007 695 4659	.568 821 0598	73.916 389 7453	.013 528 7993	97
	98	1.768 277 2293	131.704 669 6004	.007 592 7452	.565 522 1804	74.481 911 9257	.013 426 0785	98
	99	1.778 592 1899	133.472 946 8397	.007 492 1550	.562 242 4329	75.044 154 3586	.013 325 4883	99
612	100	1.788 967 3110	135.251 539 0296	.007 393 6312	.558 981 7063	75.603 136 0649	.013 226 9645	100

		XIV COMPOUND AMOUNT	XV AMOUNT OF ANNUITY	XVI SINKING FUND	XVII PRESENT VALUE	XVIII PRESENT VALUE OF ANNUITY	XIX AMORTIZATION	
RATE 2/3%	P E R I O D S	AMOUNT OF 1 *How $1 left at compound interest will grow.*	AMOUNT OF 1 PER PERIOD *How $1 deposited periodically will grow.*	SINKING FUND *Periodic deposit that will grow to $1 at future date.*	PRESENT WORTH OF 1 *What $1 due in the future is worth today.*	PRESENT WORTH OF 1 PER PERIOD *What $1 payable periodically is worth today.*	PARTIAL PAYMENT *Annuity worth $1 today. Periodic payment necessary to pay off a loan of $1.*	P E R I O D S
	1	1.006 666 6667	1.000 000 0000	1.000 000 0000	.993 377 4834	.993 377 4834	1.006 666 6667	1
	2	1.013 377 7778	2.006 666 6667	.498 338 8704	.986 798 8246	1.980 176 3081	.505 005 5371	2
	3	1.020 133 6296	3.020 044 4444	.331 120 9548	.980 263 7331	2.960 440 0411	.337 787 6215	3
	4	1.026 934 5205	4.040 178 0741	.247 513 8426	.973 771 9203	3.934 211 9614	.254 180 5093	4
	5	1.033 780 7506	5.067 112 5946	.197 351 0518	.967 323 0996	4.901 535 0610	.204 017 7184	5
	6	1.040 672 6223	6.100 893 3452	.163 910 4215	.960 916 9864	5.862 452 0473	.170 577 0882	6
	7	1.047 610 4398	7.141 565 9675	.140 025 3116	.954 553 2977	6.817 005 3450	.146 691 9783	7
	8	1.054 594 5094	8.189 176 4073	.122 112 4018	.948 231 7527	7.765 237 0977	.128 779 0685	8
	9	1.061 625 1394	9.243 770 9167	.108 180 9587	.941 952 0722	8.707 189 1699	.114 847 6254	9
	10	1.068 702 6404	10.305 396 0561	.097 036 5423	.935 713 9790	9.642 903 1489	.103 703 2089	10
	11	1.075 827 3246	11.374 098 6965	.087 919 0542	.929 517 1977	10.572 420 3466	.094 585 7209	11
	12	1.082 999 5068	12.449 926 0211	.080 321 7624	.923 361 4547	11.495 781 8013	.086 988 4291	12
	13	1.090 219 5035	13.532 925 5279	.073 893 8523	.917 246 4781	12.413 028 2794	.080 560 5190	13
	14	1.097 487 6336	14.623 146 0315	.068 384 7420	.011 171 0981	13.324 200 2775	.075 051 4087	14
	15	1.104 804 2178	15.720 632 6650	.063 610 6715	.905 137 7465	14.229 338 0240	.070 277 3382	15
	16	1.112 169 5792	16.825 436 8828	.059 433 8208	.899 143 4568	15.128 481 4808	.066 100 4874	16
	17	1.119 584 0431	17.937 606 4620	.055 748 7980	.893 188 8644	16.021 670 3452	.062 415 4647	17
	18	1.127 047 0367	19.057 190 5051	.052 473 6319	.887 273 7063	16.908 944 0515	.059 140 2986	18
	19	1.134 561 5896	20.184 238 4418	.049 543 6081	.881 397 7215	17.790 341 7730	.056 210 2748	19
	20	1.142 125 3335	21.318 800 0314	.046 906 9553	.875 560 6505	18.665 902 4236	.053 573 6220	20
	21	1.149 739 5024	22.460 925 3649	.044 521 7632	.869 762 2356	19.535 664 6592	.051 188 4299	21
	22	1.157 404 4324	23.610 664 8673	.042 353 7417	.864 002 2208	20.399 666 8800	.049 020 4083	22
	23	1.165 120 4620	24.768 069 2998	.040 374 5640	.858 280 3518	21.257 947 2317	.047 041 2307	23
	24	1.172 887 9317	25.933 189 7618	.038 560 6248	.852 596 3759	22.110 543 6077	.045 227 2915	24
	25	1.180 707 1846	27.106 077 6935	.036 892 0952	.846 950 0423	22.957 493 6500	.043 558 7619	25
	26	1.188 578 5659	28.286 784 8782	.035 352 1973	.841 341 1017	23.798 834 7517	.042 018 8640	26
	27	1.196 502 4230	29.475 363 4440	.033 926 6385	.835 769 3063	24.634 604 0580	.040 593 3052	27
	28	1.204 479 1058	30.671 865 8670	.032 603 1681	.830 234 4102	25.464 838 4682	.039 269 8348	28
	29	1.212 508 9665	31.876 344 9728	.031 371 2253	.824 736 1691	26.289 574 6373	.038 037 8920	29
	30	1.220 592 3596	33.088 853 9392	.030 221 6572	.819 274 3402	27.108 848 9774	.036 888 3238	30
	31	1.228 729 6420	34.309 446 2988	.029 146 4919	.813 848 6823	27.922 697 6597	.035 813 1586	31
	32	1.236 921 1729	35.538 175 9408	.028 138 7543	.808 458 9559	28.731 156 6156	.034 805 4209	32
	33	1.245 167 3141	36.775 097 1138	.027 192 3143	.803 104 9231	29.534 261 5387	.033 858 9810	33
	34	1.253 468 4295	38.020 264 4279	.026 301 7634	.797 786 3474	30.332 047 8861	.032 968 4301	34
	35	1.261 824 8857	39.273 732 8574	.025 462 3110	.792 502 9941	31.124 550 8802	.032 128 9777	35
	36	1.270 237 0516	40.535 557 7431	.024 669 6988	.787 254 6299	31.911 805 5101	.031 336 3655	36
	37	1.278 705 2986	41.805 794 7947	.023 920 1289	.782 041 0231	32.693 846 5333	.030 586 7956	37
	38	1.287 230 0006	43.084 500 0934	.023 210 2032	.776 861 9435	33.470 708 4767	.029 876 8698	38
	39	1.295 811 5340	44.371 730 0940	.022 536 8720	.771 717 1624	34.242 425 6392	.029 203 5386	39
	40	1.304 450 2775	45.667 541 6279	.021 897 3907	.766 606 4527	35.009 032 0919	.028 564 0573	40
	41	1.313 146 6127	46.971 991 9055	.021 289 2824	.761 529 5888	35.770 561 6807	.027 955 9491	41
	42	1.321 900 9235	48.285 138 5182	.020 710 3061	.756 486 3465	36.527 048 0272	.027 376 9728	42
	43	1.330 713 5963	49.607 039 4416	.020 158 4294	.751 476 5031	37.278 524 5303	.026 825 0960	43
	44	1.339 585 0203	50.937 753 0379	.019 631 8043	.746 499 8375	38.025 024 3678	.026 298 4710	44
	45	1.348 515 5871	52.277 338 0581	.019 128 7475	.741 556 1300	38.766 580 4978	.025 795 4142	45
	46	1.357 505 6910	53.625 853 6452	.018 647 7218	.736 645 1623	39.503 225 6601	.025 314 3885	46
	47	1.366 555 7289	54.983 359 3362	.018 187 3209	.731 766 7175	40.234 992 3776	.024 853 9876	47
	48	1.375 666 1004	56.349 915 0651	.017 746 2557	.726 920 5803	40.961 912 9579	.024 412 9223	48
	49	1.384 837 2078	57.725 581 1655	.017 323 3423	.722 106 5367	41.684 019 4946	.023 990 0089	49
	50	1.394 069 4558	59.110 418 3733	.016 917 4915	.717 324 3742	42.401 343 8688	.023 584 1582	50
	51	1.403 363 2522	60.504 487 8291	.016 527 6996	.712 573 8817	43.113 917 7505	.023 194 3663	51
	52	1.412 719 0072	61.907 851 0813	.016 153 0401	.707 854 8493	43.821 772 5998	.022 819 7068	52
	53	1.422 137 1339	63.320 570 0885	.015 792 6563	.703 167 0689	44.524 939 6687	.022 459 3230	53
	54	1.431 618 0481	64.742 707 2224	.015 445 7551	.698 510 3333	45.223 450 0020	.022 112 4218	54
	55	1.441 162 1685	66.174 325 2706	.015 111 6010	.693 884 4371	45.917 334 4391	.021 778 2677	55
	56	1.450 769 9163	67.615 487 4390	.014 789 5111	.689 289 1759	46.606 623 6150	.021 456 1777	56
	57	1.460 441 7157	69.066 257 3553	.014 478 8503	.684 724 3469	47.291 347 9679	.021 145 5170	57
	58	1.470 177 9938	70.526 699 0710	.014 179 0274	.680 189 7486	47.971 537 7105	.020 845 6941	58
	59	1.479 979 1804	71.996 877 0648	.013 889 4913	.675 685 1807	48.647 222 8912	.020 556 1580	59
	60	1.489 845 7083	73.476 856 2452	.013 609 7276	.671 210 4444	49.318 433 3356	.020 276 3943	60
	61	1.499 778 0130	74.966 701 9535	.013 339 2556	.666 765 3421	49.985 198 6778	.020 005 9223	61
	62	1.509 736 5331	76.466 479 9666	.013 077 6257	.662 349 6776	50.647 548 3554	.019 744 2923	62
	63	1.519 841 7100	77.976 256 4997	.012 824 4166	.657 963 2559	51.305 511 6113	.019 491 0833	63
	64	1.529 973 9881	79.496 098 2097	.012 579 2337	.653 605 8834	51.959 117 4947	.019 245 9004	64
	65	1.540 173 8147	81.026 072 1977	.012 341 7065	.649 277 3676	52.608 394 8623	.019 008 3731	65
	66	1.550 441 6401	82.566 246 0124	.012 111 4868	.644 977 5175	53.253 372 3798	.018 778 1535	66
	67	1.560 777 9177	84.116 687 6525	.011 888 2475	.640 706 1432	53.894 078 5229	.018 554 9141	67
	68	1.571 183 1038	85.677 465 5702	.011 671 6805	.636 463 0561	54.530 541 5791	.018 338 3471	68
	69	1.581 657 6578	87.248 648 6740	.011 461 4956	.632 248 0690	55.162 789 6481	.018 128 1622	69
	70	1.592 202 0422	88.830 306 3318	.011 257 4192	.628 060 9957	55.790 850 6438	.017 924 0859	70
	71	1.602 816 7225	90.422 508 3740	.011 059 1933	.623 901 6514	56.414 752 2952	.017 725 8600	71
	72	1.613 502 1673	92.025 325 0965	.010 866 5739	.619 769 8523	57.034 522 1475	.017 533 2406	72
	73	1.624 258 8484	93.638 827 2638	.010 679 3307	.615 665 4162	57.650 187 5637	.017 345 9973	73
	74	1.635 087 2407	95.263 086 1122	.010 497 2455	.611 588 1618	58.261 775 7256	.017 163 9121	74
	75	1.645 987 8224	96.898 173 3530	.010 320 1120	.607 537 9091	58.869 313 6347	.016 986 7787	75
	76	1.656 961 0745	98.544 161 1753	.010 147 7347	.603 514 4792	59.472 828 1139	.016 814 4013	76
	77	1.668 007 4817	100.201 122 2498	.009 979 9281	.599 517 6946	60.072 345 8085	.016 646 5948	77
	78	1.679 127 5315	101.869 129 7315	.009 816 5166	.595 547 3788	60.667 893 1873	.016 483 1832	78
	79	1.690 321 7151	103.548 257 2630	.009 657 3330	.591 603 3564	61.259 496 5437	.016 323 9996	79
	80	1.701 590 5265	105.238 578 9781	.009 502 2188	.587 685 4534	61.847 181 9970	.016 168 8854	80
	81	1.712 934 4634	106.940 169 5046	.009 351 0231	.583 793 4967	62.430 975 4937	.016 017 6898	81
	82	1.724 354 0265	108.653 103 9680	.009 203 6027	.579 927 3146	63.010 902 8083	.015 870 2694	82
	83	1.735 849 7200	110.377 457 9945	.009 059 8209	.576 086 7364	63.586 989 5447	.015 726 4876	83
	84	1.747 422 0514	112.113 307 7144	.008 919 5477	.572 271 5924	64.159 261 1371	.015 586 2144	84
	85	1.759 071 5318	113.860 729 7659	.008 782 6593	.568 481 7143	64.727 742 8514	.015 449 3260	85
	86	1.770 798 6753	115.619 801 2976	.008 649 0375	.564 716 9348	65.292 459 7862	.015 315 7042	86
	87	1.782 603 9998	117.390 599 9729	.008 518 5696	.560 977 0875	65.853 436 8737	.015 185 2363	87
	88	1.794 488 0265	119.173 203 9728	.008 391 1481	.557 262 0075	66.410 698 8812	.015 057 8147	88
	89	1.806 451 2800	120.967 691 9992	.008 266 6701	.553 571 5306	66.964 270 4118	.014 933 3367	89
	90	1.818 494 2885	122.774 143 2792	.008 145 0375	.549 905 4940	67.514 175 9057	.014 811 7042	90
	91	1.830 617 5838	124.592 637 5678	.008 026 1564	.546 263 7357	68.060 439 6414	.014 692 8231	91
	92	1.842 821 7010	126.423 255 1516	.007 909 9371	.542 646 0951	68.603 085 7365	.014 576 6038	92
	93	1.855 107 1790	128.266 076 8526	.007 796 2936	.539 052 4123	69.142 138 1489	.014 462 9603	93
	94	1.867 474 5602	130.121 184 0316	.007 685 1437	.535 482 5288	69.677 620 6777	.014 351 8104	94
	95	1.879 924 3906	131.988 658 5918	.007 576 4085	.531 936 2869	70.209 556 9646	.014 243 0752	95
	96	1.892 457 2199	133.868 582 9824	.007 470 0126	.528 413 5300	70.737 970 4946	.014 136 6793	96
	97	1.905 073 6013	135.761 040 2023	.007 365 8835	.524 914 1027	71.262 884 5973	.014 032 5501	97
	98	1.917 774 0920	137.666 113 8036	.007 263 9517	.521 437 8503	71.784 322 4477	.013 930 6184	98
	99	1.930 559 2526	139.583 887 8957	.007 164 1506	.517 984 6196	72.302 307 0672	.013 830 8173	99
	100	1.943 429 6477	141.514 447 1483	.007 066 4163	.514 554 2578	72.816 861 3250	.013 733 0830	100

	XIV COMPOUND AMOUNT	XV AMOUNT OF ANNUITY	XVI SINKING FUND	XVII PRESENT VALUE	XVIII PRESENT VALUE OF ANNUITY	XIX AMORTIZATION	
PERIODS	AMOUNT OF 1	AMOUNT OF 1 PER PERIOD	SINKING FUND	PRESENT WORTH OF 1	PRESENT WORTH OF 1 PER PERIOD	PARTIAL PAYMENT	PERIODS
	How $1 left at compound interest will grow.	How $1 deposited periodically will grow.	Periodic deposit that will grow to $1 at future date.	What $1 due in the future is worth today.	What $1 payable periodically is worth today.	Annuity worth $1 today. Periodic payment necessary to pay off a loan of $1.	
1	1.007 500 0000	1.000 000 0000	1.000 000 0000	.992 555 8313	.992 555 8313	1.007 500 0000	1
2	1.015 056 2500	2.007 500 0000	.498 132 0050	.985 167 0782	1.977 722 9094	.505 632 0050	2
3	1.022 669 1719	3.022 556 2500	.330 845 7866	.977 833 3282	2.955 556 2377	.338 345 7866	3
4	1.030 339 1907	4.045 225 4219	.247 205 0123	.970 554 1719	3.926 110 4096	.254 705 0123	4
5	1.038 066 7346	5.075 564 6125	.197 022 4155	.963 329 2029	4.889 439 6125	.204 522 4155	5
6	1.045 852 2351	6.113 631 3471	.163 568 9074	.956 158 0178	5.845 597 6303	.171 068 9074	6
7	1.053 696 1269	7.159 483 5822	.139 674 8786	.949 040 2162	6.794 637 8464	.147 174 8786	7
8	1.061 598 8478	8.213 179 7091	.121 755 5241	.941 975 4006	7.736 613 2471	.129 255 5241	8
9	1.069 560 8392	9.274 778 5569	.107 819 2858	.934 963 1768	8.671 576 4239	.115 319 2858	9
10	1.077 582 5455	10.344 339 3961	.096 671 2287	.928 003 1532	9.599 579 5771	.104 171 2287	10
11	1.085 664 4146	11.421 921 9416	.087 550 9398	.921 094 9411	10.520 674 5182	.095 050 9398	11
12	1.093 806 8977	12.507 586 3561	.079 951 4768	.914 238 1550	11.434 912 6731	.087 451 4768	12
13	1.102 010 4494	13.601 393 2538	.073 521 8798	.907 432 4119	12.342 345 0850	.081 021 8798	13
14	1.110 275 5278	14.703 403 7032	.068 011 4632	.900 677 3319	13.243 022 4169	.075 511 4632	14
15	1.118 602 5942	15.813 679 2310	.063 236 3908	.893 972 5378	14.136 994 9547	.070 736 3908	15
16	1.126 992 1137	16.932 281 8252	.059 058 7855	.887 317 6554	15.024 312 6101	.066 558 7855	16
17	1.135 444 5545	18.059 273 9389	.055 373 2118	.880 712 3131	15.905 024 9232	.062 873 2118	17
18	1.143 960 3887	19.194 718 4934	.052 097 6643	.874 156 1420	16.779 181 0652	.059 597 6643	18
19	1.152 540 0916	20.338 678 8821	.049 167 4020	.867 648 7762	17.646 829 8414	.056 667 4020	19
20	1.161 184 1423	21.491 218 9738	.046 530 6319	.861 189 8523	18.508 019 6937	.054 030 6319	20
21	1.169 893 0234	22.652 403 1161	.044 145 4266	.854 779 0097	19.362 798 7034	.051 645 4266	21
22	1.178 667 2210	23.822 296 1394	.041 977 4817	.848 415 8905	20.211 214 5940	.049 477 4817	22
23	1.187 507 2252	25.000 963 3605	.039 998 4587	.842 100 1395	21.053 314 7335	.047 498 4587	23
24	1.196 413 5294	26.188 470 5857	.038 184 7423	.835 831 4040	21.889 146 1374	.045 684 7423	24
25	1.205 386 6309	27.384 884 1151	.036 516 4956	.829 609 3340	22.718 755 4714	.044 016 4956	25
26	1.214 427 0306	28.590 270 7459	.034 976 9335	.823 433 5821	23.542 189 0535	.042 476 9335	26
27	1.223 535 2333	29.804 697 7765	.033 551 7578	.817 303 8036	24.359 492 8571	.041 051 7578	27
28	1.232 711 7476	31.028 233 0099	.032 228 7125	.811 219 6562	25.170 712 5132	.039 728 7125	28
29	1.241 957 0857	32.260 944 7574	.030 997 2323	.805 180 8001	25.975 893 3134	.038 497 2323	29
30	1.251 271 7638	33.502 901 8431	.029 848 1608	.799 186 8984	26.775 080 2118	.037 348 1608	30
31	1.260 656 3021	34.754 173 6069	.028 773 5226	.793 237 6163	27.568 317 8281	.036 273 5226	31
32	1.270 111 2243	36.014 829 9090	.027 766 3397	.787 332 6216	28.355 650 4497	.035 266 3397	32
33	1.279 637 0585	37.284 941 1333	.026 820 4795	.781 471 5847	29.137 122 0344	.034 320 4795	33
34	1.289 234 3364	38.564 578 1918	.025 930 5313	.775 654 1784	29.912 776 2128	.033 430 5313	34
35	1.298 903 5940	39.853 812 5282	.025 091 7023	.769 880 0778	30.682 656 2907	.032 591 7023	35
36	1.308 645 3709	41.152 716 1222	.024 299 7327	.764 148 9606	31.446 805 2513	.031 799 7327	36
37	1.318 460 2112	42.461 361 4931	.023 550 8228	.758 460 5068	32.205 265 7581	.031 050 8228	37
38	1.328 348 6628	43.779 821 7043	.022 841 5732	.752 814 3988	32.958 080 1569	.030 341 5732	38
39	1.338 311 2778	45.108 170 3671	.022 168 9329	.747 210 3214	33.705 290 4783	.029 668 9329	39
40	1.348 348 6123	46.446 481 6449	.021 530 1561	.741 647 9617	34.446 938 4400	.029 030 1561	40
41	1.358 461 2269	47.794 830 2572	.020 922 7650	.736 127 0091	35.183 065 4492	.028 422 7650	41
42	1.368 649 6861	49.153 291 4841	.020 344 5175	.730 647 1555	35.913 712 6046	.027 844 5175	42
43	1.378 914 5588	50.521 941 1703	.019 793 3804	.725 208 0948	36.638 920 6994	.027 293 3804	43
44	1.389 256 4180	51.900 855 7290	.019 267 5051	.719 809 5233	37.358 730 2227	.026 767 5051	44
45	1.399 675 8411	53.290 112 1470	.018 765 2073	.714 451 1398	38.073 181 3625	.026 265 2073	45
46	1.410 173 4099	54.689 787 9881	.018 284 9493	.709 132 6449	38.782 314 0074	.025 784 9493	46
47	1.420 749 7105	56.099 961 3980	.017 825 3242	.703 853 7419	39.486 167 7493	.025 325 3242	47
48	1.431 405 3333	57.520 711 1085	.017 385 0424	.698 614 1359	40.184 781 8852	.024 885 0424	48
49	1.442 140 8733	58.952 116 4418	.016 962 9194	.693 413 5344	40.878 195 4195	.024 462 9194	49
50	1.452 956 9299	60.394 257 3151	.016 557 8657	.688 251 6470	41.566 447 0665	.024 057 8657	50
51	1.463 854 1068	61.847 214 2450	.016 168 8770	.683 128 1856	42.249 575 2521	.023 668 8770	51
52	1.474 833 0126	63.311 068 3518	.015 795 0265	.678 042 8641	42.927 618 1163	.023 295 0265	52
53	1.485 894 2602	64.785 901 3645	.015 435 4571	.672 995 3986	43.600 613 5149	.022 935 4571	53
54	1.497 038 4672	66.271 795 6247	.015 089 3754	.667 985 5073	44.268 599 0222	.022 589 3754	54
55	1.508 266 2557	67.768 834 0919	.014 756 0455	.663 012 9105	44.931 611 9327	.022 256 0455	55
56	1.519 578 2526	69.277 100 3476	.014 434 7843	.658 077 3305	45.589 689 2633	.021 934 7843	56
57	1.530 975 0895	70.796 678 6002	.014 124 9564	.653 178 4918	46.242 867 7551	.021 624 9564	57
58	1.542 457 4027	72.327 653 6897	.013 825 9704	.648 316 1209	46.891 183 8760	.021 325 9704	58
59	1.554 025 8332	73.870 111 0923	.013 537 2749	.643 489 9463	47.534 673 8224	.021 037 2749	59
60	1.565 681 0269	75.424 136 9255	.013 258 3552	.638 699 6986	48.173 373 5210	.020 758 3552	60
61	1.577 423 6346	76.989 817 9525	.012 988 7305	.633 945 1103	48.807 318 6312	.020 488 7305	61
62	1.589 224 3119	78.567 241 5871	.012 727 9510	.629 225 9159	49.436 544 5471	.020 227 9510	62
63	1.601 173 7192	80.156 495 8990	.012 475 5953	.624 541 8520	50.061 086 3991	.019 975 5953	63
64	1.613 182 5221	81.757 669 6183	.012 231 2684	.619 892 6571	50.680 979 0562	.019 731 2684	64
65	1.625 281 3911	83.370 852 1404	.011 994 5997	.615 278 0715	51.296 257 1278	.019 494 5997	65
66	1.637 471 0015	84.996 133 5315	.011 765 2411	.610 697 8387	51.906 954 9655	.019 265 2411	66
67	1.649 752 0340	86.633 604 5329	.011 542 8650	.606 151 7000	52.513 106 6655	.019 042 8650	67
68	1.662 125 1743	88.283 356 5669	.011 327 1633	.601 639 4045	53.114 746 0700	.018 827 1633	68
69	1.674 591 1131	89.945 481 7412	.011 117 8458	.597 160 6992	53.711 906 7692	.018 617 8458	69
70	1.687 150 5464	91.620 072 8543	.010 914 6388	.592 715 3342	54.304 622 1035	.018 414 6388	70
71	1.699 804 1755	93.307 223 4007	.010 717 2839	.588 303 0613	54.892 925 1647	.018 217 2839	71
72	1.712 552 7068	95.007 027 5762	.010 525 5372	.583 923 6340	55.476 848 7987	.018 025 5372	72
73	1.725 396 8521	96.719 580 2830	.010 339 1681	.579 576 8079	56.056 425 6067	.017 839 1681	73
74	1.738 337 3285	98.444 977 1351	.010 157 9586	.575 262 3404	56.631 687 9471	.017 657 9586	74
75	1.751 374 8585	100.183 314 4636	.009 981 7021	.570 979 9905	57.202 667 9375	.017 481 7021	75
76	1.764 510 1699	101.934 689 3221	.009 810 2030	.566 729 5191	57.769 397 4566	.017 310 2030	76
77	1.777 743 9962	103.699 199 4920	.009 643 2760	.562 510 6889	58.331 908 1455	.017 143 2760	77
78	1.791 077 0762	105.476 943 4882	.009 480 7450	.558 323 2644	58.890 231 4099	.016 980 7450	78
79	1.804 510 1542	107.268 020 5644	.009 322 4429	.554 167 0118	59.444 398 4218	.016 822 4429	79
80	1.818 043 9804	109.072 530 7186	.009 168 2112	.550 041 6991	59.994 440 1209	.016 668 2112	80
81	1.831 679 3102	110.890 574 6990	.009 017 8990	.545 947 0959	60.540 387 2168	.016 517 8990	81
82	1.845 416 9051	112.722 254 0092	.008 871 3627	.541 882 9736	61.082 270 1903	.016 371 3627	82
83	1.859 257 5319	114.567 670 9143	.008 728 4658	.537 849 1053	61.620 119 2956	.016 228 4658	83
84	1.873 201 9633	116.426 928 4462	.008 589 0783	.533 845 2658	62.153 964 5614	.016 089 0783	84
85	1.887 250 9781	118.300 130 4095	.008 453 0761	.529 871 2316	62.683 835 7930	.015 953 0761	85
86	1.901 405 3604	120.187 381 3876	.008 320 3410	.525 926 7807	63.209 762 5736	.015 820 3410	86
87	1.915 665 9006	122.088 786 7480	.008 190 7604	.522 011 6930	63.731 774 2666	.015 690 7604	87
88	1.930 033 3949	124.004 452 6486	.008 064 2266	.518 125 7499	64.249 900 0165	.015 564 2266	88
89	1.944 508 6453	125.934 486 0435	.007 940 6367	.514 268 7344	64.764 168 7509	.015 440 6367	89
90	1.959 092 4602	127.878 994 6888	.007 819 8926	.510 440 4311	65.274 609 1820	.015 319 8926	90
91	1.973 785 6536	129.838 087 1490	.007 701 9003	.506 640 6264	65.781 249 8085	.015 201 9003	91
92	1.988 589 0460	131.811 872 8026	.007 586 5700	.502 869 1081	66.284 118 9166	.015 086 5700	92
93	2.003 503 4639	133.800 461 8486	.007 473 8158	.499 125 6656	66.783 244 5822	.014 973 8158	93
94	2.018 529 7398	135.803 965 3125	.007 363 5552	.495 410 0903	67.278 654 6722	.014 863 5552	94
95	2.033 668 7129	137.822 495 0523	.007 255 7096	.491 722 1737	67.770 376 8458	.014 755 7096	95
96	2.048 921 2282	139.856 163 7652	.007 150 2033	.488 061 7108	68.258 438 5567	.014 650 2033	96
97	2.064 288 1375	141.905 084 9934	.007 046 9638	.484 428 4971	68.742 867 0538	.014 546 9638	97
98	2.079 770 2985	143.969 373 1309	.006 945 9217	.480 822 3296	69.223 689 3834	.014 445 9217	98
99	2.095 368 5757	146.049 143 4294	.006 847 0104	.477 243 0071	69.700 932 3905	.014 347 0104	99
100	2.111 083 8400	148.144 512 0051	.006 750 1657	.473 690 3296	70.174 622 7201	.014 250 1657	100

RATE 5/6%

PERIODS	XIV COMPOUND AMOUNT — AMOUNT OF 1 — How $1 left at compound interest will grow.	XV AMOUNT OF ANNUITY — AMOUNT OF 1 PER PERIOD — How $1 deposited periodically will grow.	XVI SINKING FUND — SINKING FUND — Periodic deposit that will grow to $1 at future date.	XVII PRESENT VALUE — PRESENT WORTH OF 1 — What $1 due in the future is worth today.	XVIII PRESENT VALUE OF ANNUITY — PRESENT WORTH OF 1 PER PERIOD — What $1 payable periodically is worth today.	XIX AMORTIZATION — PARTIAL PAYMENT — Annuity worth $1 today. Periodic payment necessary to pay off a loan of $1.	PERIODS
1	1.008 333 3333	1.000 000 0000	1.000 000 0000	.991 735 5372	.991 735 5372	1.008 333 3333	1
2	1.016 736 1111	2.008 333 3333	.497 925 3112	.983 539 3757	1.975 274 9129	.506 258 6445	2
3	1.025 208 9120	3.025 069 4444	.330 570 9235	.975 410 9511	2.950 685 8640	.338 904 2569	3
4	1.033 752 3196	4.050 278 3565	.246 896 6110	.967 349 7036	3.918 035 5677	.255 229 9444	4
5	1.042 366 9223	5.084 030 6761	.196 694 3285	.959 355 0780	4.877 390 6456	.205 027 6619	5
6	1.051 053 3133	6.126 397 5984	.163 228 0609	.951 426 5296	5.828 817 1692	.171 561 3942	6
7	1.059 812 0909	7.177 450 9117	.139 325 2301	.943 563 4945	6.772 380 6637	.147 658 5635	7
8	1.068 643 6584	8.237 263 0027	.121 399 5474	.935 765 4491	7.708 146 1127	.129 732 8807	8
9	1.077 549 2238	9.305 906 8610	.107 458 6298	.928 031 8503	8.636 177 9630	.115 791 9631	9
10	1.086 528 8007	10.383 456 0849	.096 307 0477	.920 362 1656	9.556 540 1286	.104 640 3810	10
11	1.095 583 2074	11.469 984 8856	.087 184 0730	.912 755 8667	10.469 295 9953	.095 517 4064	11
12	1.104 713 0674	12.565 568 0930	.079 582 5539	.905 212 4298	11.374 508 4251	.087 915 8872	12
13	1.113 919 0097	13.670 281 1604	.073 151 3850	.897 731 3353	12.272 239 7605	.081 484 7183	13
14	1.123 201 6681	14.784 200 1701	.067 639 7768	.890 312 0001	13.162 551 8285	.075 973 1102	14
15	1.132 561 6820	15.907 401 8382	.062 863 8171	.882 954 1171	14.045 505 9457	.071 197 1505	15
16	1.141 999 6960	17.039 963 5201	.058 685 5716	.875 656 9757	14.921 162 9213	.067 018 9050	16
17	1.151 516 3601	18.181 963 2161	.054 999 5613	.868 420 1411	15.789 583 0625	.063 332 8946	17
18	1.161 112 3298	19.333 479 5763	.051 723 7467	.861 243 1152	16.650 826 1777	.060 057 0800	18
19	1.170 788 2659	20.494 591 9061	.048 793 3600	.854 125 4035	17.504 951 5811	.057 126 6933	19
20	1.180 544 8348	21.665 380 1720	.046 156 5868	.847 066 5159	18.352 018 0970	.054 489 9201	20
21	1.190 382 7084	22.845 925 0067	.043 771 4822	.840 065 9661	19.192 084 0631	.052 104 8155	21
22	1.200 302 5643	24.036 307 7151	.041 603 7277	.833 123 2722	20.025 207 3354	.049 937 0610	22
23	1.210 305 0857	25.236 610 2794	.039 624 9730	.826 237 9559	20.851 445 2913	.047 958 3063	23
24	1.220 390 9614	26.446 915 3651	.037 811 5930	.819 409 5430	21.670 854 8343	.046 144 9263	24
25	1.230 560 8861	27.667 306 3264	.036 143 7427	.812 637 5634	22.483 492 3977	.044 477 0760	25
26	1.240 815 5601	28.897 867 2125	.034 604 6299	.805 921 5504	23.289 413 9481	.042 937 9632	26
27	1.251 156 2898	30.138 682 7726	.033 179 9504	.799 261 0418	24.088 674 9898	.041 513 2837	27
28	1.261 581 9872	31.389 838 4624	.031 857 4433	.792 655 5786	24.881 330 5684	.040 190 7767	28
29	1.272 095 1704	32.651 420 4496	.030 626 5389	.786 104 7060	25.667 435 2745	.038 959 8723	29
30	1.282 695 9635	33.923 515 6200	.029 478 0768	.779 607 9729	26.447 043 2474	.037 811 4102	30
31	1.293 385 0965	35.206 211 5835	.028 404 0786	.773 164 9318	27.220 208 1793	.036 737 4119	31
32	1.304 163 3057	36.499 596 6800	.027 397 5630	.766 775 1390	27.986 983 3183	.035 730 8963	32
33	1.315 031 3332	37.803 759 9857	.026 452 3952	.760 438 1544	28.747 421 4727	.034 785 7286	33
34	1.325 989 9277	39.118 791 3189	.025 563 1620	.754 153 5415	29.501 575 0142	.033 896 4953	34
35	1.337 039 8437	40.444 781 2465	.024 725 0688	.747 920 8677	30.249 495 8819	.033 058 4022	35
36	1.348 181 8424	41.781 821 0903	.023 933 8539	.741 739 7035	30.991 235 5853	.032 267 1872	36
37	1.359 416 6911	43.130 002 9327	.023 185 7160	.735 609 6233	31.726 845 2086	.031 519 0494	37
38	1.370 745 1635	44.489 419 6238	.022 477 2543	.729 530 2049	32.456 375 4135	.030 810 5877	38
39	1.382 168 0399	45.860 164 7873	.021 805 4166	.723 501 0296	33.179 876 4431	.030 138 7500	39
40	1.393 686 1069	47.242 332 8272	.021 167 4560	.717 521 6823	33.897 398 1254	.029 500 7893	40
41	1.405 300 1578	48.636 018 9341	.020 560 8934	.711 591 7510	34.608 989 8764	.028 894 2267	41
42	1.417 010 9924	50.041 319 0919	.019 983 4860	.705 710 8275	35.314 700 7039	.028 316 8193	42
43	1.428 819 4174	51.458 330 0843	.019 433 1996	.699 878 5066	36.014 579 2105	.027 766 5329	43
44	1.440 726 2458	52.887 149 5017	.018 908 1849	.694 094 3867	36.708 673 5972	.027 241 5182	44
45	1.452 732 2979	54.327 875 7475	.018 406 7569	.688 358 0694	37.397 031 6666	.026 740 0902	45
46	1.464 838 4004	55.780 608 0454	.017 927 3772	.682 669 1598	38.079 700 8264	.026 260 7105	46
47	1.477 045 3870	57.245 446 4458	.017 468 6383	.677 027 2659	38.756 728 0923	.025 801 9717	47
48	1.489 354 0986	58.722 491 8329	.017 029 2501	.671 431 9992	39.428 160 0915	.025 362 5834	48
49	1.501 765 3828	60.211 845 9315	.016 608 0276	.665 882 9745	40.094 043 0660	.024 941 3609	49
50	1.514 280 0943	61.713 611 3142	.016 203 8808	.660 379 8094	40.754 422 8754	.024 537 2141	50
51	1.526 899 0951	63.227 891 4085	.015 815 8050	.654 922 1250	41.409 345 0003	.024 149 1383	51
52	1.539 623 2542	64.754 790 5036	.015 442 8729	.649 509 5455	42.058 854 5458	.023 776 2062	52
53	1.552 453 4480	66.294 413 7578	.015 084 2272	.644 141 6980	42.702 996 2438	.023 417 5605	53
54	1.565 390 5600	67.846 867 2058	.014 739 0741	.638 818 2129	43.341 814 4566	.023 072 4074	54
55	1.578 435 4814	69.412 257 7658	.014 406 6773	.633 538 7235	43.975 353 1801	.022 740 0107	55
56	1.591 589 1104	70.990 693 2472	.014 086 3535	.628 302 8663	44.603 656 0464	.022 419 6868	56
57	1.604 852 3530	72.582 282 3576	.013 777 4670	.623 110 2806	45.226 766 3270	.022 110 8003	57
58	1.618 226 1226	74.187 134 7106	.013 479 4261	.617 960 6089	45.844 726 9359	.021 812 7594	58
59	1.631 711 3403	75.805 360 8332	.013 191 6992	.612 853 4964	46.457 580 4323	.021 525 0125	59
60	1.645 308 9348	77.437 072 1734	.012 913 7114	.607 788 5915	47.065 369 0238	.021 247 0447	60
61	1.659 019 8426	79.082 381 1082	.012 645 0416	.602 765 5453	47.668 134 5690	.020 978 3749	61
62	1.672 845 0079	80.741 400 9508	.012 385 2198	.597 784 0118	48.265 918 5808	.020 718 5532	62
63	1.686 785 3830	82.414 245 9587	.012 133 8245	.592 843 6481	48.858 762 2289	.020 467 1579	63
64	1.700 841 9278	84.101 031 3417	.011 890 4606	.587 944 1138	49.446 706 3427	.020 223 7939	64
65	1.715 015 6106	85.801 873 2695	.011 654 7572	.583 085 0715	50.029 791 4143	.019 988 0905	65
66	1.729 307 4073	87.516 888 8801	.011 426 3660	.578 266 1867	50.608 057 6009	.019 759 6993	66
67	1.743 718 3024	89.246 196 2875	.011 204 9593	.573 487 1273	51.181 544 7282	.019 538 2927	67
68	1.758 249 2882	90.989 914 5898	.010 990 2290	.568 747 5642	51.750 292 2924	.019 323 5624	68
69	1.772 901 3657	92.748 163 8781	.010 781 8846	.564 047 1711	52.314 339 4636	.019 115 2179	69
70	1.787 675 5437	94.521 065 2437	.010 579 6522	.559 385 6243	52.873 725 0878	.018 912 9856	70
71	1.802 572 8399	96.308 740 7874	.010 383 2735	.554 762 6026	53.428 487 6904	.018 716 6069	71
72	1.817 594 2802	98.111 313 6273	.010 192 5044	.550 177 7877	53.978 665 4781 ←	.018 525 8378	72
73	1.832 740 8992	99.928 907 9076	.010 007 1143	.545 630 8638	54.524 296 3419	.018 340 4476	73
74	1.848 013 7401	101.761 648 8068	.009 826 8848	.541 121 5178	55.065 417 8598	.018 160 2181	74
75	1.863 413 8546	103.609 662 5469	.009 651 6095	.536 649 4392	55.602 067 2989	.017 984 9428	75
76	1.878 942 3033	105.473 076 4014	.009 481 0926	.532 214 3198	56.134 281 6188	.017 814 4259	76
77	1.894 600 1559	107.352 018 7048	.009 315 1485	.527 815 8544	56.662 097 4732	.017 648 4819	77
78	1.910 388 4905	109.246 618 8606	.009 153 6014	.523 453 7399	57.185 551 2131	.017 486 9347	78
79	1.926 308 3946	111.157 007 3511	.008 996 2839	.519 127 6759	57.704 678 8890	.017 329 6173	79
80	1.942 360 9645	113.083 315 7457	.008 843 0375	.514 837 3646	58.219 516 2535	.017 176 3708	80
81	1.958 547 3059	115.025 676 7103	.008 693 7111	.510 582 5103	58.730 098 7638	.017 027 0444	81
82	1.974 868 5335	116.984 224 0162	.008 548 1612	.506 362 8201	59.236 461 5840	.016 881 4945	82
83	1.991 325 7712	118.959 092 5497	.008 406 2511	.502 178 0034	59.738 639 5874	.016 739 5844	83
84	2.007 920 1527	120.950 418 3209	.008 267 8507	.498 027 7720	60.236 667 3594	.016 601 1840	84
85	2.024 652 8206	122.958 338 4736	.008 132 8360	.493 911 8400	60.730 579 1994	.016 466 1693	85
86	2.041 524 9275	124.982 991 2942	.008 001 0887	.489 829 9240	61.220 409 1234	.016 334 4220	86
87	2.058 537 6352	127.024 516 2216	.007 872 4960	.485 781 7428	61.706 190 8662	.016 205 8294	87
88	2.075 692 1155	129.083 053 8568	.007 746 9503	.481 767 0176	62.187 957 8838	.016 080 2836	88
89	2.092 989 5498	131.158 745 9723	.007 624 3486	.477 785 4720	62.665 743 3558	.015 957 6819	89
90	2.110 431 1294	133.251 735 5221	.007 504 5927	.473 836 8318	63.139 580 1876	.015 837 9260	90
91	2.128 018 0554	135.362 166 6514	.007 387 5886	.469 920 8249	63.609 501 0125	.015 720 9219	91
92	2.145 751 5392	137.490 184 7069	.007 273 2465	.466 037 1817	64.075 538 1942	.015 606 5798	92
93	2.163 632 8021	139.635 936 2461	.007 161 4803	.462 185 6348	64.537 723 8290	.015 494 8136	93
94	2.181 663 0754	141.799 569 0481	.007 052 2076	.458 365 9188	64.996 089 7478	.015 385 5409	94
95	2.199 843 6010	143.981 232 1235	.006 945 3496	.454 577 7707	65.450 667 5184	.015 278 6830	95
96	2.218 175 6310	146.181 075 7246	.006 840 8308	.450 820 9296	65.901 488 4480	.015 174 1641	96
97	2.236 660 4280	148.399 251 3556	.006 738 5785	.447 095 1368	66.348 583 5848	.015 071 9118	97
98	2.255 299 2649	150.635 911 7836	.006 638 5232	.443 400 1357	66.791 983 7205	.014 971 8566	98
99	2.274 093 4254	152.891 211 0484	.006 540 5983	.439 735 6717	67.231 719 3922	.014 873 9317	99
100	2.293 044 2039	155.165 304 4738	.006 444 7397	.436 101 4926	67.667 820 8848	.014 778 0730	100

615

	XIV COMPOUND AMOUNT	XV AMOUNT OF ANNUITY	XVI SINKING FUND	XVII PRESENT VALUE	XVIII PRESENT VALUE OF ANNUITY	XIX AMORTIZATION	
PERIODS	AMOUNT OF 1 — How $1 left at compound interest will grow.	AMOUNT OF 1 PER PERIOD — How $1 deposited periodically will grow.	SINKING FUND — Periodic deposit that will grow to $1 at future date.	PRESENT WORTH OF 1 — What $1 due in the future is worth today.	PRESENT WORTH OF 1 PER PERIOD — What $1 payable periodically is worth today.	PARTIAL PAYMENT — Annuity worth $1 today. Periodic payment necessary to pay off a loan of $1.	PERIODS
1	1.010 000 0000	1.000 000 0000	1.000 000 0000	.990 099 0099	.990 099 0099	1.010 000 0000	1
2	1.020 100 0000	2.010 000 0000	.497 512 4378	.980 296 0494	1.970 395 0593	.507 512 4378	2
3	1.030 301 0000	3.030 100 0000	.330 022 1115	.970 590 1479	2.940 985 2072	.340 022 1115	3
4	1.040 604 0100	4.060 401 0000	.246 281 0939	.960 980 3445	3.901 965 5517	.256 281 0939	4
5	1.051 010 0501	5.101 005 0100	.196 039 7996	.951 465 6876	4.853 431 2393	.206 039 7996	5
6	1.061 520 1506	6.152 015 0601	.162 548 3667	.942 045 2353	5.795 476 4746	.172 548 3667	6
7	1.072 135 3521	7.213 535 2107	.138 628 2829	.932 718 0547	6.728 194 5293	.148 628 2829	7
8	1.082 856 7056	8.285 670 5628	.120 690 2920	.923 483 2225	7.651 677 7518	.130 690 2920	8
9	1.093 685 2727	9.368 527 2684	.106 740 3628	.914 339 8242	8.566 017 5760	.116 740 3628	9
10	1.104 622 1254	10.462 212 5411	.095 582 0766	.905 286 9547	9.471 304 5307	.105 582 0766	10
11	1.115 668 3467	11.566 834 6665	.086 454 0757	.896 323 7175	10.367 628 2482	.096 454 0757	11
12	1.126 825 0301	12.682 503 0132	.078 848 7887	.887 449 2253	11.255 077 4735	.088 848 7887	12
13	1.138 093 2804	13.809 328 0433	.072 414 8197	.878 662 5993	12.133 740 0728	.082 414 8197	13
14	1.149 474 2132	14.947 421 3238	.066 901 1717	.869 962 9696	13.003 703 0423	.076 901 1717	14
15	1.160 968 9554	16.096 895 5370	.062 123 7802	.861 349 4748	13.865 052 5172	.072 123 7802	15
16	1.172 578 6449	17.257 864 4924	.057 944 5968	.852 821 2622	14.717 873 7794	.067 944 5968	16
17	1.184 304 4314	18.430 443 1373	.054 258 0551	.844 377 4873	15.562 251 2667	.064 258 0551	17
18	1.196 147 4757	19.614 747 5687	.050 982 0479	.836 017 3142	16.398 268 5809	.060 982 0479	18
19	1.208 108 9504	20.810 895 0444	.048 051 7536	.827 739 9150	17.226 008 4959	.058 051 7536	19
20	1.220 190 0399	22.019 003 9948	.045 415 3149	.819 544 4703	18.045 552 9663	.055 415 3149	20
21	1.232 391 9403	23.239 194 0347	.043 030 7522	.811 430 1687	18.856 983 1349	.053 030 7522	21
22	1.244 715 8598	24.471 585 9751	.040 863 7185	.803 396 2066	19.660 379 3415	.050 863 7185	22
23	1.257 163 0183	25.716 301 8348	.038 885 8401	.795 441 7887	20.455 821 1302	.048 885 8401	23
24	1.269 734 6485	26.973 464 8532	.037 073 4722	.787 566 1274	21.243 387 2576	.047 073 4722	24
25	1.282 431 9950	28.243 199 5017	.035 406 7534	.779 768 4430	22.023 155 7006	.045 406 7534	25
26	1.295 256 3150	29.525 631 4967	.033 868 8776	.772 047 9634	22.795 203 6640	.043 868 8776	26
27	1.308 208 8781	30.820 887 8117	.032 445 5287	.764 403 9241	23.559 607 5881	.042 445 5287	27
28	1.321 290 9669	32.129 096 6898	.031 124 4356	.756 835 5684	24.316 443 1565	.041 124 4356	28
29	1.334 503 8766	33.450 387 6567	.029 895 0198	.749 342 1470	25.065 785 3035	.039 895 0198	29
30	1.347 848 9153	34.784 891 5333	.028 748 1132	.741 922 9178	25.807 708 2213	.038 748 1132	30
31	1.361 327 4045	36.132 740 4486	.027 675 7309	.734 577 1463	26.542 285 3676	.037 675 7309	31
32	1.374 940 6785	37.494 067 8531	.026 670 8857	.727 304 1053	27.269 589 4729	.036 670 8857	32
33	1.388 690 0853	38.869 008 5316	.025 727 4378	.720 103 0745	27.989 692 5474	.035 727 4378	33
34	1.402 576 9862	40.257 698 6170	.024 839 9694	.712 973 3411	28.702 665 8885	.034 839 9694	34
35	1.416 602 7560	41.660 275 6031	.024 003 6818	.705 914 1991	29.408 580 0876	.034 003 6818	35
36	1.430 768 7836	43.076 878 3592	.023 214 3098	.698 924 9496	30.107 505 0373	.033 214 3098	36
37	1.445 076 4714	44.507 647 1427	.022 468 0491	.692 004 9006	30.799 509 9379	.032 468 0491	37
38	1.459 527 2361	45.952 723 6142	.021 761 4958	.685 153 3670	31.484 663 3048	.031 761 4958	38
39	1.474 122 5085	47.412 250 8503	.021 091 5951	.678 369 6702	32.163 032 9751	.031 091 5951	39
40	1.488 863 7336	48.886 373 3588	.020 455 5980	.671 653 1389	32.834 686 1140	.030 455 5980	40
41	1.503 752 3709	50.375 237 0924	.019 851 0232	.665 003 1078	33.499 689 2217	.029 851 0232	41
42	1.518 789 8946	51.878 989 4633	.019 275 6260	.658 418 9186	34.158 108 1403	.029 275 6260	42
43	1.533 977 7936	53.397 779 3580	.018 727 3705	.651 899 9194	34.810 008 0597	.028 727 3705	43
44	1.549 317 5715	54.931 757 1515	.018 204 4058	.645 445 4648	35.455 453 5245	.028 204 4058	44
45	1.564 810 7472	56.481 074 7231	.017 705 0455	.639 054 9156	36.094 508 4401	.027 705 0455	45
46	1.580 458 8547	58.045 885 4703	.017 227 7499	.632 727 6392	36.727 236 0793	.027 227 7499	46
47	1.596 263 4432	59.626 344 3250	.016 771 1103	.626 463 0091	37.353 699 0884	.026 771 1103	47
48	1.612 226 0777	61.222 607 7682	.016 333 8354	.620 260 4051	37.973 959 4935	.026 333 8354	48
49	1.628 348 3385	62.834 833 8459	.015 914 7393	.614 119 2129	38.588 078 7064	.025 914 7393	49
50	1.644 631 8218	64.463 182 1844	.015 512 7309	.608 038 8247	39.196 117 5311	.025 512 7309	50
51	1.661 078 1401	66.107 814 0062	.015 126 8048	.602 018 6383	39.798 136 1694	.025 126 8048	51
52	1.677 688 9215	67.768 892 1463	.014 756 0329	.596 058 0577	40.394 194 2271	.024 756 0329	52
53	1.694 465 8107	69.446 581 0678	.014 399 5570	.590 156 4928	40.984 350 7199	.024 399 5570	53
54	1.711 410 4688	71.141 046 8784	.014 056 5826	.584 313 3592	41.568 664 0791	.024 056 5826	54
55	1.728 524 5735	72.852 457 3472	.013 726 3730	.578 528 0784	42.147 192 1576	.023 726 3730	55
56	1.745 809 8192	74.580 981 9207	.013 408 2440	.572 800 0776	42.719 992 2352	.023 408 2440	56
57	1.763 267 9174	76.326 791 7399	.013 101 5595	.567 128 7898	43.287 121 0250	.023 101 5595	57
58	1.780 900 5966	78.090 059 6573	.012 805 7272	.561 513 6532	43.848 634 6782	.022 805 7272	58
59	1.798 709 6025	79.870 960 2539	.012 520 1950	.555 954 1121	44.404 588 7903	.022 520 1950	59
60	1.816 696 6986	81.669 669 8564	.012 244 4477	.550 449 6159	44.955 038 4062	.022 244 4477	60
61	1.834 863 6655	83.486 366 5550	.011 978 0036	.544 999 6197	45.500 038 0260	.021 978 0036	61
62	1.853 212 3022	85.321 230 2205	.011 720 4123	.539 603 5839	46.039 641 6099	.021 720 4123	62
63	1.871 744 4252	87.174 442 5227	.011 471 2520	.534 260 9742	46.573 902 5840	.021 471 2520	63
64	1.890 461 8695	89.046 186 9480	.011 230 1271	.528 971 2615	47.102 873 8456	.021 230 1271	64
65	1.909 366 4882	90.936 648 8174	.010 996 6665	.523 733 9223	47.626 607 7679	.020 996 6665	65
66	1.928 460 1531	92.846 015 3056	.010 770 5215	.518 548 4379	48.145 156 2058	.020 770 5215	66
67	1.947 744 7546	94.774 475 4587	.010 551 3641	.513 414 2950	48.658 570 5008	.020 551 3641	67
68	1.967 222 2021	96.722 220 2133	.010 338 8859	.508 330 9851	49.166 901 4860	.020 338 8859	68
69	1.986 894 4242	98.689 442 4154	.010 132 7961	.503 298 0051	49.670 199 4911	.020 132 7961	69
70	2.006 763 3684	100.676 336 8395	.009 932 8207	.498 314 8565	50.168 514 3476	.019 932 8207	70
71	2.026 831 0021	102.683 100 2079	.009 738 7009	.493 381 0461	50.661 895 3936	.019 738 7009	71
72	2.047 099 3121	104.709 931 2100	.009 550 1925	.488 496 0852	51.150 391 4789	.019 550 1925	72
73	2.067 570 3052	106.757 030 5221	.009 367 0646	.483 659 4903	51.634 050 9692	.019 367 0646	73
74	2.088 246 0083	108.824 600 8273	.009 189 0987	.478 870 7825	52.112 921 7516	.019 189 0987	74
75	2.109 128 4684	110.912 846 8356	.009 016 0881	.475 129 4876	52.587 051 2393	.019 016 0881	75
76	2.130 219 7530	113.021 975 3040	.008 847 8369	.469 435 1362	53.056 486 3755	.018 847 8369	76
77	2.151 521 9506	115.152 195 0570	.008 684 1593	.464 787 2636	53.521 273 6391	.018 684 1593	77
78	2.173 037 1701	117.303 717 0076	.008 524 8791	.460 185 4095	53.981 459 0486	.018 524 8791	78
79	2.194 767 5418	119.476 754 1776	.008 369 8290	.455 629 1183	54.437 088 1670	.018 369 8290	79
80	2.216 715 2172	121.671 521 7194	.008 218 8501	.451 117 9389	54.888 206 1059	.018 218 8501	80
81	2.238 882 3694	123.888 236 9366	.008 071 7914	.446 651 4247	55.334 857 5306	.018 071 7914	81
82	2.261 271 1931	126.127 119 3060	.007 928 5090	.442 229 1334	55.777 086 6639	.017 928 5090	82
83	2.283 883 9050	128.388 390 4990	.007 788 8662	.437 850 6271	56.214 937 2910	.017 788 8662	83
84	2.306 722 7440	130.672 274 4040	.007 652 7328	.433 515 4724	56.648 452 7634	.017 652 7328	84
85	2.329 789 9715	132.978 997 1481	.007 519 9845	.429 223 2400	57.077 676 0034	.017 519 9845	85
86	2.353 087 8712	135.308 787 1196	.007 390 5030	.424 973 5049	57.502 649 5083	.017 390 5030	86
87	2.376 618 7499	137.661 874 9908	.007 264 1754	.420 765 8465	57.923 415 3547	.017 264 1754	87
88	2.400 384 9374	140.038 493 7407	.007 140 8937	.416 599 8480	58.340 015 2027	.017 140 8937	88
89	2.424 388 7868	142.438 878 6781	.007 020 5551	.412 475 0970	58.752 490 2997	.017 020 5551	89
90	2.448 632 6746	144.863 267 4648	.006 903 0612	.408 391 1852	59.160 881 4849	.016 903 0612	90
91	2.473 119 0014	147.311 900 1395	.006 788 3178	.404 347 7081	59.565 229 1929	.016 788 3178	91
92	2.497 850 1914	149.785 019 1409	.006 676 2351	.400 344 2654	59.965 573 4584	.016 676 2351	92
93	2.522 828 6933	152.282 869 3323	.006 566 7268	.396 380 4608	60.361 953 9192	.016 566 7268	93
94	2.548 056 9803	154.805 698 0256	.006 459 7105	.392 455 9018	60.754 409 8210	.016 459 7105	94
95	2.573 537 5501	157.353 755 0059	.006 355 1073	.388 570 1998	61.142 980 0207	.016 355 1073	95
96	2.599 272 9256	159.927 292 5559	.006 252 8414	.384 722 9701	61.527 702 9908	.016 252 8414	96
97	2.625 265 6548	162.526 565 4815	.006 152 8403	.380 913 8318	61.908 616 8226	.016 152 8403	97
98	2.651 518 3114	165.151 831 1363	.006 055 0343	.377 142 4077	62.285 759 2303	.016 055 0343	98
99	2.678 033 4945	167.803 349 4477	.005 959 3566	.373 408 3245	62.659 167 5548	.015 959 3566	99
100	2.704 813 8294	170.481 382 9422	.005 865 7431	.369 711 2123	63.028 878 7671	.015 865 7431	100

RATE 1 1/6% PERIODS	AMOUNT OF 1 — How $1 left at compound interest will grow.	AMOUNT OF 1 PER PERIOD — How $1 deposited periodically will grow.	SINKING FUND — Periodic deposit that will grow to $1 at future date.	PRESENT WORTH OF 1 — What $1 due in the future is worth today.	PRESENT WORTH OF 1 PER PERIOD — What $1 payable periodically is worth today.	PARTIAL PAYMENT — Annuity worth $1 today. Periodic payment necessary to pay off a loan of $1.	PERIODS
1	1.011 666 6667	1.000 000 0000	1.000 000 0000	.988 467 8748	.988 467 8748	1.011 666 6667	1
2	1.023 469 4444	2.011 766 6667	.497 100 2486	.977 068 7395	1.965 536 6143	.508 766 9152	2
3	1.035 409 9213	3.035 136 1111	.329 474 5156	.965 801 0605	2.931 337 6748	.341 141 1823	3
4	1.047 489 7037	4.070 546 0324	.245 667 2869	.954 663 3217	3.886 000 9965	.257 333 9536	4
5	1.059 710 4169	5.118 035 7361	.195 387 4595	.943 654 0248	4.829 655 0212	.207 054 1261	5
6	1.072 073 7051	6.177 746 1530	.161 871 3322	.932 771 6884	5.762 426 7096	.173 537 9989	6
7	1.084 568 2317	7.249 819 8582	.137 934 4618	.922 014 8485	6.684 441 5581	.149 601 1284	7
8	1.097 234 6794	8.334 401 0898	.119 984 6263	.911 382 0578	7.595 823 6159	.131 651 2929	8
9	1.110 035 7506	9.431 635 7692	.106 026 1469	.900 871 8858	8.496 695 5017	.117 692 8136	9
10	1.122 986 1677	10.541 671 5199	.094 861 6164	.890 482 9184	9.387 178 4202	.106 528 2831	10
11	1.136 087 6730	11.664 657 6876	.085 729 0481	.880 213 7579	10.267 392 1781	.097 395 7148	11
12	1.149 342 0292	12.800 745 3606	.078 120 4509	.870 063 0227	11.137 455 2007	.089 787 1176	12
13	1.162 751 0195	13.950 087 3898	.071 684 1387	.860 029 3169	11.997 484 5477	.083 350 0054	13
14	1.176 316 4481	15.112 838 4094	.066 168 9071	.850 111 3808	12.847 595 9285	.077 835 5737	14
15	1.190 040 1400	16.289 154 8575	.061 390 5392	.840 307 7900	13.687 903 7185	.073 057 2059	15
16	1.203 923 9416	17.479 194 9975	.057 210 8727	.830 617 2553	14.518 520 9738	.068 877 5394	16
17	1.217 969 7210	18.683 118 9391	.053 524 2538	.821 038 4731	15.339 559 4469	.065 190 9205	17
18	1.232 179 3677	19.901 088 6601	.050 248 5074	.811 570 1546	16.151 129 6015	.061 915 1740	18
19	1.246 554 7937	21.133 268 0278	.047 318 7582	.802 211 0260	16.953 340 6275	.058 985 4249	19
20	1.261 097 9329	22.379 822 8214	.044 683 1062	.792 959 8280	17.746 300 4556	.056 349 7729	20
21	1.275 810 7421	23.640 920 7544	.042 299 5369	.783 815 3160	18.530 115 7716	.053 966 2036	21
22	1.290 695 2008	24.916 731 4965	.040 133 6748	.774 776 2596	19.304 892 0312	.051 800 3415	22
23	1.305 753 3115	26.207 426 6973	.038 157 1228	.765 841 4428	20.070 733 4740	.049 823 7895	23
24	1.320 987 1001	27.513 180 0087	.036 346 2166	.757 009 6634	20.827 743 1374	.048 012 8833	24
25	1.336 398 6163	28.834 167 1088	.034 681 0780	.748 279 7332	21.576 022 8706	.046 347 7447	25
26	1.351 989 9335	30.170 565 7251	.033 144 8873	.739 650 4776	22.315 673 3482	.044 811 5539	26
27	1.367 763 1494	31.522 555 6586	.031 723 3162	.731 120 7357	23.046 794 0839	.043 389 9828	27
28	1.383 720 3861	32.890 318 8079	.030 404 0835	.722 689 3598	23.769 483 4437	.042 070 7502	28
29	1.399 863 7906	34.274 039 1940	.029 176 6020	.714 355 2156	24.483 838 6593	.040 843 2687	29
30	1.416 195 5348	35.673 902 9846	.028 031 6959	.706 117 1819	25.189 955 8412	.039 698 3626	30
31	1.432 717 8161	37.090 098 5194	.026 961 3735	.697 974 1501	25.887 929 9913	.038 628 0402	31
32	1.449 432 8572	38.522 816 3355	.025 958 6420	.689 925 0248	26.577 855 0161	.037 625 3087	32
33	1.466 342 9072	39.972 249 1927	.025 017 3563	.681 968 7230	27.259 823 7391	.036 684 0230	33
34	1.483 450 2412	41.438 592 1000	.024 132 0940	.674 104 1743	27.933 927 9135	.035 798 7607	34
35	1.500 757 1606	42.922 042 3412	.023 298 0526	.666 330 3206	28.600 258 2341	.034 964 7193	35
36	1.518 265 9942	44.422 799 5018	.022 510 9631	.658 646 1159	29.258 904 3500	.034 177 6298	36
37	1.535 979 0975	45.941 065 4960	.021 767 0180	.651 050 5264	29.909 954 8764	.033 433 6847	37
38	1.553 898 8536	47.477 044 5934	.021 062 8106	.643 542 5303	30.553 497 4067	.032 729 4773	38
39	1.572 027 6735	49.030 943 4470	.020 395 2837	.636 121 1172	31.189 618 5239	.032 061 9503	39
40	1.590 367 9964	50.602 971 1206	.019 761 6855	.628 785 2889	31.818 403 8128	.031 428 3522	40
41	1.608 922 2897	52.193 339 1170	.019 159 5329	.621 534 0582	32.439 937 8709	.030 826 1996	41
42	1.627 693 0497	53.802 261 4067	.018 586 5793	.614 366 4496	33.054 304 3205	.030 253 2460	42
43	1.646 682 8020	55.429 954 4564	.018 040 7870	.607 281 4988	33.661 585 8193	.029 707 4536	43
44	1.665 894 1013	57.076 637 2584	.017 520 3034	.600 278 2525	34.261 864 0718	.029 186 9700	44
45	1.685 329 5325	58.742 531 3598	.017 023 4407	.593 355 7685	34.855 219 8403	.028 690 1074	45
46	1.704 991 7104	60.427 860 8923	.016 548 6579	.586 513 1155	35.441 732 9559	.028 215 3246	46
47	1.724 883 2804	62.132 852 6027	.016 094 5451	.579 749 3728	36.021 482 3287	.027 761 2118	47
48	1.745 006 9186	63.857 735 8831	.015 659 8098	.573 063 6305	36.594 545 9592	.027 326 4765	48
49	1.765 365 3327	65.602 742 8017	.015 243 2651	.566 454 9889	37.161 000 9481	.026 909 9318	49
50	1.785 961 2616	67.368 108 1344	.014 843 0109	.559 922 5591	37.720 923 5072	.026 510 4856	50
51	1.806 797 4763	69.154 069 3960	.014 460 4650	.553 465 4620	38.274 388 9692	.026 127 1317	51
52	1.827 876 7802	70.960 866 8723	.014 092 2743	.547 082 8290	38.821 471 7982	.025 758 9410	52
53	1.849 202 0093	72.788 743 6524	.013 738 3880	.540 773 8013	39.362 245 5996	.025 405 0546	53
54	1.870 776 0327	74.637 945 6617	.013 398 0108	.534 537 5302	39.896 783 1297	.025 064 6774	54
55	1.892 601 7531	76.508 721 6944	.013 070 4053	.528 373 1764	40.425 156 3062	.024 737 0719	55
56	1.914 682 1069	78.401 323 4475	.012 754 8867	.522 279 9108	40.947 436 2170	.024 421 5534	56
57	1.937 020 0648	80.316 005 5544	.012 450 8184	.516 256 9135	41.463 693 1304	.024 117 4851	57
58	1.959 618 6322	82.253 025 6192	.012 157 6075	.510 303 3741	41.973 996 5045	.023 824 2742	58
59	1.982 480 8496	84.212 644 2514	.011 874 7013	.504 418 4917	42.478 414 9963	.023 541 3680	59
60	2.005 609 7928	86.195 125 1010	.011 601 5842	.498 601 4745	42.977 016 4708	.023 268 2508	60
61	2.029 008 5738	88.200 734 8939	.011 337 7740	.492 851 5399	43.469 868 0106	.023 004 4407	61
62	2.052 680 3405	90.229 743 4677	.011 082 8199	.487 167 9142	43.957 035 9248	.022 749 4866	62
63	2.076 628 2778	92.282 423 8081	.010 836 7577	.481 549 8328	44.438 585 7575	.022 502 9664	63
64	2.100 855 6077	94.359 052 0859	.010 597 8174	.475 996 5399	44.914 582 2975	.022 264 4840	64
65	2.125 365 5898	96.459 907 6935	.010 367 0014	.470 507 2882	45.385 089 5857	.022 033 6681	65
66	2.150 161 5216	98.585 273 2833	.010 143 5028	.465 081 3392	45.850 170 9249	.021 810 1695	66
67	2.175 246 7394	100.735 434 8049	.009 926 9934	.459 717 9630	46.309 888 8879	.021 593 6601	67
68	2.200 624 6180	102.910 681 5443	.009 717 1643	.454 416 4379	46.764 305 3258	.021 383 8310	68
69	2.226 298 5719	105.111 306 1623	.009 513 7244	.449 176 0506	47.213 481 3764	.021 180 3911	69
70	2.252 272 0552	107.337 604 7342	.009 316 3994	.443 996 0962	47.657 477 4725	.020 983 0661	70
71	2.278 548 5625	109.589 876 7895	.009 124 9304	.438 875 8776	48.096 353 3501	.020 791 5971	71
72	2.305 131 6291	111.868 425 3520	.008 939 0728	.433 814 7060	48.530 168 0561	.020 605 7395	72
73	2.332 024 8314	114.173 556 9811	.008 758 5955	.428 811 9005	48.958 979 9566	.020 425 2621	73
74	2.359 231 7878	116.505 581 8126	.008 583 2797	.423 866 7880	49.382 846 7446	.020 249 9464	74
75	2.386 756 1587	118.864 813 6004	.008 412 9186	.418 978 7031	49.801 825 4477	.002 079 5853	75
76	2.414 601 6472	121.251 569 7591	.008 247 3159	.414 146 9882	50.215 972 4360	.019 913 9826	76
77	2.442 771 9997	123.666 171 4062	.008 086 2858	.409 370 9933	50.625 343 4293	.019 752 9524	77
78	2.471 271 0064	126.108 943 4060	.007 929 6517	.404 650 0758	51.029 993 5051	.019 596 3184	78
79	2.500 102 5015	128.580 214 4124	.007 777 2463	.399 983 6004	51.429 977 0155	.019 443 9130	79
80	2.529 270 3640	131.080 316 9139	.007 628 9105	.395 370 9395	51.825 348 0450	.019 295 5771	80
81	2.558 778 5182	133.609 587 2779	.007 484 4928	.390 811 4723	52.216 159 5173	.019 151 1595	81
82	2.588 630 9343	136.168 365 7961	.007 343 8496	.386 304 5855	52.602 464 1027	.019 010 5163	82
83	2.618 831 6285	138.756 996 7304	.007 206 8438	.381 849 6726	52.984 313 7754	.018 873 5105	83
84	2.649 384 6642	141.375 828 3589	.007 073 3449	.377 446 1344	53.361 759 9097	.018 740 0116	84
85	2.680 294 1519	144.025 213 0231	.006 943 2288	.373 093 3783	53.734 853 2880	.018 609 8954	85
86	2.711 564 2504	146.705 507 1750	.006 816 3767	.368 790 8188	54.103 644 1068	.018 483 0434	86
87	2.743 199 1666	149.417 071 4254	.006 692 6757	.364 537 8769	54.468 181 9837	.018 359 3423	87
88	2.775 203 1569	152.160 270 5920	.006 572 0178	.360 333 9804	54.828 515 9641	.018 238 6844	88
89	2.807 580 5271	154.935 473 7489	.006 454 2998	.356 178 5638	55.184 694 5279	.018 120 9665	89
90	2.840 335 6332	157.743 054 2760	.006 339 4233	.352 071 0680	55.536 765 5960	.018 006 0900	90
91	2.873 472 8823	160.583 389 9092	.006 227 2941	.348 010 9404	55.884 776 5364	.017 893 9608	91
92	2.906 996 7326	163.456 862 7915	.006 117 8221	.343 997 6347	56.228 774 1710	.017 784 4887	92
93	2.940 911 6944	166.363 859 5241	.006 010 8209	.340 030 6109	56.568 804 7819	.017 677 5876	93
94	2.975 222 3309	169.304 771 2185	.005 906 5081	.336 109 3353	56.904 914 1172	.017 573 1748	94
95	3.009 933 2581	172.279 993 5494	.005 804 5045	.332 233 2804	57.237 147 3976	.017 471 1712	95
96	3.045 049 1461	175.289 926 8075	.005 704 8344	.328 401 9246	57.565 549 3222	.017 371 5010	96
97	3.080 574 7195	178.334 975 9536	.005 607 4250	.324 614 7525	57.890 164 0746	.017 274 0917	97
98	3.116 514 7579	181.415 550 6731	.005 512 2066	.320 871 2545	58.211 035 3291	.017 178 8733	98
99	3.152 874 0967	184.532 065 4309	.005 419 1124	.317 170 9270	58.528 206 2561	.017 085 7790	99
100	3.189 657 6278	187.684 939 5276	.005 328 0780	.313 513 2722	58.841 719 5283	.016 994 7447	100

617

	XIV COMPOUND AMOUNT	XV AMOUNT OF ANNUITY	XVI SINKING FUND	XVII PRESENT VALUE	XVIII PRESENT VALUE OF ANNUITY	XIX AMORTIZATION	
PERIODS	AMOUNT OF 1	AMOUNT OF 1 PER PERIOD	SINKING FUND	PRESENT WORTH OF 1	PRESENT WORTH OF 1 PER PERIOD	PARTIAL PAYMENT	PERIODS
	How $1 left at compound interest will grow.	How $1 deposited periodically will grow.	Periodic deposit that will grow to $1 at future date.	What $1 due in the future is worth today.	What $1 payable periodically is worth today.	Annuity worth $1 today. Periodic payment necessary to pay off a loan of $1.	
1	1.012 500 0000	1.000 000 0000	1.000 000 0000	.987 654 3210	.987 654 3210	1.012 500 0000	1
2	1.025 156 2500	2.012 500 0000	.496 894 4099	.975 461 0578	1.963 115 3788	.509 394 4099	2
3	1.037 970 7031	3.037 656 2500	.329 201 1728	.963 418 3287	2.926 533 7074	.341 701 1728	3
4	1.050 945 3369	4.075 626 9531	.245 361 0233	.951 524 2752	3.878 057 9826	.257 861 0233	4
5	1.064 082 1536	5.126 572 2900	.195 062 1084	.939 777 0619	4.817 835 0446	.207 562 1084	5
6	1.077 383 1805	6.190 654 4437	.161 533 8102	.928 174 8760	5.746 009 9206	.174 033 8102	6
7	1.090 850 4703	7.268 037 6242	.137 588 7209	.916 715 9269	6.662 725 8475	.150 088 7209	7
8	1.104 486 1012	8.358 888 0945	.119 633 1365	.905 398 4463	7.568 124 2938	.132 133 1365	8
9	1.118 292 1774	9.463 374 1957	.105 670 5546	.894 220 6877	8.462 344 9815	.118 170 5546	9
10	1.132 270 8297	10.581 666 3731	.094 503 0740	.883 180 9262	9.345 525 9077	.107 003 0740	10
11	1.146 424 2150	11.713 937 2028	.085 368 3935	.872 277 4579	10.217 803 3656	.097 868 3935	11
12	1.160 754 5177	12.860 361 4178	.077 758 3123	.861 508 6004	11.079 311 9660	.090 258 3123	12
13	1.175 263 9492	14.021 115 9356	.071 320 9993	.850 872 6918	11.930 184 6578	.083 820 9993	13
14	1.189 954 7486	15.196 379 8848	.065 805 1462	.840 368 0906	12.770 552 7485	.078 305 1462	14
15	1.204 829 1829	16.386 334 6333	.061 026 4603	.829 993 1759	13.600 545 9244	.073 526 4603	15
16	1.219 889 5477	17.591 163 8162	.056 846 7221	.819 746 3466	14.420 292 2710	.069 346 7221	16
17	1.235 138 1670	18.811 053 3639	.053 160 2341	.809 626 0213	15.229 918 2924	.065 660 2341	17
18	1.250 577 3941	20.046 191 5310	.049 884 7873	.799 630 6384	16.029 548 9307	.062 384 7873	18
19	1.266 209 6116	21.296 768 9251	.046 955 4797	.789 758 6552	16.819 307 5859	.059 455 4797	19
20	1.282 037 2317	22.562 978 5367	.044 320 3896	.780 008 5483	17.599 316 1342	.056 820 3896	20
21	1.298 062 6971	23.845 015 7684	.041 937 4854	.770 378 8132	18.369 694 9474	.054 437 4854	21
22	1.314 288 4808	25.143 078 4655	.039 772 3772	.760 867 9636	19.130 562 9110	.052 272 3772	22
23	1.330 717 0868	26.457 366 9463	.037 796 6561	.751 474 5320	19.882 037 4430	.050 296 6561	23
24	1.347 351 0504	27.788 084 0331	.035 986 6480	.742 197 0686	20.624 234 5116	.048 486 6480	24
25	1.364 192 9385	29.135 435 0836	.034 322 4667	.733 034 1418	21.357 268 6533	.046 822 4667	25
26	1.381 245 3503	30.499 628 0221	.032 787 2851	.723 984 3376	22.081 252 9910	.045 287 2851	26
27	1.398 510 9172	31.880 873 3724	.031 366 7693	.715 046 2594	22.796 299 2504	.043 866 7693	27
28	1.415 992 3036	33.279 384 2895	.030 048 6329	.706 218 5278	23.502 517 7784	.042 548 6329	28
29	1.433 692 2074	34.695 376 5932	.028 822 2841	.697 499 7805	24.200 017 5587	.041 322 2841	29
30	1.451 629 0680	36.129 068 8006	.027 678 5434	.688 888 6721	24.888 906 2308	.040 178 5434	30
31	1.469 758 5270	37.580 682 1606	.026 609 4159	.680 383 8737	25.569 290 1045	.039 109 4159	31
32	1.488 130 5086	39.050 440 6876	.025 607 9056	.671 984 0728	26.241 274 1773	.038 107 9056	32
33	1.506 732 1400	40.538 571 1962	.024 667 8650	.663 687 9731	26.904 962 1504	.037 167 8650	33
34	1.525 566 2917	42.045 303 3361	.023 783 8693	.655 494 2944	27.560 456 4448	.036 283 8693	34
35	1.544 635 8703	43.570 869 6278	.022 951 1141	.647 401 7723	28.207 858 2171	.035 451 1141	35
36	1.563 943 8187	45.115 505 4982	.022 165 3285	.639 409 1578	28.847 267 3749	.034 665 3285	36
37	1.583 493 1165	46.679 449 3169	.021 422 7035	.631 515 2176	29.478 782 5925	.033 922 7035	37
38	1.603 286 7804	48.262 942 4334	.020 719 8308	.623 718 7334	30.102 501 3259	.033 219 8308	38
39	1.623 327 8652	49.866 229 2138	.020 053 6519	.616 018 5021	30.718 519 8281	.032 553 6519	39
40	1.643 619 4635	51.489 557 0790	.019 421 4139	.608 413 3355	31.326 933 1635	.031 921 4139	40
41	1.664 164 7068	53.133 176 5424	.018 820 6327	.600 902 0597	31.927 835 2233	.031 320 6327	41
42	1.684 966 7656	54.797 341 2492	.018 249 0606	.593 483 5158	32.521 318 7390	.030 749 0606	42
43	1.706 028 8502	56.482 308 0148	.017 704 6589	.586 156 5588	33.107 475 2978	.030 204 6589	43
44	1.727 354 2108	58.188 336 8650	.017 185 5745	.578 920 0581	33.686 395 3558	.029 685 5745	44
45	1.748 946 1384	59.915 691 0758	.016 690 1188	.571 772 8968	34.258 168 2527	.029 190 1188	45
46	1.770 807 9652	61.664 637 2143	.016 216 7499	.564 713 9722	34.822 882 2249	.028 716 7499	46
47	1.792 943 0647	63.435 445 1795	.015 764 0574	.557 742 1948	35.380 624 4196	.028 264 0574	47
48	1.815 354 8531	65.228 388 2442	.015 330 7483	.550 856 4886	35.931 480 9083	.027 830 7483	48
49	1.838 046 7887	67.043 743 0973	.014 915 6350	.544 055 7913	36.475 536 6995	.027 415 6350	49
50	1.861 022 3736	68.881 789 8860	.014 517 6251	.537 339 0531	37.012 875 7526	.027 017 6251	50
51	1.884 285 1532	70.742 812 2596	.014 135 7117	.530 705 2376	37.543 580 9902	.026 635 7117	51
52	1.907 838 7177	72.627 097 4128	.013 768 9655	.524 153 3211	38.067 734 3114	.026 268 9655	52
53	1.931 686 7016	74.534 936 1305	.013 416 5272	.517 682 2925	38.585 416 6038	.025 916 5272	53
54	1.955 832 7854	76.466 622 8321	.013 077 6012	.511 291 1530	39.096 707 7568	.025 577 6012	54
55	1.980 280 6952	78.422 455 6175	.012 751 4497	.504 978 9166	39.601 686 6734	.025 251 4497	55
56	2.005 034 2039	80.402 736 3127	.012 437 3877	.498 744 6090	40.100 431 2824	.024 937 3877	56
57	2.030 097 1315	82.407 770 5166	.012 134 7780	.492 587 2681	40.593 018 5505	.024 634 7780	57
58	2.055 473 3456	84.437 867 6481	.011 843 0276	.486 505 9438	41.079 524 4943	.024 343 0276	58
59	2.081 166 7624	86.493 340 9937	.011 561 5837	.480 499 6976	41.560 024 1919	.024 061 5837	59
60	2.107 181 3470	88.574 507 7561	.011 289 9301	.474 567 6026	42.034 591 7945	.023 789 9301	60
61	2.133 521 1138	90.681 689 1031	.011 027 5846	.468 708 7433	42.503 300 5378	.023 527 5846	61
62	2.160 190 1277	92.815 210 2168	.010 774 0962	.462 922 2156	42.966 222 7534	.023 274 0962	62
63	2.187 192 5043	94.975 400 3445	.010 529 0422	.457 207 1265	43.423 429 8799	.023 029 0422	63
64	2.214 532 4106	97.162 592 8489	.010 292 0267	.451 562 5941	43.874 992 4739	.022 792 0267	64
65	2.242 214 0657	99.377 125 2595	.010 062 6779	.445 987 7412	44.320 980 2012	.022 562 6779	65
66	2.270 241 7416	101.619 339 3252	.009 840 6465	.440 481 7257	44.761 461 9468	.022 340 6465	66
67	2.298 619 7633	103.889 581 0668	.009 625 6043	.435 043 6797	45.196 505 6265	.022 125 6043	67
68	2.327 352 5104	106.188 200 8301	.009 417 2421	.429 672 7700	45.626 178 3966	.021 917 2421	68
69	2.356 444 4168	108.515 553 3405	.009 215 2689	.424 368 1679	46.050 546 5645	.021 715 2689	69
70	2.385 899 9722	110.871 997 7572	.009 019 4100	.419 129 0548	46.469 675 6193	.021 519 4100	70
71	2.415 723 7216	113.257 897 7292	.008 829 4063	.413 954 6220	46.883 630 2412	.021 329 4063	71
72	2.445 920 2681	115.673 621 4508	.008 645 0133	.408 844 0711	47.292 474 3123	.021 145 0133	72
73	2.476 494 2715	118.119 541 7190	.008 456 9997	.403 796 6134	47.696 270 9258	.020 965 9997	73
74	2.507 450 4499	120.596 035 9904	.008 292 1465	.398 811 4701	48.095 082 3958	.020 792 1465	74
75	2.538 793 5805	123.103 486 4403	.008 123 2468	.393 887 8717	48.488 970 2675	.020 623 2468	75
76	2.570 528 5003	125.642 280 0208	.007 959 1042	.389 025 0584	48.877 995 3259	.020 459 1042	76
77	2.602 660 1065	128.212 808 5211	.007 799 5328	.384 222 2799	49.262 217 6058	.020 299 5328	77
78	2.635 193 3578	130.815 468 6276	.007 644 3559	.379 478 7950	49.641 696 4008	.020 144 3559	78
79	2.668 133 2748	133.450 661 9854	.007 493 4061	.374 793 8716	50.016 490 2724	.019 993 4061	79
80	2.701 484 9408	136.118 795 2603	.007 346 5240	.370 166 7868	50.386 657 0592	.019 846 5240	80
81	2.735 253 5025	138.820 280 2010	.007 203 5584	.365 596 8264	50.752 253 8856	.019 703 5584	81
82	2.769 444 1713	141.555 533 7035	.007 064 3653	.361 083 2854	51.113 337 1710	.019 564 3653	82
83	2.804 062 2234	144.324 977 8748	.006 928 8076	.356 625 4670	51.469 962 6380	.019 428 8076	83
84	2.839 113 0012	147.129 040 0983	.006 796 7547	.352 222 6835	51.822 185 3215	.019 296 7547	84
85	2.874 601 9137	149.968 153 0995	.006 668 0824	.347 874 2553	52.170 059 5768	.019 168 0824	85
86	2.910 534 4377	152.842 755 0132	.006 542 6719	.343 579 5114	52.513 639 0882	.019 042 6719	86
87	2.946 916 1181	155.753 289 4509	.006 420 4101	.339 337 7890	52.852 976 8772	.018 920 4101	87
88	2.983 752 5696	158.700 205 5690	.006 301 1891	.335 148 4336	53.188 125 3108	.018 801 1891	88
89	3.021 049 4767	161.683 958 1386	.006 184 9055	.331 010 7986	53.519 136 1094	.018 684 9055	89
90	3.058 812 5952	164.705 007 6154	.006 071 4608	.326 924 2456	53.846 060 3550	.018 571 4608	90
91	3.097 047 7526	167.763 820 2106	.005 960 7608	.322 888 1438	54.168 948 4988	.018 460 7608	91
92	3.135 760 8495	170.860 867 9632	.005 852 7152	.318 901 8704	54.487 850 3692	.018 352 7152	92
93	3.174 957 8602	173.996 628 8127	.005 747 2378	.314 964 8103	54.802 815 1795	.018 247 2378	93
94	3.214 644 8334	177.171 586 6729	.005 644 2459	.311 076 3558	55.113 891 5352	.018 144 2459	94
95	3.254 827 8938	180.386 231 5063	.005 543 6604	.307 235 9070	55.421 127 4422	.018 043 6604	95
96	3.295 513 2425	183.641 059 4001	.005 445 4053	.303 442 8711	55.724 570 3133	.017 945 4053	96
97	3.336 707 1580	186.936 572 6426	.005 349 4080	.299 696 6628	56.024 266 9761	.017 849 4080	97
98	3.378 115 9975	190.273 279 8007	.005 255 5987	.295 996 7040	56.320 263 6801	.017 755 5987	98
99	3.420 646 1975	193.651 695 7982	.005 163 9104	.292 342 4237	56.612 606 1038	.017 663 9104	99
100	3.463 404 2749	197.072 341 9957	.005 074 2788	.288 733 2580	56.901 339 3618	.017 574 2788	100

	XIV COMPOUND AMOUNT	XV AMOUNT OF ANNUITY	XVI SINKING FUND	XVII PRESENT VALUE	XVIII PRESENT VALUE OF ANNUITY	XIX AMORTIZATION	
RATE 1 1/2% PERIODS	AMOUNT OF 1 — How $1 left at compound interest will grow.	AMOUNT OF 1 PER PERIOD — How $1 deposited periodically will grow.	SINKING FUND — Periodic deposit that will grow to $1 at future date.	PRESENT WORTH OF 1 — What $1 due in the future is worth today.	PRESENT WORTH OF 1 PER PERIOD — What $1 payable periodically is worth today.	PARTIAL PAYMENT — Annuity worth $1 today. Periodic payment necessary to pay off a loan of $1.	PERIODS
1	1.015 000 0000	1.000 000 0000	1.000 000 0000	.985 221 6749	.985 221 6749	1.015 000 0000	1
2	1.030 225 0000	2.015 000 0000	.496 277 9156	.970 661 7486	1.955 883 4235	.511 277 9156	2
3	1.045 678 3750	3.045 225 0000	.328 382 9602	.956 316 9937	2.912 200 4173	.343 382 9602	3
4	1.061 363 5506	4.090 903 3750	.244 444 7860	.942 184 2303	3.854 384 6476	.259 444 7860	4
5	1.077 284 0039	5.152 266 9256	.194 089 3231	.928 260 3254	4.782 644 9730	.209 089 3131	5
6	1.093 443 2639	6.229 550 9295	.160 525 2146	.914 542 1925	5.697 187 1655	.175 525 2146	6
7	1.109 844 9129	7.322 994 1935	.136 556 1645	.901 026 7907	6.598 213 9561	.151 556 1645	7
8	1.126 492 5866	8.432 839 1064	.118 584 0246	.887 711 1238	7.485 925 0799	.133 584 0246	8
9	1.143 389 9754	9.559 331 6929	.104 609 8234	.874 592 2402	8.360 517 3201	.119 609 8234	9
10	1.160 540 8250	10.702 721 6683	.093 434 1779	.861 667 2317	9.222 184 5519	.108 434 1779	10
11	1.177 948 9374	11.863 262 4934	.084 293 8442	.848 933 2332	10.071 117 7851	.099 293 8442	11
12	1.195 618 1715	13.041 211 4308	.076 679 9929	.836 387 4219	10.907 505 2070	.091 679 9929	12
13	1.213 552 4440	14.236 829 6022	.070 240 3574	.824 027 0166	11.731 532 2236	.085 240 3574	13
14	1.231 755 7307	15.450 382 0463	.064 723 3186	.811 849 2775	12.543 381 5011	.079 723 3186	14
15	1.250 232 0667	16.682 137 7770	.059 944 3557	.799 851 5049	13.343 233 0060	.074 944 3557	15
16	1.268 985 5477	17.932 369 8436	.055 765 0778	.788 031 0393	14.131 264 0453	.070 765 0778	16
17	1.288 020 3309	19.201 355 3913	.052 079 6569	.776 385 2604	14.907 649 3057	.067 079 6569	17
18	1.307 340 6358	20.489 375 7221	.048 805 7818	.764 911 5866	15.672 560 8924	.063 805 7818	18
19	1.326 950 7454	21.796 716 3580	.045 878 4701	.753 607 4745	16.426 168 3669	.060 878 4701	19
20	1.346 855 0066	23.123 667 1033	.043 245 7359	.742 470 4182	17.168 638 7851	.058 245 7359	20
21	1.367 057 8316	24.470 522 1099	.040 865 4950	.731 497 9490	17.900 136 7341	.055 865 4950	21
22	1.387 563 6991	25.837 579 9415	.038 703 3152	.720 687 6345	18.620 824 3685	.053 703 3152	22
23	1.408 377 1546	27.225 143 6407	.036 730 7520	.710 037 0783	19.330 861 4468	.051 730 7520	23
24	1.429 502 8119	28.633 520 7953	.034 924 1020	.699 543 9195	20.030 405 3663	.049 924 1020	24
25	1.450 945 3541	30.063 023 6072	.033 263 4539	.689 205 8320	20.719 611 1984	.048 263 4539	25
26	1.472 709 5344	31.513 968 9613	.031 731 9599	.679 020 5242	21.398 631 7225	.046 731 9599	26
27	1.494 800 1774	32.986 678 4957	.030 315 2680	.668 985 7381	22.067 617 4606	.045 315 2680	27
28	1.517 222 1801	34.481 478 6732	.029 001 0765	.659 099 2494	22.726 716 7100	.044 001 0765	28
29	1.539 980 5128	35.998 700 8533	.027 778 7802	.649 358 8664	23.376 075 5763	.042 778 7802	29
30	1.563 080 2205	37.538 681 3661	.026 639 1883	.639 762 4299	24.015 838 0062	.041 639 1883	30
31	1.586 526 4238	39.101 761 5865	.025 574 2954	.630 307 8127	24.646 145 8189	.040 574 2954	31
32	1.610 324 3202	40.688 288 0103	.024 577 0970	.620 992 9189	25.267 138 7379	.039 577 0970	32
33	1.634 479 1850	42.298 612 3305	.023 641 4375	.611 815 6837	25.878 954 4216	.038 641 4375	33
34	1.658 996 3727	43.933 091 5155	.022 761 8855	.602 774 0726	26.481 728 4941	.037 761 8855	34
35	1.683 881 3183	45.592 087 8882	.021 933 6303	.593 866 0814	27.075 594 5755	.036 933 6303	35
36	1.709 139 5381	47.275 969 2065	.021 152 3955	.585 089 7353	27.660 684 3109	.036 152 3955	36
37	1.734 776 6312	48.985 108 7446	.020 414 3673	.576 443 0890	28.237 127 3999	.035 414 3673	37
38	1.760 798 2806	50.719 885 3758	.019 716 1329	.567 924 2256	28.805 051 6255	.034 716 1329	38
39	1.787 210 2548	52.480 683 6564	.019 054 6298	.559 531 2568	29.364 582 8822	.034 054 6298	39
40	1.814 018 4087	54.267 893 9113	.018 427 1017	.551 262 3219	29.915 845 2042	.033 427 1017	40
41	1.841 228 6848	56.081 912 3199	.017 831 0610	.543 115 5881	30.458 960 7923	.032 831 0610	41
42	1.868 847 1151	57.923 141 0047	.017 264 2571	.535 089 2494	30.994 050 0417	.032 264 2571	42
43	1.896 879 8218	59.791 988 1198	.016 724 6488	.527 181 5265	31.521 231 5681	.031 724 6488	43
44	1.925 333 0191	61.688 867 9416	.016 210 3801	.519 390 6665	32.040 622 2346	.031 210 3804	44
45	1.954 213 0144	63.614 200 9607	.015 719 7604	.511 714 9423	32.552 337 1770	.030 719 7604	45
46	1.983 526 2096	65.568 413 9751	.015 251 2458	.504 152 6526	33.056 489 8295	.030 251 2458	46
47	2.013 279 1028	67.551 940 1848	.014 803 4238	.496 702 1207	33.553 191 9503	.029 803 4238	47
48	2.043 478 2893	69.565 219 2875	.014 374 9996	.489 361 6953	34.042 553 6456	.029 374 9996	48
49	2.074 130 4637	71.608 697 5768	.013 964 7841	.482 129 7491	34.524 683 3947	.028 964 7841	49
50	2.105 242 4206	73.682 828 0405	.013 571 6832	.475 004 6789	34.999 688 0736	.028 571 6832	50
51	2.136 821 0569	75.788 070 4611	.013 194 6887	.467 984 9053	35.467 672 9789	.028 194 6887	51
52	2.168 873 3728	77.924 891 5180	.012 832 8700	.461 068 8722	35.928 741 8511	.027 832 8700	52
53	2.201 406 4734	80.093 764 8908	.012 485 3664	.454 255 0465	36.382 996 8977	.027 485 3664	53
54	2.234 427 5705	82.295 171 3642	.012 151 3812	.447 541 9178	36.830 538 8154	.027 151 3812	54
55	2.267 943 9840	84.529 598 9346	.011 830 1756	.440 927 9978	37.271 466 8132	.026 830 1756	55
56	2.301 963 1438	86.797 542 9186	.011 521 0635	.434 411 8205	37.705 878 6337	.026 521 0635	56
57	2.336 492 5909	89.099 506 0624	.011 223 4068	.427 991 9414	38.133 870 5751	.026 223 4068	57
58	2.371 539 9798	91.435 998 6534	.010 936 6116	.421 666 9373	38.555 537 5124	.025 936 6116	58
59	2.407 113 0795	93.807 538 6332	.010 660 1241	.415 435 4062	38.970 972 9186	.025 660 1241	59
60	2.443 219 7757	96.214 651 7126	.010 393 4274	.409 295 9667	39.380 268 8853	.025 393 4274	60
61	2.479 868 0723	98.657 871 4883	.010 136 0387	.403 247 2579	39.783 516 1432	.025 136 0387	61
62	2.517 066 0934	101.137 739 5607	.009 887 5059	.397 287 9388	40.180 804 0820	.024 887 5059	62
63	2.554 822 0848	103.654 805 6541	.009 647 4061	.391 416 6884	40.572 220 7704	.024 647 4061	63
64	2.593 144 4161	106.209 627 7389	.009 415 3423	.385 632 2054	40.957 852 9758	.024 415 3423	64
65	2.632 041 5823	108.802 772 1550	.009 190 9423	.379 933 2073	41.337 786 1831	.024 190 9423	65
66	2.671 522 2061	111.434 813 7373	.008 973 8563	.374 318 4308	41.712 104 6138	.023 973 8563	66
67	2.711 595 0392	114.106 335 9434	.008 763 7552	.368 786 6313	42.080 891 2451	.023 763 7552	67
68	2.752 268 9647	116.817 930 9825	.008 560 3297	.363 336 5826	42.444 227 8277	.023 560 3297	68
69	2.793 552 9992	119.570 199 9472	.008 363 2048	.357 967 0764	42.802 194 9042	.023 363 2878	69
70	2.835 456 2942	122.363 752 9464	.008 172 3548	.352 676 9226	43.154 871 8268	.023 172 3548	70
71	2.877 988 1386	125.199 209 2406	.007 987 2709	.347 464 9484	43.502 336 7751	.022 987 2709	71
72	2.921 157 9607	128.077 197 3793	.007 807 9011	.342 329 9984	43.844 666 7735	.022 807 7911	72
73	2.964 975 3301	130.998 355 3399	.007 633 6836	.337 270 9344	44.181 937 7079	.022 633 6836	73
74	3.009 449 9601	133.963 330 6700	.007 464 7293	.332 286 6349	44.514 224 3428	.022 464 7293	74
75	3.054 591 7095	136.972 780 6301	.007 300 7206	.327 375 9949	44.841 600 3377	.022 300 7206	75
76	3.100 410 5851	140.027 372 3395	.007 141 4609	.322 537 9260	45.164 138 2638	.022 141 4609	76
77	3.146 916 7439	143.127 782 9246	.006 986 7637	.317 771 3557	45.481 909 6195	.021 986 7637	77
78	3.194 120 4950	146.274 699 6685	.006 836 4523	.313 075 2273	45.794 984 8468	.021 836 4523	78
79	3.242 032 3025	149.468 820 1635	.006 690 9586	.308 448 4998	46.103 433 3466	.021 690 9586	79
80	3.290 662 7870	152.710 852 4660	.006 548 3231	.303 890 1476	46.407 323 4941	.021 548 3231	80
81	3.340 022 7288	156.001 515 2530	.006 410 1941	.299 399 1602	46.706 722 6543	.021 410 1941	81
82	3.390 123 0697	159.341 537 9818	.006 275 8275	.294 974 5421	47.001 697 1964	.021 275 8275	82
83	3.440 974 9158	162.731 661 0515	.006 145 0857	.290 615 3124	47.292 312 5088	.021 145 0857	83
84	3.492 589 5395	166.172 635 9673	.006 017 8380	.286 320 5048	47.578 633 0136	.021 017 8380	84
85	3.544 978 3826	169.665 225 5068	.005 893 9597	.282 089 1673	47.860 722 1808	.020 893 9597	85
86	3.598 153 0683	173.210 203 8894	.005 773 3319	.277 920 3619	48.138 642 5427	.020 773 3319	86
87	3.652 125 3542	176.808 356 9477	.005 655 8413	.273 813 1644	48.412 455 7071	.020 655 8413	87
88	3.706 907 2345	180.460 482 3019	.005 541 3795	.269 766 6644	48.682 222 3715	.020 541 3794	88
89	3.762 510 8430	184.167 389 5365	.005 429 8429	.265 779 9650	48.948 002 3365	.020 429 8429	89
90	3.818 948 5057	187.929 900 3795	.005 321 1330	.261 852 1822	49.209 854 5187	.020 321 1330	90
91	3.876 232 7333	191.748 848 8852	.005 215 1552	.257 982 4455	49.467 836 9642	.020 215 1552	91
92	3.934 376 2243	195.625 081 6185	.005 111 8190	.254 169 8971	49.722 006 8613	.020 111 8190	92
93	3.993 391 8676	199.559 457 8428	.005 011 0379	.250 413 6917	49.972 420 5530	.020 011 0379	93
94	4.053 292 7457	203.552 849 7104	.004 912 7291	.246 712 9968	50.219 133 5498	.019 912 7291	94
95	4.114 092 1368	207.606 142 4561	.004 816 8132	.243 066 9919	50.462 200 5416	.019 816 8132	95
96	4.175 803 5189	211.720 234 5929	.004 723 2141	.239 474 8688	50.701 675 4105	.019 723 2142	96
97	4.238 440 5717	215.896 038 1118	.004 631 8590	.235 935 8314	50.937 611 2419	.019 631 8590	97
98	4.302 017 1803	220.134 478 6835	.004 542 6778	.232 449 0949	51.170 060 3368	.019 542 6778	98
99	4.366 547 4380	224.436 495 8637	.004 455 6033	.229 013 8866	51.399 074 2235	.019 455 6033	99
100	4.432 045 6495	228.803 043 3017	.004 370 5712	.225 629 4450	51.624 703 6684	.019 370 5712	100

RATE 1 3/4%

PERIODS	XIV COMPOUND AMOUNT — AMOUNT OF 1 (How $1 left at compound interest will grow.)	XV AMOUNT OF ANNUITY — AMOUNT OF 1 PER PERIOD (How $1 deposited periodically will grow.)	XVI SINKING FUND (Periodic deposit that will grow to $1 at future date.)	XVII PRESENT VALUE — PRESENT WORTH OF 1 (What $1 due in the future is worth today.)	XVIII PRESENT VALUE OF ANNUITY — PRESENT WORTH OF 1 PER PERIOD (What $1 payable periodically is worth today.)	XIX AMORTIZATION — PARTIAL PAYMENT (Annuity worth $1 today. Periodic payment necessary to pay off a loan of $1.)	PERIODS
1	1.017 500 0000	1.000 000 0000	1.000 000 0000	.982 800 9828	.982 800 9828	1.017 500 0000	1
2	1.035 306 2500	2.017 500 0000	.495 662 9492	.965 897 7718	1.948 698 7546	.513 162 9492	2
3	1.053 424 1094	3.052 806 2500	.327 567 4635	.949 285 2794	2.897 984 0340	.345 067 4635	3
4	1.071 859 0313	4.106 230 3594	.243 532 3673	.932 958 5056	3.830 942 5396	.261 032 3673	4
5	1.090 616 5643	5.178 089 3907	.193 121 4246	.916 912 5362	4.747 855 0757	.210 621 4246	5
6	1.109 702 3542	6.268 705 9550	.159 522 5565	.901 142 5417	5.648 997 6174	.177 022 5565	6
7	1.129 122 1454	7.378 408 3092	.135 530 5857	.885 643 7756	6.534 641 3930	.153 030 5857	7
8	1.148 881 7830	8.507 530 4546	.117 542 9233	.870 411 5731	7.405 052 9661	.135 042 9233	8
9	1.168 987 2142	9.656 412 2376	.103 558 1306	.855 441 3495	8.260 494 3156	.121 058 1306	9
10	1.189 444 4904	10.825 399 4517	.092 375 3442	.840 728 5990	9.101 222 9146	.109 875 3442	10
11	1.210 259 7690	12.014 843 9421	.083 230 3778	.826 268 8934	9.927 491 8080	.100 730 3778	11
12	1.231 439 3149	13.225 103 7711	.075 613 7738	.812 057 8805	10.739 549 6884	.095 113 7738	12
13	1.252 989 5030	14.456 543 0261	.069 172 8305	.798 091 2830	11.537 640 9714	.086 672 8305	13
14	1.274 916 8193	15.709 532 5290	.063 655 6179	.784 364 8973	12.322 005 8687	.081 155 6179	14
15	1.297 227 8636	16.984 449 3483	.058 877 3872	.770 874 5919	13.092 880 4607	.076 377 3872	15
16	1.319 929 3512	18.281 677 2119	.054 699 5764	.757 616 3066	13.850 496 7672	.072 199 5764	16
17	1.343 028 1149	19.601 606 5631	.051 016 2265	.744 586 0507	14.595 082 8179	.068 516 2265	17
18	1.366 531 1069	20.944 634 6779	.047 744 9244	.731 779 9024	15.326 862 7203	.065 244 9244	18
19	1.390 445 4012	22.311 165 7848	.044 820 6073	.719 194 0073	16.046 056 7276	.062 320 6073	19
20	1.414 778 1958	23.701 611 1860	.042 191 2246	.706 824 5772	16.752 881 3048	.059 691 2246	20
21	1.439 536 8142	25.116 389 3818	.039 814 6399	.694 667 8891	17.447 549 1939	.057 314 6399	21
22	1.464 728 7084	26.555 926 1960	.037 656 3782	.682 720 2841	18.130 269 4780	.055 156 3782	22
23	1.490 361 4608	28.020 654 9044	.035 687 9596	.670 978 1662	18.801 247 6442	.053 187 9596	23
24	1.516 442 7864	29.511 016 3652	.033 885 6510	.659 438 0012	19.460 685 6454	.051 385 6510	24
25	1.542 980 5352	31.027 459 1516	.032 229 5163	.648 096 3157	20.108 781 9611	.049 729 5163	25
26	1.569 982 6945	32.570 439 6868	.030 702 6865	.636 949 6960	20.745 731 6571	.048 202 6865	26
27	1.597 457 3917	34.140 422 3813	.029 290 7917	.625 994 7872	21.371 726 4443	.046 790 7917	27
28	1.625 412 8960	35.737 879 7730	.027 981 5145	.615 228 2921	21.986 954 7364	.045 481 5145	28
29	1.653 857 6217	37.363 292 6690	.026 764 2365	.604 646 9701	22.591 601 7066	.044 264 2365	29
30	1.682 800 1301	39.017 150 2907	.025 629 7549	.594 247 6365	23.185 849 3431	.043 129 7549	30
31	1.712 249 1324	40.699 950 4208	.024 570 0545	.584 027 1612	23.769 876 5042	.042 070 0545	31
32	1.742 213 4922	42.412 199 5532	.023 578 1216	.573 982 4680	24.343 858 9722	.041 078 1216	32
33	1.772 702 2283	44.154 413 0453	.022 647 7928	.564 110 5336	24.907 969 5059	.040 147 7928	33
34	1.803 724 5173	45.927 115 2736	.021 773 6297	.554 408 3869	25.462 377 8928	.039 273 6297	34
35	1.835 289 6963	47.730 839 7909	.020 950 8151	.544 873 1075	26.007 251 0003	.038 450 8151	35
36	1.867 407 2660	49.566 129 4873	.020 175 0673	.535 501 8255	26.542 752 8258	.037 675 0673	36
37	1.900 086 8932	51.433 536 7533	.019 442 5673	.526 291 7204	27.069 044 5462	.036 942 5673	37
38	1.933 338 4138	53.333 623 6465	.018 749 8979	.517 240 0201	27.586 284 5663	.036 249 8979	38
39	1.967 171 8361	55.266 962 0603	.018 093 9926	.508 344 0001	28.094 628 5664	.035 593 9926	39
40	2.001 597 3432	57.234 133 8963	.017 472 0911	.499 600 9829	28.594 229 5493	.034 972 0911	40
41	2.036 625 2967	59.235 731 2395	.016 881 7026	.491 008 3370	29.085 237 8863	.034 381 7026	41
42	2.072 266 2394	61.272 356 5362	.016 320 5735	.482 563 4762	29.567 801 3625	.033 820 5735	42
43	2.108 530 8986	63.344 622 7756	.015 786 6596	.474 263 8586	30.042 065 2211	.033 286 6596	43
44	2.145 430 1893	65.453 153 6742	.015 278 1026	.466 106 9864	30.508 172 2075	.032 778 1026	44
45	2.182 975 2176	67.598 583 8635	.014 793 2093	.458 090 4043	30.966 262 6117	.032 293 2093	45
46	2.221 177 2839	69.781 559 0811	.014 330 4336	.450 211 6996	31.416 474 3113	.031 830 4336	46
47	2.260 047 8864	72.002 736 3650	.013 888 3611	.442 468 5008	31.858 942 8121	.031 388 3611	47
48	2.299 598 7244	74.262 784 2514	.013 465 6950	.434 858 4774	32.293 801 2895	.030 965 6950	48
49	2.339 841 7021	76.562 382 9758	.013 061 2445	.427 379 3390	32.721 180 6285	.030 561 2445	49
50	2.380 788 9319	78.902 224 6779	.012 673 9139	.420 028 8344	33.141 209 4629	.030 173 9139	50
51	2.422 452 7382	81.283 013 6097	.012 302 6935	.412 804 7513	33.554 014 2142	.029 802 6935	51
52	2.464 845 6611	83.705 466 3479	.011 946 6511	.405 704 9152	33.959 719 1294	.029 446 6511	52
53	2.507 980 4602	86.170 312 0090	.011 604 9249	.398 727 1894	34.358 446 3188	.029 104 9249	53
54	2.551 870 1182	88.678 292 4691	.011 276 7169	.391 869 4736	34.750 315 7925	.028 776 7169	54
55	2.596 527 8453	91.230 162 5874	.010 961 2871	.385 129 7038	35.135 445 4963	.028 461 2871	55
56	2.641 967 0826	93.826 690 4326	.010 657 9481	.378 505 8514	35.513 951 3477	.028 157 9481	56
57	2.688 201 5065	96.468 657 5152	.010 366 0611	.371 995 9228	35.885 947 2705	.027 866 0611	57
58	2.735 245 0329	99.156 859 0217	.010 085 0310	.365 597 9585	36.251 545 2290	.027 585 0310	58
59	2.783 111 8210	101.892 104 0546	.009 814 3032	.359 310 0329	36.610 855 2619	.027 314 3032	59
60	2.831 816 2778	104.675 215 8756	.009 553 3598	.353 130 2535	36.963 985 5154	.027 053 3598	60
61	2.881 373 0627	107.507 032 1534	.009 301 7171	.347 056 7602	37.311 042 2755	.026 801 7171	61
62	2.931 797 0913	110.388 405 2161	.009 058 9224	.341 087 7250	37.652 130 0005	.026 558 9224	62
63	2.983 103 5404	113.320 202 3073	.008 824 5518	.335 221 3513	37.987 351 3519	.026 324 5518	63
64	3.035 307 8523	116.303 305 8477	.008 598 2079	.329 455 8736	38.316 807 2254	.026 098 2079	64
65	3.088 425 7398	119.338 613 7001	.008 379 5175	.323 789 5563	38.640 596 7817	.025 879 5175	65
66	3.142 473 1902	122.427 039 4398	.008 168 1302	.318 220 6942	38.958 817 4759	.025 668 1302	66
67	3.197 466 4710	125.569 512 6300	.007 963 7165	.312 747 6110	39.271 565 0869	.025 463 7165	67
68	3.253 422 1343	128.766 979 1010	.007 765 9661	.307 368 6594	39.578 933 7463	.025 265 9661	68
69	3.310 357 0216	132.020 401 2353	.007 574 5869	.302 082 2206	39.881 015 9669	.025 074 5869	69
70	3.368 288 2695	135.330 758 2569	.007 389 3032	.296 886 7033	40.177 902 6702	.024 889 3032	70
71	3.427 233 3142	138.699 046 5264	.007 209 8549	.291 780 5438	40.469 683 2139	.024 709 8549	71
72	3.487 209 8972	142.126 279 8406	.007 035 9964	.286 762 2052	40.756 445 4191	.024 535 9964	72
73	3.548 236 0704	145.613 489 7378	.006 867 4956	.281 830 1771	41.038 275 5962	.024 367 4956	73
74	3.610 330 2016	149.161 725 8083	.006 704 1327	.276 982 9750	41.315 258 5712	.024 204 1327	74
75	3.673 510 9802	152.772 056 0099	.006 545 6997	.272 219 1401	41.587 477 7112	.024 045 6997	75
76	3.737 796 4223	156.445 566 9901	.006 391 9996	.267 537 2384	41.855 014 9496	.023 891 9996	76
77	3.803 208 8772	160.183 364 4124	.006 242 8455	.262 935 8608	42.117 950 8104	.023 742 8455	77
78	3.869 765 0326	163.986 573 2896	.006 098 0602	.258 410 6224	42.376 364 4329	.023 598 0602	78
79	3.937 485 9206	167.856 338 3222	.005 957 4754	.253 969 1621	42.630 333 5949	.023 457 4754	79
80	4.006 391 9242	171.793 824 2428	.005 820 9310	.249 601 1421	42.879 934 7370	.023 320 9310	80
81	4.076 503 7829	175.800 216 1671	.005 688 2751	.245 308 2478	43.125 242 9848	.023 188 2751	81
82	4.147 842 5991	179.876 719 9500	.005 559 3631	.241 089 1870	43.366 332 1718	.023 059 3631	82
83	4.220 429 8446	184.024 562 5491	.005 434 0572	.236 942 6899	43.603 274 8617	.022 934 0572	83
84	4.294 287 3669	188.244 992 3937	.005 312 2263	.232 867 5085	43.836 142 3702	.022 812 2263	84
85	4.369 437 3958	192.539 279 7606	.005 193 7454	.228 862 4162	44.065 004 7865	.022 693 7454	85
86	4.445 902 5502	196.908 717 1564	.005 078 4953	.224 926 2076	44.289 930 9941	.022 578 4953	86
87	4.523 705 8449	201.354 619 7067	.004 966 3623	.221 057 6979	44.510 988 6920	.022 466 3623	87
88	4.602 870 6972	205.878 325 5515	.004 857 2379	.217 255 7227	44.728 244 4147	.022 357 2379	88
89	4.683 420 9344	210.481 196 2487	.004 751 0182	.213 519 1378	44.941 763 5525	.022 251 0182	89
90	4.765 380 8007	215.164 617 1830	.004 647 6043	.209 846 8185	45.151 610 3711	.022 147 6043	90
91	4.848 774 9647	219.929 997 9837	.004 546 9013	.206 237 6595	45.357 848 0305	.022 046 9013	91
92	4.933 628 5266	224.778 772 9485	.004 448 8187	.202 690 5744	45.560 538 6049	.021 948 8187	92
93	5.019 967 0258	229.712 401 4751	.004 353 2695	.199 204 4957	45.759 743 1007	.021 853 2695	93
94	5.107 816 4488	234.732 368 5009	.004 260 1709	.195 778 3742	45.955 521 4749	.021 760 1709	94
95	5.197 203 2366	239.840 184 9496	.004 169 4431	.192 411 1786	46.147 932 6534	.021 669 4431	95
96	5.288 154 2933	245.037 388 1863	.004 081 0099	.189 101 8954	46.337 034 5488	.021 581 0099	96
97	5.380 696 9934	250.325 542 4795	.003 994 7981	.185 849 5286	46.522 884 0775	.021 494 7981	97
98	5.474 859 1908	255.706 239 4729	.003 910 7376	.182 653 0994	46.705 537 1769	.021 410 7376	98
99	5.570 669 2266	261.181 098 6637	.003 828 7610	.179 511 6456	46.885 048 8225	.021 328 7610	99
100	5.668 155 9381	266.751 767 8903	.003 748 8036	.176 424 2217	47.061 473 0442	.021 248 8036	100

	XIV COMPOUND AMOUNT	XV AMOUNT OF ANNUITY	XVI SINKING FUND	XVII PRESENT VALUE	XVIII PRESENT VALUE OF ANNUITY	XIX AMORTIZATION	
PERIODS	AMOUNT OF 1 *How $1 left at compound interest will grow*	AMOUNT OF 1 PER PERIOD *How $1 deposited periodically will grow.*	SINKING FUND *Periodic deposit that will grow to $1 at future date.*	PRESENT WORTH OF 1 *What $1 due in the future is worth today.*	PRESENT WORTH OF 1 PER PERIOD *What $1 payable periodically is worth today.*	PARTIAL PAYMENT *Annuity worth $1 today Periodic payment necessary to pay off a loan of $1*	PERIODS
1	1.020 000 0000	1.000 000 0000	1.000 000 0000	.980 392 1569	.980 392 1569	1.020 000 0000	1
2	1.040 400 0000	2.020 000 0000	.495 049 5050	.961 168 7812	1.941 560 9381	.515 049 5050	2
3	1.061 208 0000	3.060 400 0000	.326 754 6726	.942 322 3345	2.883 883 2726	.346 754 6726	3
4	1.082 432 1600	4.121 608 0000	.242 623 7527	.923 845 4260	3.807 728 6987	.262 623 7527	4
5	1.104 080 8032	5.204 040 1600	.192 158 3941	.905 730 8098	4.713 459 5085	.212 158 3941	5
6	1.126 162 4193	6.308 120 9632	.158 525 8123	.887 971 3822	5.601 430 8907	.178 525 8123	6
7	1.148 685 6676	7.434 283 3825	.134 511 9561	.870 560 1786	6.471 991 0693	.154 511 9561	7
8	1.171 659 3810	8.582 969 0501	.116 509 7991	.853 490 3712	7.325 481 4405	.136 509 7991	8
9	1.195 092 5686	9.754 628 4311	.102 515 4374	.836 755 2659	8.162 236 7064	.122 515 4374	9
10	1.218 994 4200	10.949 720 9997	.091 326 5279	.820 348 2999	8.982 585 0062	.111 326 5279	10
11	1.243 374 3084	12.168 715 4197	.082 177 9428	.804 263 0391	9.786 848 0453	.102 177 9428	11
12	1.268 241 7946	13.412 089 7281	.074 559 5966	.788 493 1756	10.575 341 2209	.094 559 5966	12
13	1.293 606 6305	14.680 331 5227	.068 118 3527	.773 032 5251	11.348 373 7460	.088 118 3527	13
14	1.319 478 7631	15.973 938 1531	.062 601 9702	.757 875 0246	12.106 248 7706	.082 601 9702	14
15	1.345 868 3383	17.293 416 9162	.057 825 4723	.743 014 7300	12.849 263 5006	.077 825 4723	15
16	1.372 785 7051	18.639 285 2545	.053 650 1259	.728 445 8137	13.577 709 3143	.073 650 1259	16
17	1.400 241 4192	20.012 070 9596	.049 969 8408	.714 162 5625	14.291 871 8768	.069 969 8408	17
18	1.428 246 2476	21.412 312 3788	.046 702 1022	.700 159 3750	14.992 031 2517	.066 702 1022	18
19	1.456 811 1725	22.840 558 6264	.043 781 7663	.686 430 7598	15.678 462 0115	.063 781 7663	19
20	1.485 947 3960	24.297 369 7989	.041 156 7181	.672 971 3331	16.351 433 3446	.061 156 7181	20
21	1.515 666 3439	25.783 317 1949	.038 784 7689	.659 775 8168	17.011 209 1614	.058 784 7689	21
22	1.545 979 6708	27.298 983 5388	.036 631 4005	.646 839 0361	17.658 048 1974	.056 631 4005	22
23	1.576 899 2642	28.844 963 2096	.034 668 0976	.634 155 9177	18.292 204 1151	.054 668 0976	23
24	1.608 437 2495	30.421 862 4738	.032 871 0973	.621 721 4879	18.913 925 6031	.052 871 0973	24
25	1.640 605 9945	32.030 299 7232	.031 220 4384	.609 530 8705	19.523 456 4736	.051 220 4384	25
26	1.673 418 1144	33.670 905 7177	.029 699 2308	.597 579 2848	20.121 035 7584	.049 699 2308	26
27	1.706 886 4766	35.344 323 8321	.028 293 0862	.585 862 0440	20.706 897 8024	.048 293 0862	27
28	1.741 024 2062	37.051 210 3087	.026 989 6716	.574 374 5529	21.281 272 3553	.046 989 6716	28
29	1.775 844 6903	38.792 234 5149	.025 778 3552	.563 112 3068	21.844 384 6620	.045 778 3552	29
30	1.811 361 5841	40.568 079 2052	.024 649 9223	.552 070 8890	22.396 455 5510	.044 649 9223	30
31	1.847 588 8158	42.379 440 7893	.023 596 3472	.541 245 9696	22.937 701 5206	.043 596 3472	31
32	1.884 540 5921	44.227 029 6051	.022 610 6073	.530 633 3035	23.468 334 8241	.042 610 6073	32
33	1.922 231 4039	46.111 570 1972	.021 686 5311	.520 228 7289	23.988 563 5530	.041 686 5311	33
34	1.960 676 0320	48.033 801 6011	.020 818 6728	.510 028 1656	24.498 591 7187	.040 818 6728	34
35	1.999 889 5527	49.994 477 6331	.020 002 2092	.500 027 6134	24.998 619 3320	.040 002 2092	35
36	2.039 887 3437	51.994 367 1858	.019 232 8526	.490 223 1504	25.488 842 4824	.039 232 8526	36
37	2.080 685 0906	54.034 254 5295	.018 506 7789	.480 610 9317	25.969 453 4141	.038 506 7789	37
38	2.122 298 7924	56.114 939 6201	.017 820 5663	.471 187 1880	26.440 640 6021	.037 820 5663	38
39	2.164 744 7682	58.237 238 4125	.017 171 1439	.461 948 2235	26.902 588 8256	.037 171 1439	39
40	2.208 039 6636	60.401 983 1807	.016 555 7478	.452 890 4152	27.355 479 2407	.036 555 7478	40
41	2.252 200 4569	62.610 022 8444	.015 971 8836	.444 010 2110	27.799 489 4517	.035 971 8836	41
42	2.297 244 4660	64.862 223 3012	.015 417 2945	.435 304 1284	28.234 793 5801	.035 417 2945	42
43	2.343 189 3553	67.159 467 7673	.014 889 9334	.426 768 7533	28.661 562 3334	.034 889 9334	43
44	2.390 053 1425	69.502 657 1226	.014 387 9391	.418 400 7386	29.079 963 0720	.034 387 9391	44
45	2.437 854 2053	71.892 710 2651	.013 909 6161	.410 196 8025	29.490 159 8745	.033 909 6161	45
46	2.486 611 2894	74.330 564 4704	.013 453 4159	.402 153 7280	29.892 313 6025	.033 453 4159	46
47	2.536 343 5152	76.817 175 7598	.013 017 9220	.394 268 3607	30.286 581 9632	.033 017 9220	47
48	2.587 070 3855	79.353 519 2750	.012 601 8355	.386 537 6086	30.673 119 5718	.032 601 8355	48
49	2.638 811 7932	81.940 589 6605	.012 203 9639	.378 958 4398	31.052 078 0115	.032 203 9639	49
50	2.691 588 0291	84.579 401 4537	.011 823 2097	.371 527 8821	31.423 605 8937	.031 823 2097	50
51	2.745 419 7897	87.270 989 4828	.011 458 5615	.364 243 0217	31.787 848 9153	.031 458 5615	51
52	2.800 328 1854	90.016 409 2724	.011 109 0856	.357 101 0017	32.144 949 9170	.031 109 0856	52
53	2.856 334 7492	92.816 737 4579	.010 773 9189	.350 099 0212	32.495 048 9382	.030 773 9189	53
54	2.913 461 4441	95.673 072 2070	.010 452 2618	.343 234 3345	32.838 283 2728	.030 452 2618	54
55	2.971 730 6730	98.586 533 6512	.010 143 3732	.336 504 2496	33.174 787 5223	.030 143 3732	55
56	3.031 165 2865	101.558 264 3242	.009 846 5645	.329 906 1270	33.504 693 6494	.029 846 5645	56
57	3.091 788 5922	104.589 429 6107	.009 561 1957	.323 437 3794	33.828 131 0288	.029 561 1957	57
58	3.153 624 3641	107.681 218 2029	.009 286 6706	.317 095 4700	34.145 226 4988	.029 286 6706	58
59	3.216 696 8513	110.834 842 5669	.009 022 4335	.310 877 9118	34.456 104 4106	.029 022 4335	59
60	3.281 030 7884	114.051 539 4183	.008 767 9658	.304 782 2665	34.760 886 6770	.028 767 9658	60
61	3.346 651 4041	117.332 570 2066	.008 522 7827	.298 806 1436	35.059 692 8206	.028 522 7827	61
62	3.413 584 4322	120.679 221 6108	.008 286 4306	.292 947 1996	35.352 640 0202	.028 286 4306	62
63	3.481 856 1209	124.092 806 0430	.008 058 4849	.287 203 1369	35.639 843 1571	.028 058 4849	63
64	3.551 493 2433	127.574 662 1638	.007 838 5471	.281 571 7028	35.921 414 8599	.027 838 5471	64
65	3.622 523 1081	131.126 155 4071	.007 626 2436	.276 050 6890	36.197 465 5489	.027 626 2436	65
66	3.694 973 5703	134.748 678 5153	.007 421 2231	.270 637 9304	36.468 103 4793	.027 421 2231	66
67	3.768 873 0417	138.443 652 0856	.007 223 1553	.265 331 3043	36.733 434 7837	.027 223 1553	67
68	3.844 250 5025	142.212 525 1273	.007 031 7294	.260 128 7297	36.993 563 5134	.027 031 7294	68
69	3.921 135 5126	146.056 775 6298	.006 846 6526	.255 028 1664	37.248 591 6798	.026 846 6526	69
70	3.999 558 2228	149.977 911 1424	.006 667 6485	.250 027 6141	37.498 619 2939	.026 667 6485	70
71	4.079 549 3873	153.977 469 3652	.006 494 4567	.245 125 1119	37.743 744 4058	.026 494 4567	71
72	4.161 140 3751	158.057 018 7526	.006 326 8307	.240 318 7371	37.984 063 1429	.026 326 8307	72
73	4.244 363 1826	162.218 159 1276	.006 164 5380	.235 606 6050	38.219 669 7480	.026 164 5380	73
74	4.329 250 4462	166.462 522 3102	.006 007 3582	.230 986 8677	38.450 656 6157	.026 007 3582	74
75	4.415 835 4551	170.791 772 7564	.005 855 0830	.226 457 7134	38.677 114 3291	.025 855 0830	75
76	4.504 152 1642	175.207 608 2115	.005 707 5147	.222 017 3661	38.899 131 6952	.025 707 5147	76
77	4.594 235 2075	179.711 760 3757	.005 564 4661	.217 664 0844	39.116 795 7796	.025 564 4661	77
78	4.686 119 9117	184.305 995 5833	.005 425 7595	.213 396 1612	39.330 191 9408	.025 425 7595	78
79	4.779 842 3099	188.992 115 4949	.005 291 2260	.209 211 9227	39.539 403 8635	.025 291 2260	79
80	4.875 439 1561	193.771 957 8048	.055 160 7055	.205 109 7282	39.744 513 5917	.025 160 7055	80
81	4.972 947 9392	198.647 396 9609	.005 034 0453	.201 087 9688	39.945 601 5605	.025 034 0453	81
82	5.072 406 8980	203.620 344 9001	.004 911 1006	.197 145 0674	40.142 746 6279	.024 911 1006	82
83	5.173 855 0360	208.692 751 7981	.004 791 7333	.193 279 4779	40.336 026 1058	.024 791 7333	83
84	5.277 332 1367	213.866 606 8341	.004 675 8118	.189 489 6842	40.525 515 7900	.024 675 8118	84
85	5.382 878 7794	219.143 938 9708	.004 563 2108	.185 774 2002	40.711 289 9902	.024 563 2108	85
86	5.490 536 3550	224.526 817 7502	.004 453 8110	.182 131 5688	40.893 421 5590	.024 453 8110	86
87	5.600 347 0821	230.017 354 1052	.004 347 4981	.178 560 3616	41.071 981 9206	.024 347 4981	87
88	5.712 354 0237	235.617 701 1873	.004 244 1633	.175 059 1780	41.247 041 0986	.024 244 1633	88
89	5.826 601 1042	241.330 055 2111	.004 143 7027	.171 626 6451	41.418 667 7437	.024 143 7027	89
90	5.943 133 1263	247.156 656 3153	.004 046 0169	.168 261 4168	41.586 929 1605	.024 046 0169	90
91	6.061 995 7888	253.099 789 4416	.003 951 0108	.164 962 1733	41.751 891 3339	.023 951 0108	91
92	6.183 235 7046	259.161 785 2304	.003 858 5936	.161 727 6209	41.913 618 9548	.023 858 5936	92
93	6.306 900 4187	265.345 020 9350	.003 768 6782	.158 556 4911	42.072 175 4458	.023 768 6782	93
94	6.433 038 4271	271.651 921 3537	.003 681 1814	.155 447 5403	42.227 622 9861	.023 681 1814	94
95	6.561 699 1956	278.084 959 7808	.003 596 0233	.152 399 5493	42.380 022 5354	.023 596 0233	95
96	6.692 933 1795	284.646 658 9764	.003 513 1275	.149 411 3228	42.529 433 8582	.023 513 1275	96
97	6.826 791 8431	291.339 592 1559	.003 432 4205	.146 481 6891	42.675 915 5473	.023 432 4205	97
98	6.963 327 6800	298.166 383 9991	.003 353 8321	.143 609 4991	42.819 525 0464	.023 353 8321	98
99	7.102 594 2336	305.129 711 6790	.003 277 2947	.140 793 6265	42.960 318 6729	.023 277 2947	99
100	7.244 646 1183	312.232 305 9126	.003 202 7435	.138 032 9672	43.098 351 6401	.023 202 7435	100

	XIV COMPOUND AMOUNT	XV AMOUNT OF ANNUITY	XVI SINKING FUND	XVII PRESENT VALUE	XVIII PRESENT VALUE OF ANNUITY	XIX AMORTIZATION	
PERIODS	AMOUNT OF 1	AMOUNT OF 1 PER PERIOD	SINKING FUND	PRESENT WORTH OF 1	PRESENT WORTH OF 1 PER PERIOD	PARTIAL PAYMENT	PERIODS
	How $1 left at compound interest will grow.	How $1 deposited periodically will grow.	Periodic deposit that will grow to $1 at future date.	What $1 due in the future is worth today.	What $1 payable periodically is worth today.	Annuity worth $1 today. Periodic payment necessary to pay off a loan of $1.	
1	1.022 500 0000	1.000 000 0000	1.000 000 0000	.977 995 1100	.977 995 1100	1.022 500 0000	1
2	1.045 506 2500	2.022 500 0000	.494 437 5773	.956 474 4352	1.934 469 5453	.516 937 5773	2
3	1.069 030 1406	3.068 006 2500	.325 944 5772	.935 427 3205	2.869 896 8658	.348 444 5772	3
4	1.093 083 3188	4.137 036 3906	.241 718 9277	.914 843 3453	3.784 740 2110	.264 218 9277	4
5	1.117 677 6935	5.230 119 7094	.191 200 2125	.894 712 3181	4.679 452 5291	.213 700 2125	5
6	1.142 825 4416	6.347 797 4029	.157 534 9584	.875 024 2720	5.554 476 8011	.180 034 9584	6
7	1.168 539 0140	7.490 622 8444	.133 500 2470	.855 769 4591	6.410 246 2602	.156 000 2470	7
8	1.194 831 1418	8.659 161 8584	.115 484 6181	.836 938 3464	7.247 184 6066	.137 984 6181	8
9	1.221 714 8425	9.853 993 0003	.101 481 7039	.818 521 6101	8.065 706 2167	.123 981 7039	9
10	1.249 203 4265	11.075 707 8428	.090 287 6831	.800 510 1322	8.866 216 3489	.112 787 6831	10
11	1.277 310 5036	12.324 911 2692	.081 136 4868	.782 894 9948	9.649 111 3436	.103 636 4868	11
12	1.306 049 9899	13.602 221 7728	.073 517 4015	.765 667 4765	10.414 778 8202	.096 017 4015	12
13	1.335 436 1147	14.908 271 7627	.067 076 8561	.748 819 0480	11.163 597 8681	.089 576 8561	13
14	1.365 483 4272	16.243 707 8773	.061 562 2989	.732 341 3672	11.895 939 2354	.084 062 2989	14
15	1.396 206 8044	17.609 191 3046	.056 788 5250	.716 226 2760	12.612 165 5113	.079 288 5250	15
16	1.427 621 4575	19.005 398 1089	.052 616 6300	.700 465 7956	13.312 631 3069	.075 116 6300	16
17	1.459 742 9402	20.433 019 5664	.048 940 3926	.685 052 1228	13.997 683 4298	.071 440 3926	17
18	1.492 587 1564	21.892 762 5066	.045 677 1958	.669 977 6262	14.667 661 0560	.068 177 1958	18
19	1.526 170 3674	23.385 349 6630	.042 761 8152	.665 234 8423	15.322 895 8983	.065 261 8152	19
20	1.560 509 2007	24.911 520 0304	.040 142 0708	.640 816 4177	15.963 712 3700	.062 642 0708	20
21	1.595 620 6577	26.472 029 2311	.037 775 7214	.626 715 3757	16.590 427 7457	.060 275 7214	21
22	1.631 522 1225	28.067 649 8888	.035 628 2056	.612 924 5728	17.203 352 3185	.058 128 2056	22
23	1.668 231 3703	29.699 172 0113	.033 670 9724	.599 437 2350	17.802 789 5536	.056 170 9724	23
24	1.705 766 5761	31.367 403 3816	.031 880 2289	.586 246 6846	18.389 036 2382	.054 380 2289	24
25	1.744 146 3240	33.073 169 9577	.030 235 9889	.573 346 3908	18.962 382 6290	.052 735 9889	25
26	1.783 389 6163	34.817 316 2817	.028 721 3406	.560 729 9666	19.523 112 5957	.051 221 3406	26
27	1.823 515 8827	36.600 705 8980	.027 321 8774	.548 391 1654	20.071 503 7610	.049 821 8774	27
28	1.864 544 9901	38.424 221 7807	.026 025 2506	.536 323 8781	20.607 827 6392	.048 525 2506	28
29	1.906 497 2523	40.288 766 7708	.024 820 8143	.524 522 1302	21.132 349 7693	.047 320 8143	29
30	1.949 393 4405	42.195 264 0232	.023 699 3422	.512 980 0784	21.645 329 8478	.046 199 3422	30
31	1.993 254 7929	44.144 657 4637	.022 652 7978	.501 692 0082	22.147 021 8560	.045 152 7978	31
32	2.038 103 0258	46.137 912 2566	.021 674 1493	.490 652 3308	22.637 674 1868	.044 174 1493	32
33	2.083 960 3439	48.176 015 2824	.020 757 2169	.479 855 5802	23.117 529 7670	.043 257 2169	33
34	2.130 849 4516	50.259 975 6262	.019 896 5477	.469 296 4110	23.586 826 1780	.042 396 5477	34
35	2.178 793 5643	52.390 825 0778	.019 087 3115	.458 969 5951	24.045 795 7731	.041 587 3115	35
36	2.227 816 4194	54.569 618 6421	.018 325 2151	.448 870 0197	24.494 665 7928	.040 825 2151	36
37	2.277 942 2889	56.797 435 0615	.017 606 4289	.438 992 6843	24.933 658 4771	.040 106 4289	37
38	2.329 195 9904	59.075 377 3504	.016 927 5262	.429 332 6985	25.362 991 1756	.039 427 5262	38
39	2.381 602 9002	61.404 573 3408	.016 285 4319	.419 885 2798	25.782 876 4554	.038 785 4319	39
40	2.435 188 0654	63.786 176 2410	.015 677 3781	.410 645 7504	26.193 522 2057	.038 177 3781	40
41	2.489 980 7171	66.221 365 2064	.015 100 8666	.401 609 5358	26.595 131 7416	.037 600 8666	41
42	2.546 005 2833	68.711 345 9235	.014 553 6372	.392 772 1622	26.987 903 9037	.037 053 6372	42
43	2.603 290 4022	71.257 351 2068	.014 033 6398	.384 129 2540	27.372 033 1577	.036 533 6398	43
44	2.661 864 4362	73.860 641 6090	.013 539 0105	.375 676 5320	27.747 709 6897	.036 039 0105	44
45	2.721 756 3860	76.522 506 0452	.013 068 0508	.367 409 8112	28.115 119 5009	.035 568 0508	45
46	2.782 995 9047	79.244 262 4312	.012 619 2101	.359 324 9988	28.474 444 4997	.035 119 2101	46
47	2.845 613 2126	82.027 258 3359	.012 191 0694	.351 418 0917	28.825 862 5913	.034 691 0694	47
48	2.909 639 6121	84.872 871 6484	.011 782 3279	.343 685 1753	29.169 547 7666	.034 282 3279	48
49	2.975 106 5034	87.782 511 2605	.011 391 7908	.336 122 4208	29.505 670 1874	.033 891 7908	49
50	3.042 046 3997	90.757 617 7639	.011 018 3588	.328 726 0839	29.834 396 2713	.033 518 3588	50
51	3.110 492 4437	93.799 664 1636	.010 661 0190	.321 492 5026	30.155 888 7739	.033 161 0190	51
52	3.180 478 5237	96.910 156 6073	.010 318 8359	.314 418 0954	30.470 306 8693	.032 818 8359	52
53	3.252 039 2904	100.090 635 1309	.009 990 9447	.307 499 3598	30.777 806 2291	.032 490 9447	53
54	3.325 210 1745	103.342 674 4214	.009 676 5446	.300 732 8703	31.078 539 0994	.032 176 5446	54
55	3.400 027 4034	106.667 884 5958	.009 374 8930	.294 115 2765	31.372 654 3760	.031 874 8930	55
56	3.476 528 0200	110.067 911 9993	.009 085 3000	.287 643 3022	31.660 297 6782	.031 585 3000	56
57	3.554 749 9004	113.544 440 0192	.008 807 1243	.281 313 7430	31.941 611 4212	.031 307 1243	57
58	3.634 731 7732	117.099 189 9197	.008 539 7687	.275 123 4651	32.216 734 8863	.031 039 7687	58
59	3.716 513 2381	120.733 921 6929	.008 282 6764	.269 069 4035	32.485 804 2898	.030 782 6764	59
60	3.800 534 7859	124.450 434 9310	.008 035 3275	.263 148 5609	32.748 952 8506	.030 535 3275	60
61	3.885 637 8186	128.250 569 7169	.007 797 2363	.257 358 0057	33.006 310 8563	.030 297 2363	61
62	3.973 064 6695	132.136 207 5355	.007 567 9484	.251 694 8711	33.258 005 7275	.030 067 9484	62
63	4.062 458 6246	136.109 272 2051	.007 347 0380	.246 156 3532	33.504 162 0807	.029 847 0380	63
64	4.153 863 9437	140.171 730 8297	.007 134 1061	.240 739 7097	33.744 901 7904	.029 634 1061	64
65	4.247 325 8824	144.325 594 7734	.006 928 7780	.235 442 2589	33.980 344 0493	.029 428 7780	65
66	4.342 890 7148	148.572 920 6558	.006 730 7016	.230 261 3779	34.210 605 4272	.029 230 7016	66
67	4.440 605 7558	152.915 811 3705	.006 539 5461	.225 194 5016	34.435 799 9288	.029 039 5461	67
68	4.540 519 3853	157.356 417 1264	.006 354 9998	.220 239 1214	34.656 039 0501	.028 854 9998	68
69	4.642 681 0715	161.896 936 5117	.006 176 7691	.215 392 7837	34.871 431 8339	.028 676 7691	69
70	4.747 141 3956	166.539 617 5832	.006 004 5773	.210 653 0892	35.082 084 9231	.028 504 5773	70
71	4.853 952 0770	171.286 758 9788	.005 838 1629	.206 017 6912	35.288 102 6143	.028 338 1629	71
72	4.963 165 9988	176.140 711 0559	.005 677 2792	.201 484 2946	35.489 586 9088	.028 177 2792	72
73	5.074 837 2337	181.103 877 0546	.005 521 6929	.197 050 6548	35.686 637 5637	.028 021 6929	73
74	5.189 021 0715	186.178 714 2883	.005 371 1833	.192 714 5768	35.879 352 1405	.027 871 1833	74
75	5.305 774 0456	191.367 735 3598	.005 225 5413	.188 473 9138	36.067 826 0543	.027 725 5413	75
76	5.425 153 9616	196.673 509 4054	.005 084 5689	.184 326 5660	36.252 152 6203	.027 584 5689	76
77	5.547 219 9258	202.098 663 3670	.004 948 0782	.180 270 4802	36.432 423 1006	.027 448 0782	77
78	5.672 032 3741	207.645 883 2928	.004 815 8913	.176 303 6482	36.608 726 7487	.027 315 8913	78
79	5.799 653 1025	213.317 915 6669	.004 687 8388	.172 424 1058	36.781 150 8545	.027 187 8388	79
80	5.930 145 2973	219.117 568 7694	.004 563 7600	.168 629 9323	36.949 780 7868	.027 063 7600	80
81	6.063 573 5665	225.047 714 0667	.004 443 5021	.164 919 2492	37.114 700 0360	.026 943 5021	81
82	6.200 003 9717	231.111 287 6332	.004 326 9198	.161 290 2193	37.275 990 2552	.026 826 9198	82
83	6.339 504 0611	237.311 291 6050	.004 213 8745	.157 741 0457	37.433 731 3010	.026 713 8745	83
84	6.482 142 9025	243.650 795 6661	.004 104 2345	.154 269 9714	37.588 001 2723	.026 604 2345	84
85	6.627 991 1178	250.132 938 5686	.003 997 8741	.150 875 2776	37.738 876 5500	.026 497 8741	85
86	6.777 120 9179	256.760 929 6863	.003 894 6735	.147 555 2837	37.886 431 8337	.026 394 6735	86
87	6.929 606 1386	263.538 050 6043	.003 794 5185	.144 308 3460	38.030 740 1797	.026 294 5185	87
88	7.085 522 2767	270.467 656 7429	.003 697 2998	.141 132 8567	38.171 873 0363	.026 197 2998	88
89	7.244 946 5279	277.553 179 0196	.003 602 9132	.138 027 2437	38.309 900 2800	.026 102 9132	89
90	7.407 957 8248	284.798 125 5475	.003 511 2591	.134 989 9694	38.444 890 2494	.026 011 2591	90
91	7.574 636 8759	292.206 083 3724	.003 422 2422	.132 019 5300	38.576 909 7794	.025 922 2422	91
92	7.745 066 2056	299.780 720 2482	.003 335 7716	.129 114 4547	38.706 024 2341	.025 835 7716	92
93	7.919 330 1952	307.525 786 4538	.003 251 7598	.126 273 3054	38.832 297 5395	.025 751 7598	93
94	8.097 515 1246	315.445 116 6490	.003 170 1236	.123 494 6752	38.955 792 2147	.025 670 1236	94
95	8.279 709 2149	323.542 631 7736	.003 090 7828	.120 777 1884	39.076 569 4031	.025 590 7828	95
96	8.466 002 6722	331.822 340 9885	.003 013 6609	.118 119 4997	39.194 688 9028	.025 513 6609	96
97	8.656 487 7324	340.288 343 6608	.002 938 6843	.115 520 2931	39.310 209 1959	.025 438 6843	97
98	8.851 258 7063	348.944 831 3932	.002 865 7825	.112 978 2818	39.423 187 4776	.025 365 7825	98
99	9.050 412 0272	357.796 090 0995	.002 794 8880	.110 492 2071	39.533 679 6847	.025 294 8880	99
100	9.254 046 2979	366.846 502 1267	.002 725 9358	.108 060 8382	39.641 740 5229	.025 225 9358	100

		XIV COMPOUND AMOUNT	XV AMOUNT OF ANNUITY	XVI SINKING FUND	XVII PRESENT VALUE	XVIII PRESENT VALUE OF ANNUITY	XIX AMORTIZATION		
	PERIODS	AMOUNT OF 1 How $1 left at compound interest will grow.	AMOUNT OF 1 PER PERIOD How $1 deposited periodically will grow.	SINKING FUND Periodic deposit that will grow to $1 at future date.	PRESENT WORTH OF 1 What $1 due in the future is worth today.	PRESENT WORTH OF 1 PER PERIOD What $1 payable periodically is worth today.	PARTIAL PAYMENT Annuity worth $1 today Periodic payment necessary to pay off a loan of $1.	PERIODS	
	1	1.025 000 0000	1.000 000 0000	1.000 000 0000	.975 609 7561	.975 609 7561	1.025 000 0000	1
	2	1.050 625 0000	2.025 000 0000	.493 827 1605	.951 814 3962	1.927 424 1523	.518 827 1605	2
	3	1.076 890 6250	3.075 625 0000	.325 137 1672	.928 599 4109	2.856 023 5632	.350 137 1672	3
	4	1.103 812 8906	4.152 515 6250	.240 817 8777	.905 950 6448	3.761 974 2080	.265 817 8777	4
	5	1.131 408 2129	5.256 328 5156	.190 246 8609	.883 854 2876	4.645 828 4956	.215 246 8609	5
	6	1.159 693 4182	6.387 736 7285	.156 549 9711	.862 296 8660	5.508 125 3616	.181 549 9711	6
	7	1.188 685 7537	7.547 430 1467	.132 495 4296	.841 265 2351	6.349 390 5967	.157 495 4296	7
	8	1.218 402 8975	8.736 115 9004	.114 467 3458	.820 746 5708	7.170 137 1675	.139 467 3458	8
	9	1.248 862 9699	9.954 518 7979	.100 456 8900	.800 728 3618	7.970 865 5292	.125 456 8900	9
	10	1.280 084 5442	11.203 381 7679	.089 258 7632	.781 198 4017	8.752 063 9310	.114 258 7632	10
	11	1.312 086 6578	12.483 466 3121	.080 105 9558	.762 144 7822	9.514 208 7131	.105 105 9558	11
	12	1.344 888 8242	13.795 552 9699	.072 487 1270	.743 555 8850	10.257 764 5982	.097 487 1270	12
	13	1.378 511 0449	15.140 441 7941	.066 048 2708	.725 420 3757	10.983 184 9738	.091 048 2708	13
	14	1.412 973 8210	16.518 952 8390	.060 536 5249	.707 727 1958	11.690 912 1696	.085 536 5249	14
	15	1.448 298 1665	17.931 926 6599	.055 766 4561	.690 465 5568	12.381 377 7264	.080 766 4561	15
	16	1.484 505 6207	19.380 224 8264	.051 598 9886	.673 624 9335	13.055 002 6599	.076 598 9886	16
	17	1.521 618 2612	20.864 730 4471	.047 927 7699	.657 195 0571	13.712 197 7170	.072 927 7699	17
	18	1.559 658 7177	22.386 348 7083	.044 670 0805	.641 165 9093	14.353 363 6264	.069 670 0805	18
	19	1.598 650 1856	23.946 007 4260	.041 760 6151	.625 527 7164	14.978 891 3428	.066 760 6151	19
	20	1.638 616 4403	25.544 657 6116	.039 147 1287	.610 270 9429	15.589 162 2856	.064 147 1287	20
	21	1.679 581 8513	27.183 274 0519	.036 787 3273	.595 386 2857	16.184 548 5714	.061 787 3273	21
	22	1.721 571 3976	28.862 855 9032	.034 646 6061	.580 864 6690	16.765 413 2404	.059 646 6061	22
	23	1.764 610 6825	30.584 427 3008	.032 696 3781	.566 697 2380	17.332 110 4784	.057 696 3781	23
	24	1.808 725 9496	32.349 037 9833	.030 912 8204	.552 875 3542	17.884 985 8326	.055 912 8204	24
	25	1.853 944 0983	34.157 763 9329	.029 275 9210	.539 390 5894	18.424 376 4220	.054 275 9210	25
	26	1.900 292 7008	36.011 708 0312	.027 768 7467	.526 234 7214	18.950 611 1434	.052 768 7467	26
	27	1.947 800 0183	37.912 000 7320	.026 376 8722	.513 399 7282	19.464 010 8717	.051 376 8722	27
	28	1.996 495 0188	39.859 800 7503	.025 087 9327	.500 877 7836	19.964 888 6553	.050 087 9327	28
	29	2.046 407 3942	41.856 295 7690	.023 891 2685	.488 661 2523	20.453 549 9076	.048 891 2685	29
	30	2.097 567 5791	43.902 703 1633	.022 777 6407	.476 742 6852	20.930 292 5928	.047 777 6407	30
	31	2.150 006 7686	46.000 270 7424	.021 739 0025	.465 114 8148	21.395 407 4076	.046 739 0025	31
	32	2.203 756 9378	48.150 277 5109	.020 768 3123	.453 770 5510	21.849 177 9586	.045 768 3123	32
	33	2.258 850 8612	50.354 034 4487	.019 859 3819	.442 702 9766	22.291 880 9352	.044 859 3819	33
	34	2.315 322 1327	52.612 885 3099	.019 006 7508	.431 905 3430	22.723 786 2782	.044 006 7508	34
	35	2.375 205 1861	54.928 207 4426	.018 205 5823	.421 371 0664	23.145 157 3447	.043 205 5823	35
	36	2.432 535 3157	57.301 412 6287	.017 451 5767	.411 093 7233	23.556 251 0680	.042 451 5767	36
	37	2.493 348 6986	59.733 947 9444	.016 740 8992	.401 067 0471	23.957 318 1151	.041 740 8992	37
	38	2.555 682 4161	62.227 296 6430	.016 070 1180	.391 284 9240	24.348 603 0391	.041 070 1180	38
	39	2.619 574 4765	64.782 979 0591	.015 436 1534	.381 741 3893	24.730 344 4284	.040 436 1534	39
	40	2.685 063 8384	67.402 553 5356	.014 836 2332	.372 430 6237	25.102 775 0521	.039 836 2332	40
	41	2.752 190 4343	70.087 617 3740	.014 267 8555	.363 346 9499	25.466 122 0020	.039 267 8555	41
	42	2.820 995 1952	72.839 807 8083	.013 728 7567	.354 484 8292	25.820 606 8313	.038 728 7567	42
	43	2.891 520 0751	75.660 803 0035	.013 216 8833	.345 838 8578	26.166 445 6890	.038 216 8833	43
	44	2.963 808 0770	78.552 323 0786	.012 730 3683	.337 403 7637	26.503 849 4527	.037 730 3683	44
	45	3.037 903 2789	81.516 131 1556	.012 267 5106	.329 174 4036	26.833 023 8563	.037 267 5106	45
	46	3.113 850 8609	84.554 034 4345	.011 826 7568	.321 145 7596	27.154 169 6159	.036 826 7568	46
	47	3.191 697 1324	87.667 885 2954	.011 406 6855	.313 312 9362	27.467 482 5521	.036 406 6855	47
	48	3.271 489 5607	90.859 582 4277	.011 005 9938	.305 671 1573	27.773 153 7094	.036 005 9938	48
	49	3.353 276 7997	94.131 071 9884	.010 623 4847	.298 215 7632	28.071 369 4726	.035 623 4847	49
	50	3.437 108 7197	97.484 348 7881	.010 258 0569	.290 942 2080	28.362 311 6805	.035 258 0569	60
	51	3.523 036 4377	100.921 457 5078	.009 908 6956	.283 846 0566	28.646 157 7371	.034 908 6956	51
	52	3.611 112 3486	104.444 493 9455	.009 574 4635	.276 922 9820	28.923 080 7191	.034 574 4635	52
	53	3.701 390 1574	108.055 606 2942	.009 254 4944	.270 168 7629	29.193 249 4821	.034 254 4944	53
	54	3.793 924 9113	111.756 996 4515	.008 947 9856	.263 579 2809	29.456 828 7630	.033 947 9856	54
	55	3.888 773 0341	115.550 921 3628	.008 654 1932	.257 150 5180	29.713 979 2810	.033 654 1932	55
	56	3.985 992 3599	119.439 694 3969	.008 372 4260	.250 878 5541	29.964 857 8351	.033 372 4260	56
	57	4.085 642 1689	123.425 686 7568	.008 102 0412	.244 759 5650	30.209 617 4001	.033 102 0412	57
	58	4.187 783 2231	127.511 328 9257	.007 842 4404	.238 789 8195	30.448 407 2196	.032 842 4404	58
	59	4.292 477 8037	131.699 112 1489	.007 593 0656	.232 965 6776	30.681 372 8972	.032 593 0656	59
	60	4.399 789 7488	135.991 589 9526	.007 353 3959	.227 283 5879	30.908 656 4851	.032 353 3959	60
	61	4.509 784 4925	140.391 379 7014	.007 122 9445	.221 740 0857	31.130 396 5708	.032 122 9445	61
	62	4.622 529 1048	144.901 164 1940	.006 901 2558	.216 331 7910	31.346 728 3617	.031 901 2558	62
	63	4.738 092 3325	149.523 693 2988	.006 687 9033	.211 055 4058	31.557 783 7676	.031 687 9033	63
	64	4.856 544 6408	154.261 785 6313	.006 482 4869	.205 907 7130	31.763 691 4805	.031 482 4869	64
	65	4.977 958 2568	159.118 330 2721	.006 284 6311	.200 885 5736	31.964 577 0542	.031 284 6311	65
	66	5.102 407 2132	164.096 288 5289	.006 093 9830	.195 985 9255	32.160 562 9797	.031 093 9830	66
	67	5.229 967 3936	169.198 695 7421	.005 910 2110	.191 205 7810	32.351 768 7607	.030 910 2110	67
	68	5.360 716 5784	174.428 663 1356	.005 733 0027	.186 542 2253	32.538 310 9860	.030 733 0027	68
	69	5.494 734 4929	179.789 379 7140	.005 562 0638	.181 992 4150	32.720 303 4010	.030 562 0638	69
	70	5.632 102 8552	185.284 114 2069	.005 397 1168	.177 553 5756	32.897 856 9766	.030 397 1168	70
	71	5.772 905 4266	190.916 217 0620	.005 237 8997	.173 223 0006	33.071 079 9772	.030 237 8997	71
	72	5.917 228 0622	196.689 122 4886	.005 084 1652	.168 998 0543	33.240 078 0265	.030 084 1652	72
	73	6.065 158 7638	202.606 350 5508	.004 935 6794	.164 876 1457	33.404 954 1722	.029 935 6794	73
	74	6.216 787 7329	208.671 509 3146	.004 792 2211	.160 854 7763	33.565 808 9485	.029 792 2211	74
	75	6.372 207 4262	214.888 297 0474	.004 653 5806	.156 931 4891	33.722 740 4375	.029 653 5806	75
	76	6.531 512 6118	221.260 504 4736	.004 519 5594	.153 103 8918	33.875 844 3293	.029 519 5594	76
	77	6.694 800 4271	227.792 017 0855	.004 389 9695	.149 369 6505	34.025 213 9798	.029 389 9695	77
	78	6.862 170 4378	234.486 817 5126	.004 264 6321	.145 726 4883	34.170 940 4681	.029 264 6321	78
	79	7.033 724 6988	241.348 987 9504	.004 143 3776	.142 172 1837	34.313 112 6518	.029 143 3776	79
	80	7.209 567 8162	248.382 712 6492	.004 026 0451	.138 704 5695	34.451 817 2213	.029 026 0451	80
	81	7.389 807 0116	255.592 280 4654	.003 912 4812	.135 321 5312	34.587 138 7525	.028 912 4812	81
	82	7.574 552 1869	262.982 087 4770	.003 802 5404	.132 021 0060	34.719 159 7585	.028 802 5404	82
	83	7.763 915 9916	270.556 639 6640	.003 696 0838	.128 800 9815	34.847 960 7400	.028 696 0838	83
	84	7.958 013 8914	278.320 555 6556	.003 592 9793	.125 659 4941	34.973 620 2342	.028 592 9793	84
	85	8.156 964 2387	286.278 569 5470	.003 493 1011	.122 594 6284	35.096 214 8626	.028 493 1011	85
	86	8.360 888 3446	294.435 533 7856	.003 396 3292	.119 604 5155	35.215 819 3781	.028 396 3292	86
	87	8.569 910 5533	302.796 422 1303	.003 302 5489	.116 687 3322	35.332 506 7104	.028 302 5489	87
	88	8.784 158 3171	311.366 332 6835	.003 211 6510	.113 841 2997	35.446 348 0101	.028 211 6510	88
	89	9.003 762 2750	320.150 491 0006	.003 123 5311	.111 064 6827	35.557 412 6928	.028 123 5311	89
	90	9.228 856 3319	329.154 253 2756	.003 038 0893	.108 355 7880	35.665 768 4808	.028 038 0893	90
	91	9.459 577 7402	338.383 109 6075	.002 955 2302	.105 712 9639	35.771 481 4447	.027 955 2302	91
	92	9.696 067 1837	347.842 687 3477	.002 874 8628	.103 134 5989	35.874 616 0436	.027 874 8628	92
	93	9.938 468 8633	357.538 754 5314	.002 796 8996	.100 619 1209	35.975 235 1645	.027 796 8996	93
	94	10.186 930 5849	367.477 223 3947	.002 721 2571	.098 164 9960	36.073 400 1605	.027 721 2571	94
	95	10.441 603 8495	377.664 153 9796	.002 647 8552	.095 770 7278	36.169 170 8882	.027 647 8552	95
	96	10.702 643 9457	388.105 757 8290	.002 576 6173	.093 434 8564	36.262 605 7446	.027 576 6173	96
	97	10.970 210 0444	398.808 401 7748	.002 507 4697	.091 155 9574	36.353 761 7021	.027 507 4697	97
	98	11.244 465 2955	409.778 611 8191	.002 440 3421	.088 932 6414	36.442 694 3435	.027 440 3421	98
	99	11.525 576 9279	421.023 077 1146	.002 375 1667	.086 763 5526	36.529 457 8961	.027 375 1667	99
	100	11.813 716 3511	432.548 654 0425	.002 311 8787	.084 647 3684	36.614 105 2645	.027 311 8787	100

	PERIODS	XIV COMPOUND AMOUNT	XV AMOUNT OF ANNUITY	XVI SINKING FUND	XVII PRESENT VALUE	XVIII PRESENT VALUE OF ANNUITY	XIX AMORTIZATION	PERIODS
RATE 3%		AMOUNT OF 1	AMOUNT OF 1 PER PERIOD	SINKING FUND	PRESENT WORTH OF 1	PRESENT WORTH OF 1 PER PERIOD	PARTIAL PAYMENT	
		How $1 left at compound interest will grow.	How $1 deposited periodically will grow.	Periodic deposit that will grow to $1 at future date.	What $1 due in the future is worth today.	What $1 payable periodically is worth today.	Annuity worth $1 today. Periodic payment necessary to pay off a loan of $1.	
	1	1.030 000 0000	1.000 000 0000	1.000 000 0000	.970 873 7864	.970 873 7864	1.030 000 0000	1
	2	1.060 900 0000	2.030 000 0000	.492 610 8374	.942 595 9091	1.913 469 6955	.522 610 8374	2
	3	1.092 727 0000	3.090 900 0000	.323 530 3633	.915 141 6594	2.828 611 3549	.353 530 3633	3
	4	1.125 508 8100	4.183 627 0000	.239 027 0452	.888 487 0479	3.717 098 4028	.269 027 0452	4
	5	1.159 274 0743	5.309 135 8100	.188 354 5714	.862 608 7844	4.579 707 1872	.218 354 5714	5
	6	1.194 052 2965	6.468 409 8843	.154 597 5005	.837 484 2567	5.417 191 4439	.184 597 5005	6
	7	1.229 873 8654	7.662 462 1808	.130 506 3538	.813 091 5113	6.230 282 9552	.160 506 3538	7
	8	1.266 770 0814	8.892 336 0463	.112 456 3888	.789 409 2343	7.019 692 1895	.142 456 3888	8
	9	1.304 773 1838	10.159 106 1276	.098 433 8570	.766 416 7323	7.786 108 9219	.128 433 8570	9
	10	1.343 916 3793	11.463 879 3115	.087 230 5066	.744 093 9149	8.530 202 8368	.117 230 5066	10
	11	1.384 233 8707	12.807 795 6908	.078 077 4478	.722 421 2766	9.252 624 1134	.108 077 4478	11
	12	1.425 760 8868	14.192 029 5615	.070 462 0855	.701 379 8802	9.954 003 9936	.100 462 0855	12
	13	1.468 533 7135	15.617 790 4484	.064 029 5440	.680 951 3400	10.634 955 3336	.094 029 5440	13
	14	1.512 589 7249	17.086 324 1618	.058 526 3390	.661 117 8058	11.296 073 1394	.088 526 3390	14
	15	1.557 967 4166	18.598 913 8867	.053 766 5805	.641 861 9474	11.937 935 0868	.083 766 5805	15
	16	1.604 706 4391	20.156 881 3033	.049 610 8493	.623 166 9392	12.561 102 0260	.079 610 8493	16
	17	1.652 847 6323	21.761 587 7424	.045 952 5294	.605 016 4458	13.166 118 4718	.075 952 5294	17
	18	1.702 433 0612	23.414 435 3747	.042 708 6959	.587 394 6076	13.753 513 0795	.072 708 6959	18
	19	1.753 506 0531	25.116 868 4359	.039 813 8806	.570 286 0268	14.323 799 1063	.069 813 8806	19
	20	1.806 111 2347	26.870 374 4890	.037 215 7076	.553 675 7542	14.877 474 8605	.067 215 7076	20
	21	1.860 294 5717	28.676 485 7236	.034 871 7765	.537 549 2759	15.415 024 1364	.064 871 7765	21
	22	1.916 103 4089	30.536 780 2954	.032 747 3948	.521 892 5009	15.936 916 6372	.062 747 3948	22
	23	1.973 586 5111	32.452 883 7042	.030 813 9027	.506 691 7484	16.443 608 3857	.060 813 9027	23
	24	2.032 794 1065	34.426 470 2153	.029 047 4159	.491 933 7363	16.935 542 1220	.059 047 4159	24
	25	2.093 777 9297	36.459 264 3218	.027 427 8710	.477 605 5693	17.413 147 6913	.057 427 8710	25
	26	2.156 591 2675	38.553 042 2515	.025 938 2903	.463 694 7274	17.876 842 4187	.055 938 2903	26
	27	2.221 289 0056	40.709 633 5190	.024 564 2103	.450 189 0558	18.327 031 4745	.054 564 2103	27
	28	2.287 927 6757	42.930 922 5246	.023 293 2334	.437 076 7532	18.764 108 2277	.053 293 2334	28
	29	2.356 565 5060	45.218 850 2003	.022 114 6711	.424 346 3623	19.188 454 5900	.052 114 6711	29
	30	2.427 262 4712	47.575 415 7063	.021 019 2593	.411 986 7595	19.600 441 3495	.051 019 2593	30
	31	2.500 080 3453	50.002 678 1775	.019 998 9288	.399 987 1452	20.000 428 4946	.049 998 9288	31
	32	2.575 082 7557	52.502 758 5228	.019 046 6183	.388 337 0341	20.388 765 5288	.049 046 6183	32
	33	2.652 335 2384	55.077 841 2785	.018 156 1219	.377 026 2467	20.765 791 7755	.048 156 1219	33
	34	2.731 905 2955	57.730 176 5169	.017 321 9633	.366 044 8997	21.131 836 6752	.047 321 9633	34
	35	2.813 862 4544	60.462 081 8124	.016 539 2916	.355 383 3978	21.487 220 0731	.046 539 2916	35
	36	2.898 278 3280	63.275 944 2668	.015 803 7942	.345 032 4251	21.832 252 4981	.045 803 7942	36
	37	2.985 226 6778	66.174 222 5948	.015 111 6244	.334 982 9369	22.167 235 4351	.045 111 6244	37
	38	3.074 783 4782	69.159 449 2726	.014 459 3401	.325 226 1524	22.492 461 5874	.044 459 3401	38
	39	3.167 026 9825	72.234 232 7508	.013 843 8516	.315 753 5460	22.808 215 1334	.043 843 8516	39
	40	3.262 037 7920	75.401 259 7333	.013 262 3779	.306 556 8408	23.114 771 9742	.043 262 3779	40
	41	3.359 898 9258	78.663 297 5253	.012 712 4089	.297 628 0008	23.412 399 9750	.042 712 4089	41
	42	3.460 695 8935	82.023 196 4511	.012 191 6731	.288 959 2240	23.701 359 1990	.042 191 6731	42
	43	3.564 516 7703	85.483 892 3446	.011 698 1103	.280 542 9360	23.981 902 1349	.041 698 1103	43
	44	3.671 452 2734	89.048 409 1149	.011 229 8469	.272 371 7825	24.254 273 9174	.041 229 8469	44
	45	3.781 595 8417	92.719 861 3884	.010 785 1757	.264 438 6238	24.518 712 5412	.040 785 1757	45
	46	3.895 043 7169	96.501 457 2300	.010 362 5378	.256 736 5279	24.775 449 0691	.040 362 5378	46
	47	4.011 895 0284	100.396 500 9469	.009 960 5065	.249 258 7650	25.024 707 8341	.039 960 5065	47
	48	4.132 251 8793	104.408 395 9753	.009 577 7738	.241 998 8009	25.266 706 6350	.039 577 7738	48
	49	4.256 219 4356	108.540 647 8546	.009 213 1383	.234 950 2922	25.501 656 9272	.039 213 1383	49
	50	4.383 906 0187	112.796 867 2902	.008 865 4944	.228 107 0798	25.729 764 0070	.038 865 4944	50
	51	4.515 423 1993	117.180 773 3089	.008 533 8232	.221 463 1843	25.951 227 1913	.038 533 8232	51
	52	4.650 885 8952	121.696 196 5082	.008 217 1837	.215 012 8003	26.166 239 9915	.038 217 1837	52
	53	4.790 412 4721	126.347 082 4035	.007 914 7059	.208 750 2915	26.374 990 2830	.037 914 7059	53
	54	4.934 124 8463	131.137 494 8756	.007 625 5841	.202 670 1859	26.577 660 4690	.037 625 5841	54
	55	5.082 148 5917	136.071 619 7218	.007 349 0710	.196 767 1708	26.774 427 6398	.037 349 0710	55
	56	5.234 613 0494	141.153 768 3135	.007 084 4726	.191 036 0882	26.965 463 7279	.037 084 4726	56
	57	5.391 651 4409	146.388 381 3629	.006 831 1432	.185 471 9303	27.150 935 6582	.036 831 1432	57
	58	5.553 400 9841	151.780 032 8038	.006 588 4819	.180 069 8352	27.331 005 4934	.036 588 4819	58
	59	5.720 003 0136	157.333 433 7879	.006 355 9281	.174 825 0827	27.505 830 5761	.036 355 9281	59
	60	5.891 603 1040	163.053 436 8015	.006 132 9587	.169 733 0900	27.675 563 6661	.036 132 9587	60
	61	6.068 351 1972	168.945 039 9056	.005 919 0847	.164 789 4978	27.840 353 0739	.035 919 0847	61
	62	6.250 401 7331	175.013 391 1027	.005 713 8485	.159 989 7163	28.000 342 7902	.035 713 8485	62
	63	6.437 913 7851	181.263 792 8358	.005 516 8216	.155 329 8216	28.155 672 6118	.035 516 8216	63
	64	6.631 051 1986	187.701 706 6209	.005 327 6021	.150 805 6521	28.306 478 2639	.035 327 6021	64
	65	6.829 982 7346	194.332 757 8195	.005 145 8128	.146 413 2544	28.452 891 5184	.035 145 8128	65
	66	7.034 882 2166	201.162 740 5541	.004 971 0995	.142 148 7907	28.595 040 3091	.034 971 0995	66
	67	7.245 928 6831	208.197 622 7707	.004 803 1288	.138 008 5347	28.733 048 8438	.034 803 1288	67
	68	7.463 306 5436	215.443 551 4539	.004 641 5871	.133 988 8686	28.867 037 7124	.034 641 5871	68
	69	7.687 205 7399	222.906 857 9975	.004 486 1787	.130 086 2802	28.997 123 9926	.034 486 1787	69
	70	7.917 821 9121	230.594 063 7374	.004 336 6251	.126 297 3594	29.123 421 3521	.034 336 6251	70
	71	8.155 356 5695	238.511 885 6495	.004 192 6632	.122 618 7956	29.246 040 1476	.034 192 6632	71
	72	8.400 017 2666	246.667 242 2190	.004 054 0446	.119 047 3743	29.365 087 5220	.034 054 0446	72
	73	8.652 017 7846	255.067 259 4856	.003 920 5345	.115 579 9751	29.480 667 4971	.033 920 5345	73
	74	8.911 578 3181	263.719 277 2701	.003 791 9109	.112 213 5680	29.592 881 0651	.033 791 9109	74
	75	9.178 925 6676	272.630 855 5882	.003 667 9634	.108 945 2117	29.701 826 2768	.033 667 9634	75
	76	9.454 293 4377	281.809 781 2559	.003 548 4929	.105 772 0502	29.807 598 3270	.033 548 4929	76
	77	9.737 922 2408	291.264 074 6936	.003 433 3105	.102 691 3109	29.910 289 6379	.033 433 3105	77
	78	10.030 059 9080	301.001 996 9344	.003 322 2371	.099 700 3018	30.009 989 9397	.033 322 2371	78
	79	10.330 961 7053	311.032 056 8424	.003 215 1027	.096 796 4095	30.106 786 3492	.033 215 1027	79
	80	10.640 890 5564	321.363 018 5477	.003 111 7457	.093 977 0966	30.200 763 4458	.033 111 7457	80
	81	10.960 117 2731	332.003 909 1041	.003 012 0127	.091 239 8996	30.292 003 3455	.033 012 0127	81
	82	11.288 920 7913	342.964 026 3772	.002 915 7577	.088 582 4268	30.380 585 7723	.032 915 7577	82
	83	11.627 588 4151	354.252 947 1685	.002 822 8417	.086 002 3561	30.466 588 1284	.032 822 8417	83
	84	11.976 416 0675	365.880 535 5836	.002 733 1325	.083 497 4332	30.550 085 5616	.032 733 1325	84
	85	12.335 708 5495	377.856 951 6511	.002 646 5042	.081 065 4691	30.631 151 0307	.032 646 5042	85
	86	12.705 779 8060	390.192 660 2006	.002 562 8365	.078 704 3389	30.709 855 3696	.032 562 8365	86
	87	13.086 953 2002	402.898 440 0067	.002 482 0151	.076 411 9795	30.786 267 3491	.032 482 0151	87
	88	13.479 561 7962	415.985 393 2069	.002 403 9306	.074 186 3879	30.860 453 7370	.032 403 9306	88
	89	13.883 948 6501	429.464 955 0031	.002 328 4787	.072 025 6193	30.932 479 3563	.032 328 4787	89
	90	14.300 467 1096	443.348 903 6532	.002 255 5599	.069 927 7857	31.002 407 1421	.032 255 5599	90
	91	14.729 481 1229	457.649 370 7628	.002 185 0789	.067 891 0541	31.070 298 1962	.032 185 0789	91
	92	15.171 365 5566	472.378 851 8856	.002 116 9449	.065 913 6448	31.136 211 8409	.032 116 9449	92
	93	15.626 506 5233	487.550 217 4422	.002 051 0708	.063 993 8299	31.200 205 6708	.032 051 0708	93
	94	16.095 301 7190	503.176 723 9655	.001 987 3733	.062 129 9319	31.262 335 6027	.031 987 3733	94
	95	16.578 160 7705	519.272 025 6844	.001 925 7729	.060 320 3223	31.322 655 9250	.031 925 7729	95
	96	17.075 505 5936	535.850 186 4550	.001 866 1932	.058 563 4197	31.381 219 3446	.031 866 1932	96
	97	17.587 770 7615	552.925 692 0486	.001 808 5613	.056 857 6890	31.438 077 0336	.031 808 5613	97
	98	18.115 403 8843	570.513 462 8101	.001 752 8070	.055 201 6398	31.493 278 6734	.031 752 8070	98
	99	18.658 866 0008	588.628 866 6944	.001 698 8633	.053 593 8250	31.546 872 4985	.031 698 8633	99
	100	19.218 631 9809	607.287 732 6952	.001 646 6659	.052 032 8399	31.598 905 3383	.031 646 6659	100

	XIV COMPOUND AMOUNT	XV AMOUNT OF ANNUITY	XVI SINKING FUND	XVII PRESENT VALUE	XVIII PRESENT VALUE OF ANNUITY	XIX AMORTIZATION	
PERIODS	AMOUNT OF 1 — How $1 left at compound interest will grow.	AMOUNT OF 1 PER PERIOD — How $1 deposited periodically will grow.	SINKING FUND — Periodic deposit that will grow to $1 at future date.	PRESENT WORTH OF 1 — What $1 due in the future is worth today.	PRESENT WORTH OF 1 PER PERIOD — What $1 payable periodically is worth today.	PARTIAL PAYMENT — Annuity worth $1 today. Periodic payment necessary to pay off a loan of $1.	PERIODS
1	1.035 000 0000	1.000 000 0000	1.000 000 0000	.966 183 5749	.966 183 5749	1.035 000 0000	1
2	1.071 225 0000	2.035 000 0000	.491 400 4914	.933 510 7004	1.899 694 2752	.526 400 4914	2
3	1.108 717 8750	3.106 225 0000	.321 934 1806	.901 942 7057	2.801 636 9809	.356 934 1806	3
4	1.147 523 0006	4.214 942 8750	.237 251 1395	.871 442 2277	3.673 079 2086	.272 251 1395	4
5	1.187 686 3056	5.362 465 8756	.186 481 3732	.841 973 1669	4.515 052 3755	.221 481 3732	5
6	1.229 255 3263	6.550 152 1813	.152 668 2087	.813 500 6443	5.328 553 0198	.187 668 2087	6
7	1.272 279 2628	7.779 407 5076	.128 544 4938	.785 990 9607	6.114 543 9805	.163 544 4938	7
8	1.316 809 0370	9.051 686 7704	.110 476 6465	.759 411 5562	6.873 955 5367	.145 476 6465	8
9	1.362 897 3533	10.368 495 8073	.096 446 0051	.733 730 9722	7.607 686 5089	.131 446 0051	9
10	1.410 598 7606	11.731 393 1606	.085 241 3679	.708 918 8137	8.316 605 3226	.120 241 3679	10
11	1.459 969 7172	13.141 991 9212	.076 091 9658	.684 945 7137	9.001 551 0363	.111 091 9658	11
12	1.511 068 6573	14.601 961 6385	.068 483 9493	.661 783 2983	9.663 334 3346	.103 483 9493	12
13	1.563 956 0604	16.113 030 2958	.062 061 5726	.639 404 1529	10.302 738 4875	.097 061 5726	13
14	1.618 694 5275	17.676 986 3562	.056 570 7287	.617 781 7903	10.920 520 2778	.091 570 7287	14
15	1.675 348 8308	19.295 680 8786	.051 825 0694	.596 890 6186	11.517 410 8964	.086 825 0694	15
16	1.733 986 0398	20.971 029 7094	.047 684 8306	.576 705 9117	12.094 116 8081	.082 684 8306	16
17	1.794 675 5512	22.705 015 7492	.044 043 1317	.557 203 7794	12.651 320 5876	.079 043 1317	17
18	1.857 489 1955	24.499 691 3004	.040 816 8408	.538 361 1396	13.189 681 7271	.075 816 8408	18
19	1.922 501 3174	26.357 180 4960	.037 940 3252	.520 155 6904	13.709 837 4175	.072 940 3252	19
20	1.989 788 8635	28.279 681 8133	.035 361 0768	.502 565 8844	14.212 403 3020	.070 361 0768	20
21	2.059 431 4737	30.269 470 6768	.033 036 5870	.485 570 9028	14.697 974 2048	.068 036 5870	21
22	2.131 511 5753	32.328 902 1505	.030 932 0742	.469 150 6308	15.167 124 8355	.065 932 0742	22
23	2.206 114 4804	34.460 413 7257	.029 018 8042	.453 285 6336	15.620 410 4691	.064 018 8042	23
24	2.283 228 4872	36.666 528 2061	.027 272 8303	.437 957 1339	16.058 367 6030	.062 272 8303	24
25	2.363 244 9843	38.949 856 6933	.025 674 0354	.423 146 9893	16.481 514 5923	.060 674 0354	25
26	2.445 958 5587	41.313 101 6776	.024 205 3963	.408 837 6708	16.890 352 2631	.059 205 3963	26
27	2.531 567 1083	43.759 060 2363	.022 852 4103	.395 012 2423	17.285 364 5054	.057 852 4103	27
28	2.620 171 9571	46.290 627 3446	.021 602 6452	.381 654 3404	17.667 018 8458	.056 602 6452	28
29	2.711 877 9756	48.910 799 3017	.020 445 3825	.368 748 1550	18.035 767 0008	.055 445 3825	29
30	2.806 793 7047	51.622 677 2772	.019 371 3316	.356 278 4106	18.392 045 4114	.054 371 3316	30
31	2.905 031 4844	54.429 470 9819	.018 372 3998	.344 230 3484	18.736 275 7598	.053 372 3998	31
32	3.006 707 5863	57.334 502 4663	.017 441 5048	.332 589 7086	19.068 865 4684	.052 441 5048	32
33	3.111 942 3518	60.341 210 0526	.016 572 4221	.321 342 7136	19.390 208 1820	.051 572 4221	33
34	3.220 860 3342	63.453 152 4044	.015 759 6583	.310 476 0518	19.700 684 2338	.050 759 6583	34
35	3.333 590 4459	66.674 012 7386	.014 998 3473	.299 976 8617	20.000 661 0955	.049 998 3473	35
36	3.450 266 1115	70.007 603 1845	.014 284 1628	.289 832 7166	20.290 493 8121	.049 284 1628	36
37	3.571 025 4254	73.457 869 2959	.013 613 2454	.280 031 6102	20.570 525 4223	.048 613 2454	37
38	3.696 011 3152	77.028 894 7213	.012 982 1414	.270 561 9422	20.841 087 3645	.047 982 1414	38
39	3.825 371 7113	80.724 906 0365	.012 387 7506	.261 412 5046	21.102 499 8691	.047 387 7506	39
40	3.959 259 7212	84.550 277 7478	.011 827 2823	.252 572 4682	21.355 072 3373	.046 827 2823	40
41	4.097 833 8114	88.509 537 4690	.011 298 2174	.244 031 3702	21.599 103 7075	.046 298 2174	41
42	4.241 257 9948	92.607 371 2804	.010 798 2765	.235 779 1017	21.834 882 8092	.045 798 2765	42
43	4.389 702 0246	96.848 629 2752	.010 325 3914	.227 805 8953	22.062 688 7046	.045 325 3914	43
44	4.543 341 5955	101.238 331 2998	.009 877 6816	.220 102 3143	22.282 791 0189	.044 877 6816	44
45	4.702 358 5513	105.781 672 8953	.009 453 4334	.212 659 2409	22.495 450 2598	.044 453 4334	45
46	4.866 941 1006	110.484 031 4467	.009 051 0817	.205 467 8656	22.700 918 1254	.044 051 0817	46
47	5.037 284 0392	115.350 972 5473	.008 669 1944	.198 519 6769	22.899 437 8023	.043 669 1944	47
48	5.213 588 9805	120.388 256 5864	.008 306 4580	.191 806 4511	23.091 244 2534	.043 306 4580	48
49	5.396 264 5948	125.601 845 5670	.007 961 6665	.185 320 2426	23.276 564 4961	.042 961 6665	49
50	5.584 926 8557	130.997 910 1618	.007 633 7096	.179 053 3745	23.455 617 8706	.042 633 7096	50
51	5.780 399 2956	136.582 837 0175	.007 321 5641	.172 998 4295	23.628 616 3001	.042 321 5641	51
52	5.982 713 2710	142.363 236 3131	.007 024 2854	.167 148 2411	23.795 764 5412	.042 024 2854	52
53	6.192 108 2354	148.345 949 5840	.006 740 9997	.161 495 8851	23.957 260 4263	.041 740 9997	53
54	6.408 832 0237	154.538 057 8195	.006 470 8979	.156 034 6716	24.113 295 0978	.041 470 8979	54
55	6.633 141 1445	160.946 889 8432	.006 213 2297	.150 758 1368	24.264 053 2346	.041 213 2297	55
56	6.865 301 0846	167.580 030 9877	.005 967 2981	.145 660 0355	24.409 713 2702	.040 967 2981	56
57	7.105 586 6225	174.445 332 0722	.005 732 4549	.140 734 3339	24.550 447 6040	.040 732 4549	57
58	7.354 282 1543	181.550 918 6948	.005 508 0966	.135 975 2018	24.686 422 8058	.040 508 0966	58
59	7.611 682 0297	188.905 200 8491	.005 293 6605	.131 377 0066	24.817 799 8124	.040 293 6605	59
60	7.878 090 9008	196.516 882 8788	.005 088 6213	.126 934 3059	24.944 734 1182	.040 088 6213	60
61	8.153 824 0823	204.394 973 7796	.004 892 4882	.122 641 8414	25.067 375 9597	.039 892 4882	61
62	8.439 207 9252	212.548 797 8619	.004 704 8020	.118 494 5328	25.185 870 4924	.039 704 8020	62
63	8.734 580 2025	220.988 005 7870	.004 525 1325	.114 487 4713	25.300 357 9637	.039 525 1325	63
64	9.040 290 5096	229.722 585 9896	.004 353 0765	.110 615 9143	25.410 973 8780	.039 353 0765	64
65	9.356 700 6775	238.762 876 4992	.004 188 2558	.106 875 2795	25.517 849 1575	.039 188 2558	65
66	9.684 185 2012	248.119 577 1767	.004 030 3148	.103 261 1396	25.621 110 2971	.039 030 3148	66
67	10.023 131 6832	257.803 762 3779	.003 878 9193	.099 769 2170	25.720 879 5141	.038 878 9193	67
68	10.373 941 2921	267.826 894 0611	.003 733 7550	.096 395 3788	25.817 274 8928	.038 733 7550	68
69	10.737 029 2374	278.200 835 3532	.003 594 5255	.093 135 6316	25.910 410 5245	.038 594 5255	69
70	11.112 825 2607	288.937 864 5906	.003 460 9517	.089 986 1175	26.000 396 6420	.038 460 9517	70
71	11.501 774 1448	300.050 689 8512	.003 332 7702	.086 943 1087	26.087 339 7507	.038 332 7702	71
72	11.904 336 2399	311.552 463 9960	.003 209 7323	.084 003 0036	26.171 342 7543	.038 209 7323	72
73	12.320 988 0083	323.456 800 2359	.003 091 6030	.081 162 3223	26.252 505 0766	.038 091 6030	73
74	12.752 222 5885	335.777 788 2442	.002 978 1601	.078 417 7027	26.330 922 7794	.037 978 1601	74
75	13.198 550 3791	348.530 010 8327	.002 869 1934	.075 765 8964	26.406 688 6757	.037 869 1934	75
76	13.660 499 6424	361.728 561 2119	.002 764 5038	.073 203 7646	26.479 892 4403	.037 764 5038	76
77	14.138 617 1299	375.389 060 8543	.002 663 9029	.070 728 2750	26.550 620 7153	.037 663 9029	77
78	14.633 468 7294	389.527 677 9842	.002 567 2117	.068 336 4976	26.618 957 2128	.037 567 2117	78
79	15.145 640 1350	404.161 146 7136	.002 474 2606	.066 025 6015	26.684 982 8143	.037 474 2606	79
80	15.675 737 5397	419.306 786 8486	.002 384 8887	.063 792 8517	26.748 775 6660	.037 384 8887	80
81	16.224 388 3536	434.982 524 3883	.002 298 9429	.061 635 6055	26.810 411 2715	.037 298 9429	81
82	16.792 241 9460	451.206 912 7419	.002 216 2781	.059 551 3097	26.869 962 5812	.037 216 2781	82
83	17.379 970 4141	467.999 154 6878	.002 136 7560	.057 537 4973	26.927 500 0784	.037 136 7560	83
84	17.988 269 3786	485.379 125 1019	.002 060 2452	.055 591 7848	26.983 091 8632	.037 060 2452	84
85	18.617 858 8068	503.367 394 4805	.001 986 6205	.053 711 8694	27.036 803 7326	.036 986 6205	85
86	19.269 483 8651	521.985 253 2873	.001 915 7629	.051 895 5260	27.088 699 2585	.036 915 7629	86
87	19.943 915 8003	541.254 737 1524	.001 847 5589	.050 140 6048	27.138 839 8633	.036 847 5589	87
88	20.641 952 8533	561.198 652 9527	.001 781 9002	.048 445 0288	27.187 284 8921	.036 781 9002	88
89	21.364 421 2032	581.840 605 8060	.001 718 6838	.046 806 7911	27.234 091 6832	.036 718 6838	89
90	22.112 175 9453	603.205 027 0092	.001 657 8111	.045 223 9527	27.279 315 6359	.036 657 8111	90
91	22.886 102 1034	625.317 202 9546	.001 599 1884	.043 694 6403	27.323 010 2762	.036 599 1884	91
92	23.687 115 6770	648.203 305 0580	.001 542 7259	.042 217 0438	27.365 227 3200	.036 542 7259	92
93	24.516 164 7257	671.890 420 7350	.001 488 3379	.040 789 4143	27.406 016 7343	.036 488 3379	93
94	25.374 230 4911	696.406 585 4607	.001 435 9428	.039 410 0621	27.445 426 7965	.036 435 9428	94
95	26.262 328 5583	721.780 815 9519	.001 385 4621	.038 077 3547	27.483 504 1512	.036 385 4621	95
96	27.181 510 0579	748.043 144 5102	.001 336 8213	.036 789 7147	27.520 293 8659	.036 336 8213	96
97	28.132 862 9099	775.224 654 5680	.001 289 9487	.035 545 6181	27.555 839 4839	.036 289 9487	97
98	29.117 513 1117	803.357 517 4779	.001 244 7758	.034 343 5923	27.590 183 0763	.036 244 7758	98
99	30.136 626 0706	832.475 030 5896	.001 201 2372	.033 182 2148	27.623 365 2911	.036 201 2372	99
100	31.191 407 9831	862.611 656 6603	.001 159 2702	.032 060 1109	27.655 425 4020	.036 159 2702	100

	XIV COMPOUND AMOUNT	XV AMOUNT OF ANNUITY	XVI SINKING FUND	XVII PRESENT VALUE	XVIII PRESENT VALUE OF ANNUITY	XIX AMORTIZATION	
RATE 3 3/4% PERIODS	AMOUNT OF 1 — How $1 left at compound interest will grow.	AMOUNT OF 1 PER PERIOD — How $1 deposited periodically will grow.	SINKING FUND — Periodic deposit that will grow to $1 at future date.	PRESENT WORTH OF 1 — What $1 due in the future is worth today.	PRESENT WORTH OF 1 PER PERIOD — What $1 payable periodically is worth today.	PARTIAL PAYMENT — Annuity worth $1 today. Periodic payment necessary to pay off a loan of $1.	PERIODS
1	1.037 500 0000	1.000 000 0000	1.000 000 0000	.963 855 4217	.963 855 4217	1.037 500 0000	1
2	1.076 406 2500	2.037 500 0000	.490 797 5460	.929 017 2739	1.892 872 6956	.528 297 5460	2
3	1.116 771 4844	3.113 906 2500	.321 140 0472	.895 438 3363	2.788 311 0319	.358 640 0472	3
4	1.158 650 4150	4.230 677 7344	.236 368 7482	.863 073 0952	3.651 384 1271	.273 868 7482	4
5	1.202 099 8056	5.389 328 1494	.185 551 8856	.831 877 6822	4.483 261 8093	.223 051 8856	5
6	1.247 178 5483	6.591 427 9550	.151 712 1945	.801 809 8141	5.285 071 6234	.189 212 1945	6
7	1.293 947 7439	7.838 606 5033	.127 573 6956	.772 828 7365	6.057 900 3599	.165 073 6956	7
8	1.342 470 7843	9.132 554 2472	.109 498 3915	.744 895 1677	6.802 795 5276	.146 998 3915	8
9	1.392 813 4387	10.475 025 0315	.095 465 1657	.717 971 2460	7.520 766 7736	.132 965 1657	9
10	1.445 043 9426	11.867 838 4702	.084 261 3423	.692 020 4781	8.212 787 2517	.121 761 3423	10
11	1.499 233 0905	13.312 882 4128	.075 115 2131	.667 007 6897	8.879 794 9414	.112 615 2131	11
12	1.555 454 3314	14.812 115 5033	.067 512 3010	.642 898 9780	9.522 693 9194	.105 012 3010	12
13	1.613 783 8688	16.367 569 8346	.061 096 4248	.619 661 6656	10.142 355 5850	.098 596 4248	13
14	1.674 300 7639	17.981 353 7034	.055 613 1655	.597 264 2560	10.739 619 8409	.093 113 1655	14
15	1.737 087 0425	19.655 654 4673	.050 875 9452	.575 676 3913	11.315 296 2322	.088 375 9452	15
16	1.802 227 8066	21.392 741 5098	.046 744 8270	.554 868 8109	11.870 165 0431	.084 244 8270	16
17	1.869 811 3494	23.194 969 3165	.043 112 7968	.534 813 3117	12.404 978 3548	.080 612 7968	17
18	1.939 929 2750	25.064 780 6658	.039 896 6188	.515 482 7101	12.920 461 0649	.077 396 6188	18
19	2.012 676 6228	27.004 709 9408	.037 030 5773	.496 850 8049	13.417 311 8698	.074 530 5773	19
20	2.088 151 9961	29.017 386 5636	.034 462 0973	.478 892 3421	13.896 204 2118	.071 962 0973	20
21	2.166 457 6960	31.105 538 5597	.032 148 6155	.461 582 9803	14.357 787 1921	.069 648 6155	21
22	2.247 699 8596	33.271 996 2557	.030 055 3051	.444 899 2581	14.802 686 4502	.067 555 3051	22
23	2.331 988 6043	35.519 696 1153	.028 153 3940	.428 818 5620	15.231 505 0123	.065 653 3940	23
24	2.419 438 1770	37.851 684 7196	.026 418 9033	.413 319 0959	15.644 824 1082	.063 918 9033	24
25	2.510 167 1086	40.271 122 8966	.024 831 6890	.398 379 8515	16.043 203 9597	.062 331 6890	25
26	2.604 298 3752	42.781 290 0052	.023 374 7042	.383 980 5798	16.427 184 5395	.060 874 7042	26
27	2.701 959 5643	45.385 588 3804	.022 033 4259	.370 101 7636	16.797 286 3031	.059 533 4259	27
28	2.803 283 0479	48.087 547 9447	.020 795 4043	.356 724 5915	17.154 010 8946	.058 295 4043	28
29	2.908 406 1622	50.890 830 9926	.019 649 9051	.343 830 9315	17.497 841 8261	.057 149 9051	29
30	3.017 471 3933	53.799 237 1548	.018 587 6242	.331 403 3075	17.829 245 1336	.056 087 6242	30
31	3.130 626 5706	56.816 708 5481	.017 600 4564	.319 424 8747	18.148 670 0083	.055 100 4564	31
32	3.248 025 0670	59.947 335 1187	.016 681 3087	.307 879 3973	18.456 549 4056	.054 181 3087	32
33	3.369 826 0070	63.195 360 1856	.015 823 9465	.296 751 2263	18.753 300 6319	.053 323 9465	33
34	3.496 194 4822	66.565 186 1926	.015 022 8679	.286 025 2784	19.039 325 9102	.051 522 8679	34
35	3.627 301 7753	70.061 380 6748	.014 273 1986	.275 687 0153	19.315 012 9255	.051 773 1986	35
36	3.763 325 5919	73.688 682 4501	.013 570 6050	.265 722 4244	19.580 735 3499	.051 070 6050	36
37	3.904 450 3016	77.452 008 0420	.012 911 2211	.256 117 9994	19.836 853 3493	.050 411 2211	37
38	4.050 867 1879	81.356 458 3436	.012 291 5872	.246 860 7223	20.083 714 0716	.049 791 5872	38
39	4.202 774 7074	85.407 325 5315	.011 708 5975	.237 938 0456	20.321 652 1172	.049 208 5975	39
40	4.360 378 7590	89.610 100 2389	.011 159 4563	.229 337 8753	20.550 989 9925	.048 659 4563	40
41	4.523 892 9624	93.970 478 9979	.010 641 6399	.221 048 5545	20.772 038 5470	.048 141 6399	41
42	4.693 538 9485	98.494 371 9603	.010 152 8644	.213 058 8477	20.985 097 3947	.047 652 8644	42
43	4.869 546 6591	103.187 910 9088	.009 691 0577	.205 357 9255	21.190 455 3202	.047 191 0577	43
44	5.052 154 6588	108.057 457 5679	.009 254 3358	.197 935 3499	21.388 390 6701	.046 754 3358	44
45	5.241 610 4585	113.109 612 2267	.008 840 9816	.190 781 0601	21.579 171 7302	.046 340 9816	45
46	5.438 170 8507	118.351 222 6852	.008 449 4269	.183 885 3592	21.763 057 0893	.045 949 4269	46
47	5.642 102 2576	123.789 393 5359	.008 078 2365	.177 238 9004	21.940 295 9897	.045 578 2365	47
48	5.853 681 0923	129.431 495 7935	.007 726 0947	.170 832 6751	22.111 128 6648	.045 226 0947	48
49	6.073 194 1332	135.285 176 8857	.007 391 7928	.164 658 0001	22.275 786 6649	.044 891 7928	49
50	6.300 938 9132	141.358 371 0189	.007 074 2185	.158 706 5061	22.434 493 1709	.044 574 2185	50
51	6.537 224 1225	147.659 309 9321	.006 772 3464	.152 970 1264	22.587 463 2973	.044 272 3464	51
52	6.782 310 0270	154.196 534 0546	.006 485 2301	.147 441 0856	22.734 904 3829	.043 985 2301	52
53	7.036 708 9031	160.978 904 0816	.006 211 9941	.142 111 8898	22.877 016 2727	.043 711 9941	53
54	7.300 585 4869	168.015 612 9847	.005 951 8278	.136 975 3154	23.013 991 5882	.043 451 8278	54
55	7.574 357 4427	175.316 198 4716	.005 703 9795	.132 024 4004	23.146 015 9886	.043 203 9795	55
56	7.858 395 8468	182.890 555 9143	.005 467 7509	.127 252 4341	23.273 268 4227	.042 967 7509	56
57	8.153 085 6910	190.748 951 7611	.005 242 4928	.122 652 9486	23.395 921 3713	.042 742 4928	57
58	8.458 826 4045	198.902 037 4521	.005 027 6006	.118 219 7095	23.514 141 0808	.042 527 6006	58
59	8.776 032 3946	207.360 863 8566	.004 822 5108	.113 946 7079	23.628 087 7887	.042 322 5108	59
60	9.105 133 6094	216.136 896 2512	.004 626 6973	.109 828 1522	23.737 915 9409	.042 126 6973	60
61	9.446 576 1198	225.242 029 8606	.004 439 6687	.105 858 4600	23.843 774 4009	.041 939 6687	61
62	9.800 822 7243	234.688 605 9804	.004 260 9653	.102 032 2506	23.945 806 6515	.041 760 9653	62
63	10.168 353 5764	244.489 428 7047	.004 090 1564	.098 344 3379	24.044 150 9894	.041 590 1564	63
64	10.549 666 8355	254.657 782 2811	.003 926 8386	.094 789 7233	24.138 940 7126	.041 426 8386	64
65	10.945 279 3419	265.207 449 1166	.003 770 6332	.091 363 5887	24.230 304 3013	.041 270 6332	65
66	11.355 727 3172	276.152 728 4585	.003 621 1846	.088 061 2903	24.318 365 5916	.041 121 1846	66
67	11.781 567 0916	287.508 455 7757	.003 478 1586	.084 878 3521	24.403 243 9438	.040 978 1586	67
68	12.223 375 8575	299.290 022 8673	.003 341 2407	.081 810 4599	24.485 054 4036	.040 841 2407	68
69	12.681 752 4522	311.513 398 7248	.003 210 1348	.078 853 4553	24.563 907 8589	.040 710 1348	69
70	13.157 318 1691	324.195 151 1770	.003 084 5619	.076 003 3304	24.639 911 1893	.040 584 5619	70
71	13.650 717 6005	337.352 469 3461	.002 964 2587	.073 256 2221	24.713 167 4114	.040 464 2587	71
72	14.162 619 5105	351.003 186 9466	.002 848 9770	.070 608 4068	24.783 775 8182	.040 348 9770	72
73	14.693 717 7421	365.165 806 4571	.002 738 4820	.068 056 2957	24.851 832 1139	.040 238 4820	73
74	15.244 732 1575	379.859 524 1992	.002 632 5521	.065 596 4296	24.917 428 5435	.040 132 5521	74
75	15.816 409 6134	395.104 256 3567	.002 530 9775	.063 225 4743	24.980 654 0179	.040 030 9775	75
76	16.409 524 9739	410.920 665 9701	.002 433 5598	.060 940 2162	25.041 594 2341	.039 933 5598	76
77	17.024 882 1604	427.330 190 9440	.002 340 1108	.058 737 5578	25.100 331 7919	.039 840 1108	77
78	17.663 315 2414	444.355 073 1044	.002 250 4525	.056 614 5135	25.156 946 3054	.039 750 4525	78
79	18.325 689 5630	462.018 388 3458	.002 164 4160	.054 568 2058	25.211 514 5113	.039 664 4160	79
80	19.012 902 9216	480.344 077 9088	.002 081 8410	.052 595 8610	25.264 110 3723	.039 581 8410	80
81	19.725 886 7811	499.356 980 8303	.002 002 5754	.050 694 8058	25.314 805 1781	.039 502 5754	81
82	20.465 607 5354	519.082 867 6115	.001 926 4747	.048 862 4634	25.363 667 6416	.039 426 4747	82
83	21.233 067 8180	539.548 475 1469	.001 853 4016	.047 096 3503	25.410 763 9919	.039 353 4016	83
84	22.029 307 8612	560.781 542 9649	.001 783 2256	.045 394 0726	25.456 158 0645	.039 283 2256	84
85	22.855 406 9060	582.810 850 8261	.001 715 8225	.043 753 3230	25.499 911 3874	.039 215 8225	85
86	23.712 484 6650	605.666 257 7321	.001 651 0743	.042 171 8776	25.542 083 2650	.039 151 0743	86
87	24.601 702 8399	629.378 742 3970	.001 588 8684	.040 647 5928	25.582 730 8578	.039 088 8684	87
88	25.524 266 6964	653.980 445 2369	.001 529 0977	.039 178 4027	25.621 909 2606	.039 029 0977	88
89	26.481 426 6975	679.504 711 9333	.001 471 6601	.037 762 3159	25.659 671 5764	.038 971 6601	89
90	27.474 480 1987	705.986 138 6308	.001 416 4584	.036 397 4129	25.696 068 9893	.038 916 4584	90
91	28.504 773 2061	733.460 618 8294	.001 363 3997	.035 081 8438	25.731 150 8331	.038 863 3997	91
92	29.573 702 2013	761.965 392 0356	.001 312 3956	.033 813 8253	25.764 964 6584	.038 812 3956	92
93	30.682 716 0339	791.539 094 2369	.001 263 3615	.032 591 6389	25.797 556 2973	.038 763 3615	93
94	31.833 317 8852	822.221 810 2708	.001 216 2168	.031 413 6278	25.828 969 9251	.038 716 2168	94
95	33.027 067 3058	854.055 128 1559	.001 170 8846	.030 278 1955	25.859 248 1205	.038 670 8846	95
96	34.265 582 3298	887.082 195 4618	.001 127 2913	.029 183 8029	25.888 431 9234	.038 627 2913	96
97	35.550 541 6672	921.347 777 7916	.001 085 3665	.028 128 9666	25.916 560 8900	.038 585 3665	97
98	36.883 686 9797	956.898 319 4588	.001 045 0431	.027 112 2570	25.943 673 1470	.038 545 0431	98
99	38.266 825 2414	993.782 006 4385	.001 006 2569	.026 132 2959	25.969 805 4429	.038 506 2569	99
100	39.701 831 1880	1032.048 831 6799	.000 968 9464	.025 187 7551	25.994 993 1980	.038 468 9464	100

	XIV COMPOUND AMOUNT	XV AMOUNT OF ANNUITY	XVI SINKING FUND	XVII PRESENT VALUE	XVIII PRESENT VALUE OF ANNUITY	XIX AMORTIZATION	
RATE 4% PERIODS	AMOUNT OF 1 — How $1 left at compound interest will grow.	AMOUNT OF 1 PER PERIOD — How $1 deposited periodically will grow.	SINKING FUND — Periodic deposit that will grow to $1 at future date.	PRESENT WORTH OF 1 — What $1 due in the future is worth today.	PRESENT WORTH OF 1 PER PERIOD — What $1 payable periodically is worth today.	PARTIAL PAYMENT — Annuity worth $1 today. Periodic payment necessary to pay off a loan of $1.	PERIODS
1	1.040 000 0000	1.000 000 0000	1.000 000 0000	.961 538 4615	.961 538 4615	1.040 000 0000	1
2	1.081 600 0000	2.040 000 0000	.490 196 0784	.924 556 2130	1.886 094 6746	.530 196 0784	2
3	1.124 864 0000	3.121 600 0000	.320 348 5392	.888 996 3587	2.775 091 0332	.360 348 5392	3
4	1.169 858 5600	4.246 464 0000	.235 490 0454	.854 804 1910	3.629 895 2243	.275 490 0454	4
5	1.216 652 9024	5.416 322 5600	.184 627 1135	.821 927 1068	4.451 822 3310	.224 627 1135	5
6	1.265 319 0185	6.632 975 4624	.150 761 9025	.790 314 5257	5 242 136 8567	.190 761 9025	6
7	1.315 931 7792	7.898 294 4809	.126 609 6120	.759 917 8132	6.002 054 6699	.166 609 6120	7
8	1.368 569 0504	9.214 226 2601	.108 527 8320	.730 690 2050	6.732 744 8750	.148 527 8320	8
9	1.423 311 8124	10.582 795 3105	.094 492 9927	.702 586 7356	7.435 331 6105	.134 492 9927	9
10	1.480 244 2849	12.006 107 1230	.083 290 9443	.675 564 1688	8.110 895 7794	.123 290 9443	10
11	1.539 454 0563	13.486 351 4079	.074 149 0393	.649 580 9316	8.760 476 7109	.114 149 0393	11
12	1.601 032 2186	15.025 805 4642	.066 552 1727	.624 597 0496	9.385 073 7605	.106 552 1727	12
13	1.665 073 5073	16.626 837 6828	.060 143 7278	.600 574 0861	9.985 647 8466	.100 143 7278	13
14	1.731 676 4476	18.291 911 1901	.054 668 9731	.577 475 0828	10.563 122 9295	.094 668 9731	14
15	1.800 943 5055	20.023 587 6377	.049 941 1004	.555 264 5027	11.118 387 4322	.089 941 1004	15
16	1.872 981 2457	21.824 531 1432	.045 819 9992	.533 908 1757	11.652 295 6079	.085 819 9992	16
17	1.947 900 4956	23.697 512 3889	.042 198 5221	.513 373 2459	12.165 668 8537	.082 198 5221	17
18	2.025 816 5154	25.645 412 8845	.038 993 3281	.493 628 1210	12.659 296 9747	.078 993 3281	18
19	2.106 849 1760	27.671 229 3998	.036 138 6184	.474 642 4240	13.133 939 3988	.076 138 6184	19
20	2.191 123 1430	29.778 078 5758	.033 581 7503	.456 386 9462	13.590 326 3450	.073 581 7503	20
21	2.278 768 0688	31.969 201 7189	.031 280 1054	.438 833 6021	14.029 159 9471	.071 280 1054	21
22	2.369 918 7915	34.247 969 7876	.029 198 8111	.421 955 3867	14.451 115 3337	.069 198 8111	22
23	2.464 715 5432	36.617 888 5791	.027 309 0568	.405 726 3333	14.856 841 6671	.067 309 0568	23
24	2.563 304 1649	39.082 604 1223	.025 586 8313	.390 121 4743	15.246 963 1414	.065 586 8313	24
25	2.665 836 3315	41.645 908 2872	.024 011 9628	.375 116 8023	15.622 079 9437	.064 011 9628	25
26	2.772 469 7847	44.311 744 6187	.022 567 3805	.360 689 2329	15.982 769 1766	.062 567 3805	26
27	2.883 368 5761	47.084 214 4034	.021 238 5406	.346 816 5701	16.329 585 7467	.061 238 5406	27
28	2.998 703 3192	49.967 582 9796	.020 012 9752	.333 477 4713	16.663 063 2180	.060 012 9752	28
29	3.118 651 4519	52.966 286 2987	.018 879 9342	.320 651 4147	16.983 714 6327	.058 879 9342	29
30	3.243 397 5100	56.084 937 7507	.017 830 0991	.308 318 6680	17.292 033 3007	.057 830 0991	30
31	3.373 133 4104	59.328 335 2607	.016 855 3524	.296 460 2577	17.588 493 5583	.056 855 3524	31
32	3.508 058 7468	62.701 468 6711	.015 948 5897	.285 057 9401	17.873 551 4984	.055 948 5897	32
33	3.648 381 0967	66.209 527 4180	.015 103 5665	.274 094 1731	18.147 645 6715	.055 103 5665	33
34	3.794 316 3406	69.857 908 5147	.014 314 7715	.263 552 0896	18.411 197 7611	.054 314 7715	34
35	3.946 088 9942	73.652 224 8553	.013 577 3224	.253 415 4707	18.664 613 2318	.053 577 3224	35
36	4.103 932 5540	77.598 313 8495	.012 886 8780	.243 668 7219	18.908 281 9537	.052 886 8780	36
37	4.268 089 8561	81.702 246 4035	.012 239 5655	.234 296 8479	19.142 578 8016	.052 239 5655	37
38	4.438 813 4504	85.970 336 2596	.011 631 9191	.225 285 4307	19.367 864 2323	.051 631 9191	38
39	4.616 365 9884	90.409 149 7100	.011 060 8274	.216 620 6064	19.584 484 8388	.051 060 8274	39
40	4.801 020 6279	95.025 515 6984	.010 523 4893	.208 289 0447	19.792 773 8834	.050 523 4893	40
41	4.993 061 4531	99.826 536 3264	.010 017 3765	.200 277 9276	19.993 051 8110	.050 017 3765	41
42	5.192 783 9112	104.819 597 7794	.009 540 2007	.192 574 9303	20.185 626 7413	.049 540 2007	42
43	5.400 495 2676	110.012 381 6906	.009 089 8859	.185 168 2023	20.370 794 9436	.049 089 8859	43
44	5.616 515 0783	115.412 876 9582	.008 664 5444	.178 046 3483	20.548 841 2919	.048 664 5444	44
45	5.841 175 6815	121.029 392 0365	.008 262 4558	.171 198 4118	20.720 039 7038	.048 262 4558	45
46	6.074 822 7087	126.870 567 7180	.007 882 0488	.164 613 8575	20.884 653 5613	.047 882 0488	46
47	6.317 815 6171	132.945 390 4267	.007 521 8855	.158 282 5553	21.042 936 1166	.047 521 8855	47
48	6.570 528 2418	139.263 206 0438	.007 180 6476	.152 194 7647	21.195 130 8814	.047 180 6476	48
49	6.833 349 3714	145.833 734 2855	.006 857 1240	.146 341 1199	21.341 472 0013	.046 857 1240	49
50	7.106 683 3463	152.667 083 6570	.006 550 2004	.140 712 6153	21.482 184 6167	.046 550 2004	50
51	7.390 950 6801	159.773 767 0032	.006 258 8497	.135 300 5917	21.617 485 2083	.046 258 8497	51
52	7.686 588 7073	167.164 717 6834	.005 982 1236	.130 096 7228	21.747 581 9311	.045 982 1236	52
53	7.994 052 2556	174.851 306 3907	.005 719 1451	.125 093 0027	21.872 674 9337	.045 719 1451	53
54	8.313 814 3459	182.845 358 6463	.005 469 1025	.120 281 7333	21.992 956 6671	.045 469 1025	54
55	8.646 366 9197	191.159 172 9922	.005 231 2426	.115 655 5128	22.108 612 1799	.045 231 2426	55
56	8.992 221 5965	199.805 539 9119	.005 004 8662	.111 207 2239	22.219 819 4037	.045 004 8662	56
57	9.351 910 4603	208.797 761 5083	.004 789 3234	.106 930 0229	22.326 749 4267	.044 789 3234	57
58	9.725 986 8787	218.149 671 9687	.004 584 0087	.102 817 3297	22.429 566 7564	.044 584 0087	58
59	10.115 026 3539	227.875 658 8474	.004 388 3581	.098 862 8171	22.528 429 5735	.044 388 3581	59
60	10.519 627 4081	237.990 685 2013	.004 201 8451	.095 060 4010	22.623 489 9745	.044 201 8451	60
61	10.940 412 5044	248.510 312 6094	.004 023 9779	.091 404 2318	22.714 894 2062	.044 023 9779	61
62	11.378 029 0045	259.450 725 1137	.003 854 2964	.087 888 6844	22.802 782 8906	.043 854 2964	62
63	11.833 150 1647	270.828 754 1183	.003 692 3701	.084 508 3504	22.887 291 2410	.043 692 3701	63
64	12.306 476 1713	282.661 904 2830	.003 537 7955	.081 258 0292	22.968 549 2702	.043 537 7955	64
65	12.798 735 2182	294.968 380 4544	.003 390 1939	.078 132 7204	23.046 681 9905	.043 390 1939	65
66	13.310 684 6269	307.767 115 6725	.003 249 2100	.075 127 6157	23.121 809 6063	.043 249 2100	66
67	13.843 112 0120	321.077 800 2994	.003 114 5099	.072 238 0921	23.194 047 6984	.043 114 5099	67
68	14.396 836 4925	334.920 912 3114	.002 985 7795	.069 459 7039	23.263 507 4023	.042 985 7795	68
69	14.972 709 9522	349.317 748 8039	.002 862 7231	.066 788 1768	23.330 295 5791	.042 862 7231	69
70	15.571 618 3502	364.290 458 7560	.002 745 0623	.064 219 4008	23.394 514 9799	.042 745 0623	70
71	16.194 483 0843	379.862 077 1063	.002 632 5344	.061 749 4238	23.456 264 4038	.042 632 5344	71
72	16.842 262 4076	396.056 560 1905	.002 524 8919	.059 374 4460	23.515 638 8498	.042 524 8919	72
73	17.515 952 9039	412.898 822 5981	.002 421 9008	.057 090 8135	23.572 729 6632	.042 421 9008	73
74	18.216 591 0201	430.414 775 5021	.002 323 3403	.054 895 0130	23.627 624 6762	.042 323 3403	74
75	18.945 254 6609	448.631 366 5221	.002 229 0015	.052 783 6663	23.680 408 3425	.042 229 0015	75
76	19.703 064 8473	467.576 621 1830	.002 138 6869	.050 753 5253	23.731 161 8678	.042 138 6869	76
77	20.491 187 4412	487.279 686 0303	.002 052 2095	.048 801 4666	23.779 963 3344	.042 052 2095	77
78	21.310 834 9389	507.770 873 4716	.001 969 3922	.046 924 4871	23.826 887 8215	.041 969 3922	78
79	22.163 268 3364	529.081 708 4104	.001 890 0672	.045 119 6992	23.872 007 5207	.041 890 0672	79
80	23.049 799 0699	551.244 976 7468	.001 814 0755	.043 384 3261	23.915 391 8468	.041 814 0755	80
81	23.971 791 0327	574.294 775 8167	.001 741 2661	.041 715 6982	23.957 107 5450	.041 741 2661	81
82	24.930 662 6740	598.266 566 8494	.001 671 4957	.040 111 2483	23.997 218 7933	.041 671 4957	82
83	25.927 889 1809	623.197 229 5233	.001 604 6284	.038 568 5079	24.035 787 3013	.041 604 6284	83
84	26.965 004 7482	649.125 118 7043	.001 540 5351	.037 085 1038	24.072 872 4050	.041 540 5351	84
85	28.043 604 9381	676.090 123 4525	.001 479 0928	.035 658 7537	24.108 531 1587	.041 479 0928	85
86	29.165 349 1356	704.133 728 3906	.001 420 1848	.034 287 2631	24.142 818 4218	.041 420 1848	86
87	30.331 963 1010	733.299 077 5262	.001 363 7001	.032 968 5222	24.175 786 9441	.041 363 7001	87
88	31.545 241 6251	763.631 040 6272	.001 309 5329	.031 700 5022	24.207 487 4462	.041 309 5329	88
89	32.807 051 2901	795.176 282 2523	.001 257 5828	.030 481 2521	24.237 968 6983	.041 257 5828	89
90	34.119 333 3417	827.983 333 5424	.001 207 7538	.029 308 8962	24.267 277 5945	.041 207 7538	90
91	35.484 106 6754	862.102 666 8841	.001 159 9547	.028 181 6310	24.295 459 2255	.041 159 9547	91
92	36.903 470 9424	897.586 773 5595	.001 114 0984	.027 097 7221	24.322 556 9476	.041 114 0984	92
93	38.379 609 7801	934.490 244 5018	.001 070 1021	.026 055 5020	24.348 612 4496	.041 070 1021	93
94	39.914 794 1713	972.869 854 2819	.001 027 8867	.025 053 3673	24.373 665 8169	.041 027 8867	94
95	41.511 385 9381	1012.784 648 4532	.000 987 3767	.024 089 7763	24.397 755 5932	.040 987 3767	95
96	43.171 841 3757	1054.296 034 3913	.000 948 5002	.023 163 2464	24.420 918 8396	.040 948 5002	96
97	44.898 715 0307	1097.467 875 7670	.000 911 1884	.022 272 3523	24.443 191 1919	.040 911 1884	97
98	46.694 663 6319	1142.366 590 7976	.000 875 3757	.021 415 7234	24.464 606 9153	.040 875 3757	98
99	48.562 450 1772	1189.061 254 4296	.000 840 9996	.020 592 0417	24.485 198 9570	.040 840 9996	99
100	50.504 948 1843	1237.623 704 6067	.000 808 0000	.019 800 0401	24.504 998 9972	.040 808 0000	100

RATE 4 1/2%

	XIV COMPOUND AMOUNT	XV AMOUNT OF ANNUITY	XVI SINKING FUND	XVII PRESENT VALUE	XVIII PRESENT VALUE OF ANNUITY	XIX AMORTIZATION	
PERIODS	AMOUNT OF 1	AMOUNT OF 1 PER PERIOD	SINKING FUND	PRESENT WORTH OF 1	PRESENT WORTH OF 1 PER PERIOD	PARTIAL PAYMENT	PERIODS
	How $1 left at compound interest will grow.	How $1 deposited periodically will grow.	Periodic deposit that will grow to $1 at future date.	What $1 due in the future is worth today.	What $1 payable periodically is worth today.	Annuity worth $1 today. Periodic payment necessary to pay off a loan of $1.	
1	1.045 000 0000	1.000 000 0000	1.000 000 0000	.956 937 7990	.956 937 7990	1.045 000 0000	1
2	1.092 025 0000	2.045 000 0000	.488 997 5550	.915 729 9512	1.872 667 7503	.533 997 5550	2
3	1.141 166 1250	3.137 025 0000	.318 773 3601	.876 296 6041	2.748 964 3543	.363 773 3601	3
4	1.192 518 6006	4.278 191 1250	.233 743 6479	.838 561 3436	3.587 525 6979	.278 743 6479	4
5	1.246 181 9377	5.470 709 7256	.182 791 6395	.802 451 0465	4.389 976 7444	.227 791 6395	5
6	1.302 260 1248	6.716 891 6633	.148 878 3875	.767 895 7383	5.157 872 4827	.193 878 3875	6
7	1.360 861 8305	8.019 151 7881	.124 701 4680	.734 828 4577	5.892 700 9404	.169 701 4680	7
8	1.422 100 6128	9.380 013 6186	.106 609 6533	.703 185 1270	6.595 886 0674	.151 609 6533	8
9	1.486 095 1404	10.802 114 2314	.092 574 4700	.672 904 4277	7.268 790 4951	.137 574 4700	9
10	1.552 969 4217	12.288 209 3718	.081 378 8217	.643 927 6820	7.912 718 1771	.126 378 8217	10
11	1.622 853 0457	13.841 178 7936	.072 248 1817	.616 198 7388	8.528 916 9159	.117 248 1817	11
12	1.695 881 4328	15.464 031 8393	.064 666 1886	.589 663 8649	9.118 580 7808	.109 666 1886	12
13	1.772 196 0972	17.159 913 2721	.058 275 3528	.564 271 6410	9.682 852 4218	.103 275 3528	13
14	1.851 944 9216	18.932 109 3693	.052 820 3160	.539 972 8622	10.222 825 2840	.097 820 3160	14
15	1.935 282 4431	20.784 054 2909	.048 113 8061	.516 720 4423	10.739 545 7263	.093 113 8061	15
16	2.022 370 1530	22.719 336 7340	.044 015 3694	.494 469 3228	11.234 015 0491	.089 015 3694	16
17	2.113 376 8099	24.741 706 8870	.040 417 5833	.473 176 3854	11.707 191 4346	.085 417 5833	17
18	2.208 478 7664	26.855 083 6970	.037 236 8975	.452 800 3688	12.159 991 8034	.082 236 8975	18
19	2.307 860 3108	29.063 562 4633	.034 407 3443	.433 301 7884	12.593 293 5918	.079 407 3443	19
20	2.411 714 0248	31.371 422 7742	.031 876 1443	.414 642 8597	13.007 936 4515	.076 876 1443	20
21	2.520 241 1560	33.783 136 7990	.029 600 5669	.396 787 4255	13.404 723 8770	.074 600 5669	21
22	2.633 652 0080	36.303 377 9550	.027 545 6461	.379 700 8857	13.784 424 7627	.072 545 6461	22
23	2.752 166 3483	38.937 029 9629	.025 682 4930	.363 350 1298	14.147 774 8925	.070 682 4930	23
24	2.876 013 8340	41.689 196 3113	.023 987 0299	.347 703 4735	14.495 478 3660	.068 987 0299	24
25	3.005 434 4565	44.565 210 1453	.022 439 0280	.332 730 5967	14.828 208 9627	.067 439 0280	25
26	3.140 679 0071	47.570 644 6018	.021 021 3674	.318 402 4849	15.146 611 4476	.066 021 3674	26
27	3.282 009 5624	50.711 323 6089	.019 719 4616	.304 691 3731	15.451 302 8206	.064 719 4616	27
28	3.429 699 9927	53.993 333 1713	.018 520 8051	.291 570 6919	15.742 873 5126	.063 520 8051	28
29	3.584 036 4924	57.423 033 1640	.017 414 6147	.279 015 0162	16.021 888 5288	.062 414 6147	29
30	3.745 318 1345	61.007 069 6564	.016 391 5429	.267 000 0155	16.288 888 5443	.061 391 5429	30
31	3.913 857 4506	64.752 387 7909	.015 443 4459	.255 502 4072	16.544 390 9515	.060 443 4459	31
32	4.089 981 0359	68.666 245 2415	.014 563 1962	.244 499 9112	16.788 890 8627	.059 563 1962	32
33	4.274 030 1825	72.756 226 2774	.013 744 5281	.233 971 2069	17.022 862 0695	.058 744 5281	33
34	4.466 361 5407	77.030 256 4599	.012 981 9119	.223 895 8917	17.246 757 9613	.057 981 9119	34
35	4.667 347 8100	81.496 618 0005	.012 270 4478	.214 254 4419	17.461 012 4031	.057 270 4478	35
36	4.877 378 4615	86.163 965 8106	.011 605 7796	.205 028 1740	17.666 040 5772	.056 605 7796	36
37	5.096 860 4922	91.041 344 2720	.010 984 0206	.196 199 2096	17.862 239 7868	.055 984 0206	37
38	5.326 219 2144	96.138 204 7643	.010 401 6920	.187 750 4398	18.049 990 2266	.055 401 6920	38
39	5.565 899 0790	101.464 423 9787	.009 855 6712	.179 665 4926	18.229 655 7192	.054 855 6712	39
40	5.816 364 5376	107.030 323 0577	.009 343 1466	.171 928 7011	18.401 584 4203	.054 343 1466	40
41	6.078 100 9418	112.846 687 5953	.008 861 5804	.164 525 0728	18.566 109 4931	.053 861 5804	41
42	6.351 615 4842	118.924 788 5371	.008 408 6759	.157 440 2611	18.723 549 7542	.053 408 6759	42
43	6.637 438 1810	125.276 404 0213	.007 982 3492	.150 660 5369	18.874 210 2911	.052 982 3492	43
44	6.936 122 8991	131.913 842 2022	.007 580 7056	.144 172 7626	19.018 383 0536	.052 580 7056	44
45	7.248 248 4296	138.849 965 1013	.007 202 0184	.137 964 3661	19.156 347 4198	.052 202 0184	45
46	7.574 419 6089	146.098 213 5309	.006 844 7107	.132 023 3169	19.288 370 7366	.051 844 7107	46
47	7.915 268 4913	153.672 633 1398	.006 507 3395	.126 338 1023	19.414 708 8389	.051 507 3395	47
48	8.271 455 5734	161.587 901 6311	.006 188 5821	.120 897 7055	19.535 606 5444	.051 188 5821	48
49	8.643 671 0742	169.859 357 2045	.005 887 2235	.115 691 5842	19.651 298 1286	.050 887 2235	49
50	9.032 636 2725	178.503 028 2787	.005 602 1459	.110 709 6500	19.762 007 7785	.050 602 1459	50
51	9.439 104 9048	187.535 664 5512	.005 332 3191	.105 942 2488	19.867 950 0273	.050 332 3191	51
52	9.863 864 6255	196.974 769 4560	.005 076 7923	.101 380 1424	19.969 330 1697	.050 076 7923	52
53	10.307 738 5337	206.838 634 0815	.004 834 6867	.097 014 4903	20.066 344 6600	.049 834 6867	53
54	10.771 586 7677	217.146 372 6152	.004 605 1886	.092 836 8328	20.159 181 4928	.049 605 1886	54
55	11.256 308 1722	227.917 959 3829	.004 387 5437	.088 839 0745	20.248 020 5673	.049 387 5437	55
56	11.762 842 0400	239.174 267 5551	.004 181 0518	.085 013 4684	20.333 034 0357	.049 181 0518	56
57	12.292 169 9318	250.937 109 5951	.003 985 0622	.081 352 6013	20.414 386 6370	.048 985 0622	57
58	12.845 317 5787	263.229 279 5269	.003 798 9695	.077 849 3793	20.492 236 0163	.048 798 9695	58
59	13.423 356 8698	276.074 597 1056	.003 454 2558	.074 497 0137	20.566 733 0299	.048 622 2094	59
60	14.027 407 9289	289.497 953 9753	.003 454 2558	.071 289 0083	20.638 022 0382	.048 454 2558	60
61	14.658 641 2857	303.525 361 9042	.003 294 6176	.068 219 1467	20.706 241 1849	.048 294 6176	61
62	15.318 280 1435	318.184 003 1899	.003 142 8356	.065 281 4801	20.771 522 6650	.048 142 8356	62
63	16.007 602 7500	333.502 283 3335	.002 998 4802	.062 470 3159	20.833 992 9808	.047 998 4802	63
64	16.727 944 8738	349.509 886 0835	.002 861 1494	.059 780 2066	20.893 773 1874	.047 861 1494	64
65	17.480 702 3931	366.237 830 9572	.002 730 4661	.057 205 9393	20.950 979 1267	.047 730 4661	65
66	18.267 334 0008	383.718 533 3503	.002 606 0769	.054 742 5256	21.005 721 6523	.047 606 0769	66
67	19.089 364 0308	401.985 867 3511	.002 487 6496	.052 385 1920	21.058 106 8443	.047 487 6496	67
68	19.948 385 4122	421.075 231 3819	.002 374 8725	.050 129 3703	21.108 236 2147	.047 374 8725	68
69	20.846 062 7557	441.023 616 7941	.002 267 4523	.047 970 6893	21.156 206 9040	.047 267 4523	69
70	21.784 135 5797	461.869 679 5498	.002 165 1129	.045 904 9659	21.202 111 8699	.047 165 1129	70
71	22.764 421 6808	483.653 815 1295	.002 067 5946	.043 928 1970	21.246 040 0668	.047 067 5946	71
72	23.788 820 6565	506.418 236 8104	.001 974 6524	.042 036 5521	21.288 076 6190	.046 974 6524	72
73	24.859 317 5860	530.207 057 4668	.001 886 0556	.040 226 3657	21.328 302 9847	.046 886 0556	73
74	25.977 986 8774	555.066 375 0528	.001 801 5863	.038 494 1298	21.366 797 1145	.046 801 5863	74
75	27.146 996 2869	581.044 361 9302	.001 721 0390	.036 836 4879	21.403 633 6024	.046 721 0390	75
76	28.368 611 1198	608.191 358 2171	.001 644 2194	.035 250 2276	21.438 883 8301	.046 644 2194	76
77	29.645 198 6202	636.559 969 3368	.001 570 9439	.033 732 2753	21.472 616 1053	.046 570 9439	77
78	30.979 232 5581	666.205 167 9570	.001 501 0391	.032 279 6892	21.504 895 7946	.046 501 0391	78
79	32.373 298 0232	697.184 400 5151	.001 434 3408	.030 889 6548	21.535 785 4494	.046 434 3408	79
80	33.830 096 4342	729.557 698 5382	.001 370 6935	.029 559 4783	21.565 344 9276	.046 370 6935	80
81	35.352 450 7738	763.387 794 9725	.001 309 9502	.028 286 5821	21.593 631 5097	.046 309 9502	81
82	36.943 311 0586	798.740 245 7462	.001 251 9715	.027 068 4996	21.620 700 0093	.046 251 9715	82
83	38.605 760 0562	835.683 556 8048	.001 196 6252	.025 902 8704	21.646 602 8797	.046 196 6252	83
84	40.343 019 2587	874.289 316 8610	.001 143 7861	.024 787 4358	21.671 390 3155	.046 143 7861	84
85	42.158 455 1254	914.632 336 1198	.001 093 3355	.023 720 0343	21.695 110 3497	.046 093 3355	85
86	44.055 585 6060	956.790 791 2452	.001 045 1606	.022 698 5974	21.717 808 9471	.046 045 1606	86
87	46.038 086 9583	1000.846 376 8512	.000 999 1543	.021 721 1458	21.739 530 0929	.045 999 1543	87
88	48.109 800 8714	1046.884 463 8095	.000 955 2152	.020 785 7855	21.760 315 8784	.045 955 2152	88
89	50.274 741 9106	1094.994 264 6809	.000 913 2468	.019 890 7038	21.780 206 5822	.045 913 2468	89
90	52.537 105 2966	1145.269 006 5916	.000 873 1573	.019 034 1663	21.799 240 7485	.045 873 1573	90
91	54.901 275 0350	1197.806 111 8882	.000 834 8597	.018 214 5132	21.817 455 2617	.045 834 8597	91
92	57.371 832 4115	1252.707 386 9232	.000 798 2710	.017 430 1562	21.834 885 4179	.045 798 2710	92
93	59.953 564 8701	1310.079 219 3347	.000 763 3126	.016 679 5753	21.851 564 9932	.045 763 3126	93
94	62.651 475 2892	1370.032 784 2048	.000 729 9095	.015 961 3161	21.867 526 3093	.045 729 9095	94
95	65.470 791 6772	1432.684 259 4940	.000 697 9905	.015 273 9867	21.882 800 2960	.045 697 9905	95
96	68.416 977 3027	1498.155 051 1712	.000 667 4877	.014 616 2552	21.897 416 5512	.045 667 4877	96
97	71.495 741 2813	1566.572 028 4739	.000 638 3364	.013 986 8471	21.911 403 3983	.045 638 3364	97
98	74.713 049 6390	1638.067 769 7552	.000 610 4754	.013 384 5427	21.924 787 9409	.045 610 4754	98
99	78.075 136 8727	1712.780 819 3942	.000 583 8459	.012 808 1748	21.937 596 1157	.045 583 8459	99
100	81.588 518 0320	1790.855 956 2670	.000 558 3922	.012 256 6266	21.949 852 7423	.045 558 3922	100

628

RATE 5%	PERIODS	XIV COMPOUND AMOUNT	XV AMOUNT OF ANNUITY	XVI SINKING FUND	XVII PRESENT VALUE	XVIII PRESENT VALUE OF ANNUITY	XIX AMORTIZATION	PERIODS
		AMOUNT OF 1	AMOUNT OF 1 PER PERIOD	SINKING FUND	PRESENT WORTH OF 1	PRESENT WORTH OF 1 PER PERIOD	PARTIAL PAYMENT	
		How $1 left at compound interest will grow.	How $1 deposited periodically will grow.	Periodic deposit that will grow to $1 at future date.	What $1 due in the future is worth today.	What $1 payable periodically is worth today.	Annuity worth $1 today. Periodic payment necessary to pay off a loan of $1.	
	1	1.050 000 0000	1.000 000 0000	1.000 000 0000	.952 380 9524	.952 380 9524	1.050 000 0000	1
	2	1.102 500 0000	2.050 000 0000	.487 804 8780	.907 029 4785	1.859 410 4308	.537 804 8780	2
	3	1.157 625 0000	3.152 500 0000	.317 208 5646	.863 837 5985	2.723 248 0294	.367 208 5646	3
	4	1.215 506 2500	4.310 125 0000	.232 011 8326	.822 702 4748	3.545 950 5042	.282 011 8326	4
	5	1.276 281 5625	5.525 631 2500	.180 974 7981	.783 526 1665	4.329 476 6706	.230 974 7981	5
	6	1.340 095 6406	6.801 912 8125	.147 017 4681	.746 215 3966	5.075 692 0673	.197 017 4681	6
	7	1.407 100 4227	8.142 008 4531	.122 819 8184	.710 681 3301	5.786 373 3974	.172 819 8184	7
	8	1.477 455 4438	9.549 108 8758	.104 721 8136	.676 839 3620	6.463 212 7594	.154 721 8136	8
	9	1.551 328 2160	11.026 564 3196	.090 690 0800	.644 608 9162	7.107 821 6756	.140 690 0800	9
	10	1.628 894 6268	12.577 892 5355	.079 504 5750	.613 913 2535	7.721 734 9292	.129 504 5750	10
	11	1.710 339 3581	14.206 787 1623	.070 388 8915	.584 679 2891	8.306 414 2183	.120 388 8915	11
	12	1.795 856 3260	15.917 126 5204	.062 825 4100	.556 837 4182	8.863 251 6364	.112 825 4100	12
	13	1.885 649 1423	17.712 982 8465	.056 455 7652	.530 321 3506	9.393 572 9871	.106 455 7652	13
	14	1.979 931 5994	19.598 631 9888	.051 023 9695	.505 067 9530	9.898 640 9401	.101 023 9695	14
	15	2.078 928 1794	21.578 563 5882	.046 342 2876	.481 017 0981	10.379 658 0382	.096 342 2876	15
	16	2.182 874 5884	23.657 491 7676	.042 269 9080	.458 111 5220	10.837 769 5602	.092 269 9080	16
	17	2.292 018 3178	25.840 366 3560	.038 699 1417	.436 296 6876	11.274 066 2478	.088 699 1417	17
	18	2.406 619 2337	28.132 384 6738	.035 546 2223	.415 520 6549	11.689 586 9027	.085 546 2223	18
	19	2.526 950 1954	30.539 003 9075	.032 745 0104	.395 733 9570	12.085 320 8597	.082 745 0104	19
	20	2.653 297 7051	33.065 954 1029	.030 242 5872	.376 889 4829	12.462 210 3425	.080 242 5872	20
	21	2.785 962 5904	35.719 251 8080	.027 996 1071	.358 942 3646	12.821 152 7072	.077 996 1071	21
	22	2.925 260 7199	38.505 214 3984	.025 970 5086	.341 849 8711	13.163 002 5783	.075 970 5086	22
	23	3.071 523 7559	41.430 475 1184	.024 136 8219	.325 571 3058	13.488 573 8841	.074 136 8219	23
	24	3.225 099 9437	44.501 998 8743	.022 470 9008	.310 067 9103	13.798 641 7943	.072 470 9008	24
	25	3.386 354 9409	47.727 098 8180	.020 952 4573	.295 302 7717	14.093 944 5660	.070 952 4573	25
	26	3.555 672 6879	51.113 453 7589	.019 564 3207	.281 240 7350	14.375 185 3010	.069 564 3207	26
	27	3.733 456 3223	54.669 126 4468	.018 291 8599	.267 848 3190	14.643 033 6200	.068 291 8599	27
	28	3.920 129 1385	58.402 582 7692	.017 122 5304	.255 093 6371	14.898 127 2571	.067 122 5304	28
	29	4.116 135 5954	62.322 711 9076	.016 045 5149	.242 946 3211	15.141 073 5782	.066 045 5149	29
	30	4.321 942 3752	66.438 847 5030	.015 051 4351	.231 377 4487	15.372 451 0269	.065 051 4351	30
	31	4.538 039 4939	70.760 789 8782	.014 132 1204	.220 359 4749	15.592 810 5018	.064 132 1204	31
	32	4.764 941 4686	75.298 829 3721	.013 280 4189	.209 866 1666	15.802 676 6684	.063 280 4189	32
	33	5.003 188 5420	80.063 770 8407	.012 490 0437	.199 872 5396	16.002 549 2080	.062 490 0437	33
	34	5.253 347 9691	85.066 959 3827	.011 755 4454	.190 354 7996	16.192 904 0076	.061 755 4454	34
	35	5.516 015 3676	90.320 307 3518	.011 071 7072	.181 290 2854	16.374 194 2929	.061 071 7072	35
	36	5.791 816 1360	95.836 322 7194	.010 434 4571	.172 657 4146	16.546 851 7076	.060 434 4571	36
	37	6.081 406 9428	101.628 138 8554	.009 839 7945	.164 435 6330	16.711 287 3405	.059 839 7945	37
	38	6.385 477 2899	107.709 545 7982	.009 284 2282	.156 605 3647	16.867 892 7053	.059 284 2282	38
	39	6.704 751 1544	114.095 023 0881	.008 764 6242	.149 147 9664	17.017 040 6717	.058 764 6242	39
	40	7.039 988 7121	120.799 774 2425	.008 278 1612	.142 045 6823	17.159 086 3540	.058 278 1612	40
	41	7.391 988 1477	127.839 762 9546	.007 822 2924	.135 281 6022	17.294 367 9562	.057 822 2924	41
	42	7.761 587 5551	135.231 751 1023	.007 394 7131	.128 839 6211	17.423 207 5773	.057 394 7131	42
	43	8.149 666 9329	142.993 338 6575	.006 993 3328	.122 704 4011	17.545 911 9784	.056 993 3328	43
	44	8.557 150 2795	151.143 005 5903	.006 616 2506	.116 861 3344	17.662 773 3128	.056 616 2506	44
	45	8.985 007 7935	159.700 155 8699	.006 261 7347	.111 296 5089	17.774 069 8217	.056 261 7347	45
	46	9.434 258 1832	168.685 163 6633	.005 928 2036	.105 996 6752	17.880 066 4968	.055 928 2036	46
	47	9.905 971 0923	178.119 421 8465	.005 614 2109	.100 949 2144	17.981 015 7113	.055 614 2109	47
	48	10.401 269 6469	188.025 392 9388	.005 318 4306	.096 142 1090	18.077 157 8203	.055 318 4306	48
	49	10.921 333 1293	198.426 662 5858	.005 039 6453	.091 563 9133	18.168 721 7336	.055 039 6453	49
	50	11.467 399 7858	209.347 995 7151	.004 776 7355	.087 203 7270	18.255 925 4606	.054 776 7355	50
	51	12.040 769 7750	220.815 395 5008	.004 528 6697	.083 051 1685	18.338 976 6291	.054 528 6697	51
	52	12.642 808 2638	232.856 165 2759	.004 294 4966	.079 096 3510	18.418 072 9801	.054 294 4966	52
	53	13.274 948 6770	245.498 973 5397	.004 073 3368	.075 329 8581	18.493 402 8382	.054 073 3368	53
	54	13.938 696 1108	258.773 922 2166	.003 864 3770	.071 742 7220	18.565 145 5602	.053 864 3770	54
	55	14.635 630 9164	272.712 618 3275	.003 666 8637	.068 326 4019	18.633 471 9621	.053 666 8637	55
	56	15.367 412 4622	287.348 249 2439	.003 480 0978	.065 072 7637	18.698 544 7258	.053 480 0978	56
	57	16.135 783 0853	302.715 661 7060	.003 303 4300	.061 974 0607	18.760 518 7865	.053 303 4300	57
	58	16.942 572 2396	318.851 444 7913	.003 136 2568	.059 022 9149	18.819 541 7014	.053 136 2568	58
	59	17.789 700 8515	335.794 017 0309	.002 978 0161	.056 212 2999	18.875 754 0013	.052 978 0161	59
	60	18.679 185 8941	353.583 717 8825	.002 828 1845	.053 535 5237	18.929 289 5251	.052 828 1845	60
	61	19.613 145 1888	372.262 903 7766	.002 686 2736	.050 986 2131	18.980 275 7382	.052 686 2736	61
	62	20.593 802 4483	391.876 048 9654	.002 551 8273	.048 558 2982	19.028 834 0363	.052 551 8273	62
	63	21.623 492 5707	412.469 851 4137	.002 424 4196	.046 245 9983	19.075 080 0346	.052 424 4196	63
	64	22.704 667 1992	434.093 343 9844	.002 303 6520	.044 043 8079	19.119 123 8425	.052 303 6520	64
	65	23.839 900 5592	456.798 011 1836	.002 189 1514	.041 946 4837	19.161 070 3262	.052 189 1514	65
	66	25.031 895 5871	480.637 911 7428	.002 080 5683	.039 949 0321	19.201 019 3583	.052 080 5683	66
	67	26.283 490 3665	505.669 807 3299	.001 977 5751	.038 046 6972	19.239 066 0555	.051 977 5751	67
	68	27.597 664 8848	531.953 297 6964	.001 879 8643	.036 234 9497	19.275 301 0052	.051 879 8643	68
	69	28.977 548 1291	559.550 962 5812	.001 787 1473	.034 509 4759	19.309 810 4812	.051 787 1473	69
	70	30.426 425 5355	588.528 510 7103	.001 699 1530	.032 866 1676	19.342 676 6487	.051 699 1530	70
	71	31.947 746 8123	618.954 936 2458	.001 615 6265	.031 301 1120	19.373 977 7607	.051 615 6265	71
	72	33.545 134 1529	650.902 683 0581	.001 536 3280	.029 810 5828	19.403 788 3435	.051 536 3280	72
	73	35.222 390 8605	684.447 817 2110	.001 461 0318	.028 391 0313	19.432 179 3748	.051 461 0318	73
	74	36.983 510 4036	719.670 208 0715	.001 389 5254	.027 039 0774	19.459 218 4522	.051 389 5254	74
	75	38.832 685 9238	756.653 718 4751	.001 321 6085	.025 751 5023	19.484 969 9545	.051 321 6085	75
	76	40.774 320 2209	795.486 404 3989	.001 257 0925	.024 525 2403	19.509 495 1947	.051 257 0925	76
	77	42.813 036 2309	836.260 724 6188	.001 195 7993	.023 357 3717	19.532 852 5664	.051 195 7993	77
	78	44.953 688 0425	879.073 760 8497	.001 137 5610	.022 245 1159	19.555 097 6823	.051 137 5610	78
	79	47.201 372 4446	924.027 448 8922	.001 082 2189	.021 185 8247	19.576 283 5069	.051 082 2189	79
	80	49.561 441 0668	971.228 821 3368	.001 029 6235	.020 176 9759	19.596 460 4828	.051 029 6235	80
	81	52.039 513 1202	1020.790 262 4037	.000 979 6332	.019 216 1675	19.615 676 6503	.050 979 6332	81
	82	54.641 488 7762	1072.829 775 5239	.000 932 1143	.018 301 1119	19.633 977 7622	.050 932 1143	82
	83	57.373 563 2150	1127.471 264 3001	.000 886 9406	.017 429 6304	19.651 407 3925	.050 886 9406	83
	84	60.242 241 3758	1184.844 827 5151	.000 843 9924	.016 599 6480	19.668 007 0405	.050 843 9924	84
	85	63.254 353 4445	1245.087 068 8908	.000 803 1567	.015 809 1885	19.683 816 2291	.050 803 1567	85
	86	66.417 071 1168	1308.341 422 3354	.000 764 3265	.015 056 3700	19.698 872 5991	.050 764 3265	86
	87	69.737 924 6726	1374.758 493 4521	.000 727 4005	.014 339 4000	19.713 211 9992	.050 727 4005	87
	88	73.224 820 9062	1444.496 418 1247	.000 692 2828	.013 656 5715	19.726 868 5706	.050 692 2828	88
	89	76.886 061 9515	1517.721 239 0310	.000 658 8825	.013 006 2585	19.739 874 8292	.050 658 8825	89
	90	80.730 365 0491	1594.607 300 9825	.000 627 1136	.012 386 9129	19.752 261 7421	.050 627 1136	90
	91	84.766 883 3016	1675.337 666 0317	.000 596 8946	.011 797 0599	19.764 058 8020	.050 596 8946	91
	92	89.005 227 4667	1760.104 549 3332	.000 568 1481	.011 235 2951	19.775 294 0971	.050 568 1481	92
	93	93.455 488 8400	1849.109 776 7999	.000 540 8008	.010 700 2811	19.785 994 3782	.050 540 8008	93
	94	98.128 263 2820	1942.565 265 6399	.000 514 7832	.010 190 7439	19.796 185 1221	.050 514 7832	94
	95	103.034 676 4461	2040.693 528 9219	.000 490 0295	.009 705 4704	19.805 890 5925	.050 490 0295	95
	96	108.186 410 2684	2143.728 205 3680	.000 466 4770	.009 243 3051	19.815 133 8976	.050 466 4770	96
	97	113.595 730 7818	2251.914 615 6364	.000 444 0666	.008 803 1477	19.823 937 0453	.050 444 0666	97
	98	119.275 517 3209	2365.510 346 4182	.000 422 7418	.008 383 9502	19.832 320 9955	.050 422 7418	98
	99	125.239 293 1870	2484.785 863 7391	.000 402 4492	.007 984 7145	19.840 305 7100	.050 402 4492	99
	100	131.501 257 8463	2610.025 156 9261	.000 383 1381	.007 604 4900	19.847 910 2000	.050 383 1381	100

		XIV COMPOUND AMOUNT	XV AMOUNT OF ANNUITY	XVI SINKING FUND	XVII PRESENT VALUE	XVIII PRESENT VALUE OF ANNUITY	XIX AMORTIZATION		
RATE 6%	P E R I O D S	AMOUNT OF 1 *How $1 left at compound interest will grow.*	AMOUNT OF 1 PER PERIOD *How $1 deposited periodically will grow.*	SINKING FUND *Periodic deposit that will grow to $1 at future date.*	PRESENT WORTH OF 1 *What $1 due in the future is worth today.*	PRESENT WORTH OF 1 PER PERIOD *What $1 payable periodically is worth today.*	PARTIAL PAYMENT *Annuity worth $1 today. Periodic payment necessary to pay off a loan of $1.*	P E R I O D S	
	1	1.060 000 0000	1.000 000 0000	1.000 000 0000	.943 396 2264	.943 396 2264	1.060 000 0000	1	
	2	1.123 600 0000	2.060 000 0000	.485 436 8932	.889 996 4400	1.833 392 6664	.545 436 8932	2	
	3	1.191 016 0000	3.183 600 0000	.314 109 8128	.839 619 2830	2.673 011 9495	.374 109 8128	3	
	4	1.262 476 9600	4.374 616 0000	.228 591 4924	.792 093 6632	3.465 105 6127	.288 591 4924	4	
	5	1.338 225 5776	5.637 092 9600	.177 396 4004	.747 258 1729	4.212 363 7856	.237 396 4004	5	
	6	1.418 519 1123	6.975 318 5376	.143 362 6285	.704 960 5404	4.917 324 3260	.203 362 6285	6	
	7	1.503 630 2590	8.393 837 6499	.119 135 0181	.665 057 1136	5.582 381 4396	.179 135 0181	7	
	8	1.593 848 0745	9.897 467 9088	.101 035 9426	.627 412 3713	6.209 793 8110	.161 035 9426	8	
	9	1.689 478 9590	11.491 315 9834	.087 022 2350	.591 898 4635	6.801 692 2745	.147 022 2350	9	
	10	1.790 847 6965	13.180 794 9424	.075 867 9582	.558 394 7769	7.360 087 0514	.135 867 9582	10	
	11	1.898 298 5583	14.971 642 6389	.066 792 9381	.526 787 5254	7.886 874 5768	.126 792 9381	11	
	12	2.012 196 4718	16.869 941 1973	.059 277 0294	.496 969 3636	8.383 843 9404	.119 277 0294	12	
	13	2.132 928 2601	18.882 137 6691	.052 960 1053	.468 839 0222	8.852 682 9626	.112 960 1053	13	
	14	2.260 903 9558	21.015 065 9292	.047 584 9090	.442 300 9644	9.294 983 9270	.107 584 9090	14	
	15	2.396 558 1931	23.275 969 8850	.042 962 7640	.417 265 0607	9.712 248 9877	.102 962 7640	15	
	16	2.540 351 6847	25.672 528 0781	.038 952 1436	.393 646 2837	10.105 895 2715	.098 952 1436	16	
	17	2.692 772 7858	28.212 879 7628	.035 444 8042	.371 364 4186	10.477 259 6901	.095 444 8042	17	
	18	2.854 339 1529	30.905 652 5485	.032 356 5406	.350 343 7911	10.827 603 4812	.092 356 5406	18	
	19	3.025 599 5021	33.759 991 7015	.029 620 8604	.330 513 0105	11.158 116 4917	.089 620 8604	19	
	20	3.207 135 4722	36.785 591 2035	.027 184 5570	.311 804 7269	11.469 921 2186	.087 184 5570	20	
	21	3.399 563 6005	39.992 726 6758	.025 004 5467	.294 155 4027	11.764 076 6213	.085 004 5467	21	
	22	3.603 537 4166	43.392 290 2763	.023 045 5685	.277 505 0969	12.041 581 7182	.083 045 5685	22	
	23	3.819 749 6616	46.995 827 6929	.021 278 4847	.261 797 2612	12.303 378 9794	.081 278 4847	23	
	24	4.048 934 6413	50.815 577 3545	.019 679 0050	.246 978 5483	12.550 357 5278	.079 679 0050	24	
	25	4.291 870 7197	54.864 511 9957	.018 226 7182	.232 998 6305	12.783 356 1583	.078 226 7182	25	
	26	4.549 382 9629	59.156 382 7155	.016 904 3467	.219 810 0288	13.003 166 1870	.076 904 3467	26	
	27	4.822 345 9407	63.705 765 6784	.015 697 1663	.207 367 9517	13.210 534 1387	.075 697 1663	27	
	28	5.111 686 6971	68.528 111 6191	.014 592 5515	.195 630 1431	13.406 164 2818	.074 592 5515	28	
	29	5.418 387 8990	73.639 798 3162	.013 579 6135	.184 556 7388	13.590 721 0206	.073 579 6135	29	
	30	5.743 491 1729	79.058 186 2152	.012 648 9115	.174 110 1309	13.764 831 1515	.072 648 9115	30	
	31	6.088 100 6433	84.801 677 3881	.011 792 2196	.164 254 8405	13.929 085 9920	.071 792 2196	31	
	32	6.453 386 6819	90.889 778 0314	.011 002 3374	.154 957 3967	14.084 043 3887	.071 002 3374	32	
	33	6.840 589 8828	97.343 164 7133	.010 272 9350	.146 186 2233	14.230 229 6119	.070 272 9350	33	
	34	7.251 025 2758	104.183 754 5961	.009 598 4254	.137 911 5314	14.368 141 1433	.069 598 4254	34	
	35	7.686 086 7923	111.434 779 8719	.008 973 8590	.130 105 2183	14.498 246 3616	.068 973 8590	35	
	36	8.147 251 9999	119.120 866 6642	.008 394 8348	.122 740 7720	14.620 987 1336	.068 394 8348	36	
	37	8.636 087 1198	127.268 118 6640	.007 857 4274	.115 793 1811	14.736 780 3147	.067 857 4274	37	
	38	9.154 252 3470	135.904 205 7839	.007 358 1240	.109 238 8501	14.846 019 1648	.067 358 1240	38	
	39	9.703 507 4879	145.058 458 1309	.006 893 7724	.103 055 5190	14.949 074 6838	.066 893 7724	39	
	40	10.285 717 9371	154.761 965 6188	.006 461 5359	.097 222 1877	15.046 296 8715	.066 461 5359	40	
	41	10.902 861 0134	165.047 683 5559	.006 058 8551	.091 719 0450	15.138 015 9165	.066 058 8551	41	
	42	11.557 032 6742	175.950 544 5692	.005 683 4152	.086 527 4010	15.224 543 3175	.065 683 4152	42	
	43	12.250 454 6346	187.507 577 2434	.005 333 1178	.081 629 6235	15.306 172 9410	.065 333 1178	43	
	44	12.985 481 9127	199.758 031 8780	.005 006 0565	.077 009 0788	15.383 182 0198	.065 006 0565	44	
	45	13.764 610 8274	212.743 513 7907	.004 700 4958	.072 650 0743	15.455 832 0942	.064 700 4958	45	
	46	14.590 487 4771	226.508 124 6181	.004 414 8527	.068 537 8060	15.524 369 9002	.064 414 8527	46	
	47	15.465 916 7257	241.098 612 0952	.004 147 6805	.064 658 3075	15.589 028 2077	.064 147 6805	47	
	48	16.393 871 7293	256.564 528 8209	.003 897 6549	.060 998 4033	15.650 026 6110	.063 897 6549	48	
	49	17.377 504 0330	272.958 400 5502	.003 663 5619	.057 545 6635	15.707 572 2746	.063 663 5619	49	
	50	18.420 154 2750	290.335 904 5832	.003 444 2864	.054 288 3618	15.761 860 6364	.063 444 2864	50	
	51	19.525 363 5315	308.756 058 8582	.003 238 8028	.051 215 4357	15.813 076 0721	.063 238 8028	51	
	52	20.696 885 3434	328.281 422 3897	.003 046 1669	.048 316 4488	15.861 392 5208	.063 046 1669	52	
	53	21.938 698 4640	348.978 307 7331	.002 865 5076	.045 581 5554	15.906 974 0762	.062 865 5076	53	
	54	23.255 020 3718	370.917 006 1970	.002 696 0209	.043 001 4674	15.949 975 5436	.062 696 0209	54	
	55	24.650 321 5941	394.172 026 5689	.002 536 9634	.040 567 4221	15.990 542 9657	.062 536 9634	55	
	56	26.129 340 8898	418.822 348 1630	.002 387 6472	.038 271 1529	16.028 814 1186	.062 387 6472	56	
	57	27.697 101 3432	444.951 689 0528	.002 247 4350	.036 104 8612	16.064 918 9798	.062 247 4350	57	
	58	29.358 927 4238	472.648 790 3959	.002 115 7359	.034 061 1898	16.098 980 1696	.062 115 7359	58	
	59	31.120 463 0692	502.007 717 8197	.001 992 0012	.032 133 1979	16.131 113 3676	.061 992 0012	59	
	60	32.987 690 8533	533.128 180 8889	.001 875 7215	.030 314 3377	16.161 427 7052	.061 875 7215	60	
	61	34.966 952 3045	566.115 871 7422	.001 766 4228	.028 598 4318	16.190 026 1370	.061 766 4228	61	
	62	37.064 969 4428	601.082 824 0467	.001 663 6642	.026 979 6526	16.217 005 7896	.061 663 6642	62	
	63	39.288 867 6094	638.147 793 4895	.001 567 0351	.025 452 5025	16.242 458 2921	.061 567 0351	63	
	64	41.646 199 6659	677.436 661 0989	.001 476 1528	.024 011 7948	16.266 470 0869	.061 476 1528	64	
	65	44.144 971 6459	719.082 860 7649	.001 390 6603	.022 652 6366	16.289 122 7235	.061 390 6603	65	
	66	46.793 669 9446	763.227 832 4107	.001 310 2248	.021 370 4119	16.310 493 1354	.061 310 2248	66	
	67	49.601 290 1413	810.021 502 3554	.001 234 5351	.020 160 7659	16.330 653 9013	.061 234 5351	67	
	68	52.577 367 5498	859.622 792 4967	.001 163 3009	.019 019 5905	16.349 673 4918	.061 163 3009	68	
	69	55.732 009 6028	912.200 160 0465	.001 096 2506	.017 943 0099	16.367 616 5017	.061 096 2506	69	
	70	59.075 930 1790	967.932 169 6493	.001 033 1302	.016 927 3678	16.384 543 8695	.061 033 1302	70	
	71	62.620 485 9897	1027.008 099 8283	.000 973 7022	.015 969 2149	16.400 513 0844	.060 973 7022	71	
	72	66.377 715 1491	1089.628 585 8180	.000 917 7439	.015 065 2971	16.415 578 3816	.060 917 7439	72	
	73	70.360 378 0580	1156.006 300 9670	.000 865 0472	.014 212 5444	16.429 790 9260	.060 865 0472	73	
	74	74.582 000 7415	1226.366 679 0251	.000 815 4168	.013 408 0608	16.443 198 9868	.060 815 4168	74	
	75	79.056 920 7860	1300.948 679 7666	.000 768 6698	.012 649 1140	16.455 848 1007	.060 768 6698	75	
	76	83.800 336 0332	1380.005 600 5526	.000 724 6347	.011 933 1264	16.467 781 2271	.060 724 6347	76	
	77	88.828 356 1951	1463.805 936 5857	.000 683 1507	.011 257 6664	16.479 038 8935	.060 683 1507	77	
	78	94.158 057 5669	1552.634 292 7808	.000 644 0667	.010 620 4400	16.489 659 3335	.060 644 0667	78	
	79	99.807 541 0209	1646.792 350 3477	.000 607 2411	.010 019 2830	16.499 678 6165	.060 607 2411	79	
	80	105.795 993 4821	1746.599 891 3686	.000 572 5410	.009 452 1538	16.509 130 7703	.060 572 5410	80	
	81	112.143 753 0910	1852.395 884 8507	.000 539 8414	.008 917 1262	16.518 047 8965	.060 539 8414	81	
	82	118.872 378 2765	1964.539 637 9417	.000 509 0251	.008 412 3832	16.526 460 2797	.060 509 0251	82	
	83	126.004 720 9731	2083.412 016 2182	.000 479 9819	.007 936 2106	16.534 396 4903	.060 479 9819	83	
	84	133.565 004 2315	2209.416 737 1913	.000 452 6081	.007 486 9911	16.541 883 4814	.060 452 6081	84	
	85	141.578 904 4854	2342.981 741 4228	.000 426 8066	.007 063 1992	16.548 946 6806	.060 426 8066	85	
	86	150.073 638 7545	2484.560 645 9082	.000 402 4856	.006 663 3954	16.555 610 0760	.060 402 4856	86	
	87	159.078 057 0798	2634.634 284 6626	.000 379 5593	.006 286 2221	16.561 896 2981	.060 379 5593	87	
	88	168.622 740 5045	2793.712 341 7424	.000 357 9467	.005 930 3982	16.567 826 6963	.060 357 9467	88	
	89	178.740 104 9348	2962.335 082 2469	.000 337 5715	.005 594 7153	16.573 421 4116	.060 337 5715	89	
	90	189.464 511 2309	3141.075 187 1818	.000 318 3623	.005 278 0333	16.578 699 4450	.060 318 3623	90	
	91	200.832 381 9048	3330.539 698 4127	.000 300 2516	.004 979 2767	16.583 678 7217	.060 300 2516	91	
	92	212.882 324 8190	3531.372 080 3174	.000 283 1761	.004 697 4308	16.588 376 1525	.060 283 1761	92	
	93	225.655 264 3082	3744.254 405 1365	.000 267 0759	.004 431 5385	16.592 807 6910	.060 267 0759	93	
	94	239.194 580 1667	3969.909 669 4447	.000 251 8949	.004 180 6967	16.596 988 3878	.060 251 8949	94	
	95	253.546 254 9767	4209.104 249 6113	.000 237 5802	.003 944 0535	16.600 932 4413	.060 237 5802	95	
	96	268.759 030 2753	4462.650 504 5880	.000 224 0821	.003 720 8052	16.604 653 2465	.060 224 0821	96	
	97	284.884 572 0918	4731.409 534 8633	.000 211 3535	.003 510 1936	16.608 163 4401	.060 211 3535	97	
	98	301.977 646 4173	5016.294 106 9551	.000 199 3504	.003 311 5034	16.611 474 9435	.060 199 3504	98	
	99	320.096 305 2023	5318.271 753 3724	.000 188 0310	.003 124 0598	16.614 599 0033	.060 188 0310	99	
	100	339.302 083 5145	5638.368 058 5748	.000 177 3563	.002 947 2262	16.617 546 2295	.060 177 3563	100	

	P E R I O D S	XIV COMPOUND AMOUNT	XV AMOUNT OF ANNUITY	XVI SINKING FUND	XVII PRESENT VALUE	XVIII PRESENT VALUE OF ANNUITY	XIX AMORTIZATION	P E R I O D S
		AMOUNT OF 1	AMOUNT OF 1 PER PERIOD	SINKING FUND	PRESENT WORTH OF 1	PRESENT WORTH OF 1 PER PERIOD	PARTIAL PAYMENT	
		How $1 left at compound interest will grow.	*How $1 deposited periodically will grow.*	*Periodic deposit that will grow to $1 at future date.*	*What $1 due in the future is worth today.*	*What $1 payable periodically is worth today.*	*Annuity worth $1 today. Periodic payment necessary to pay off a loan of $1.*	
	1	1.070 000 0000	1.000 000 0000	1.000 000 0000	.934 579 4393	.934 579 4393	1.070 000 0000	1
	2	1.144 900 0000	2.070 000 0000	.483 091 7874	.873 438 7283	1.808 018 1675	.553 091 7874	2
	3	1.225 043 0000	3.214 900 0000	.311 051 6657	.816 297 8769	2.624 316 0444	.381 051 6657	3
	4	1.310 796 0100	4.439 943 0000	.225 228 1167	.762 895 2120	3.387 211 2565	.295 228 1167	4
	5	1.402 551 7207	5.750 739 0100	.173 890 6944	.712 986 1795	4.100 197 4359	.243 890 6944	5
	6	1.500 730 3518	7.153 290 7407	.139 795 7998	.666 342 2238	4.766 539 6598	.209 795 7998	6
	7	1.605 781 4765	8.654 021 0925	.115 553 2196	.622 749 7419	5.389 289 4016	.185 553 2196	7
	8	1.718 186 1798	10.259 802 5690	.097 467 7625	.582 009 1046	5.971 298 5062	.167 467 7625	8
	9	1.838 459 2124	11.977 988 7489	.083 486 4701	.543 933 7426	6.515 232 2488	.153 486 4701	9
	10	1.967 151 3573	13.816 447 9613	.072 377 5027	.508 349 2921	7.023 581 5409	.142 377 5027	10
	11	2.104 851 9523	15.783 599 3186	.063 356 9048	.475 092 7964	7.498 674 3373	.133 356 9048	11
	12	2.252 191 5890	17.888 451 2709	.055 901 9887	.444 011 9592	7.942 686 2966	.125 901 9887	12
	13	2.409 845 0002	20.140 642 8598	.049 650 8481	.414 964 4479	8.357 650 7444	.119 650 8481	13
	14	2.578 534 1502	22.550 487 8600	.044 344 9386	.387 817 2410	8.745 467 9855	.114 344 9386	14
	15	2.759 031 5407	25.129 022 0102	.039 794 6247	.362 446 0196	9.107 914 0051	.109 794 6247	15
	16	2.952 163 7486	27.888 053 5509	.035 857 6477	.338 734 5978	9.446 648 6029	.105 857 6477	16
	17	3.158 815 2110	30.840 217 2995	.032 425 1931	.316 574 3905	9.763 222 9934	.102 425 1931	17
	18	3.379 932 2757	33.999 032 5105	.029 412 6017	.295 863 9163	10.059 086 9097	.099 412 6017	18
	19	3.616 627 5350	37.378 964 7862	.026 753 0148	.276 508 3330	10.335 595 2427	.096 753 0148	19
	20	3.869 684 4625	40.995 492 3212	.024 392 9257	.258 419 0028	10.594 014 2455	.094 392 9257	20
	21	4.140 562 3749	44.865 176 7837	.022 289 0017	.241 513 0867	10.835 527 3323	.092 289 0017	21
	22	4.430 401 7411	49.005 739 1586	.020 405 7732	.225 713 1652	11.061 240 4974	.090 405 7732	22
	23	4.740 529 8630	53.436 140 8997	.018 713 9263	.210 946 8833	11.272 187 3808	.088 713 9263	23
	24	5.072 366 9534	58.176 670 7627	.017 189 0207	.197 146 6199	11.469 334 0007	.087 189 0207	24
	25	5.427 432 6401	63.249 037 7160	.015 810 5172	.184 249 1775	11.653 583 1783	.085 810 5172	25
	26	5.807 352 9249	68.676 470 3562	.014 561 0279	.172 195 4930	11.825 778 6713	.084 561 0279	26
	27	6.213 867 6297	74.483 823 2811	.013 425 7340	.160 930 3673	11.986 709 0386	.083 425 7340	27
	28	6.648 838 3638	80.697 690 9108	.012 391 9283	.150 402 2124	12.137 111 2510	.082 391 9283	28
	29	7.114 257 0492	87.346 529 2745	.011 448 6518	.140 562 8154	12.277 674 0664	.081 448 6518	29
	30	7.612 255 0427	94.460 786 3237	.010 586 4035	.131 367 1172	12.409 041 1835	.080 586 4035	30
	31	8.145 112 8956	102.073 041 3664	.009 796 9061	.122 773 0067	12.531 814 1902	.079 796 9061	31
	32	8.715 270 7983	110.218 154 2621	.009 072 9155	.114 741 1277	12.646 555 3179	.079 072 9155	32
	33	9.325 339 7542	118.933 425 0604	.008 408 0653	.107 234 6988	12.753 790 0168	.078 408 0653	33
	34	9.978 113 5370	128.258 764 8146	.007 796 7381	.100 219 3447	12.854 009 3615	.077 796 7381	34
	35	10.676 581 4846	138.236 878 3516	.007 233 9596	.093 662 9390	12.947 672 3004	.077 233 9596	35
	36	11.423 942 1885	148.913 459 8363	.006 715 3097	.087 535 4570	13.035 207 7574	.076 715 3097	36
	37	12.223 618 1417	160.337 402 0248	.006 236 8480	.081 808 8383	13.117 016 5957	.076 236 8480	37
	38	13.079 271 4117	172.561 020 1665	.005 795 0515	.076 456 8582	13.193 473 4539	.075 795 0515	38
	39	13.994 820 4105	185.640 291 5782	.005 386 7616	.071 455 0077	13.264 928 4616	.075 386 7616	39
	40	14.974 457 8392	199.635 111 9887	.005 009 1389	.066 780 3810	13.331 708 8426	.075 009 1389	40
	41	16.022 669 8880	214.609 569 8279	.004 659 6245	.062 411 5710	13.394 120 4137	.074 659 6245	41
	42	17.144 256 7801	230.632 239 7158	.004 335 9072	.058 328 5711	13.452 448 9847	.074 335 9072	42
	43	18.344 354 7547	247.776 496 4959	.004 035 8953	.054 512 6832	13.506 961 6680	.074 035 8953	43
	44	19.628 459 5875	266.120 851 2507	.003 757 6913	.050 946 4329	13.557 908 1009	.073 757 6913	44
	45	21.002 451 7587	285.749 310 8382	.003 499 5710	.047 613 4887	13.605 521 5896	.073 499 5710	45
	46	22.472 623 3818	306.751 762 5969	.003 259 9650	.044 498 5876	13.650 020 1772	.073 259 9650	46
	47	24.045 707 0185	329.224 385 9787	.003 037 4421	.041 587 4650	13.691 607 6423	.073 037 4421	47
	48	25.728 906 5098	353.270 092 9972	.002 830 6953	.038 866 7898	13.730 474 4320	.072 830 6953	48
	49	27.529 929 9655	378.998 999 5070	.002 638 5294	.036 324 1026	13.766 798 5346	.072 638 5294	49
	50	29.457 025 0631	406.528 929 4724	.002 459 8495	.033 947 7594	13.800 746 2940	.072 459 8495	50
	51	31.519 016 8175	435.985 954 5355	.002 293 6519	.031 726 8780	13.832 473 1720	.072 293 6519	51
	52	33.725 347 9947	467.504 971 3530	.002 139 0147	.029 651 2878	13.862 124 4598	.072 139 0147	52
	53	36.086 122 3543	501.230 319 3477	.001 995 0908	.027 711 4839	13.889 835 9437	.071 995 0908	53
	54	38.612 150 9191	537.316 441 7021	.001 861 1007	.025 898 5831	13.915 734 5269	.071 861 1007	54
	55	41.315 001 4835	575.928 592 6212	.001 736 3264	.024 204 2833	13.939 938 8102	.071 736 3264	55
	56	44.207 051 5873	617.243 594 1047	.001 620 1059	.022 620 8255	13.962 559 6357	.071 620 1059	56
	57	47.301 545 1984	661.450 645 6920	.001 511 8286	.021 140 9584	13.983 700 5941	.071 511 8286	57
	58	50.612 653 3623	708.752 190 8905	.001 410 9304	.019 757 9051	14.003 458 4991	.071 410 9304	58
	59	54.155 739 0977	759.364 844 2528	.001 316 8900	.018 465 3318	14.021 923 8310	.071 316 8900	59
	60	57.946 426 8345	813.520 383 3505	.001 229 2255	.017 257 3195	14.039 181 1504	.071 229 2255	60
	61	62.002 676 7130	871.466 810 1850	.001 147 4906	.016 128 3360	14.055 309 4864	.071 147 4906	61
	62	66.342 864 0829	933.469 486 8980	.001 071 2723	.015 073 2112	14.070 382 6976	.071 071 2723	62
	63	70.986 864 5687	999.812 350 9808	.001 000 1877	.014 087 1132	14.084 469 8108	.071 000 1877	63
	64	75.955 945 0885	1070.799 215 5495	.000 933 8819	.013 165 5264	14.097 635 3372	.070 933 8819	64
	65	81.272 861 2447	1146.755 160 6379	.000 872 0257	.012 304 2303	14.109 939 5675	.070 872 0257	65
	66	86.961 961 5318	1228.028 021 8826	.000 814 3137	.011 499 2806	14.121 438 8481	.070 814 3137	66
	67	93.049 298 8390	1314.989 983 4144	.000 760 4621	.010 746 9912	14.132 185 8394	.070 760 4621	67
	68	99.562 749 7577	1408.039 282 2534	.000 710 2075	.010 043 9171	14.142 229 7564	.070 710 2075	68
	69	106.532 142 2408	1507.602 032 0111	.000 663 3050	.009 386 8384	14.151 616 5948	.070 663 3050	69
	70	113.989 392 1976	1614.134 174 2519	.000 619 5272	.008 772 7461	14.160 389 3409	.070 619 5272	70
	71	121.968 649 6515	1728.123 566 4495	.000 578 6623	.008 198 8282	14.168 588 1691	.070 578 6623	71
	72	130.506 455 1271	1850.092 216 1010	.000 540 5136	.007 662 4562	14.176 250 6253	.070 540 5136	72
	73	139.641 906 9860	1980.598 671 2281	.000 504 8978	.007 161 1740	14.183 411 7993	.070 504 8978	73
	74	149.416 840 4750	2120.240 578 2140	.000 471 6446	.006 692 6860	14.190 104 4854	.070 471 6446	74
	75	159.876 019 3082	2269.657 418 6890	.000 440 5951	.006 254 8468	14.196 359 3321	.070 440 5951	75
	76	171.067 340 6598	2429.533 437 9973	.000 411 6017	.005 845 6512	14.202 204 9833	.070 411 6017	76
	77	183.042 054 5060	2600.600 778 6571	.000 384 5265	.005 462 2254	14.207 668 2087	.070 384 5265	77
	78	195.854 998 3214	2783.642 833 1631	.000 359 2415	.005 105 8181	14.212 774 0268	.070 359 2415	78
	79	209.564 848 2039	2979.497 831 4845	.000 335 6270	.004 771 7926	14.217 545 8194	.070 335 6270	79
	80	224.234 387 5782	3189.062 679 6884	.000 313 5718	.004 459 6193	14.222 005 4387	.070 313 5718	80
	81	239.930 794 7087	3413.297 067 2666	.000 292 9719	.004 167 8685	14.226 173 3072	.070 292 9719	81
	82	256.725 950 3383	3653.227 861 9752	.000 273 7305	.003 895 2042	14.230 068 5114	.070 273 7305	82
	83	274.696 766 8619	3909.953 812 3135	.000 255 7575	.003 640 3778	14.233 708 8892	.070 255 7575	83
	84	293.925 540 5423	4184.650 579 1754	.000 238 9686	.003 403 2222	14.237 111 1114	.070 238 9686	84
	85	314.500 328 3802	4478.576 119 7177	.000 223 2853	.003 179 6469	14.240 290 7583	.070 223 2853	85
	86	336.515 351 3669	4793.076 448 0980	.000 208 6343	.002 971 6326	14.243 262 3909	.070 208 6343	86
	87	360.071 425 9625	5129.591 799 4648	.000 194 9473	.002 777 2268	14.246 039 6177	.070 194 9473	87
	88	385.276 425 7799	5489.663 225 4273	.000 182 1605	.002 595 5390	14.248 635 1567	.070 182 1605	88
	89	412.245 775 5845	5874.939 651 2073	.000 170 2145	.002 425 7374	14.251 060 8941	.070 170 2145	89
	90	441.102 979 8754	6287.185 426 7918	.000 159 0537	.002 267 0443	14.253 327 9384	.070 159 0537	90
	91	471.980 188 4667	6728.288 406 6672	.000 148 6262	.002 118 7330	14.255 446 6714	.070 148 6262	91
	92	505.018 801 6594	7200.268 595 1339	.000 138 8837	.001 980 1243	14.257 426 7957	.070 138 8837	92
	93	540.370 117 7755	7705.287 396 7933	.000 129 7810	.001 850 5835	14.259 277 3792	.070 129 7810	93
	94	578.196 026 0198	8245.657 514 5688	.000 121 2760	.001 729 5172	14.261 006 8965	.070 121 2760	94
	95	618.669 747 8412	8823.853 540 5886	.000 113 3292	.001 616 3713	14.262 623 2677	.070 113 3292	95
	96	661.976 630 1901	9442.523 288 4298	.000 105 9039	.001 510 6273	14.264 133 8951	.070 105 9039	96
	97	708.314 994 3034	10104.499 918 6199	.000 098 9658	.001 411 8013	14.265 545 6963	.070 098 9658	97
	98	757.897 043 9046	10812.814 912 9233	.000 092 4829	.001 319 4404	14.266 865 1367	.070 092 4829	98
	99	810.949 836 9780	11570.711 956 8279	.000 086 4251	.001 233 1219	14.268 098 2586	.070 086 4251	99
	100	867.716 325 5664	12381.661 793 8059	.000 080 7646	.001 152 4504	14.269 250 7090	.070 080 7646	100

631

		XIV COMPOUND AMOUNT	XV AMOUNT OF ANNUITY	XVI SINKING FUND	XVII PRESENT VALUE	XVIII PRESENT VALUE OF ANNUITY	XIX AMORTIZATION	
RATE 7 1/2%	PERIODS	AMOUNT OF 1 — How $1 left at compound interest will grow.	AMOUNT OF 1 PER PERIOD — How $1 deposited periodically will grow.	SINKING FUND — Periodic deposit that will grow to $1 at future date.	PRESENT WORTH OF 1 — What $1 due in the future is worth today.	PRESENT WORTH OF 1 PER PERIOD — What $1 payable periodically is worth today.	PARTIAL PAYMENT — Annuity worth $1 today. Periodic payment necessary to pay off a loan of $1.	PERIODS
	1	1.075 000 0000	1.000 000 0000	1.000 000 0000	.930 232 5581	.930 232 5581	1.075 000 0000	1
	2	1.155 625 0000	2.075 000 0000	.481 927 7108	.865 332 6122	1.795 565 1704	.556 927 7108	2
	3	1.242 296 8750	3.230 625 0000	.309 537 6282	.804 960 5695	2.600 525 7399	.384 537 6282	3
	4	1.335 469 1406	4.472 921 8750	.223 567 5087	.748 800 5298	3.349 326 2696	.298 567 5087	4
	5	1.435 629 3262	5.808 391 0156	.172 164 7178	.696 558 6324	4.045 884 9020	.247 164 7178	5
	6	1.543 301 5256	7.244 020 3418	.138 044 8912	.647 961 5185	4.693 846 4205	.213 044 8912	6
	7	1.659 049 1401	8.787 321 8674	.113 800 3154	.602 754 9009	5.296 601 3214	.188 800 3154	7
	8	1.783 477 8256	10.446 371 0075	.095 727 0232	.560 702 2334	5.857 303 5548	.170 727 0232	8
	9	1.917 238 6625	12.229 848 8331	.081 767 1595	.521 583 4729	6.378 887 0277	.156 767 1595	9
	10	2.061 031 5622	14.147 087 4955	.070 685 9274	.485 193 9283	6.864 080 9560	.145 685 9274	10
	11	2.215 608 9293	16.208 119 0577	.061 697 4737	.451 343 1891	7.315 424 1451	.136 697 4737	11
	12	2.381 779 5990	18.423 727 9870	.054 277 8313	.419 854 1294	7.735 278 2745	.129 277 8313	12
	13	2.560 413 0690	20.805 507 5860	.048 064 1963	.390 561 9808	8.125 840 2554	.123 064 1963	13
	14	2.752 444 0491	23.365 920 6550	.042 797 3721	.363 313 4706	8.489 153 7259	.117 797 3721	14
	15	2.958 877 3528	26.118 364 7041	.038 287 2363	.337 966 0191	8.827 119 7450	.113 287 2363	15
	16	3.180 793 1543	29.077 242 0569	.034 391 1571	.314 386 9945	9.141 506 7396	.109 391 1571	16
	17	3.419 352 6408	32.258 035 2112	.031 000 0282	.292 453 0182	9.433 959 7577	.106 000 0282	17
	18	3.675 804 0889	35.677 387 8520	.028 028 9578	.272 049 3192	9.706 009 0770	.103 028 9578	18
	19	3.951 489 3956	39.353 191 9410	.025 410 8994	.253 069 1342	9.959 078 2111	.100 410 8994	19
	20	4.247 851 1002	43.304 681 3365	.023 092 1916	.235 413 1481	10.194 491 3592	.098 092 1916	20
	21	4.566 439 9328	47.552 532 4368	.021 029 3742	.218 988 9749	10.413 480 3341	.096 029 3742	21
	22	4.908 922 9277	52.118 972 3695	.019 186 8710	.203 710 6744	10.617 191 0085	.094 186 8710	22
	23	5.277 092 1473	57.027 895 2972	.017 535 2780	.189 498 3017	10.806 689 3102	.092 535 2780	23
	24	5.672 874 0583	62.304 987 4445	.016 050 0795	.176 277 4900	10.982 966 8002	.091 050 0795	24
	25	6.098 339 6127	67.977 861 5029	.014 710 6716	.163 979 0605	11.146 945 8607	.089 710 6716	25
	26	6.555 715 0837	74.076 201 1156	.013 499 6124	.152 538 6609	11.299 484 5215	.088 499 6124	26
	27	7.047 393 7149	80.631 916 1992	.012 402 0369	.141 896 4287	11.441 380 9503	.087 402 0369	27
	28	7.575 948 2436	87.679 309 9142	.011 405 1993	.131 996 6779	11.573 377 6282	.086 405 1993	28
	29	8.144 144 3618	95.255 258 1578	.010 498 1081	.122 787 6073	11.696 165 2355	.085 498 1081	29
	30	8.754 955 1890	103.399 402 5196	.009 671 2358	.114 221 0301	11.810 386 2656	.084 671 2358	30
	31	9.411 576 8281	112.154 357 7086	.008 916 2831	.106 252 1210	11.916 638 3866	.083 916 2831	31
	32	10.117 445 0903	121.565 934 5367	.008 225 9887	.098 839 1823	12.015 477 5689	.083 225 9887	32
	33	10.876 253 4720	131.683 379 6269	.007 593 9728	.091 943 4254	12.107 420 9943	.082 593 9728	33
	34	11.691 972 4824	142.559 633 0990	.007 014 6084	.085 528 7678	12.192 949 7622	.082 014 6084	34
	35	12.568 870 4186	154.251 605 5814	.006 482 9147	.079 561 6445	12.272 511 4067	.081 482 9147	35
	36	13.511 535 7000	166.820 476 0000	.005 994 4680	.074 010 8321	12.346 522 2388	.080 994 4680	36
	37	14.524 900 8775	180.332 011 7000	.005 545 3271	.068 847 2857	12.415 369 5244	.080 545 3271	37
	38	15.614 268 4433	194.856 912 5775	.005 131 9709	.064 043 9867	12.479 413 5111	.080 131 9709	38
	39	16.785 338 5766	210.471 181 0208	.004 751 2443	.059 575 8016	12.538 989 3127	.079 751 2443	39
	40	18.044 238 9698	227.256 519 5974	.004 400 3138	.055 419 3503	12.594 408 6629	.079 400 3138	40
	41	19.397 556 8925	245.300 758 5672	.004 076 6282	.051 552 8840	12.645 961 5469	.079 076 6282	41
	42	20.852 373 6595	264.698 315 4597	.003 777 8858	.047 956 1711	12.693 917 7181	.078 777 8858	42
	43	22.416 301 6839	285.550 689 1192	.003 502 0052	.044 610 3918	12.738 528 1098	.078 502 0052	43
	44	24.097 524 3102	307.966 990 8031	.003 247 1012	.041 498 0388	12.780 026 1487	.078 247 1012	44
	45	25.904 838 6335	332.064 515 1134	.003 011 4630	.038 602 8268	12.818 628 9755	.078 011 4630	45
	46	27.847 701 5310	357.969 353 7469	.002 793 5352	.035 909 6064	12.854 538 5819	.077 793 5352	46
	47	29.936 279 1458	385.817 055 2779	.002 591 9020	.033 404 2850	12.887 942 8669	.077 591 9020	47
	48	32.181 500 0818	415.753 334 4237	.002 405 2724	.031 073 7535	12.919 016 6203	.077 405 2724	48
	49	34.595 112 5879	447.934 834 5055	.002 232 4676	.028 905 8172	12.947 922 4375	.077 232 4676	49
	50	37.189 746 0320	482.529 947 0934	.002 072 4102	.026 889 1323	12.974 811 5698	.077 072 4102	50
	51	39.978 976 9844	519.719 693 1254	.001 924 1141	.025 013 1463	12.999 824 7161	.076 924 1141	51
	52	42.977 400 2582	559.698 670 1098	.001 786 6757	.023 268 0431	13.023 092 7591	.076 786 6757	52
	53	46.200 705 2776	602.676 070 3681	.001 659 2661	.021 644 6912	13.044 737 4504	.076 659 2661	53
	54	49.665 758 1734	648.876 775 6457	.001 541 1247	.020 134 5965	13.064 872 0469	.076 541 1247	54
	55	53.390 690 0364	698.542 533 8191	.001 431 5521	.018 729 8572	13.083 601 9040	.076 431 5521	55
	56	57.394 991 7892	751.933 223 8555	.001 329 9053	.017 423 1230	13.101 025 0270	.076 329 9053	56
	57	62.699 616 1734	809.328 215 6447	.001 235 5927	.016 207 5563	13.117 232 5833	.076 235 5927	57
	58	66.327 087 3864	871.027 831 8180	.001 148 0689	.015 076 7965	13.132 309 3798	.076 148 0689	58
	59	71.301 618 9403	937.354 919 2044	.001 066 8318	.014 024 9270	13.146 334 3068	.076 066 8318	59
	60	76.649 240 3609	1008.656 538 1447	.000 991 4178	.013 046 4437	13.159 380 7505	.075 991 4178	60
	61	82.397 933 3879	1085.305 778 5056	.000 921 3993	.012 136 2267	13.171 516 9772	.075 921 3993	61
	62	88.577 778 3920	1167.703 711 8935	.000 856 3816	.011 289 5132	13.182 806 4904	.075 856 3816	62
	63	95.221 111 7714	1256.281 490 2855	.000 795 9999	.010 501 8728	13.193 308 3632	.075 795 9999	63
	64	102.362 695 1543	1351.502 602 0569	.000 739 9172	.009 769 1840	13.203 077 5471	.075 739 9172	64
	65	110.039 897 2908	1453.865 297 2112	.000 687 8216	.009 087 6130	13.212 165 1601	.075 687 8216	65
	66	118.292 889 5877	1563.905 194 5020	.000 639 4249	.008 453 5935	13.220 618 7536	.075 639 4249	66
	67	127.164 856 3067	1682.198 084 0897	.000 594 4603	.007 863 8079	13.228 482 5615	.075 594 4603	67
	68	136.702 220 5297	1809.362 940 3964	.000 552 6807	.007 315 1701	13.235 797 7316	.075 552 6807	68
	69	146.954 887 0695	1946.065 160 9261	.000 513 8574	.006 804 8094	13.242 602 5411	.075 513 8574	69
	70	157.976 503 5997	2093.020 047 9956	.000 477 7785	.006 330 0553	13.248 932 5963	.075 477 7785	70
	71	169.824 741 3696	2250.996 551 5952	.000 444 2477	.005 888 4235	13.254 821 0198	.075 444 2477	71
	72	182.561 596 9724	2420.821 292 9649	.000 413 0829	.005 477 6033	13.260 298 6231	.075 413 0829	72
	73	196.253 716 7453	2603.382 889 9373	.000 384 1156	.005 095 4449	13.265 394 0680	.075 384 1156	73
	74	210.972 745 5012	2799.636 606 6826	.000 357 1892	.004 739 9487	13.270 134 0168	.075 357 1892	74
	75	226.795 701 4138	3010.609 352 1837	.000 332 1587	.004 409 2546	13.274 543 2714	.075 332 1587	75
	76	243.805 379 0198	3237.405 053 5975	.000 308 8894	.004 101 6322	13.278 644 9036	.075 308 8894	76
	77	262.090 782 4463	3481.210 432 6173	.000 287 2564	.003 815 4718	13.282 460 3755	.075 287 2564	77
	78	281.747 591 1298	3743.301 215 0636	.000 267 1439	.003 549 2761	13.286 009 6516	.075 267 1439	78
	79	302.878 660 4645	4025.048 806 1934	.000 248 4442	.003 301 6522	13.289 311 3038	.075 248 4442	79
	80	325.594 559 9993	4327.927 466 6579	.000 231 0575	.003 071 3044	13.292 382 6082	.075 231 0575	80
	81	350.014 151 9993	4653.522 026 6573	.000 214 8910	.002 857 0273	13.295 239 6355	.075 214 8910	81
	82	376.265 213 3992	5003.536 178 6566	.000 199 8587	.002 657 6998	13.297 897 3354	.075 199 8587	82
	83	404.485 104 4042	5379.801 392 0558	.000 185 8805	.002 472 2789	13.300 369 6143	.075 185 8805	83
	84	434.821 487 2345	5784.286 496 4600	.000 172 8822	.002 299 7944	13.302 669 4087	.075 172 8822	84
	85	467.433 098 7771	6219.107 983 6945	.000 160 7948	.002 139 3436	13.304 808 7522	.075 160 7948	85
	86	502.490 581 1854	6686.541 082 4716	.000 149 5542	.001 990 0871	13.306 798 8393	.075 149 5542	86
	87	540.177 374 7743	7189.031 663 6569	.000 139 1008	.001 851 2438	13.308 650 0831	.075 139 1008	87
	88	580.690 677 8823	7729.209 038 4312	.000 129 3793	.001 722 0872	13.310 372 1703	.075 129 3793	88
	89	624.242 478 7235	8309.899 716 3136	.000 120 3384	.001 601 9416	13.311 974 1119	.075 120 3384	89
	90	671.060 664 6278	8934.142 195 0371	.000 111 9302	.001 490 1782	13.313 464 2901	.075 111 9302	90
	91	721.390 214 4749	9605.202 859 6649	.000 104 1102	.001 386 2123	13.314 850 5025	.075 104 1102	91
	92	775.494 480 5605	10326.593 074 1397	.000 096 8374	.001 289 4998	13.316 140 0023	.075 096 8374	92
	93	833.656 566 6025	11102.087 554 7002	.000 090 0732	.001 199 5347	13.317 339 5370	.075 090 0732	93
	94	896.180 809 0977	11935.744 121 3027	.000 083 7820	.001 115 8463	13.318 455 3833	.075 083 7820	94
	95	963.394 369 7800	12831.924 930 4004	.000 077 9306	.001 037 9965	13.319 493 3798	.075 077 9306	95
	96	1035.648 947 5135	13795.319 300 1804	.000 072 4884	.000 965 5782	13.320 458 9579	.075 072 4884	96
	97	1113.322 618 5770	14830.968 247 6940	.000 067 4265	.000 898 2122	13.321 357 1702	.075 067 4265	97
	98	1196.821 814 9703	15944.290 866 2710	.000 062 7184	.000 835 5463	13.322 192 7164	.075 062 7184	98
	99	1286.583 451 0931	17141.112 681 2414	.000 058 3393	.000 777 2523	13.322 969 9688	.075 058 3393	99
	100	1383.077 209 9251	18427.696 132 3345	.000 054 2661	.000 723 0254	13.323 692 9942	.075 054 2661	100

RATE 8%

PERIODS	AMOUNT OF 1 — How $1 left at compound interest will grow.	AMOUNT OF 1 PER PERIOD — How $1 deposited periodically will grow.	SINKING FUND — Periodic deposit that will grow to $1 at future date	PRESENT WORTH OF 1 — What $1 due in the future is worth today.	PRESENT WORTH OF 1 PER PERIOD — What $1 payable periodically is worth today.	PARTIAL PAYMENT — Annuity worth $1 today. Periodic payment necessary to pay off a loan of $1	PERIODS
1	1.080 000 0000	1.000 000 0000	1.000 000 0000	.925 925 9259	.925 925 9259	1.080 000 0000	1
2	1.166 400 0000	2.080 000 0000	.480 769 2308	.857 338 8203	1.783 264 7462	.560 769 2308	2
3	1.259 712 0000	3.246 400 0000	.308 033 5140	.793 832 2410	2.577 096 9872	.388 033 5140	3
4	1.360 488 9600	4.506 112 0000	.221 920 8045	.735 029 8528	3.312 126 8400	.301 920 8045	4
5	1.469 328 0768	5.866 600 9600	.170 456 4546	.680 583 1970	3.992 710 0371	.250 456 4546	5
6	1.586 874 3229	7.335 929 0368	.136 315 3862	.630 169 6269	4.622 879 6640	.216 315 3862	6
7	1.713 824 2688	8.922 803 3597	.112 072 4014	.583 490 3953	5.206 370 0592	.192 072 4014	7
8	1.850 930 2103	10.636 627 6285	.094 014 7606	.540 268 8845	5.746 638 9437	.174 014 7606	8
9	1.999 004 6271	12.487 557 8388	.080 079 7092	.500 248 9671	6.246 887 9109	.160 079 7092	9
10	2.158 924 9973	14.486 562 4659	.069 029 4887	.463 193 4881	6.710 081 3989	.149 029 4887	10
11	2.331 638 9971	16.645 487 4632	.060 076 3421	.428 882 8592	7.138 964 2583	.140 076 3421	11
12	2.518 170 1168	18.977 126 4602	.052 695 0169	.397 113 7586	7.536 078 0169	.132 695 0169	12
13	2.719 623 7262	21.495 296 5771	.046 521 8052	.367 697 9247	7.903 775 9416	.126 521 8052	13
14	2.937 193 6243	24.214 920 3032	.041 296 8528	.340 461 0414	8.244 236 9830	.121 296 8528	14
15	3.172 169 1142	27.152 113 9275	.036 829 5449	.315 241 7050	8.559 478 6879	.116 829 5449	15
16	3.425 942 6433	30.324 283 0417	.032 976 8720	.291 890 4676	8.851 369 1555	.112 976 8720	16
17	3.700 018 0548	33.750 225 6850	.029 629 4315	.270 268 9514	9.121 638 1069	.109 629 4315	17
18	3.996 019 4992	37.450 243 7398	.026 702 0959	.250 249 0291	9.371 887 1360	.106 702 0959	18
19	4.315 701 0591	41.446 263 2390	.024 127 6275	.231 712 0640	9.603 599 2000	.104 127 6275	19
20	4.660 957 1438	45.761 964 2981	.021 852 2088	.214 548 2074	9.818 147 4074	.101 852 2088	20
21	5.033 833 7154	50.422 921 4420	.019 832 2503	.198 655 7476	10.016 803 1550	.099 832 2503	21
22	5.436 540 4126	55.456 755 1573	.018 032 0684	.183 940 5070	10.200 743 6621	.098 032 0684	22
23	5.871 463 6456	60.893 295 5699	.016 422 1692	.170 315 2843	10.371 058 9464	.096 422 1692	23
24	6.341 180 7372	66.764 759 2155	.014 977 9616	.157 699 3373	10.528 758 2837	.094 977 9616	24
25	6.848 475 1962	73.105 939 9527	.013 678 7791	.146 017 9049	10.674 776 1886	.093 678 7791	25
26	7.396 353 2119	79.954 415 1490	.012 507 1267	.135 201 7638	10.809 977 9524	.092 507 1267	26
27	7.988 061 4689	87.350 768 3609	.011 448 0962	.125 186 8183	10.935 164 7707	.091 448 0962	27
28	8.627 106 3864	95.338 829 8297	.010 488 9057	.115 913 7207	11.051 078 4914	.090 488 9057	28
29	9.317 274 8973	103.965 936 2161	.009 618 5350	.107 327 5192	11.158 406 0106	.089 618 5350	29
30	10.062 656 8891	113.283 211 1134	.008 827 4334	.099 377 3325	11.257 783 3431	.088 827 4334	30
31	10.867 669 4402	123.345 868 0025	.008 107 2841	.092 016 0487	11.349 799 3918	.088 107 2841	31
32	11.737 082 9954	134.213 537 4427	.007 450 8132	.085 200 0451	11.434 999 4368	.087 450 8132	32
33	12.676 049 6350	145.950 620 4381	.006 851 6324	.078 888 9306	11.513 888 3674	.086 851 6324	33
34	13.690 133 6059	158.626 670 0732	.006 304 1101	.073 045 3061	11.586 933 6736	.086 304 1101	34
35	14.785 344 2943	172.316 803 6790	.005 803 2646	.067 634 5427	11.654 568 2163	.085 803 2646	35
36	15.968 171 8379	187.102 147 9733	.005 344 6741	.062 624 5766	11.717 192 7928	.085 344 6741	36
37	17.245 625 5849	203.070 319 8112	.004 924 4025	.057 985 7190	11.775 178 5119	.084 924 4025	37
38	18.625 275 6317	220.315 945 3961	.004 538 9361	.053 690 4806	11.828 868 9925	.084 538 9361	38
39	20.115 297 6822	238.941 221 0278	.004 185 1297	.049 713 4080	11.878 582 4004	.084 185 1297	39
40	21.724 521 4968	259.056 518 7100	.003 860 1615	.046 030 9333	11.924 613 3337	.083 860 1615	40
41	23.462 483 2165	280.781 040 2068	.003 561 4940	.042 621 2345	11.967 234 5683	.083 561 4940	41
42	25.339 481 8739	304.243 523 4233	.003 286 8407	.039 464 1061	12.006 698 6743	.083 286 8407	42
43	27.366 640 4238	329.583 005 2972	.003 034 1370	.036 540 8389	12.043 239 5133	.083 034 1370	43
44	29.555 971 6577	356.949 645 7210	.002 801 5156	.033 834 1101	12.077 073 6234	.082 801 5156	44
45	31.920 449 3903	386.505 617 3787	.002 587 2845	.031 327 8797	12.108 401 5032	.082 587 2845	45
46	34.474 085 3415	418.426 066 7690	.002 389 9085	.029 007 2961	12.137 408 7992	.082 389 9085	46
47	37.232 012 1688	452.900 152 1105	.002 207 9922	.026 858 6075	12.164 267 4067	.082 207 9922	47
48	40.210 573 1423	490.132 164 2793	.002 040 2660	.024 869 0810	12.189 136 4877	.082 040 2660	48
49	43.427 418 9937	530.342 737 4217	.001 885 5731	.023 026 9268	12.212 163 4145	.081 885 5731	49
50	46.901 612 5132	573.770 156 4154	.001 742 8582	.021 321 2286	12.233 484 6431	.081 742 8582	50
51	50.653 741 5143	620.671 768 9286	.001 611 1575	.019 741 8783	12.253 226 5214	.081 611 1575	51
52	54.706 040 8354	671.325 510 4429	.001 489 5903	.018 279 5169	12.271 506 0383	.081 489 5903	52
53	59.082 524 1023	726.031 551 2783	.001 377 3506	.016 925 4786	12.288 431 5169	.081 377 3506	53
54	63.809 126 0304	785.114 075 3806	.001 273 7003	.015 671 7395	12.304 103 2564	.081 273 7003	54
55	68.913 856 1129	848.923 201 4111	.001 177 9629	.014 510 8699	12.318 614 1263	.081 177 9629	55
56	74.426 964 6019	917.837 057 5239	.001 089 5180	.013 435 9906	12.332 050 1170	.081 089 5180	56
57	80.381 121 7701	992.264 022 1259	.001 007 7963	.012 440 7321	12.344 490 8490	.081 007 7963	57
58	86.811 611 5117	1072.645 143 8959	.000 932 2748	.011 519 1964	12.356 010 0454	.080 932 2748	58
59	93.756 540 4326	1159.456 755 4076	.000 862 4729	.010 665 9226	12.366 675 9680	.080 862 4729	59
60	101.257 063 6672	1253.213 295 8402	.000 797 9488	.009 875 8542	12.376 551 8222	.080 797 9488	60
61	109.357 628 7606	1354.470 359 5074	.000 738 2960	.009 144 3095	12.385 696 1317	.080 738 2960	61
62	118.106 239 0614	1463.827 988 2680	.000 683 1404	.008 466 9532	12.394 163 0849	.080 683 1404	62
63	127.554 738 1864	1581.934 227 3295	.000 632 1375	.007 839 7715	12.402 002 8564	.080 632 1375	63
64	137.759 117 2413	1709.488 965 5158	.000 584 9701	.007 259 0477	12.409 261 9040	.080 584 9701	64
65	148.779 846 6206	1847.248 082 7571	.000 541 3458	.006 721 3404	12.415 983 2445	.080 541 3458	65
66	160.682 234 3502	1996.027 929 3777	.000 500 9950	.006 223 4634	12.422 206 7079	.080 500 9950	66
67	173.536 813 0982	2156.710 163 7279	.000 463 6692	.005 762 4661	12.427 969 1739	.080 463 6692	67
68	187.419 758 1461	2330.246 976 8261	.000 429 1391	.005 335 6167	12.433 304 7907	.080 429 1391	68
69	202.413 338 7978	2517.666 734 9722	.000 397 1932	.004 940 3859	12.438 245 1766	.080 397 1932	69
70	218.606 405 9016	2720.080 073 7700	.000 367 6362	.004 574 4314	12.442 819 6079	.080 367 6362	70
71	236.094 918 3737	2938.686 479 6716	.000 340 2881	.004 235 5846	12.447 055 1925	.080 340 2881	71
72	254.982 511 8436	3174.781 398 0453	.000 314 9823	.003 921 8376	12.450 977 0301	.080 314 9823	72
73	275.381 112 7911	3429.763 909 8889	.000 291 5653	.003 631 3311	12.454 608 3612	.080 291 5653	73
74	297.411 601 8144	3705.145 022 6800	.000 269 8950	.003 362 3436	12.457 970 7048	.080 269 8950	74
75	321.204 529 9596	4002.556 624 4944	.000 249 8403	.003 113 2811	12.461 083 9860	.080 249 8403	75
76	346.900 892 3563	4323.761 154 4540	.000 231 2801	.002 882 6677	12.463 966 6537	.080 231 2801	76
77	374.652 963 7448	4670.662 046 8103	.000 214 1024	.002 669 1368	12.466 635 7904	.080 214 1024	77
78	404.625 200 8444	5045.315 010 5551	.000 198 2037	.002 471 4229	12.469 107 2134	.080 198 2037	78
79	436.995 216 9120	5449.940 211 3995	.000 183 4883	.002 288 3546	12.471 395 5679	.080 183 4883	79
80	471.954 834 2649	5886.935 428 3115	.000 169 8677	.002 118 8468	12.473 514 4147	.080 169 8677	80
81	509.711 221 0061	6358.890 262 5764	.000 157 2601	.001 961 8952	12.475 476 3099	.080 157 2601	81
82	550.488 118 6866	6868.601 483 5825	.000 145 5900	.001 816 5696	12.477 292 8796	.080 145 5900	82
83	594.527 168 1815	7419.089 602 2691	.000 134 7874	.001 682 0089	12.478 974 8885	.080 134 7874	83
84	642.089 341 6361	8013.616 770 4506	.000 124 7876	.001 557 4157	12.480 532 3042	.080 124 7876	84
85	693.456 488 9669	8655.706 112 0867	.000 115 5307	.001 442 0515	12.481 974 3557	.080 115 5307	85
86	748.933 008 0843	9349.162 601 0536	.000 106 9615	.001 335 2329	12.483 309 5886	.080 106 9615	86
87	808.847 648 7310	10098.095 609 1379	.000 099 0286	.001 236 3268	12.484 545 9154	.080 099 0286	87
88	873.555 460 6295	10906.943 257 8690	.000 091 6847	.001 144 7470	12.485 690 6624	.080 091 6847	88
89	943.439 897 4799	11780.498 718 4985	.000 084 8860	.001 059 9509	12.486 750 6133	.080 084 8860	89
90	1018.915 089 2783	12723.938 615 9783	.000 078 5920	.000 981 4360	12.487 732 0494	.080 078 5920	90
91	1100.428 296 4205	13742.853 705 2566	.000 072 7651	.000 908 7371	12.488 640 7865	.080 072 7651	91
92	1188.462 560 1342	14843.282 001 6771	.000 067 3705	.000 841 4232	12.489 482 2097	.080 067 3705	92
93	1283.539 564 9449	16031.744 561 8113	.000 062 3762	.000 779 0956	12.490 261 3053	.080 062 3762	93
94	1386.222 730 1405	17315.284 126 7562	.000 057 7524	.000 721 3848	12.490 982 6901	.080 057 7524	94
95	1497.120 548 5517	18701.506 856 8967	.000 053 4716	.000 667 9489	12.491 650 6389	.080 053 4716	95
96	1616.890 192 4359	20198.627 405 4485	.000 049 5083	.000 618 4712	12.492 269 1101	.080 049 5083	96
97	1746.241 407 8307	21815.517 597 8843	.000 045 8389	.000 572 6585	12.492 841 7686	.080 045 8389	97
98	1885.940 720 4572	23561.759 005 7151	.000 042 4417	.000 530 2394	12.493 372 0080	.080 042 4417	98
99	2036.815 978 0938	25447.699 726 1723	.000 039 2963	.000 490 9624	12.493 862 9704	.080 039 2963	99
100	2199.761 256 3413	27484.515 704 2661	.000 036 3841	.000 454 5948	12.494 317 5652	.080 036 3841	100

	XIV COMPOUND AMOUNT	XV AMOUNT OF ANNUITY	XVI SINKING FUND	XVII PRESENT VALUE	XVIII PRESENT VALUE OF ANNUITY	XIX AMORTIZATION	
RATE 9% / PERIODS	AMOUNT OF 1 — How $1 left at compound interest will grow.	AMOUNT OF 1 PER PERIOD — How $1 deposited periodically will grow.	SINKING FUND — Periodic deposit that will grow to $1 at future date.	PRESENT WORTH OF 1 — What $1 due in the future is worth today.	PRESENT WORTH OF 1 PER PERIOD — What $1 payable periodically is worth today.	PARTIAL PAYMENT — Annuity worth $1 today. Periodic payment necessary to pay off a loan of $1.	PERIODS
1	1.090 000 0000	1.000 000 0000	1.000 000 0000	.917 431 1927	.917 431 1927	1.090 000 0000	1
2	1.188 100 0000	2.090 000 0000	.478 468 8995	.841 679 9933	1.759 111 1859	.568 468 8995	2
3	1.295 029 0000	3.278 100 0000	.305 054 7573	.772 183 4801	2.531 294 6660	.395 054 7573	3
4	1.411 581 6100	4.573 129 0000	.218 668 6621	708 425 2111	3.239 719 8771	.308 668 6621	4
5	1.538 623 9549	5.984 710 6100	.167 092 4570	.649 931 3863	3.889 651 2634	.257 092 4570	5
6	1.677 100 1108	7.523 334 5649	.132 919 7833	.596 267 3265	4.485 918 5902	.222 919 7833	6
7	1.828 039 1208	9.200 434 6757	.108 690 5168	.547 034 2448	5.032 952 8351	.198 690 5168	7
8	1.992 562 6417	11.028 473 7966	.090 674 3778	.501 866 2797	5.534 819 1147	.180 674 3778	8
9	2.171 893 2794	13.021 036 4382	.076 798 8021	.460 427 7795	5.995 246 8943	.166 798 8021	9
10	2.367 363 6746	15.192 929 7177	.065 820 0899	.422 410 8069	6.417 657 7012	.155 820 0899	10
11	2.580 826 4053	17.560 293 3923	.056 946 6567	.387 532 8504	6.805 190 5515	.146 946 6567	11
12	2.812 664 7818	20.140 719 7976	.049 650 6585	.355 534 7251	7.160 725 2766	.139 650 6585	12
13	3.065 804 6121	22.953 384 5794	.043 566 5597	.326 178 6469	7.486 903 9235	.133 566 5597	13
14	3.341 727 0272	26.019 189 1915	.038 433 1730	.299 246 4650	7.786 150 3885	.128 433 1730	14
15	3.642 482 4597	29.360 916 2188	.034 058 8827	.274 538 0413	8.060 688 4299	.124 058 8827	15
16	3.970 305 8811	33.003 398 6784	.030 299 9097	.251 869 7627	8.312 558 1925	.120 299 9097	16
17	4.327 633 4104	36.973 704 5595	.027 046 2485	.231 073 1768	8.543 631 3693	.117 046 2485	17
18	4.717 120 4173	41.301 337 9699	.024 212 2907	.211 993 7402	8.755 625 1094	.114 212 2907	18
19	5.141 661 2548	46.018 458 3871	.021 730 4107	.194 489 6699	8.950 114 7793	.111 730 4107	19
20	5.604 410 7678	51.160 119 6420	.019 546 4750	.178 430 8898	9.128 545 6691	.109 546 4750	20
21	6.108 807 7369	56.764 530 4098	.017 616 6348	.163 698 0640	9.292 243 7331	.107 616 6348	21
22	6.658 600 4332	62.873 338 1466	.015 904 9930	.150 181 7101	9.442 425 4432	.105 904 9930	22
23	7.257 874 4722	69.531 938 5798	.014 381 8800	.137 781 3854	9.580 206 8286	.104 381 8800	23
24	7.911 083 1747	76.789 813 0520	.013 022 5607	.126 404 9408	9.706 611 7694	.103 022 5607	24
25	8.623 080 6604	84.700 896 2267	.011 806 2505	.115 967 8356	9.822 579 6049	.101 806 2505	25
26	9.399 157 9198	93.323 976 8871	.010 715 3599	.106 392 5097	9.928 972 1146	.100 715 3599	26
27	10.245 082 1326	102.723 134 8069	.009 734 9054	.097 607 8070	10.026 579 9217	.099 734 9054	27
28	11.167 139 5246	112.968 216 9396	.008 852 0473	.089 548 4468	10.116 128 3685	.098 852 0473	28
29	12.172 182 0818	124.135 356 4641	.008 055 7226	.082 154 5384	10.198 282 9069	.098 055 7226	29
30	13.267 678 4691	136.307 538 5459	.007 336 3514	.075 371 1361	10.273 654 0430	.097 336 3514	30
31	14.461 769 5314	149.575 217 0150	.006 685 5995	.069 147 8313	10.342 801 8743	.096 685 5995	31
32	15.763 328 7892	164.036 986 5464	.006 096 1861	.063 438 3773	10.406 240 2517	.096 096 1861	32
33	17.182 028 3802	179.800 315 3356	.005 561 7255	.058 200 3462	10.464 440 5979	.095 561 7255	33
34	18.728 410 9344	196.982 343 7158	.005 076 5971	.053 394 8130	10.517 835 4109	.095 076 5971	34
35	20.413 967 9185	215.710 754 6502	.004 635 8375	.048 986 0670	10.566 821 4779	.094 635 8375	35
36	22.251 225 0312	236.124 722 5687	.004 235 0500	.044 941 3459	10.611 762 8237	.094 235 0500	36
37	24.253 835 2840	258.375 947 5999	.003 870 3293	.041 230 5925	10.652 993 4163	.093 870 3293	37
38	26.436 680 4595	282.629 782 8839	.003 538 1975	.037 826 2317	10.690 819 6480	.093 538 1975	38
39	28.815 981 7009	309.066 463 3434	.003 235 5500	.034 702 9648	10.725 522 6128	.093 235 5500	39
40	31.409 420 0540	337.882 445 0443	.002 959 6092	.031 837 5824	10.757 360 1952	.092 959 6092	40
41	34.236 267 8588	369.291 865 0983	.002 707 8853	.029 208 7912	10.786 568 9865	.092 707 8853	41
42	37.317 531 9661	403.528 132 9572	.002 478 1420	.026 797 0562	10.813 366 0426	.092 478 1420	42
43	40.676 109 8431	440.845 664 9233	.002 268 3675	.024 584 4552	10.837 950 4978	.092 268 3675	43
44	44.336 959 7290	481.521 774 7664	.002 076 7493	.022 554 5461	10.860 505 0439	.092 076 7493	44
45	48.327 286 1046	525.858 734 4954	.001 901 6514	.020 692 2441	10.881 197 2880	.091 901 6514	45
46	52.676 741 8540	574.186 020 6000	.001 741 5959	.018 983 7102	10.900 180 9981	.091 741 5959	46
47	57.417 648 6209	626.862 762 4540	.001 595 2455	.017 416 2479	10.917 597 2460	.091 595 2455	47
48	62.585 236 9967	684.280 411 0748	.001 461 3892	.015 978 2090	10.933 575 4550	.091 461 3892	48
49	68.217 908 3264	746.865 648 0716	.001 338 9289	.014 658 9074	10.948 234 3624	.091 338 9289	49
50	74.357 520 0758	815.083 556 3980	.001 226 8681	.013 448 5389	10.961 682 9013	.091 226 8681	50
51	81.049 696 8826	889.441 076 4738	.001 124 3016	.012 338 1091	10.974 021 0104	.091 124 3016	51
52	88.344 169 6021	970.490 773 3565	.001 030 4065	.011 319 3661	10.985 340 3765	.091 030 4065	52
53	96.295 144 8663	1058.834 942 9585	.000 944 4343	.010 384 7396	10.995 725 1160	.090 944 4343	53
54	104.961 707 9042	1155.130 087 8248	.000 865 7034	.009 527 2840	11.005 252 4000	.090 865 7034	54
55	114.408 261 6156	1260.091 795 7290	.000 793 5930	.008 740 6275	11.013 993 0276	.090 793 5930	55
56	124.705 005 1610	1374.500 057 3447	.000 727 5373	.008 018 9243	11.022 011 9519	.090 727 5373	56
57	135.928 455 6255	1499.205 062 5057	.000 667 0202	.007 356 8113	11.029 368 7632	.090 667 0202	57
58	148.162 016 6318	1635.133 518 1312	.000 611 5709	.006 749 3682	11.036 118 1314	.090 611 5709	58
59	161.496 598 1287	1783.295 534 7630	.000 560 7595	.006 192 0809	11.042 310 2123	.090 560 7595	59
60	176.031 291 9602	1944.792 132 8917	.000 514 1938	.005 680 8082	11.047 991 0204	.090 514 1938	60
61	191.874 108 2367	2120.823 424 8519	.000 471 5150	.005 211 7506	11.053 202 7710	.090 471 5150	61
62	209.142 777 9780	2312.697 533 0886	.000 432 3955	.004 781 4226	11.057 984 1936	.090 432 3955	62
63	227.965 627 9960	2521.840 311 0665	.000 396 5358	.004 386 6262	11.062 370 8198	.090 396 5358	63
64	248.482 534 5156	2749.805 939 0625	.000 363 6620	.004 024 4277	11.066 395 2475	.090 363 6620	64
65	270.845 962 6220	2998.288 473 5782	.000 333 5236	.003 692 1355	11.070 087 3831	.090 333 5236	65
66	295.222 099 2580	3269.134 436 2002	.000 305 8914	.003 387 2803	11.073 474 6634	.090 305 8914	66
67	321.792 088 1912	3564.356 535 4582	.000 280 5555	.003 107 5966	11.076 582 2600	.090 280 5555	67
68	350.753 376 1285	3886.148 623 6495	.000 257 3242	.002 851 0061	11.079 433 2660	.090 257 3242	68
69	382.321 179 9800	4236.901 999 7779	.000 236 0215	.002 615 6019	11.082 048 8679	.090 236 0215	69
70	416.730 086 1782	4619.223 179 7579	.000 216 4866	.002 399 6348	11.084 448 5027	.090 216 4866	70
71	454.235 793 9343	5035.953 265 9361	.000 198 5721	.002 201 4998	11.086 650 0025	.090 198 5721	71
72	495.117 015 3883	5490.189 059 8704	.000 182 1431	.002 019 7246	11.088 669 7270	.090 182 1431	72
73	539.677 546 7733	5985.306 075 2587	.000 167 0758	.001 852 9583	11.090 522 6853	.090 167 0758	73
74	588.248 525 9829	6524.983 622 0220	.000 153 2571	.001 699 9618	11.092 222 6471	.090 153 2571	74
75	641.190 893 3213	7113.232 148 0149	.000 140 5831	.001 559 5979	11.093 782 2450	.090 140 5831	75
76	698.898 073 7203	7754.423 041 3362	.000 128 9587	.001 430 8238	11.095 213 0689	.090 128 9587	76
77	761.798 900 3551	8453.321 115 0565	.000 118 2967	.001 312 6824	11.096 525 7512	.090 118 2967	77
78	830.360 801 3870	9215.120 015 4116	.000 108 5173	.001 204 2958	11.097 730 0470	.090 108 5173	78
79	905.093 273 5119	10045.480 816 7986	.000 099 5473	.001 104 8585	11.098 834 9055	.090 099 5473	79
80	986.551 668 1279	10950.574 090 3105	.000 091 3194	.001 013 6317	11.099 848 5372	.090 091 3194	80
81	1075.341 318 2595	11937.125 758 4384	.000 083 7723	.000 929 9373	11.100 778 4745	.090 083 7723	81
82	1172.122 036 9028	13012.467 076 6979	.000 076 8494	.000 853 1535	11.101 631 6280	.090 076 8494	82
83	1277.613 020 2241	14184.589 113 6007	.000 070 4990	.000 782 7096	11.102 414 3376	.090 070 4990	83
84	1392.598 192 0442	15462.202 133 8247	.000 064 6738	.000 718 0822	11.103 132 4198	.090 064 6738	84
85	1517.932 029 3282	16854.800 325 8690	.000 059 3303	.000 658 7910	11.103 791 2108	.090 059 3303	85
86	1654.545 911 9677	18372.732 355 1972	.000 054 4285	.000 604 3954	11.104 395 6063	.090 054 4285	86
87	1803.455 044 0448	20027.278 267 1649	.000 049 9319	.000 554 4912	11.104 950 0975	.090 049 9319	87
88	1965.765 998 0089	21830.733 311 2098	.000 045 8070	.000 508 7075	11.105 458 8050	.090 045 8070	88
89	2142.684 937 8297	23796.499 309 2187	.000 042 0230	.000 466 7042	11.105 925 5092	.090 042 0230	89
90	2335.526 582 2343	25939.184 247 0483	.000 038 5517	.000 428 1690	11.106 353 6782	.090 038 5517	90
91	2545.723 974 6354	28274.710 829 2827	.000 035 3673	.000 392 8156	11.106 746 4937	.090 035 3673	91
92	2774.839 132 3526	30820.434 803 9181	.000 032 4460	.000 360 3813	11.107 106 8750	.090 032 4460	92
93	3024.574 654 2644	33595.273 936 2708	.000 029 7661	.000 330 6250	11.107 437 5000	.090 029 7661	93
94	3296.786 373 1482	36619.848 590 6231	.000 027 3076	.000 303 3257	11.107 740 8257	.090 027 3076	94
95	3593.497 146 7315	39916.634 963 6833	.000 025 0522	.000 278 2804	11.108 019 1060	.090 025 0522	95
96	3916.911 889 9373	43510.132 110 4148	.000 022 9832	.000 255 3032	11.108 274 4093	.090 022 9832	96
97	4269.433 960 0317	47427.044 000 3521	.000 021 0850	.000 234 2231	11.108 508 6324	.090 021 0850	97
98	4653.683 016 4345	51696.477 960 3838	.000 019 3437	.000 214 8836	11.108 723 5159	.090 019 3437	98
99	5072.514 487 9137	56350.160 976 8183	.000 017 7462	.000 197 1409	11.108 920 6568	.090 017 7462	99
100	5529.040 791 8259	61422.675 464 7320	.000 016 2806	.000 180 8632	11.109 101 5200	.090 016 2806	100

634

10%

| XIV COMPOUND AMOUNT | XV AMOUNT OF ANNUITY | XVI SINKING FUND | XVII PRESENT VALUE | XVIII PRESENT VALUE OF ANNUITY | XIX AMORTIZATION |
| AMOUNT OF 1 | AMOUNT OF 1 PER PERIOD | SINKING FUND | PRESENT WORTH OF 1 | PRESENT WORTH OF 1 PER PERIOD | PARTIAL PAYMENT |

RATE 10%

PERIODS	Amount of 1	Amount of 1 per period	Sinking Fund	Present Worth of 1	Present Worth of 1 per period	Partial Payment	PERIODS
1	1.100 000 0000	1.000 000 0000	1.000 000 0000	.909 090 9091	.909 090 9091	1.100 000 0000	1
2	1.210 000 0000	2.100 000 0000	.476 190 4762	.826 446 2810	1.735 537 1901	.576 190 4762	2
3	1.331 000 0000	3.310 000 0000	.302 114 8036	.751 314 8009	2.486 851 9910	.402 114 8036	3
4	1.464 100 0000	4.641 000 0000	.215 470 8037	.683 013 4554	3.169 865 4463	.315 470 8037	4
5	1.610 510 0000	6.105 100 0000	.163 797 4808	.620 921 3231	3.790 786 7694	.263 797 4808	5
6	1.771 561 0000	7.715 610 0000	.129 607 3804	.564 473 9301	4.355 260 6995	.229 607 3804	6
7	1.948 717 1000	9.487 171 0000	.105 405 4997	.513 158 1182	4.868 418 8177	.205 405 4997	7
8	2.143 588 8100	11.435 888 1000	.087 444 0176	.466 507 3802	5.334 926 1979	.187 444 0176	8
9	2.357 947 6910	13.579 476 9100	.073 640 5391	.424 097 6184	5.759 023 8163	.173 640 5391	9
10	2.593 742 4601	15.937 424 6010	.062 745 3949	.385 543 2894	6.144 567 1057	.162 745 3949	10
11	2.853 116 7061	18.531 167 0611	.053 963 1420	.350 493 8995	6.495 061 0052	.153 963 1420	11
12	3.138 428 3767	21.384 283 7672	.046 763 3151	.318 630 8177	6.813 691 8229	.146 763 3151	12
13	3.452 271 2144	24.522 712 1439	.040 770 5230	.289 664 3797	7.103 356 2026	.140 778 5238	13
14	3.797 498 3358	27.974 983 3583	.035 746 2232	.263 331 2543	7.366 687 4569	.135 746 2232	14
15	4.177 248 1694	31.772 481 6942	.031 473 7769	.239 392 0494	7.606 079 5063	.131 473 7769	15
16	4.594 972 9864	35.949 729 8636	.027 816 6207	.217 629 1358	7.823 708 6421	.127 816 6207	16
17	5.054 470 2850	40.544 702 8499	.024 664 1344	.197 844 6689	8.021 553 3110	.124 664 1344	17
18	5.559 917 3135	45.599 173 1349	.021 930 2222	.179 858 7899	8.201 412 1009	.121 930 2222	18
19	6.115 909 0448	51.159 090 4484	.019 546 8682	.163 507 9908	8.364 920 0917	.119 546 8682	19
20	6.727 499 9493	57.274 999 4933	.017 459 6248	.148 643 6280	8.513 563 7198	.117 459 6248	20
21	7.400 249 9443	64.002 499 4426	.015 624 3898	.135 130 5709	8.648 694 2907	.115 624 3898	21
22	8.140 274 9387	71.402 749 3868	.014 005 0630	.122 845 9736	8.771 540 2643	.114 005 0630	22
23	8.954 302 4326	79.543 024 3255	.012 571 8127	.111 678 1578	8.883 218 4221	.112 571 8127	23
24	9.849 732 6758	88.497 326 7581	.011 299 7764	.101 525 5980	8.984 744 0201	.111 299 7764	24
25	10.834 705 9434	98.347 059 4339	.010 168 0722	.092 295 9982	9.077 040 0182	.110 168 0722	25
26	11.918 176 5377	109.181 765 3773	.009 159 0386	.083 905 4529	9.160 945 4711	.109 159 0386	26
27	13.109 994 1915	121.099 941 9150	.008 257 6423	.076 277 6844	9.237 223 1556	.108 257 6423	27
28	14.420 993 6106	134.209 936 1065	.007 451 0132	.069 343 3495	9.306 566 5051	.107 451 0132	28
29	15.863 092 9717	148.630 929 7171	.006 728 0747	.063 039 4086	9.369 605 9137	.106 728 0747	29
30	17.449 402 2689	164.494 022 6889	.006 079 2483	.057 308 5533	9.426 914 4670	.106 079 2483	30
31	19.194 342 4958	181.943 424 9578	.005 496 2140	.052 098 6848	9.479 013 1518	.105 496 2140	31
32	21.113 776 7454	201.137 767 4535	.004 971 7167	.047 362 4407	9.526 375 5926	.104 971 7167	32
33	23.225 154 4199	222.251 544 1989	.004 499 4063	.043 056 7643	9.569 432 3569	.104 499 4063	33
34	25.547 669 8619	245.476 698 6188	.004 073 7064	.039 142 5130	9.608 574 8699	.104 073 7064	34
35	28.102 436 8481	271.024 368 4806	.003 689 7051	.035 584 1027	9.644 158 9726	.103 689 7051	35
36	30.912 680 5329	299.126 805 3287	.003 343 0638	.032 349 1843	9.676 508 1569	.103 343 0638	36
37	34.003 948 5862	330.039 485 8616	.003 029 9405	.029 408 3494	9.705 916 5063	.103 029 9405	37
38	37.404 343 4448	364.043 434 4477	.002 746 9250	.026 734 8631	9.732 651 3694	.102 746 9250	38
39	41.144 777 7893	401.447 777 8925	.002 490 9840	.024 304 4210	9.756 955 7903	.102 490 9840	39
40	45.259 255 5682	442.592 555 6818	.002 259 4144	.022 094 9282	9.779 050 7185	.102 259 4144	40
41	49.785 181 1250	487.851 811 2499	.002 049 8028	.020 086 2983	9.799 137 0168	.102 049 8028	41
42	54.763 699 2375	537.636 992 3749	.001 859 9911	.018 260 2712	9.817 397 2880	.101 859 9911	42
43	60.240 069 1612	592.400 691 6124	.001 688 0466	.016 600 2465	9.833 997 5345	.101 688 0466	43
44	66.264 076 0774	652.640 760 7737	.001 532 2365	.015 091 1332	9.849 088 6678	.101 532 2365	44
45	72.890 483 6851	718.904 836 8510	.001 391 0047	.013 719 2120	9.862 807 8798	.101 391 0047	45

RATE 12%

PERIODS	Amount of 1	Amount of 1 per period	Sinking Fund	Present Worth of 1	Present Worth of 1 per period	Partial Payment	PERIODS
1	1.120 000 0000	1.000 000 0000	1.000 000 0000	.892 857 1429	.892 857 1429	1.120 000 0000	1
2	1.254 400 0000	2.120 000 0000	.471 698 1132	.797 193 8776	1.690 051 0204	.591 698 1132	2
3	1.404 928 0000	3.374 400 0000	.296 348 9806	.711 780 2478	2.401 831 2682	.416 348 9806	3
4	1.573 519 3600	4.779 328 0000	.209 234 4363	.635 518 0784	3.037 349 3466	.329 234 4363	4
5	1.762 341 6832	6.352 847 3600	.157 409 7319	.567 426 8557	3.604 776 2023	.277 409 7319	5
6	1.973 822 6852	8.115 189 0432	.123 225 7184	.506 631 1212	4.111 407 3235	.243 225 7184	6
7	2.210 681 4074	10.089 011 7284	.099 117 7359	.452 349 2153	4.563 756 5389	.219 117 7359	7
8	2.475 963 1763	12.299 693 1358	.081 302 8414	.403 883 2280	4.967 639 7668	.201 302 8414	8
9	2.773 078 7575	14.775 656 3121	.067 678 8888	.360 610 0250	5.328 249 7918	.187 678 8888	9
10	3.105 848 2083	17.548 735 0695	.056 984 1642	.321 973 2366	5.650 223 0284	.176 984 1642	10
11	3.478 549 9933	20.654 583 2779	.048 415 4043	.287 476 1041	5.937 699 1325	.168 415 4043	11
12	3.895 975 9925	24.133 133 2712	.041 436 8076	.256 675 0929	6.194 374 2255	.161 436 8076	12
13	4.363 493 1117	28.029 109 2638	.035 677 1951	.229 174 1901	6.423 548 4156	.155 677 1951	13
14	4.887 112 2851	32.392 602 3754	.030 871 2461	.204 619 8126	6.628 168 2282	.150 871 2461	14
15	5.473 565 7593	37.279 714 6605	.026 824 2396	.182 696 2613	6.810 864 4895	.146 824 2396	15
16	6.130 393 6504	42.753 280 4197	.023 390 0180	.163 121 6618	6.973 986 1513	.143 390 0180	16
17	6.866 040 8884	48.883 674 0701	.020 456 7275	.145 644 3409	7.119 630 4922	.140 456 7275	17
18	7.689 965 7950	55.749 714 9585	.017 937 3114	.130 039 5901	7.249 670 0824	.137 937 3114	18
19	8.612 761 6904	63.439 680 7535	.015 763 0049	.116 106 7769	7.365 776 8592	.135 763 0049	19
20	9.646 293 0933	72.052 442 4440	.013 878 9800	.103 666 7651	7.469 443 6243	.333 878 9800	20
21	10.803 848 8645	81.698 735 5372	.012 240 0915	.092 559 6117	7.562 003 2360	.132 240 0915	21
22	12.100 310 0562	92.502 583 8017	.010 810 5088	.082 642 5104	7.644 645 7464	.130 810 5088	22
23	13.552 347 2629	104.602 893 8579	.009 559 9650	.073 787 9557	7.718 433 7022	.129 559 9650	23
24	15.178 628 9345	118.155 241 1209	.008 463 4417	.065 882 1033	7.784 315 8055	.128 463 4417	24
25	17.000 064 4066	133.333 870 0554	.007 499 9698	.058 823 3066	7.843 139 1121	.127 499 9698	25
26	19.040 072 1354	150.333 934 4620	.006 651 8581	.052 520 8094	7.895 659 9215	.126 651 8581	26
27	21.324 880 7917	169.374 006 5974	.005 904 0937	.046 893 5798	7.942 553 5013	.125 904 0937	27
28	23.883 866 4867	190.698 887 3891	.005 243 8691	.041 869 2677	7.984 422 7690	.125 243 8691	28
29	26.749 930 4651	214.582 753 8758	.004 660 2068	.037 383 2747	8.021 806 0438	.124 660 2068	29
30	29.959 922 1209	241.332 684 3409	.004 143 6576	.033 377 9239	8.055 183 9677	.124 143 6576	30
31	33.555 112 7754	271.292 606 4618	.003 686 0570	.029 801 7177	8.084 985 6854	.123 686 0570	31
32	37.581 726 3085	304.847 719 2373	.003 280 3263	.026 608 6766	8.111 594 3620	.123 280 3263	32
33	42.091 533 4655	342.429 445 5457	.002 920 3096	.023 757 7469	8.135 352 1089	.122 920 3096	33
34	47.142 517 4813	384.520 979 0112	.002 600 6383	.021 212 2740	8.156 564 3830	.122 600 6383	34
35	52.799 619 5791	431.663 496 4926	.002 316 6193	.018 939 5304	8.175 503 9134	.122 316 6193	35
36	59.135 573 9286	484.463 116 0717	.002 064 1406	.016 910 2950	8.192 414 2084	.122 064 1406	36
37	66.231 842 8000	543.598 690 0003	.001 839 5924	.015 098 4777	8.207 512 6860	.121 839 5924	37
38	74.179 663 9360	609.830 532 8003	.001 639 7998	.013 480 7836	8.220 993 4697	.121 639 7998	38
39	83.081 223 6084	684.010 196 7363	.001 461 9665	.012 036 4140	8.233 029 8836	.121 461 9665	39
40	93.050 970 4414	767.091 420 3447	.001 303 6256	.010 746 7982	8.243 776 6818	.121 303 6256	40
41	104.217 086 8943	860.142 390 7861	.001 162 5982	.009 595 3555	8.253 372 0373	.121 162 5982	41
42	116.723 137 3216	964.359 477 6804	.001 036 9577	.008 567 2817	8.261 939 3190	.121 036 9577	42
43	130.729 913 8002	1081.082 615 0020	.000 924 9987	.007 649 3587	8.269 588 6777	.120 924 9987	43
44	146.417 503 4563	1211.812 528 8023	.000 825 2102	.006 829 7845	8.276 418 4623	.120 825 2102	44
45	163.987 603 8710	1358.230 032 2586	.000 736 2523	.006 098 0219	8.282 516 4842	.120 736 2523	45

Column descriptions: AMOUNT OF 1 — How $1 left at compound interest will grow. AMOUNT OF 1 PER PERIOD — How $1 deposited periodically will grow. SINKING FUND — Periodic deposit that will grow to $1 at future date. PRESENT WORTH OF 1 — What $1 due in the future is worth today. PRESENT WORTH OF 1 PER PERIOD — What $1 payable periodically is worth today. PARTIAL PAYMENT — Annuity worth $1 today. Periodic payment necessary to pay off a loan of $1.

RATE 14%

PERIODS	AMOUNT OF 1 — How $1 left at compound interest will grow.	AMOUNT OF 1 PER PERIOD — How $1 deposited periodically will grow.	SINKING FUND — Periodic deposit that will grow to $1 at future date.	PRESENT WORTH OF 1 — What $1 due in the future is worth today.	PRESENT WORTH OF 1 PER PERIOD — What $1 payable periodically is worth today.	PARTIAL PAYMENT — Annuity worth $1 today. Periodic payment necessary to pay off a loan of $1.	PERIODS
1	1.140 000 0000	1.000 000 0000	1.000 000 0000	.877 192 9825	.877 192 9825	1.140 000 0000	1
2	1.299 600 0000	2.140 000 0000	.467 289 7196	.769 467 5285	1.646 660 5109	.607 289 7196	2
3	1.481 544 0000	3.439 600 0000	.290 731 4804	.674 971 5162	2.321 632 0271	.430 731 4804	3
4	1.688 960 1600	4.921 144 0000	.203 204 7833	.592 080 2774	2.913 712 3045	.343 204 7833	4
5	1.925 414 5824	6.610 104 1600	.151 283 5465	.519 368 6644	3.433 080 9689	.291 283 5465	5
6	2.194 972 6239	8.535 518 7424	.117 157 4957	.455 586 5477	3.888 667 5165	.257 157 4957	6
7	2.502 268 7913	10.730 491 3663	.093 192 3773	.399 637 3225	4.288 304 8391	.233 192 3773	7
8	2.852 586 4221	13.232 760 1576	.075 570 0238	.350 559 0549	4.638 863 8939	.215 570 0238	8
9	3.251 948 5212	16.085 346 5797	.062 168 3838	.307 507 9429	4.946 371 8368	.202 168 3838	9
10	3.707 221 3141	19.337 295 1008	.051 713 5408	.269 743 8095	5.216 115 6463	.191 713 5408	10
11	4.226 232 2981	23.044 516 4150	.043 394 2714	.236 617 3768	5.452 733 0231	.183 394 2714	11
12	4.817 904 8198	27.270 748 7131	.036 669 3269	.207 559 1024	5.660 292 1255	.176 669 3269	12
13	5.492 411 4946	32.088 653 5329	.031 163 6635	.182 069 3881	5.842 361 5136	.171 163 6635	13
14	6.261 349 1038	37.581 065 0275	.026 609 1448	.159 709 9696	6.002 071 5032	.166 609 1448	14
15	7.137 937 9784	43.842 414 1313	.022 808 9630	.140 096 4821	6.142 167 9852	.162 808 9630	15
16	8.137 249 2954	50.980 352 1097	.019 615 4000	.122 891 6509	6.265 059 6362	.159 615 4000	16
17	9.276 464 1967	59.117 601 4051	.016 915 4359	.107 799 6938	6.372 859 3300	.156 915 4359	17
18	10.575 169 1843	68.394 065 6018	.014 621 1516	.094 561 1349	6.467 420 4649	.154 621 1516	18
19	12.055 692 8700	78.969 234 7861	.012 663 1593	.082 948 3640	6.550 368 8288	.152 663 1593	19
20	13.743 489 8719	91.024 927 6561	.010 986 0016	.072 761 7228	6.623 130 5516	.150 986 0016	20
21	15.667 578 4539	104.768 417 5280	.009 544 8612	.063 826 0726	6.686 956 6242	.149 544 8612	21
22	17.861 039 4375	120.435 995 9819	.008 303 1654	.055 987 7830	6.742 944 4072	.148 303 1654	22
23	20.361 584 9587	138.297 035 4193	.007 230 8130	.049 112 0903	6.792 056 4976	.147 230 8130	23
24	23.212 206 8529	158.658 620 3780	.006 302 8406	.043 080 7810	6.835 137 2786	.146 302 8406	24
25	26.461 915 8123	181.870 827 2310	.005 498 4079	.037 790 1588	6.872 927 4373	.145 498 4079	25
26	30.166 584 0261	208.332 743 0433	.004 800 0136	.033 149 2621	6.906 076 6994	.144 800 0136	26
27	34.389 905 7897	238.499 327 0694	.004 192 8839	.029 078 3001	6.935 154 9995	.144 192 8839	27
28	39.204 492 6003	272.889 232 8591	.003 664 4905	.025 507 2808	6.960 662 2803	.143 664 4905	28
29	44.693 121 5643	312.093 725 4594	.003 204 1657	.022 374 8077	6.983 037 0879	.143 204 1657	29
30	50.950 158 5833	356.786 847 0237	.002 802 7939	.019 627 0243	7.002 664 1122	.142 802 7939	30
31	58.083 180 7850	407.737 005 6070	.002 452 5613	.017 216 6880	7.019 880 8002	.142 452 5613	31
32	66.214 826 0949	465.820 186 3920	.002 146 7511	.015 102 3579	7.034 983 1581	.142 146 7511	32
33	75.484 901 7482	532.035 012 4868	.001 879 5755	.013 247 6823	7.048 230 8404	.141 879 5755	33
34	86.052 787 9929	607.519 914 2350	.001 646 0366	.011 620 7740	7.059 851 6144	.141 646 0366	34
35	98.100 178 3119	693.572 702 2279	.001 441 8099	.010 193 6614	7.070 045 2758	.141 441 8099	35
36	111.834 203 2756	791.672 880 5398	.001 263 1480	.008 941 8082	7.078 987 0840	.141 263 1480	36
37	127.490 991 7342	903.507 083 8154	.001 106 7982	.007 843 6914	7.086 830 7755	.141 106 7982	37
38	145.339 730 5769	1030.998 075 5495	.000 969 9339	.006 880 4311	7.093 711 2065	.140 969 9339	38
39	165.687 292 8577	1176.337 806 1264	.000 850 0959	.006 035 4659	7.099 746 6724	.140 850 0959	39
40	188.883 513 8578	1342.025 098 9841	.000 745 1425	.005 294 2683	7.105 040 9407	.140 745 1425	40
41	215.327 205 7979	1530.908 612 8419	.000 653 2069	.004 644 0950	7.109 685 0357	.140 653 2069	41
42	245.473 014 6096	1746.235 818 6398	.000 572 6603	.004 073 7675	7.113 758 8033	.140 572 6603	42
43	279.839 236 6549	1991.708 833 2494	.000 502 0814	.003 573 4803	7.117 332 2836	.140 502 0814	43
44	319.016 729 7866	2271.548 069 9043	.000 440 2284	.003 134 6318	7.120 466 9154	.140 440 2284	44
45	363.679 071 9567	2590.564 799 6909	.000 386 0162	.002 749 6771	7.123 216 5925	.140 386 0162	45

RATE 15%

PERIODS	AMOUNT OF 1	AMOUNT OF 1 PER PERIOD	SINKING FUND	PRESENT WORTH OF 1	PRESENT WORTH OF 1 PER PERIOD	PARTIAL PAYMENT	PERIODS
1	1.150 000 0000	1.000 000 0000	1.000 000 0000	.869 565 2174	.869 565 2174	1.150 000 0000	1
2	1.322 500 0000	2.150 000 0000	.465 116 2791	.756 143 6673	1.625 708 8847	.615 116 2791	2
3	1.520 875 0000	3.472 500 0000	.287 976 9618	.657 516 2324	2.283 225 1171	.437 976 9618	3
4	1.749 006 2500	4.993 375 0000	.200 265 3516	.571 753 2456	2.854 978 3627	.350 265 3516	4
5	2.011 357 1875	6.742 381 2500	.148 315 5525	.497 176 7353	3.352 155 0980	.298 315 5525	5
6	2.313 060 7656	8.753 738 4375	.114 236 9066	.432 327 5959	3.784 482 6939	.264 236 9066	6
7	2.660 019 8805	11.066 799 2031	.090 360 3636	.375 937 0399	4.160 419 7338	.240 360 3636	7
8	3.059 022 8625	13.726 819 0836	.072 850 0896	.326 901 7738	4.487 321 5077	.222 850 0896	8
9	3.517 876 2919	16.785 841 9461	.059 574 0150	.284 262 4120	4.771 583 9197	.209 574 0150	9
10	4.045 557 7357	20.303 718 2381	.049 252 0625	.247 184 7061	5.018 768 6259	.199 252 0625	10
11	4.652 391 3961	24.349 275 9738	.041 068 9830	.214 943 2227	5.233 711 8486	.191 068 9830	11
12	5.350 250 1055	29.001 667 3698	.034 480 7761	.186 907 1502	5.420 618 9988	.184 480 7761	12
13	6.152 787 6213	34.351 917 4753	.029 110 4565	.162 527 9567	5.583 146 9554	.179 110 4565	13
14	7.075 705 7645	40.504 705 0966	.024 688 4898	.141 328 6580	5.724 475 6134	.174 688 4898	14
15	8.137 061 6292	47.580 410 8611	.021 017 0526	.122 894 4852	5.847 370 0986	.171 017 0526	15
16	9.357 620 8735	55.717 472 4902	.017 947 6914	.106 864 7697	5.954 234 8684	.167 947 6914	16
17	10.761 264 0046	65.075 093 3638	.015 366 8623	.092 925 8867	6.047 160 7551	.165 366 8623	17
18	12.375 453 6053	75.836 357 3683	.013 186 2874	.080 805 1189	6.127 965 8740	.163 186 2874	18
19	14.231 771 6460	88.211 810 9736	.011 336 3504	.070 265 3208	6.198 231 1948	.161 336 3504	19
20	16.366 537 3929	102.443 582 6196	.009 761 4704	.061 100 2789	6.259 331 4737	.159 761 4704	20
21	18.821 518 0019	118.810 120 0126	.008 416 7914	.053 130 6773	6.312 462 1511	.158 416 7914	21
22	21.644 745 7022	137.631 638 0145	.007 265 7713	.046 200 5890	6.358 662 7401	.157 265 7713	22
23	24.891 457 5575	159.276 383 7166	.006 278 3947	.040 174 4252	6.398 837 1653	.156 278 3947	23
24	28.625 176 1911	184.167 841 2741	.005 429 8286	.034 934 2828	6.433 771 4481	.155 429 8296	24
25	32.918 952 6198	212.793 017 4653	.004 699 4023	.030 377 6372	6.464 149 0853	.154 699 4023	25
26	37.856 795 5128	245.711 970 0851	.004 069 8058	.026 415 3367	6.490 564 4220	.154 069 8058	26
27	43.535 314 8397	283.568 765 5978	.003 526 4815	.022 969 8580	6.513 534 2800	.153 526 4815	27
28	50.065 612 0656	327.104 080 4375	.003 057 1309	.019 973 7896	6.533 508 0695	.153 057 1309	28
29	57.575 453 8755	377.169 692 5031	.002 651 3265	.017 368 5127	6.550 876 5822	.152 651 3265	29
30	66.211 771 9568	434.745 146 3786	.002 300 1982	.015 103 0545	6.565 979 6367	.152 300 1982	30
31	76.143 537 7503	500.956 918 3354	.001 996 1796	.013 133 0909	6.579 112 7276	.151 996 1796	31
32	87.565 068 4128	577.100 456 0857	.001 732 8006	.011 420 0790	6.590 532 8066	.151 732 8006	32
33	100.699 828 6748	664.665 524 4985	.001 504 5161	.009 930 5035	6.600 463 3101	.151 504 5161	33
34	115.804 802 9760	765.365 353 1733	.001 306 5655	.008 635 2204	6.609 098 5305	.151 306 5655	34
35	133.175 523 4224	881.170 156 1493	.001 134 8546	.007 508 8873	6.616 607 4178	.151 134 8546	35
36	153.151 851 9358	1014.345 679 5717	.000 985 8572	.006 529 4672	6.623 136 8851	.150 985 8572	36
37	176.124 629 7261	1167.497 531 5074	.000 856 5329	.005 677 7976	6.628 814 6827	.150 856 5329	37
38	202.543 324 1850	1343.622 161 2335	.000 744 2569	.004 937 2153	6.633 751 8980	.150 744 2569	38
39	232.924 822 8128	1546.165 485 4186	.000 646 7613	.004 293 2307	6.638 045 1287	.150 646 7613	39
40	267.863 546 2347	1779.090 308 2314	.000 562 0850	.003 733 2441	6.641 778 3728	.150 562 0850	40
41	308.043 078 1699	2046.953 854 4661	.000 488 5308	.003 246 2992	6.645 024 6720	.150 488 5308	41
42	354.249 539 8954	2354.996 932 6360	.000 424 6290	.002 822 8689	6.647 847 5408	.150 424 6290	42
43	407.386 970 8797	2709.246 472 5314	.000 369 1063	.002 454 6686	6.650 302 2094	.150 369 1063	43
44	468.495 016 5117	3116.633 443 4111	.000 320 8590	.002 134 4944	6.652 436 7038	.150 320 8590	44
45	538.769 268 9884	3585.128 459 9227	.000 278 9300	.001 856 0821	6.654 292 7860	.150 278 9300	45

		XIV COMPOUND AMOUNT	XV AMOUNT OF ANNUITY	XVI SINKING FUND	XVII PRESENT VALUE	XVIII PRESENT VALUE OF ANNUITY	XIX AMORTIZATION	
RATE 16%	P E R I O D S	AMOUNT OF 1 *How $1 left at compound interest will grow.*	AMOUNT OF 1 PER PERIOD *How $1 deposited periodically will grow.*	SINKING FUND *Periodic deposit that will grow to $1 at future date.*	PRESENT WORTH OF 1 *What $1 due in the future is worth today.*	PRESENT WORTH OF 1 PER PERIOD *What $1 payable periodically is worth today.*	PARTIAL PAYMENT *Annuity worth $1 today. Periodic payment necessary to pay off a loan of $1.*	P E R I O D S
	1	1.160 000 0000	1.000 000 0000	1.000 000 0000	.862 068 9655	.862 068 9655	1.160 000 0000	1
	2	1.345 600 0000	2.160 000 0000	.462 962 9630	.743 162 9013	1.605 231 8668	.622 962 9630	2
	3	1.560 896 0000	3.505 600 0000	.285 257 8731	.640 657 6735	2.245 889 5404	.445 257 8731	3
	4	1.810 639 3600	5.066 496 0000	.197 375 0695	.552 291 0979	2.798 180 6382	.357 375 0695	4
	5	2.100 341 6576	6.877 135 3600	.145 409 3816	.476 113 0154	3.274 293 6537	.305 409 3816	5
	6	2.436 396 3228	8.977 477 0176	.111 389 8702	.410 442 2547	3.684 735 9083	.271 389 8702	6
	7	2.826 219 7345	11.413 873 3404	.087 612 6771	.353 829 5299	4.038 565 4382	.247 612 6771	7
	8	3.278 414 8920	14.240 093 0749	.070 224 2601	.305 025 4568	4.343 590 8950	.230 224 2601	8
	9	3.802 961 2747	17.518 507 9669	.057 082 4868	.262 952 9800	4.606 543 8750	.217 082 4868	9
	10	4.411 435 0786	21.321 469 2416	.046 901 0831	.226 683 6034	4.833 227 4785	.206 901 0831	10
	11	5.117 264 6912	25.732 904 3202	.038 860 7515	.195 416 8995	5.028 644 3780	.198 860 7515	11
	12	5.936 027 0418	30.850 169 0114	.032 414 7333	.168 462 8444	5.197 107 2224	.192 414 7333	12
	13	6.885 791 3685	36.786 196 0533	.027 184 1100	.145 226 5900	5.342 333 8124	.187 184 1100	13
	14	7.987 517 9875	43.671 987 4218	.022 897 9733	.125 195 3362	5.467 529 1486	.182 897 9733	14
	15	9.265 520 8655	51.659 505 4093	.019 357 5218	.107 927 0140	5.575 456 1626	.179 357 5218	15
	16	10.748 004 2040	60.925 026 2748	.016 413 6162	.093 040 5293	5.668 496 6919	.176 413 6162	16
	17	12.467 684 8766	71.673 030 4787	.013 952 2494	.080 207 3528	5.748 704 0447	.173 952 2494	17
	18	14.462 514 4569	84.140 715 3553	.011 884 8526	.069 144 2697	5.817 848 3144	.171 884 8526	18
	19	16.776 516 7700	98.603 229 8122	.010 141 6556	.059 607 1290	5.877 455 4435	.170 141 6556	19
	20	19.460 759 4531	115.379 746 5821	.008 667 0324	.051 385 4561	5.928 840 8996	.168 667 0324	20
	21	22.574 480 9656	134.840 506 0353	.007 416 1691	.044 297 8070	5.973 138 7065	.167 416 1691	21
	22	26.186 397 9201	157.414 987 0009	.006 352 6353	.038 187 7646	6.011 326 4711	.166 352 6353	22
	23	30.376 221 5874	183.601 384 9211	.005 446 5820	.032 920 4867	6.044 246 9579	.165 446 5820	23
	24	35.236 417 0414	213.977 606 5085	.004 673 3862	.028 379 7299	6.072 626 6878	.164 673 3862	24
	25	40.874 243 7680	249.214 023 5498	.004 012 6153	.024 465 2844	6.097 091 9723	.164 012 6153	25
	26	47.414 122 7708	290.088 267 3178	.003 447 2266	.021 090 7624	6.118 182 7347	.163 447 2266	26
	27	55.000 382 4142	337.502 390 0886	.002 962 9420	.018 181 6918	6.136 364 4265	.162 962 9420	27
	28	63.800 443 6004	392.502 772 5028	.002 547 7527	.015 673 8722	6.152 038 2987	.162 547 7527	28
	29	74.008 514 5765	456.303 216 1032	.002 191 5252	.013 511 9588	6.165 550 2575	.162 191 5252	29
	30	85.849 876 9088	530.311 730 6798	.001 885 6833	.011 648 2403	6.177 198 4978	.161 885 6833	30
	31	99.585 857 2142	616.161 607 5885	.001 622 9508	.010 041 5865	6.187 240 0843	.161 622 9508	31
	32	115.519 594 3684	715.747 464 8027	.001 397 1408	.008 656 5401	6.195 896 6244	.161 397 1408	32
	33	134.002 729 4674	831.267 059 1711	.001 202 9828	.007 462 5346	6.203 359 1590	.161 202 9828	33
	34	155.443 166 1822	965.269 788 6385	.001 035 9798	.006 433 2194	6.209 792 3784	.161 035 9798	34
	35	180.314 072 7713	1120.712 954 8207	.000 892 2891	.005 545 8788	6.215 338 2573	.160 892 2891	35
	36	209.164 324 4147	1301.027 027 5920	.000 768 6235	.004 780 9300	6.220 119 1873	.160 768 6235	36
	37	242.630 616 3211	1510.191 352 0067	.000 662 1677	.004 121 4914	6.224 240 6787	.160 662 1677	37
	38	281.451 514 9324	1752.821 968 3278	.000 570 5086	.003 553 0098	6.227 793 6885	.160 570 5086	38
	39	326.483 757 3216	2034.273 483 2602	.000 491 5760	.003 062 9395	6.230 856 6281	.160 491 5760	39
	40	378.721 158 4931	2360.757 240 5818	.000 423 5929	.002 640 4651	6.233 497 0932	.160 423 5929	40
	41	439.316 543 8520	2739.478 399 0749	.000 365 0330	.002 276 2630	6.235 773 3562	.160 365 0330	41
	42	509.607 190 8683	3178.794 942 9269	.000 314 5846	.001 962 2957	6.237 735 6519	.160 314 5846	42
	43	591.144 341 4072	3688.402 133 7952	.000 271 1201	.001 691 6342	6.239 427 2861	.160 271 1201	43
	44	685.727 436 0324	4279.546 475 2025	.000 233 6696	.001 458 3054	6.240 885 5915	.160 233 6696	44
	45	795.443 825 7976	4965.273 911 2349	.000 201 3988	.001 257 1598	6.242 142 7513	.160 201 3988	45
RATE 18%	1	1.180 000 0000	1.000 000 0000	1.000 000 0000	.847 457 6271	.847 457 6271	1.180 000 0000	1
	2	1.392 400 0000	2.180 000 0000	.458 715 5963	.718 184 4298	1.565 642 0569	.638 715 5963	2
	3	1.643 032 0000	3.572 400 0000	.279 923 8607	.608 630 8727	2.174 272 9296	.459 923 8607	3
	4	1.938 777 7600	5.215 432 0000	.191 738 6709	.515 788 8752	2.690 061 8047	.371 738 6709	4
	5	2.287 757 7568	7.154 209 7600	.139 777 8418	.437 109 2162	3.127 171 0209	.319 777 8418	5
	6	2.699 554 1530	9.441 967 5168	.105 910 1292	.370 431 5392	3.497 602 5601	.285 910 1292	6
	7	3.185 473 9006	12.141 521 6698	.082 361 0332	.313 925 0332	3.811 527 5933	.262 361 0332	7
	8	3.758 859 2027	15.326 995 5704	.065 244 3589	.266 038 1637	4.077 565 7571	.245 244 3589	8
	9	4.435 453 8592	19.085 854 7731	.052 394 8239	.225 456 0710	4.303 021 8280	.232 394 8239	9
	10	5.233 835 5538	23.521 308 6322	.042 514 6413	.191 064 4669	4.494 086 2949	.222 514 6413	10
	11	6.175 925 9535	28.755 144 1860	.034 776 3862	.161 919 0398	4.656 005 3347	.214 776 3862	11
	12	7.287 592 6251	34.931 070 1395	.028 627 8089	.137 219 5252	4.793 224 8599	.208 627 8089	12
	13	8.599 359 2976	42.218 662 7646	.023 686 2073	.116 287 7332	4.909 512 5931	.203 686 2073	13
	14	10.147 243 9712	50.818 022 0622	.019 678 0583	.098 548 9265	5.008 061 5196	.199 678 0583	14
	15	11.973 747 8860	60.965 266 0334	.016 402 7825	.083 516 0394	5.091 577 5590	.196 402 7825	15
	16	14.129 022 5055	72.939 013 9195	.013 710 0839	.070 776 3046	5.162 353 8635	.193 710 0839	16
	17	16.672 246 5565	87.068 036 4250	.011 485 2711	.059 979 9191	5.222 333 7827	.191 485 2711	17
	18	19.673 250 9367	103.740 282 9814	.009 639 4570	.050 830 4399	5.273 164 2226	.189 639 4570	18
	19	23.214 436 1053	123.413 533 9181	.008 102 8390	.043 076 6440	5.316 240 8666	.188 102 8390	19
	20	27.393 034 6042	146.627 970 0234	.006 819 9812	.036 505 6305	5.352 746 4971	.186 819 9812	20
	21	32.323 780 8330	174.021 004 6276	.005 746 4327	.030 936 9750	5.383 683 4721	.185 746 4327	21
	22	38.142 061 3829	206.344 785 4605	.004 846 2577	.026 217 7754	5.409 901 2476	.184 846 2577	22
	23	45.007 632 4318	244.486 846 8434	.004 090 1996	.022 218 4538	5.432 119 7013	.184 090 1996	23
	24	53.109 006 2695	289.494 479 2752	.003 454 2973	.018 829 1981	5.450 948 8994	.183 454 2973	24
	25	62.668 627 3981	342.603 485 5448	.002 918 8261	.015 956 9475	5.466 905 8470	.182 918 8261	25
	26	73.948 980 3297	405.272 112 9429	.002 467 4779	.013 522 8369	5.480 428 6839	.182 467 4779	26
	27	87.259 796 7891	479.221 093 2726	.002 086 7195	.011 460 0313	5.491 888 7152	.182 086 7195	27
	28	102.966 560 2111	566.480 890 0616	.001 765 2846	.009 711 8909	5.501 600 6061	.181 765 2846	28
	29	121.500 541 0491	669.447 450 2727	.001 493 7692	.008 230 4160	5.509 831 0221	.181 493 7692	29
	30	143.370 638 4379	790.947 991 3218	.001 264 3056	.006 974 9288	5.516 805 9509	.181 264 3056	30
	31	169.177 353 3568	934.318 629 7597	.001 070 2987	.005 910 9566	5.522 716 9076	.181 070 2987	31
	32	199.629 276 9610	1103.495 983 1165	.000 906 2108	.005 009 2853	5.527 726 1928	.180 906 2108	32
	33	235.562 546 8139	1303.125 260 0775	.000 767 3859	.004 245 1570	5.531 971 3499	.180 767 3859	33
	34	277.963 805 2405	1538.687 806 8914	.000 649 9044	.003 597 5907	5.535 568 9406	.180 649 9044	34
	35	327.997 290 1837	1816.651 612 1319	.000 550 4633	.003 048 8057	5.538 617 7462	.180 550 4633	35
	36	387.036 802 4168	2144.648 902 3156	.000 466 2768	.002 583 7336	5.541 201 4799	.180 466 2768	36
	37	456.703 426 8518	2531.685 704 7324	.000 394 9937	.002 189 6048	5.543 391 0846	.180 394 9937	37
	38	538.910 043 6852	2988.389 131 5843	.000 334 6284	.001 855 5973	5.545 246 6819	.180 334 6284	38
	39	635.913 851 5485	3527.299 175 2694	.000 283 5030	.001 572 5401	5.546 819 2219	.180 283 5030	39
	40	750.378 344 8272	4163.213 026 8179	.000 240 1991	.001 332 6611	5.548 151 8830	.180 240 1991	40
	41	885.446 446 8961	4913.591 371 6451	.000 203 5171	.001 129 3738	5.549 281 2568	.180 203 5171	41
	42	1044.826 807 3374	5799.037 818 5413	.000 172 4424	.000 957 0964	5.550 238 3532	.180 172 4424	42
	43	1232.895 632 6582	6843.864 625 8787	.000 146 1163	.000 811 0987	5.551 049 4519	.180 146 1163	43
	44	1454.816 846 5366	8076.760 258 5369	.000 123 8120	.000 687 3717	5.551 736 8236	.180 123 8120	44
	45	1716.683 878 9132	9531.577 105 0735	.000 104 9144	.000 582 5184	5.552 319 3420	.180 104 9144	45

637

Table C-24 Sum of months' digits

Number of months	Sum: 1 through largest month
6	21
9	45
10	55
12	78
15	120
18	171
24	300
36	666

Table C-25 Monthly payment per $1,000 of mortgage[a]

Rate	20 years	25 years	30 years
8%	$8.37	$7.72	$7.34
$8\frac{1}{4}$	8.53	7.89	7.52
$8\frac{1}{2}$	8.68	8.06	7.69
$8\frac{3}{4}$	8.84	8.23	7.87
9	9.00	8.40	8.05
$9\frac{1}{4}$	9.16	8.57	8.23
$9\frac{1}{2}$	9.33	8.74	8.41
$9\frac{3}{4}$	9.49	8.92	8.60
10	9.66	9.09	8.78
11	10.33	9.81	9.53

[a] Monthly payments including principal and interest.

Table C-26 Metric measures with U.S. equivalents

Length	Weight	Liquid capacity
1 millimeter = 0.039 inch	1 gram[a] = 0.035 ounce	1 liter[b] = 33.8 ounces
1 centimeter = 0.394 inch	1 kilogram = 2.20 pounds	= 2.11 pints
1 decimeter = 3.94 inches		= 1.06 quarts
1 meter = 39.37 inches	**Temperature**	= 0.264 gallon
= 3.28 feet		
= 1.09 yards	$C = \frac{5}{9}(F - 32)$	
1 kilometer = 0.621 mile	or $C = (F - 32) \div 1.8$	

[a] 1 gram is defined as the weight of 1 cubic centimeter of pure water at 4° Celsius.
[b] 1 liter is defined as the capacity of a cube that is 1 decimeter long on each edge.

TABLE C-27 U.S. MEASURES WITH METRIC EQUIVALENTS

LENGTH	WEIGHT	LIQUID CAPACITY
1 inch = 2.54 centimeters	1 ounce = 28.4 grams	1 ounce = 0.030 liter
1 foot = 0.305 meter	1 pound = 454 grams	1 pint = 0.473 liter
1 yard = 0.914 meter	= 0.454 kilograms	1 quart = 0.946 liter
1 mile = 1.61 kilometers		1 gallon = 3.79 liters

TEMPERATURE

$$F = \tfrac{9}{5}C + 32$$
$$\text{or} \quad F = 1.8C + 32$$

TABLE C-28 SAMPLE FOREIGN CURRENCY EXCHANGE RATES

COUNTRY AND CURRENCY	$1 U.S. IN FOREIGN CURRENCY	FOREIGN CURRENCY IN U.S. DOLLARS
Britain (£)	0.6371	1.5695
Canada (C$)	1.3570	0.7369
France (F̄)	5.0800	0.1969
Germany (DM)	1.4380	0.6954
Japan (¥)	86.5500	0.011554
Mexico (N$)	5.9200	0.168919

Answers to Odd-Numbered Problems

## CHAPTER 1

REVIEW OF OPERATIONS

SECTION 1, PAGE 11

1. a. 900   b. 6,240   c. 14,800
   d. 37,800   e. 285,000   f. 718,200
   g. 151,632   h. 24,449,824   i. 725,700
   j. 65.1   k. 25.20   l. 3.69
   m. 51.2   n. 3.33   o. 4,096
   p. 243   q. 216   r. 1.0816

3. a. 15   b. 6   c. 70
   d. 70   e. 3.9   f. −$57.81
   g. −$12,167   h. −$185   i. −$2,225

5. a. $\dfrac{607}{600}$ or $1\dfrac{7}{600}$ b. $\dfrac{103}{100}$ or $1\dfrac{3}{100}$ c. $\dfrac{116}{125}$

   d. $\dfrac{173}{175}$ e. 1,270   f. 2,500

   g. 740   h. $200 - 14y$ i. $400 - 12a$

   j. $cde + cf$ k. $s - stu$

7. a.   66.7   b.   84.67   c.   86.455
      368.0   488.93   410.015
     1,542.3   591.55   217.675
   d. 8.2;   8.19;   8.188
      14.5;   14.53;   14.527

9. a. 116.2;   116   b. 6.0;   6.0
   c. 24.16;   24.2   d. 16.2;   16

11. a. $901.51   b. $1,801.44   c. $806.49
    d. $643.88

SECTION 2, PAGE 17

1. a. 1,500   b. 337   c. 240

d. 80   e. 272   f. 160

g. 45   h. 400

3. a. 223.6   b. 768   c. 56

d. 720

5. a. $40.50; $72; $135   b. 3.12; 6.24; 8.645

c. 2,125   d. 20.102

e. 1,580   f. 1,395

g. 850.75   h. 32

CHAPTER 2

USING EQUATIONS

SECTION 1, PAGE 26

1.	47	**3.**	12	**5.**	9	**7.**	20	**9.**	10	**11.**	36
13.	10	**15.**	8	**17.**	63	**19.**	6	**21.**	9	**23.**	19
25.	5	**27.**	1	**29.**	3	**31.**	75	**33.**	63	**35.**	81

SECTION 2, PAGE 30

1. a. $n - 12$ b. $D + J$ c. $b + 3$

d. $x - \$15$ e. $\frac{1}{4}c \ \ or \ \ \frac{c}{4}$ f. $\frac{2}{3}k - 10$

g. $4(c + d)$ h. $w = z - \$18$ i. $b = 8(j + k)$

j. $d = 3.8e$ k. $h = \frac{1}{2}m - 5$ l. $\$30v$

3. 65   **5.** $43   **7.** $26

9. $28,000   **11.** $108,000   **13.** 147

15. 5,000   **17.** 18¢   **19.** OH: $6,250
                                             WC: $18,750

21. F = 13 gal.   **23.** C = $80   **25.** C = 40
S = 8 gal.   F = $320   G = 184
                        E = $100

27. S = $2.25   **29.** CD = 20   **31.** S = 80
C = $4.50   T = 30   B = 30
H = $1.25

33. J = 460   **35.** G = 7
G = 320   S = 5

Section 3, page 36

1. a. 1 to 6   b. 3 to 4   c. 1 to 3   d. 7 to 6

    1:6   3:4   1:3   7:6

    $\dfrac{1}{6}$ $\dfrac{3}{4}$ $\dfrac{1}{3}$ $\dfrac{7}{6}$

    e. 6 to 1

    6:1

    $\dfrac{6}{1}$

3. a. 27   b. 20   c. 21   d. 8

5. $\dfrac{5}{9}$ 7. $\dfrac{5}{6}$ 9. 1,536

11. 30   13. 15   15. 208

17. 366   19. 344   21. 18

23. 28   25. 960   27. 14 hours

SEVENTH EDITION

BUSINESS
MATHEMATICS

A COLLEGIATE APPROACH

NELDA W. ROUECHE

VIRGINIA H. GRAVES

PRENTICE HALL

UPPER SADDLE RIVER, NEW JERSEY 07458

Library of Congress Cataloging-in-Publication Data

Roueche, Nelda W.
    Business mathematics : a collegiate approach / Nelda W. Roueche,
Virginia H. Graves.—7th ed.
      p. cm.
    Includes index.
    ISBN 0-13-500000-9
    1. Business mathematics.   I. Graves, Virginia H., (date)
II. Title.
HF5691.R68 1997
650′.01′513—dc20   96-10831
                                                  CIP

Director of Production and Manufacturing: **Bruce Johnson**
Managing Editor: **Mary Carnis**
Editorial/production supervision: **WordCrafters Editorial Services, Inc.**
Interior design: **Amy Rosen**
Cover design: **Amy Rosen**
Manufacturing buyer: **Ilene Sanford**
Acquisition editor: **Stephen Helba**
Editorial Assistant: **Craig Campanella**
Marketing manager: **Frank Mortimer, Jr.**
Supplements editor: **Judy Casillo**
Cover illustration: Jasper Johns, **Numbers in Color,** 1958–59. Encaustic and collage on
canvas, 5′7″ × 4′12″. Albright-Knox Gallery, Buffalo, NY.

 1997, 1993, 1988, 1983, 1978, 1973, 1969 by Prentice-Hall, Inc.
A Simon & Schuster Company
Upper Saddle River, New Jersey 07458

Printed in the United States of America

10 9 8 7 6 5 4 3 2

ISBN 0-13-500000-9

Prentice-Hall International (UK) Limited, *London*
Prentice-Hall of Australia Pty. Limited, *Sydney*
Prentice-Hall Canada Inc., *Toronto*
Prentice-Hall Hispanoamericana, S.A., *Mexico*
Prentice-Hall of India Private Limited, *New Delhi*
Prentice-Hall of Japan, Inc., *Tokyo*
Simon & Schuster Asia Pte. Ltd., *Singapore*
Editora Prentice-Hall do Brasil, Ltda., *Rio de Janeiro*